FLORA ZAMBESIACA

Flora terrarum Zambesii aquis conjunctarum

T0293720

VOLUME FOURTEEN

CYPERACEAE

FLORA ZAMBESIACA

MOZAMBIQUE, MALAWI, ZAMBIA,

ZIMBABWE, BOTSWANA

VOLUME FOURTEEN
CYPERACEAE

Editors

M.A. GARCÍA & J.R. TIMBERLAKE
on behalf of the Editorial Board:

FRANCES CHASE
National Botanical Research Institute, Windhoek, Namibia

MARTIN CHEEK
Royal Botanic Gardens, Kew, U.K.

DAVID CHUBA
School of Natural Sciences, UNZA, Lusaka, Zambia

IAIN DARBYSHIRE
Royal Botanic Gardens, Kew, U.K.

DAVID GOYDER
Royal Botanic Gardens, Kew, U.K.

SHAKKIE KATIVU
University of Zimbabwe, Harare, Zimbabwe

JAMESON SEYANI
Herbarium and National Botanical Gardens, Malawi

Authors

J. Browning, K.D. Gordon-Gray†, M. Lock,
H. Beentje, K. Vollesen, K. Bauters, C. Archer,
I. Larridon, M. Xanthos, P. Vorster, J. Bruhl,
K. Wilson and X. Zhang

Published by the Royal Botanic Gardens, Kew
for the Flora Zambesiaca Managing Committee
2020

First published in 2020 by
Royal Botanic Gardens, Kew,
Richmond, Surrey, TW9 3AB, UK
www.kew.org

Distributed on behalf of the Royal Botanic Gardens, Kew in North America by the University of Chicago Press, 1427 East 60th Street, Chicago, IL 60637, USA

ISBN 978-1-84246-707-7
eISBN 978-1-84246-708-4

British Library Cataloguing in Publication Data
A catalogue record for this book is available from the British Library

Copy-editing: Ruth Linklater
Design and page layout: Christine Beard
Production management: Jo Pillai

Printed in the UK by Short Run Press Limited

For information or to purchase all Kew titles please visit shop.kew.org/kewbooksonline or email publishing@kew.org

Kew's mission is to inspire and deliver science-based plant conservation worldwide, enhancing the quality of life.

Kew receives approximately one third of its running costs from Government through the Department for Environment, Food and Rural Affairs (Defra). All other funding needed to support Kew's vital work comes from members, foundations, donors and commercial activities including book sales.

CONTENTS

NEW TAXA AND COMBINATIONS PUBLISHED IN THIS VOLUME

14. CYPERACEAE[1]

by J. Browning, K.D. Gordon-Gray†, M. Lock, H. Beentje, K. Vollesen, K. Bauters, C. Archer, I. Larridon, M. Xanthos, P. Vorster, J. Bruhl, K. Wilson and X. Zhang

Annual or perennial herbs, usually tufted with rhizomes or stolons; monoecious or hermaphrodite, rarely dioecious (in *Scirpoides*). Culms solid or hollow, triangular or less often rounded, 4–6 angular or flattened, sometimes with transverse septa. Leaves basal or cauline, usually 3-ranked with a closed sheath, rarely 2-ranked with an open sheath (in *Coloeochloa*); blade linear when present. Inflorescence terminal or pseudolateral, umbellate, anthelate, capitate, spicate, paniculate, corymbose (or combinations thereof); with few to many spikelets, sometimes reduced to a solitary spikelet, or with spikelets aggregating into pseudo-spikelets often surrounded by conspicuous leafy bracts. Spikelets consisting of few to many inconspicuous unisexual or bisexual flowers each subtended by a glume (bract). Glumes spirally or distichously arranged, usually all fertile (occasionally only uppermost glume fertile). Stamens 1–3(6); filaments free, sometimes strongly elongating after anthesis; anthers basifixed, opening lengthwise by a slit. Ovary solitary and superior, 1-locular, of 2–3 joined carpels; hypogynous scales or bristles absent or 3–6; style 2–3-branched, rarely unbranched. Fruit an indehiscent 1-seeded nutlet, sessile or nearly so, sometimes (in *Carex*, *Schoenoxiphium*) surrounded by a sac-like utricle.

A family of 104 genera and 5000 species, particularly in the tropics and subtropics but with the large genus *Carex* well-represented in temperate zones. Often found in moist conditions and can be the dominant plants in wetlands.

The order of genera below follows that in F.T.E.A., based on Goetghebeur in Kubitzki (1998).

Recently the genera *Alinula*, *Ascolepis*, *Kyllinga*, *Lipocarpha*, *Pycreus*, *Sphaerocyperus* and *Volkiella* have been subsumed into the genus *Cyperus* sensu lato (Larridon *et al.* in Phytotaxa **166**: 33–48, 2014; in Pl. Ecol. Evol. **144**: 327–356, 2011; Bauters *et al.* in Phytotaxa **166**: 1–32, 2014). In light of the recent nature of this change, and to ensure consistency with F.T.E.A., we are retaining the original genera here. Larridon *et al.* (2011, 2014) and Bauters *et al.* (2014) provided the necessary taxonomic changes to subsume the listed genera into Cyperus.

KEY TO THE GENERA

[adapted by M. Xanthos from Goetghebeur in Kubitzki, Fam. Pl. Gen. Pl. **4**: 154–159 (1998)]

1. Plants dioecious, with separate male and female individual plants . . **13. Scirpoides**
– Plants monoecious i.e. separate male and female flowers present on same plant, or flowers bisexual 2
2. All flowers unisexual i.e. glumes subtending either stamens or ovary 3
– At least some flowers bisexual 7
3. Female flowers enclosed in a sac-like utriculiform prophyll (commonly known as a utricle), these flowers usually aggregating together to form a spike or basal portion of a bisexual spike 4
– Female flowers not enclosed in a utricle 5

---

[1] *Fuirena* by J. Browning and M. Xanthos; *Nemum* by I. Larridon; *Cyperus*, *Scleria* and *Schoenoxiphium* by M. Lock; *Courtoisina* by P. Vorster; *Alinula*, *Ascolepis* and *Lipocarpha* by K. Bauters; *Pycreus* by K. Vollesen; *Kyllinga* by H. Beentje; *Carpha* by J. Bruhl, K. Wilson and X. Zhang; *Coleochloa* and *Queenslandiella* by M. Xanthos; *Carex* by C. Archer. All other genera by J. Browning and K. Gordon-Gray†.

4. Rachilla of female spikelet usually small and not protruding out of the utricle; utricles closed except for small apical orifice . . . . **37. Carex**
- Rachilla of female spikelet protruding in at least a few spikelets; usually some of the utricles partly open . . . . **36. Schoenoxiphium**
5. Leaves distichously arranged with deciduous blades, leaf sheaths opening on the ventral side with ciliate ligule . . . . **33. Coleochloa**
– Leaves not with this combination of characters. 6
6. Inflorescence a sessile to shortly stalked cluster of spikelets arising in the leaf axils (except lowest 1–2 leaves) . . . . **35. Diplacrum**
– Inflorescence paniculate or spicate arising on the upper part of the culm. . . . .**34. Scleria**
7. Bisexual flowers subtended by hypogynous bristles or scales 8
– Bisexual flowers not subtended by hypogynous bristles or scales. 16
8. Inflorescence reduced to a single terminal spikelet 9
– Inflorescence a cluster of few to many spikelets 10
9. Plants aquatic, floating or submerged with many pseudoverticillate vegetative branchlets; style base usually indistinct, not thickened. . . . . **6. Websteria**
– Plants rarely completely aquatic, ± unbranched (but see *Eleocharis naumanniana*); style base usually distinct, thickened and persistent on nutlet apex .**5. Eleocharis**
10. Style base persistent as a short-conical or long-conical cap 11
– Style base not persistent 14
11. Style unbranched or 2-branched . . . . **26. Rhynchospora**
– Style 3-branched 12
12. Ligules tubular and hairy; glumes in partially or clearly defined ranks **1. Fuirena**
– Not with this combination of characters 13
13. Anthers conspicuously greenish-yellow; hypogynous bristles antrorsely scabrous . . . . **30. Carpha**
– Anthers not greenish-yellow; hypogynous bristles plumose on top-half . . . .**32. Costularia**
14. Leaves reduced to sheaths without blades (rarely with blades) or with very short blades; inflorescence pseudolateral or apparently so . . . . **3. Schoenoplectus**/**4. Schoenoplectiella**
– Leaves with well-developed blades 15
15. Leaves eligulate . . . . **2. Bolboschoenus**
– Leaves ligulate; rachilla internodes elongated, conspicuously zigzag . . . . **28. Schoenus**
16. Number of glumes per spikelet reduced to 1 with supporting bract which may be larger or smaller than glume; spikelet prophyll present or absent 17
– Glumes more than 1 per spikelet, spirally or distichously arranged 20
17. Spikelet prophyll absent . . . . **20. Ascolepis**
– Spikelet prophyll present 18
18. Spikelet glume larger than spikelet bract, floret enclosed within glume . . . . **19. Alinula**
– Spikelet glume smaller than spikelet bract, floret enclosed between glume and prophyll 19
19. Prophyll thin and hyaline . . . .**25. Lipocarpha**
– Prophyll thick and coriaceous; dwarf annual buried in sand . . . . **24. Volkiella**
20. Glumes distichously arranged. 21
– Glumes, or at least upper ones, spirally arranged 27
21. Only uppermost glume of spikelet fertile; glumes gradually increasing in size towards apex 22
– More than one glume within a spikelet fertile. 23

22. Plants stoloniferous; inflorescence of 3–4 spikelets, glumes deciduous; leaves ligulate . . . . **27. Actinoschoenus**
– Plants with an elongated rhizome; inflorescence a globose cluster of many greenish-white spikelets, which are shed as units; leaves eligulate . . . . **18. Sphaerocyperus**

23. Nutlet usually trigonous with flat side appressed to rachilla axis 24
– Nutlet usually biconvex appressed edgeways to rachilla axis, or nutlet laterally compressed 25

24. Plants annual and smelling of curry; inflorescence anthelate with digitately arranged spikelets; spikelets deciduous as a unit; glumes with conspicuously winged keel . . . . **17. Courtoisina**
– Plants without this combination of characters. . . . . **16. Cyperus**

25. Spikelets reduced to 2–3 glumes folded within one another; style 2-branched; nutlet oblong or ellipsoid, laterally compressed . . . .**23. Kyllinga**
– Spikelets with many glumes; style 2 or 3-branched 26

26. Plants annual with a pungent smell of curry or fenugreek; glumes persistent on a deciduous rachilla, the spikelets falling as a unit . . . .**22. Queenslandiella**
– Plants without this combination of characters. . . . . **21. Pycreus**

27. Leaf blades absent; inflorescence reduced to a single terminal spikelet. . . . .**5. Eleocharis**
– Plants without this combination of characters. 28

28. Ligules tubular, completely enclosing the culm; usually hairy . . . . **1. Fuirena**
– Ligules not as above. 29

29. Inflorescence paniculate or shortly corymbose. 30
– Inflorescence umbellate, anthelate, capitate, terminal or pseudolateral 31

30. Robust perennials with thick rhizomes and occasionally stolons; leaves V-shaped in cross-section or ± flat with scabrid margins, eligulate. . . . . **29. Cladium**
– Slender tufted perennials without rhizomes or stolons; leaves, when present, inrolled without scabrid margins, ligulate; sheaths conspicuously dark reddish-brown . . . .**31. Tetraria**

31. Style base distinct and thickened, persistent on the nutlet or deciduous with the style. 32
– Style base neither distinct nor thickened. 34

32. Style glabrous to villous, leaf sheath orifice usually with long white hairs, style base usually persistent as a conspicuous knob on the nutlet . . . . **8. Bulbostylis**
– Style flattened and fimbriate, leaf sheath orifice glabrous, style base deciduous 33

33. Nutlet >1 mm long, conspicuously clavate-stipitate; glumes usually coriaceous; inflorescence always with 1–few spikelets. . . . .**9. Abildgaardia**
– Nutlet usually <1 mm long, rarely stipitate; glumes usually more delicate; inflorescence often with many spikelets. . . . .**7. Fimbristylis**

34. Floating aquatic with inflorescence an open umbel of globose heads subtended by foliose bracts; nutlet with a corky base, margin and tip . . . .**14. Oxycaryum**
– Plants without this combination of characters. 35

35. Glumes long persistent, dark reddish brown to black, ± scabrous; nutlet blackish, rarely brownish to greyish, nutlet surface smooth and shiny . . . .**10. Nemum**
– Plants without this combination of characters. 36

36. Inflorescence a terminal compact head of several rounded or ovoid sessile spikelets, involucral bracts spreading or recurved . . . . **15. Kyllingiella**
– Inflorescence pseudolateral or if terminal consisting of a single solitary spikelet 37

1. **FUIRENA** Rottb.[2]

Fuirena Rottb., Descr. Icon. Rar. Pl.: 70 (1773). —Forbes, Revision *Fuirena*: thesis, Univ. Witwatersrand, South Africa (1980).
Scirpus L. subgen. *Fuirena* Kuntze, Rev. Gen. Pl. **3**: 331 (1898).
Scirpus L. sect. *Vaginaria* Koyama in J. Fac. Sci. Univ. Tokyo, Bot., Sect.3, Bot. **7**: 286 (1958).

Annuals or rhizomatous perennials. Culms erect or decumbent, 3–5 angled, glabrous or hairy. Leaf sheaths well-developed, closed, hairy. Leaf blades linear or lanceolate, glabrous or hairy; ligules tubular, completely enclosing culm, usually hairy. Inflorescences paniculate; a terminal, partial inflorescence accompanied by 1–3(7) lateral, spikelet-bearing branches at 1–3(6) nodes below, or these lateral branches lacking when inflorescence is terminal of loosely arranged spikelets or these grouped into heads. Bracts to terminal and lateral inflorescence branches with or without well-defined tubular sheaths, branches usually hairy. Spikelets ovoid, ellipsoid, terete or 5-angled, of many imbricate floral scales in drier habitats, partially or clearly ranked. Glumes glabrous or densely pubescent, trinervate, nerves converging distally becoming excurrent into a subterminal or terminal awn, straight or markedly outward curving, giving a prickle-covered appearance to the spikelet. Perianth extremely variable, in 2 whorls, each of 3 segments alternating with stamens (inner whorl) and with lower (outer) segments, or reduced to microscopic remnants, or entirely absent; outer whorl composed of bristles or bristle-like structures (expanded below, narrowed above), inner whorl composed mostly of scales on short or long stalks, Stamens (2)3, Style short, thickened, remaining as beak on nutlet; style branches 3. Nutlet small, 3-angled, stipitate or sessile, surface topography variable, cells on 3 faces differing from more uniform cells on angles.

A cosmopolitan genus, essentially of warm temperate to tropical areas; estimates of species vary from 30 to 59, with c. 19 in south tropical Africa. Species grow in permanent or seasonal lakes, streams and wet depressions in grassland, often shaded by taller vegetation.

Owing to morphological variation (especially of perianth segments and general plant pubescence), taxonomic opinion on species limits has differed. Pubescence, commonly present, is characteristic of the genus, but variable in type and distribution on individual plants.

Inner perianth segments are used in the key; in one species they are represented by bristles, but in the majority they are scales consisting of claw (stalk), blade and often an arista (small projection). Illustrations of these scales are given in Fig. 14.1 and are denoted in the key by square brackets.

1. Glumes glabrous . . . .**1.** *abnormalis*
– Glumes hairy with long to short hairs on abaxial side 2
2. Glumes with short hairs of uniform length 3

---

[2] by J. Browning and M. Xanthos

– Glumes with intermingled short and long hairs, on all, many, or on a few basal glumes only 9

3. Glumes pubescent, except distally where a glabrous band with ciliate edge borders the margin . . . . **2.** *coerulescens*

– Glumes uniformly pubescent, lacking distal glabrous band 4

4. Inner perianth of 3 well-developed bristles as long as nutlet; rhizome absent (occasionally poorly developed) . . . . **3.** *stricta*

– Inner perianth of scales or these reduced or absent; rhizome present, elongate or contracted 5

5. Inner perianth scales with well-developed hairy claws; culms glabrous; rhizome well developed [F] . . . . **4.** *hirsuta*

– Inner perianth scales lacking claws, or claws glabrous 6

6. Inner perianth scales well-developed, equalling or slightly shorter than nutlet; claws glabrous, lobed blade with subterminal arista developed from inner (adaxial) face. Leaves with sheaths covering culm so that internodes often not visible [E] . . . . **5.** *ecklonii*

– Inner perianth scales usually absent, if present vestigial. 7

7. Plant with short hairs, 0.1–0.2 mm long on all parts of the plant, forming a dense velvet-like covering, except towards base . . . . **6.** *pachyrrhiza*

– Plant with short hairs not forming a dense covering, restricted or absent from certain parts of plant 8

8. Hairs restricted to small portions of culm, nodes, midribs, leaf blade margins or inflorescence branches; leaf blades 5–7 mm wide . . . . **7.** *pubescens*

– Hairs infrequently present, if present restricted to leaf margins and apices; leaf blades 2–3(4) mm wide . . . . **8.** *welwitschii*

9. Culms and leaf sheath 5-angled; leaf blades with 5 well-marked nerves [L] . . . . **9.** *umbellata*

– Culms and leaf sheath 3-angled to terete; leaf blades with visible nerves, but not regularly 5 in number 10

10. Inner perianth scales obcordate and subsessile, equalling mature nutlet in length. Rhizome well-developed [H] . . . . **10.** *obcordata*

– Inner perianth scales absent or present and variously shaped, never obcordate, always with claw 11

11. Claw of inner perianth scale hairy; rhizome well-developed 12

– Claw of inner perianth scale glabrous, or scales absent; rhizome usually absent or poorly developed 13

12. Inner perianth scales with large crest; leaf blades many nerved; culms ribbed [J] . . . . **11.** *ochreata*

– Inner perianth scales not crested; leaf blades with midrib; culm smooth [F] . . . . **4.** *hirsuta*

13. Awns of glumes straight, ⅕ of total glume length or shorter 14

– Awns of glumes recurved, ⅓–¼ of total glume length 15

14. Glumes elliptical, pilose; inner perianth segments lunate, inflated with outer edges produced into long downward pointing lobes [D] . . . . **12.** *claviseta*

– Glumes oblong, scabrid at least on margin and awn; inner perianth segment broadly-winged (rhomboid) and aristate [M] . . . . **13.** *zambesiaca*

15. Stamens 3; mature nutlet (0.7)0.8–1.3 mm long 16

– Stamens 2; nutlet 0.4–0.6 mm long 19

16. Inner perianth scale 1 (rarely 2), claw shorter than cup-shaped blade [B] . . . . **14.** *bullifera*

– Inner perianth scales 3, scale claw equal or longer than blade; blade variously shaped, not cup-shaped 17

17. Blade of inner perianth scale quadrate with 3 nerves [C] . . . . **15.** *ciliaris*
– Blade of inner perianth scale subquadrate (seldom quadrate), uninervate, enervate or lunate, or scales anchor-shaped 18
18. Inner perianth scales subquadrate, lacking nerves, convex on abaxial side, sometimes inflated [A] . . . .**16.** *angolensis*
– Inner perianth scales with anchor-shaped blade, lacking nerves, not convex on abaxial side, not inflated [K] . . . .**17.** *sagittata*
19. Inner and outer perianth present, inner one of 3 anchor-shaped scales [G] . . . .
. . . . **18.** *leptostachya*
– Inner and outer perianth lacking. 20
20. Nutlet convex, greenish-white colour, semi-transparent, faintly patterned by large bullate cells at maturity . . . . **19.** *microcarpa*
– Nutlet with flat or concave sides, brown, opaque, smooth or faintly transversely striate . . . . **18.** *leptostachya* forma *nudiflora*

1. **Fuirena abnormalis** C.B. Clarke in F.T.A. **8**: 462 (1902). —Napper in J. E. Africa Nat. Hist. Soc. Natl. Mus. **25**(110): 20 (1965). —Haines & Lye, Sedges & Rushes E. Afr.: 25, fig.63 (1983). —Muasya in Kew Bull. **53**: 193 (1998); in F.T.E.A., Cyperaceae: 10 (2010). Type: Mozambique, Zambezi R., Boroma (Boruma), ii.1892, Menyharth 1060 (K holotype).

Tufted annual. Culms 20–65 cm tall, 2–3 mm wide, weak, erect or decumbent and rooting at nodes, terete to obscurely trigonous, ridged, glabrous. Leaf sheath loose, ribbed, membranous, glabrous; ligule membranous, glabrous, mouth oblique; leaf blade 7–14 cm long, 6 mm wide, flat, lanceolate to linear, scabrid on margin, upper surface with sharp transparent teeth, auriculate encircling sheath mouth with a membrous ridge. Inflorescence an elongated compound panicle 13–20 cm long, lateral panicles of 2–5 unequal branches subtended by leafy bracts; peduncles compressed, scabrid; pedicellate spikelets in corymbose clusters. Spikelets, 4–6 × 1–2 mm, ellipsoid at flowering, ovoid in fruit, glumes tardily deciduous, golden-brown. Glumes not ranked, glabrous, 2–3 mm long (awn included), c. 1.5 mm wide, tardily persistent after nutlet fall, ovate, midrib stout, green, excurrent into straight or curved awn 0.5–0.9 mm long. Perianth segments absent. Stamens 3; anthers c. 0.4 mm long with red crest. Nutlet dark brown to black at maturity, 1.2–1.4 × 0.9–1 mm (including beak and base), obovoid, trigonous with pale raised angles, circled by a median rounded ridge; surface of continuous or interrupted lines of hexagonal cells in transverse lines with large central papillae.

Zambia. N: Mbala Dist., Mwamba, 22.v.1962, *Robinson* 5419 (K). **Zimbabwe**. S: Mwenezi Dist., Mateke Hills, S slopes, 9.v.1958, *Boughey* 2824 (SRGH). **Mozambique**. T: Changara Dist., Zambezi R., Boroma (Boruma), ii.1892, *Menyharth* 1060 (K).

Also in Kenya and Tanzania. Edges of rivers and streams and perennially wet grassland; 300–1500 m.

Conservation notes: Widespread; not threatened.

Recognised by the glabrous glumes, stems and leaves, and dark black nutlets lacking perianth segments.

2. **Fuirena coerulescens** Steud. in Flora **12**: 153 (1829). —Gordon-Gray in Strelitzia **2**: 99 (1995). —Forbes in S. African J. Bot. **63**: 514 (1997). Type: South Africa, Cape Province, Cape Flats, xii.1830, *Ecklon* 868 (B† holotype, K lectotype, M, PRE), lectotypified by Forbes (1997).

Fuirena gracilis Kunth, Enum. Pl. **2**: 181(1837), invalid name. Type: South Africa, Cape Province, between Gekau and Bashee R., 24.i.1932, Drège 4341 (K lectotype, P), lectotypified by Forbes (1997).

Fuirena enodis C.B. Clarke in Fl. Cap. **7**: 263 (1898). Type: South Africa, Cape Province (Griquatown), 1.xii.1811, *Burchell* 1865 (K holotype, P).

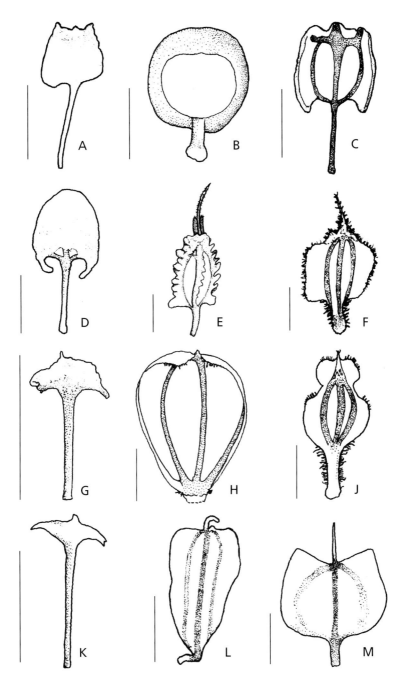

Fig. 14.1. FUIRENA inner perianth scales. A. —FUIRENA ANGOLENSIS, from *Gilliland* 88. B. —FUIRENA BULLIFERA, from *Du Preez & Steenkamp* 55. C. —FUIRENA CILIARIS, from *Golding et al.* P76. D. —FUIRENA CLAVISETA, from *Ward* 7887. E. —FUIRENA ECKLONII, from *Robinson* 6703. F. —FUIRENA HIRSUTA, from *Balsinhas* 505. G. —FUIRENA LEPTOSTACHYA, from *Heery* 19. H. —FUIRENA OBCORDATA, from *Merrett* 1381. J. —FUIRENA OCHREATA, from *Bullock* 1325. K. —FUIRENA SAGITTATA, from *P.A. Smith* 1079. L. —FUIRENA UMBELLATA, from *Pope & Müller* 1302. M. —FUIRENA ZAMBESIACA, from *Gomes e Sousa* 4063. Scale bars = 0.5 mm. Drawn by Jane Browning.

Fuirena subdigitata C.B. Clarke in J. Linn. Soc., Bot. **37**: 477 (1906). —Brain in Proc. & Trans. Rhodesia Sci. Assoc. **33**: 80 (1934). Type: Zimbabwe, Matopo Hills, ix.1905, *Gibbs* 196 (BM holotype).

Fuirena reticulata Kük. in Repert. Spec. Nov. Regni Veg. **41**: 271 (1937). Type: Namibia, Grootfontein, 20.i.1934, *Schoenfelder* 472 (PRE holotype, K, M).

Perennial, culms contiguous or to 20 mm apart; rhizome woody and stout, horizontal, 3–9 mm diameter. Culms 25–60(80) cm tall, 1–3 mm wide, sharply triangular, glabrous apart from short distance below inflorescence. Leaf sheath glabrous, occasionally puberulous; ligule glabrous or puberulous on upper margin; leaf blade 8–20(31) cm long, 2–5 mm wide, flat or infolded, glabrous, sometimes puberulous adaxially in region of midrib adaxially, ciliate on parts of margin. Inflorescence a terminal or pseudolateral cluster of up to 12 sessile to shortly stalked spikelets, sometimes with an added cluster, branches puberulous. Bracts 2, unequal, erect, puberulous. Spikelets 6–15 × 3–4 mm (awns excluded), terete, narrowly ovoid, dark silver-grey. Glumes 3.5–5 mm long (1–1.5 mm long awn included), puberulous abaxially with a glabrous band (in some collections coloured blue-black) with ciliate edge on distal margin of body, awn subterminal, straight or slightly recurved. Perianth segments 6 in 2 whorls, outer 3 segments of scabrous bristles, as short or longer than nutlet, or absent, inner 3 segments very variable, of scales (with claw and obtriangular blade bearing a scabrid arista) or bristles, or a mixture of these, or reduced or absent. Stamens 3; anthers 1.25 mm (orange-reddish in pressed material). Nutlet pale to dark brown, 1.3–1.6 (including stalk and short conical papillate beak) × 0.7–0.9 mm, dark to pale brown, trigonous, obovoid, substipitate, subhexagonal surface cells arranged in somewhat longitudinal rows may be visible (× 10).

Zambia. C: Lusaka Dist., University Campus, 25.viii.1993, *Bingham* 9634 (K). **Zimbabwe**. N: Mazoe Dist., Chiweshe communal land (TTL), Mwenje Dam, 12.ix.1970, *Mitchell* (K). C: Gweru, Fletcher High School, 22.xi.1962, *Loveridge* 497 (K, SRGH). E: Mutasa Dist., Honde Dip, 26.ix.1947 *Whellan* 255 (SRGH). S: Gwanda Dist., near Sisilala R., c. 14.4 km due W of Fort Tuli in old road to Macloutse (Maclautsi), 13.v.1959, *Drummond* 6129 (SRGH).

Also in South Africa (Eastern Cape, Free State, Gauteng, KwaZulu-Natal, Limpopo, Mpumulanga, Northern Cape and Western Cape) and Madagascar. In temporary wet situations in seasonal pans and damp grassland where vegetation is sparse; 400–1500 m.

Conservation notes: Widespread; not threatened.

This polymorphic species is widespread in southern Africa. *Fuirena coerulescens* can be recognised by the glabrous band, sometimes a different colour, and a ciliate margin to the glume. The rhizome is markedly wide and the culm triangular. The hypogynous whorls in particular show a wide range of variation throughout the distributional range. No evidence was seen of development of either the inner or outer perianth segments in the limited amount of material from Zambia and Zimbabwe. Inner perianth segments resemble those of *F. ecklonii* Nees but the arista arises from the blade apex.

3. **Fuirena stricta** Steud., Syn. Pl. Glumac. **2**: 128 (1855). —Clarke in Durand & Schinz, Consp. Fl. Afr. **5**: 648 (1894); in F.T.A. **8**: 465 (1902). —Gordon-Gray in Strelitzia **2**: 103, fig.41 (1995). Type: Madagascar, Sainte Marie, n.d., *Boivin* 1656 (P holotype, K).

Tufted annual or perennial; rhizome if present, short. Culms 30–100 cm tall, 1–2(4) mm diameter, trigonous, glabrous, rarely pubescent except below inflorescence. Leaf sheath glabrous; ligule puberulous to pubescent, mouth oblique; leaf blade 2–4(7) cm long, 2–4(5) mm wide, erect or reflexed, flat, short, linear-triangular, glabrous. Inflorescence a narrow panicle with 1–2 erect lateral branches with spikelet clusters, sometimes reduced to a terminal spikelet cluster. Spikelets 5–7(15) × 2–4 mm (awns included), distinctly or indistinctly 5–angled (ranked), less often unranked, blackish to red-brown. Glumes 2–3.5 mm long (including awn 0.5–0.6(0.8) mm), with very short hairs, appearing almost glabrous, 3-nerved, often only midnerve conspicuous. Perianth segments 6, in 2 whorls, both outer and inner segments of retrorsely scabrous bristles, as long or overtopping nutlet. Stamens 3. Nutlets brown, dark green or almost black, 1.3–1.5 × 0.6–0.8 mm (including stalk and conical beak), obovoid to ellipsoid, triangular-triquetrous, angles ridged, base cuneate, surface smooth.

Spikelets (ranked) 5-angled; mature nutlets usually brown. . . . . .a) subsp. *stricta*
Spikelets (not ranked) terete; mature nutlets usually dark green. . . . .
. . . . .b) subsp. *chlorocarpa*

a) Subsp. **stricta**. —Haines & Lye, Sedges & Rushes E. Afr.: 42, fig.37 (1983). —Muasya
    in Kew Bull. **53**: 192 (1998); in F.T.E.A., Cyperaceae: 10 (2010).

    Fuirena friesii Kük. in Fries, Wiss. Ergebn. Schwed. Rhodesia-Kongo-Exped., 1911–1912: 8
    (1921). Type: Zambia, near Mansa (Fort Roseberry), 1911–1912, *Fries* 614 (UPS holotype).

Botswana. N: Okavango Delta, near Gwetsa, Moanachira R., N of Gcobega lagoon, 31.xii.2009, *A. & R. Heath* 1799 (K). **Zambia**. N: Chinsali Dist., Shiwa Ngandu (Ishiba Nganda), 1524 m, 1.v.1956, *Robinson* 1516 (K); Isoka Dist., Nyika Plateau on path to N Rukuru waterfall, 2150 m, 27.x.1958, *Robson & Angus* 400 (K, LISC). W: Mwinilunga Dist., Mujileshi R., c. 9.6 km from Angolan border, 1290 m, 7.xi.1962, *Richards* 16962 (K).. S: Choma Dist., Mochipapa, 10 km SE of Choma, 20.vi.1978, *Heery* 21 (K). **Zimbabwe**. N: Guruve Dist., Nyamnyetsi Estate, 0.5 km S of Mawire Hill, 24.viii.1978, *Nyariri* 308 (K, SRGH). W: Matobo Dist., Besna Kobila, x.1956, *Miller* 3726 (K, LISC, SRGH). C: Harare (Salisbury), 29.xi.1931, *Brain* 7215 (K, LISC, SRGH). E: Mutare Dist., Bvumba, Eastern Beacon, 17.iv.1993, *Browning* 553 (GENT, K, NU, PRE). **Malawi**. N: Mzimba Dist., Vipya Plateau, 16 km SW of Mzuzu, c. 1370 m, 23.xii.1973, *Pawek* 7645 (K). S: Mangochi Dist., Cape Maclear, Chemba village, 13.viii.1987, *Salubeni & Patel* 5102 (K). **Mozambique**. Z: Gurué Dist., Mt Namuli, 1893 m, 31.v.2007, *Patel* 7398 (K). MS: Sussundenga Dist., S slopes of Serra Macuta, 700 m, 1.vi.1971, *Pope* 430 (K, LISC).

Also in West Africa, Sudan, Congo, Rwanda, D.R. Congo, Kenya, Uganda, Tanzania, South Africa (Gauteng, KwaZulu-Natal, Limpopo, Mpumalanga, North West) and Swaziland, Comores and Madagascar. Seasonally wet grassland, edges of permanent swamps, often in shallow water; 500–2200 m.

Conservation notes: Widespread; not threatened.

b) Subsp. **chlorocarpa** (Ridl.) Lye in Nordic J. Bot. **3**: 241 (1983). —Haines & Lye,
    Sedges & Rushes E. Afr.: 43, fig.38 (1983). —Muasya in Kew Bull. **53**: 192 (1998);
    in F.T.E.A., Cyperaceae: 10 (2010). Type: Angola, Huíla, Lopollo, iv.1860, *Welwitsch*
    7113 (?LISU holotype, BM).

    Fuirena chlorocarpa Ridl. in Trans. Linn. Soc. London, Bot. **2**: 159 (1884). —Clarke in
    F.T.A. **8**: 465 (1902).

    Fuirena stricta Steud. var. *chlorocarpa* (Ridl.) Kük. in Notizbl. Bot. Gart. Berlin-Dahlem **9**:
    310 (1925). —Hooper in F.W.T.A., ed.2 **3**(2): 326 (1972).

    Fuirena nyasensis Nelmes in Mem. New York Bot. Gard. **9**: 98 (1954). Type: Malawi,
    Nkotakota Dist., Ntchisi Mt, 27.vii.1946, *Brass* 16980 (K holotype).

Zambia. B: Kalabo Dist., Kalabo, 14.x.1962, *Robinson* 5485 (K). W: Mwinilunga Dist., Kalenda Plain, 12.xii.1937, *Milne-Redhead* 3655 (K). N: Mbala Dist., Chilongowelo, old Mpulungu road, c. 1460 m, 24.xii.1951, *Richards* 128b (K). C: Lusaka Dist., Munali, 8 km E of Lusaka, Little Dambo, 15.vi.1955, *Robinson* 1300 (K). S: Choma Dist., 35 km N of Choma, 17.v.1954, *Robinson* 753 (K). **Zimbabwe**. W: Matobo Dist., Besna Kobila, 1460 m, vi.1957, *Miller* 4405 (K). C: Marondera Dist., Grasslands Research Station, Grasslands vlei, 5.i.1949, *Corby* 337 (K, SRGH). E: Mutare Dist., S of Penhalonga, 1934, *Gilliland* K1209 (K). **Malawi**. N: Chitipa Dist., above Chisenga village towards foot of Mafinga Hills, 1585 m, 12.vii.1970, *Brummitt* 12047 (K, LISC). C: Chipata Dist.,

Chipata Mt (Chipata Hill), 4.v.1963, *Verboom* 122S (K). S: Zomba Dist., Domasi R., between Zomba and Ncheu, 20.viii.1950, *Jackson* 118 (K). **Mozambique**. N: Marrupa Dist., Okoewangoe, 16 km on road from Marrupa to Nungo, 6.viii.1981, *Jansen et al.* 98 (K). MS: Chimoio Dist., Bandula, 697 m, 6.iv.1954, *Chase* 4539 (K, SRGH).

Also in West Africa, Ethiopia, D.R. Congo, Kenya, Uganda, Tanzania, Angola and Madagascar. Seasonally wet grassland, edges of permanent swamps and streams; 600–1600 m.

Conservation notes: Widespread; not threatened.

Following F.T.E.A. (2010), infraspecific ranking has been given here, but may not be accepted elsewhere where collections are all placed under *Fuirena stricta sensu lato*.

Within the species there is sometimes difficulty in assigning collections to subspecies due to the number of intermediates and the characters used to define the subspecies (and previously varieties). Infraspecific categories are based mainly on the ranking of spikelets; those of typical subspecies being clearly ranked, while those in subsp. *chlorocarpa* are unranked. Colour of nutlets is used, but this is variable, possibly due to degree of maturity at collection time.

Fuirena stricta is easily distinguished from other *Fuirena* species by the usual presence of 6 well-developed bristles, which frequently exceed nutlet length and may be visible within the glumes of mature spikelets.

4. **Fuirena hirsuta** (P.J. Bergius) P.L. Forbes in J. S. Afr. Bot. **35**: 83 (1969). —Gordon-Gray in Strelitzia **2**: 99 (1995). Type: South Africa, Cape of Good Hope ('e Cap. b. sp.') sent to Bergius by M. Grubb, n.d., *Bergius Bot. Garden Herb.* (S). FIGURE 14.1**E**.

    Cyperus hirsutus P.J. Bergius, Descr. Pl. Cap.: 11 (1767).

    Scirpus hottentotus L., Mant. Pl. Alt. **2**: 182 (1771). Type: South Africa, 'Cap. b. spei', n.d., *Koenig* s.n. (LINN).

    Fuirena hirta Vahl, Enum. Pl. **2**: 387 (1805). Type: Uncertain, from South Africa, W Cape "Cap. b. spei".

    Fuirena glabra Kunth, Enum. Pl. **2**: 182 (1837). Type: South Africa, Cape, n.d., *Ecklon* 882 (K syntype, M, PRE).

Perennial, rhizome woody, 3–8 mm diameter, with culms up to 20 mm apart. Culms trigonous, 40–60(100) cm tall, glabrous, with short hirsute zone immediately below inflorescence. Leaf sheath shorter than internode, glabrous to hirsute; ligule a membranous collar, mouth oblique, usually hirsute, sometimes puberulous or glabrous; leaf blade up to 15 cm long and 4 mm wide, more or less erect, hirsute to subglabrous, apex acuminate. Inflorescence terminal with 1–3 globose, echinate heads 1.5 mm across, rarely with 1–2 long hirsute-stemmed lateral heads from node below inflorescence. Spikelets squarrose, 6–8 × 3–4 mm (excluding awns), ovoid to oblong. Glumes not ranked, 3.2 mm long (including 0.9–1 mm subterminal awn), oblong-obovoid to oblong-elliptic, puberulous, with or without long stiff hairs, keel rounded, excurrent into straight or recurved awn. Perianth segments 6, in 2 whorls, outer 3 segments of bristles with barbs mainly antrorse, inner 3 segments of scales as long or exceeding nutlet and beak, with markedly hairy stalks (claws); blade (lamina) square to oblong, 3-nerved, lateral margins ciliate, arista antrorsely and retrorsely hairy. Stamens 3. Nutlet dark yellow, 1.2–1.3 mm long (including beak and stalk, each 0.3 mm), 0.6 mm wide, broadly obovate, triquetrous, surface smooth (stalk and beak may be papillate).

Mozambique. M: Namaacha Dist., Namaacha (Namahacha), 30.vi.1961, *Balsinhas* 505 (K, LISC, PRE).

Also in South Africa (Eastern and Western Cape, KwaZulu-Natal, Mpumalanga) and Swaziland. Wet sandy areas along the coast and on riverine fringes; 5–100 m.

Conservation notes: Widespread, but localised within the Flora area; not threatened.

In N KwaZulu-Natal and Mozambique the glumes generally lack the long stiff hairs [Forbes in J. S. Afr. Bot. **35**: 90–91 (1969)].

Less well known synonyms have been omitted.

5. **Fuirena ecklonii** Nees in Linnaea **10**: 143 (1835). —Gordon-Gray in Strelitzia **2**: 99, fig.38C,F (1995). Types: South Africa, Cape Province, Olifantshoek by Bushman R., Uitenhage, n.d., *Ecklon* s.n. (PRE syntype, P); Eastern Cape, near Grahamstown (Albany), *Ecklon* s.n. (PRE? syntype); Mpumalanga, Quaggas vlei, *Ecklon* s.n. (PRE? syntype). FIGURE 14.1F.

    Fuirena mollicula Kunth, Enum. Pl. **2**: 182 (1837), also including var. B. Types: South Africa, Cape Province, Omtando, 25.ii.1832, *Drège* 4340 (P syntype); Cape Province, Koratra, 29.ix.1831, *Drège* 2039 (P syntype).

    Fuirena coerulescens Steud. var. *glabrescens* Schönl. in Mem. Bot. Surv. S. Afr. **3**: 51 (1922). Types: South Africa, N Cape, Olifantshoek, *Ecklon & Zeyher* s.n. (P, PRE?); E Cape, Bethelsdorp, *Paterson* 347; E Cape (Transkei), Ibisi R., *Wood* 3144; KwaZulu-Natal, Clairmont, *Wood* 9969; KwaZulu-Natal, Kokstad, *Tyson* 1469, herbaria uncertain.

Perennial, rhizome woody, 2–3.5 mm diameter, scales hirsute or hirsute-pubescent, culms contiguous, 2–15 mm apart. Culms trigonous, 30–60 cm tall, pilose or pilose distally and sparsely pilose to glabrescent proximally, or hairs present mainly or only in distal part of each internode and/or at angles. Leaf sheath covering much of culm, glabrous or with short hairs near mouth only; ligule with very short hairs, mouth ciliate on side adjacent to blade; leaf blade suberect to spreading, 8–20 cm long, 4–7 mm wide, glabrous adaxial side, midrib and margins abaxial side pilose (hairs long), apex attenuate. Inflorescence usually terminal with spikelets shortly pedicellate in corymbose cluster or sessile in a head, or paniculate with 1–2 lateral branches at nodes below terminal portion; inflorescence stems and bracts pilose to glabrescent. Spikelets squarrose, (6)9–13 × 3–5 mm (awns excluded), ovoid to ellipsoid, apex acute or obtuse. Glumes usually unranked, occasionally 5-ranked, pubescent, 4.6–6 mm long including (1.1)1.5–2.5(3) mm awn, recurved. Perianth segments in 2 whorls, outer 3 segments as long as scale claw, reduced, extrorsely to retrorsely scabrous; inner 3 segments well-developed, equal or slightly shorter than nutlet, rarely smaller; claw ¼–½ as long as blade, obovate, margins irregularly lobed and unevenly thickened with few papillae on lobes; blade apex thickened with extrorsely scabrous arista arising on inner surface. Stamens 3. Nutlet white or brownish at maturity, 1.6–2.5 mm long including stalk and beak, obovoid, trigonous with longitudinal angles ridged, beak conical papillate, up to 0.5 mm long; surface cells appearing polygonal, irregularly arranged, becoming hexagonal with age.

There has been confusion between *Fuirena ecklonii* and *F. coerulescens*; the difference can only convincingly be decided by examination of the nutlets. Such confusion is most likely to occur in southern African collections and not in those present in the Flora area (see note under *F. coerulescens*).

Subsp. **barotsia** Xanthos & Browning, subsp. nov. Distinguished from the typical subspecies by the relatively lax, less squarrose inflorescence; longer, narrower, sometime pedicellate spikelets with more numerous florets; glumes with shorter, straight awns and better developed perianth segments. Type: Zambia, Mongu Dist., Lake Lutende, E of Mongu, 6.xi.1965, *Robinson* 6703 (K holotype, NU).

Perennial, rhizome woody, 6–8 mm diameter, scales glabrous, culms contiguous, 2–15 mm apart. Culms triangular, 30–60 cm tall, glabrous to pilose or pilose distally. Leaf sheaths numerous, covering much of culm, glabrous, with short hairs only near mouth; ligule with very short hairs; leaf blade suberect to spreading, 40–140 mm long, 3–7 mm wide, glabrous adaxially, pilose on midrib and margins abaxially, apex attenuate. Inflorescence paniculate with 1–2 lateral branches at nodes below terminal portion, seldom capitate; inflorescence stems and bracts pilose to glabrescent. Spikelets 15–20 × 3–4 mm, oblong-oblanceolate, apex acute or obtuse. Glumes unranked, pubescent, 3.3–4 mm long including 1–1.2 mm long straight awn. Perianth segments 6, in 2 whorls, outer 3 segments represented by bristles, extrorsely to retrorsely scabrous, as long or longer than nutlet, inner 3 segments well-developed, equal or slightly shorter than nutlet, rarely smaller; claw ¼ length of blade, blade obovate, margins irregularly lobed and unevenly thickened, with few papillae on lobes, apex thickened with arista arising on inner side, arista extrorsely scabrous. Stamens 3. Nutlet white or brownish at maturity, 1.4–1.5 × 0.8 mm including stalk and beak, obovoid, trigonous with longitudinal angles ridged, beak conical papillate.

Zambia. B: Mongu Dist., 40 km NE of Mongu, 12.xi.1965, *Robinson* 6735 (K); Lake Lutende c. 22.5 km E of Mongu, 18.xi.1959, *Drummond & Cookson* 6601 (K, SRGH). W: Mwinilunga Dist., Matonchi dambo, 26.x.1937, *Milne-Rehead* 2966 (K).

Only in W Zambia. Very wet situations, margins of lakes, ponds and water bodies; 1000–1300 m.

Conservation notes: Only known from W Zambia; possibly Near Threatened.

This subspecies is found at higher elevations than subsp. *ecklonii*, which only occurs in South Africa. It is distinguished by the paniculate inflorescence, with ± 1 lateral branch below the terminal pedicellate cluster of loosely arranged spikelets, glumes with straight awns c. 1 mm long, and both outer and inner perianth segments present and well-developed.

6. **Fuirena pachyrrhiza** Ridl. in Trans. Linn. Soc. London, Bot. **2**: 161 (1884). —
     Clarke in Durand & Schinz, Consp. Fl. Afr. **5**: 647 (1894); in Fl. Cap. **7**: 262
     (1897); in F.T.A. **8**: 464 (1902). —Gordon-Gray in Strelitzia **2**: 101, fig.39B,E
     (1995). —Muasya in Kew Bull. **53**: 204 (1998); in F.T.E.A., Cyperaceae: 13 (2010).
     Types: Angola, Pungo Andongo, near Muta Locala, iii.1857, *Welwitsch* 7117 (LISU
     syntype, BM); between Cagui and R. Cuanza, iv.1857, *Welwitsch* 7118 (BM syntype,
     LISU).

   Fuirena macrostachya Boeckeler in Bot. Jahrb. Syst. **5**: 507 (1884). Type: Tanzania, Tabora
   Dist., Igonda (Gonda), i.1882, *Böhm* 73a (B† holotype, CORD, K).

   Fuirena pubescens (Poir.) Kunth var. *major* Lye in Bot. Not. **127**: 112 (1974). —Haines &
   Lye, Sedges & Rushes E. Afr.: 51, fig.60 (1983). Type as for *F. pachyrrhiza*.

Perennial, rhizome woody, 4–6 mm diameter, culms contiguous or 5–10 mm apart. Culms sharply triangular, 30–80 cm tall, 1–2 mm wide, velutinous-pubescent or hispidulous, or glabrescent proximally where hairs often present only on angles. Leaf sheath velutinous-glabrescent, proximal sheaths often sparsely hairy; ligule hairy; leaf blade 5–14 cm long, 5–8 mm wide, velutinous-pubescent or hispidulous, margins revolute in dried material, midrib projecting abaxially. Inflorescence paniculate, consisting of terminal partial inflorescence and 1–3 lateral branches at 1–4 nodes below terminal portion, spikelets in heads or loosely arranged, inflorescence stems and bracts velutinous-pubescent or hispidulous. Spikelets 7–15 × 3–5 mm excluding awns, ellipsoid to terete. Glumes usually not ranked or indistinctly or distinctly 5-ranked, 3.5–7 mm long (including 0.1–2 mm awn), pubescent with short hairs, awn arising 0.1–0.2 mm below apex, straight or slightly recurved. Perianth segments absent (if present, variable even within a spikelet of 3 slender vestigial outgrowths, of either or both outer and inner hypogynous whorls). Stamens 3. Nutlet body darker (brown to black) than the spongy and often shrunken white beak, 1.3–1.6 × 0.8–0.9 mm long (beak and stalk included), obovoid, trigonous, longitudinally ridged at angles with a few small scattered tubercles.

Botswana. N: North East Dist., 26 km S of Francistown on main road to Palapye, 16.iv.1978, *P.A. Smith* 2382 (PRE, PSUB). **Zambia**. N: Mbala Dist., Saisi R. marsh, 1500 m, 27.ii.1957, *Richards* 8348 (K). C: Lusaka Dist., 9.6 km E of Lusaka, 20.xii.1957, *King* 394 (K). S: Choma Dist., SW of Choma, c. 1310 m, 2.i.1955, *Robinson* 1041 (K). **Zimbabwe**. W: Hwange Dist., Dete vlei, 14.xi.1956, *Lovemore* 501 (K, LISC, SRGH). C: Kwekwe Dist., 50 km NE of Kwekwe, Iwaba Estate, 1195 m, 9.ii.2000, *Lye* 23776 (K, SRGH). E: Mutare Dist., Quagga's Hoek commonage, 22.iii.1953, *Chase* 4884 (K, SRGH). S: Mwenezi Dist., Malangwe R., SW of Mateke Hills, 6.v.1958, *Drummond* 5658 (K, SRGH). **Malawi**. C: Dedza Dist., Kachera Village, 8.i.1959, *Jackson* 2269 (K). S: Blantyre Dist., Namadzi (Namasi) R., 20.xi.1899, *Cameron* 11 (K). **Mozambique**. MS: Sussundenga Dist., 2 km from Dombe, 28.x.1953, *Gomes Pedro* 4501 (LISC, PRE).

Also in Uganda, Kenya, Tanzania, Angola, South Africa (KwaZulu-Natal, Limpopo, Mpumalanga) and Swaziland. In damp situations such as margins of streams and dams; 100–1600 m.

Conservation notes: Widespread; not threatened.

Forbes (1980) recognised but did not publish two varieties of *Fuirena pachyrrhiza*; of these only the northern one with less echinate spikelets is present in the Flora region.

Fuirena pachyrrhiza may be distinguished from *F. pubescens* and *F. welwitschii* by the velvety (velutinous) texture of the culms and leaves (seen at × 40) that is lacking from only small areas of the plant.

7. **Fuirena pubescens** (Poir.) Kunth, Enum. Pl. **2**: 182 (1837). —Clarke in Durand & Schinz, Consp. Fl. Afr. **5**: 648 (1894); in F.T.A. **8**: 463 (1902). —Haines & Lye, Sedges & Rushes E. Afr.: 50, figs.58,59 (1983). —Gordon-Gray in Strelitzia **2**: 102, figs.39C,F, 40 (1995). —Muasya in Kew Bull. **53**: 193 (1998); in F.T.E.A., Cyperaceae: 11, 12, fig.2, 3 (2010). Type: NE Algeria (Numidia), near La Calle, n.d., *Poiret* s.n. (P holotype). FIGURE 14.**2**.

    Carex pubescens Poir., Voy. Barbarie **2**: 254 (1789).
    Scirpus pubescens (Poir.) Lam., Tabl. Encycl. **1**: 139 (1791).
    Isolepis pubescens (Poir.) Roem. & Schult., Syst. Veg. **2**: 118 (1817).

Perennial, rhizome woody 2–3(4) mm diameter, culms contiguous, 20 mm apart. Culms triangular, 30–100 cm tall, 1–2 mm wide; pilose when hairs sparse or glabrous, except sometimes below inflorescence. Leaf sheath glabrous or with short hairs near mouth only; ligule hairy (short or long hairs) or completely glabrous; leaf blade suberect to spreading, 5–17 cm long, 5–7 mm wide; flat or V-shaped, glabrous or with short soft hairs on margin and midrib of upper surface, or long hairs present on both surfaces or lower surface only. Inflorescence terminal with spikelets shortly pedicellate in corymbose clusters or sessile in a head, or paniculate with 1–2 lateral branches at nodes below terminal portion; inflorescence stems and bracts puberulous to pilose. Spikelets 6–10 × 3–5 mm, ovoid, ellipsoid or terete, apex acute or obtuse. Glumes not ranked or distinctly or indistinctly 5-ranked, 3–5 mm long including 0.8–2 mm awn, pubescent with short hairs. Perianth segments absent, or more rarely outer 3 segments and inner 3 segments vestigial. Stamens 3. Nutlet brown to greenish-black at maturity, 1.2–1.6(1.8) mm long (including stalk and conical, white beak up to 0.2 mm long), 0.8–1 mm wide, obovoid, trigonous with ridged longitudinal angles; surface cells appearing polygonal, becoming hexagonal with age.

Botswana. N: Okavango swamps, Chief's Is., c. 2 km SW of Chief's Camp, 14°18.583'S 22°54.171'E, 975 m, 12.ix.2011, *A. & R. Heath* 2165 (K). **Zambia**. B: Senanga Dist., Mashi R., in salt pan, 30.x.1964, *Verboom* 1766 (K). W: Ndola Dist., Itawa dambo, 25.v.1950, *Jackson* 35 (K). N: Kasama Dist., path from Kalolo village to Chambeshi Flats, 4 km from Kalolo, 9.xii.1964, *Richards* 19320 (K). C: Chongwe Dist., Chakwenga headwaters, 100–129 km E of Lusaka, 5.iii.1965, *Robinson* 6401 (K). S: Mazabuka Dist., Kafue Flats, 18.vi.1955, *Robinson* 1312 (K). **Zimbabwe**. N: Guruve Dist., Nyamunyeche (Nyamnyetsi) Estate, E of station, 7.ix.1978, *Nyariri* 333 (K, SRGH). C: Harare, Borrowdale, source of Umwindsi R, c. 1430 m, 7.xii.1974, *Burrows* 631 (K, SRGH). E: Mutare Dist., Nyamakwarara Valley, Stapleford, 1.xi.1967, *Mavi* 350 (K, SRGH). S: Masvingo Dist., Makoholi Exp. Station, 18.xi.1976, *Senderayi* 59 (K, SRGH). **Malawi**. N: Mzimba Dist., 5 km W of Mzuzu, grounds of Bishop's house at Katoto, 1310 m, 23.v.1970, *Brummit & Pawek* 11079 (K). C: Dowa Dist., Mwera Hill Station, 31.x.1950, *Jackson* 224 (K). S: Shire Highlands, n.d., *Adamson* 106 (K). **Mozambique**. MS: Gondola Dist., 38 km from camp along track from Inchope to R. Revuè, 3.xi.1953, *Gomes Pedro* 4557 (LISC, PRE).

Also in NW India, Mediterranean, the Mascarenes, and across N & W and Central Africa, Uganda, Kenya, Tanzania, Namibia, South Africa (Eastern Cape, Free State, Gauteng, KwaZulu-Natal, Limpopo, Mpumalanga, Northern Cape, North West), and Swaziland. Seasonally wet marshy areas, margins of lake and swamps; 700–1500 m.

Conservation notes: Widespread; not threatened.

Fuirena pubescens is a polymorphic species well represented in the Flora area. There appears to be a continuous range of variation, an absence of reliable characters for separation, and intermediates that are difficult to place. Some authors have distinguished three separate species and varieties within them, namely *F. pubescens, F. pachyrrhiza* and *F. welwitschii* (Muasya in F.T.E.A. 2010). Others, such as Lye (Fl. Ethiopia **6**: 395, 1997), have *F. pachyrrhiza* and *F. welwitschii* as synonyms of *F. pubescens*. In this account the species are kept separate and varieties within them are not accounted for.

Fig. 14.**2**. FUIRENA PUBESCENS. 1, habit; 2, leaf sheath apex; 3, inflorescence; 4, glume lateral view; 5, floret; 6, nutlet. 1–3 from *Taylor* 36; 4–6 from *Browning* 165. Scale bars: 1 = 40 mm; 2, 3 = 5 mm; 4–6 = 1 mm. Drawn by Jane Browning. Reproduced from Strelitzia (1995) by kind permission of the South African National Biodiversity Institute, Pretoria.

Gordon-Gray (1995) suggests the following may be used to separate *F. pubescens* from *F. pachyrrhiza* – "Plants may be recognised by: the well-developed elongate rhizome and indumentums of uniform, short, patent hairs limited to small portions such as culms, or nodes, or midribs and margins of leaf blades or inflorescences; reduced hypogynous whorl consisting of minute outgrowths, or these may be entirely absent."

8. **Fuirena welwitschii** Ridl. in Trans. Linn. Soc. London, Bot. **2**: 161 (1884). —Clarke in Trans. Linn. Soc. London, Bot. **4**: 54 (1894); in Durand & Schinz, Consp. Fl. Afr. **5**: 648 (1894); in F.T.A. **8**: 463 (1902). —Muasya in Kew Bull. **53**: 195 (1998); in F.T.E.A., Cyperaceae: 13, 14 (2010). Types: Angola, Pungo Andongo near Quibanga, i.1857, *Welwitsch* 7108 (BM syntype); Huíla, Morro de Monino, iv.1860, *Welwitsch* 7109 (BM syntype); near Lopollo, i.1860, *Welwitsch* 7114 (BM syntype); near Eme, ii.1860, *Welwitsch* 7115 (BM syntype).

   Fuirena buchananii Boeckeler, Beitr. Cyper. **1**: 20 (1888). Type: Malawi, Shire R., n.d., *Buchanan* s.n. (BM holotype, K).
   Fuirena pubescens (Poir.) Kunth var. *buchananii* (Boeckeler) C.B. Clarke in Durand & Schinz, Consp. Fl. Afr. **5**: 648 (1894). —Haines & Lye, Sedges & Rushes E. Afr.: 51 (1983).

Perennial, rhizome woody, 2.5–4 mm diameter, culms arising 4–15 mm apart or almost contiguous. Culms triangular, 20–60 cm tall, 1–2 mm wide, glabrous except below inflorescence. Leaf sheath 3-angled, glabrous or pubescent; ligule membranous with a rim of short hairs; leaf blade 14–24 cm long, 2–3(4) mm wide, very short hairs sometimes present in angle between sheath and abaxial side of blade, on blade margins and midrib and on ligule, otherwise leaves glabrous. Inflorescence a subdigitate cluster of spikelets. Spikelets 8–20 mm long, ovoid. Glumes distinctly or indistinctly 5-ranked, 4–5 mm long (including 1.5–1.75(2) mm awn, straight or slightly recurved). Perianth segments generally absent, if present both outer 3 and inner 3 segments usually rudimentary. Stamens 3. Nutlet white to light brown, 1.5–1.6 mm long (including beak and base) × 0.7–0.8 mm wide, trigonous, surface with a few tubercles, otherwise smooth.

Zambia. W: Mwinilunga Dist., SW of Dobeka (Jobeka) Bridge, 13.x.1937, *Milne-Redhead* 2745 (K). N: Mbala Dist., 80 km S of Mbala (Abercorn), 12.xi.1960, *Robinson* 4061(K). C: Kabwe Dist., 19 km N of Kabwe (Broken Hill), 23.ix.1947, *Brenan & Greenway* 7944 (K). E: Chipata Dist., Ngoni area, 10.iii.1963, *Verboom* 88 (K). S: Choma Dist., Muckle Neuk, 19 km N of Choma, 14.i.1954, *Robinson* 469 (K). **Zimbabwe**. N: Makonde Dist., Banket, 23.xi.1944, *Hopkins* GH 13063 (K, SRGH). W: Matobo Dist., Besna Kobila, ix.1957, *Miller* 4549 (K, LISC, SRGH). C: Harare, Arundel vlei, 9.6 km N of Harare, 10.x.1957, *Seagrief* 3079 (K, P, SRGH). E: Chimanimani Dist., Stonehenge plateau, 1680 m, 2.ii.1957, *Phipps* 410 (K, SRGH). **Malawi**. N: Nkhata Dist., Vipya Plateau, 56 km SW of Mzuzu, Elephant Rock dambo, c. 1680 m, 27.iii.1976, *Pawek* 10949 (K, MO, SRGH). **Mozambique**. MS: Beira, n.d., *Rogers* 5463 (K).

Also in Angola, Uganda, Kenya and Tanzania, possibly in South Africa (former Transvaal and KwaZulu-Natal) but not listed due to possible inclusion with *F. pubescens*. Seasonally wet grassland, frequently in black mud at edges of permanent pools and streams and seepage areas; 10–1700 m.

Conservation notes: Widespread; not threatened.

Plants of *Fuirena welwitschii* can be recognised by the narrow leaves compared to *F. pubescens* and *F. pachyrrhiza*, to which it is very closely related. In comparison to the latter species the short hairs are infrequently present on ligule and leaf blade margins.

9. **Fuirena umbellata** Rottb., Descr. Icon. Rar. Pl.: 70, t.19 (1773). —Kunth, Enum. Pl. **2**: 185 (1837). —Clarke in Durand & Schinz, Consp. Fl. Afr. **5**: 648 (1894). —Haines & Lye, Sedges & Rushes E. Afr.: 49, fig.55 (1983). —Gordon-Gray in Strelitzia **2**: 104, fig.41B,D (1995). —Muasya in Kew Bull. **53**: 201 (1998); in F.T.E.A., Cyperaceae: 20 (2010). Type: Surinam, no locality, *Rowlander* s.n.

(C lectotype); Rottbøll (1773) cites several pre-Linnean syntypes. FIGURE 14.**1L**.

Fuirena oedipus C.B. Clarke in J. Linn. Soc., Bot. **37**: 478 (1906). Type: Zimbabwe, Victoria Falls rain forest, ix.1905, *Gibbs* 125 (BM holotype).

Perennial rhizome woody, 4–5 mm diameter, culms contiguous to 5 mm apart. Culms 5-angled, 50–170 cm tall, 5–10 mm wide, glabrous to hirsute distally, bases sometimes swollen and corm-like. Leaf sheath 5-angled, glabrous to hirsute (reduced leaf sheaths covered in short white hairs are not obviously 5-angled); ligule membranous, mouth oblique, cilate, sometimes with long hairs; leaf blades 9–26 cm long, 3–20 mm wide, with 5 large veins, glabrous or upper surface minutely pubescent, margins with translucent hairs. Inflorescence rarely terminal, usually a panicle 7–30 cm long with 2–5 densely pubescent branches subtended by leafy bracts, each further branched ending in digitate clusters of spikelets in globose heads to 15 mm across, sometimes reduced to one head. Spikelets 5–8 × 1.5–3 mm, ovoid, terete, apex acute to sometimes obtuse, green turning brown. Glumes 2.5–4 mm long including 0.5–1(2) mm awn, with short and long hairs, prominently 3-nerved. Perianth segments 3, in 1 whorl, outer 3 segments absent or vestigial, inner 3 segments subsessile (claw very short); blade oblong-obcordate, 3-veined, apical lobes fimbriate, short arista arising from apical notch. Stamens 3. Nutlet brown, 0.9–1.5 (including papillate stalk and beak) × 0.5–0.8 mm, triangular-triquetrous, angles ridged, surface smooth.

Botswana. N: Ngamiland Dist., Kwando R. floodplain, 27.vii.1973, *P.A. Smith* 681 (K). **Zambia**. B: Mongu Dist., Mongu, 14.i.,1960, *Gilges* 907 (K). N: Chinsali Dist., Shiwa Ngandu (Ishiba Ngandu), 5.vi.1956, *Robinson* 1606 (K, SRGH). C: Mkushi Dist., David Moffat's Farm, Munchiwemba dambo, 20.ix.1993, *Bingham & Nkhoma* 9712 (K). **Zimbabwe**. W: Hwange Dist., Victoria Falls rainforest, 6.i.1954, *Robinson* 1075 (K, SRGH). C: Kadoma Dist., 'Chengiri Native Purchase Area', Gadze R., 5.ix.1963, *Bingham* 812 (K, SRGH). E: Mutare Dist., Nyamkwara (Nyumquara) Valley, ii.1935, *Gilliland* K1499 (K). **Malawi**. N: Mzimbe Dist., Mzuzu, Marymount dambo, c. 1370 m, 20.iii.1974, *Pawek* 8233 (K). C: Mchinji Dist., Tembwe, banks of Bua R. on Lilongwe–Mchinji road, 1075 m, 30.iii.1970, *Brummitt* 9552 (K). **Mozambique**. MS: Sussundenga Dist., 28 km W of Dombe, base of SE slopes of Chimanimani, 25.iv.1974, *Pope & Müller* 1302 (K, LISC). GI: Chibuto Dist., Chongoéne, experimental fields (campo de ensaios), 9.vii.1948, *Myre* 58 (K). M: Matutuine Dist., Zitundo, 15. xi.1979, *de Koning* 7624 (K).

Widespread in tropical Africa from West Africa, Uganda, Kenya, Tanzania to Namibia, South Africa (KwaZulu-Natal); also in India, SE Asia, Polynesia, tropical South and Central America. Common on riverbanks and other moist places; 5–1600 m.

Conservation notes: Widespread; not threatened.

A full list of African synonyms that apply to this taxon is extensive and can be found in Forbes (1980).

The species can generally be identified by the 5-angled culm and leaf blade with 5 prominent veins (sometimes 7). In unfavourable growing conditions the culm bases may become swollen and corm-like. If growing with *F. obcordata* the two taxa may be confused – dissection will show the presence of basal and apical hairs on the scales of the inner hypogynous whorl of *F. obcordata*.

Depauperate examples of *F. umbellata* may also be confused with *F. ciliaris* but examination of *F. umbellata* scales shows the inner perianth segments have blades that are subsessile (very short claws).

10. **Fuirena obcordata** P.L. Forbes in S. African J. Bot. **53**: 185 (1987); in S. African J. Bot. **52**: 237–240 (1986), invalidly published. —Gordon-Gray in Strelitzia **2**: 100, fig.39A,D (1995). Type: South Africa, Durban, 4.iv.1832, *Drège* 4339 (B† holotype, P lectotype), lectotypified by Forbes (1987). FIGURE 14.**1H**.

Fuirena glabra Krauss in Flora **28**: 757 (1845), invalid name.

Perennial, rhizome woody, 4–6 mm diameter, culms contiguous or up to 15 mm apart. Culms 0.5–1.4 m tall, 3–4 mm wide, 3-angled, glabrous, bases sometimes swollen and corm-like. Leaf sheath shorter than internode, glabrous but proximal sheaths maybe hirsute; ligule a tubular membranous collar, 1 mm high, glabrous apart from short hairs abaxially and adaxially forming an apical rim; leaf blade to 20 cm long, 2–3 mm wide, glabrous, erect or suberect, flat with midrib projecting abaxially, margin often revolute. Inflorescence a terminal globose head, c. 8 mm wide of numerous sessile, subsessile and pedunculate spikelets; subtending bracts as long or longer than inflorescence, with long white hairs. Spikelets 5–8 mm long excluding awn, ovoid, light-brown when dry, green and white when fresh. Glumes elliptic to obovate-elliptic, 3–4 mm long including 1–1.2 mm awn, short hairs interspersed with stiff bristles at base of awn, 3-nerved. Outer 3 perianth segments vestigial, represented by small outgrowths on receptacle below each stamen, inner 3 segments of substipitate, aristate scales, equal to nutlet in length; claw short with few hairs, scale blade obcordate, 3-veined, midvein curving inwards and uniting with marginal veins, apical margins incurved and fimbriate, arista short, scabridulous, incurved over nutlet. Stamens 3. Nutlet dull brown, 1–1.3 mm long (including beak and stalk), 0.6–0.8 mm wide, trigonous, obovoid, beak cylindrical papillate, surface appearing tessellated.

Caprivi. Kwando R., Singalamwe, 1006 m, 31.xii.1958, *Killick & Leistner* 3221 (K, PRE). **Zambia**. C: Mpika Dist., Mutinondo Wilderness Area, main bridge, 1424 m, 1.iv.2013, *Merrett* 1381 (K). **Mozambique**. GI: Inhassoro Dist., Bazaruto Is, 7.xi.1958, *Mogg* 28847 (K). M: Maputo, Inhaca Is, 4.iii.1958, *Mogg* 31409 (K, J).

Also in Namibia and South Africa (Eastern Cape, KwaZulu-Natal and Mpumalanga). Very wet places, such as rivers and swamp margins; 5–1500 m.

Conservation notes: not threatened.

In South Africa and Zambia may be sympatric with *F. umbellata*, from which it is distinguished from *F. obcordata* by the 5-veined leaf blade.

11. **Fuirena ochreata** Kunth, Enum. Pl. **2**: 184 (1837). —Raynal & Roessler in Mitt. Bot. Staatssamml. München **13**: 357 (1977). —Haines & Lye, Sedges & Rushes E. Afr.: 48, fig.53, 5 (1983). —Muasya in Kew Bull. **53**: 200 (1998); in F.T.E.A., Cyperaceae: 20 (2010) Type: Zanzibar, n.d., *Bojer* s.n. (?B† holotype, K, M, P). FIGURE 14.**1J**.

    Fuirena cinerascens Ridl. in Trans. Linn. Soc. London, Bot. **2**: 161 (1884). —Clarke in F.T.A. **8**: 467 (1902). Types: Angola, Huíla, Catumba, n.d., *Welwitsch* 7110 (BM); Zanzibar, ix.1873, *Hildebrandt* 1057 (K syntype, BM, P); Tanzania, Usambara Mts, Kibafuta swamp, ii.1893, *Holst* 2133 (K syntype); Tanzania, Usarmo, 7.iv.1894, *Kuntze* 283 (K syntype); Zanzibar, n.d., *Stulmann* 52 (K syntype, BM); Usarmo, n.d., *Kirk* 62 (K lectotype), lectotypified by Verdcourt on sheet.

    Fuirena calolepis K. Schum. in Abh. Königl. Akad. Wiss. Berlin **1894**: 20 (1894). Types: Tanzania, Usambara Mts, Kibafuta swamp, ii.1893, *Holst* 2133 (B† lectotype, K) from four syntypes, lectotypfied on sheet by Verdcourt.

    Fuirena cristata Turrill in Bull. Misc. Inform. Kew **1914**: 170 (1914). Type: Angola, Benguella, n.d., *Gossweiler* 2204 (K syntype); Benguella, n.d., *Gossweiler* 2166 (K syntype).

Perennial, rhizome woody, 5–8 mm diameter, culms up to 10 mm apart. Culms terete-trigonous, 20–70 cm tall, longitudinally ribbed, densely pubescent or glabrous, except below inflorescence. Leaf sheath usually hirsute, sometimes glabrescent; ligule 0.5 mm long with stiff straight hairs of varying length on mouth; leaf blade 8–16 cm long, 7–12 mm wide, linear lancelolate, densely pubescent, 9–12 veins projecting slightly on abaxial surface, midrib absent. Inflorescence a panicle to 14 cm long, of 1–2 lateral branches bearing several spikelet heads. Spikelets squarrose, 7–11 × 2–4 mm, ellipsoid to ovoid. Glumes not ranked, 4.5–5.3 mm long including 1.9–2.5 mm subterminal awn, with short and long hairs, prominently 3-nerved. Perianth segments 6, in 2 whorls; outer 3 segments short, bristle-like, incurved (difficult to see), inner 3 segments 1.2–1.4 mm long, hairy stalk (claw) 0.4 mm long, blade trinervate, oblong to oval, lateral margins broadly winged, apex crested, crest funnel-shaped with mouth oblique on inner side and with a small barbed appendage at centre of outer edge. Margins and wings papillate. Stamens 3, anthers c. 1.5 mm long. Nutlet yellow-brown, 1–1.1 mm long (including beak and stalk) × 0.6 mm wide, triquetrous in cross-section, base cuneate, surface smooth.

Zambia. N: Mbala Dist., Kaputa, 18.x.1949, *Bullock* 1325 (K). **Mozambique**. MS: Muanza Dist., Cheringoma, 60 km from Beira on road to Inhaminga, 27.iv.1942, *Torre* 4010 (BR, LISC, LMA).

Also in Angola, Kenya and Tanzania. In seasonally wet ground; sea-level to c. 250 m
Conservation notes: Scattered distribution; probably not threatened.

Fuirena ochreata and *F. bernieri* Cherm. from Madagascar closely resemble each other; Forbes (1980: 147 & pers. comm.) has pointed out that it is questionable if they should be kept as two separate species.

12. **Fuirena claviseta** Peter in Abh. Königl. Ges. Wiss. Göttingen, Math.-Phys. Kl. ser.2 **13**(2): 50 (1928). —Raynal & Roessler in Mitt. Bot. Staatssamml. München **13**: 356 (1977). —Haines & Lye, Sedges & Rushes E. Afr.: 48, fig.51 (1983). —Muasya in Kew Bull. **53**: 198 (1998); in F.T.E.A., Cyperaceae: 16 (2010). Type: Tanzania, Uzaramo Dist., Dar es Salaam, Geresane, 2.iv.1926, *Peter* 39372 (K lectotype, B, P), lectotypified by Haines & Lye (1983). In original reference Peter gives 'Usaramo' but cites no specimens. FIGURE 14.**1D**.

Sparsely tufted annual or short-lived perennial; sometimes with a contracted rhizome. Culms 2–10 per plant, 30–90 cm tall, 1.5–3 mm wide, ribbed, glabrous, except below inflorescence, sometimes becoming woody at base. Lower leaf sheath pubescent, upper glabrous except at mouth; ligule with sparse long or short hairs and/or ciliate on mouth; leaf blade 7–11 cm long, 3–10 mm wide, lanceolate, many-nerved, but without distinct midrib, glabrous, but margins with long translucent hairs. Inflorescence sometimes terminal, usually paniculate, with terminal cluster and 1–2(3) lateral branches, each with a head of 1–7 spikelets; inflorescence stems pubescent distally. Spikelets 7–10 × 3–4 mm (including short awn), cylindrical to narrowly ovoid, apex rounded. Glumes not ranked, 2.4–2.9 mm long including 0.2–0.3 mm straight awn, broadly elliptical or oblong-ovoid with short and long hairs, nerves converging in a scarcely excurrent awn. Perianth segments 6, in 2 whorls, outer 3 segments, retrorsely scabrid bristles half nutlet length, inner 3 segments overtopped by nutlet, smooth claw equalling blade in length; blade lunate, inflated except for narrow band on proximal side with 3 inconspicuous nerves in centre, disappearing into swollen portion above and with the outer edges produced into long downwardly pointing lobes. Stamens 3, anthers, linear 0.5 mm long. Nutlet light brown, elliptical, sharply trigonous, 1–3 (including beak and stalk) × 0.6–0.7 mm, surface smooth, sometimes faintly lineolate.

Mozambique. MS: Gorongosa Dist., Gorongosa Nat. Park (Game Reserve), W slopes Cheringoma Plateau, Mussambidzi Falls, 100 m, 12.xi.1971, *Ward* 7443 (K, NU); Muanza Dist., Cheringoma Section, lower Chimizíua R., 1 km from Gana, 40–45 m, 13.vii.1972, *Ward* 7887 (K, J).

Also in Kenya, Tanzania, Zanzibar Is. and Madagascar. Near to and on the margins of open water and on damp ground; 40–100 m.

Conservation notes: Widespread, although local in the Flora area; not threatened.

Spikelets of *Fuirena claviseta* are generally not squarrose as the species has very short, straight awns. It could be confused with *F. angolensis*, but the leaves have long translucent hairs along the margins.

13. **Fuirena zambesiaca** Lye in Bot. Not. **127**: 109 (1974). —Raynal & Roessler in Mitt. Bot. Staatssamml. München **13**: 357 (1977). —Haines & Lye, Sedges & Rushes E. Afr.: 43, figs.14b,39 (1983). —Muasya in Kew Bull. **53**: 200 (1998); in F.T.E.A., Cyperaceae: 19 (2010). Type: Mozambique, Marrupa, 4 km from Maua, n.d., *Pedro & Pedrogão* 4222 (EA holotype). FIGURE 14.**1M**.

Annual. Culms 11–60 cm tall, 1–1.5 mm wide, trigonous-terete, ribbed, hirsute. Leaf sheath hirsute, rarely glabrescent; ligule hispid with intermingled long hairs; leaf blade 40–100 × 2–5 mm, densely pubescent. Inflorescence terminal or paniculate, 7–30 cm long, with 1–2 stalked spikelet clusters from upper leaf sheaths; peduncles to 50 mm long with short hairs. Spikelets with a smooth outline as awns straight, 4–8 × 2–3 mm, ovoid, apex acute to sometimes obtuse, green turning brown. Glumes not ranked, oblong, 2.5–2.8 mm long including 0.4–0.5 mm awn, appearing glabrous as hairs extremely short on glume flanks; a few long, straight stiff hairs on 3-nerved excurrent midrib. Perianth segments 6, in 2 whorls; outer 3 segments with antrorsely scabridulous bristles, shorter than nutlet, inner 3 segments large, longer than nutlet, claw short, smooth and narrowly winged, blade rhomboid, broadly winged and ciliated with deeply notched apex and slender antrorsely scabridulous arista 0.3 mm long. Stamens 3; anthers 0.3 mm long. Nutlet off-white, brown centrally, 1–1.1 mm (including beak and stalk, both 0.1 mm) × 0.4–0.5 mm, triangular-triquetrous, angles ridged, surface smooth.

Mozambique. N: Ribáuè Dist., near Ribáuè, 14°50'S 38°21'E, 600 m, viii.1931, *Gomes e Sousa* 770 (K). MS: Manica Dist., Bandula, iv.1952, *Schweickerdt* 2333a (M).

Also in Kenya and Tanzania. Seasonally wet grassland, pools on sandy ground and in crevices of rock outcrops; 50–800 m.

Conservation notes: Not widespread; not threatened.

The nutlet is distinctive in comparison to others in the *F. ciliaris* group in having both a long stalk and beak and a comparatively small part containing the seed.

14. **Fuirena bullifera** J. Raynal & Roessler in Mitt. Bot. Staatssamml. München **13**: 355 (1977). Type: Maputo, Matola, v.1893, *Quintas* 187 (P holotype, EA, LISC). FIGURE 14.**1B**.

Annual, with tufted roots. Culms trigonous, 20–50 cm tall, pubescent. Leaf sheath 10–30 mm long, hirsute; ligule tubular, 2 mm long, membranous; leaf blade 30–110 mm long, 3–5 mm wide, pubescent. Inflorescence paniculate with 2–3 branches subtended by leafy bracts, each further branched ending in digitate clusters of spikelets in subglobose heads, sometimes reduced to one head. Spikelets squarrose, 6–10 × 2.1–2.5 mm, conical, with extended awns, green turning brown. Glumes 1.5 mm long excluding recurved 1 mm awn, with short and long hairs particularly on awns, prominently 3-nerved. Perianth segments 6, in 2 whorls, outer 3 segments with retrorsely scabrous bristles, very short or absent, inner 3 segments 1(2) on lower side of flower overtopped by nutlet, claw short, blade cup-shaped with opening against nutlet. Stamens 3. Nutlet brown, 0.9–1.1 mm long including stalk and papillate beak, 0.5 mm wide; triangular-triquetrous, angles ridged, base cuneate, surface smooth.

Botswana. N: Central Dist., 29 km from Nata (direction not stated, 2026AA), 1300 m, 24.iii.1976, *Du Preez & Steenkamp* 55 (PRE). **Mozambique**. M: Maputo (Lourenco Marques), Matola (Vila de Salazar), v.1893, *Quintas* 187 (P holotype, LISC).

Also in Namibia and South Africa (former Transvaal). In wet soils; 10–1300 m.

Conservation notes: Said to be rare or rarely found.

This rare and unusual *Fuirena* for many years was apparently incorrectly identified until published by Raynal & Roessler (1977). Forbes (1980) in her revision suggested that it was a variety of *F. ciliaris* (L.) Roxb. and named it *F. ciliaris* var. *calpolepis*, but did not validly publish it.

The presence often of only one unusually shaped inner perianth segment, asymmetrically placed, is distinctive and separates *F. bullifera* from other *Fuirena* species in the Flora area.

15. **Fuirena ciliaris** (L.) Roxb., Fl. Ind. **1**: 184 (1820). —Hooper in F.W.T.A., ed.2 **3**(2): 326 (1972). —Haines & Lye, Sedges & Rushes E. Afr.: 46, fig.46 (1983). — Gordon-Gray in Strelitzia **2**: 96 (1995).—Muasya in Kew Bull. **53**: 199 (1998); in F.T.E.A., Cyperaceae: 17, fig.3 (2010). Type: India ("India orientalis"), *König* s.n. (BM holotype). FIGURES 14.**1C**, 14.**3**.

Scirpus ciliaris L., Mant. Pl. Altera **2**: 182 (1771).
Fuirena glomerata sensu C.B. Clarke in F.T.A. **8**: 465 (1902). —Hutchinson in F.W.T.A. **2**: 470 (1936), non Lam.

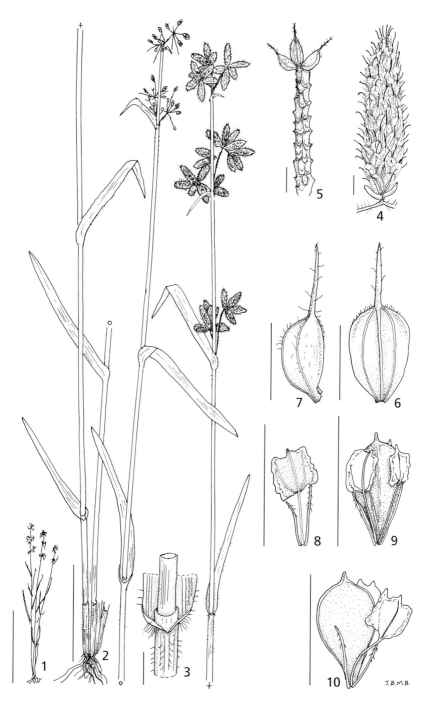

Fig. 14.**3**. FUIRENA CILIARIS. 1, habit; 2, partial habit; 3, leaf sheath apex; 4, spikelet; 5, rachilla; 6, 7, glume, adaxial and lateral views; 8, two bristles and one scale; 9, 10, nutlet with bristles and scales, adaxial and abaxial views. All from *Drummond* 5574. Scale bars: 1 = 250 mm; 2 = 40 mm; 3 = 5 mm; 4–10 = 1 mm. Drawn by Jane Browning.

Fuirena ciliaris (L.) Roxb. var. *ciliaris*. —Gordon-Gray in Strelitzia **2**: 97 (1995). —Mapaura & Timberlake, Checkl. Zimbabwe Pl.: 88 (2004).

Tufted annual; rhizome lacking. Culms 25–55 cm tall, 1–3 mm wide, trigonous-terete, usually densely hairy. Leaf sheath hirsute or glabrescent; ligule glabrous or hairy; leaf blade 4–12 cm long, 4–7 mm wide, erect or spreading, hirsute or glabrescent, hairs more numerous on abaxial side of nerves, on margins and towards apex on adaxial side than on other parts, large veins present at leaf base. Inflorescence terminal, more usually paniculate 7–30 cm long, of a terminal partial inflorescence and 1–3 lateral branches subtended by leafy pubescent bracts, each hirsute branch ending in digitate clusters of spikelets in globose heads, sometimes reduced to one head, rarely to one spikelet. Spikelets, squarrose, 5–7(12) × 2–4 mm (excluding awns), ovoid, apex acute to sometimes obtuse, green turning brown. Glumes not ranked, 2–3.4 mm long (including recurved awn 0.8–1.5 mm long), with short hairs and few long stiff bristles, prominently 3-nerved. Perianth segments 6, in 2 whorls; outer 3 segments retrorsely scabrid bristles equalling scale claw in length, sometimes short or absent, inner 3 segments of claw equal to or longer than 3-veined quadrate flat blade with apical fimbriate lobes and short arista. Stamens 3. Nutlet brown, 0.7–1.1 mm (including beak to 0.1 mm long and stalk) × 0.4–0.7 mm, triangular-triquetrous, often papillate near base, beak smooth or papillate; surface smooth.

Caprivi. Kwando R., Singalamwe, iii.1946, *Kruger* s.n. (PRE cited in Forbes, 1980). **Botswana**. N: Okavango Delta, 3 km NW of Duba Plains Camp, 19°0.053'S 22°40.669'E, 9.iv.2013, *A. & R. Heath* 2642 (K). SE: Central Dist., Selebi-Pikwe, iv.1978, *Kerfoot* 8036a (J). **Zambia**. B: Senanga Dist., Sioma (Ngonge) Falls, W bank, 23.vi.1964, *Verboom* 1750 (K). S: Namwala Dist., Kabulamwanda, c. 120 km N of Choma, 1006 m, 21.iv.1955, *Robinson* 1244 (K, SRGH). **Zimbabwe**. W: Hwange Dist., Hwange (Wankie), c. 730 m, v.1915, *Rogers* 13216 (K). S: Mwenezi Dist., Malangwe R., SW of Mateke Hills, 700 m, 5.v.1958, *Drummond* 5574 (K, LISC, PRE, SRGH). **Malawi**. S: Mulanje Dist., 16 km NW of Likabwe Forestry Dept., 700 m, 15.vi.1962, *Robinson* 5361(K, SRGH). **Mozambique**. N: Ribáuè Dist., Ribáuè, 600 m, viii.1931, *Gomes e Sousa* 774 (K). Z: Namacurra Dist., 14 km from Namacurra towards Macuze, 28.viii.1949, *Barbosa & Carvalho* 3851 (K). MS: Beira Dist., Manga, 6.4 km N of Macuti beach & 0.8 km inland, 8.ix.1962, *Noel* 2458 (K, LISC). GI: Inhassoro Dist., Bazaruto Is, 1–92 m, 28.x.1958, *Mogg* 28699 (K, LISC). M: Moamba Dist., Pessene, v.1893, *Quintas* 89 (LISC, not seen).

Widespread in tropical and subtropical Africa; Kenya, Tanzania, Angola, South Africa (KwaZulu-Natal, Limpopo, Mpumalanga) and Swaziland, extending to China, SE Asia and Australia. Seasonally wet grassland, freshwater swamps, stream margins of organically rich sand, and as a weed in fields; 5–1100 m.

Conservation notes: Widespread; not threatened.

The species could be confused with *Fuirena umbellata* if basal parts are lacking and dissection is not carried out to determine the nature of the perianth segments.

Forbes (1980) recognised a number of varieties (two unpublished) based on variations in the perianth segments. Of these, two were published as species – *Fuirena bullifera* and *F. angolensis*.

Muasya in F.T.E.A. (2010) considered *Fuirena ciliaris* to have two forms, forma *ciliaris* and forma *apetala* (Wingf.) Lye. The latter lacks perianth segments and is not represented in the Flora Zambesiaca area.

16. **Fuirena angolensis** (C.B. Clarke) J. Raynal & Roessler in Mitt. Bot. Staatssamml. München **13**: 354 (1977). —Haines & Lye, Sedges & Rushes E. Afr.: 47, figs.49,50 (1983). —Muasya in Kew Bull. **53**: 199 (1998); in F.T.E.A., Cyperaceae: 17 (2010). Type: Angola, Huíla to Humpata, ix.1883, *Johnston* s.n. (K holotype). FIGURE 14.**1A**.
    Fuirena glomerata Lam. var. *angolensis* C.B. Clarke in Durand & Schinz, Consp. Fl. Afr. **5**: 647 (1894); in F.T.A. **8**: 466 (1902).

Fuirena ciliaris (L.) Roxb. var. *angolensis* C.B. Clarke in Bull. Herb. Boissier **4**, app.III: 31 (1896). —Mapaura & Timberlake, Checkl. Zimbabwe Pl.: 88 (2004), not (C.B. Clarke) Podlech.

Tufted annual; rhizome lacking. Culms 15–50 cm tall, 1–2 mm wide, trigonous-terete, hirsute, glabrous to hirsute distally. Leaf sheath with long colourless hairs, more numerous or present near mouth; ligule membranous, margin with rim of short stiff hairs; leaf blade 6–10 cm long, 2–5 mm wide, erect or spreading, densely hairy, 7–11(17) large veins present at leaf base. Inflorescence terminal, more usually paniculate consisting of a terminal partial inflorescence with 1–3 lateral branches subtended by pubescent leafy bracts, each hirsute branch ending in globose heads of clusters of digitate spikelets, sometimes reduced to one head, rarely to one spikelet. Spikelets squarrose, 5–10 × 2–3 mm excluding awns, ovoid, apex acute to sometimes obtuse, green turning brown. Glumes 2–3.6 mm long including 1–1.5 mm recurved awn, with short hairs and few long stiff bristles, prominently 3-nerved. Perianth segments 6, in 2 whorls; outer 3 segments with retrosely scabrid bristles, equal to scale claw in length, sometimes short or absent, inner 3 segments of claw equal or longer than quadrate or subquadrate to lunate blade, apex 3-lobed, or 2-lobed with central lobe absent, sides infolded, sometimes inflated, lacking veins. Stamens (2)3. Nutlet brown, 0.9–1.3 mm (including beak to 0.1 mm long and stalk) × 0.4–0.6 mm, triangular-triquetrous, papillate near base, beak smooth, surface smooth.

Zambia. E: Chama Dist., Lundazi–Chama, 1350 m, 19.x.1958, *Robson & Angus* 170 (K, LISC, SRGH). **Zimbabwe**. N: Gokwe Dist., Sengwa Research Station, 2.v.1969, *Jacobsen* 639 (K, SRGH). C: Kwekwe Dist., 50 km NE of Kwekwe town, Iwaba Estate, 10 km SE of Camp, 1200 m, 13.ii.2001, *Lye* 24552 (K). S: Gwanda Dist., Doddieburn Ranch, Makoli kopje, 11.v.1972, *Pope* 753 (K, SRGH).

Also in Kenya, Tanzania, Angola, Namibia, and South Africa (Limpopo). Seasonally wet grassland on seepage areas and in *Brachystegia* woodland; 600–1400 m.

Conservation notes: Widespread; not threatened.

This species may be confused with *Fuirena ciliaris*, with which there is great similarity. It is distinguished mainly on the degree and development of hairs and on the shape of the scales in the inner perianth segments which have a long claw and lack the veins present on *F. ciliaris*. Forbes (unpublished) considered it as a variety of *F. ciliaris*.

Clarke's var. *angolensis* was not validly published until 1902. He cited four specimens, which are syntypes (*Gregory* 102, *Newton* s.n., *Johnston* s.n. and *Rautanen* s.n.).

17. **Fuirena sagittata** Lye in Bot. Not. **127**: 110 (1974). —Haines & Lye, Sedges & Rushes E. Afr.: 46, fig.45 (1983). —Muasya in Kew Bull. **53**: 197 (1998); in F.T.E.A., Cyperaceae: 16 (2010). Type: Tanzania, Mwitikera, c. 47 km S of Dodoma, 16.viii.1928, *Greenway* 780 (EA holotype, K, PRE). FIGURE 14.**1K**.

Robust, tufted annual. Culms terete, 8–45 cm tall, 1–2 mm wide but c. 3 mm wide across sheath, densely pubescent. Leaf sheath and ligule hairy; blade 4–11 cm by 3–6 mm wide, densely pubescent. Inflorescence an irregular terminal cluster of spikelets, 1–2 lateral pubescent branches may be present below main spikelet group. Spikelets, squarrose, 4–12 × 2–3 mm, many-flowered. Glumes, ovoid, 2–2.3 mm long including 0.4–0.7 mm awn, with short and long hairs. Perianth segments 6, in 2 whorls; outer 3 segments with retrosely scabrous bristles, shorter than or equal to scale claw in length, inner 3 segments as long as nutlet, anchor-shaped, claws smooth, longer than blade; blade lacking veins, transversely elongate, thickened distally, produced into downward pointing lobes on each side, upper margin obtuse with small central lobe. Stamens 3. Nutlet dark brown, 0.9–1 (including style-base) × 0.5–0.6 mm, triangular, papillae on cuneate base, surface smooth.

Botswana. N: Okavango Swamps, Lopis Is., 19°09.7'S 23°08.9'E, 23.viii.1974, *P.A. Smith* 1079 (K, LISC); Chobe Dist., Zwezwe (Zweizewe) Flats, 18.v.1977, *P.A. Smith* 2043 (K, PSUB).

Also in Tanzania. Periodically flooded grassland in moist white sand; 900–1000 m.

Conservation notes: Apparently scarce; possibly Vulnerable.

The arrow-shaped blade of the inner perianth is variable even within one collection. *Fuirena trilobites* C.B. Clarke from India has scales of the inner perianth whorl most closely resembling those of *F. sagittata*. Forbes suggested (1980) that these two species should be combined, in which case the earlier name *F. trilobites* would take preference.

18. **Fuirena leptostachya** Oliv. in Trans. Linn. Soc. London **29**: 168, t.108 (1875). — Clarke in Durand & Schinz, Consp. Fl. Afr. **5**: 647 (1894); in F.T.A. **8**: 466 (1902). —Hooper in F.W.T.A., ed.2 **3**(2): 326 (1972). Type: Uganda, West Nile Dist., Madi, xii.1862, *Grant* s.n. (K holotype). FIGURE 14.**1G**.

Annual. Culms 15–35 cm tall, 1–2 mm wide, trigonous-terete, densely hairy in parts or glabrous. Leaf sheath and ligule hairy, with long and short hairs; leaf blade suberect to spreading, 5–13 cm long, 2–4 mm wide, pubescent, 5–10 nerves at base. Inflorescence sometimes terminal, usually paniculate, with 1–2(3) lateral branches each with several heads of numerous spikelets; inflorescence stems with long and short hairs. Spikelets squarrose, 4–7(10) × 1.3–2.3 mm (awns excluded), terete-ovoid, apex acute. Glumes not ranked, 1.5–2(3) mm long (including 0.5–0.8 mm awns), hispidulous with intermingled stiff bristles. Perianth segments absent or present in 2 whorls; outer 3 segments present as small retrorsely scabrid bristles in forma *leptostachya*, inner 3 segments with smooth claw and crescent-shaped blade, lacking nerves, upper margin with small apiculum. Stamens 2. Nutlet pale brown, 0.4–0.6 × 0.3–0.4 mm (including 0.1 mm style base), obovoid, smooth, triangular in cross-section (stipitate or substipitate and papillate near base in forma *leptostachya*, sessile and smooth in forma *nudiflora*).

a) Forma **leptostachya**. —Haines & Lye, Sedges & Rushes E. Afr.: 44, fig.42 (1983). —Muasya in Kew Bull. **53**: 197 (1998); in F.T.E.A., Cyperaceae: 15 (2010).

    Fuirena leptostachya var. *leptostachya*. —Forbes, Rev. *Fuirena*: 183 (1980). —Mapaura & Timberlake, Checkl. Zimbabwe Pl.: 88 (2004).

Perianth segments present.

Botswana. N: Central Dist., Makgadikgadi depression, main Maun–Francistown road, 31.2 km E of Gweta turnoff, 9.iv.1978, *P.A. Smith* 2372 (K). **Zambia**. W: Mfumbwe Dist., Mfumbwe (Chizera), 100 km WNW of Kasempa, 7.vii.1963, *Robinson* 5554 (K). C: Lusaka Dist., Mt Makulu Pasture Research Station, 1250 m, 15.iv.1956, *Robinson* 1490 (K). S: Choma Dist., Mochipapa, 10 km SE of Choma, 20.vi.1978, *Heery* 19 (K). **Zimbabwe**. W: Matobo Dist., Besna Kobila, 1460 m, iii.1957, *Miller* 4222 (K, SRGH). C: Chegutu Dist., Poole Farm, 3.v.1945, *Hornby* 2392 (K, SRGH). **Malawi**. N: Chitipa Dist., Chitipa (Fort Hill), 1220 m, vii.1896, *Whyte* s.n. (K).

Widespread in tropical and subtropical Africa, in Uganda, Kenya, Tanzania, and in South Africa (Gauteng). Seasonally wet margins of dams and rivers and a weed of disturbed areas; 1000–1500 m.

Conservation notes: Widespread; not threatened.

Forma *leptostachya* has similarly shaped inner perianth segments to *F. sagittata*, but those in forma *leptostachya* are smaller, as is the nutlet. In addition, forma *leptostachya* only has 2 stamens per floret whereas *F. sagittata* has 3.

b) Forma **nudiflora** Lye in Nordic J. Bot. **3**: 241 (1983). —Haines & Lye, Sedges & Rushes E. Afr.: 45, fig.43 (1983). —Gordon-Gray in Strelitzia **2**: 100 (1995). — Muasya in Kew Bull. **53**: 197 (1998); in F.T.E.A., Cyperaceae: 15 (2010). Type: Uganda, Masaka Dist., Bugabo, 1.ii.1969, *Lye* 1825 (MHU holotype, EA, K).

    Fuirena schweinfurthiana Boeckeler in Flora **63**: 438 (1880). Type: Sudan, Djur, 20.x.1869, *Schweinfurth* III 190 (B† holotype, K, P).

    Fuirena pygmaea Ridl. in Trans. Linn. Soc. London, Bot. **2**: 160 (1884). Types: Angola, Pungo Andongo, near Sansamanda, ii.1857, *Welwitsch* 7111 (BM syntype, LISU); between

Mopopo & Sansamanda, v.1857, *Welwitsch* 7171 (BM syntype, LISU); Huíla, Monino, iv.1860, *Welwitsch* 7112 (BM syntype, LISU).

    Fuirena leptostachya var. *nudiflora* C.B. Clarke in Durand & Schinz, Consp. Fl. Afr. **5**: 647 (1894); in F.T.A. **8**: 466 (1902), invalid name. —Mapaura & Timberlake, Checkl. Zimbabwe Pl.: 88 (2004).

    Fuirena moiseri Turrill in Bull. Misc. Inform. Kew **1925**: 71 (1925). Type: Nigeria, Fodama, 13.xii.1921, *Moiser* 157 (K holotype).

    Fuirena glomerata sensu Boeckeler in Flora **62**: 566 (1879), not Lam.

Perianth segments absent.

Zambia. B: Mongu Dist., plain 56 km N of Mongu, 19.vii.1961, *Angus* 2997 (K). W: Ndola Dist., 15 km S of Ndola, 1300 m, 29.iv.1960, *Robinson* 3680 (K). N: Mbala Dist., Ndundu, 1740 m, 20.viii.1960, *Richards* 13139 (K). S: Choma Dist., 4.8 km NE of Mapanza, 1070 m, 27.iv.1955, *Robinson* 1253 (K). **Zimbabwe**. C: Chivhu Dist., near Sebakwe Bridge, 5.iv.1993, *Micho* 59 (SRGH). **Malawi**. C: Nkhotakota Dist., 6 km N of Nkhotakota, 16.vi.1970, *Brummitt* 11440a (K). S: Zomba Dist., Zomba neighbourhood, c. 900 m, 1936, *Cormack* 302 (K). **Mozambique**. MS: Cheringoma Dist., Inhaminga, tando de Inhauga, 15.vii.1946, *Simão* 801 (LISC, not seen).

Widespread in tropical and subtropical Africa; in Uganda, Tanzania, South Africa (KwaZulu-Natal, Limpopo, Mpumalanga) and Swaziland. In roadside ditches and sandy soils at edge of temporary waterbodies; 500–1800 m.

Conservation notes: Widespread; not threatened.

Forma *nudiflora* has also been reported from S Mozambique (Forbes 1980: 151), but no specimens have been seen. The exact identity of the Mozambique specimen cited has not been confirmed.

19. **Fuirena microcarpa** Lye in Bot. Not. **127**: 111 (1974). —Haines & Lye, Sedges & Rushes E. Afr.: 44, fig.40 (1983). —Muasya in Kew Bull. **53**: 196 (1998); in F.T.E.A., Cyperaceae: 14 (2010). Type: Tanzania, Uzaramo Dist., Dar es Salaam, 11.viii.1972, *Wingfield* 2100 (DSM holotype, EA, K, NU).

Annual, slender and weak with fibrous roots. Culms clustered, to 5–20 cm tall, 1 mm wide, obscurely trigonous, finely ridged, pubescent. Leaf sheath pubescent; ligule brown, sparsely hairy; leaf blade 2–8 cm long, 2–3 mm wide, narrowly lanceolate, a few long stiff hairs on both surfaces, ciliate on margin. Inflorescence of 1–2 head-like clusters of digitate spikelets. Spikelets squarrose, narrowly ovoid, 4–5 × 1.5–2 mm (excluding awns). Glumes broadly elliptical, 1–1.9 mm (including subterminal awn 0.7 mm long), with short and few long hairs on flanks, awn recurved with some bristle-like hairs. Perianth segments absent. Stamens 2, anthers 0.1–0.15 mm. Nutlet off-white to greenish, semi-transparent, elliptical, 0.5–0.6 × 0.3 mm with a minute stalk and short beak, trigonous, faces smooth and convex, obscurely patterned by large bullate cells when ripe.

Botswana. SE: Gaborone Dist., N of Kgale Siding, 1080 m, 8.iv.1978, *Hansen* 3403 (K, PRE, SRGH). **Zimbabwe**. C: no locality, 1931, *Chase* 4548 (SRGH).

Also in Tanzania. In seasonally wet sandy areas; ?50–1100 m.

Conservation notes: Localised but probably not threatened.

Fuirena microcarpa has also been reported in coastal Mozambique (Forbes 1980), but no specimens were seen for the Flora treatment.

Forbes (1980) considered that the difference between *Fuirena leptostachya* var. *nudiflora* and *F. microcarpa* did not justify upholding two distinct species. The nutlets of the latter are within the range of the former but are slightly narrower. With more collections and assessment it may be that *F. microcarpa* should be considered as a synonym of *F. leptostachya* var. *nudiflora*.

2. **BOLBOSCHOENUS** (Asch.) Palla[3]

Bolboschoenus (Asch.) Palla in Koch, Syn. Deutsch. Schweiz. Fl., ed.3 **3**: 2531 (1905). —Goetghebeur & Simpson in Kew Bull. **46**: 169–178 (1991).

Scirpus [no rank] *Bolboschoenus* Asch., Fl. Brandenburg **1**: 753 (1864).

Rhizomatous perennial herbs; culms leafy, noded, trigonous. Leaves eligulate, blades V-shaped in cross section. Inflorescence variable, bracteated, few to many spikelets. Bracts, several, leaf-like, erect or spreading. Spikelets, cylindric, many flowered, sessile or pedunculate. Glumes spiral, imbricate, awned, each subtending a bisexual floret. Hypogynous bristles present. Stamens 3, crested. Style long, branches 2 or 3. Nutlet trigonous or lenticular, slightly beaked, surface smooth, exocarp of isodiametric with central papilla or radially elongate cells.

A genus of 14 species with 4 species in Africa. The generic description applies only to the African species.

Bolboschoenus glaucus (Lam.) S.G. Sm. in Novon **5**: 101 (1995). Type: Senegal, no locality, n.d., *Roussillon* s.n. (P-Lam 673/14 holotype). FIGURE 14.4.

Scirpus maritimus L., Sp. Pl.: 51 (1753). —Clarke in F.T.A. **8**: 455 (1902).
Scirpus glaucus Lam., Tabl. Encycl. **1**: 142 (1791).
Bolboschoenus maritimus (L.) Palla in Koch, Syn. Deut. Schweiz. Fl. ed. 3 **3**: 2531 (1904). —Gordon-Gray in Strelitzia **2**: 25, 209, fig.8 (1995). —Mapaura & Timberlake, Checkl. Zimbabwe Pl.: 87 (2004). —Verdcourt in F.T.E.A., Cyperaceae: 23, fig.4 (2010).
Schoenoplectus maritimus (L.) Lye in Blyttia **29**: 145 (1971). —Haines & Lye, Sedges & Rushes E. Afr.: 53, fig.64 (1983).

Perennial, 1–1.5(2) m tall, swollen stem bases and rhizomes exceeding 150 mm long, 5–8 mm wide, with membranous scale leaves soon disintegrating; corms at base of culm or on rhizome from which roots arise. Culms erect, leaf-bearing, trigonous, smooth to scabridulous on angles, occasionally so well-developed that leaves appear basal. Leaves up to 10 per culm, to 570 × 4–10 mm, V-shaped, margins scabridulous especially towards apex. Inflorescence a simple or compound anthelodium 42–65 × 55–98 mm long carrying numerous sessile or peduncled spikelets. Bracts leaf-like 1–2(3), 90–100 mm long, overtopping inflorescence. Spikelets solitary or clustered (12)15–20(85) × 3–4(5) mm, elongate, 'catkin-like' with age. Glumes 5–6 × 2.5–4 mm (excluding recurved, smooth or scabrid awn), mid- to light brown, apex emarginate. Perianth bristles 4–6, unequal, shorter than or equal to nutlet length, densely retrorsely spinulose. Stamens 3. Style branches 3, ± half style length. Nutlet 2.4–3.3 × 1.6–2.2 mm, obtusely trigonous, obovate to elliptic, beak poorly defined, light to dark brown, surface smooth, non-cellular (× 20 magnification); pericarp with narrow exocarp of ± isodiametric cells.

Botswana. N: Chobe Dist., Savuti R, just above Savuti Marsh, 24°15'E 18°30'S, 25.x.1972, *Gibbs Russell* 2326 (K, PRE). **Zambia**. S: Mumbwa Dist., Kafue floodplain below Narubamba village, 26.vi.1963, *van Rensburg* 2313 (K, MRSC, PRE). **Zimbabwe**. S: Bikita Dist., 48 km S of Devuli Ranch Halt, 24.vii.1958, *Chase* 6961 (K, LISC, SRGH). **Malawi**. S: Machinga Dist., Kasupe, sand bar between Lakes Chilwa and Chiuta, 31.vii.1969, *Howard-Williams* 173 (K). **Mozambique**. Z: Chinde Dist., Chinde, 19.viii.1974, *Bond* W530 (LISC). MS: Beira Dist., Beira, rice field, 12.i.1980, *Roelofsen* 8078 (K, LMU). GI: Xai Xai Dist., Xai Xai town, 7.x.1978, *de Koning* 7217 (K). M: Namaacha Dist., Changalane, Estação do C.F.M., 17.x.1983, *Zunguse et al.* 631 (K, LMU).

Also in sub-Saharan and Mediterranean Africa, extending into Mediterranean Europe, the Middle East and India. Reported as an introduction in North America. Mainly a freshwater plant but also common on estuarine inlets, in shallows of rivers, streams and around margins of pans; 10–1200 m.

---

[3] by J. Browning and K.D. Gordon-Gray†

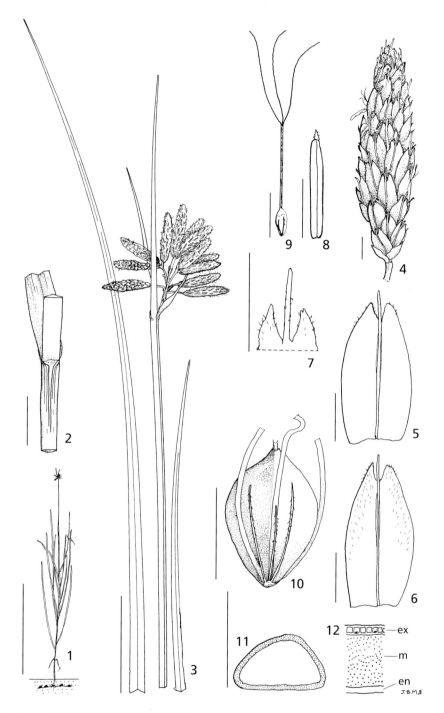

Fig. 14.4. BOLBOSCHOENUS GLAUCUS. 1, habit; 2, leaf sheath apex; 3, inflorescence and leaves; 4, spikelet; 5, 6, glume, adaxial and abaxial surface; 7, glume apex; 8, anther; 9, ovary, style and branches; 10, nutlet, abaxial view; 11, 12, nutlet sections. All from *van Rensburg* 2404. Scale bars: 1 = 250 mm; 2 = 10 mm; 3 = 40 mm; 4–11 = 2 mm; ex = exocarp, m = mesocarp, en = endocarp. Drawn by Jane Browning.

Conservation notes: Widespread; not threatened.

This species is common along the sea coasts and estuarine inlets of the W and SW coasts of the Cape Province of South Africa. In tropical areas it grows in dense stands; high temperatures are known to promote flowering and good seed-set, while poorer habitats result in elongated, abnormally catkin-like spikelets without viable seed-set. Rhizomes may survive underground for several years, and flowering shoots are relished by cattle and other large grazers.

Most African plants of the genus were previously named *Bolboschoenus maritimus* (Gordon-Gray 1995, Verdcourt 2010). This is now understood to be a halophytic species with its main centre of distribution in Scandinavia.

3. **SCHOENOPLECTUS** (Rchb.) Palla[4]

Schoenoplectus (Rchb.) Palla in Verh. K.K. Zool.-Bot. Ges. Wien **38**: 49 (1888), conserved name.
Scirpus L. subgen. *Schoenoplectus* Rchb., Icon. Fl. Germ. Helv. **8**: 40 (1846)

and

4. **SCHOENOPLECTIELLA** Lye

Schoenoplectiella Lye in Lidia **6**: 20 (2003).

Annual or perennial herbs, leaves usually reduced to a sheath, sometimes with a ligulate blade. Culms scapose or almost so, with or without a node above base. Inflorescence appearing pseudolateral, anthelate or capitate with few to many spikelets. Lowermost involucral bract stem-like, erect surpassing inflorescence. Spikelets with spirally arranged glumes each subtending a bisexual floret. Perianth 0–6. Stamens usually 3, with persistent filaments. Style 2–3 branched. Nutlet obovoid, trigonous, sometimes biconvex, often beaked, smooth or rugulose.

The genus *Schoenoplectus* incorporates perennial plants, and Lye (2003) separated the genus *Schoenoplectiella* from *Schoenoplectus* to include annual taxa that in addition to aerial flowers have basal flowers in the leaf sheaths. Hayasaka (2012) expanded the circumscription of the genus *Schoenoplectiella*, but this has not been used in this account. Lye's separation is applied to the key below, which includes both genera.

In the Flora area the species may be sympatric and are frequent inhabitants of margins of freshwater pools, dams and damp areas. Characters delimiting species are neither clear-cut nor concise; features of the nutlet are perhaps the most consistent for diagnosing species, but considerable variation occurs and intermediates are present. Heterogeneity has been noted in herbarium specimens and efforts have been made to apply various criteria by numerous authors, none of which are wholly satisfactory.

Keys have been adapted from Beentje in F.T.E.A., Cyperaceae: 24–25 (2010).

1. Aquatic, floating or submerged, more rarely terrestrial and tufted. . . **1**. *rhodesicus*
 – Terrestrial perennial or annual herb, bases sometimes in water 2
2. Perennial with rhizomes; basal pistillate flowers absent; culms 1.5–12 mm thick above leaf sheaths, except in *S. muricinux* and *S. muriculatus*; inflorescence bract much shorter than culm; *Schoenoplectus* sensu stricto 3
 – Annual or short-lived perennials lacking obvious rhizomes; basal pistillate flowers usually present in lowest leaf sheath; culms 0.3–2 mm wide, (except in *S. articulata* where stem base is 1–8 mm thick, culm septate and inflorescence bract as long or longer than culm); previously under *Schoenoplectiella*. 8

[4] by J. Browning

3. Spikelets sessile or shortly stalked, catkin-like, to 22 mm long; bristles plumose . . . . **2.** *scirpoides*
− Spikelets sessile or stalked, not catkin-like; bristles if present not plumose 4
4. Culms terete; perianth present or absent; nutlets smooth or transversely wavy. . .5
− Culms triangular, perianth of 6 retrorsely scabrid bristles; nutlet faintly transversely wavy. . . . . **3.** *mucronatus*
5. Perianth absent; nutlet smooth or rarely transversely wavy . . . . **4.** *corymbosus*
− Perianth absent or present; nutlet smooth or transversely wavy. 6
6. Culms terete, pith-filled, longitudinally ridged, 1.5–5 mm wide; spikelets generally pallid, 5–10 mm long. . . . .**5.** *confusus*
− Culms firm, not or rarely pith-filled, ridged, 1–3 mm wide; spikelets usually dark brown or pale with inverted brown V; less than 10 mm long 7
7. Culms compressible when dry; nutlets markedly transversely wavy, 12–16 transverse ridges per face; ligule 1–1.8 mm high . . . . **6.** *muricinux*
− Culms firm when dry; nutlets markedly transversely wavy, 8–10 transverse ridges per face; ligule 0.3–0.5 mm high . . . . **7.** *muriculatus*
8. Culms and inflorescence bract not septate 11
− Culms and inflorescence bract markedly or faintly septate 9
9. Septa obvious; nutlets smooth . . . . **8.** *articulata*
− Septa not always obvious; nutlets transversely wavy. 10
10. Glumes broadly ovate, concave 3–4 mm long; nutlet, 1.3–1.4 mm, waves not on shoulders (angles) . . . . **9.** *senegalensis*
− Glumes ovate, slightly concave 2–2.75 mm, nutlet, 0.8–1.2 mm, waves even on angles . . . .**10.** *roylei*
11. Glumes 2–4 mm long 12
− Glu mes usually 1–2 mm long. 15
12. Nutlets smooth, with remnants of bristles . . . . **11.** *hooperae*
− Nutlets transversely wavy, lacking bristle remnants 13
13. Spikelets angular, glumes in longitudinal rows . . . . **12.** *juncea*
− Spikelets not angular, glumes not in longitudinal rows 14
14. Style branches 2; spikelets 3–18 × 2–3.5 mm . . . . **13.** *erecta*
− Style branches 3; spikelets 4–10 × 2–2.5 mm . . . . **14.** *lateriflora*
15. Spikelets 'bristly' with glume mucros; glumes 1.3–1.7 mm long . . **15.** *microglumis*
− Spikelets not-'bristly' 16
16. Plant dwarf, 2–8 cm tall; inflorescence usually unispicate; nutlet dark brown-black, transversely wavy . . . . **16.** *proxima*
− Plant 5–20 cm tall; inflorescence of 1–3 obtuse spikelets; nutlet often bright orange to orange brown, with many transverse waves. . . . . **17.** *leucantha*

1. **Schoenoplectus rhodesicus** (Podlech) Lye in Nordic J. Bot. **3**: 242 (1983). —Haines & Lye, Sedges & Rushes E. Afr.: 54, figs.67,68 (1983). —Beentje in F.T.E.A., Cyperaceae: 25 (2010). —Browning in Kew Bull. 67: 59 (2012). Type: Zambia, Kasama, 20.vi.1960, *Robinson* 3758 (M holotype, K).

Scirpus rhodesicus Podlech in Mitt. Bot. Staatssamml. München **4**: 117 (1961).

Tufted and/or creeping perennial up to c. 30 cm tall; rhizome condensed at plant base, inconspicuous in tufted examples, or slender, green, stoloniferous, nodose, leafy shoots occasional from nodes in some terrestrial and occasional apparently aquatic examples (see note). Culms occasionally contiguous, more often slightly to more widely distanced in inundated examples, terete, 1–1.5 mm thick, scapose in terrestrial plants. Leaves at lower nodes reduced to sheaths, mouths truncate or oblique; at upper nodes sheaths membranous (in terrestrial or partially temporarily inundated specimens), red spotted; blade 6–10 cm long, 1 mm wide, canaliculate, glabrous. Inflorescence terminal of 1–2(3) sessile spikelets, 9–11 × 5 mm, appearing pseudolateral, usually aerial. Bracts 2, lower 2–5 × 0.8 mm continuing line of culm; upper shorter, glume-like, mucronate, or inconspicuous. Spikelets compact, narrowly cylindric. Glumes spiral, 4–5 × 2–3 mm, light brown, midrib green, obscurely 3-nerved, apex with shortly reflexed, barbellate mucro. Perianth of 5–6 slender bristles, mostly as long as or slightly exceeding nutlet; barbellate at apex. Stamens 3, anther crest round in outline with marginal hairs. Style trifid, bifid in lowest florets. Nutlet 1.8–2.2 × 1.5–1.6 mm, obtusely trigonous, broadly ovate in outline, minutely apiculate, surface smooth to obscurely wavy.

Zambia. N: Mbala Dist., Lake Chila (Shila), 15.iv.1963, 1650 m, *Richards* 18081 (K). **Zimbabwe**. C: Marondera Dist., Digglefold, 25.ix.1949, *Corby* 482 (K, SRGH).

Also in D.R. Congo (Katanga) and Tanzania (Rungwe Dist.). Growing submerged as trailing stems with inflorescences above the surface in waterways and lakes, or on bare mud by waterbodies; 1600–1750 m.

Conservation notes: Rare and infrequently collected; possibly threatened.

The species is found growing in damp mud of lake and dam margins; water levels and habitat conditions vary from terrestrial to temporarily submerged. Plants are sometimes uprooted by floods and carried into shallows from which they appear as floating aquatics. What were stolons now look like branches producing shoots from stem nodes. Two ecological forms are recognised – terrestrial and aquatic, the latter is more frequently collected.

2. **Schoenoplectus scirpoides** (Schrad.) Browning in S. African J. Bot. **60**: 172 (1994) as *scirpoideus*. —Beentje in F.T.E.A., Cyperaceae: 29 (2010). Types: South Africa, Cape Province, n.d., *Hesse* s.n. (GOET syntype); Uitenhage Dist., channel of Swartkoprivier, n.d., *Zeyher* 13 (NBG syntype).

    Pterolepis scirpoides Schrad. in Gött. Gel. Anz. **3**: 2071 (1821).
    Scirpus littoralis Schrad. var. *pterolepis* (Kunth) C.B. Clarke in Durand & Schinz, Consp. Fl. Afr. **5**: 625 (1894), illegitimate name.
    Schoenoplectus littoralis (Schrad.) Palla var. *pterolepis* (Nees) C.C. Towns. in Kew Bull. **15**: 417 (1962).
    Scirpus littoralis sensu C.B. Clarke in F.T.A. **8**: 456 (1902), non Schrad.
    Schoenoplectus subulatus sensu Lye in Bot. Not. **124**: 290 (1971). —Haines & Lye, Sedges & Rushes E. Afr.: 54 (1983), non Vahl.

Perennial; rhizome stout, to 11 mm wide when dry, horizontal, stoloniferous, bearing overlapping scale leaves. Culms erect, nodeless, 1–2 m long, 2–12 mm thick, terete and spongy in middle, tapering, becoming stiffer and ± trigonous up to inflorescence (never sharply triangular throughout), glabrous. Leaves usually reduced to membranous, greyish-brown sheaths, 2–3 per culm; sheath mouth obliquely truncate, margin hyaline; ligule a membranous flap; lamina occasionally developed, 5–140 cm, flexuous, thin, almost translucent when under water. Inflorescence pseudolateral, anthelate, with single spikelets or umbels of spikelets on stalks of unequal length, 1 cm long; peduncle including primary rays stoutish; usually overtopped by terete sharply pointed bract, 2–9 cm long continuing line of culm. Spikelet sessile to shortly pedicelled, ovate-oblong to oblong, 8–22 × 2–4 mm, lengthening with age when proximal rachilla becomes bare, rust-brown. Glumes extreme proximal, 1–2 sometimes sterile, otherwise fertile, ovate to oblong-ovate, 3–4 × 2.5–3 mm, concave, opaque, thin, rust brown, apex acute, glume tissue continuous adaxially, not notched; mucro minute, recurved, flanked by margins minutely ciliate distally, glabrous proximally. Perianth 4–5, abaxial (rarely + 1 adaxial), slightly exceeding mature nutlet, pale brown, narrow to medium-slender proximally, plumose-fimbriate distally for c. 2/3 of total length. Stamens 3, usually persistent; anthers with fan-shaped apex. Style 2-branched. Nutlet broadly obovoid, planoconvex (lacking abaxial ridge) apiculate, 2.3–2.7 × 1.3–1.7 mm (including apiculus), greyish-brown to greyish-black at maturity, surface smooth.

Botswana. N: Chobe Dist., Boteti (Botletle) riverbanks, 21°14.5'S 24°47.7'E, 5.xii.1978, *P.A. Smith* 2556 (K). SE: Central Dist., SE Lake Xau (Dow), 1.xii.1968, *Child* s.n. in SRGH 129,159 (K). **Zambia**. N: Nchelenge Dist., Lake Mweru, Puta village, 945 m,

27.v.1961, *Astle* 736 (K). **Malawi**. S: Mangochi Dist., Lake Malombe (Pamolombe), Upper Shire Valley, ix-x.1861, *Kirk* s.n. (K). **Mozambique**. MS: Cheringoma Dist.?, Inhamissembe R. near fishing village, 14.vii.1972, *Ward* 7944 (K, NU). GI: Chibuto Dist., Chibuto–Baixo Changana–Rio Tzengu, 23.viii.1963, *Macedo & Macuácua* 1133 (K). M: Maputo Dist., Costa do Sol towards Marracuene, 19.ii.1981, *de Koning & Boane* 8648 (K).

Also in Ethiopia, Somalia, Uganda, Kenya, Tanzania, Namibia, South Africa (Eastern, Western and Northern Cape, KwaZulu-Natal, Limpopo) and Madagascar. In estuarine environments where it may be locally dominant and an indicator of freshwater; inland in places in South Africa, in water up to 3 m deep; 5–1000 m.

Conservation notes: Widespread; not threatened.

3. **Schoenoplectus mucronatus** (L.) A. Kern., Sched. Fl. Exs. Austro-Hung. **5**: 91 (1888). —Haines & Lye, Sedges & Rushes E. Afr.: 55, fig.69 (1983). —Beentje in F.T.E.A., Cyperaceae: 26 (2010). Type: "Habitat in Angliae, Italiae, Helvetiae, Virginiae stagnis maritimis", *Rathgeb* s.n. in Herb. Linn. no.71.31 (LINN lectotype), lectotypified by Kukkonen in Taxon **53**: 181 (2004).

   Scirpus mucronatus L., Sp. Pl.: 50 (1753). —Clarke in F.T.A. **8**: 454 (1902).

Perennial, 60–100 cm tall; rhizome horizontal. Leaves absent, reduced to sheaths, 9–15 cm, light brown with short triangular apex. Culms contiguous, 60–100 cm long, 4–8 mm thick, sharply triangular. Inflorescence appearing lateral, head-like with 5–20 sessile spikelets. Involucral bract stem-like, 15–20 mm, ± twice as long as inflorescence, erect or angled, triangular, acute. Spikelets 5–15 × 4–6 mm (increasing in length at fruiting), oblong-ovoid, apex acute or obtuse. Glumes 3.3–3.5 × 2.5 mm, glabrous, greenish brown, strongly concave, ridged, apex acute, margin ciliate. Perianth 5 or 6, bristles retrorsely spinulose, unequal, longer than nutlet. Stamens 3. Style 2–3-branched. Nutlet 2 × 1.5–1.6 mm, obvoid, lenticular to obscurely trigonous, dark brown at maturity; surface faintly transversely wavy to smooth.

Zambia. N: Mbala Dist., Mbala (Abercorn) sandpits, 1500 m, 5.ii.1964, *Richards* 18936 (K). W: Mwinilunga Dist., between Kalene and Sakatwala, 1400 m, 22.ii.1975, *Hooper & Townsend* 316 (K). C: Chongwe Dist., Chakwenga headwaters, 100–129 km E of Lusaka, 16.xi.1963, *Robinson* 5820 (K). **Malawi**. N: Mzimba Dist., Vipya Plateau, Mbowe Dam, 16 km SW of Mzuzu, 1370 m, 23.xii.1973, *Pawek* 7638 (K).

Also in Guinea, Liberia, Nigeria, D.R. Congo, Burundi, Uganda, Tanzania and Angola; widespread in temperate and tropical areas of Europe and Asia. Possibly introduced into Africa as a weed of rice-fields. In beds of streams and pools with roots submerged at 15–20 cm depth; 1000–1500 m.

Conservation notes: Infrequently collected but not threatened.

The species is distinguished by the very sharply triangular culm with a rosette-like inflorescence close to the apex of green-brown spikelets with boat-shaped overlapping glumes.

In recent revisions this species has been included in *Schoenoplectiella* as *S. mucronata* (L.) J. Jung & H.K. Choi (J. Pl. Biol. **53**: 230, 2010).

4. **Schoenoplectus corymbosus** (Roem. & Schult.) J. Raynal in Fabregues & Lebrun, Cat. Pl. Vasc. Niger: 343 (1976). —Haines & Lye, Sedges & Rushes E. Afr.: 56, fig.70 (1983). —Gordon-Gray in Strelitzia **2**: 156 (1995). —Beentje in F.T.E.A., Cyperaceae: 25, fig.5 (2010). Type: India: "in India orientali", unknown collector (Z or BM holotype). FIGURE 14.**5**.

   Isolepis corymbosa Roem. & Schult., Syst. Veg., ed.15 **2**: 110 (1817).
   Scirpus brachyceras A. Rich., Tent. Fl. Abyss. **2**: 496 (1851). Type: Ethiopia, Tigray, Adua, 16.vi.1837, *Schimper* 288 (P holotype, BR, G, K).

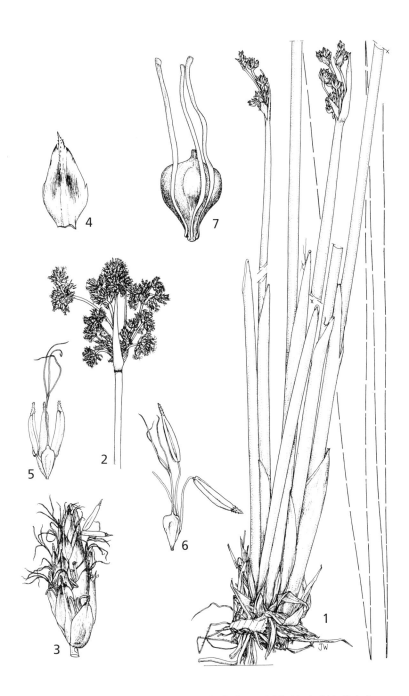

Fig. 14.**5**. SCHOENOPLECTUS CORYMBOSUS. 1, partial habit (× ²/₃); 2, inflorescence (× ²/₃); 3, spikelet (× 6); 4, glume (× 10); 5, young floret (× 10); 6, floret (× 10); 7, nutlet (× 20). 1 from *Richards* 6616; 2, 7 from Greenway & Kanuri 12546; 3–6 from *Grimshaw* 93. Drawn by Juliet Williamson. Reproduced from Flora of Tropical East Africa (2010).

Schoenoplectus brachyceras (A. Rich.) Lye in Bot. Not. **124**: 290 (1971).
Schoenoplectus inclinatus (Delile) Lye in Bot. Not. **124**: 290 (1971).
Schoenoplectus corymbosus var. *brachyceras* (A. Rich.) Lye in Nordic J. Bot. **3**: 242 (1983). —
Haines & Lye, Sedges & Rushes E. Afr.: 56, fig.70 (1983).

Perennial; rhizome abbreviated, 3–4 mm wide when dry; roots numerous, 2 mm wide when dry. Culms contiguous, to 400 cm tall, 2–10 mm thick near base, somewhat terete, glabrous. Leaves reduced to 2–3 sheaths obliquely truncate at top, sometimes with scabrid-margined reduced leaf blade. Inflorescence a pseudolateral head of clusters of sessile spikelets on 1–10 slender, very unequal scabridulous branches, 20–110 mm long (length greater if proliferation occurring); inflorescence bract 20–100 mm, tapering to a sharp point. Spikelets sessile, 5–8(10) × 2–3 mm, ovoid to cylindric, many-flowered, pale straw-coloured, overlaid by streaks of chestnut-brown. Glumes all fertile, 3–4 (including mucro, 0.12–0.20 mm) × 2–2.5 mm, ovate, keeled, convex; glabrous, lower glume keels and mucros may be hispidulous; margins slightly ciliated on either side of mucro. Perianth 0; rudimentary when present, unequal. Stamens 3, filaments, 0.12–0.7 wide, persistent on mature nutlets. Style branches 3. Nutlet (including beak) 1.3–2 × 0.9–1.4 mm, ovoid to obovoid in outline, trigonous, clearly or faintly rugulose (faces and margins), or smooth, dark-brown or black at maturity.

Botswana. N: Chobe Dist., Dassakao Is. off Moanachira R., 19°17.2'S 29°19.7'E, 23.viii.1976, *P.A. Smith* 1764 (K, PRE, SRGH). **Zambia**. B: Kalabo Dist., 15 km S of Kalabo, 2.viii.1962, *Robinson* 5446 (K). W: Sesheke Dist., Masese, Loanja dambo, 950 m, 12.ii.1991, *Bingham* 7318 (K, NU). N: Mbala Dist., Isanya (Isanga) road, 1520 m, 18.i.1955, *Richards* 4149a (K). C: Mpongwe Dist., (Ndola) Sacred Lake near St. Anthony's Mission, c. 48 km SW of Luanshya, 1200 m, 14.ii.1975, *Hooper & Townsend* 43 (K). S: Choma Dist., Muckle Neuk, 13.i.1954, *Robinson* 453 (K). **Zimbabwe**. W: Matobo Dist., Sandy Spruit Dam, 20.ii.1965, *Simon* 152 (K, SRGH). C: Marondera Dist., Marondera (Marandellas), 1520 m, i-ii.1948, *Colville* 59 (K, SRGH). E: Nyanga Dist., Nyanga, Mororo R., 1830 m, 23.x.1996, *Rattray* 985 (K). **Malawi**. N: Rumphi Dist., Nyika Plateau, Chelinda, 2300 m, 4.i.1977, *Pawek* 12261 (K). C: Lilongwe Dist., 1–5 km S of Lilongwe, 2.ii.1989, *Thompson & Rawlins* 6331 (K). S: Zomba Dist., Zomba Plateau, by Chagwa Dam, 1710 m, 19.iv.1970, *Brummitt* 9982 (K). **Mozambique**. T: Angónia Dist., Ulónguè veterinary centre, R. Capambadzi, 26.xi. 1980, *Macuácua* 1322 (K, LISC).

Also in W and N Africa, Namibia, South Africa (Free State, Gauteng, KwaZulu-Natal, Limpopo, Mpumalanga, North West, Northern Cape, Western Cape); Indian subcontinent and introduced into Spain. Forms extensive stands in dams where rhizomes generally submerged under 0.5 m of standing water; proliferation common especially where flowering culms dip into water; 900–2200 m.

Conservation notes: Widespread; not threatened.

The species has been differentiated into two varieties, the typical var. *corymbosus* and var. *brachyceras* (A. Rich.) Lye, both reportedly occuring in eastern and southern Africa. A detailed study of the two varieties was made by Browning (S. African J. Bot. **57**: 335–343, 1991). However, later Browning (1992) reverted var. *brachyceras* to specific status as *Schoenoplectus brachyceras*. Differences between the two species (or varieties) are difficult to assess unless a number of specimens can be examined; variation is considerable and overlap may occur. Intermediates may also be present. Beentje in F.T.E.A. (2010) united the two varieties, a position that is supported here.

5. **Schoenoplectus confusus** (N.E. Br.) Lye in Bot. Not. **124**: 290 (1971). —Haines
& Lye, Sedges & Rushes E. Afr.: 57, fig.72 (1983). —Gordon-Gray in Strelitzia **2**:
154, fig.66b (1995). —Beentje in F.T.E.A., Cyperaceae: 28 (2010). Type: Ethiopia,
Amogai, 1863, *Schimper* 253 (K holotype).

Scirpus confusus N.E. Br. in Bull. Misc. Inform. Kew **1921**: 300 (1921).

Perennial; rhizome abbreviated, woody, short; roots numerous. Culms 43–110 cm tall, 1.5–5 mm thick, somewhat terete, glabrous. Leaves reduced to 2–3 sheaths obliquely truncate at top, with or without a glabrous or scabrid-margined reduced leaf blade 2–5 mm long; uppermost leaf sheath sharply tapering to 1 mm wide at origin of reduced leaf blade, margin narrow (0.25–1 mm), with or without red spots, showing marked tendency to strip forming a ladder around culm; ligule (if present) 0.5–0.8 mm high. Inflorescence a pseudolateral head or contracted anthela of clusters of spikelets on 1–8 slender, unequal, scabridulous branches to 40 mm long, overtopped by 20–90 mm tapering bract appearing as continuation of culm. Spikelets 5–10 × 2–3 mm, pale yellow, with or without dark brown markings. Glumes all fertile, 2.4–3.5 mm long, ovate, convex, glabrous, pale to dark brown, margins narrow, transparent, mucro pale or dark. Perianth absent or present. Stamens 3, anthers with small crest. Style linear, 3-branched, papillate. Nutlet 1.1–1.5 × 1–1.3 mm (from apex of beak to nutlet base), ovate to obovate in outline, trigonous, shining dark brown to black at maturity, surface transversely wavy with 12–16 waves extending over nutlet shoulders; markings visible on immature nutlets.

Two subspecies of *Schoenoplectus confusus* have been described. All material from our area is subsp. *confusus*. Subsp. *natalitius* Browning is confined to KwaZulu-Natal.

Subsp. **confusus** —Browning in S. African J. Bot. **57**: 258 (1991).

a) Var. **confusus**

Perianth absent.

Zambia. C: Lusaka Dist., Mt Makulu Research Station, 13.ix.1957, *Angus* 1778 (K). S: Choma Dist., Mapanza, 1070 m, 17.v.1953, *Robinson* 243 (K). **Zimbabwe**. N: Hurungwe Dist., Mwami, Miami Expt. Farm, 1370 m, 6.iii.1947, *Wild* 1751 (K, SRGH). C: Gweru Dist., Gweru Teachers College dam, 1402 m, 3.ii.1967, *Biegel* 1851 (K, SRGH).

Also in Kenya, Uganda, Tanzania, Ethiopia and Angola. Wet grassland, seasonal pools, dambos, river margins in either sand or clay soils, or in standing water; 1000–1500 m.

Conservation notes: Fairly widespread; not threatened.

b) Var. **rogersii** (N.E. Br.) Lye in Nordic J. Bot. **3**: 242 (1983). —Haines & Lye, Sedges & Rushes E. Afr.: 57 (1983). —Gordon-Gray in Strelitzia **2**:154 (1995). —Beentje in F.T.E.A., Cyperaceae: 29 (2010). Type: Zimbabwe, Matopos, v.1915, *Rogers* 7914 (K holotype).

> *Scirpus rogersii* N.E. Br. in Bull. Misc. Inform., Kew **1921**: 301 (1921).
> *Schoenoplectus rogersii* (N.E. Br.) Lye in Bot. Not. **124**: 290 (1971).
> *Schoenoplectus confusus* subsp. *natalitius* Browning sensu Mapuara & Timberlake, Checkl. Zimbabwe Vasc. Pl.: 89 (2004), in error.

Perianth of well-developed bristles.

Zambia. N: Mbala Dist., Kawimbe road, 1500 m, 29.ix.1963, *Richards* 18217 (K). C: Chongwe Dist., Chakwenga headwaters, 100–129 km E of Lusaka, 8.ix.1963, *Robinson* 5641 (K). E: Katete Dist., 11 km E of Katete, 1100 m, 8.x.1958, *Robson & Angus* 16 (K). S: Choma Dist., Muckle Neuk, 19 km N of Choma, 1280 m, 11.x.1954, *Robinson* 916 (K). **Zimbabwe**. W: Matobo Dist., Besna Kobila, 1460 m, x.1958, *Miller* 5436 (K, SRGH). C: Harare Dist., Kaola Estate, 11 km S of Harare (Salisbury), 1460 m, fl. 14.x.1955, *Drummond* 4899 (K, SRGH). S: Masvingo Dist., Mushandike (Umshondige) Nat. Park, 30.xi.1975, *Bezuidenhout* 272 (K). **Malawi**. N: Mzimba Dist., Perekzi Forest Reserve, 30 km S from Mzimba, 1510 m, 6.vii.1970, *Brummitt* 11829 (K). C: Kasungu Dist., Kasungu, 1000 m, 28.viii.1946, *Brass* 174448 (K).

Also in Kenya and Tanzania. Damp grassland, pools, swamps, dambos and margins of streams; 1000–1500 m.

Conservation notes: Fairly widespread; not threatened.

The culms of both varieties are rounded, bright green and appear to be full of pith. Generally the spikelets are pale in colour with patches of brown, but in some cases are almost uniformly brown.

Schoenoplectus confusus var. *confusus* is sometimes confused with *S. corymbosus*, but the latter is generally a larger plant with a smooth or faintly rugulose nutlet lacking the transverse waves over the nutlet shoulders. The distinction between var. *confusus* and *S. muricinux* is often not clear; determinations are difficult and may differ.

Schoenoplectus confusus var. *rogersii* is well-represented in Zimbabwe. The presence of bristles, usually of unequal length, distinguishes this from the typical variety and from *S. muricinux*, which has dark brown to black nutlets that are markedly transversely ridged and lack bristles.

6. **Schoenoplectus muricinux** (C.B. Clarke) J. Raynal in Adansonia, n.s. **15**: 538 (1976). —Haines & Lye, Sedge Rushes E. Afr.: 57 (1983). —Gordon-Gray in Strelitzia **2**: 158 (1995). Types: South Africa, Orange Free State, iv.1883, *Buchanan* 163 (K syntype); Zimbabwe, Bulawayo, xii.1902, *Eyles* 1202 (K lectotype, P, SRGH), lectotypified by Browning (S. African J. Bot. **57**: 249, 1991).

    Scirpus muricinux C.B. Clarke in Bot. Jahrb. Syst. **38**: 135 (1906).

Perennial; rhizome abbreviated, woody, 3–4 mm wide; roots numerous. Culms 20–60 cm long, 1.5–3.6 mm thick, somewhat terete, glabrous. Leaves reduced to 2–3 sheaths, obliquely truncate at top, with a scabrid-margined reduced leaf blade 1–20 mm long; mouth of uppermost leaf sheath at origin of reduced leaf blade 1.25–2.5 mm across, broad hyaline margin with tendency to strip forming a ladder around culm; ligule 1–1.8 mm high. Inflorescence a pseudolateral head or a contracted anthela of clusters of congested spikelets, on 1–5 slender, unequal scabridulous branches 50 mm (more commonly 10–20 mm) long, overtopped by a 80–100 mm tapering bract appearing as continuation of culm. Spikelets sessile, 4–8 × 2–3 mm, shortly conical, many-flowered. Glumes all fertile, 1.8–2.6 × 1.5–2.3 mm, ovate, convex, glabrous, pale whitish-yellow, frequently green along keel, flanked by dark brown to blackish inverted V, margins narrow, transparent, apex acute to mucronate. Lower glumes and bracts usually with spine-like projections distally on keels and mucros. Perianth absent. Stamens 3. Style linear, 3-branched. Nutlet 1.2–1.6 (from apex of beak to nutlet base) × 0.9–1.2 mm, ovoid to obovoid, trigonous with short beak, shining black at maturity, surface markedly transversely wavy (rugose), 12–16 waves extending over nutlet shoulders; markings visible on immature nutlets.

Botswana. N: Chobe Dist., Segxebe Pan in Mababe Depression, 10.vi. 1978, *P.A. Smith* 2423 (K). SW: Ghanzi Dist., Groot Laagte (W), fossil river valley, 20°58.25'S 21°11.3'E, 20.iii.1980, *P.A. Smith* 3274 (K). SE: Kweneng Dist., Thamaga, 25.iii.1977, *Camerik* 137 (K). **Zambia**. N: Mbala Dist., 129 km from Mbala (Abercorn) on road to Tunduma, 1520 m, 16.ii.1961, *Vesey-FitzGerald* 2968 (K). S: Choma Dist., Mapanza, 1070 m, 9.v.1954, *Robinson* 739 (K). **Zimbabwe**. N: Gokwe Dist., 8 km N of Gokwe on road to Chinyemigetu (Nhongo), 25.iii.1963, *Bingham* 592 (K, SRGH). W: Bulawayo, Lakeside Dam, 24.i.1976, *Cross* 343(K, PRE, SRGH). S: Zvishavane Dist., Tweefontein on Great Dyke, 17.iii.1964, *Wild* 6417 (K). **Mozambique**. T: Angónia Dist., Ulónguè Veterinária, Capambadzi R., 11.xii.1980, *Macuácua* 1427 (K, LISC).

Also in D.R. Congo, Rwanda, Namibia, South Africa (Free State, Gauteng, KwaZulu-Natal, Limpopo, Mpumalanga, Northern Cape, North West) and Swaziland. Wet grassland, in black clay soils of dambos and edges of shallow dams; 800–1600 m.

Conservation notes: Widespread; not threatened.

Plants from Botswana tend to be taller with paler inflorescences and more compressible culms than those with a more easterly distribution into Zimbabwe.

Schoenoplectus muricinux and *S. muriculatus* can be easily confused; the differences are summarised under the latter species. There is also confusion with *S. confusus* and *S. corymbosus*. Examination of nutlets may clarify distinctions.

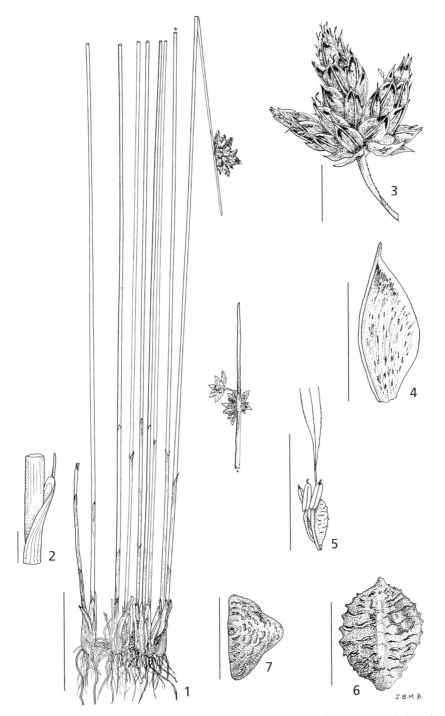

Fig. 14.**6**. SCHOENOPLECTUS MURICULATUS. 1, habit; 2, leaf sheath; 3, spikelets; 4, glume, lateral view; 5, floret; 6, 7, nutlet, abaxial and apical view. All from *Robinson* 1802. Sale bars: 1 = 40 mm; 2–5 = 2 mm; 6, 7 = 1 mm. Drawn by Jane Browning.

7. **Schoenoplectus muriculatus** (Kük.) Browning in S. African J. Bot. **57**: 251 (1991).
—Gordon-Gray in Strelitzia **2**: 160 (1995). Type: Zimbabwe, Masvingo, by Mtilikwe
R., 20.x.1930, *Fries, Norlindh & Weimark* 2137 (LD holotype, B, S). FIGURE 14.**6**.
   Scirpus muriculatus Kük. in Bot. Not. **1934**: 75 (1934).

Perennial; rhizome abbreviated, woody, 2–3 mm wide. Culms 20–75 (usually 30–40) cm long,
1–2 mm thick, somewhat terete, glabrous. Leaves reduced to 2–3 sheaths, obliquely truncate
at top, with glabrous or scabrid-margined leaf blade 0.5–2 mm (occasionally to 10 mm) long;
uppermost leaf sheath at origin of reduced blade 0.6–1.5 mm across with narrow hyaline margin,
with or without brown spots and streaks; ligule 0.3–0.5 mm. Inflorescence a pseudolateral head
or a contracted anthela of clusters of spikelets on 1–5 slender, unequal, scabridulous branches
(to 10 mm long), overtopped by 20–80 mm tapering bract appearing as continuation of culm.
Spikelets sessile, 4–6 (longest 8–10) × 2–3 mm, ovoid to cylindric, many-flowered, chestnut-
brown or rust coloured. Glumes all fertile, 2.3–2.6 × 1.5–2.3 mm, ovate, keel pallid, convex,
glabrous, streaked reddish-brown, margins narrow transparent, apex acute to mucronate; lower
glumes and bracts with spine-like projections on keels and mucros. Perianth absent. Stamens 3,
anthers with minute crest, lacking marginal microscopic projections. Style 3-branched. Nutlet
1.2–1.6 (from apex of beak to nutlet base) × 0.9–1.2 mm, ovate to obovate in outline, trigonous,
shining dark brown or black at maturity, surface with 8–10 transverse ridges which may extend
over nutlet shoulders, markings visible on young immature nutlets.

Botswana. N: Ngamiland Dist., Maun Camp No. 9, across floodplain to east of
Thalamakane R., 14.i.1974, *P.A. Smith* 776 (K, SRGH), possibly an intermediate.
SE: Kanye Dist., Mmathethe (Mathethe), 18.xi.1948, *Hillary & Robertson* 596 (PRE).
Zimbabwe. W: Matobo Dist., Matopos, 16.i.1931, *Brain* 7639 (SRGH). C: Harare,
Rainham Dam, 8.xi.1956, *Robinson* 1802 (K, LISC, SRGH). S: Masvingo Dist., Masvingo
(Victoria) near Mtilikwe (Mitilinwe) R., 20.x.1930, *Fries et al.* 2137 (B, LD, S).

Also present in South Africa (Eastern Cape, Free State, Gauteng, KwaZulu-Natal,
Limpopo, Mpumulanga, Northern Cape, North West), Lesotho and Swaziland. Wet
places on margins of dams and streams; 800–1500 m.

Conservation notes: Widespread; not threatened.

Schoenoplectus muriculatus is morphologically close to *S. muricinux* – the types of
both come from Zimbabwe, and they are sympatric in the Harare area. Observable
differences are as follows:

Schoenoplectus muricinux culm c. 2 mm thick halfway up; inflorescence well-branched;
glumes straw-coloured with dark brown-black inverted V either side of midvein.

Schoenoplectus muriculatus culm c. 1 mm thick; inflorescence basically digitate
with 1 or 2 additional short branches; glumes uniformly medium reddish-brown.
Intermediates may occur.

8. **Schoenoplectiella articulata** (L.) Lye in Lidia **6**: 20 (2003). —Beentje in F.T.E.A.,
   Cyperaceae: 30 (2010). Type: "Habitat in Malabariae aquosis arenosis"; lectotype:
   India, "Tsjeli" in Rheede, Hort. Malab. **12**: 135, t.71(1693), lectotypified by
   Simpson in Taxon **53**: 180 (2004).
   Scirpus articulatus L., Sp. Pl.: 47 (1753).
   Schoenoplectus articulatus (L.) Palla in Engler, Bot. Jarb. Syst. **10**: 299 (1888). —Haines &
   Lye, Sedges & Rushes E. Afr.: 58, fig.73 (1983). —Gordon-Gray in Strelitzia **2**: 154 (1995).
   —Mapaura & Timberlake, Checkl. Zimbabwe Pl.: 89 (2004).
   Schoenoplectus patentiglumis Hayas. in J. Jap. Bot. **78**: 65 (2003). Type: South Africa,
   KwaZulu-Natal, Umfolozi Game Reserve, 16.viii.1957, *Ward* 3137 (NU holotype).
   Schoenoplectiella patentiglumis (Hayas.) Hayas. in J. Jap. Bot. **87**: 181 (2012).

Tufted annual or short-lived perennial to 120 cm tall, including c. 40 cm inflorescence bract
often longer than culm. Culm terete, spongy, compressible, shining, 1–8 mm thick, ridged, septate
(seen best when dry). Leaves reduced to sheaths to 9 cm long ending in broad lobe; may enclose

basal female floret. Inflorescence apparently pseudolateral, globose cluster of c. 25 brown-tinged spikelets, 6–18 × 4–10 mm. Glumes 3–6 mm long, ovate-triangular, firm, midrib green, slightly concave, acute to mucronate. Perianth absent. Stamens 3. Style 3-branched. Nutlet 1.8–2 × 1.5–1.7 mm, black at maturity, trigonous, obovoid; surface smooth; basal nutlet c. 3.5 mm long

Caprivi. Reported present in PRE, but no specimen seen. **Botswana**. N: Ngmailand Dist., 100 km NNE of Maun on road to Moremi, 16.v.1977, *Ellis* 3039 (K, PRE). SE: Gaborone, N of Content Farm, 24o33'S 25o57'E, 1050 m, 10.iii.1978, *Hansen* 3367 (K). **Zambia**. E: Chipata Dist., dambo in Chipata (Fort Jameson) area, 7.v.1963, *Verboom* 135S (K, LISC). S: Choma Dist., 7 km S of Mapanza, 1070 m, 20.iii.1055, *Robinson* 1137 (K). **Zimbabwe**. N: Binga Dist., Lake Kariba, Sengwa West cleared area, 490 m, 19.xii.1964, *Mitchell* 938 (K). E: Chipinge Dist., pan near Chibuwe 32°18'E 20°30'S, 1.vi.1972, *Gibbs Russell* 2075 (K, SRGH). S: Mwenezi Dist., Urumbo pan, between Fishan and Kapateni, 25.v.1962, *Drummond* 7709 (K). **Malawi**. S: Chikwanwa Dist., Lengwe Nat. Park (Game Reserve), 100 m, 8.iii.1970, *Brummitt & Hall-Martin* 8957 (K, LISC). **Mozambique**. T: Changara Dist. (Baroma), Sisitso Camp, Zambezi R., 275 m, 11.vii.1950, *Chase* 2672 (K, SRGH). GI: Xai Xai Dist., Xai Xai town, small lake by town centre, 7.x.1978, *de Koning* 7262 (K).

Also in N and W Africa, Sudan, Ethiopia, Kenya, Uganda, Tanzania, Namibia, South Africa (Limpopo, KwaZulu-Natal, Mpumalanga) and Swaziland; and Mascarene Is., India, Far East and Australia. In seasonal pools, waterholes, wet grassland and coastal margins; 5–1100 m.

Conservation notes: Widespread; not threatened.

The inflorescence may be just above water level or ⅓ to ½ plant height. Sheaths often 2, the lower shorter at 3 cm, sometimes enclosing a basal female floret. The septa, seen as transverse markings on the culm, and an 'overtopping' involucral bract distinguish this species, the type species of the genus *Schoenoplectiella*.

The status of this species, which is close to *S. senegalensis*, is unclear as according to some authors intermediates occur and distinguishing characters cannot always be satisfactorily applied.

Raynal (1976) noted on a Zambian collection: "Rhodesian form with pale glumes despite the mostly 2-sided nuts a few normal 3-sided can be seen". There is considerable variation in plant size, colour and appearance of the glumes and in the nutlet size and surface. This has lead to separation of species such as *Schoenoplectiella patentiglumis* (Hayas.) Hayas., which differs from *S. articulata* in the narrower culm, less numerous spikelets and glumes which spread on fruiting. Variation within *S. articulata* has not been studied adequately throughout its distributional range. Until further studies are done it is considered prudent that *S. patentiglumis* is placed in synonymy.

The status of *Schoenoplectiella articulata*, which is close to *S. senegalensis*, is unclear as according to some authors intermediates occur and distinguishing characters cannot always be satisfactorily applied.

9. **Schoenoplectiella senegalensis** (Steud.) Lye in Lidia **6**: 27 (2003). —Beentje in F.T.E.A., Cyperaceae: 31, fig.6 (2010). Type: Ethiopia, Gafta, 15.ix.1838, *Schimper* 1194 (B holotype, not found, BR, K, P, S, WAG). FIGURE 14.**7**.

Isolepis senegalensis Steud., Syn. Pl. Glumac. **2**: 96 (1855).

Schoenoplectus senegalensis (Steud.) Palla in Bot. Jahrb. Syst. **10**: 299 (1888). —Haines & Lye, Sedges & Rushes E. Afr.: 58, fig.74 (1983). —Gordon-Gray in Strelitzia **2**: 160 (1995). —Mapaura & Timberlake, Checkl. Zimbabwe Pl.: 89 (2004).

Scirpus jacobii C.E.C. Fisch. in Bull. Misc. Inform., Kew **1931**: 103 (1931); in F.W.T.A., ed.2 **3**: 310 (1972), new name for *Isolepis senegalensis*, non *Scirpus senegalensis* Lam. (1791).

Scirpus praelongatus sensu Cufod. in Bull. Jard. Bot. Natl. Belg. **40**, suppl.: 1472 (1970), non Poir.

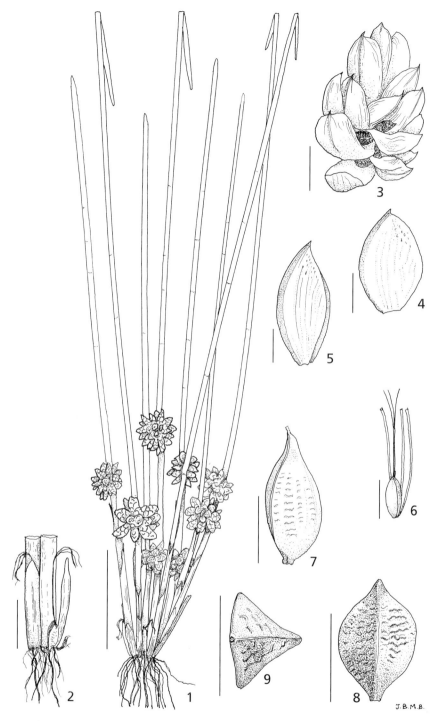

Fig. 14.**7**. SCHOENOPLECTIELLA SENEGALENSIS. 1, habit; 2, detail showing position of cleistogamous florets; 3, spikelet; 4, 5, glume from spikelet apex and base, lateral view; 6, floret; 7, basal nutlet; 8, 9, aerial nutlets, lateral and apical view. All from *Gibbs-Russell* 1588. Scale bars: 1 = 40 mm; 2 = 10 mm; 3, 7 = 2 mm; 4–6, 8, 9 = 1 mm. Drawn by Jane Browning.

Tufted annual. Culms 1–60 cm (including overtopping bract of c. 23–30 cm), round to angular, light green, hollow with transverse septa developed or not; roots shallow. Leaves absent, reduced to sheaths with short triangular lobes, basal female florets sometimes present, styles extending through sheath orifice. Inflorescence ± third along stem, a pseudolateral dense head-like cluster 17–19 mm across of 1–25 sessile spikelets; spikelets 4–9 × 2–4 mm, shortly conical or rounded (obtuse), many-flowered, golden yellow. Glumes 3–4 mm long, broadly ovate, strongly concave, midrib pronounced, mucro short. Perianth absent. Stamens 3; anthers 0.3–0.4 mm, oblong, crest minute. Style 3-branched. Nutlet 1.3–1.4 × 0.9–1.2 mm, dark brown, obovoid, sharply 3-angled, surface transversely wavy, waves not extending over nutlet shoulders; basal nutlet 4 × 2.3 mm long, surface slightly wavy; basal nutlet 4 x 2.3 mm long, surface slightly wavy.

Botswana. N: Ngamiland Dist., Samedupe Drift, 20°07'S 23°3.05'E, 11.i.1975, *P.A. Smith* 1232 (K). SE: Gaborone, Content Farm North, 24°33'S 25°57'E, 1050 m, 10.iii.1978, *Hansen* 3368 (K). **Zimbabwe**. N: Binga Dist., Sebungwe, along valley S side of Binga Hill, 460 m, 10.xi.1958, *Phipps* 1434 (K). W: Hwange Dist., Hwange Nat. Park, Makasa Pan, 18°20'S 27°03'E, 17.iv.1973, *Gibbs Russell* 1588 (K). C: Kwekwe Dist., c. 80 km NE of Kwekwe town, Iwaba Estate, 3 km SE of Camp, 1200 m, 30°05'E 18°49'S, 13.ii.2000, *Lye* 23827 (K). E: Chipinge Dist., pan near Chibuwe, 32°18'E 20°30'S, 700 m, 1.vi.1972, *Gibbs Russell* 2080 (K). S: Mwenezi Dist., Malangwe R., SW Mateke Hills, 620 m, 6.v.1958, *Drummond* 5629 (K, LISC). **Malawi**. S: Chiradzulu Dist., near Chiradzulu (Chiradzura), 21.x.1905, *Cameron* 102 (K). **Mozambique**. N: Eráti Dist., Namapa, R. Namapa, 230 m, 5.iv.1964, *Torre* 11646 (BR, LD, LISC, LMU, UPS, WAG). T: Lower Zambezi, between Tete and the coast, N'Kweza, Livingstone's Zambezi Exped., iii–iv.1860, *Kirk* s.n. (K).

Widespread in tropical Africa from Egypt, Ethiopia, Nigeria, Senegal, Rwanda, Kenya, Uganda and Tanzania to Angola, Namibia and South Africa (Free State, KwaZulu-Natal, Limpopo, Mpumalanga) and Swaziland; also in India. Margins of clay pans and dams, seasonally flooded pools and ditches; 50–1500 m.

Conservation notes: Widespread; not threatened.

May be confused with *Schoenoplectiella articulata*. Limits between *S. senegalensis* and *S. roylei* are not clearly defined, but differences can be seen in the nutlets.

10. **Schoenoplectiella roylei** (Nees) Lye in Lidia **6**: 26 (2003). —Beentje in F.T.E.A., Cyperaceae: 30 (2010). Type: Nepal, no locality, n.d., *Royle* 48 (B holotype, not found).

  Isolepis roylei Nees in Wight, Contr. Bot. India: 107 (1834).

  Scirpus quinquefarius Boeckeler in Linnaea **36**: 701 (1870). —Clarke in F.T.A. **8**: 454 (1902). Type: India, Nawanggunj, n.d., *Wallich* 3465 (K holotype, BM).

  Schoenoplectus roylei (Nees) Ovcz. & Czukav. in Fl. Tadjikist. **2**: 40 (1963). —Haines & Lye in Sedges & Rushes E. Afr.: 59, fig.76 (1983). —Mapaura & Timberlake, Checkl. Zimbabwe Pl.: 89 (2004).

Tufted annual. Culms 2–20 cm long (excluding overtopping bract which may add 5–22 cm), rounded, transverse septa present but not easily seen. Leaves absent; leaf-sheaths may contain basal female florets. Inflorescence 1/2–1/3 from base, pseudolateral, dense head-like cluster of c. 10 sessile spikelets, 4–6 × 2–3 mm, major inflorescence bract stem-like or flattened 6–15 cm long. Glumes subdistichous, ovate (boat-shaped), c. 2–2.75 mm long, straw-coloured with golden tinge, midrib broad, c. 5 pairs lateral veins, margins incurved or not, apex mucronate, reflexed. Perianth absent. Stamens 3(2). Style 3-branched. Nutlet obovoid, trigonous, 0.8–1.2 × 0.6–0.8 mm, transversely wrinkled, c. 8 wrinkles extending over shoulders, dark brown to black at maturity; basal nutlet 2–2.5 × 1–1.2 mm, transversely wrinkled.

Botswana. N: Chobe Dist., Zweizwe (Zwezwe) flats, 18°39.1'S 24°21.7'E, 18.v.1977, *P.A. Smith* 2042 (K). **Zambia**. N: Kasama Dist., 100 km E of Kasama, 6.v.1962, *Robinson* 5138 (K). S: Namwala Dist., Kabulamwanda, 121 km N of Choma, 1000 m, 21.iv.1955,

Robinson 1246 (K). **Zimbabwe**. N: Gokwe, 8 km N on road to Chinyenyetu Nhongo, 25.iii.1963, *Bingham* 590 (K). W: Hwange Dist., Hwange Nat. Park, Shumba Pan, 80 km W of Main Camp, 18 iv.1972, *Gibbs Russell* 1671 (K). C: Chegutu Dist., Chegutu (Hartley) Poole Farm, 5.iii.1946, *Hornby* 2432 (K). **Malawi**. S: Zomba Dist., grasslands W of Lake Chilwa (Chirwa), 600 m, 21.vi.1962, *Robinson* 5400 (K).

Also in Senegal, Mali, Ghana, Niger, Nigeria, D.R. Congo, Ethiopia, Somalia, Kenya, Uganda, Tanzania, Namibia and Angola; also in Iran to the Indian subcontinent. Seasonally wet grasslands and margins of pans; 1000–1100 m.

Conservation notes: Widespread; not threatened.

The golden heads with boat-shaped glumes with lateral veins on the flanks are distinctive. Long overtopping bracts may be grazed by animals.

11. **Schoenoplectiella hooperae** (J. Raynal) Lye in Lidia **6**: 25 (2003) as *hooperiae*. —Beentje in F.T.E.A., Cyperaceae: 33 (2010). Type: Tanzania, Iringa Dist., Kinyantupa, 25 km from Msembi, 2.v.1970, *Greenway & Kanuri* 14441 (K holotype, NY, P, PRE).

> *Schoenoplectus hooperae* J. Raynal in Adansonia, n.s. **16**: 146 (1976). —Haines & Lye, Sedges & Rushes E. Afr.: 60, fig.78 (1983).

Annual or short lived tufted perennial, 4–60 cm tall (including inflorescence bract of 15–40 cm); roots numerous, shallow. Culms rounded to triangular, scapose, 4–15 cm long. Leaves absent, sheaths wider than culms with short triangular lobes; basal solitary female florets often frequently present, with long styles extruded through the sheath orifice. Inflorescence a compact, pseudolateral, globose head, 6–12 mm wide, spikelets numerous, sessile. Glumes 2–3 × 1–2 mm, lanceolate, with slightly recurved apex, brown to greenish-brown, 4–5 pairs of veins, midrib distinct, green. Perianth c. 5 of very reduced bristles. Stamens 3. Style trifid. Nutlet 1 × 0.6 mm, dark brown to grey-brown, obovoid-triangular, ± smooth; basal nutlet ovoid, 1.4 × 1.4 mm, lacking bristles.

Zambia. S: Kazangula Dist., banks of Zambezi R, Katambora–Kazangula, 5.viii.1947, *Greenway & Brenan* 7980, in part (K).

Also in Tanzania, but not in South Africa. Seasonally flooded pans; 900–970 m.

Conservation notes: Infrequently collected in the Flora area; possibly threatened.

The only collection (one sheet) from the Flora area is a mixed one that includes a plant of *Scheonoplectiella proxima* and several of *S. hooperae*; the latter are atypical as the characteristic extended long overtopping bract is not present. It was possibly grazed as the collectors note the plant to be 10 cm tall. Some collections, including the type material from Tanzania, have overtopping bracts to 40 cm above the inflorescence.

The World Checklist of Cyperaceae (2007) has the specific name spelt '*hooperae*', as does IPNI. However, the spelling given by Raynal (1976) in his description of the species (and in F.T.E.A.) is '*hooperiae*'; the former spelling is correct. Raynal also mentions there are species with hypogynous bristles that are constant, although not very developed, especially in the aerial flowers. This statement refers to *S. hooperae* where the abaxial face in some nutlets has two larger bristles with a curled apex (see Haines & Lye 1983: 60, fig.78).

12. **Schoenoplectiella juncea** (Willd.) Lye in Lidia **6**: 25 (2003). —Beentje in F.T.E.A., Cyperaceae: 34 (2010). Type: Ghana, no locality, n.d., *Isert* s.n. (C holotype).
> *Schoenus junceus* Willd. in Phytographia **1**: 2, t.14 (1794).
> *Scirpus aureiglumis* S.S. Hooper in Kew Bull. **26**: 581 (1972); in F.W.T.A., ed.2 **3**: 310 (1972). Type as for *S. juncea*; replacement name because of an existing *Scirpus junceus* G. Forst.
> *Schoenoplectus junceus* (Willd.) J. Raynal in Adansonia, n.s. **16**: 139 (1976). —Haines & Lye, Sedges & Rushes E. Afr.: 63, fig.88 (1983).

Tufted annual, 5–45 cm tall excluding stem-like inflorescence bract of c. 20 cm. Culms numerous, 0.6–1.5 mm thick, terete, soft, pith-filled, ridged, lacking transverse septa. Leaves absent, reduced to pale green sheaths ending in 5 mm linear lobes; lowest sheath may enclose basal female florets with a 3-branched style. Inflorescence pseudolateral, a dense head-like cluster overtopped by stem-like inflorescence bract. Spikelets 10–20, clustered together, sessile, 4–10 × 3–4 mm, sometimes appearing stalked when glumes have fallen. Glumes 2.5–3 mm long, arranged in rows, ovate, concave, golden or orange-green, flanks veinless, midrib green, mucronate. Perianth absent. Stamens 3. Style 3-branched. Nutlet dark brown to black at maturity, c. 1 × 0.8–0.9 mm, subtrigonous, apiculate, strongly and sharply transversely wrinkled, very obvious over nutlet shoulders; basal nutlet c. 3 × 2 mm, transversely wavy.

Zimbabwe. S: Chiredzi Dist., Gonarezhou, between Chuhonja Range (Chuanja Hills) and Chipinda Pools, 27.v.1971, *Grosvenor* 552 (K, SRGH).

Also in West Africa, Uganda, Kenya and Tanzania. Margins of pools, waterholes and swamps; 350–600 m.

Conservation notes: Undercollected, but perhaps infrequent; not threatened.

The infloresence appears about half-way up the stem. The golden sessile spikelets with glumes in angular rows are distinctive.

13. **Schoenoplectiella erecta** (Poir.) Lye in Lidia **6**: 25 (2003). —Beentje in F.T.E.A., Cyperaceae: 34 (2010). Type: Mauritius, no locality, *du Petit Thouars* 13 (P holotype).

    Scirpus erectus Poir. in Lamarck, Encycl. **6**: 761 (1804).

Annual or short-lived tufted perennial, 5–30 cm tall including inflorescence bract. Leaves absent or less often present; basal sheaths ending in short subulate lobe, may enclose basal female florets with a 3-branched style and nutlets c. 2.2 × 1.4 mm, sometimes with 2 bristles. Inflorecence pseudolateral, overtopped by lower bract, upper bract absent or present, short; spikelets clustered, subsessile or stalked; reddish-brown to cream, ovoid 3–18 × 2–3 mm. Glumes (2.5)3–4 mm long, triangular, lanceolate, glabrous, margin ciliate or not, mucronate. Perianth absent. Stamens 3. Style 2-branched (rarely 3 as in S. lateriflora). Nutlets dark brown to black, lenticular, ovoid, 1.4–1.6 × 1.1–1.4 mm, transversely (12–15 waves) or hardly wrinkled; basal nutlet c. 2.2 × 1.4 mm, sometimes with 2 bristles.

a) Subsp. **erecta**

    Isolepis uninodis Delile, Descr. Egypte, Hist. Nat. **2**: 152, t.6, fig.1 (1813). Type: Egypt, n.d., *Delile* s.n. (P, MPU).

    Scirpus uninodis (Delile) Coss. & Durieu in Bory & Durieu, Expl. Sci. Algérie, Bot. **2**: 310 (1868).

    Scirpus uninodis (Delile) Boiss., Fl. Orient. **5**: 380 (1884), superfluous combination. —Hooper in F.W.T.A., ed.2 **3**: 310 (1972).

    Scirpus supinus L. var. *uninodis* (Delile) Asch. & Schweinf., Ill. Fl. Égypte: 157 (1887); in F.T.A. **8**: 453 (1902).

    Schoenoplectus erectus (Poir.) J. Raynal subsp. *erectus* in Adansonia, n.s. **16**: 141 (1976). —Haines & Lye, Sedges & Rushes E. Afr.: 61 (1983). —Mapaura & Timberlake, Checkl. Zimbabwe Pl.: 89 (2004).

    Schoenoplectus lateriflorus (J.F. Gmel.) Lye subsp. *laevinux* Lye in Nordic J. Bot. **3**: 242 (1983). Type: Tanzania, near Dar es Salaam, 15.vii.1972, *Wingfield* 2036 (DSM holotype, EA, K).

    Schoenoplectiella lateriflora (J.F. Gmel.) Lye subsp. *laevinux* (Lye) Beentje in F.T.E.A., Cyperaceae: 36 (2010).

Nutlets slightly to hardly wrinkled.

Zambia. B: Kalabo Dist., 2.viii.1962, *Robinson* 5448a, 5448b (K).

Also in Algeria, Egypt, Libya West Africa (Senegal, Mali, N Nigeria), Tanzania, Mauritius, Madagascar; possibly in Asia and Australia. Swampy ground on edges of water bodies: 1000–1100 m.

Conservation notes: Uncommon in the Flora area, a possible introduction with rice seed; not threatened.

b) Subsp. **raynalii** (Schuyler) Beentje in F.T.E.A., Cyperaceae: 34 (2010). Type: Botswana, 23 km on Maun-Shorobe road, 6.ii.1964, *Yalala* 425 (P holotype, K).

 Scirpus raynalii Schuyler in Notul. Nat. Acad. Nat. Sci. Philadelphia **438**: 1 (1971).

 Schoenoplectus erectus subsp. *raynalii* (Schuyler) Lye in Nordic. J. Bot. **3**: 243 (1983). — Haines & Lye in Sedges & Rushes E. Afr.: 62, fig.83 (1983).

Nutlcts strongly wrinkled.

Botswana. N: Ngamiland Dist., Selinda spillway, Makwegana R., 23 km south of the Tsetse Fly Control road, 18°44.779'S 23°06.260'E, 970 m, 25.xi.2004, *Heath* 703 (K). **Zambia**. B: Sesheke Dist., NE corner of Sioma Ngwezi Nat. Park, 9.viii. 2006, *Bingham & Vestergaard* 13096 (K). **Zimbabwe**. E: Chipinge Dist., pan near Chibuwe 30°55'E 20°15'S, 1.vi.1972, *Gibbs Russell* 2085 (K). **Mozambique**. M: 'Magaia', 15 m, 18.i.1898, *Schlechter* 12054 (K).

In USA (Texas), Mexico, Argentina and Paraguay; also in Uganda, Tanzania, Namibia and South Africa (KwaZulu-Natal, Limpopo, North West, Northern Cape). River banks and dambo margins; 20–1400 m.

Conservation notes: Widespread; not threatened.

There is confusion in the covers at K between *Schoenoplectiella erecta* subsp. *raynalii* and *S. lateriflora*. Haines & Lye (1983) distinguish the two taxa on the basis of style branch number and ciliated glumes, but these criteria are not always applicable to the material available; both 2 and 3 style branches may be observed within the same collection. In citing the above I have placed those that obviously have 2 style branches in *S. erecta* and those with 3 style branches in *S. lateriflora*.

According to Gordon-Gray (1995), typical subsp. *erecta* (as *Schoenoplectus erectus* subsp. *erectus*) was not confirmed on the African continent but was only known from Indian Ocean Islands. However, recent examination of collections at K show it is in fact present in West Africa, and a collection from Kalabo (*Robinson* 5448) extends its distribution to Zambia.

Schoenoplectiella erecta subsp. *raynalii* is well represented in N Botswana. However, it is very variable morphologically and nodes on the culm are not always evident in collections.

14. **Schoenoplectiella lateriflora** (J.F. Gmel.) Lye in Lidia **6**: 25 (2003). —Beentje in F.T.E.A., Cyperaceae: 35 (2010). Type: Sri Lanka, no locality, *König* s.n. (LD holotype).

 Scirpus lateriflorus J.F. Gmel., Syst. Nat. **2**: 127 (1791).

 Schoenoplectus lateriflorus (J.F. Gmel.) Lye in Bot. Not. **124**: 290 (1971). —Haines & Lye, Sedges & Rushes E. Afr.: 62, fig.84 (1983). —Gordon-Gray in Strelitzia **2**: 158 (1995). — Mapaura & Timberlake, Checkl. Zimbabwe Pl.: 89 (2004).

Tufted annual tufted, few to many closely placed culms 5–50 cm long (including overtopping erect, inflorescence bract), 0.4–1.8 mm thick, ridged, triangular to almost terete. Leaves absent, if present to 30 cm long, flat or folded; leaf sheaths 1–2 cm long, often lowest enclosing basal female florets with a 3-branched style 15–20 mm long. Inflorescence a cluster of spikelets, appearing pseudolateral, overtopped by inflorescence bract, shorter than culm; spikelets 1–20, sessile or shortly pedunculate, 4–10 × 2–2.5 mm, ovoid, acute. Glumes 2–3 mm long, lowest to 3.5 mm, ovate to broadly boat-shaped, pale brown with reddish streaks, green midrib, mucronate, margin ciliate apically. Perianth absent. Stamens 3. Style 3-branched (rarely 2-branched within same plant, see Gordon-Gray 1995). Nutlet broadly ovoid, trigonous, dark brown, black at maturity, 1.1–1.4 × 0.8–1.3 mm, surface strongly transversely wrinkled; basal nutlet 2–2.5 × 1.4–2 mm.

Zambia. B: Kalabo Dist., Kalabo, 2.viii.1962, *Robinson* 5448 sheet 1, B (K, see note below). N: Kasama Dist., 96 kms E of Kasama, 6.v.1962, *Robinson* 5153 (K). S: Namwala Dist., Kabulamwanda, c. 120 km N of Choma, 21.iv.1955, *Robinson* 1248 (K).

Zimbabwe. W: Bulilimamangwe Dist., stream flowing into Mananda R., 21 km NW of Marula, 20°20'S 28°05'E, 1300 m, 20.iv.1972, *Gibbs Russell* 1732 (K). S: Mwenezi Dist., Malangwe R., SW Mateke hills, 640 m, 7.v.1958, *Drummond* 5678 (K, LISC, SRGH). **Malawi**. C: Nhakotakota Dist., Chia area, 480 m, 5.ix.1946, *Brass* 17541 (K). S: Zomba Dist., W of L. Chilwa, 600 m, 21.vi.1962, *Robinson* 5399 (K). **Mozambique**. T: Changara Dist., "Baroma", Sisitso, Zambezi R., 275 m, 11.vii.1950, *Chase* 2673 (K). GI: Xai-Xai Dist., Xai-Xai, small lake in town centre, 7.x.1978, *de Koning* 7261 (K).

Also in Ghana, Senegal, Mali, Nigeria, Central African Republic, Equatorial Guinea, Ethiopia, Somalia, Kenya, Uganda, Tanzania, Angola and Namibia; and in tropical Asia and Australia. Waterholes, swamps, lagoon margins and lake margins; 10–1300 m.

Conservation notes: Widespread; not threatened.

The *Robinson* 5448 collection at Kew comprises two mounted sheets – sheet 1 is a mixed collection of *Schoenoplectiella erecta* subsp. *erecta* and *S. lateriflora*.

On the covers of *Schoenoplectus lateriflorus* at Kew, Hooper mentions an 'African form' without defining it.

Schoenoplectiella lateriflora could be confused with *Schoenoplectus muricinux*, but the culms of the latter lack the 1–2 nodes above the base, which can sometimes be seen in the former. See also note under *Schoenoplectiella erecta*.

15. **Schoenoplectiella microglumis** (Lye) Lye in Lidia **6**: 26 (2003). —Beentje in F.T.E.A., Cyperaceae: 33 (2010). Type: Uganda, Busoga Dist., Ndolwa, 12.viii.1957, *Langdale-Brown* 2323 (KAW holotype, MHU).

    Scirpus tenerrimus Peter, Repert. Spec. Nov. Regni Veg. Beih. **40**(1): 124 (1936), illegitimate name. Type: Tanzania, Tabora Dist., Unyanyembe, E of Makongwe, 8.i.1926, *Peter* 45850 (GOET holotype, B).

    Schoenoplectus microglumis Lye in Bot. Not. **124**: 287 (1971). —Haines & Lye in Sedges & Rushes E. Afr.: 60, fig.81 (1983).

Tufted annual, 5–25 cm tall (excluding overtopping inflorescence bract 5–15 cm), slender-stemmed, stem round or angular, 0.2–0.8 mm thick, ridged, 1-noded c. 20 mm above base. Leaves absent, if present short; leaf sheaths ending in a linear lobe, 1–5 mm long or leaf-like; lowest sheath may contain basal female florets with 3–branched style. Inflorescence pseudolateral, a dense cluster of 3–15 sessile or shortly stalked spikelets, 3–5 × 1.5–2 mm, lengthening with age to 6–9 mm. Glumes 1.3–1.7 mm, ovate, mucronate, margins not ciliate, midrib green, flanks dark reddish brown. Perianth absent. Stamens 3. Style 3-branched. Nutlet 0.7–0.9 × 0.5–0.7 mm, brown to black at maturity, obovoid, subtrigonous, transversely wrinkled; basal nutlet 1.3–1.5 × 0.9–1 mm.

Botswana. N: Ngamiland Dist., Dassako Is. off Moanachira R., 23.viii.1976, *P.A. Smith* 1759 (K). **Zambia**. N: Kasama Dist., 100 km E of Kasama, 17.iv.1962, *Robinson* 5081 (K). S: Mazabuka Dist., between Mazabuka (Masabuka) and Kaleya, 15.iv.1963, *van Rensburg* 1910 (K). **Zimbabwe**. S: Mwenezi Dist., near Malipati, 2.v.1961, *Drummond & Rutherford-Smith* 7679 (K, NU).

Also in Ethiopia, Rwanda, Kenya, Uganda and Tanzania. Seasonally wet grassland and river banks; 900–1100 m.

Conservation notes: Locally common; not threatened.

As noted by Hooper (1968), *Robinson* 1247 has a 2-branched style and lenticular nutlets, while the typical species has a 3-branched style and trigonous nutlets.

Very mature inflorescences may carry spikelets on elongated stalks (rays) up to 65 mm long, as in *Robinson* 5233 (K) from E Zambia (45 km NE of Isoka). With long rachilla and few glumes this looks different from specimens collected soon after flowering. In common with *Schoenoplectiella lateriflora* subsp. *lateriflora* it has a node on the culm just above the uppermost leaf sheath.

This species may be confused with *Schoenoplectiella lateriflora*, *S. erecta* subsp. *raynalii* and *Schoenoplectus muricinux*, but is recognised and distinguished by the very small glumes.

16. **Schoenoplectiella proxima** (Steud.) Lye in Lidia **6**: 26 (2003). —Beentje in
F.T.E.A., Cyperaceae: 36 (2010). Type: Egypt, "prope Abu-Zabel", 1835, *Schimper*
31 (B† holotype, P, M).

> *Isolepis proxima* Steud., Syn. Pl. Glumac. **2**: 95 (1855).
> *Schoenoplectus proximus* (Steud.) J. Raynal in Adansonia, n.s. **16**: 144 (1976). —Haines &
> Lye, Sedges & Rushes E. Afr.: 60, fig.79 (1983).
> *Schoenoplectus proximus* var. *botswanensis* Hayas. in J. Jap. Bot. **80**: 161 (2005). Type:
> Botswana, Tsau–Xai Xai road, 19.iv.1980, *P.A. Smith* 3360 (NU holotype, K, PRE).
> *Schoenoplectiella proxima* var. *botswanensis* (Hayas.) Hayas. in J. Jap. Bot. **87**: 181 (2012).

Tufted annual, 1–6 cm tall (including inflorescence bract). Leaves absent; leaf sheath
with extended lobe to 5 mm, sometimes enclosing basal female florets with 3-branched style.
Inflorescence pseudolateral, overtopping bract up to 4 cm long; spikelets sessile, 2–4 × 1.5–
2 mm. Glumes ovate, 1.5–1.8 mm long, light brown with green midrib, margin entire, apex
mucronate. Perianth absent. Stamens 3. Style 3-branched, sometimes 2-branched. Nutlet dark
brown, obovoid, trigonous, 0.8–1 × 0.5–0.7 mm, transversely wrinkled; basal nutlet dark brown,
c. 1.7 mm long.

Botswana. N: Ngamiland Dist., Tsau (Tsao)–Xai Xai (Nxainxai) road, 19°55.7'S
21°27.4'E, 19.iv.1980, *P.A. Smith* 3360 (K, PRE, NU). **Zambia**. S: Kazungula Dist., banks
of Zambezi R., Katambora–Kazangula, 915 m, 15.viii.1947, *Greenway & Brenan* 7980 in
part (K).

Also in Yemen, Chad, Egypt, Ethiopia and Tanzania. In seasonally wet grassland; c.
900 m.

Conservation notes: Although few collections, probably not threatened.

Greenway & Brenan 7980 (K) was a mixed collection including mainly *Schoenoplectiella
hooperae*, but with one plant determined by Raynal as *Schoenoplectus proximus*. Further
collections have not been located. However, *P.A. Smith* 3360 from Botswana was named
Schoenoplectus proximus var. *botswanensis* on the basis of 2-branched style and plano-
convex nutlet with fewer transverse waves. Other species have been differentiated on
the basis of number of style branches which may be an unsatisfactory criterion. Until
further material is found it is regarded as a synonym of *S. proxima*.

17. **Schoenoplectiella leucantha** (Boeckeler) Lye in Lidia **6**: 25 (2003). Type: Namibia,
"Gross-Namaland", Kleiner Fischfluss, iv.1885, *Schinz* 379 (K holotype).

> *Scirpus leucanthus* Boeckeler in Verh. Bot. Vereins Prov. Brandenburg **29**: 46 (1888).
> *Scirpus supinus* L. var. *leucosperma* C.B. Clarke in Fl. Cap. **7**: 228 (1898). Type: South Africa,
> Western Cape, Ebenezer, by Olifants R., xi.1833, *Drège* 7414 (P holotype, BM, CGE, K, P).
> *Schoenoplectus leucanthus* (Boeckeler) J. Raynal in Adansonia, n.s. **16**: 143 (1976).

Annual tufted with many ridged, triangular bright green culms 5–10 cm long (excluding 15–30
cm long overtopping inflorescence bract); 0.5–0.8 mm thick. Leaves absent, or if present short,
flat or folded; lowest leaf sheaths often with basal female floret. Inflorescence pseudolateral, a
digitate cluster of few spikelets, rarely a solitary spikelet, obtuse (in the Cape form), brown and
green, 2–3.5 mm long. Glumes 1.2–1.5 mm long, translucent, membranous, keel bright green,
flanks with dark brown patches, ovate to hemispherical, neither pointed nor keeled. Perianth
absent. Stamens 3. Style 3-branched. Nutlet to 1.5 mm long (including beak) × 0.8–0.9 mm
wide, subequally trigonous, dark brown, almost black at maturity, surface with very many close
transverse wrinkles.

Botswana. N: Ngamiland Dist., c. 30 km S of Mopipi, pans next to road, 1080 m,
7.iii.1996, *Burgoyne & Snow* 5403 (PRE).

Also in Namibia and South Africa (Northern Cape, Free State, Limpopo, a few
records in E & W Cape, C. Archer, pers. comm.). Riverbanks and drying clay pans
frequently surrounded by sand; c. 1000 m.

Conservation notes: Not threatened.

The above description is based on information supplied by C. Archer (PRE) and supplemented by measurements from *Drège* 7414 (K). Nutlets from *Giess* 13312 (K) from Namibia are bright orange in colour, ovoid, trigonous, 0.9 × 0.8 mm, with many fine rugae on the surface.

Archer notes that there are specimens of *Schoenoplectiella leucantha* that have 2-styled branches but are otherwise identical to those with 3 style branches.

Difficulty is experienced in distinguishing the following *Schoenoplectiella* species – *S. lateriflora*, *S. erecta* subsp. *raynalii*, *S. leucantha* and *S. proxima*, where the criteria for separation are based on plant size, cilia present or absent on glumes and number of style branches.

5. **ELEOCHARIS** R. Br.[5]

Eleocharis R. Br., Prodr. Fl. Nov. Holland.: 224 (1810). —Svenson in Rhodora **31**: 121–242 (1929).

Annual or perennial herbs, robust to medium sized; rhizome stout to slender, horizontally creeping; or compactly uniseriate. Culms erect, 0.5–1 m tall, usually numerous, hollow or pith-filled, occasionally septate; leaves reduced to inconspicuous sheaths basal to culms. Short-lived perennials appearing leafless; tufted with numerous spikelet-terminated, slender nodeless culms mostly under 0.5 m tall, often producing slender, white rhizomes branching and sometimes with small tubers. These are inconspicuous, easily broken and often do not persist in herbarium material. Annuals similar but lacking rhizomes or above ground stolons, aerial organs delicate and sometimes very small. Inflorescence throughout a solitary terminal spike (spikelet) of one, few or many bisexual florets, florets spirally placed when numerous. Glumes imbricate to fairly loose, lowest 2 sometimes sterile, resembling greenish enveloping bracts, or bracts undeveloped. Perianth of 3–9 barbellate, marginally toothed or smooth bristles, sometimes absent, or reduced to a narrow rim underlying ovary. Stamens 1–3. Style base enlarged, usually persistent as conical or flattened appendage on nutlet; style 2–3 branched, variable even in individual spikelets; upper portion deciduous, base persistent as appendage to nutlet. Nutlets obovoid, biconvex or trigonous, longitudinally 3-ridged, smooth, variously patterned, or with pits placed in longitudinal rows.

A genus of about 250 species, mostly American, in permanently to periodically wet habitats.

The genus is most easily recognised by its inflorescence – a terminal spike, usually quite short and visibly ebracteate (a spikelet), sometimes pseudoviviparous. Culms are nodeless and photosynthetic, replacing the reduced leaves each of prophyll and sheath at culm base. Detail of persistent style base is not reliable in species identification as the rim curls upward on drying and the body shrinks; ideally both stages should be described. Nutlet measurements given in the descriptions exclude the style base.

1. Plant a submerged aquatic or semi-aquatic with capillary culms, more rarely amphibious and present as a tufted terrestrials with wider stems; spikelets somewhat flattened with few (up to 8) glumes; nutlet surfaces markedly pitted16
   – Plant normally terrestrial in damp situations with erect culms; spikelets terete with many glumes; nutlet surfaces variously patterned but not markedly pitted 2
2. Culms hollow, septate, septa marked by darker transverse lines . . . .**1.** *dulcis*
   – Culms pith-filled or hollow, non-septate 3
3. Culms acutely 3-angled 4
   – Culms round to slightly angled, sometimes compressed distally 5

---

5 by J. Browning and K.D. Gordon-Gray†

4. Glumes clearly veined, oblong . . . .**2.** *acutangula*
–   Glumes obscurely veined or lacking veins, obovate to orbicular . . . .**3.** *mutata*
5. Plant perennial, of variable height (20–150 cm or more); rhizomes horizontal, usually robust. 6
–   Plant appearing annual or a short-lived perennial; sometimes with delicate rhizomes or stolons bearing easily-overlooked small tubers, not enduring in herbarium specimens 9
6. Plant with slender, usually uniseriate rhizomes clothed in persistent culm bases; culms terete; vegetative organs often glaucous; spikelets often exhibiting pseudovivipary. . . . .**4.** *limosa*
–   Plant with thick horizontal rhizomes densely clothed in persistent culm bases; culms ridged, flattened distally; aerial organs not evidently glaucous; pseudovivipary not reported. 7
7. Plants widespread in wet or inundated habitats, morphologically extremely variable; stolons usually present, sometimes shoot bases crowded, thickened; perianth bristles brown, variable when present from exceeding nutlet length to appearing absent (reduced to a rim underlying nutlet), marginal retrorse barbs few to many; spikelet exceeding 10 mm at maturity . . . . **5.** *variegata*
–   Plants differing in some respects 8
8. Tall plants to 80 cm, spikelets 8–20 mm long; glumes c. 4 mm long. .**6.** *marginulata*
–   Short plants from 8–25 cm, spikelets 6–10 mm long; glumes 2.5–3 mm long . . . . **7.** *welwitschii*
9. Glumes with distinct 0.2–0.5 mm wide dark marginal bands . . . . . . **8.** *decoriglumis*
–   Glumes without distinct wide dark marginal bands 10
10. Perianth bristles present, brown, retrorsely scabrid 11
–   Perianth bristles absent, if present not brown, reduced and much shorter than nutlet 13
11. Perianth bristles much longer than nutlet. . . . .**9.** *minuta*
–   Perianth bristles as long or slightly longer than nutlet. 12
12. Spikelet ± twice as long as wide; nutlet appendage triangular . . . . **10.** *caduca*
–   Spikelet ± as long as wide; nutlet appendage conical . . . .**11.** *geniculata*
13. Glumes 1–1.3 mm, pale brown; nutlet yellowish-brown, 0.5–0.7 mm long . . . .**12.** *setifolia*
–   Not this combination of characters 14
14. Tufted annual with filiform dark green culms; style branches 2; nutlet biconvex, 0.5–0.4 mm long, black, shining, smooth at maturity, with a minute, vertically-flattened style base; perianth bristles white, shorter than nutlet or absent. . . . .**13.** *atropurpurea*
–   Tufted annual, culms filiform, not dark green; style branches 3 15
15. Spikelet 2–5 mm long; glumes dark purple with paler margins; nutlet 3-ribbed longitudinally, pale with darker small appendage. . . . . **14.** *nigrescens*
–   Spikelet 1–2 mm long; glumes reddish brown, basal glume uncoloured; nutlet 3-ribbed with conspicuous large pits in longitudinal rows . . . . **15.** *brainii*
16. Aquatic, usually submerged; stems septate; spikelets with 2–3 glumes . . . . **16.** *naumanniana*
–   Semi-aquatic; stems not septate; spikelets with 2–8 glumes . . . . **17.** *retroflexa*

1. **Eleocharis dulcis** (Burm.f.) Hensch., Vita Rumphii: 186 (1833). —Haines & Lye, Sedges & Rushes E. Afr.: 66, fig.89 (1983). —Gordon-Gray in Strelitzia **2**: 80 fig.28E (1995). —Browning *et al.* in S. African J. Bot. **63**: 173, fig.1 (1997). —Beentje in F.T.E.A., Cyperaceae: 37, fig.7 (2010). Type: East Indies, based on *Cyperus dulcis* Rumph., Herb. Amb. **6**: 7 (1750). FIGURES 14.**8**, 14.**10A**.

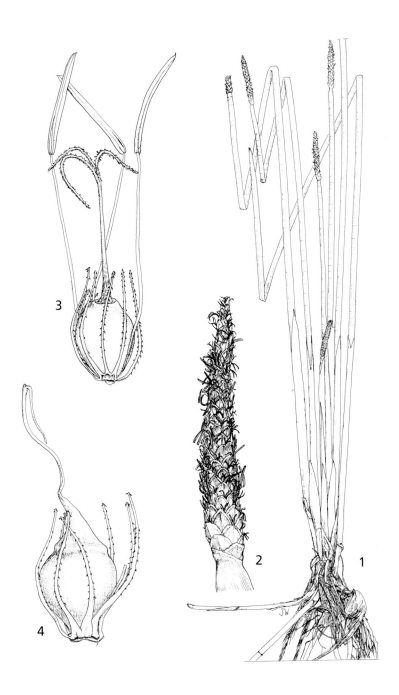

Fig. 14.**8**. ELEOCHARIS DULCIS. 1, habit (× ⅓); 2, inflorescence (× 1½); 3, floret (× 10); 4, nutlet (× 12). 1 from *Kirika et al.* NMK 778; 2 from *Vesey-Fitzgerald* 401; 3 from *Milne-Redhead & Taylor* 9164; 4 from *Faden et al.* 96/468. Drawn by Juliet Williamson. Reproduced from Flora of Tropical East Africa (2010).

Andropogon dulcis Burm. f., Fl. Ind.: 219 (1768).

Eleocharis plantaginea (Retz.) Roem. & Schult., Syst. Veg. **2**: 150 (1817). Type: India, no locality, *Koenig* s.n.

Perennial; rhizome abbreviated linking shoot bases into tufts; stolons elongate, nodose, lacking tubers, terminating in erect, rounded, hollow culms to 150 cm tall with septa 3–10 cm apart (contracted upwards to 2–3 cm). Leaves reduced to nodal sheaths with sloping mouth, blade a small triangular outgrowth. Inflorescence a terminal cylindrical spike (a solitary spikelet), 25–50 × 2–6 mm, narrower than its culm. Glumes closely imbricate, 4–6 mm long, not keeled, veins numerous (× 20 magnification needed to observe distally), ovate with dark brown submarginal band and a membranous ± colourless margin. Perianth usually of 7 strong bristles exceeding mature nutlet, margins barbellate, retrorse. Stamens 3. Style bifid or trifid. Nutlet 2–2.4 mm long, ± round, ovate in outline, surface smooth, persistent style base long-triangular, shrinking on drying.

Botswana. N: Ngamiland Dist., Okavango swamps, Khwai (Kwikhai) R., upstream of Txatxanika Camp, 23°25'E 19°10'S, 1000 m, 3.iii.1972, *Gibbs Russell & Biegel* 1500 (K). **Zambia**. B: Mongu Dist., Barotse floodplain, Mongu, 8.iv.1964, *Verboom* 1719 (K). N: Luapula Dist., Lake Bangweulu, S of Ncheta Is., 1060 m, 6.ii.1996, *Renvoize* 5526 (K). C: Chongwe Dist., Chakwenga headwaters, 100–129 km E of Lusaka, 27.iii.1965, *Robinson* 6471 (K). **Zimbabwe**. C: Harare, Cleveland Dam, 15.ii.1946, *Wild* 792 (SRGH). **Mozambique**. M: Maputo Dist., Inhaca Is., Hlanginini freshwater swamp, 28.ix.1959, *Mogg* 31360 (K).

In Uganda, Kenya, Tanzania and South Africa (KwaZulu-Natal); also in Madagascar, Asia, Australia and Polynesia. In freshwater pans or seepage areas and margins of lakes; 10–1400 m.

Conservation notes: Widespread; not threatened.

Plants have septate culms which can be noticed on breakage.

The cultivated form of this species is known as Chinese Water Chestnut. Robinson notes the vernacular name in Zambia is *nasunga* and that the plant is burnt and used as a salt, while the stems may be used in basket making.

2. **Eleocharis acutangula** (Roxb.) Schult., Mant. **2**: 91 (1824). —Haines & Lye, Sedges Rushes E. Afr.: 67, figs.92,93 (1983). —Gordon-Gray in Strelitzia **2**: 77 (1995). —Browning *et al.* in S. African J. Bot. **63**: 175, fig.7 (1997). Type: India, *Roxburgh* s.n. (BM000847992 lectotype), lectotypified by Rosen (on sheet, 2007). FIGURE 14.**10B**.

Scirpus fistulosus Poir. in Lamarck, Encycl. **6**(2): 749 (1805), illegitimate name.

Scirpus acutangulus Roxb., Fl. Ind. **1**: 216 (1820).

Eleocharis fistulosa (Poir.) Schult. in Roem. & Schult., Mant. **2**: 89 (1824). —Clarke in F.T.A. **8**: 406 (1902). —Beentje in F.T.E.A., Cyperaceae: 39 (2010). Type: Madagascar, 'dans les marais', *Du Petit-Thouars* s.n. (P holotype).

Perennial to 120 cm tall, tufted at intervals by abbreviated, closely placed shoot bases; stolons long, nodose, terminating in erect, pith-filled, sharply 3-angled culms (triangular in transverse section). Leaves reduced to sheaths bearing a small triangular outgrowth at mouth, eventually blackening and decaying (especially if in water). Inflorescence a terminal cylindrical spike (a spikelet) 10–60 × 3–5 mm. Lowest glumes green, becoming brown, firm, persistent, acting as bracts; flowering glumes green to yellow-brown, imbricate, 4–5 mm long, ± ovate, not keeled but with many ridges (veins?), margin narrow, translucent, bounding a narrow darker zone, soon fading. Perianth of few short or long bristles, barbellate or smooth margined. Stamens 3. Style trifid. Nutlet 1.4–2.2 × 1.2–1.6 mm, surface longitudinally ridged with transverse bars between ridges, (almost trabeculate). Appendage (persistent style base) eventually drying, margin curling up as hollow triangular style base eventually shrivels and implodes, leaving narrow space between nutlet and still attached style base.

Botswana. N: Okavango swamps, Txatxanika Lagoon, 23°25'E 19°10'S, 1.iii.1972, *Gibbs Russell & Biegel* 1484 (K, LISC). **Zambia**. N: Mwense Dist., Chipili, 1220 m, 27.vi.1956, *Robinson* 1761 (K). C: Mkushi Dist., David Moffat's Farm, Munchiwemba dambo, 13°45'S 29°40'E, 1400 m, 20.ix.1993, *Bingham & Nkhoma* 9711 (K). S: Choma Dist., Mapanza, 1070 m, 21.iii.1954, *Robinson* 628 (K). **Zimbabwe**. W: Matobo Dist., Farm Besna Kobila, 1460 m, iii.1951, *Miller* 4245 (K, SRGH). C: Harare (Salisbury), 1640 m, fl. 17.ii.1927, *Eyles* 4742 (K, SRGH). E: Chimanimani Dist., Tarka Forest Reserve, 1000 m, xi.1970, *Goldsmith* 32/70 (K, SRGH). **Mozambique**. M: Matutuíne Dist., near Zitundo, 15.xi.1979, *de Koning* 7629 (K, LMU).

In Uganda, Kenya, Tanzania, Namibia, South Africa (Eastern Cape, KwaZulu-Natal, Limpopo, Mpumalanga) and Swaziland; also in Madagascar, Asia, Australia, South and Central America. In shallow water bodies, flooded grasslands and margins of lakes and streams; 10–1500 m.

Conservation notes: Widespread; not threatened.

3. **Eleocharis mutata** (L.) Roem. & Schult., Syst. Veg. **2**: 155 (1817). —Hooper in F.W.T.A.: 314 (1972). —Haines & Lye, Sedges & Rushes E. Afr.: 66 (1983). — Browning *et al.* in S. African J. Bot. **63**: 172 (1997). Type: Jamaica, no locality, *Patrick Browne* s.n. (LINN HL 71-2). FIGURE 14.**10C**.

    Scirpus mutatus L., Syst. Nat., ed.10 **2**: 867 (1759).
    Eleocharis fistulosa sensu C.B. Clarke, F.T.A. **8**: 406 (1902) in part, non (Poir.) Schult.

Tufted perennial; rhizomes long, 2–5 mm thick, tubers generally absent; roots coarse fibrous. Culms sharply 3-angled. Leaves reduced to sheaths with small triangular outgrowth at mouth. Inflorescence a terminal, cylindrical spike (a spikelet), tapering slightly distally, 22–40 × 3–4 mm. Bract 1, hard, without obvious venation, 4–6 mm deep, edges folding over and enclosing spikelet base as collar. Glumes closely spirally imbricate, appearing wider than long (much of length hidden by overlap), obovate to almost square when detached and flattened (depending upon truncate base width), 4–4.8 × (2.5)3–3.5(4.5) mm, apex broadly obtuse with central area slightly peaked, firm to hard, lacking obvious veins except for central midvein forming slightly projecting keel terminating below membranous apical edge (clearly visible above), grey green with dark brown to black submarginal line below membranous pallid edge, becoming uniform yellowish to pallid brown in age, lacking gland dots. Perianth usually of 6 bristles, exceeding nutlet in length, retrorsely barbed. Stamens 3; anthers 1.65–2.25 mm, crest 0.3 mm. Style trifid. Nutlet obpyriform to obovoid, biconvex, constricted subapically into clearly-defined thick rim just wider than ½ nutlet width, 1.9–2.2 × 1.4–1.6 mm excluding style base; surface finely trabeculate, marked by numerous closely placed longitudinal ridges linked by faint cross lines, cells slightly wider than deep, some almost isodiametric. Appendage (persistent style base) eventually drying, margin curling up as hollow triangular style base eventually shrivels and implodes, leaving narrow space between nutlet and style base.

Zambia. N: Mbala Dist., Saisi valley, 1525 m, 22.i.1970, *Sanane* 1049 (K).

Also Burkino Faso, Guinea, Liberia, Senegal, Togo, Congo, Gabon, Equatorial Guinea, Tanzania, Angola and South Africa (KwaZulu-Natal); also in West Indies, central and northern South America. In coastal habitats (the Zambian freshwater record may represent an introduction); c. 1500 m.

Conservation notes: Data deficient.

In its perennial stoloniferous growth form and sharply trigonous culms, *Eleocharis mutata* is indistinguishable from *E. acutangula*. It differs in its floral features in that there is resemblance to *E. dulcis* in the perianth, nutlet and style base. The shape of the glumes, nutlet apex and style base have been relied upon to distinguish it from *E. acutangula* (Hooper 1972: 314; Haines & Lye 1983: 66), but these characters are difficult to assess owing to the range variability encountered. The trigonous, nonseptate culms set it apart from *E. dulcis*.

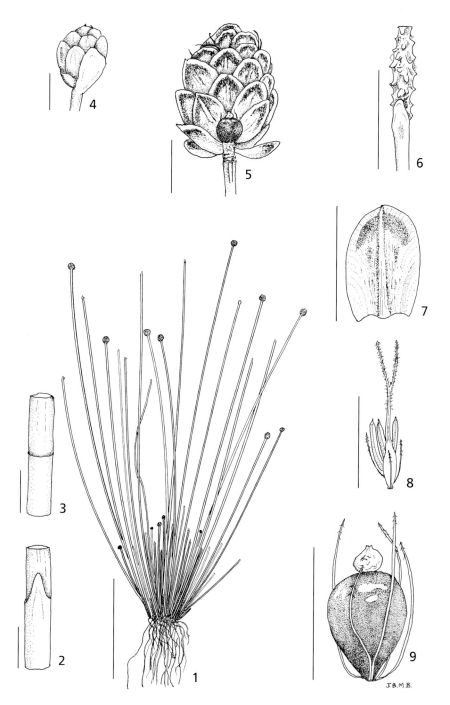

Fig. 14.**9**. ELEOCHARIS GENICULATA. 1, habit; 2, 3, leaf sheath; 4, 5, spikelet, young and mature; 6, spikelet after glumes shed; 7, glume, abaxial view; 8, floret; 9, nutlet. All from *Ward* 7945. Scale bars: 1 = 40 mm; 2–7 = 2 mm; 8, 9 = 1 mm. Drawn by Jane Browning.

4. **Eleocharis limosa** (Schrad.) Schult. in Roem. & Schult., Mant. **2**: 87 (1824). — Gordon-Gray in Strelitzia **2**: 80 (1995). Type: South Africa, Cape Province, *Hesse s.n.* (K, not located). FIGURE 14.**10D**.

Scirpus limosus Schrad. in Gött. Gel. Anz. **3**: 2069 (1821).
Limnochloa limosa (Schrad.) Nees in Linnaea **9**: 294 (1834).

Perennial; rhizome usually uniseriate; shoot bases mostly contiguous. Culms terete, 30–50 cm tall, width variable, (2)3–5(10) mm; vegetative organs often glaucous. Culm bases persistent on rhizome, not always conspicuous; sheath apex acute. Inflorescence a spike (spikelet) (15)36 mm long, mostly slightly broader than culm, often exhibiting pseudovivipary. Glumes c. 4 mm long, dark centrally, margins widely membranous, apex acute to obtuse. Perianth of (5)6 bristles, usually shorter than mature nutlet, margins toothed (scabrid). Stamens 3. Style trifid, occasionally bifid, variable within some florets of a single spikelet. Nutlet (excluding style base) 1.3–1.5 × 0.85–1 mm, dark brown, trigonous, obovoid; persistent style base triangular in outline; nutlet surface slightly rough, not otherwise marked.

Mozambique. GI: Xai-Xai Dist., Limpopo valley between Xai-Xai (João Belo) and Macia, 21.ix.1948, *Myre & Carvalho* 261 (K, LMA).

Also in Namibia, South Africa (Eastern Cape, KwaZulu-Natal, Limpopo, Mpumalanga, Northern Cape, North West, Western Cape), Lesotho and Madagascar. Usually in sandy substrates at margins of swamps and permanent water bodies where levels may fluctuate; 10–100 m.

Conservation notes: Widespread outside the Flora area; not threatened.

This species and its more tropical close associate, *Eleocharis marginulata*, are extremely variable morphologically. Study across the distributional area of both species in sub-Saharan Africa may indicate synonomy or infraspecific recognition.

5. **Eleocharis variegata** (Poir.) C. Presl in Isis (Oken) **21**: 269 (1828). —Svenson in Rhodora **41**: 8 (1939). —Haines & Lye, Sedges & Rushes E. Afr.: 68, figs.95,96 (1983). —Browning *et al.* in S. African J. Bot. **63**: 179, fig.8 (1997). —Beentje in F.T.E.A., Cyperaceae : 40 (2010). Type: Madagascar, "dans les marais", *Petit-Thouars* s.n. (P holotype). FIGURE 14.**10G1,2**.

Scirpus variegatus Poir. in Lamarck, Encycl. **6**: 749 (1805).

Tussocky perennial; stolons to 30 cm long, usually with black scales on internodes and adventitious roots from nodes; rhizome connecting shoot bases short and thickened. Culms 30(90) cm tall, variable, either rounded, triangular or quadrangular; leaf sheaths 3. Inflorescence a cylindric spike 10–55 mm long. Bracts inconspicuous; glumes 3–5 mm long, dark, apex obtuse, often frayed, margins translucent, central zone appearing ribbed (veined). Perianth bristles 7–8, longer than nutlet, reddish brown, margins densely set with retrorse teeth, or depauperate to absent except occasionally for a rim not easily observed. Stamens 3. Style branches 2–3, variable. Nutlet 1.5–1.9 × 1–1.5 mm, biconvex to rounded (not 3-ribbed longitudinally), surface finely ridged, eventually reticulated between ridges; style base appendage conical.

Botswana. N: Ngamiland Dist., Khianadiandavhu R. floodplain, 19°01.9'S 22°55.5'E, 16.v.1979, *P.A. Smith* 2748 (K). **Zambia**. B: Mongu Dist., Mongu, 5.iii.1956, *Robinson* 6861 (K). N: Luapula Dist., S part of Lake Bangweulu, Chishi Ngombe, S of Chafye Is., 1060 m, 14.ii.1996, *Renvoize* 5596a (K). C: Chongwe Dist., Chakwenga headwaters, 100–129 km E of Lusaka, 10.i.1964, *Robinson* 6201 (K). **Zimbabwe**. W: Matobo Dist., Besna Kobila, 12.iv.2015, *Browning* 950 (K). C: Marondera Dist., Grasslands Research Station, 3.i.1949, *Corby* 335 (K, SRGH).

Also in West Africa (Guinea, Ivory Coast, Nigeria), D.R. Congo, Angola, Uganda, Tanzania, South Africa (Eastern Cape, KwaZulu-Natal, Limpopo), Namibia, Madagascar and Indian Ocean Is. Floodplains and grasslands, stagnant shallow waters of pans, dambos and near hot springs; 1000–1400 m.

Conservation notes: Widespread; not threatened.

6. **Eleocharis marginulata** Steud., Syn. Pl. Glumac. **2**: 78 (1854). —Haines & Lye, Sedges & Rushes E. Afr.: 69, fig.98 (1983). —Beentje in F.T.E.A., Cyperaceae: 40 (2010). Type: Ethiopia, near Adoa, 20.xii.1838, *Schimper* II: 915 (P holotype, BR, K not located, M). FIGURE 14.**10E**.

    Scirpus marginulatus (Steud.) Kuntze, Revis. Gen. Pl. **2**: 757 (1891).

Robust perennial; culm bases persistent; rhizome creeping, horizontal. Culms numerous, closely placed, 20–80 cm × 1.5–2 mm, ridged, becoming compressed upwards. Leaves limited to sheaths, mouths truncate, blades reduced to minute outgrowths, or absent. Spikelet ovoid 8–20 × 3–4 mm. Glumes c. 4 mm long, margins narrow translucent; bracts equalling 2 basal sterile glumes, or all glumes fertile and bracts inconspicuous. Perianth of c. 6 minutely barbellate bristles, slightly shorter than mature nutlet. Stamens 3. Style trifid, occasionally bifid. Nutlet dull yellow to olive green at maturity, obovoid, 1.3–1.6 × 0.8–1.2 mm, faintly trigonous, apex narrowed, surface smooth to slightly rough; persistent style base conical, smooth.

Botswana. N: Okavango Swamps, Chief's Is., 19°15.5'S 22°50.8'E, fl. 22.iii.1979, *P.A. Smith* 2736 (K). **Zimbabwe**. C: Harare (Salisbury), Rainham Dam, 1460 m, 16.ix.1956, *Robinson* 1803 (K, LISC, NU, SRGH). E: Mutare Dist., Gairesi Ranch, 9.6 km N of Troutbeck, 1830 m, 16.xi.1956, *Robinson* 1911 (K). **Malawi**. N: Nyika Plateau, Lake Kaulime, 2285 m, 4.i.1959, *Robinson* 3051 (K).

Also in the Arabian Peninsula, Eritrea, Ethiopia, Kenya and Tanzania. In freshwater swamps, often in grassland, seldom coastal; 1000–2300 m.

Conservation notes: Widespread; not threatened.

7. **Eleocharis welwitschii** Nelmes in Kew Bull. **13**: 282 (1958). Type: Angola, Huíla, 1.iv.1860, *Welwitsch* 6969 (LISU holotype, K).

Perennial; rhizomes horizontal, creeping, robust with culm bases persistent. Culms numerous, closely placed, 8–25 cm tall, 0.75–1.5 mm thick, smooth, becoming compressed upwards. Leaves limited to sheaths, mouths truncate, blades reduced to minute outgrowths or absent. Spikelet ovoid or cylindric ovoid, 6–10 × 1.5–3 mm. Glumes 2.5–3 mm oblong-ovate, chestnut-coloured, margins wide, translucent; bracts equalling 2 basal sterile glumes, or all glumes fertile and bracts inconspicuous. Bristles 0 to few, barbellate, very short to slightly shorter than mature nutlet. Stamens 3. Style trifid, occasionally bifid. Nutlet dull yellow to olive green at maturity, obovoid, 1.2–1.5 (including style-base) × 0.8–1.1 mm, faintly trigonous, apex narrowed, surface smooth to slightly rough; persistent style base conical, smooth.

Botswana. N: Okavango Delta, 1.2 km W of Chief's Island, 19°17.898'S 22°53.954'E, 965 m, *A. & R. Heath* 1759 (K).

Also in Angola and Namibia. Freshwater swamps, often in grassland, seldom coastal; 500–1000 m.

Conservation notes: Moderately widespread; not threatened

Eleocharis welwitschii is a dwarf plant in comparison to *E. marginulata*, being c. 9 cm tall when in flower.

8. **Eleocharis decoriglumis** Berhaut in Bull. Soc. Bot. France **100**: 174 (1953). — Haines & Lye, Sedges & Rushes E. Afr.: 70, fig.99 (1983). —Browning *et al.* in S. African J. Bot. **63**: 181, fig.10 (1997). —Beentje in F.T.E.A., Cyperaceae: 41 (2010). Type: Senegal, no locality, n.d., *Perrotet* 839 (P holotype). FIGURE 14.**10F**.

Tufted annual. Culms (at flowering) 2–5 cm long, to 50 cm when fruiting, 2–4 mm thick, triangular with rather indistinct longitudinal ribs. Leaves few, each reduced to prophyll and sheath at culm base, blade only a small lobe. Inflorescence a rounded, terminal spike (spikelet) (1)2–3(5) cm long. Glumes greenish with distinct dark sub-marginal 0.2–0.5 mm band, apex obtuse, keel softly rounded. Perianth bristles 7–9, approximately nutlet length, margins with recurved hooks. Stamen 1. Style bifid. Nutlet 1.5–1.6 × 1.2–1.3 mm, lustrous cream when young,

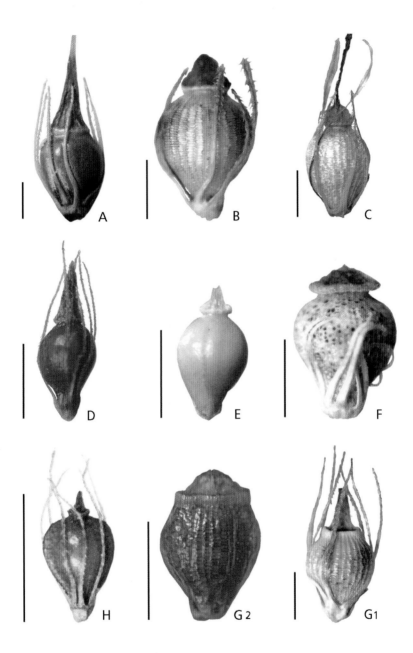

Fig. 14.**10**. ELEOCHARIS nutlets (1). A. —ELEOCHARIS DULCIS, from *Robinson* 1695.
B. —ELEOCHARIS ACUTANGULA, from *Eyles* 4742. C. —ELEOCHARIS MUTATA, from
Sanane 1049. D. —ELEOCHARIS LIMOSA, from *Myre & Carvalho* 261. E. —ELEOCHARIS
MARGINATULA, from *Heath* 1681. F. —ELEOCHARIS DECORIGLUMIS, from *Davey* 86.
G .—ELEOCHARIS VARIEGATA, G1, from *Robinson* 4288; G2, from *Robinson* 6861. H. —
ELEOCHARIS MINUTA, from *Dummer* 3163. Scale bar = 1 mm. Photomicrographs by Jane
Browning.

turning brown, broadly obovoid, base narrowed, upper section widened, at maturity surface marked by 18 or more longitudinal rows of isodiametric ± 6-sided cells; style base a swollen appendage c. 0.3 mm long, divided from nutlet by a visible constriction.

Botswana. N: Ngamiland Dist., Boteti R. floodplain below Samedupe Bridge, 5.ii.1977, *P.A. Smith* 1900 (NU, PSUB, SRGH).

In Senegal, Mali, Chad, Uganda and Tanzania. In seasonally wet depressions; 900–1000 m.

Conservaton notes: Rare, possibly threatened.

Eleocharis decoriglumis is a distinctive and attractive species with the very dark black bands near the glume margins. The surface of the nutlet shows shallow depressions rather than pits.

9. **Eleocharis minuta** Boeckeler in Bot. Jahrb. Syst. **5**: 503 (1884) as *Heleocharis*. — Svenson in Rhodora **41**: 54 (1939). —Haines & Lye, Sedges & Rushes E. Afr.: 71, fig.102 (1983). —Beentje in F.T.E.A., Cyperaceae: 43 (2010). Type: Madagascar, Imerina, vii.1880, *Hildebrandt* 3527 (B holotype, G, K, M). FIGURE 14.**10H**.

Annual or short lived perennial, densely tufted; stolons c. 2 cm long, inconspicuous, easily broken. Culms 1–15 cm × 0.2–0.7 mm, slightly flattened to 4-angled; leaf sheaths purple tinged, blade reduced to small triangular lobes. Spikelet ovoid, 2–6 × 1–2 mm elongating to c. 7 mm in fruit, 3–7-flowered. Bracts similar to glumes; glumes 1–2 mm long, reddish brown, margin paler, midvein greenish. Perianth of 5–7 unequal barbellate bristles, longer than nutlet. Stamens 3(2). Style bifid. Nutlet greenish when immature, turning dark olive brown, obovoid, 0.7–0.8 × 0.5–0.6 mm, surface smooth to minutely reticulate; persistent style base conical, set in conspicuous rim.

Zimbabwe. E: Chipinge Dist., Mt Selinda, 1160 m, fl. 14.ix.1947, *Whellan* 169 (K, SRGH).

Also in Burundi, Uganda, Tanzania, Madagascar, Mascarenes and E Australia. Fringes of permanent water bodies; 900–1200 m.

Conservation notes: Data Deficient.

Eleocharis minuta differs from *E. caduca* (Delile) Schult. only in the detail of style base. Both species have glumes 2 mm long and cannot be convincingly distinguished morphologically except on style base size and structure; both criteria are unsatisfactory as the style base (length of persistent portion and its basal form) varies on drying out when the basal rim turns upward and the main body may shrink and implode.

Svenson (in Rhodora **20**: 239, 1929) mentions that the type *Hildebrandt* 3527 is a dwarf plant 1–2 cm tall.

10. **Eleocharis caduca** (Delile) Schult. in Roem. & Schult., Mant. **2**: 88 (1824). — Haines & Lye, Sedges & Rushes E. Afr.: 71 fig.103 (1983). —Gordon-Gray in Strelitzia **2**: 78 (1995). Type: Egypt, Damietta, *Delile* s.n. (G lectotype, P), lectotypified by Menapace (on sheet, 1990). FIGURE 14.**11A**.

Scirpus caducus Delile, Descr. Egypte, Hist. Nat. **2**: 153 t.6, fig.2 (1813).
Eleocharis intricata Kük. in Repert. Spec. Nov. Regni Veg. **13**: 135 (1914). —Beentje in F.T.E.A., Cyperaceae: 42 (2010). Type: Tanzania, Rungwe Dist., Kyimbila, 17.ii.1912, *Stolz* 1132 (B holotype, K, P, PRE).
Eleocharis madagascariensis Cherm. in Bull. Soc. Bot. France **75**: 284 (1928). Types: Madagascar, Tamatave, 27.ix.1912, *Viguier* 410 (B syntype, P); Sambirano, Andranomafana, xi.1909, *Perrier de la Bâthie* 2646 (P syntype); Mahatringo, iii.1926, *Perrier de la Bâthie* 17630 (B syntype).
Eleocharis intricata var. *peteri* Schultze-Motel in Willdenowia **2**: 507 (1960). Type: Zimbabwe, Victoria Falls, 8.ix.1925, *Peter* 30806 (B holotype).

Annual or short-lived perennial; stolons horizontal, slender, easily overlooked and damaged. Culms 3–15 cm × 0.4–0.8 mm, angular; leaf sheaths membranous, often pallid, ending in a transparent triangular lobe. Spikelet 2–5 × 1–2.5 mm, ovate, apex acute. Glumes 1.8–2.5 mm long, light brown with green midrib (only 4–8 reported to bear florets), lowest glume bract-like, sterile. Perianth bristles 6–7, dark brown at maturity, margins finely toothed, exceeding nutlet length. Stamens 3. Style bifid. Nutlet dark brown to blackish, glossy, obovoid and flattened, 0.8–0.9 mm long, surface smooth; persistent style base triangular c. 0.3 mm long.

Botswana. N: Okavango swamps, Zibadianja lagoon, 23°30'E 18°30'S, 970 m, 19.x.1972, *Gibbs Russell* 2139 (K, SRGH). **Zambia**. C: Chibombo Dist., Chipembi, 1220 m, 23 km E of Chisamba, 16.vi.1955, *Robinson* 1303 (K). **Zimbabwe**. W: Hwange Dist., Victoria Falls, rainforest, 4.iv.1956, *Robinson* 1416 (K). **Mozambique**. MS: Gorongosa Dist., Gorongosa Nat. Park (Game Reserve), W slopes Cheringoma Plateau, Mussambidze Falls, c. 100 m, 12.xi.1971, *Ward* 7442 (K).

Also in North Africa (Algeria, Egypt, Libya), the Mediterranean and Arabian Peninsula, West Africa, Sudan, Somalia, Tanzania to South Africa (Eastern Cape, KwaZulu-Natal, Mpumalanga, North West Province), Mauritius and Madagascar. Fresh to saline water seepages in sand along coast, particularly adjacent to streamlets; 100–1300 m.

Conservation notes: Widespread; not threatened.

This species is similar in appearance to *Eleocharis geniculata* with a rounded inflorescence, but the inflorescence of *E. caduca* is approximately twice as long as wide. The nutlets of both species are also similar, but *E. caduca* may have a larger style base.

11. **Eleocharis geniculata** (L.) Roem. & Schult., Syst. Veg. **2**: 150 (1817). —Haines & Lye, Sedges & Rushes E. Afr.: 70, fig.101 (1983). —Gordon-Gray in Strelitzia **2**: 80 (1995). —Beentje in F.T.E.A., Cyperaceae: 43 (2010). Type: Jamaica, *Herb. Clifford* 21, Scirpus 1 (BM-000557653 lectotype), lectotypified by Furtado in Gard. Bull. Straits Settlem. **9**: 299 (1937). FIGURES 14.**9**, 14.**11B**.

    Scirpus geniculatus L., Sp. Pl.: 48 (1753), in part.

    Eleocharis capitata R. Br., Prodr. Fl. Nov. Holland.: 225 (1810). —Clarke in F.T.A. **8**: 407 (1902). Type: Australia, North Is., Gulf of Carpentaria, 16.xii.1802, *Brown* 5930 (BM holotype, E).

    Eleocharis kirkii C.B. Clarke in F.T.A. **8**: 410 (1902). Type: Zimbabwe, Victoria Falls, 1860, *Kirk s.n.* (K holotype).

Annual or short-lived perennial, closely tufted; often with inconspicuous fragile white stolons; adventitious roots shallow. Culms 12–15 cm × 0.6–0.9 mm, lengthening to c. 22 cm when fruiting, ridged; leaves few, reduced to small lobes terminating sheaths tinted reddish-purple. Inflorescence a solitary ovoid spikelet, 3–4 × 2.5–3.5 mm, partially covered when young by 2 greenish bracts resembling glumes. Glumes greyish below to brown apically, 1.8–2 × 1.4–1.5 mm, imbricate, loosening as fruits mature, midribs obscure, apex rounded. Perianth bristles c. 7, pink, softly glabrous, exceeding nutlet length. Stamens 3, filaments persisting on fruit until after dissemination. Style bifid. Nutlet biconvex, obovoid, 0.6–0.9 × 0.6–0.7 mm, smooth, shining purple/black at maturity; persistent style base conical, small.

Zambia. E: Mambwe Dist., Msoro (Mushoro), c. 80 km W of Chipata (Fort Jameson), Luangwa valley, 730 m, 9.vi.1954, *Robinson* 847 (K). **Zimbabwe**. N: Binga Dist., Binga, Chikwatata Hot Springs, 480 m, 7.xi.1958, *Phipps* 1392 (K, SRGH). W: Hwange Dist., Victoria Falls rainforest, 30.viii.1947, *Greenway & Brenan* 8030 (K). **Mozambique**. MS: Cheringoma Dist., Inhamissene R. flats, 8 m, 14.vii.1972, *Ward* 7945 (K). GI: Vilankulo Dist., San Sebastian Peninsula, 40 m, 27.vi.2002, *Jacobsen* 6014 (K, PRE).

Also in West Africa, Somalia, D.R. Congo, Uganda, Kenya, Tanzania, Namibia, South Africa (KwaZulu-Natal), and widely distributed in temperate and subtropical zones. Predominantly coastal along subtropical to temperate continental margins or adjacent to permanent waters of mangrove swamps and other temporary water bodies, rarely on inland lake margins; 5–1000 m.

Conservation notes: Widespread; not threatened.

Fig. 14.**11**. ELEOCHARIS nutlets (2). A. —ELEOCHARIS CADUCA, from *Ward* 7442. B. —ELEOCHARIS GENICULATA, from *Ward* 7945. C. —ELEOCHARIS SETIFOLIA, from *van Rensburg* 1912. D. —ELEOCHARIS COMPLANATA, from *Hooper & Townsend* 1995. E. —ELEOCHARIS ATROPURPUREA, from *Heath* 728; F. —ELEOCHARIS NIGRESCENS, from *Robinson* 3678. G. —ELEOCHARIS BRAINII, from *P.A. Smith* 1770. H. —ELEOCHARIS NAUMANNIANA, from *Jordan* 814. J. —ELEOCHARIS RETROFLEXA subsp. CHAETARIA, from *Verboom* 1784. Scale bar = 0.5 mm. Photomicrographs by Jane Browning.

Similar in appearance to *Eleocharis caduca*. Bristles are reported to be glabrous but in the nutlet illustrated they are barbellate.

12. **Eleocharis setifolia** (A. Rich.) J. Raynal in Adansonia, n.s. **7**: 318 (1967) as *Heleocharis*. —Haines & Lye, Sedges & Rushes E. Afr.: 73, fig.109 (1983). — Beentje in F.T.E.A., Cyperaceae: 42 (2010). Type: Ethiopia, Tacazze R., 1844, *Quartin Dillon & Petit* s.n. (P holotype). FIGURE 14.**11C**.
   Isolepis setifolia A. Rich., Tent. Fl. Abyss. **2**: 498 (1851).

Subsp. **setifolia**

Perennial or annual with annual aerial organs; small tubers on scale-bearing, branching, white, delicate, easily-overlooked, underground rhizomes, twisted among adventitious roots. Culms filiform, erect, crowded, 4–20 cm × 0.3–0.4 mm; leaves few, reduced to pale sheaths each tipped by a minute triangular outgrowth. Spikelet ovoid to obovoid, 2–5 × 1–2.5 mm, lengthening to 5 mm in fruit. Bracts 2, greenish, glume-like; glumes pale brown or pale greenish brown, 1–1.3 mm long, apex obtuse to emarginate. Perianth lacking. Stamen 1. Style trifid, white. Nutlet yellow to light brown, 0.5–0.7 × 0.3–0.4 mm, ureolate-obovoid, strongly 3-ribbed longitudinally, surface smooth to minutely tuberculate; style base conical to 3-sided, set in conspicuous rim.

Zambia. W: Mwinilunga Dist., c. 2 km E of Matonchi Farm, c. 1350 m, 18.ii.1975, *Hooper & Townsend* 204 (K). S: Mazabuka Dist., between Mazabuka and Kaleya, 15.iv.1963, *van Rensburg* KBS 1912 (K); Choma Dist., 47 km N of Choma, 1220 m, 5.i.1957, *Robinson* 2024 (K).

Found from Senegal to Nigeria, Ethiopia, Sudan, D.R. Congo, Kenya and Tanzania; also in Phillipines, Australia, Brazil and the Carribean. In seasonally wet grassland and fringes of localised water bodies; 850–1700 m.

Conservation notes: Widespread; not threatened.

Another subspecies, subsp. *schweinfurthiana* (Boeckeler) D.A. Simpson, is found from Mali to Nigeria, Senegal, Sudan and D.R. Congo. In this subspecies the bristles are at least half the nutlet length.

The species is close to *Eleocharis nigrescens* but differences can be seen in the nutlets.

13. **Eleocharis atropurpurea** (Retz.) J. Presl & C. Presl, Reliq. Haenk. **1**: 196 (1828). —Haines & Lye, Sedges & Rushes E. Afr.: 72, fig.104 (1983). —Gordon-Gray in Strelitzia **2**: 77 (1995). —Beentje in F.T.E.A., Cyperaceae: 44, fig.8 (2010). Type: India, no locality, n.d., *Konig* s.n. (LD holotype). FIGURE 14.**11E**.
   Scirpus atropurpureus Retz., Observ. Bot. **5**: 14 (1788).

Annual with shallow rooting system. Culms numerous, clustered, filiform, 1–12(30) cm × 0.2–0.3 mm. Leaves reduced to 2 sheaths above prophyll at culm base, sometimes purple tinged; blade a minute outgrowth at sheath mouth or absent. Spikelet ovoid, 2–8 × 1–2 mm. Bracts lacking; glumes all fertile, dark reddish brown or purple with green midrib, basal ones early deciduous, upper ones loosening to expose mature fruits, c. 1 mm long, apex narrowly obtuse to broadly acute. Perianth of 4–5 barbellate bristles, shorter than nutlet, or lacking. Stamens 1–3. Style bifid. Nutlet blackish brown to shining black at full maturity, surface smooth, biconvex, broadly ovate in outline, 0.5–0.6 × 0.3–0.4 mm; persistent style base small, round, button-like.

Botswana. N: Okavango Delta, Chief's Is., 19°18'26.7S 22°54'16.38E, 960 m, 5.ix.2009, *A. & R. Heath* 1710 (K). **Zambia**. W: Solwezi Dist., c. 5 km E of Kabompo R. on Mwinilunga–Solwezi road, 1400 m, 27.ii.1975, *Hooper & Townsend* 421 (K). S: Choma Dist., Mapanza, 1070 m, 9.v.1954, *Robinson* 740 (K). **Zimbabwe**. W: Bulilamamangwe Dist., 21 km NW of Marula, 20°20'S 28°02'E, 20.iv.1972, *Gibbs Russell* 1739 (K, SRGH). C: Chegutu Dist., Chegutu (Hartley), Poole Farm, 1219 m, fl. 5.iii.1946, *Hornby* 2433 (K, SRGH). E: Chipinge Dist., pan near Chibuwe (Chibuya), 32°18'E 20°30'S, 700 m, fl. 1.vi.1972, *Gibbs Russell* 2077 (K, SRGH). S: Beitbridge Dist., near Tshiturapadsi (Chiturupadzi) store, 40 km NNW of Bubi (Bubye)–Limpopo confluence, 12.v.1958, *Drummond* 5763 (K, LISC, LMA, SRGH). **Malawi**. C: Nkhotakota Dist., 6 km N of Nkhotakota (Nkhota Kota), 490 m, 16.vi.1970, *Brummitt* 11449 (K).

Also in West Africa, Sudan, Ethiopia, D.R. Congo, Uganda, Kenya, Tanzania, Namibia and South Africa (Gauteng, KwaZulu-Natal, Limpopo, Mpumalanga) and Swaziland; widespread in the tropics and subtropics of Old and New Worlds. Margins of seasonal water bodies (pools, rice paddies, seepage areas), usually in alluvium or in fallows; 300–1400 m.

Conservation notes: Widespread; not threatened.

The species is most easily recognised by the small, smooth and shiny black nutlets, which may have short perianth bristles.

14. **Eleocharis nigrescens** (Nees) Kunth, Enum. Pl. **2**: 157 (1837). —Haines & Lye, Sedges & Rushes E. Afr.: 72, fig.106 (1983). —Beentje in F.T.E.A., Cyperaceae: 46 (2010). Type: Brazil, Bahia ("in maritimis"), n.d., *Salzmann* s.n. (CGE holotype, K, P). FIGURE 14.**11F**.

    Scirpidium nigrescens Nees in Martius, Fl. Bras. **2**(1): 97 (1842).
    Isolepis nigrescens Steud., Syn. Pl. Glumac. **2**: 91 (1855).
    Eleocharis hildebrandtii Boeckeler in Flora **61**: 34 (1878). —Clarke in Durand & Schinz, Consp. Fl. Afr. **5**: 598 (1894); in F.T.A. **8**: 409 (1902), as *Heleocharis*. Type: Tanzania, Zanzibar, x.1873, *Hildebrandt* 1063 (B holotype, K).

Annual or short-lived perennial; loosely tufted with slender shortly branched rhizomes and shallow roots. Culms filiform, 3–12 cm × 0.2–0.5 mm, obscurely 3–4 angled, sometimes flattened; leaf sheaths reddish to purple near base, uppermost sheath terminating in lobe 1–2 mm long. Spikelet ovoid, 2–5 × 1–2.5 mm. Inflorescence bracts (2 basal glumes) fertile, slightly larger than upper glumes; glumes 0.8–1.3 × 0.4–1 mm, apex obtuse, body usually dark, margins and midrib paler. Perianth lacking. Stamen 1. Style trifid. Nutlet pale brownish yellow, semi-translucent, obovoid-triangular, 0.4–0.6 × 0.2–0.4 mm, with 3 longitudinal pale ribs, almost winged on drying, surface smooth; style base flattened or minutely conical, persistent.

Zambia. N: Kasama Dist., 100 km E of Kasama, 18.ii.1962, *Robinson* 4941 (K, excl. 4941a). S: Choma Dist., 8 km E of Choma, 1310 m, 27.iii.1955, *Robinson* 1195 (K). **Zimbabwe**. C: Harare (Salisbury), 1460 m, 1.v.1932, *Brain* 8971 (K, SRGH). **Malawi**. N: Chitipa Dist., Namwila Valley, Chisenga, 10°3'42.36"S 33°22'25.87"E, 3.viii.2007, *Chapama et al.* 690 (K).

Also in Senegal, Mali, Ivory Coast, Nigeria, Chad, Sudan, Kenya, Uganda, Tanzania; also in Madagascar and tropical America. Margins of temporary water bodies such as pool edges, depressions in disused cultivated land and rice paddies; 800–1500 m.

Conservation notes: Widespread; not threatened.

Similar to *Eleocharis atropurpurea* except for lack of perianth and the triangular, not biconvex, nutlet. Nutlet most similar to *E. setifolia* in colour, shape and size but not in shape and form of the style base.

15. **Eleocharis brainii** Svenson in Rhodora **39**: 251 (1937). —Hooper in F.W.T.A., ed.2 **3**: 313 (1972). —Raynal in Adansonia, n.s. **7**: 318 (1967). —Haines & Lye, Sedges & Rushes E. Afr.: 74, fig.112 (1983). —Beentje in F.T.E.A., Cyperaceae: 44 (2010). Type: Zimbabwe, Harare (Salisbury), 1.v.1932, *Brain* 8963 (K holotype, G, LISC, LMA, SRGH). FIGURE 14.**11G**.

Annual with white slender roots, sometimes with inconspicuous rhizomes making it a short-lived perennial. Culms numerous, tufted, 1–5(13) cm tall, filiform or obscurely 3-angled; leaves reduced to sheaths, reddish purple near base, each topped by minute triangular lobe. Spikelet ovoid, 1–2 × 0.5–1 mm of 4–5 fertile glumes; lower glume almost colourless, upper reddish-brown with green mid-nerve, 1–1.5 mm long, apex obtuse, spreading widely at nutlet maturity. Perianth absent or with a few short colourless barbed bristles. Stamens 1. Style trifid. Nutlet greyish-white, urn-shaped, 0.5–0.6 × 0.4–0.5 mm, 3-ribbed longitudinally, surface with longitudinal rows of rounded pits; style base low-pyramidal or merely a shallow rim.

Botswana. N: Okavango swamps, Gidiba Is., 19°02.1'S 22°31.8'E, 14.ix.1976, *P.A. Smith* 1770 (K). **Zambia**. W: Solwezi Dist., Solwezi, 1350 m, 10.iv.1960, *Robinson* 3510 (K). N: Mbala Dist., Kali dambo near Kawembi Mission, 1550 m, 16.v.1952, *Richards* 1664 (K). S: Choma Dist., Choma–Namwala road, 31 km N of Choma, 1040 m, 20.ii.1956, *Robinson* 1351 (K). **Zimbabwe**. W: Hwange Dist., Victoria Falls, rainforest, 30.viii.1947, *Greenway & Brenan* 8028 (K). C: Harare (Salisbury), no locality, 1460 m, 1.v.1932, *Brain* 8963 (GH, K, LISC, LMA, SRGH). **Malawi**. N: Chitipa Dist., above Chisenga Village towards foot of Mafinga Hills, 1580 m, 12.vii.1970, *Brummitt* 12048 (K). **Mozambique**. MS: Dondo Dist., 25 Miles [40 km] Station, 60 m, 10.iv.1898, *Schlechter* 12232 (K).

Also in Ghana, Nigeria, D.R. Congo, Sudan, Uganda and Tanzania. Plants often partly submerged in swamps in grassland, rock pools and fringes of permanent water bodies; 50–1600 m.

Conservation notes: Widespread; not threatened.

16. **Eleocharis naumanniana** Boeckeler in Bot. Jahrb. Syst. **5**: 92 (1884), as *Heleocharis*. —Clarke in C.F.A. **5**: 599 (1895); in F.T.A. **8**: 411 (1902). Type: Liberia, Monrovia, viii.1874, *Naumann* 20 (B holotype, K). FIGURE 14.**11H**.

  Heleocharis naumanniana Boeckeler in Bot. Jahrb. Syst. **5**: 92 (1884).
  Eleocharis testui Cherm. in Bull. Soc. Bot. France **77**: 276 (1930). Type: Gabon, Haute Ngounye, Les Echiras, 15.xii.1925, *Le Testui* 5816 (P holotype).
  Eleocharis naumanniana var. *naumanniana* Nelmes & Baldwin in Amer. J. Bot. **39**: 373 (1952). —Hooper in F.W.T.A., ed.2 **3**: 312 (1972).

Aquatic, forming submerged mats in freshwater. Stems extremely fine, 0.2–0.4 mm wide, rounded or angular, with distinct septa, branching, forming a tangled mass with roots at nodes. Leaves reduced to sheaths. Inflorescence a single spikelet, 4–5 mm long, acuminate. Glumes c. 3, lanceolate, overlapping, streaked maroon, enclosing a single floret. Perianth absent or very reduced. Stamens 3. Style trifid. Nutlet 1.3 × 0.9 mm, trigonous, ovoid to obovoid, yellow to brownish, surface longitudinally ridged with shallow transverse bars over pits; style base triangular, persistent.

Botswana. N: Okavango swamps, mouth of Dxherega Lediba on Moanachira R., 17.i.1980, *P.A. Smith* 2988 (PRE, SRGH). **Zambia**. N: Chinsali Dist., Shiwa Ngandu (Ishiba Ngandu), 1525 m, 4.vi.1956, *Robinson* 1567 (K).

Also in West Africa (Guinea, Ivory Coast, Senegal), Gabon and D.R. Congo. Submerged in water bodies; 900–1500 m.

Conservation notes: Widespread; not threatened.

Plants may be terrestrial, appearing tufted in damp mud beside water bodies, or aquatic and submerged with spreading mat-forming capillary stems (culms).

Two varieties are mentioned by Hooper (1972: 311) – the typical variety and var. *caillei* Nelmes (in Bull. Misc. Inform. Kew **1951**: 165, 1951). Differences between the two are in the length of the spikelets and degree of development of septa.

Eleocharis naumanniana resembles *Websteria confervoides* (Poir.) S.S. Hooper in its habit and habitat, but has 3 style branches and a differing nutlet. *Websteria* has been included in *Eleocharis* but is here kept separate.

Although well-represented in West Tropical Africa, *Eleocharis naumanniana* does not appear to have been collected in lakes and dams in other parts of Zambia.

17. **Eleocharis retroflexa** (Poir.) Urb., Symb. Antill. **2**: 165 (1900). Type: not known.

  Scirpus retroflexus Poir. in Lamarck, Encycl. **6**: 753 (1805).

Subsp. **chaetaria** (Roem. & Schult.) T. Koyama in Bull. Natl. Sci. Mus. Tokyo **17**: 68
(1974). —Haines & Lye, Sedges & Rushes E. Afr.: 75, fig.114 (1983). —Beentje in
F.T.E.A., Cyperaceae: 44 (2010). Type: India, "in humid grassy places of Calcutta",
König s.n. FIGURE 14.**11J**.
> *Eleocharis chaetaria* Roem. & Schult., Syst. Veg., ed.15 **2**: 154 (1817).
> *Scirpus chaetarius* Spreng., Syst. Veg., ed.16 **1**: 203 (1824).

Annual, closely tufted, often with inconspicuous fragile white stolons; adventitious roots
shallow. Culms 5–17 cm × 0.5–0.7 mm, sometimes elongating in fruit to 25 cm, angular, ridged;
leaves reduced to small lobes terminating membranous sheaths, sometimes tinted reddish-brown.
Inflorescence a solitary ovate spikelet, 4–6 × 2.5–3.5 mm. Glumes c. 3.5 mm long, streaked red-
brown, with wide transparent margin and apex, minutely emarginate, midnerve green, ending
prior to obtuse apex. Perianth bristles 3 + 3, with recurved spines, sometimes exceeding nutlet
length. Stamens 3. Style trifid. Nutlet 0.8–1 (excluding appendage) × 0.7–0.8 mm, broadly ovoid,
trigonous, angles prominent, greenish when young turning almost black at maturity, surface
pitted in longitudinal rows; appendage large triangular, pale yellow-white.

Zambia. N: Mongu Dist., Mongu, 20.ii.1966, *Robinson* 6847 (K, LMA); Mongu,
6.xi.1964, *Verboom* 1784 (K).
In Angola and possibly W Tanzania. In seasonally wet grassland; c. 1000 m.
Conservation notes: Data Deficient.
One problem that requires investigation is that *Eleocharis retroflexa* subsp. *chaetaria* is
perhaps amphibious and the other subspecies, subsp. *subtilissima* (Nelmes) Lye may
be the aquatic 'form' that has been given a separate ranking. C. Archer (pers. comm.)
writes that in 1988 J. Bruhl identified a specimen from the Okavango swamps in N
Botswana (*Ellery* 15 (J, PRE), backwaters W of Xaxanika lagoon, 19°11'S 23°23'E, on
floating mats with other sedges and grasses, 6.iv.1984) as subsp. *subtilissima*. Claire
Archer wrote: "I noted that the fruit had bristles. The leaves and culms are filiform, but
I would not call them thread-like. On this plant some inflorescences were proliferous."
In addition she matched this with another Okavango specimen (*Ellery* 392 (J, PRE),
backwater just S of Godikwe lagoon, 19°09'S 23°14'E, on floating sudd, 10.v.1985) and
stated "These are quite neat little tufts. Leaves and culms as above but a bit shorter."

General note on aquatic species of *Eleocharis*.

Little material of aquatic *Eleocharis* species is available, but further study especially
under field conditions is required for fragile underwater species which up to now
have been linked by vague morphological similarities. In our opinion the following
species need a detailed examination and assessment across the African continent –
E. brainii, *E. retroflexa* subsp. *chaetaria* and subsp. *subtilissima*, *E. naumanniana* subsp.
caillei and *E. cubangensis* H.E. Hess. More complete descriptions and confirmation of
distributions are also required.

Excluded taxa.

Eleocharis complanata Boeckeler in Flora **62**: 562 (1879). —Haines & Lye, Sedges &
Rushes E. Afr.: 73, fig.110 (1983). —Mapaura & Timberlake, Checkl. Zimbabwe
Pl.: 88 (2004), in error? —Beentje in F.T.E.A., Cyperaceae: 41 (2010). Type: Sudan,
Bongo, Gir, 25.x.1869, *Schweinfurth* 2576 (B holotype, S, K). FIGURE 14.**11D**.
> *Eleocharis anceps* Ridl. in Trans. Linn. Soc. London, Bot. **2**: 148 (1884). Type: Angola,
> Cuanza Norte (Pungo Andongo), v.1857, *Welwitsch* 7170 (BM, LISU).

Densely tufted annual, lacking a rhizome. Many culms per plant, 5–20 cm × 1–2 mm, markedly
flattened; leaf sheaths usually purplish towards base, ending in a triangular lobe. Spikelets mostly
narrowly ovoid, 3–15 × 1.5–3 mm. Glumes pale with dark-red midsection divided by pale midrib,
1.7–2 × 0.9–1 mm, apex obtuse, becoming frayed. Perianth lacking. Nutlet ovate, trigonous,
3-ridged longitudinally, otherwise surface smooth.

Not yet recorded from the Flora area, but likely to be found here. Otherwise known from Nigeria, Senegal, Sudan, Tanzania and Angola. In sandy substrates in seepages or fringing other small temporary water bodies.

Close to *Eleocharis nigrescens*, but with flattened, taller culms, longer glumes and nutlets.

Eleocharis cubangensis H.E. Hess in Ber. Schweiz. Bot. Ges. **62**: 339 (1953). Type: Angola, Cuchi, 12.ii.1952, *Hess* 52/674 (Z holotype).

This species, described from a collection in Angola, was reported to be morphologically close to *E. brainii*. No material was available for study. There have been reports that it may be "endemic to the Okavango River and in Namibia" (Golding 2002), but this requires further investigation.

6. **WEBSTERIA** S.H. Wright[6]

Websteria S.H. Wright in Bull. Torrey Bot. Club **14**: 135 (1887).

Perennial herbs, submerged or floating. Culms elongate, many-noded, frequently branched, with whorls of vegetative branchlets subtended by small bract scales; occasionally intermingled with more robust fertile whorls with single terminal spikelets. Leaves basal to branchlets, reduced to tubular sheaths. Inflorescences often emergent, a single spikelet terminal on each branch, 1-flowered, bisexual. Glumes 2, distichous, elongate, eventually deciduous; lower barren, upper subtending floret. Perianth of bristles disseminated with nutlet. Stamens 3. Style branches 2. Nutlets ovoid, crowned by flattened persistent style base.

A monotypic genus widely distributed in tropical Asia, Africa and Americas.

Websteria confervoides (Poir.) S.S. Hooper in Kew Bull. **26**: 582 (1972); in F.W.T.A., ed.2 **3**: 314 (1972). —Haines & Lye, Sedges & Rushes E. Afr.: 76, fig.117 (1983). —Beentje in F.T.E.A., Cyperaceae: 47, fig.9 (2010). Type: Madagascar, no locality, n.d., *Petit-Thouars* s.n. (P holotype). FIGURE 14.**12**.
    Scirpus confervoides Poir. in Lamarck, Encycl. **6**: 755 (1805).
    Eleocharis confervoides (Poir.) Steud., Syn. Pl. Glumac. **2**: 82 (1855).

Perennial, submerged or floating, delicate elongate noded stems with numerous whorls of lateral branches subtended by pink to purplish bract scales, each branch base enveloped by a tubular prophyll. Inflorescence a series of spikelets at ends of most whorl branches, many emergent from water. Spikelets 8–12 × 1.5–2 mm, 1-flowered, bisexual; glumes 2, lanceolate, 8–12 mm long, closely enveloping floret, green to reddish brown. Perianth segments up to 11, filiform with marginal retrorse spines, longer than surrounding nutlet, persistent during dispersal (air enclosed among bristles possibly aiding flotation). Stamens 3, filaments lengthening at anthesis to extrude anthers beyond enveloping glumes. Style branches 2; style base persistent on nutlet, elongate, flattened. Nutlet ovoid, pale brown, rounded not angled, c. 2 × 1.5 mm, with tapering beak, surface smooth to faintly cellular.

Botswana. N: Okavango swamps, Hamokae Lediba, 19°07.5'S 23°07.6'E, 25.vi.1973, *P.A. Smith* 649 (K); Okavango, backwater of Moanachira R., 3.v.1973, *P.A. Smith* 571 (K). **Zambia**. N: Luwingu Dist., Luwingu, i.1961, *Robinson* 4245 (NY).

In W Africa, D.R. Congo and Tanzania; also in Madagascar, S.E. Asia, tropical America and southern USA. In tropical, warm fresh or stagnant permanent waters, water depth 20–100 cm; 800–1000 m.

Conservation notes: Widespread; not threatened.

---

[6] by J. Browning and K.D. Gordon-Gray†

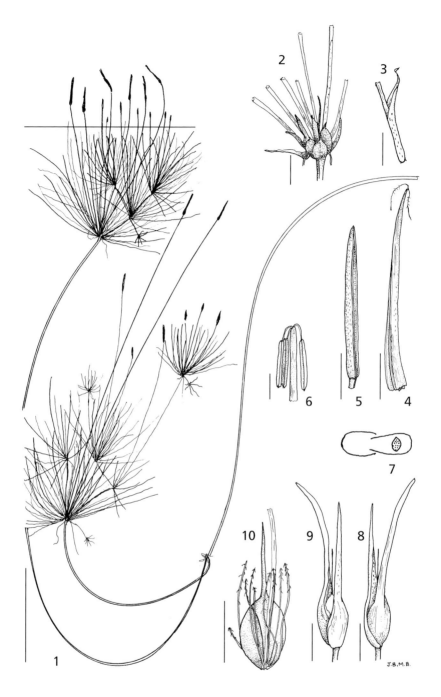

Fig. 14.**12**. WEBSTERIA CONFERVOIDES. 1, habit; 2, 'node' subtending branches; 3, leaf sheath apex; 4, 5, spikelet with upper and lower scale, respectively; 6, spikelet apex; 7, diagram of spikelet; 8, 9, spikelets, opposing views; 10, nutlet with bristles. All from *P.A. Smith* 571. Scale bars: 1 = 40 mm; 2–10 (except 7) = 2 mm. Drawn by Jane Browning.

As a submerged perennial hydrophyte, sexual organs are usually sporadic and sparse. Its main relationship is with *Eleocharis*, although embryos of the two genera differ slightly, that of *Websteria* having an enlarged cotyledon and third leaf.

7. **FIMBRISTYLIS** Vahl[7]

Fimbristylis Vahl, Enum. Pl. **2**: 285 (1805), conserved name.

Perennial and annual, occasionally biennial herbs, densely to sparsely tufted; rhizome sympodial, woody, or hardly developed. Leaves variable, sheaths usually closely enclosing culms, splitting with age; mouth truncate or sloping, glabrous, or ciliate; ligule 0 or a line of short hairs; blades well-developed or obsolete (both can be found in same population or on same plant, *F. complanata*), width variable, occasionally filiform, apex often asymmetric. Culms scapose, terete or ridged by 2–6 angles, variously pubescent to glabrous. Inflorescences terminal, variable on individual plants, anthelate of c. 1–200 spikelet(s) in 3, 2, 1 or 0 branching systems. Spikelets compact, mostly cylindric, many-flowered, basal glumes early deciduous. Glumes basal, 1–2, sterile, remainder (except most apical) subtending one bisexual floret. Perianth 0. Stamens 3, 2 or 1. Style 2(3)-angled, ribbon-like or filiform, villous, ciliate, glabrescent or glabrous, base thickened, distinct, occasionally of style width. Nutlet biconvex or trigonous, obovate, ovate-orbicular in outline; gynophore sometimes present; surface usually longitudinally ridged, trabeculate, occasionally smooth, seldom tuberculate; epidermal cells subquadrate or transversely elongate, generally lacking silica deposits.

A large genus of c. 300 species, cosmopolitan with its maximum diversity in Malaysia. Species are not easily identified by a few simple characters. Several sub-genera have been established, with some having been given generic status.

Key to Fimbristylis

In couplet 5 there may be some exceptions in which style branch number is not uniform in all florets. Use the majority number, or try the alternative entry. In couplet 6 note that *F. madagascariensis* closely resembles *F. ferruginea* and should also key out here. However, the cells of the nutlet may be somewhat transversely oblong not strictly isodiametric.

1. Style base with many to one pendant outgrowth(s) that fringe developing nutlet
   . . . . **14.** *squarrosa*
- Style base lacking outgrowths 2
2. Caespitose perennial with coarse, spongy culms mostly 1.5–1.8 m tall, nodeless between base and inflorescence; leaves reduced to bladeless sheaths. . .**6.** *bivalvis*
- Plants differing markedly from above 3
3. Plant short; tufted leafy annual or 1–2 year perennials; culms slender, unbranched, bearing terminally 1(3) sessile or almost sessile spikelets. 4
- Plant differing in one or more respects 5
4. Inflorescence of 1(3) spikelets; stamens 3 per floret (occasionally number reduced, usually to one); wet grassland habitats . . . .**13.** *schoenoides*
- Plants not as above. 5
5. Style branches mostly 2; nutlets biconvex to rounded 6
- Style branches mostly 3; nutlets trigonous or rounded. 9
6. Glumes bearing fine, short white hairs on upper section abaxially (need ×10 magnification; hairs wearing away with age); nutlets 1.2–1.7 × 0.7–1.3 mm, clearly biconvex, round in outline with well-developed persistent stalk, surface almost smooth, reticulated by small isodiametric cells . . **7.** *ferruginea* /**8.** *madagascariensis*

─────────────────────

[7] by K.D. Gordon-Gray† and J. Browning

–   Plants not as above. 7
7. Spikelets 1.2–1.6 mm wide, clearly angled by projecting glume keels; nutlets obovoid, 0.4–0.6 mm at widest, surface with 5(6) longitudinal ridges to each face; plants of wet habitats, often stream margins in coastal situations. . **10.** *bisumbellata*
–   Spikelets often exceeding 2 mm width, not markedly angled by projecting glume keels; nutlets c. 1.2 mm at widest breadth; plants often scattered in grassland or as undergrowth in woodland, wet, damp or dry 8
8. Leaves with scabrid hairs marginally (if margins pubescent then with hairs soft not scabrid); Nutlets lacking marked longitudinal ridges (avoid young examples) . .
    . . . . **12.** *scabrida*
–   Leaves variously glabrous to pubescent to inconspicuously pubescent; nutlets with defined (usually well marked) longitudinal ridges 9
9. Nutlet (6)7–10-ridged, plants ± glabrous to inconspicuously pubescent; inflorescence with many spikelets in (1)2–3 orders of branching, main bract usually exceeding inflorescence . . . .**9.** *dichotoma*
–   Plants differing from above in being clearly pubescent; nutlets with more than 10 longitudinal ridges 10
10. Nutlets with 12–20 longitudinal ridges (avoid young examples as ridges are late in development); plants pilose pubescent . . . . **11.** *pilosa*
–   Plants differing from above. 11
11. Leaf blades numerous, often radiating, thickened; culms often, but not always, flat and twisted. 12
–   Leaf blades reduced to filiform outgrowths up to 4 mm long 13
12. Inflorescence a central compact head of small spikelets, 3–8 × 1.2–2.5 mm, with or lacking one to many lateral pedicels with terminal reduced heads or a solitary spikelet; leaf apices truncate; plants coastal. . . . . **3.** *cymosa*
–   Inflorescence an open compound anthela of central spikelet with few to many lateral pedicels, often branching and bearing mostly solitary spikelets terminally; culms often flat and twisted; leaf apices with blunt apiculate apices not strictly truncate; plants coastal and inland in permanently wet or damp habitats . . . . . .
    . . . .**1.** *complanata*
13. Leaf blade reduced to filiform tips up to 4 mm long, white ciliate on margin surrounding reduced blade, but no formal ligule; culms 4–5 angled, spongy, glabrous, conspicuous. . . . .**2.** *aphylla*
–   Plants differing in one or more criteria. 14
14. Ligule a clearly defined fringe of dense white hairs; culms 4–6 angled sufficiently pronounced to give a slightly flattened appearance . . . . **5.** *microcarya*
–   Ligule absent; leaf blades with prominent midrib. . . . .**4.** *quinquangularis*

1. **Fimbristylis complanata** (Retz.) Link, Hort. Berol. **1**: 292 (1827). —Haines & Lye in Sedges & Rushes E. Afr.: 78, fig.119 (1983). —Goetghebeur & Coudijzer in Bull. Jard. Bot. Belg. **54**: 82 (1984). —Gordon-Gray in Strelitzia **2**: 90 (1995). —Verdcourt in F.T.E.A., Cyperaceae: 50 (2010). Type: India, no locality, n.d., *König* s.n. (LD holotype, C).

    Scirpus complanatus Retz., Observ. Bot. **5**: 14 (1788).
    Fimbristylis consanguinea Kunth, Enum. Pl. **2**: 228 (1837). Types: South Africa, Cape Province, 1840, *Drège* 4414, 4418, 7404 (B† syntypes).
    Fimbristylis bequaertii De Wild., Pl. Bequaert. **4**: 200 (1927). Type: D.R. Congo, Kibimbi, 2.ii.1911, *Bequaert* 126 (BR holotype).
    Fimbristylis complanata subsp. *keniaeensis* (Kük.) Lye in Nordic. J. Bot. **2**: 334 (1982). —Haines & Lye in Sedges & Rushes E. Afr.: 79, fig.122 (1983). —Verdcourt in F.T.E.A.,

Cyperaceae: 51 (2010). Types: Kenya, Mt Kenya, W foothills, Cole's Mill, 18.i.1922, *R.E. &
T.C.E. Fries* 1072 (B† syntype, K, UPS); W Kenya Forest Station, 18.i.1922, *R.E. & T.C.E. Fries*
728 (B† syntype, K, UPS).

Perennial, sparsely to densely tufted, to 80 cm tall; rhizome creeping, uniseriate or irregular,
woody, up to 140 × 5–7 mm, clothed in imbricate scale leaves and leaf sheaths. Leaves with
flattened sheaths, angles glabrous or scabrid pubescent, mouth markedly sloping, white ciliate
to glabrous; ligule a fringe of dense white short hairs; blade variable in development, 16–320 ×
0.9–4.5 mm, or reduced to absent, occasionally falcate or twisted, especially in slender examples,
densely to sparsely scabrid pubescent to glabrous, apex often scabrid. Culms flattened, ribbon-
like, 1–3 mm wide, sometimes twisted. Inflorescence terminal, anthelate, decompound, simple
or up to 40 solitary (occasionally paired) spikelets or reduced to a fairly compact head of 3–4
spikelets; bracts 2–4, shorter to slightly exceeding inflorescence, apex abruptly acuminate.
Spikelets many-flowered, compact, 4–9 × 2 mm, polygonal-terete, rachilla ragged. Glumes 2.2–
3.6 × 1.6–3.6 mm, breaking away ± mid-length, bases persistent to form rachilla wings. Stamens
3. Style trigonous with thickened pyramidal base, mostly glabrous, 3-branched, usually exceeding
style length. Nutlet 3-angled, obovoid, 0.7–1.15 × 0.4–0.75 mm, apex mostly rounded; surface
reticulate, obscurely longitudinally ridged with scattered warts; epidermal cells subquadrate to
transversely oblong arranged in longitudinal series, eventually velate.

Botswana. N: Ngamiland Dist., Okavango Delta, Kaporota Camp, 19°00.472'S
22°55.624'E, 8.iii.2010, *Heath* 1955 (K). **Zambia**. B: Mongu Dist., Lake Lutende, c.
23 km E of Mongu, 18.ix.1959, *Drummond & Cookson* 6606 (K, LISC). N: Mbala Dist.,
Namkole Marsh, 780 m, 20.ii.1964, *Richards* 19062 (K). W: Ndola Dist., Itawa Dam,
25.v.1950, *Jackson* 29 (K). C: Mkushi Dist., Fiwila dambo, 1220 m, 6.i.1958, *Robinson*
2659 (K). S: Choma Dist., 5 km E of Choma, 28.v.1955, *Robinson* 1264 (K). **Zimbabwe**.
N: Chegutu Dist., 6 km S of Darwendale near Rhochrome Mine, Great Dyke, 3.iv.1964,
Wild 6485 (K, SRGH). W: Matobo Dist., Maleme Dam, 6.i.1963, *Wild* 5936 (K, SRGH).
C: Harare Dist., University College, edge of hockey pitch, 29.iii.1963, *Loveridge* 644 (K,
SRGH). S: Masvingo Dist., Mutirikwe (Kyle) Dam Recreational Park (Game Reserve),
i-ii.1969, *Junor* 4 (K, SRGH). **Malawi**. N: Mzimba Dist., foot hills of Viphya, 1220 m,
31.i.1938, *Fenner* 248 (K). S: Zomba Dist., Namadzi (Namasi) R., Zomba, xi-xii.1899,
Cameron 96 (K). **Mozambique**. T: Angónia Dist., Ulónguè, 16.xii.1980, *Macuácua* 1450
(K, LISC). GI: Jangamo Dist., Nhálué, near Chacane vlei, 76 m, 26.x.1935, *Lea* 174
(K). M: Maputo Dist., Inhaca Is., 28.vii.1957, *Mogg* 31983 (K).

Also in Nigeria, Gabon, Rwanda, D.R. Congo, Kenya, Uganda, Tanzania, Angola,
Namibia, South Africa (Eastern Cape, Free State, North West, Gauteng, KwaZulu-Natal,
Limpopo, Mpumalanga) and Swaziland. Damp grassveld and swampy areas; 80–1300 m.

Conservation notes: Common and widespread; not threatened.

Infraspecific status (subsp. *keniaeensis*) is given to specimens with clustered spikelets
(Haines & Lye 1983, Verdcourt 2010), but the variation is such that subspecific
recognition is not helpful. *Fimbristylis consanguinea* has been described as distinct
from *F. complanata* based on a Drège gathering from the Eastern Cape, but the
differentiating criteria are not reliable.

2. **Fimbristylis aphylla** Steud., Syn. Pl. Glumac. **2**: 114 (1855). —Kern in Blumea **8**:
117 (1955). —Goetghebeur & Coudijzer in Bull. Jard. Bot. Belg. **54**: 83 (1984).
—Gordon-Gray in Strelitzia **2**: 89 (1995). Types: Indonesia, Java, *Goering* sect.
II, 149 (Herb. Goering lectotype); W Java, n.d., *Zollinger* 1609 (P syntype, F, G);
Java, Prabaksi, n.d., *Zollinger* 3524 (P syntype, BM, F, G, GH), lectotypified by
Kern (1955).

> *Fimbristylis globulosa* (Retz.) Kunth var. *aphylla* (Steud.) Miq., Fl. Ned. Ind. **3**: 322 (1856).
> *Fimbristylis vanderystii* De Wild., Pl. Bequaert. **4**: 205 (1927). Type: D.R. Congo, Kisantu,
> n.d., *Vanderyst* s.n. (BR holotype).

Fimbristylis testui Cherm. in Arch. Bot. Mém. **4**(7): 33 (1931). Types: Central African Republic, Oubangui-Chari, 23.viii.1921, *Le Testu* 3129 (P syntype); Oubangui, 20.viii.1921, *Tisserant* 2765 (P syntype); Oubangui-Chari, 22.v.1924, *Tisserant* 1523 (P syntype).

Perennial tufted herb up to 120 cm tall; rhizome woody, contracted, less often elongate then up to 6 mm wide, often obliquely vertical in soil, clothed in yellow to grey brown scale leaves when young, soon decaying. Leaves reduced to sheaths, closely surrounding culms, mouth sloping, white ciliate adjacent to reduced blade; ligule 0; blades obsolete, filiform apiculate with 4 mm extension of main sheath nerves, margins densely white pubescent. Culms 4–5-angled, glabrous. Inflorescence anthelate with up to 50(80) spikelets, mostly pedicellate; bracts 2–5, not exceeding inflorescence. Spikelets 4–7 × 2.2–3 mm, brown, compact, basal glumes early deciduous, wings on rachilla minute. Glumes 2–2.4 × 1.8–2 mm, keel 3-veined, terminating slightly below apex as minute recurved mucron, apex rounded or less often emarginate. Stamens (2)3. Style 3-angled to faintly 3-winged, angles finely fringed above or glabrous, base long pyramidal, branches 3. Nutlet 0.9–1 × 0.65–0.7 mm, trigonous, gynophore minute, surface with longitudinal ridges faint to indistinguishable, usually markedly warty.

Zambia. N: Kawambwa Dist., Mbereshi, permanent swamp, 910 m, 24.vi.1957, *Robinson* 2394 (K). W: Ndola Dist., 16 km S of Ndola, 1300 m, 3.iii.1960, *Robinson* 3374 (K). **Mozambique**. MS: Sussundenga Dist., Dombe, 27.x.1953, *Gomes Pedro* 4449 (K).

Also in West and Central Tropical Africa, South Africa (KwaZulu-Natal) and Asia. Permanently wet swamps and dambos; 900–1300 m.

Conservation notes: Infrequently collected; possibly threatened.

Haines & Lye (1983) and F.T.E.A give an account of *Fimbristylis subaphylla* Boeckeler. This taxon is not yet recorded for the Flora area but is likely to be present in swamps in N Zambia.

3. **Fimbristylis cymosa** R. Br., Prodr. Fl. Nov. Holland.: 228 (1810). —Haines & Lye, Sedges & Rushes E. Afr.: 80, fig.125 (1983). —Goetghebeur & Coudijzer in Bull. Jard. Bot. Belg. **54**: 85 (1984). —Verdcourt in F.T.E.A., Cyperaceae: 51 (2010). Type: Australia, Gulf of Carpentaria Is. & Prince of Wales Is., 1802–1805, *R. Brown* s.n. (BM syntype, K). FIGURE 14.**13**.

Fimbristylis obtusifolia (Lam.) Kunth, Enum. Pl. **2**: 240 (1837). —Clarke in F.T.A. **8**: 423 (1902). —Napper in J. E. Africa Nat. Hist. Soc. Natl. Mus. **25**(110): 10 (1965). —Gordon-Gray in Strelitzia **2**: 93 figs.35C,F, 36 (1995). Type: India, no further details.

Perennial herb up to 70 cm tall, tufted; rhizome woody, up to 14 mm wide, almost vertical, clothed in coarse, hard, markedly striate leaf bases, eventually fibrous. Leaves stiff, forming radiating rosette to each tuft, sheaths short, mouths sloping, glabrous or sparsely ciliate; ligule 0; blade to 210 × 2.7–4.2 mm, stiff, one margin usually involute, often scabrid pubescent, apex obtuse, almost truncate, usually apiculate. Culms slightly flattened, glabrous. Inflorescence anthelate, variable from open to contracted, a central head with up to 8 branches, each with a single terminal head, or much reduced to sessile heads, sometimes to single head only. Bracts 2–5 with 3 well-developed, leaf-like, mostly shorter than inflorescence. Spikelets close-packed, within heads, 3–8 × 1.2–2.5 mm, polygonal-cylindric due to projecting glume keels. Glumes 1.6–2.4 × 1.2–1.7 mm, dark beige, glabrous, apex rounded to minutely mucronate. Stamens (2)3, variable number in single spikelets. Style 3-angled, glabrous, base long pyramidal; style branches variable, in Flora area mostly 3, equalling or exceeding style length. Nutlet 0.6–0.85 × 0.5–0.7 mm, trigonous tending to round, dark brown-black, marked white by deciduous outermost cells at full maturity; surface reticulate, faintly longitudinally striated, indistinctly tuberculate to smooth, outermost cells often irregular, subquadrate to slightly transversely elongate.

Botswana. N: Chobe Dist., near Safari Co. boat harbour on Kwando R., 18°22–23'S 23°33–34'E, 12.xi.1980, *P.A. Smith* 3568 (K). **Zambia**. C: Mpika Dist., Mutinondo Wilderness Area, 12.iv.2013, *Merrett* 1434 (K). **Mozambique**. MS: Gorongosa Dist., Gorongosa Nat. Park (Game Reserve), W slopes of Cheringoma Plateau, 19°01'S 34°40'E, 12.xi.1971, *Ward* 7441 (K, NU). GI: Govuro Dist., c. 4 km E of Nova Mambone

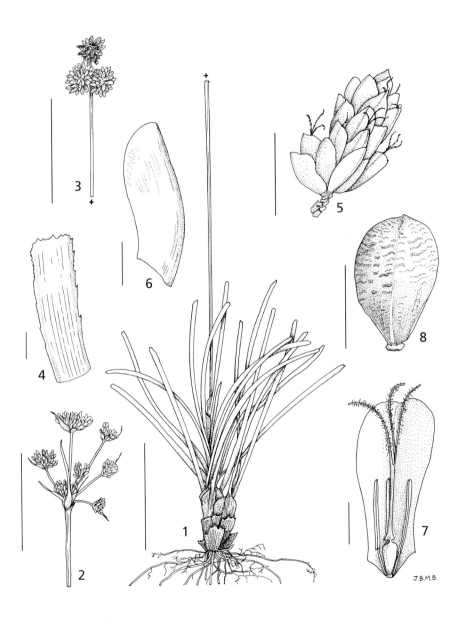

Fig. 14.**13**. FIMBRISTYLIS CYMOSA. 1, habit; 2, 3, inflorescence, branched and contracted; 4, leaf blade apex; 5, spikelet; 6, glume, lateral view; 7, floret and glume; 8, nutlet. 1, 3–8 from *Huntley* 727; 2 from *Ward* 663. Scale bars: 1, 3 = 40 mm; 2 = 25 mm; 4, 5 = 2 mm; 6–8 = 0.5 mm. Drawn by Jane Browning. Reproduced from Strelitzia (1995) by kind permission of the South African National Biodiversity Institute, Pretoria.

ferry landing, 18.vii.1972, *Ward* 7997 (K); Inhassoro Dist., Bazaruto Is., Ponta Estone mud flats, 21°30–47'S 35°30'E, 1–92 m, 28.x.1958, *Mogg* 28698 (K). M: Maputo Dist., Inhaca Is., 0–30 m, 31.xii.1956, *Mogg* 27086 (K, LISC).

In Kenya, Tanzania and South Africa (KwaZulu-Natal, Eastern Cape), also on Indian and Pacific Ocean islands, in the Americas, Asia and Australia. On sand with freshwater seeps or in thin soil overlying sandstone rocks; at the coast dwarf specimens under 5 cm tall are seen in the spray zone, often inundated at high tide; 0–1000 m.

Conservation notes: Widespread; not threatened.

Apart from the Mozambique coast, where *Fimbristylis cymosa* seems to have been rarely collected, there are few records for the Flora area. It is very occasional inland near hot springs.

The type locality is said to be in India, where the name is *Fimbristylis spathacea* and the styles 2-branched. Haines & Lye give the locality as Australia, without details, while Verdcourt in F.T.E.A. (2010: 51) states Australia, Gulf of Carpentaria Is. and Prince of Wales Is., *R. Brown* s.n. (BM syntype, K) with *R. Brown* 5959 as an isosyntype.

4. **Fimbristylis quinquangularis** (Vahl) Kunth, Enum. Pl. **2**: 229 (1837). —Verdcourt in F.T.E.A., Cyperaceae: 53 (2010). Type: E India, no locality, n.d., *Konig* s.n. (C holotype).

  Scirpus miliaceus L., Syst. Nat. ed.10: 868 (1759), rejected name. Type: E India, unknown collector, Herb. Linn. 71/40 (LINN lectotype), lectotypified by Blake (J. Arnold Arbor. **35**: 216, 1954).
  Scirpus quinquangularis Vahl, Enum. Pl. **2**: 279 (1805).
  Fimbristylis miliacea (L.) Vahl, Enum. Pl. **2**: 287 (1805), rejected name. —Clarke in F.T.A. **8**: 421 (1902). —Vollesen in Opera Bot. **59**: 94 (1980). —Goetghebeur & Coudijzer in Bull. Jard. Bot. Belg. **54**: 84 (1984). —Mapaura & Timberlake, Checkl. Zimbabwe Pl.: 88 (2004).
  Fimbristylis miliacea subsp. *miliacea* (L.) Vahl in Haines & Lye, Sedge Rushes E. Afr.: 81, fig.128 (1983).

Annual slender herb to 60(125) cm tall, (plants occasionally persisting longer than one year). Leaf blade 14–25 cm × 1.5–4 mm, with prominent midrib; ligule absent (lacking). Culms 4–5-angled, equalling or exceeding leaf length. Inflorescence anthelate, open, mostly 2-branched, spikelets many; involucral bracts slender, to 30 mm long. Spikelets compact, 2–5 × 1–2 mm, mostly small for the genus. Glumes 1.5–2 mm long, obovoid, light brown, glabrous, shortly mucronate. Stamens 1(2). Style triangular, 3-branched. Nutlets obscurely 3-angled, 0.4–0.6 × 0.3–0.5 mm, crystal white to pale brown; epidermal cells forming raised transverse ridges in 2–8 vertical and 15–30 horizontal rows on each face, papillae mostly not chambered.

Subsp. **quinquangularis**.

Zambia. N: Mpika Dist., Mfuwe, 610 m, 12.iii.1969, *Astle* 5589 (K). E: Mambwe Dist., Luangwa Valley, Lusangazi, 14.iv.1963, *Verboom* 895 (K). S: Namwala Dist., Kabulamwanda, fl. 21.iv.1965, *Robinson* 1245 (K). **Malawi**. S: Mulanje Dist., Phalombe, Ngolozi dambo, 770 m, 15°42'33.8"S 35°47'45.1"E, 15.iv.2008, *Chapama et al.* 836 (K).

In tropical Africa, Kenya, Uganda and Tanzania; also in Asia and N Australia. Wet grassland, waterlogged soils and edge of dambos in mopane woodland; 600–1100 m.

Conservation notes: Widespread; not threatened.

Reported to be a weed of rice fields.

Only subsp. *quinquangularis* is recorded from the Flora area, while two additional subspecies are recorded in the F.T.E.A. These subspecies differ in nutlet conformation and glume length but both criteria can vary with spikelet age of development. Care must be taken to consider only fully mature spikelets, which is not reliably achieved with herbarium specimens.

In his key Verdcourt (2010) noted that *Fimbristylis littoralis* and *F. quinquangularis* differ in glume apex. This needs checking and authenticated specimens are needed; *F. littoralis* is not recorded from the Flora Zambesiaca area.

5. **Fimbristylis microcarya** F. Muell., Fragm. **1**: 200 (1859). —Goetghebeur & Coudijzer in Bull. Jard. Bot. Belg. **54**: 83 (1984). Type: Australia, Depot Creek, iv.1856, *Müller* s.n. (MEL).

> *Fimbristylis thonningiana* Boeckeler in Linnaea **38**: 395 (1874). Type: Guinea, *Thonning* s.n. (herb?).
>
> *Fimbristylis complanata* (Retz.) Link var. *microcarya* (F. Muell.) C.B. Clarke in Hooker, Fl. Brit. India **6**: 646 (1894).
>
> *Fimbristylis autumnalis* (L.) Roem. & Schult. var. *microcarya* (F. Muell.) Kük. in Bot. Jahrb. Syst. **69**: 258 (1939).

Annual herb up to 30 cm tall, slender to sparsely tufted; rhizome inconspicuous, slightly woody. Leaves numerous, half to equal to culm length, sheaths flattened, mouth markedly sloping, white ciliate; ligule a fringe of dense white hairs; blade to 150 × 2.8–3 mm, pilose above ligule, otherwise glabrous, margins thickened, scabrid to glabrous, apex broadly acute, scabrid. Culms 4–6-angled, 2 of them pronounced causing flattening, scabrid pubescent especially upwards. Inflorescence anthelate or up to 50 spikelets, pedicels flattened; bracts (2)3(4), usually well-developed. Spikelets 2.5–7 × 1–2, compact, polygonal, acute. Glumes 1.7–1.9 × 1.4–1.6 mm, keel prominent, of 3 nerves excurrent into short, recurved mucro. Stamens (1)2. Style faintly 3-angled, base long pyramidal, exceeding style width, branches 3, shorter than style, shortly fringed. Nutlet 0.65–0.8 × 0.4–0.75 mm, 3-angled, obovoid, apex minutely projecting; surface faintly longitudinally striate, trabeculate, prominently tuberculate; outermost cells transversely elongate, in 4–6 longitudinal series on each face.

Zambia. S: Mazabuka Dist., 1070 m, 8.iii.1958, *Robinson* 2788 (K); Choma Dist., Muckle Neuk, 19 km N of Choma, 1280 m, 28.ii.1954, *Robinson* 583 (K).

Also in West and Central Africa (Burundi), Namibia and Asia. In wet grasslands; 1000–1300 m.

Conservation notes: Widespread outside the Flora area; not threatened.

The taxonomy of this species is confusing. *Fimbristylis microcarya* is not recorded for the F.T.E.A. area, but Goetghebeur & Coudijzer (1984: 83) record it for Central Africa, giving the name *F. thonningiana* Boeckeler as a synonym – "Annual with flattened stem and flat, ligulate leaves; spikelets small, up to 4 × 1.2 mm; fruit small, up to 0.7 mm long, pale, warty." Cited is Burundi: 1X, plaines Rusizi, km 14, *Reekams* 4397 (BR).

In Verdcourt (F.T.E.A., Cyperaceae: 54, 2010) it is listed as *Fimbristylis quinquangularis* subsp. *pallescens* (Lye) Verdc., comb. nov. Type: Tanzania, Usaramo Dist.: Dar es Salaam Univ. Campus *Wingfield* 2087 (DSM holotype, K). Synonyms: *F. miliacea* (L.) Vahl subsp. *pallescens* Lye in Nordic J. Bot **2**: 333 (1982). —Haines & Lye, Sedges & Rushes E. Afr.: 81, figs.129,130 (1983). There is also a note that Gordon-Gray had previously identified *Wingfield* 2087 as *F. thonningiana* Boeckeler.

The above indicates *Fimbristylis thonningiana* as a synonym of both *F. microcarya* subsp. *pallescens* and *F. quinquangularis*. Note that Haines & Lye figs. 129 and 130 (1983) are both taken from *Wingfield* 2087.

We do not think that *Fimbristylis microcarya* and *F. quinquangularis* subsp. *pallescens* are synonymous, although the latter was described from a specimen (*Wingfield* 2087) previously identified as *F. thonningiana* Boeckeler. However, the differences are slight. Most importantly the ligule is a fringe of short hairs in *F. microcarya* but absent in subsp. *pallescens*. All descriptions of *F. littoralis* and *F. quinquangularis* seen indicate there is no ligule. *F. microcarya* also lacks the bladeless cauline leaves (sheaths?) present in subsp. *pallescens*.

6. **Fimbristylis bivalvis** (Lam.) Lye in Lidia **3**: 144 (1995). Type: Madagascar, Ile Saint Marie, Nosy Boraha, i.1848, *Bovin* 1657 (P holotype, K).

    Scirpus bivalvis Lam., Tabl. Encycl. **1**: 141 (1791).

    Fimbristylis longiculmis Steud., Syn. Pl. Glumac. **2**: 110 (1855). —Clarke in F.T.A. **8**: 417 (1902). —Haines & Lye, Sedges & Rushes E. Afr.: 82, fig.132 (1983). —Goetghebeur & Coudijzer in Bull. Jard. Bot. Belg. **54**: 78 (1984). —Gordon-Gray in Strelitzia **2**: 93 (1995). —Verdcourt in F.T.E.A., Cyperaceae: 54 (2010).

    Fimbristylis sansibariensis Boeckeler in Flora **63**: 437 (1880). Type: Zanzibar Is., x.1873, *Hildebrandt* 1058b (B† holotype, K, L).

Perennial herb up to 135(180) cm tall, tufted; rhizome woody, clothed in hard outer scale leaves. Leaves limited to sheaths, flattened, glabrescent, mouth truncate, white ciliate; ligule a fringe of dense white hairs; blades obsolete, up to 7 mm long, stiff, coriaceous. Culms erect, compressed, nodeless, 2-angled when young and below inflorescence, glabrous. Inflorescence anthelate, of c. 30 solitary pedicelled spikelets; bracts 2–4, poorly developed, shorter than inflorescence. Spikelets 13–15 × 3.2–5 mm, compact, cylindric. Glumes 3.4–4.4 mm, keel poorly defined, excurrent into minute mucro, apex obtuse. Stamens 3. Style ribbon-like, margins long villous especially upwards, base dark brown slightly broadened, branches 2, shorter than style. Nutlet 1.1–1.2 × 0.9–1.1 mm, biconvex, obovoid, gynophore poorly developed; surface reticulate to faintly longitudinally ridged and trabeculate, markedly to sparsely warted especially near ridges, occasionally almost etuberculate, outermost cells wider than deep, in longitudinal series or irregular.

Botswana. N: Ngamiland Dist., Nxabega Is., 19°28'S 22°45'E, 14.iii.1982, *P.A. Smith* 3799 (K). **Zambia**. W: Mwinilunga Dist., lake c. 2 km on road to Kanyama from Mwinilunga–Solwezi, 1400 m, 25.ii.1975, *Hooper & Townsend* 387 (K). N: Luwingu Dist., Nsombo, 72 km S of Luwingu on Lake Bangweulu, 22.v.1961, *Robinson* 4697 (K). **Mozambique**. GI: Inhassoro Dist., Bazaruto Is., W coast, Ponta Estone marsh, 21°30–47'S 35°25–30'E, 28.x.1958, *Mogg* 28696 (K).

Also in Kenya, Tanzania, D.R. Congo, South Africa (KwaZulu-Natal) and Madagascar. Margins of freshwater swamps and fringes of coastal islands, usually partially submerged; 5–1400 m.

Conservation notes: Widespread; not threatened.

This closest affinity of this species appears to be with *Fimbristylis ferruginea*. It is known from inland river banks or hotsprings, as is occasionally *F. ferruginea*.

7. **Fimbristylis ferruginea** (L.) Vahl, Enum. Pl. **2**: 291 (1805). —Goetghebeur & Coudijzer in Bull. Jard. Bot. Belg. **54**: 80 (1984). —Gordon-Gray in Strelitzia **2**: 92 (1995). Type: Jamaica, *Herb. van Royen* 902.77–420 (L lectotype), lectotypified by Adams in Taxon **53**: 180 (2004).

    Scirpus ferrugineus L., Sp. Pl.: 50 (1753).

    Fimbristylis sieberiana Kunth, Enum. Pl. **2**: 237 (1837).

    Fimbristylis ferruginea subsp. *ferruginea*. —Haines & Lye, Sedges & Rushes E. Afr.: 83, fig.134 (1983). —Verdcourt in F.T.E.A., Cyperaceae: 55, fig.10 (2010).

    Fimbristylis ferruginea subsp. *sieberiana* (Kunth) Lye in Nordic J. Bot. **2**: 335 (1982). —Haines & Lye, Sedges & Rushes E. Afr.: 83, fig.135 (1983). —Verdcourt in F.T.E.A., Cyperaceae: 55 (2010). Type: Mauritius, no locality, n.d., *Sieber* 201 (B† holotype, AWH, BR).

Biennial herb, sometimes annual or a short-lived perennial up to 80 cm tall, sparsely to densely tufted; rhizome inconspicuous, persistent tunic on leaf bases when present dark or pale brown, not readily breaking into fibres. Leaves numerous, half to equalling culm length, sheaths glabrous to sparsely white pilose, mouth truncate, white ciliate; ligule a fringe of dense white hairs; blade to 310 × 1.2–2.3 mm, linear, glabrous, almost glaucous when fresh, margins thickened or ridged, scabrid pubescent, apex often asymmetric. Culms terete, slightly flattened below inflorescence, usually glabrous. Inflorescence anthelate, compound or simple, spikelets few to many, shortly stalked (pedicels 0–35 mm) to sessile or subsessile; bracts 2–4, usually much exceeding inflorescence. Spikelets 6–17 × 3–5 mm, compact, markedly cylindric, apex obtuse, becoming acute when mature. Lowest glumes bract-like, sterile, remainder each subtending a single floret, 3.2–4.9 × 2.7–4.9 mm, broadly ovate-orbicular, closely grey-pubescent abaxially in upper half, apex rounded, mucronate, keel hardly projecting. Stamens 3. Style ribbon-like, flattened in plane of nut, 1–3 nerved, margins villous in upper 2/3, glabrous below; base seldom exceeding style width, otherwise undistinguished; branches 2, shorter than style, margins fringed. Nutlet 1.3–1.7 × 0.7–1.27 mm, biconvex, obovoid to almost orbicular, apex slightly projecting, gynophore well developed; surface faintly reticulate, almost smooth, outermost cells subquadrate or hexagonal.

Botswana. N: Ngamiland Dist., Qangwa watercourse, 19°31.07'S 21°10.25'E, 9.iv.1980, *P.A. Smith* 3533 (K, LISC). **Zambia**. C: Chibombo Dist., Chipembi, 23 km E of Chisamba, 1220 m, 16.vi.1955, *Robinson* 1304 (K). E: Chipata Dist., Mushoro (Msoro), 80 km W of Chipata (Fort Jameson), 370 m, 9.vi.1954, *Robinson* 848 (K). S: Mazabuka Dist., Lochinvar Nat. Park, Gwishe hot spring, 1000 m, 15°59'S 27°15'E, 29.vii.1972, *Kornaś* 1969 (K). **Zimbabwe**. N: Binga Dist., Binga, Chiwata hot springs, fl. 7.xi.1958, *Phipps* 1390 (K, SRGH). E: Chipinge Dist., E Sabi, Upper Rupembi, 400 m, fl. 21.i.1957, *Phipps* 81 (K, SRGH). S: Gwanda, 14 km W of Fort Tuli, 13.v.1959, *Drummond* 6132 (K, SRGH). **Mozambique**. MS: Dondo Dist., Nhangau, 19°44'S 35°02'E, 6 m, 16.vii.1972, *Ward* 7974 (K). M: Maputo Dist., Delagoa Bay, 29.iii.1894, *Kuntze* 221 (K).

Widespread in the tropics and subtropics, in Kenya, Tanzania, Namibia, South Africa (Eastern Cape, Gauteng, KwaZulu-Natal, Limpopo, Mpumalanga, North West) and Swaziland. Common in low-lying, marshy, estuarine grassland periodically inundated by fresh or sea-water, also within the salt-spray zone, tolerant of saline conditions; also found inland in muddy verges of permanent watercourses, occasionally fringing hot springs; 10–1300 m.

Two subspecies, subsp. *ferruginea* and subsp. *sieberiana* (Kunth) Lye, have been recognised. In general African plants link more convincingly with subsp. *sieberiana*, but individual specimens are often difficult to place and the value of infraspecific distinction is questionable.

As a species *Fimbristylis ferruginea* is often superficially identified by the short, fine, grey pubescence abaxially present on the upper half of glume surfaces. Exceptions from this easily visible feature are rare, despite Clarke's (1898) comment "but the glumes are occasionally absolutely glabrous".

8. **Fimbristylis madagascariensis** Boeckeler in Abh. Naturwiss. Vereine Bremen **7**: 38 (1880). —Haines & Lye, Sedges & Rushes E. Afr.: 88, fig.146 (1983). —Goetghebeur & Coudijzer in Bull. Jard. Bot. Belg. **54**: 79 (1984). —Verdcourt in F.T.E.A., Cyperaceae: 61 (2010). Type: Madagascar, Antananarivo, *Rustenberg* s.n. (B holotype).

Appears to differ from *Fimbristylis ferruginea* only in the presence of lateral rhizome (stolons) and in the absence of glume hairs.

Zambia. N: Chinsali Dist., Shiwa Ngandu (Ishiba Ngandu), dambo, 29.i.1961, *Robinson* 4318 (K).

Also in Kenya, Uganda, Tanzania, Rwanda, Burundi and Madagascar. Fringing freshwater swamps and seasonally wet grassland; c. 1500 m.

Conservation notes: Widespread; not threatened.

As most *Fimbristylis* species (except true annuals) will produce lateral rhizomes, especially when growing in ± permanent water, it is not clear if the differences between *F. madagascariensis* and *F. ferruginea* are taxonomically valid. If not, the latter name should prevail.

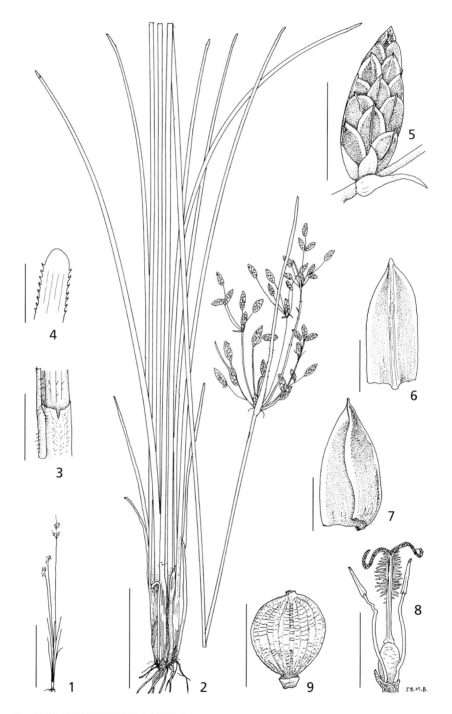

Fig. 14.**14**. FIMBRISTYLIS DICHOTOMA. 1, 2, habit; 3, leaf sheath apex; 4, leaf apex; 5, spikelet; 6, 7, glume, abaxial and lateral view; 8, floret; 9, nutlet. All from *Brummitt* 9546. Scale bars: 1 = 250 mm; 2 = 40 mm; 3, 5 = 5 mm; 4 = 2 mm; 6–9 = 1 mm. Drawn by Jane Browning.

9. **Fimbristylis dichotoma** (L.) Vahl, Enum. Pl. **2**: 287 (1805), excl. description and reference. —Haines & Lye, Sedges & Rushes E. Afr.: 85, fig.140 (1983). —Goetghebeur & Coudijzer in Bull. Jard. Bot. Belg. **54**: 77 (1984). —Gordon-Gray in Strelitzia **2**: 91, fig.34 H,K (1995). —Verdcourt in F.T.E.A., Cyperaceae: 57, fig.11 (2010). Type: Sri Lanka, no locality, *Herb. Hermann* 2, fol.63 bottom left (BM lectotype), lectotypified by Koyama (1979). FIGURE 14.**14**.

   Scirpus dichotomus L., Sp. Pl.: 50 (1753).
   Scirpus diphyllus Retz., Observ. Bot. **5**: 15 (1788). Type: India, Tranquebar, n.d., *König* s.n. (LD holotype).
   Fimbristylis diphylla (Retz) Vahl, Enum. Pl. **2**: 289 (1805). —Clarke in F.T.A. **8**: 415 (1902).

Annual or short-lived perennial herb, sparsely tufted, tunic of persistent leaf bases usually lacking; rhizome inconspicuous connecting shoot bases only. Leaves numerous, up to   culm length, sheaths glabrous to white pilose in upper half, mouth truncate, ciliate; ligule a fringe of dense, short white hairs, blade to 360 × 1.3–4.1 mm, linear, margins smooth or closely scabrid pubescent, apex often obtuse, scabrid. Culms 60–70(126) cm long, terete to faintly ridged, scattered white pilose on angles and below inflorescence. Inflorescence anthelate, decompound, compound or simple with from c. 75 solitary spikelets, or reduced to a head of c. 3 shortly pedicelled or sessile units; bracts (2)3(4), leaf-like, usually white pilose to shortly pubescent. Spikelets compact, cylindric 4–12 × 2–3 mm, basal glumes early deciduous. Glumes 2–2.5 × 1.7–3.3 mm, shining chestnut red with greenish well-defined keel, produced apically into a short mucro. Stamens 2(3). Style 1.6–2.1 mm long, flattened in plane of nut, 3-nerved, margins villous, base widened, style branches 2, shorter than style. Nutlet 0.9–1.24 × 0.65–0.95 mm, biconvex, ovate to almost orbicular in outine, apex minutely projecting; surface prominently (6)7–10 ridged longitudinally, cells between ridges transversely elongate.

Botswana. N: Ngamiland Dist., Boteti R., 20°06.8'S 23°30.4'E, 15.xii.1976, *P.A. Smith* 1841 (K). **Zambia**. B: Kaoma Dist., 14.6 km NE of Kaoma (Mankoya)–Kasempe road, 14°46'03"S 24°48'07"E, 1080 m, 28.ii.1996, *Harder et al.* 3585 (K, MO). N: Mbala Dist., Chilongowelo, N of Mbala (Abercorn), 1460 m, 20.vi.1956, *Robinson* 1703 (K). W: Ndola Dist., Ndola, 1220 m, 1.i.1960, *Robinson* 3264 (K). C: Serenje Dist., Kalwa Farm, 13°13'35"S 30°21'30"E, 1540 m, 3.xii.2006, *Bingham* 13168 (K). S: Choma Dist., 8 km E of Choma, 1310 m, 26.i.1955, *Robinson* 1180 (K). **Zimbabwe**. W: Nyamandhlovu Dist., Umguza spur, Amandundumella vlei, 1370 m, 10.iii.1931, *Pardy* 3757 (K, SRGH). C: Harare Dist., Hatfield, 8.v.1934, *Gilliland* 89 (K, SRGH). E: Chimanimani Dist., Chimanimani (Melsetter), 11.i.1949, *Fisher & Schweickerdt* 405 (K, NU). **Malawi**. N: Chitipa Dist., 14 km E of crossroads towards Karonga, 1100 m, 19.iv.1969, *Pawek* 2261 (K). C: Mchinji Dist., Mchinji, riverbank by road to Zambia, 1190 m, 30.iii.1970, *Brummitt* 9546 (K). S: Mulanje Dist., Likubula gorge, river bed, 840 m, 20.vi.1946, *Brass* 16386 (K, MO). **Mozambique**. N: Angoche Dist., Angoche (Antonio Enes), swamp by R.C. Mission, 16°14'S 39°54'E, 16–31.x.1965, *Mogg* 32490 (K). Z: Lugela Dist., Namagoa, n.d., *Faulkner* K20 (K). MS: Cheringoma Dist., Cheringoma Plateau, source of R. Mueredzi, 11.vii.1972, *Ward* 7815 (K, NU).

Widespread across the tropics and subtropics, also in Kenya, Uganda, Tanzania, D.R. Congo, Rwanda, Burundi, Namibia, South Africa (Eastern Cape, Free State, Gauteng, KwaZulu-Natal, Limpopo, Mpumalanga, North West) and Swaziland. Common in damp mud on streambanks, dambos, seepages and drainage ditches by roadsides, also in grassland as isolated plants. Flowering period from November to March; 10–1600 m.

Conservation notes: Widespread; not threatened.

Fimbristylis dichotoma does not have the tolerance for saline conditions and hot springs as does *F. ferruginea*.

In South Africa in early spring, in areas where from mid to late summer it is abundant, *F. dichotoma* can only be detected by the previous year's inflorescences persistent on old plants, some of which develop new leaves at this time, while others appear dead. Among old plants nutlets may be found germinating, sometimes in quantity. Young plants produce their first inflorescences within 4 months.

In Zimbabwe it is common as a weed of dambos (vleis) that have been cultivated.

The type of *Scirpus dichotomus* is based on "Gramen Cyperoid. Maderaspatanum, Juncelli Gesneris…" in Plukenet, Phytographia: tab.119 fig.3 (1691) and "Gramen parvum…" in Hermann, Mus. Zeylan.: 26 (1717). Hermann's types are at the Natural History Museum, London (BM).

10. **Fimbristylis bisumbellata** (Forssk.) Bubani, Dodecanthea: 30 (1850). —Haines & Lye, Sedges & Rushes E. Afr.: 86, fig.141 (1983). —Gordon-Gray in Strelitzia **2**: 90, fig.3B,E (1995). —Verdcourt in F.T.E.A., Cyperaceae: 58 (2010). Type: Egypt, Rashid and Cairo, 1761, *Forsskål* s.n. (C holotype, BM).

> *Scirpus bisumbellatus* Forssk., Fl. Aegypt.-Arab.: 15 (1775).
> *Fimbristylis dichotoma* sensu Vahl, Enum. Pl.: **2**: 287 (1805), description. —Clarke in F.T.A. **8**: 414 (1902).
> *Fimbristylis dichotoma* (L.) Vahl subsp. *bisumbellata* (Forssk.) Luceño in Anales Jard. Bot. Madrid **57**: 176 (1999).

Annual herb 10–26(46) cm tall, tufted; rhizome inconspicuous, tunic of leaf bases lacking. Leaves numerous, half to equalling culm length; sheaths usually flattened, white velutinous, mouth sloping; ligule 0; blade 30–70(150) × 1–2.8 mm, margins thickened, smooth to scabrid pubescent, apex obtuse or acute. Culms slightly flattened, faintly ridged, glabrous. Inflorescence anthelate, of up to 175 solitary spikelets, decompound, compound, or simple when reduced to 30–40 spikelets; pedicels markedly flattened, glabrous; bracts (2)3(4), to 120 mm long. Spikelets 3–11 × 1.2–1.6 mm, compact, polygonal-angled by projecting glume keels. Glumes 1.5–2 × 0.95–1.3 mm, glabrous, pallid to golden yellow with delicate narrow membranous margin, apex mucronate. Stamens 1(2–3), sometimes within single spikelets on an individual plant or within florets of a single spikelet. Style thread-like, slightly flattened in plane of nut, margins villous in upper half, base widened, deciduous with style, branches 2, shorter than style, margins fringed. Nutlet 0.65–0.8 × 0.4–0.6 mm, biconvex, obovoid, apex slightly projecting, gynophore small; surface 5–6 ridged longitudinally, trabeculate, outermost cells transversely elongate in longitudinal series.

Botswana. N: Ngamiland Dist., Kwando (Mashi) R. floodplain, 18°19'S 23°29'30E, 27.vii.1973, *P.A. Smith* 678 (K, LISC). **Zambia**. C: Kabwe Dist., Kabwe (Broken Hill), Mwomboshi R., 12 km N of Chipenbi, 14°50'S 28°36'E, 26.viii.1972, *Kornaś* 2009 (K). E: Lundazi Dist., Lundazi R., above dam, 1200 m, 19.xi.1958, *Robson* 672 (K, LISC). S: Choma Dist., Mapanza, 1070 m, 24.x.1953, *Robinson* 354 (K, NU, SRGH). **Zimbabwe**. N: Hurungwe Dist., Sanyati R., junction with Fulechi R., 760 m, 11.x.1957, *Phipps* 743 (K, PRE, SRGH). E: Mutare Dist., Mutare (Umtali), Chimedzi R., 3 km from junction with Sungwizi (Tsungwesi) R., 990 m, 8.ix.1955, *Drummond* 4865 (K, PRE, SRGH). S: Beitbridge Dist., Shashe R., 10 km upstream of Limpopo junction, 6.v.1989, *Drummond* 6098 (K, SRGH). **Malawi**. C: Dedza Dist., SE bank of Linthipe R. on road from Nkhoma to Linthipe, 24.vi.1970, *Brummitt* 11678 (K, LISC). S: Mwanza Dist., lower Mwanza R., 3.x.1946, *Brass* 17936 (K). **Mozambique**. MS: Nhamatanda Dist., pontoon bridge on Chimoio road, lower floodplain of Pungwe S bank, 50 m, 11.xi.1971, *Ward* 7412 (K, NU).

Found in Kenya, Tanzania, Botswana, Namibia and South Africa (KwaZulu-Natal, Limpopo, Mpumalanga, Northern Cape); also in Madagascar, S Europe, Asia, Malaysia, Phillipines and Australia. A riverine species common on alluvial soils, along stream margins, in warm localities with permanent water during the growing season; sometimes a weed; 50–1200 m.

Conservation notes: Widespread; not threatened.

Gordon-Gray (pers. comm.) notes that for the ligule "there may, or may not, be some short white hairs at the sheath mouth, but there is no recognisable ligule of hairs; some examples are glabrous."

11. **Fimbristylis pilosa** Vahl, Enum. Pl. **2**: 290 (1805). —Haines & Lye, Sedges & Rushes E. Afr.: 86, figs.143a,b (1983). —Goetghebeur & Coudijzer in Bull. Jard. Bot. Belg. **54**: 82 (1984). —Verdcourt in F.T.E.A., Cyperaceae: 60 (2010). Type: Guinea, no locality, *Thonning* 393 (C syntype, P).

    Fimbristylis castanea Vahl var. *thonningiana* Boeckeler in Linnaea **37**: 19 (1871). Type as for *F. pilosa*.

    Fimbristylis madagascariensis sensu Vollesen in Opera Bot. **59**: 94 (1980), non Boeckeler.

Perennial (usually 2–3 years) or occasionally annual, up to c. 60 cm tall; base slightly thickened, surrounded by leaf sheath remnants. Leaves 120–200 × 0.5–3 mm, margins densely hairy; ligule a fringe of short hairs. Culms 25–70 cm tall, angular to compressed above, densely villous, hairs sometimes stiff (spine-like). Inflorescence a simple or compound anthela with 4–9 spikelets or consisting of one sessile and 1–2 stalked spikelets; bracts leafy, to 6 cm long, hairy. Spikelets ovoid, 3.5–7(12) × 2–4 mm, compact. Glumes 2.5–3 mm long, dark chestnut with black patches, keel green, strongly concave, margins shortly ciliate, apex obtuse to shortly acuminate. Stamens 2. Style upper half ciliate, branches 2. Nutlet biconvex, 1.2–1.7 × 1.4 mm, gynophore present; surface smooth in young examples, developing usually ± 20 longitudinal ridges with age, ridges transversely elongated; tubercles present or absent.

Zambia. S: Choma Dist., 11 km S of Mapanza, 1070 m, 20.iii.1955, *Robinson* 1145 (K); Mapanza, SE, 26.i.1965, *Robinson* 1333 (K). **Zimbabwe**. N: Hurungwe Dist., c. 6 km N of Manora in E, 610 m, 1.iii.1958, *Phipps* 1001 (K).

Also in West Africa, Kenya, Tanzania, Uganda, D.R. Congo, Burundi and Angola. In seasonally wet grasslands and swamp edges; 600–1100 m.

Conservation notes: Widespread; not threatened.

Fimbristylis pilosa and *F. tomentosa* Vahl were described in 1805 but it is not clear how they were differentiated; the specific names imply *F. tomentosa* has denser, longer hairs than *F. pilosa*. Haines & Lye (1983) consider only the latter for E Tropical Africa, while Goetghebeur (1984) includes both for Central Africa. Goetghebeur's descriptions are brief and unsatisfactory, namely *F. tomentosa*: annual, spikelets 5–7 mm long, fruit conspicuously reticulate with 16–24 vertical rows of cells (Fig 2B); *F. pilosa*: caespitose perennial; spikelets 6.5 × 4.5 mm, fruit smooth. Both are probably variants of *F. dichotoma* with the increased pubescence related to growth in more tropical, extreme habitats.

The terms annual or perennial are not possible to apply accurately, either in field or herbarium, unless there is definite evidence of perennation. However, true annuals begin to yellow and shrivel immediately following reduction in daylength in December. Many sedges are not annual, but shortly perennial with the aerial parts decaying, with the underground organs regenerating for another 1–3 years.

We are unsure how reliable fruit wall topography is – Haines & Lye (1983) say that in *F. pilosa* young fruits are ± smooth (the ribs become clear only at full maturity) and the presence/absence of warts is also variable.

Both *Fimbristylis pilosa* and *F. tomentosa* are probably variants of *F. dichotoma*; increased pubescence is related to growth in more tropical extreme habitats. It is noticeable that in sub-Saharan Africa, there is gradual desiccation from E to W of the continent, showing in most of the sedges.

The fruits of *Fimbristylis pilosa* are longitudinally ribbed like those of *F. dichotoma*, but in *F. pilosa* they have many more closely-spaced ribs.

12. **Fimbristylis scabrida** Schumach., Beskr. Guin. Pl.: 32 (1827). —Haines & Lye, Sedges & Rushes E. Afr.: 87, fig.145 (1983). —Goetghebeur & Coudijzer in Bull. Jard. Bot. Belg. **54**: 82 (1984). —Verdcourt in F.T.E.A., Cyperaceae: 61 (2010). Type: Guinea, no locality, *Thonning* 394 (C holotype).

Fimbristylis muriculata Benth. in Hooker, Niger Fl.: 554 (1849). Type: Ghana, Accra, *Don* s.n. (K syntype).

Perennial tufted herb up to 50 cm tall; rhizome contracted, swollen, covered in fibrous leaf base fragments. Leaves c. 100 × 1–2.5 mm, margins scabrid, apex acuminate, surface scabrid between ridges throughout; ligule 0. Culm 0.7–1 mm wide, rounded, or slightly compressed, deeply ridged, scabrid. Inflorescence anthelate, spikelets less than 40 per culm; bracts leaf-like, 1–3 cm long. Spikelets 6–7(14) mm long, c. 2 mm wide, rachilla shortly winged after glume dehiscence. Glumes brown with pale margins, c. 3 mm long, apex acuminate, scabrid, with long scabrid mucro, surface mostly glabrous shining with some short hairs adjacent to midrib. Style 3-branched. Nutlet ± trigonous, lacking distinct ribs, 1.2–1.3 × 1.2 mm; surface smooth, but densely warty at maturity.

Zambia. N: Kasama Dist., 10 km E of Kasama, 19.i.1961, *Robinson* 4278 (K). S: Choma Dist., Simansunda, 3.2 km E of Mapanza, 1070 m, 5.i.1954, *Robinson* 428 (K).

Also in West Africa, Central African Republic, Burundi, D.R. Congo, Uganda and Madagascar. Dry grassland and edges of dambos; 1000–1300 m.

Conservation notes: Widespread; not threatened.

According to Napper (F.W.T.A. 1972: 323) this species is very variable in size with stem-bases forming coarse matted fibrous tufts and with a panicle of lanceolate-cylindric spikelets with acute glumes.

Gordon-Gray (pers. comm.) notes that the long mucro of *Fimbristylis scabrida* is distinctive; probably the result of drier habitat conditions. It is closely related to *F. dichotoma*.

13. **Fimbristylis schoenoides** (Retz.) Vahl, Enum. Pl. **2**: 286 (1805). —Haines & Lye, Sedges & Rushes E. Afr.: 89, figs.150,151 (1983). —Verdcourt in F.T.E.A., Cyperaceae: 62 (2010). Type: India, no locality, *König* s.n. (LD holotype).

Scirpus schoenoides Retz., Observ. Bot. **5**: 14 (1788).

Short-lived perennial, sometimes annual, herb up to 50 cm, tufted, glabrous; rhizome short if developed. Leaves shorter than culms, 0.6–1 mm wide, upper margins scabrous-spinulose, apex abruptly acuminate, sheaths sometimes ferruginous; ligule a fringe of dense hairs. Culms compressed, smooth. Inflorescence a solitary terminal spikelet, or with 1–2 additional pedicelled spikelets; lowest bracts to 40 mm long, or glume-like as remainder. Spikelets 5–6 × 3–4 mm, lengthening to 15 mm with age, many-flowered, rachilla narrowly winged. Glumes spiral, ovate, usually broader than long, 2–3 × 3–3.5 mm, whitish-pink to pale brown, several-nerved in addition to keel (laterals not always easily observed), occasionally almost transparent, apex obtuse to shortly apiculate. Stamens 3. Style flattened, dilated basally, ciliate in upper half, style branches 2, shorter than style. Nutlet biconvex, obovoid, 1.25–1.5 × 1.2 mm, distinctly stipitate, gynophore c. 0.3mm deep, apex slightly raised where style base fallen; surface smooth to faintly reticulate, epidermal cells isodiametric.

Zambia. N: Mporokoso Dist., Mporokoso, N end of Mweru Wantipa, 1040 m, 17.iv.1961, *Phipps & Vesey-Fizgerald* 3278 (K, SRGH). S: Namwala Dist., Namwala, Baambwe, 17.iv.1963, *van Rensburg* KBS 2023 (K).

In Senegal, Ghana and Tanzania; also in India, Thailand, Indo-China, S China to tropical Australia, introduced in America. Open wet grasslands, mixed woodland and swampy ground; 1000–1100 m.

Conservation notes: Possibly introduced.

Uncommon in Africa and probably introduced from Asia as a contaminant of rice.

14. **Fimbristylis squarrosa** Vahl, Enum. Pl. **2**: 289 (1805). —Haines & Lye, Sedges & Rushes E. Afr.: 90, fig.152 (1983). —Goetghebeur & Coudijzer in Bull. Jard. Bot. Belg. **54**: 85 (1984). —Gordon-Gray in Strelitzia **2**: 95 (1995). —Verdcourt in F.T.E.A., Cyperaceae: 63 (2010). Type: South America, no locality, possibly *Loefling* s.n. (C holotype).

    Scirpus squarrosus (Vahl) Poir. in Lamarck, Encycl. Suppl. **5**: 100 (1817), illegitimate name, non L.

    Fimbristylis aestivalis (Retz.) Vahl var. *squarrosa* (Vahl) T. Koyama in J. Fac. Sci. Univ. Tokyo, Sect.3, Bot. **8**: 116 (1961).

Annual herb up to 35 cm tall, generally densely tufted; rhizome inconspicuous, persistent leaf bases to shoots lacking. Leaves numerous, 0.3 mm wide, half to equalling culm length; ligule 0; blade to 180 × 0.4–1.2 mm, margins white pilose to scabrid pubescent; sheath mouth sloping or almost truncate. Culms slightly flattened, faintly ridged and furrowed, glabrous. Inflorescence anthelate, of up to 53 solitary spikelets, decompound or simple, pedicels flattened; bracts 3(4), to 65 mm long. Spikelets 5–7(13) × 1–2 mm, compact, polygonal-angled by projecting glume keels. Glumes lowest sterile, remainder fertile, 1.9–2.2 × 0.8–1.3 mm, abaxial surface pilose becoming glabrescent upwards in spikelet sequence, apices with mucros up to 1 mm giving somewhat spiky outline to spikelets. Stamen 1. Style slightly flattened, in plane of nutlet, 1-nerved, margin in upper half fringed with conspicuous pendulous hairs, base widened capping nutlet with fringe of drooping white hairs usually of unequal length, branches 2, shorter than style. Nutlet 0.6–0.8 × 0.3–0.6 mm, biconvex, obovoid, apex slightly projecting; surface finely reticulate, faintly longitudinally striated, sometimes almost smooth, outermost cells subquadrate to narrowly transversely elongate, placed in irregular longitudinal lines.

Botswana. N: Ngamiland Dist., Boteti (Botletle) R. bank, S of Toromoja School, 1000 m, 24°30'E 21°05'S, 22.iv.1971, *Pope* 362 (K, LISC). **Zambia**. B: Kalabo Dist., Sandaula pontoon, 12.xi.1959, *Drummond & Cookson* 6369 (K, LISC). N: Mansa Dist., Mansa (Fort Roseberry), 1180 m, 14.xii.1960, *Symoens* 7975 (K). W: Solwezi Dist., Solwezi dambo, 10.vi.1930, *Milne-Redhead* 467 (K). S: Victoria Falls, Palm Is., 910 m, 20.xi.1949, *Wild* 3118 (K, SRGH). **Zimbabwe**. W: Bulawayo Dist., Bulawayo, Hillside Dams, 1370 m, xi.1958, *Miller* 5367 (K, SRGH). C: Chegutu Dist., Chegutu (Hartley), Nkuti Farm, 24.ii.1969, *Mavi* 951 (K, SRGH, NU). **Malawi**. C: Nkhotakota Dist., Nkhotakota (Kota Kota), Benga, 470 m, 2.ix.1946, *Brass* 17490 (K). **Mozambique**. MS: Gorongosa Dist., Gorongosa Nat. Park (Game Reserve), banks of Urema R., 610 m, 25.viii.1958, *Chase* 6982 (K).

Widespread in the tropics and subtropics in Asia, Central and South America (introduced in North America), also in Senegal, D.R. Congo, Burundi, Tanzania, Angola, Namibia, South Africa (KwaZulu-Natal, Limpopo, Mpumalanga, Western Cape) and Madagascar. Sandbanks, mudflats, shallow pools and seasonally wet habitats; 500–1400 m.

Conservation notes: Widespread; not threatened.

Fimbristylis squarrosa is distinguished from all other species in the genus by the long white pendant hairs developed from the margins of the style base. If spikelets are not dissected, it is easily confused with plants of *F. bisumbellata*. Both species favour similar habitats and may grow in close proximity. Detailed study shows differences – *F. squarrosa* has a longer glume length, the apex is markedly mucronate, basal glumes are pilose along keel as are the leaf margins in the basal section of the blade.

Excluded taxa.

Fimbristylis polytrichoides (Retz.) Vahl, Enum. Pl. **2**: 248 (1805). —Haines & Lye, Sedges & Rushes E. Afr.: 89, fig.148 (1983). —Verdcourt in F.T.E.A., Cyperaceae: 62 (2010). Type: Sri Lanka, no locality, *König* s.n. (LD syntype, BM).

Not recorded in the Flora area, but present in Kenya, Tanzania, Madagascar, tropical Asia and Australia. Like *Fimbristylis schoenoides*, it may be an introduction from Asia Seed probably carried with rice.

8. **BULBOSTYLIS** Kunth[8]

Bulbostylis Kunth, Enum. Pl. **2**: 205 (1837), conserved name.

Perennial or annual herbs, mostly sparsely tufted; rhizome woody, variable, usually compact with swollen confluent shoot bases, occasionally elongate in uniseriate or multiseriate rows, less often of uniform thickness throughout. Leaves variable, mostly grass-like, occasionally reduced to sheaths forming a persistent clothing to shoot bases; sheath mouth conspicuously white-bearded to woolly; ligule mostly 0, seldom a ring of short hairs; blades well-developed or reduced to apiculate projections. Culms scapose, terete, ridged and furrowed, glabrous to pilose or densely velutinous apically. Inflorescence a supra-decompound, compound or simple anthela (of up to 90 spikelets) or reduced to a terminal head or single spikelet (very variable within a clone or population). Bracts (2)3(4), mostly only one elongate, others hardly exceeding inflorescence; glabrous to white woolly. Spikelets cylindric/polygonal or compressed, many-flowered, basal glumes mostly early deciduous. Glumes imbricate, fertile (except lowest 1–2 in perennials), pubescent to sub-glabrous abaxially, apex emarginate to fringed when old or mucronate, florets bisexual, bristles 0. Stamens 3(2)–1. Style filiform, glabrous to villous, swollen basally into a pyramidal, more often rounded, vertically compressed, hardened base mostly persistent on nutlet post-dispersal. Nutlets trigonous (perennials) or dorsiventrally convex (annuals), elliptic to oblanceolate in outline; surface reticulate, transversely lineolate and/or ridged, etuberculate; epidermal cells mostly longitudinally (seldom transversely) elongate, sometimes becoming puncticulate.

A genus of about 100 species found throughout the tropics; abundant in the grasslands of sub-Saharan Africa.

In this treatment of *Bulbostylis* a broad view has been taken for the perennial species. Much synonymy exists – the outcome of sometimes personal interpretation and experience. Because descriptions of some of the annual species have been derived from examination of a single type collection, descriptions may be incomplete or misleading.

Morphological species limits are difficult to determine as gradations and differences in minutiae exist, particularly in size and pubescence. Features of nutlets, such as colour (even at maturity), surface conformation of the pericarp and dimensions inclusive of the persistent style-base should be treated with caution as many annual species are at present unreliably distinguished one from another, and only by very slight morphological differences.

Morphological features such as inflorescence type may be very variable and unreliable in some species where up to three types are encountered. However, these have been used along with dimensional data for both annual and perennial species.

Although differences between similar species are often slight, they may provide important evidence of active speciation occurring in the Flora area. The arbitrary placement of similar species in larger groups, although convenient, might hide the significance of small differences, and is not undertaken here.

The key for annual species has been adapted from Goetghebeur & Coudijzer (1985).

Plants mostly perennial, shoot bases thickened, closely placed, on elongate rhizomes or soboles (rhizomes bearing early-deciduous white to brown scale leaves). . **Key 1**

Plants mostly annual, roots fibrous, underground perennating organs lacking (except water plants that may persist by fragmentation) . . . . **Key 2**

---

[8] by K.D. Gordon-Gray† and J. Browning

Note: If in doubt, follow both alternatives; variable taxa, such as the annual/perennial *B. hispidula*, *B. densa* and *B. viridecarinata*, are included in both keys.

Key 1 – Mostly perennial species

1. Inflorescence a solitary spikelet 2
   – Inflorescence of 2 or more spikelets 6
2. Rhizome elongated; culms isolated 3
   – Rhizome seriate; culms closely placed or plants caespitose 4
3. Leaf blade to 1.5 mm wide; plants of dry *Brachystegia* woodland . . . **1.** *rhizomatosa*
   – Leaf blade generally absent; plants of inundated marshy ground . . . . **2.** *schlechteri*
4. Leaf blade usually developed 5
   – Leaf blade generally absent, reduced to leaf sheaths purple to reddish in colour; growing in permanently waterlogged soil . . . . **3.** *clarkeana*
5. Glumes 2.4–3.6 mm long; on mountain grassland and slopes . . . . **4.** *oritrephes*
   – Glumes 3.8–6 mm; in dry and often stony *Brachystegia* woodland on rocky hilltops and slopes. . . . .**9.** *macra*
6. Inflorescence compact and head-like, of (2)3 to many sessile spikelets, rarely with an additional stalked spikelet or spikelet cluster. 7
   – Inflorescence open and anthelate, of 2–3 or many mostly pedicellate spikelets 15
7. Spikelets flattened and compressed 8
   – Spikelets not flattened or compressed 9
8. Glumes 7–10 mm long; involucral bracts generally inconspicuous, sometimes exceeding inflorescence; nutlets 1.8–2 mm long, rugose . . . .**5.** *pilosa*
   – Glumes 4–5 mm long; involucral bracts generally short, not easily seen in mature inflorescences; nutlets 0.8–1.2 mm, not rugose. . . . . **6.** *parvinux*
9. Flowering head often spherical, spikelets reflexed, closely packed; bracts with long white hairs; lower glumes with a short or obsolete mucro . . . . **7.** *laniceps*
   – Not this combination of characters 10
10. Glumes bifid, young glume apex emarginate or rounded, later splitting to become lacerate with acute apex . . . . **10.** *schoenoides*
    – Glumes not bifid 11
11. Glumes ranging within 1.5–3 mm in length 12
    – Glumes ranging within 4–9 mm in length 14
12. Culms mostly completely scabrid with upward directed hairs . . . . .**11.** *scabricaulis*
    – Culms smooth, rarely slightly scabridulous near top. 13
13. Glumes c. 2 mm long, broadly boat-shaped, midrib reaching apex. . . . .**12.** *cardiocarpoides*
    – Glumes 2.4–3 mm long, narrowly boat-shaped, midvein excurrent into a distinct mucro . . . .**13.** *filamentosa*
14. Glumes 4–6 mm long; inflorescence very variable, usually a terminal head, but sometimes with secondary stalked heads; nutlet 1.7–1.9 mm long . . . .**14.** *boeckeleriana*
    – Glumes 8–9 mm long; nutlets 2.1–2.3 mm long; plant bases frequently burnt . . . . **15.** *igneotonsa*
15. Largest inflorescence usually with 2–6 spikelets only 16
    – Largest inflorescence usually with 7 to many spikelets 20
16. Rhizome elongate, culms isolated; leaf blade to 1.5 mm wide; in dry *Brachystegia* woodland . . . . **1.** *rhizomatosa*
    – Not this combination of characters 17
17. Leaf blades absent; leaf sheath purple to reddish; glumes 3–5 mm long; in permanently water-logged soils. . . . . **3.** *clarkeana*

–   Leaf blades present; leaf-sheath generally not purple; plants not found in permanently inundated situations 18
18. Rhizome of swollen shoot bases confluent in a uniseriate, sometimes biseriate row, or occasionally irregular; glumes 2.4–3.6 mm long. . . . . **4.** *oritrephes*
–   Rhizome not of compact confluent swollen shoot bases, not in a seriate row . . . 19
19. Inflorescence branches flattened, erect or spreading, not noticeably of equal length; glumes 6–8 mm long; nutlets 2–2.3 mm long. . . . . **16.** *macrostachya*
–   Inflorescence branches not flattened, erect and noticeably of equal length; glumes 2.8–3.5 mm long; nutlets 0.8–1 mm long . . . . **17.** *quaternella*
20. Style base deciduous 21
–   Style base persistent. 22
21. Culms, inflorescence branches and leaves usually scabrid; glumes 2–3.5 mm, mostly dark brown and mucronulate. . . . . **21.** *hispidula*
–   Culms glabrous, scabridulous in upper part; glumes 3.5–6 mm, pale brown, midnerve of glume green when fresh . . . . **22.** *viridecarinata*
22. Glumes less than 2 mm long, gaping at fruit maturity; nutlet surface punctate . . . . **47.** *densa*
–   Glumes more than 2 mm long, not gaping at fruit maturity; nutlet surface not puncticulate 23
23. Spikelet slender, 1–2.5 mm wide. 24
–   Spikelet wider, 2–4 mm wide. 25
24. Glumes 2.2–3.6 mm long, pale brown with white hairs on margin; leaves filiform, well developed forming a conspicuous feature . . . . **19.** *burchellii*
–   Not this combination of characters 25
25. Inflorescence variable, often a hemispherical head with 2–4 branches; glumes 4–6 mm long; nutlets 1.7–1.9 mm long; plants sprawling, robust, coarse-textured . . . . **14.** *boeckeleriana*
–   Inflorescence variable, a dense head, or frequently more open with 1 or more sessile spikelets and 2–5 stalked-spikelet groups or individual spikelets; glumes 3.2–5.3 mm long; nutlet 1.3–1.6 mm long, style base temporarily persistent; plants erect, not sprawling and more finely-textured. **20.** *contexta*

Key 2 – Mostly annual species

1. Inflorescence consisting of a single spikelet 2
–   At least some inflorescences with 2 or more spikelets. 8
2. Involucral bracts entirely glume-like, generally shorter than inflorescence 3
–   Involucral bracts not entirely glume-like, shorter or longer than inflorescence. 4
3. Glumes 30–60; spikelet 6–8 mm long, apex obtuse. . . . . **23.** *densiflora*
–   Glumes less than 30; spikelet 7–8 mm long (sometimes spikelets twinned), apex acute. . . . . **24.** *melanocephala*
4. Culm glabrous except below inflorescence, culm base conspicuously ascending, rooting at lower nodes; inflorescence sometimes anthelate; spikelet 5–8(10) mm; nutlet 1–1.2 × 0.8–0.9 mm, rugose, style base deciduous, occasionally persistent. . . . . **25.** *wombaliensis*
–   Not this combination of characters 5
5. Glumes with a noticeable recurved mucro c. 0.5 mm long . . . . **27.** *mucronata*
–   Glumes without a pronounced recurved mucro. 6
6. Spikelet 4–8 mm long; glumes 2.5–3 mm; nutlets 0.8–1 × 0.6–0.7 mm, lacking persistent style base, obpyriform, base pedunculate, surface pitted . . **28.** *lacunosa*
–   Not this combination of characters 7

7. Culms hispid; spikelet 3–8 mm long, sometimes with 2–3 sessile spikelet clusters; nutlets 0.9–1.3 × 0.8–1 mm, surface rugose without obvious papillae . . **29.** *humilis*
– Culms glabrous, except below inflorescence; spikelet 4–9 mm long, sometimes not unispicate but anthelate; nutlet 0.7–0.9 × 0.5–0.6 mm, rugose with papillae on angles . . . . **30.** *acutispicata*
8. Inflorescence head-like of (2)3 to many sessile spikelets, occasionally with an additional stalked spikelet or cluster of spikelets 9
– Inflorescence not head-like, an umbel-like anthela of 2–3 or many individually pedicelled spikelets, rarely all spikelets sessile in a few inflorescences 17
9. Inflorescence a single compact head 10
– Inflorescence occasionally a single compact head of sessile spikelets, more often with 1–3 added stalked sessile spikelet clusters; glumes 1.3–1.8 mm long; nutlet 0.7–0.8 × 0.5–0.6 mm. . . . . **31.** *pluricephala*
10. One or more involucral bracts exceeding head-like inflorescence 11
– Most involucral bracts as long or shorter than inflorescence, but 1 or 2 may surpass the inflorescence 13
11. Bracts with many conspicuous long white hairs; glumes 2.5–4 mm long; nutlet 0.9–1 × 0.6–0.8 mm, rugose, angles with prominent raised papillae . . . .
. . . .**32.** *buchananii*
– Bracts with few long white hairs, or these absent 12
12. Plants favouring metalliferous soils; stems and leaves scabrid; spikelets 4–7 mm long; glumes 2–3.5 mm long, mucro 0.5–1.5 mm . . . .**26.** *abbreviata*
– Plants infrequent on metalliferous soils; stems glabrous, except below inflorescence; spikelets 4–5(6) mm long; glumes 2–2.5 mm long, mucro 0.3–0.5 mm . . . .
. . . . **27.** *mucronata*
13. Inflorescence a compact spherical head 6 × 6 mm; involucral bracts with long white hairs; lower glumes with conspicuous dark aristae; nutlet 0.5–0.6 × 0.3 mm, elliptical to oblong, surface smooth to puncticulate. . . . .**8.** *fimbristyloides*
– Not this combination of characters 14
14. Nutlet surface trabeculate (striate lengthwise and transversely barred), 0.5–0.6 × 0.4–0.5 mm; inflorescence 3–6 mm across; glumes 1.5–1.7 mm long, mucronate
. . . . **33.** *trabeculata*
– Nutlet surface not trabeculate 15
15. Culm hispid, spikelets 3–8 mm long, sometimes with 2–3 sessile spikelet clusters; nutlet 0.9–1.3 × 0.8–1 mm, surface rugose without obvious papillae. . .**29.** *humilis*
– Culm glabrous, sometimes scabrid below inflorescence. 16
16. Spikelet 3–6 mm long; glumes 1.5–2 mm long, apex obtuse to subacute; nutlet 0.4–0.5 mm long, smooth to faintly rugose, papillae indistinct . . . .
. . . .**34.** *afromicrocephala*
– Spikelet 6–7 mm long; glumes 2–2.5 mm long, apex rounded, more or less emarginate; nutlet 0.5–0.6 mm long, rugose, angles with papillae . . . **35.** *capitata*
17. Style base deciduous 18
– Style base persistent. 24
18. Culms and inflorescence branches glabrous, sometimes scabridulous near top .
. . . . 19
– Culms (especially near base) and inflorescence branches with short spreading hairs, rarely glabrous. 23
19. Glumes 3.5–7 mm long 20
– Glumes less than 3.5 mm long 21
20. Leaf blades absent; spikelets 10–12 mm long; glumes 5–7 mm long; nutlet obovoid, 1–1.2 × 0.9–1.1 mm . . . .**36.** *nudiuscula*

– Leaf blades present; spikelets 8–15 mm long; glumes 3.5–6 mm long; nutlet 0.9–1 × 0.7–0.8 mm . . . . **22.** *viridecarinata*

21. Stem base ascending, often rooting at lower nodes; spikelets 5–8(10) mm long; glumes 2.5–3 mm; style base usually not persistent. . . . .**25.** *wombaliensis*
– Not this combination of characters 22

22. Spikelets ovoid; nutlets 0.9–1 × 0.8–0.9 mm, white at maturity, with cuneate base and expanded distally with 3 'lobes' visible on abaxial face . . . . **37.** *rotundata*
– Spikelets not ovoid; nutlets with or without cuneate base, not expanded distally with 3 obvious 'lobes' visible on abaxial face 23

23. Culm with spreading hairs; glumes 2–3.5 mm, boat-shaped, mucronulate, dark brown to almost black; nutlet 0.8–1.3 × 0.7–1.1 mm, obpyriform, rugose, angles smooth or papillate . . . . **21.** *hispidula*
– Not this combination of characters 24

24. Inflorescence with 6 or fewer spikelets. 25
– Inflorescence with more than 6 spikelets. 27

25. Nutlets c. 1 mm long, surface with large quadrate cells in longitudinal rows; maybe present at high altitude in Malawi . . . .**38.** *johnstonii*
– Not this combination of characters 26

26. Glumes longer than 3 mm; plant hairy; style base persistent . . . . . **18.** *andongensis*
– Glumes less than 3 mm . . . . **30.** *acutispicata*

27. Involucral bracts lacking long white hairs; plant hispid; spikelets 3–5 mm long; glumes 1.8–2.5 mm long . . . . **39.** *congolensis*
– Involucral bracts with few to many long white hairs 28

28. Culm robust, 1.5–1.8 mm wide; spikelets sometimes compacted together, appearing head-like; glumes 2.5–3.2 mm long . . . . **40.** *flexuosa*
– Culm slender, less than 1.5 mm wide 29

29. Small plant up to 15 cm tall; on metalliferous soils; nutlet narrowly obovoid, dark rusty-red, 0.7 × 0.4 mm . . . .**41.** *cupricola*
– Not this combination of characters 30

30. Spikelets 5–7 mm long, rarely longer; glumes 1.8–2 mm, nutlet 0.7–0.8 mm long 31
– Not this combination of characters 33

31. Glumes 1.8–2 mm long . . . . **42.** *laxispicata*
– Glumes more than 2 mm long 32

32. Anthela with a central group of spikelets. . . . . **44.** *capitulatoradiata*
– Anthela lacking a central group of spikelets . . . . **43.** *longiradiata*

33. Glumes markedly fimbriolate, with or without a hyaline margin. 34
– Not this combination of characters 35

34. Glumes glabrous, markedly fimbriolate with a broad hyaline margin; nutlets faintly transversely rugose, 0.5–0.6 × 0.4–0.5 mm . . . . **45.** *tanzaniae*
– Glumes glabrous, fimbriolate, with or without a narrow hyaline margin; nutlets puncticulate, 0.7–0.8 × 0.4–0.5 mm . . . . **46.** *abortiva*

35. Spikelets ovate; glumes gaping at nutlet maturity; nutlet surface with many rows of small cones (punctate) . . . . **47.** *densa*
– Spikelets ovate to lanceolate; glumes not or slightly gaping at maturity; nutlet surface without many rows of small cones 36

36. Spikelets 2–5 mm; glumes 1–2.5(3) mm, rounded with midnerve not reaching apex; nutlet 0.7–0.9 × 0.6–0.7 mm, transversely rugose with raised papillae on folds . . . . **48.** *pusilla*
– Spikelets 3–4 mm long; glumes 1.3–1.6 mm long, margin fimbriolate; nutlet 0.5–0.6 × 0.4–0.5 mm, folds shallow, banded white in mature nutlets . . **49.** *scrobiculata*

1. **Bulbostylis rhizomatosa** (Lye) R.W. Haines in Haines & Lye, Sedges & Rushes E.
Afr., app.3: 1 (1983). —Goetghebeur & Coudijzer in Bull. Jard. Bot. Belg. **55**: 229
(1985). Type: D.R. Congo, Katanga, 1952, *Schmitz* s.n. (BR holotype).

Abildgaardia rhizomatosa Lye in Nordic J. Bot. **1**: 749 (1982). —Haines & Lye, Sedges &
Rushes E. Afr.: 97, figs.165,166 (1983).

Perennial; rhizome elongated, horizontal, conspicuous, c. 3 mm wide when dry, fleshy and soft
when live, covered in dark acuminate scales. Leaf sheath mouth sloping, orifice hairs 5–7 mm
long, light reddish-brown; leaf blades 120–200 × 1–1.5 mm wide, margins scabid. Culms isolated,
glabrous, triangular, 20–70 cm high by c. 1 mm thick. Inflorescence an anthela, one sessile and
2–4 stalked spikelets, or reduced to a single spikelet; bracts 5–7 mm long, glume-like basally.
Spikelets 4–10 × 2–3 mm, ovoid, 5–12 flowered. Glumes 4–5 mm long, reddish-brown, flanks
glabrous to finely scabrid, long marginal hairs may be present, midnerve pronounced ending
before the apex. Stamens 3. Style 3-branched, base dark, persistent. Nutlet 1.8 × 1.3 mm, broadly
obovoid, trigonous, transversely rugose, light brown to darker at maturity.

Zambia. N: Kasama Dist., 10 km E of Kasama, 19.i.1961, *Robinson* 4277 (K). W:
Mwinilunga Dist., Mwinilunga, source of Zambezi R., 13.xii.1963, *Robinson* 5999 (K).
Malawi. N: Nyika Plateau, Western Valley, 2010 m, 6.i.1959, *Robinson* 3093 (K).

Also from southern D.R. Congo and Tanzania. In open dry *Brachystegia* woodland;
1300–2100 m.

Conservation notes: Not widespread, maybe infrequently collected; possibly
threatened.

Some descriptions note there is just one stamen, but 3 stamens were found in all
florets of *Robinson* 4277 (K).

The only species in the genus producing well-developed horizontal rhizomes and
scattered culms (Lye 1982). However, *B. schlechteri* and some collections of *B. clarkeana*
also have horizontal rhizomes with scattered culms.

2. **Bulbostylis schlechteri** C.B. Clarke in Bull. Herb. Boissier ser.2 **4**: 995 (1904). —
Clarke in Kew Bull., add. ser. **8**: 26 (1908). Type: South Africa, Aapies R., 1600 m,
5.xi.1893, *Schlechter* 3634 (K holotype).

Abildgaardia schlechteri (C.B. Clarke) Lye in Nordic J. Bot. **1**: 758 (1982).

Perennial to 50 cm tall; rhizome conspicuous, woody, sympodial. Leaf sheaths frequently
becoming red, with short leaf-like tips; leaf blades absent. Culms erect, 20–50 cm tall by 0.4–0.8 mm
wide, 1–10 mm apart, angular, glabrous but minutely scabrid below inflorescence. Inflorescence
a solitary chestnut-red spikelet; bracts 1–2, shorter than, equal to or slightly overtopping spikelet.
Spikelet 5–8 × 1–2 mm, ovoid to oblong lanceolate, compressed. Glumes 2.9–3.7 mm long,
densely packed, ovate-lanceolate, minutely scabrid, light to dark reddish-brown, keel prominent,
green when fresh, excurrent, margins ciliated. Stamens 3. Style 3-branched, style base persistent.
Nutlet 0.8–1 × 0.5–0.8 mm, obovoid, trigonous, surface faintly transversely wavy.

Zambia. B: Mongu Dist., 90 km E of Mongu, 15.xi.1965, *Robinson* 6710 (LMA, PRE);
Mongu Dist., 10 km E of Mongu, 31.x.1965, *Robinson* 6699 (LMA).

Also in Angola and South Africa (Gauteng, Limpopo, Mpumalanga). Marshy
ground, grassland near streams and seasonal seepage areas; 1000–1200 m.

Conservation notes: Local in distribution in Flora area; possibly threatened.

Drummond & *Cookson* 6494 (K) from just over the border in E Angola has leaves, and
may be incorrectly assigned to *B. schlechteri*; further investigation is needed.

3. **Bulbostylis clarkeana** M. Bodard in Bull. Soc. Bot. France **108**: 308 (1962). —
Goetghebeur & Coudijzer in Bull. Jard. Bot. Belg. **55**: 228 (1985). —Verdcourt in
F.T.E.A., Cyperaceae: 78 (2010). Type: Guinea, Sulimania, Erimakuna, 20.iii.1892,
Scott Elliot 5244 (K holotype).

Abildgaardia clarkeana (Bodard) Lye in Bot. Not. **127**: 495 (1974). —Haines & Lye, Sedges & Rushes E. Afr.: 102, figs.175,176 (1983).

Perennial to 50 cm tall; seriate rhizome conspicuous, woody. Leaf blades absent; leaf sheaths dark purple to reddish, glabrous with short leaf-like tips. Culms erect, 30–50 cm by 0.4–0.8 mm wide, angular, glabrous, minutely scabrid below inflorescence; culm bases ± thickened. Inflorescence a terminal spikelet, or with 1–2 additional stalked spikelets; bracts 1–2, shorter than or equal to inflorescence, bases glume-like, apices subulate. Spikelets 8–12 mm long, ovate to oblong-lanceolate, compressed. Glumes 3–5(6) mm long, ovate-lanceolate, minutely scabrid, light to dark brown with darker patches, keel prominent, shortly excurrent, green when fresh. Stamens 3. Style 3-branched, base persistent on nutlet. Nutlet 1.4–1.5 × 0.8–0.9 mm, obovoid, trigonous, light to darker brown at maturity, surface rugose.

Zambia. N: Mbala Dist., Chila pans, 12 km from Mbala (Abercorn), fl. 17.iii.1961, *Robinson* 4541 (K). C: Mkushi Dist., Fiwila, Chalobeti dambo, fl. 4.i.1958, *Robinson* 2622 (K).

Also in Guinea, Burundi, Uganda and Tanzania. In perennial wet bogs and swamps; 1200–1800 m.

Conservation notes: Scattered distribution; probably threatened.

Bulbostylis clarkeana has been collected infrequently although there are a number of Robinson collections from perennial bogs in N Zambia; Goetghebeur & Coudijzer (1985) record its distribution as scattered. *Robinson* 2985 (K) from N Malawi (Chisenga, 64 km W of Karonga) was determined by Lye as *B. clarkeana*, but leaves are present and this collection is possibly *B. oritrephes*.

Hooper (F.W.T.A.: 317, 1972) included *Bulbostylis clarkeana* as a synonym of *B. oritrephes*, while Gordon-Gray (pers. comm.) considered it a synonym of *B. schlechteri*. Claire Archer (pers. comm.) noted that *B. schlechteri* in southern Africa has a single spikelet while in *B. clarkeana* the inflorescence may be anthelate or a single terminal spikelet. The type, *Scott Elliot* 5244 (K), has solitary spikelets. In this account these three species are kept separate as further study is needed of all African material with conspicuous and well-developed rhizomes.

4. **Bulbostylis oritrephes** (Ridl.) C.B. Clarke in Trans. Linn. Soc. London, Bot. **4**: 54 (1894). —Goetghebeur & Coudijzer in Bull. Jard. Bot. Belg. **55**: 228 (1985). —Gordon-Gray in Strelitzia **2**: 33, fig.12A,D (1995). —Verdcourt in F.T.E.A., Cyperaceae: 77 (2010). Types: Angola, Golungo Alto, mountains E of Quilombo, Sobato Quilombo, i.1855, *Welwitsch* 7016 (LISU syntype, BM); slope of Queta Mt, xi.1855, *Welwitsch* 7020 (LISU syntype, BM, K).

   Fimbristylis oritrephes Ridl. in Trans. Linn. Soc. London, Bot. **2**: 155 (1884).
   Bulbostylis trichobasis (Baker) C.B. Clarke var. *uniseriata* C.B. Clarke in Bull. Soc. Roy. Bot. Belg **39**(3): 37 (1900). Type: D.R. Congo, Kisantu, 1899, *J. Gillet* 333 (G?).
   Bulbostylis trichobasis var. *leptocaulis* C.B. Clarke in Bull. Herb. Boissier ser.2 **1**: 58 (1901). Type: D.R. Congo, Kisantu, 1899, *J. Gillet* 179 (G?).
   Bulbostylis caespitosa Peter in Repert. Spec. Nov. Regni Veg. Beih. **40**(Anhang): 127 (1936). Type: Tanzania, Lushoto, Dist., E Usambaras, Monga, 7.iii.1918, *Peter* 22665 (B† holotype, GOET, K, S).
   Bulbostylis trichobasis var. *caespitosa* (Peter) Kük. in Repert. Spec. Nov. Regni Veg. Beih. **40**: 417 (1937).
   Bulbostylis oritrephes var. *major* Meneses in Garcia de Orta **4**: 254 (1956). Type: Angola, Menongue, Tiengo Valley, 4.xii.1906, *Gossweiler* 3759 (LISC holotype).
   Abildgaardia oritrephes (Ridl.) Lye in Bot. Not. **127**: 497 (1974). —Haines & Lye, Sedges & Rushes E. Afr.: 101, fig.173 (1983).

Perennial to 90 cm tall; rhizome conspicuous, woody, sympodial, up to 90 × 3–4 mm, of swollen shoot bases confluent in a uniseriate or sometimes biseriate row, clothed in golden to blackish closely imbricate scale leaves to 10 mm long. Leaves variable, sometimes appearing undeveloped, usually 1/20th to half culm length; leaf bases sheathing, mouths sloping, conspicuously long white-bearded, ligule 0; leaf blades 4–200 × 0.2–0.5 mm, filiform to setaceous, densely to sparsely white villous, hairs mostly confined to nerves (often inconspicuous unless plant held against a light source). Culms slender, developed from only most recent 4–6 rhizome sections, stiffly erect when short (early in growing season following fire), elongating remarkably to c. 100 cm long by fruit maturation, densely villous below to glabrescent/glabrous above. Inflorescence variable, often on individual clones, up to 27 × 24 mm, anthelate of central sessile spikelet with 1–5 glabrous inflorescence branches, or less often of 2–3 sessile clustered units; bracts 2(4), to 22 mm long, shorter than or equal to inflorescence; bases sheathing, blades pilose. Spikelets 4–12 × 1.5–3 mm, ovate oblong to lanceolate oblong. Glumes 2.4–3.6 × 2.3 mm, ovate, dark chestnut red to black, short white pubescent abaxially, keel prominent, apex rounded to mucronate, margin densely shortly white-fringed. Stamens 3. Style 3-branched, 1.4–2.4 mm long, 3-angled, base round, vertically flattened, persistent on nutlet. Nutlet 0.9–1.3 × 0.6–1 mm, rounded obovoid, grey to brown at maturity, surface finely wrinkled.

Zambia. N: Isoka Dist., Mafinga Mts, NW slopes, 2100 m, 18.xii.1964, *Robinson* 6303 (K). W: Mwinilunga Dist., Dobeka bridge near Matonchi, 1350 m, 17.ii.1975, *Hooper & Townsend* 147 (K). C: Serenje Dist., Kundalila Falls, S of Kanona, 1400 m, 14.iii.1975, *Hooper & Townsend* 749 (K). E: Nyika Plateau, E side, 2500 m, 5.i.1959, *Robinson* 3073 (K). **Zimbabwe**. E: Nyanga Dist., Gairesi Ranch, N of Troutbeck, 13.xi.1956, *Robinson* 1877 (K, NU, SRGH). **Malawi**. N: Chitipa Dist., Misuku Mts, 1520 m, 10.i.1959, *Robinson* 3143 (K). C/S: Kirk Range, 14.xi.1950, *Jackson* 270 (K). S: Mt Mulanje, Chambe Mt, 1850 m, 14.vi.1962, *Robinson* 5345 (K).

Also in Guinea to Cameroon, Rwanda, Burundi, Kenya, Uganda, Tanzania, D.R. Congo, Angola, South Africa (Eastern Cape, Free State, North West, Gauteng, Limpopo, Mpumalanga, KwaZulu-Natal), Lesotho and Swaziland. Montane grasslands, burnt or unburnt, and montane marshland; 1300–2500 m.

Conservation notes: Widespread; not threatened.

The majority of specimens of *Bulbostylis oritrephes* from sub-Saharan Africa have an indumentum of fine villous hairs, particularly evident on culms when held against a light source (otherwise microscopic enlargement is necessary). Some examples from tropical localities are more densely villous (not necessarily uniformly), approaching *B. trichobasis* (Baker) C.B. Clarke in Madagascar. In the past, southern African plants have been named *B. trichobasis* but do not precisely match those from Madagascar. Hence it is prudent to accept *B. oritrephes* for all sub-Saharan specimens.

5. **Bulbostylis pilosa** (Willd.) Cherm. in Bull. Soc. Bot. France **81**: 266 (1934); **82**: 341 (1935). —Goetghebeur & Coudijzer in Bull. Jard. Bot. Belg. **55**: 229 (1985). —Verdcourt in F.T.E.A., Cyperaceae: 70 (2010). Type: West Africa, Guinea, 1784, *Isert* s.n. (B-W 1095 holotype, C).

  Schoenus pilosus Willd., Phytographia **1**: 3, t.1, fig.3 (1794).
  Abildgaardia pilosa (Willd.) Nees in Linnaea **9**: 289 (1834). —Haines & Lye, Sedges & Rushes E. Afr.: 96, fig.161 (1983).
  Bulbostylis aphyllanthoides (Ridl.) C.B. Clarke in Durand & Schinz, Consp. Fl. Afr. **5**: 611 (1894); in F.T.A. **8**: 436 (1902). —Napper in J. E. Africa Nat. Hist. Soc. Natl. Mus. **25**(110): 6 (1965). Type: Angola, Pungo Andongo, low hills near Conde by R. Cuanza, also Quissande, iii.1857, *Welwitsch* 6837 (LISU holotype, BM, K).
  Fimbristylis africana C.B. Clarke in Durand & Schinz, Consp. Fl. Afr. **5**: 601 (1894); in F.T.A. **8**: 425 (1902), invalid name. Type as for *B. pilosa*.
  Fimbristylis pilosa (Willd.) K. Schum. in Engler, Pflanzenw. Ost-Afrikas **C**: 124 (1895), invalid name.

Perennial to 75 cm tall, loosely tufted; rhizome variable, of slightly thickened shoot bases irregularly linked, less often uniseriate, or conspicuously elongate (c. 8 mm wide), clothed with markedly striate scale-leaves, soon decaying. Leaves variable, outermost reduced to short sheaths, inner mostly bladed, occasionally obsolete; sheath mouth conspicuously long white hairy; leaf blades (when present) to $^1/_3$ culm length, 0.8–3.5 mm wide, glabrous, short hairy or white pilose. Culms terete, ridged and furrowed, white-pilose to glabrescent, generally pubescent at least apically. Inflorescence terminal, a head to 18 × 23 mm with (1)10 spikelets, long hairy when immature; bracts 2, generally inconspicuous, occasionally exceeding inflorescence. Spikelets laterally compressed throughout, 10–18 × 3–6 mm, greenish, buff to golden brown or chestnut red. Glumes 7–10 × 3.2–4.4 mm, distichous, persistent or some deciduous, apex shortly mucronate. Stamens 3. Style 3-angled, 3-branched, base c. 4 mm deep, persistent but occasionally shed before nutlet dissemination. Nutlet 1.8–2 × 1.2–1.5 mm, trigonous, obovoid with tapering base, eventually greyish, surface faintly lineate becoming transversely rugose, outermost cells longitudinally oblong.

Zambia. B: Mongu Dist., Looma, Nanganda 15°06'S 23°41'E, 1060 m, 21.xii.1993, *Bingham* 9885 (K). N: Chinsali Dist., Shiwa Ngandu (Tshiba Ngandu), 1520 m, 2.vi.1956, *Robinson* 1547 (K). W: Mwinilunga Dist., Dobeka bridge near Matonchi, 1350 m, 17.ii.1975, *Hooper & Townsend* 136 (K). C: Mkushi Dist., Fiwila, 7.i.1958, *Robinson* 2678 (K). E: Lundazi Dist., Tigone dam, 2.4 km along Lundazi–Chama road, 1200 m, 18.x.1958, *Robson & Angus* 156 (K). S: Namwala Dist., Namwala, edge of dambo, 9.i.1957, *Robinson* 2073 (K, NU, SRGH). **Zimbabwe**. N: Gokwe Dist., Gokwe, 27.ii.1962, *Bingham* 366 (K, NU). E: Chipinge Dist., 1.5 km E of Rusango, 500 m, 1.ii.1975, *Gibbs Russell* 2735 (K, SRGH). **Malawi**. N: Mzimba Dist., Mzuzu, Marymount, 2.iv.1971, *Pawek* 4563 (K). S: Blantyre Dist., Ndiranda Mt, S shoulder, 1390 m, 3.vii.1970, *Brummitt* 11787 (K, LISC). **Mozambique**. Z: Milange Dist., Chiperone Mt, 35.72915°E 16.50967°S, 1045 m, 28.xi.2006, *Harris* 58 (K). T: Angónia Dist., Ulónguè, 7.xii.1980, *Macuácua* 1462 (K, LISC, LMA). MS: Cheringoma Dist., Gorongosa Nat. Park, Cheringoma Plateau, headwaters of Muaredzi stream, 230 m, i.1972, *Tinley* 2346 (K, LISC, SRGH). GI: Vilankulo Dist., San Sebastian peninsula, 22°7.7'–8.7 E 35°26.9–27.5'S, 10.xi.1958, *Mogg* 2915 B (K).

Also in Senegal, Nigeria, Burundi, Kenya, Tanzania, D.R. Congo and Angola. Open *Brachystegia* woodland, seasonally wet grassland, dambos, often on white sand; 5–1400 m.

Conservation notes: Widespread; not threatened.

6. **Bulbostylis parvinux** C.B. Clarke in Fl. Cap. **7**: 207 (1898); in Bull. Misc. Inform. Kew, add. ser. **8**: 110 (1908), name only. —Gordon-Gray in Strelitzia **2**: 35 fig.12B,E (1995). Type: Mozambique, Delagoa Bay area, 29.iii.1894, *Kuntze* 217 (K holotype).

Scirpus parvinux (C.B. Clarke) K. Schum. in Just's Bot. Jahresb. **26**(1): 329 (1900).

Abildgaardia parvinux (C.B. Clarke) Lye in Mitt. Bot. Staatssamml. München **10**: 545 (1971).

Bulbostylis mozambica M. Raymond in Naturaliste Canad. **99**: 30 (1972). Type: Mozambique, Marrucuene, 1958, *Da Costa* 23 (MTJB holotype, COI).

Tufted perennial up to 60 cm tall; rhizome woody, uniseriate, shoot bases slightly swollen, mostly clothed in white woolly scale leaves. Photosynthetic leaves usually appearing undeveloped; leaf sheaths with mouth sloping, often long white-bearded, hairs curled (woolly), ligule a dense line of white wool; leaf blades often obsolete, setaceous when present, to 26 × 0.2–0.4 mm, margins thickened. Culms terete, ridged and furrowed, glabrous. Inflorescence a single terminal head up to 25 × 23 mm, almost spherical at maturity, of up to 16 spikelets, usually densely white woolly during development; bracts 3–6, visible when inflorescence young, later masked by reflexing spikelets, margins of sheaths conspicuously white bearded. Spikelets compressed throughout, 6–9 × 2.1–4 mm, ovate to oblong in age. Glumes distichous, most fertile, 4–5 × 2.7–3.9 mm, finely pubescent below or glabrous, apex mucronate. Stamens 3. Style 2.6–4 mm long, 3-angled, 3-branched, glabrous, base round, vertically compressed, c. 0.1 mm deep, dark brown, persistent. Nutlet 0.8–1.2 × 0.5–1 mm, ovoid, trigonous, eventually golden brown, surface reticulate, not transversely ridged, outermost cells predominantly sub-quadrate or slightly longer than wide.

Mozambique. M: Marracuene Dist., Maputo (Lourenço Marques), 30.xi.1897, *Schlechter* 11543 (BOL, BR, COI, GRA, K, PRE).

Also in South Africa (KwaZulu-Natal). Locally common on white and pale red sandy soils among grasses in *Terminalia* and *Acacia burkei* woodland in between sand forest strips; 10–120 m.

Conservation notes: Restricted to the Maputaland area; possibly Near Threatened.

Clare Archer (pers. comm.) reports limited distribution both in South Africa (2632CC & 2732AB) and in S. Mozambique (2532DC & DD).

In first describing *Bulbostylis parvinux*, Clarke (1898) made reference to its relationship with *B. cinnamomea* (Boeckeler) C.B. Clarke with "ciliate-hairy leaf sheaths and very much longer nuts". However, *B. cinnamomea sensu* C.B. Clarke is *B. boeckeleriana* (Schweinf.) Beetle.

7. **Bulbostylis laniceps** C.B. Clarke in Mém. Couronnés Autres Mém. Acad. Roy. Sci. Belgique **53**: 306 (1896). —Hooper in F.W.T.A. **3**: 316 (1972). —Goetghebeur & Coudijzer in Bull. Jard. Bot. Belg. **55**: 234 (1985). Type: Equatorial Guinea, Grand Corisco Is., x.1862, *Mann* 1885 (K holotype).

   Bulbostylis wittei Cherm. in Rev. Zool. Bot. Africaines **22**: 69 (1932).

Perennial up to 50 cm tall, densely tufted, somewhat swollen at base. Leaf sheath mouth sloping, sparsely white bearded to more densely woolly; leaf blades ¼ to ⅓ culm length, 0.25–0.3 mm wide, filiform, channelled, minutely setaceous. Culms scapose, 0.3–0.9 mm wide, glabrous. Inflorescence terminal, capitate, c. 10 × 10 mm, with many closely-packed sessile spikelets interspersed by long white hairs; bracts 3–4, unequal, narrow, longest c. 20 mm, with many long white hairs. Spikelets 4–6 mm long, not flattened, reflexed. Glumes 3–4 mm long, distichous, ovate, acute to acuminate, lower with short or obsolete mucro, flank 1-nerved with brown patch, margin ciliate. Stamens 3; anther 1.5 mm long, apiculate. Style 3-branched, base small, persistent. Nutlet 0.8–1 × 0.6–0.7 mm, obovoid, trigonous, smooth to colliculate, white when immature, brown to red-brown at maturity.

Zambia. B: Mongu Dist., 25 km NE of Mongu, 12.xii.1965, *Robinson* 6734 (K). N: Mansa Dist., 90 km S of Mansa (Fort Roseberry), 4.i.1961, *Robinson* 4232 (K, LISC). W: Mwinilunga Dist., source of Matonchi dambo, 16.xi.1937, *Milne-Redhead* 3263 (K).

Also in West Africa, Congo, D.R. Congo and Angola. In grassland or seasonal dambos on sandy or peaty damp soil; 800–1400 m.

Conservation notes: Widespread; not threatened.

Bulbostylis laniceps closely resembles *B. fimbristyloides*. Apart from size, few morphological differences are seen, *B. fimbristyloides* being approximately half the size of *B. laniceps* with a tendency to appear to be an annual and grow on damp lateritic outcrops in the Mwinilunga area of NW Zambia.

8. **Bulbostylis fimbristyloides** C.B. Clarke in Bull. Soc. Bot. France **54**(8): 28 (1907). —Hooper in F.W.T.A., ed.2 **3**: 316 (1972). —Goetghebeur & Coudijzer in Bull. Jard. Bot. Belg. **55**: 235 (1985) as *fimbristylidoides*. Types: Mali, Kouli Koro, 6.x.1899, *Chevalier* 2457 (P syntype); *Chevalier* 2458 (P syntype, K); *Chevalier* 2462 (P syntype).

   Bulbostylis cyrtathera Cherm. in Arch. Bot. Mém. **4**(7): 35 (1931). Types: Central African Republic, Oubangui-Chari, 23.vii.1921, *Le Testu* 3004 (P syntype, BR); Central African Republic, Oubangui, 103.ix.1922, *Tisserant* 662 (P syntype).

Tufted erect annual, 15–25 cm tall; lacking rhizome and tunic of leaf bases. Leaves numerous, ½ to ⅓ or occasionally equalling culm length, scabrid; leaf sheath mouth conspicuously fringed with long white hairs, ligule 0; leaf blades to 80 × 0.2–0.4 mm, filiform, margins scabrous. Culms 0.2–0.4 mm wide, ridged and furrowed, glabrous. Inflorescence capitate, with long white hairs, often spherical, c. 6 × 6 mm, dark brown, spikelets often reflexed; bracts 3–4, short or twice as long as head, filiform, scabrid with numerous long white hairs. Spikelets 2–3 mm long, interspersed by scabrid dark aristae (subulae) of lower glumes. Glumes ± distichous, 2–3 mm long, margins with very fine hairs, flanks minutely hispid; lower glume apex extended into hispid subula 0.5 mm long, upper glume mucronate to shortly apiculate. Stamens (1)2–3. Style glabrous, branches 3, usually shorter than style, base minute, persistent on nutlet. Nutlet 0.5–0.6 × 0.3 mm, elliptical to oblong in outline, subtrigonous, white to grey-white at maturity, surface smooth to somewhat puncticulate.

Zambia. W: Mwinilunga Dist., 7 km N of Kalene Hill, 17.iv.1965, *Robinson* 6618 (K); Mufumbwa Dist., 7 km E of Mufumbwa (Chizera), 27.iii.1961, *Drummond & Rutherford-Smith* 7439 (K, LISC, SRGH).

Also in Burkina Faso, Guinea, Chad, Mali, Central African Republic, D.R. Congo and Angola. Damp soil over rocky outcrops and edges of laterite pans; 1000–1400 m.

Conservation notes: Widespread, although local within the Flora area; not threatened.

There is close relationship to *Bulbostylis laniceps*, a perennial larger in all its parts, and also with *B. afromicrocephala*.

9. **Bulbostylis macra** (Ridl.) C.B. Clarke in Durand & Schinz, Consp. Fl. Afr. **5**: 614 (1894). —Goetghebeur & Coudijzer in Bull. Jard. Bot. Belg. **55**: 227 (1985). —Verdcourt in F.T.E.A., Cyperaceae: 72 (2010). Type: Angola, Huíla, Lopollo, ii.1860, *Welwitsch* 6955 (LISU holotype, BM, K).

Fimbristylis macra Ridl. in Trans. Linn. Soc. London, Bot. **2**: 150 (1884).

Bulbostylis zambesica C.B. Clarke in F.T.A. **8**: 430 (1902). Types: Malawi, Shire Highlands, Kampala, i.1896, *Scott Elliot* 8464 (BM syntype); Malawi, Mt Sochi, x.1859, *Kirk* s.n (K syntype); Malawi, between Blantyre and Matope, 24.x.1887, *Scott* s.n. (K syntype).

Abildgaardia macra (Ridl.) Lye in Bot. Not. **127**: 497 (1974). —Haines & Lye, Sedges & Rushes E. Afr.: 97, fig.164 (1983).

Perennial to 40 cm, densely tufted with a slightly swollen woody base; persistent tunic of leaf sheaths. Leaves numerous, 1/3 to 1/2 culm length; leaf sheaths densely white pubescent to glabrescent below, mouth truncate, occasionally auriculate laterally, white-bearded; leaf blades 40–70(110) × 0.3 mm, filiform, margins pubescent to pilose. Culms scapose, terete, ridged and furrowed, white pilose to pubescent below inflorescence, occasionally glabrescent or glabrous. Inflorescence a solitary sessile spikelet, sometimes with an additional pedicelled spikelet; bracts 2–3, to 20 mm long, glume-like. Spikelet cylindric, compact, 8–12(16) × 2–4 mm, elliptic to oblong, golden to chestnut brown, basal glumes tardily deciduous. Glumes 3.8–6 × 1.5–2.5 mm, finely pubescent abaxially, veins terminating below margin, apex acute to rounded when young, margins long ciliate. Stamens 3. Style 20–30 mm long, trigonous, 3-branched, densely fringed, base rounded, vertically depressed, persistent on nutlet. Nutlet 1.1–1.5 × 0.8–1 mm, trigonous, occasionally biconvex, obovoid, surface closely transversely rugose, outermost cells longitudinally oblong.

Zambia. B: Kaoma Dist., 14 km ESE of Kaoma (Mankoya), near Luena R., 21.xi.1959, *Drummond & Cookson* 6713 (K). N: Kasama Dist., Kasama, 11.xii.1960, *Robinson* 4171 (K). W: Solwezi Dist., 1250 m, 1.x.1947, *Greenway & Brenan* 8141 (K). C: Mkushi Dist., Fiwila, 1220 m, 7.i.1958, *Robinson* 2662 (K). **Zimbabwe**. N: Makonde Dist., Silverside Mine, 4.xi.1965, *Wild* 7466 (K, SRGH). C: Harare (Salisbury), NW shore of Lake Chivero (McIlwaine), 1400 m, 27.x.1956, *Robinson* 1822 (K, NU, SRGH). E: Nyanga Dist., Gairesi Ranch, 19.xi.1956, *Robinson* 1951 (K, LISC, NU). **Malawi**. N: Nyika Plateau, SW slopes, 1680 m, 7.i.1959, *Robinson* 3102 (K). C: Dedza Dist., Chongoni Forest, 16.ii.1969, *Chapman* 1091 (K). S: Mt Mulanje, W slope of Mchese Mt, near Bwaibwai forest village, 950 m, 17.x.1987, *Chapman & Chapman* 8918 (K). **Mozambique**. MS: Gorongosa Dist., Gorongosa Mt, Mt Gogogo summit area, iii.1972, *Tinley* 2405 (K).

Also in Burundi, Tanzania, D.R. Congo and Angola. Often in dry stony areas in miombo woodland; 1000–1700 m.

Conservation notes: Widespread; not threatened.

This species is closely related to *Bulbostylis schoenoides* and *B. macrostachya*. It differs from the latter which has an anthelate inflorescence of 3–5 branches with spikelets surrounding the solitary terminal one.

Particularly in N Zambia, plants with wider leaves may occur. These are unispicate but with larger spikelets, longer narrower glumes and larger nutlets than in typical *Bulbostylis macra*. Intermediates have also been noted in other parts of the distributional range (Goetghebeur & Coudijzer 1985).

Plants are resistant to burning, developing tunics of persistent leaf sheaths that are shorter, finer and more slender than those of *Bulbostylis schoenoides*.

10. **Bulbostylis schoenoides** (Kunth) C.B. Clarke in Trans. Linn. Soc. London, Bot. **4**: 54 (1894); in Durand & Schinz, Consp. Fl. Afr. **5**: 616 (1894). —Gordon-Gray in S. Afr. J. Bot. **54**: 565–570 (1988); in Strelitzia **2**: 36, fig.12H,K (1995). —Verdcourt in F.T.E.A., Cyperaceae: 76 (2010). Type: South Africa, Cape Province, between Yellow R. and Zandplaat, 1838, *Drège* 1040 (P lectotype, K), lectotypified by Haines & Lye (1983). Other Drège syntypes from the E Cape.

Isolepis schoenoides Kunth, Enum. Pl. **2**: 208 (1837), non *Abildgaardia schoenoides* R. Br.

Schoenus erraticus Hook. f. in J. Linn. Soc., Bot. **6**: 22 (1862). Type: Bioko (Fernando Po), Clarence Peak, xii.1860, *Mann* 655 (K holotype).

Scirpus cinnamomeus Boeckeler in Bot. Jahrb. Syst. **5**: 505 (1884). Type: Madagascar, Andrangolaoka, xi.1880, *Hildebrandt* 3737 (BM, GOET, JE, K).

Bulbostylis cinnamomea (Boeckeler) C.B. Clarke in Durand & Schinz, Consp. Fl. Afr. **5**: 612 (1894); in F.T.A. **8**: 432 (1902).

Bulbostylis megastachys (Ridl.) C.B. Clarke in Durand & Schinz, Consp. Fl. Afr. **5**: 614 (1894). —Mapaura & Timberlake, Checkl. Zimbabwe Pl.: 87 (2004). Type: Angola, Huíla Dist., v.1860, *Welwitsch* 6952 (BM holotype).

Fimbristylis schoenoides (Kunth) K. Schum. in Engler, Pflanzenw. Ost-Afrikas **C**: 125 (1895), in part.

Bulbostylis scleropus C.B. Clarke in Fl. Cap. **7**: 207 (1898). Type: South Africa, former Transvaal, Aapies R., n.d., *Burke* s.n. (K lectotype), lectotypified by Burtt (1986).

Bulbostylis erratica (Hook. f.) C.B. Clarke in F.T.A. **8**: 434 (1902).

Bulbostylis stricta Turrill in Bull. Misc. Inform. Kew **1925**: 70 (1925). Type: Zimbabwe, Harare (Salisbury) Dist., xi.1919, *Eyles* 1890 (K holotype, PRE).

Abildgaardia erratica (Hook. f.) Lye in Nordic J. Bot. **3**: 239 (1983). —Haines & Lye, Sedges & Rushes E. Afr.: 101, fig.172 (1983).

Perennial up to 80 cm tall; rhizome woody, with compactly confluent swollen stem bases clothed in leaf bases varing from soft dull brown to coarse hard shining black, forming persistent tunics to aerial shoots, especially prominent following fire. Leaves numerous, half to equalling culm length; leaf sheath mouths truncate to sloping, long white-bearded when young, hairs soon drying and breaking away, ligule 0; leaf blades filiform to linear, to 210 × 0.2–1.5 mm, glabrous except for densely scabrid to ciliate margins. Culms terete, several-ridged, glabrous, occasionally sparsely pubescent apically. Inflorescence a terminal head of 4–6 spikelets (occasionally one below the rest), or anthelate with 1 central sessile spikelet with (1)5 additional pedicelled ones, total dimensions up to 28 × 25 mm; bracts 2–4, to 40 mm long, glabrous to finely white pubescent. Spikelets 7–17 × 2–4.5 mm, elliptic to lanceolate, dark brown to chestnut-red to black. Glumes 4–7 × 1.9–3.5 mm, glabrous, less often finely white pubescent below, apex emarginate to rounded with veins of keel terminating before margin, soon splitting to appear acute or bifid (important to examine young undamaged examples). Stamens 3. Style 1.4–4.7 mm long, 3-angled, glabrous, branches 3, densely fringed, base 3-angled, persistent. Nutlet 1.5–1.9 × 1.1–1.3 mm, brown to slate grey, trigonous, surface closely transversely lineolate, outermost cells longitudinally oblong, eventually velate.

Zambia. N: Mbala Dist., escarpment above Chilongowelo, 1520 m, 15.xii.1951, *Richards* 69 (K). W: Mufilira Dist., 8 km S of Mufilira, 1300 m, 27.iii.1960, *Robinson* 3433 (K). C: Chongwe Dist., Chakwenga headwaters, 100–129 km E of Lusaka, 27.x.1963, *Robinson* 5768 (K). S: Choma Dist., Muckle Neuk, 19 km N of Choma, 1280 m, 28.xi.1954, *Robinson* 991 (K). **Zimbabwe**. W: Matobo Dist., Besna Kobila, 1460 m, x.1956, *Miller* 3676 (K). C: Gweru Dist., Watershed Block, 1400 m, 13.xi.1976, *Biegel* 2318 (K, LISC, SRGH). E: Nyanga Dist., Mare R., 27.x.1946, *Wild* 1563 (GRA, K, PRE). **Malawi**. N: Nkhata Dist., Vipya Link road, 24 km from S end, 11.i.1975, *Pawek* 8947 (K, MO, SRGH). **Mozambique**. Z: Gurué Dist., Mt Namuli, 15°23'14.7S 37°02'40.7E, 1880 m, 19.xi.2007, *Mphamba* 327 (K). T: Tete Dist., Angonia, 16.xii.1980, *Macuácua* 1449a (K, LISC). MS: Gorongosa Dist., Mt Gorongosa summit, c. 1700 m, *Ward* 7381 (NU).

Also in Ethiopia, Kenya, Uganda, Tanzania, Angola, South Africa (Eastern Cape, Gauteng, KwaZulu-Natal), Lesotho and Madagascar. Dry or damp grasslands and upland grassland with shrubs; 1200–1700 m.

Conservation notes: Widespread; not threatened.

Goetghebeur & Coudijzer (1985) suggest that *Bulbostylis schoenoides* is a complex and a thorough study is required. A full list of synonyms and types is not provided here, only those most likely to be known in the Flora area. *Schoenus erraticus* (=*B. erratica*) from West Africa has been included as it may not be morphologically distinct from *B. scleropus*.

Gordon-Gray's recommendation of indicating synonyms rather than imposing them could not be upheld for this flora account as no further morphological and distributional data is presently available from detailed studies. The taxa outlined in the key are here treated as synonyms of a polymorphic *Bulbostylis schoenoides*. In some parts of the Flora area it is possible to recognise some taxa mentioned here as synonyms of *B. schoenoides*; this applies particularly to *B. cinnamomea*, which is sometimes given specific rank.

Collections of intermediate specimens are difficult to determine, which has led to a variety of names being applied in checklists for duplicates of certain collections.

Bulbostylis schoenoides shows a close morphological relationship with *B. macra* and *B. macrostachya*.

11. **Bulbostylis scabricaulis** Cherm. in Bull. Soc. Bot. France **68**: 419 (1922). — Goetghebeur & Coudijzer in Bull. Jard. Bot. Belg. **55**: 233 (1985). —Gordon-Gray in Strelitzia **2**: 35, figs.12G,J (1995). —Verdcourt in F.T.E.A., Cyperaceae: 103 (2010). Types: Madagascar, Mevatanana, i.1900, *Perrier de la Bathie* 477 (P syntype); Ampombo, ii.1907, *Perrier de la Bathie* 4578 (P syntype, K).

    Fimbristylis collina Ridl. in Trans. Linn. Soc. London, Bot. **2**:154 (1884), non *B. collina* (Kunth) C.B. Clarke. Type: Angola, mountains E of Quilombo, Quiacatubia, i.1855, *Welwitsch* 7004b (LISU holotype, BM, K).

    Bulbostylis cardiocarpa (Ridl.) C.B. Clarke var. *holubii* C.B. Clarke in F.T.A. **8**: 434 (1902). Type: Botswana, Leshumo Valley, n.d., *Holub* s.n. (K holotype).

    Abildgaardia filamentosa (Vahl) Lye var. *holubii* (C.B. Clarke) Lye in Bot. Not. **127**: 496 (1974).

    Abildgaardia collina (Ridl.) Lye in Nordic J. Bot. **1**: 757 (1982). —Haines & Lye, Sedges & Rushes E. Afr.: 121, fig.228 (1983).

Perennial up to 70 cm tall, tufted, with a tunic of dark brown leaf bases sometimes sparsely developed, seldom becoming loose as in *B. filamentosa*. Leaf sheath orifice with long flexuous hairs; leaf blades 80–180 × 0.4–0.6 mm, abaxially densely white pilose below to glabrescent with age, hairs mostly confined to nerves. Culms scapose, generally not conspicuous among leaves, terete to slightly angled below inflorescence, ridged and furrowed, densely white hispid (scabrid) but velutinous below inflorescence, occasionally glabrescent to glabrous. Inflorescence 6–14 × 9–17 mm, of 12–15 spikelets closely packed when young, becoming distinct at maturity; bracts with sheathing part short, remainder narrowed, as long as or exceeding inflorescence. Spikelets

4–12 × 1.5–2.3 mm, polygonal-cylindric, sometimes flattened in upper half. Glumes 1.8–2.7 × 2.1–3.3 mm, broadly boat-shaped, acute to mucronulate, margins shortly ciliate. Style base c. 0.1 mm deep, persistent, same colour as nutlet, 3-angled, not vertically flattened, more conspicuous than in *B. filamentosa*. Nutlet 1–1.4 × 0.7–1 mm, golden becoming slate grey at full maturity, apex never depressed, surface reticulate, finely and closely transversely rugulose, not punctulate.

Botswana. N: Chobe Dist., Leshumo (Lyshuma) Valley, 1883?, *Holub* s.n. (K). **Zambia**. W: Ndola, 1400 m, 14.ii.1960, *Robinson* 3342 (LISC). N: Mbala Dist., Simenwe Farm, 1520 m, 16.iv.1959, *Webster* 622 (K). S: Choma Dist., Muckle Neuk, 19 km N of Choma, 1260 m, 13.i.1954, *Robinson* 458 (K). **Zimbabwe**. E: Mutare Dist., Nyamkwarara (Nyumquarara) Valley, 1070 m, ii.1935, *Gilliland* Q1631 (K). **Malawi**. N: Mzimba Dist., Kasitu Valley, 1040 m, 27.i.1938, *Fenner* 227 (K). S: Zomba Dist., country round Zombwe Farm, iii.1939, *Wilson* 227 (K).

Also Senegal to Angola, Uganda, Kenya, Tanzania, South Africa (Mpumalanga, North West, KwaZulu-Natal), Lesotho and Swaziland. Grassland, savanna and *Brachystegia* woodland; 1000–1600 m.

Conservation notes: Very scattered, never common, rare in South Africa; not threatened.

Distinguished from *Bulbostylis filamentosa* by the scabrid culm and smaller glumes.

12. **Bulbostylis cardiocarpoides** Cherm. in Rev. Zool. Bot. Afric. **24**: 298 (1934). — Goetghebeur & Coudijzer in Bull. Jard. Bot. Belg. **55**: 234 (1985). —Verdcourt in F.T.E.A., Cyperaceae: 103 (2010). Type: D.R. Congo, Lower Congo, Kisantu, 4.ii.1932, *Vanderyst* 28175, 28178, 28182 (BR syntypes); Kisantu, 2.xii.1930, *Vanderyst* 28044 (BR syntype); between Kasinde & Lubango (Kibale–Ituri), i.1932, *Lebrun* 4765 (BR syntype, P); Kisantu, 4.ii.1932, *Vanderyst* 28181 (BR lectotype, P), lectotypified by Huntley on specimen (1953).

    Abildgaardia cardiocarpoides (Cherm.) Lye in Bot. Not. **127**: 495 (1974). —Haines & Lye, Sedges & Rushes E. Afr.: 122, fig.231 (1983).

Small tufted perennial, sometimes an annual, 6–20 cm tall; new shoots appearing amongst old which are usually rotted or burnt. Leaf sheath mouth with many hairs to 5 mm long, ligule 0; leaf blades 40–80 × 0.2–0.4 mm, flat or channeled, glabrous or with a few spine-like hairs particularly near apex. Culms ridged and furrowed, glabrous, 0.4–0.5 mm thick. Inflorescence head-like, of 3 to many sessile spreading spikelets; bracts 3–12 mm long, ± erect, lacking long white hairs. Spikelets 3–5 × 1–1.5 mm, acute. Glumes 2 mm long, reddish brown, boat-shaped, acute with paler margins and midrib reaching apex. Stamens 3. Style fimbriate, 3-branched, base persistent as brown or blackish knob. Nutlet narrowly obovoid, pale brown with 3 light yellow angles, 1–1.1 × 0.6–7 mm, surface smooth to slightly papillose-reticulate.

Zambia. W: Mwinilunga Dist., lake c. 2 km along road to Kanyama from Mwinilunga–Solwezi road, 1400 m, 25.ii.1975, *Hooper & Townsend* 391 (K).

Also in Burundi, Uganda and D.R. Congo. Sandy soil in open grassland and near lake margins; 1200–1400 m.

Conservation notes: Localised in the Flora area, but probably not threatened.

It is likely that this species has been undercollected in the Flora area. A fimbriate style is uncommon in the genus. The capitate inflorescence with short bracts resembles the type of *Bulbostylis capitata*.

13. **Bulbostylis filamentosa** (Vahl) C.B. Clarke in Durand & Schinz, Consp. Fl. Afr. **5**: 613 (1894); in Fl. Cap. **7**: 206 (1898). —Goetghebeur & Coudijzer in Bull. Jard. Bot. Belg. **55**: 231 (1985). —Verdcourt in F.T.E.A., Cyperaceae: 101, fig.16 (2010). Type: Ghana, *Thonning* s.n. (C holotype, P-JU). FIGURE 14.**15**.

    Scirpus filamentosus Vahl, Enum. Pl. **2**: 262 (1805).

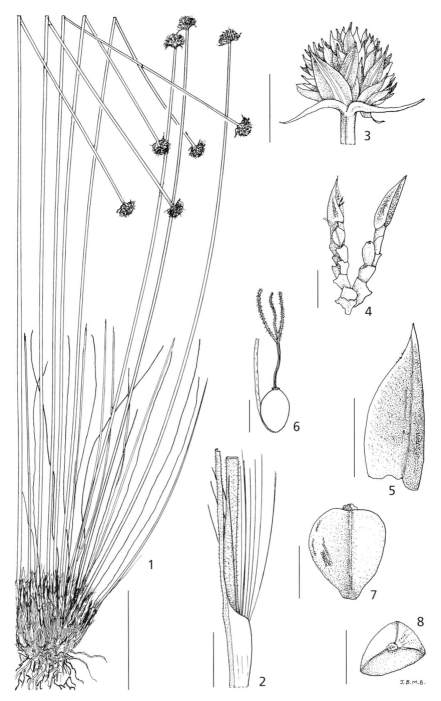

Fig. 14.**15**. BULBOSTYLIS FILAMENTOSA. 1, habit; 2, leaf sheath apex; 3 inflorescence; 4, two spikelets, lower glumes removed; 5, glume; 6, floret; 7, 8, nutlet abaxial and apical views. 1–6 from *Brummitt* 11305; 7, 8 from *Pawek* 13802. Scale bars: 1 = 40 mm; 2 = 3 mm; 3 = 4 mm; 4, 5 = 1 mm; 6–8 = 0.5 mm. Drawn by Jane Browning.

Fimbristylis cardiocarpa Ridl. in Trans. Linn. Soc. London, Bot. **2**: 154 (1884), illegitimate name.

Bulbostylis cardiocarpa (Ridl.) C.B. Clarke in Durand & Schinz, Consp. Fl. Afr. **5**: 612 (1894), invalid name.

Bulbostylis metralis Cherm. in Arch. Bot. Mém. **4**(7): 34 (1931). Type: Central African Republic, Haut-Ubangi between Wadda & Kotto, 15.viii.1922, *Le Testu* 4122 (P holotype).

Abildgaardia filamentosa (Vahl)Lye var. *filamentosa* in Bot. Not. **127**: 496 (1974). —Haines & Lye, Sedges & Rushes E. Afr.: 120, fig.227 (1983).

Abildgaardia filamentosa var. *metralis* (Cherm.) Lye in Bot. Not. **127**: 496 (1974). —Haines & Lye, Sedges & Rushes E. Afr.: 121 (1983).

Bulbostylis filamentosa var. *metralis* (Cherm.) R.W. Haines in Haines & Lye, Sedges & Rushes E. Afr., app.3: 1 (1983).

Perennial up to 90 cm tall, densely tufted with tunic of dark brown leaf bases; rhizome compact, inconspicuous, woody, up to 3 mm wide. Leaves ± half culm length, densely to sparsely pubescent abaxially; outer leaf sheaths brown, shorter than inner darker ones, sheath-mouth sloping, white bearded; leaf blades to 260 × 0.2–0.6 mm, filiform to setaceous, scabrid. Culms terete, glabrous, occasionally minutely hairy, conspicuous among leaves. Inflorescence terminal, capitate, 8–16 × 11–19 mm, of 12–20 closely packed, sessile spikelets; bracts 2–3, glume-like, as long as head, lacking white hairs. Spikelets 4–7 × 1.3–2 mm, rachilla straight, massive. Glumes 2.4–3 × 1.4–2.2 mm, oval to elliptic, pubescent below, with 3–5 yellowish veins, keel prominent, excurrent into a short mucro, flanks dark brown, margins paler. Stamens 3. Style and base 1–1.2 mm long, 3-angled, branches 3, densely fimbriate when young, base round, persistent, vertically depressed, becoming inconspicuous following style dehiscence. Nutlet 0.8–0.9 × 0.6–0.8 mm, oblanceolate to obovate in outline, trigonous, eventually dark brown, surface finely reticulate becoming papillose.

Zambia. W: Mwinilunga Dist., lake c. 2 km along road to Kanyama from Mwinilungu–Solwezi road, 25.ii.1975, *Hooper & Townsend* 392 (K). N: Mbala Dist., Lake Chila, 1585 m, 25.i.1952, *Richards* 584 (K). C: Serenje Dist., Kundalila Falls, 13 km SE of Kanona, 1400 m, 15.x.1967, *Simon & Williamson* 1028 (K, SRGH). S: Kalomo Dist., Kalomo, sandy riverside, 1220 m, 1.i.1958, *Robinson* 2548 (K, SRGH). **Zimbabwe**. E: Nyanga Dist., Nyanga Nat. Park, Alt., 1830 m, 4.iii.1969, *Jacobsen* 3726 (K, SRGH). **Malawi**. N: Mzimba Dist., Viphya Plateau, 48 km SW of Mzuzu, 1.6 km N of Lumono Hills road, 19.ii.1978, *Pawek* 13802 (K). C: Ntchisi Dist., Ntchisi Forest Reserve above resthouse, 1590 m, 26.iii.1970, *Brummitt* 9436 (K). S: Mulanje Mt, Lichenya Plateau, firebreaks, 6.vi.1962, *Robinson* 5270 (K). **Mozambique**. MS: Manica Dist., Manica (Macequece), i.1949, *Fisher & Schweickerdt* 447 (K, NU).

Also in Guinea, Mali, Nigeria, Rwanda, Burundi, Kenya, Uganda, Tanzania, D.R. Congo, Angola and South Africa (Limpopo). Dry or seasonally wet grassland, sandy river banks and dambos, shallow soil over rocky outcrops; 500–1900 m.

Conservation notes: Widespread; not threatened.

14. **Bulbostylis boeckeleriana** (Schweinf.) Beetle in Amer. Midl. Naturalist **41**: 458 (1949). —Goetghebeur & Coudijzer in Bull. Jard. Bot. Belg. **55**: 225 (1985). —Gordon-Gray in Strelitzia **2**: 30, fig.10B,E (1995). —Verdcourt in F.T.E.A., Cyperaceae: 73, fig.12 (2010). Types: Eritrea, below summit of Lalamba, near Keren, 12.iii.1891, *Schweinfurth* 837 (K syntype); Eritrea, Mt Oualid, N of Mt Bizen, 10.v.1892, *Schweinfurth & Riva* 1873 (FT syntype); Mt Bizen, n.d., *Schweinfurth & Riva* 1851 (G lectotype), lectotypified by Lye. FIGURE 14.**16**.

Scirpus boeckelerianus Schweinf. in Bull. Herb. Boissier **2**, app.2: 50 (1894).

Scirpus collinus Boeckeler var. *boeckelerianus* (Schweinf.) Schweinf. in Bull. Herb. Boissier **2**, app.2: 104 (1894).

Abildgaardia boeckeleriana (Schweinf.) Lye in Bot. Not. **126**: 327 (1973).

Abildgaardia boeckeleriana (Schweinf.) Lye var. *boeckeleriana*. —Haines & Lye, Sedges & Rushes E. Afr.: 98, fig.167,168 (1983).

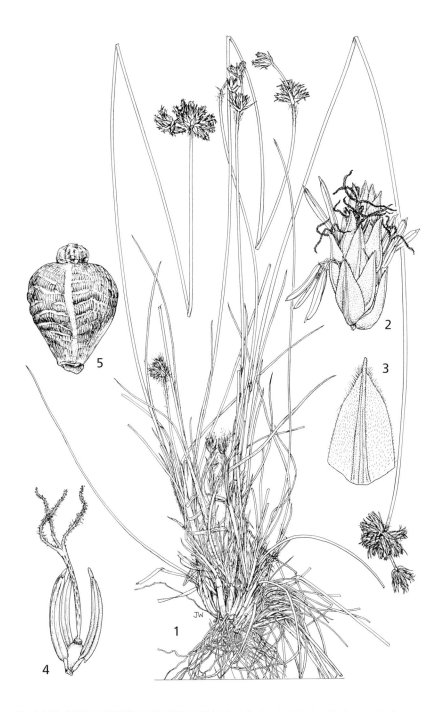

Fig. 14.**16**. BULBOSTYLIS BOECKELERIANA. 1, habit (× ²/₃); 2, spikelets (× 5); 3, glume (× 8); 4, floret (× 10); 5, nutlet (× 24). All from *Conrads in EAH* 10542. Drawn by Juliet Williamson. Reproduced from Flora of Tropical East Africa (2010).

Perennial up to 80 cm tall; rhizome woody of slightly swollen shoot bases confluent irregularly, or in bi-or multi-seriate rows, up to 70 × 7–9 mm, clothed in imbricate scale leaves, occasionally upper scales bearded with white hairs. Leaves variable, usually up to 1/3 culm length, occasionally obsolete; leaf sheath mouths sloping, conspicuously white fringed (except when growing in lush conditions), ligule 0; leaf blades 4–280 × 0.3–1 mm, linear to setaceous, glabrous to densely pilose. Culms stout, terete, ridged and furrowed, white pubescent below inflorescence, otherwise glabrous. Inflorescence variable, usually a hemispherical terminal head of up to 16 spikelets, often (0)2–4 pedicels carrying additional heads, occasionally reduced to single spikelet (all forms can be represented in a clone); bracts 2(3), usually shorter than inflorescence, with few to many long white hairs. Spikelets 7–12 × 2–4 mm. Glumes spiral, 4–6 × 2.4–4 mm, golden brown, white pubescent below, margin ciliate, apex obtuse or minutely muronate. Stamens 3. Style 2.1–3.7 mm long, 3-angled, glabrous, 3-branched 3. Nutlet (including style-base) 1.7–1.9 × 1.1–1.2 mm, obovoid, trigonous, crowned by persistent pyramidal base, surface transversely wavy and faintly lineate, ridged, outermost cells longitudinally oblong.

Zambia. N: Kasama Dist., 25 km W of Kasama, 11.xii.1960, *Robinson* 4178b (K). **Zimbabwe**. S: Masvingo Dist., Masvingo (Fort Victoria), 10.i.1949, *Fisher & Schweickerdt* 376 (K, NU). **Malawi**. N: Chitipa Dist., Mafinga Hills, 1850 m, 12.iii.1961, *Robinson* 4465 (K). **Mozambique**. MS: Dondo Dist., forest next to railway track near Dondo, 5.x.1925, *Peter* 31191 (K).

Also in Ethiopia, Sudan, Rwanda, Uganda, Kenya, Tanzania, South Africa (Gauteng, KwaZulu-Natal, Limpopo, Mpumalanga). Dry open grassland, damp *Brachystegia* woodland and shallow soil over rock outcrops; 600–1800 m.

Conservation notes: Widespread; not threatened.

In East Africa *Bulbostylis boeckeleriana* appears to be well-represented, while for the Flora area few collections have been seen. It would appear that the closely affiliated *B. igneotonsa* is more frequent.

Bulbostylis boeckeleriana has been much confused in the literature with *B. contexta* and also with *B. cinnamomea* (Boeckeler) C.B. Clarke – see Gordon-Gray (1995).

The World Cyperaceae Checklist lists two varieties, typical var. *boeckeleriana* and var. *transiens* (K. Schum.) R.W. Haines & Lye based on difference in the inflorescence. According to Gordon-Gray this is an unreliable character throughout the genus.

15. **Bulbostylis igneotonsa** Raymond in Naturaliste Canad. **99**: 29, fig.2 (1972). — Verdcourt in F.T.E.A., Cyperaceae: 71 (2010). Type: Zambia, 15 km E of Kasama, 20.xii.1961, *Robinson* 4732 (Herb. Raymond holotype, K, MTJB).

Abildgaardia igneotonsa (Raymond) Kornaś in Flora **176**: 64 (1985).

Perennial 15–60 cm tall; rhizome woody with slightly swollen shoot bases, burnt down to leave stubs of ± equal length, growth arising above these. Leaves variable, usually to 1/3 culm length, occasionally obsolete; leaf sheath mouth sloping, conspicuously long, white-fringed, ligule 0; leaf blades 4–200 × 0.3–1 mm, linear to setaceous, glabrous to densely pilose. Culms c. 1 mm thick, stout, terete, ridged and furrowed, glabrous or shortly pilose. Inflorescence a pale to dark brown terminal head of 2–5 sessile spikelets, occasionally reduced to a single spikelet; bracts 2–4, equalling inflorescence, few to many long white hairs, long-awned. Spikelets 8–12 × 2–4 mm, oblong, acute. Glumes spiral, 8–9 mm long, golden to dark brown, long-acuminate, glabrous to shortly pubescent, margin ciliate. Stamens 3. Style 3-angled, glabrous, branches 3. Nutlet 2.1–2.3 (including style base) × 1.3–1.6 mm, triangular-rhomboid, angles blunt, crowned by persistent pyramidal base, surface transversely rugose.

Zambia. W: Mwinilunga Dist., Camp 25, 35 km E of Boma,10.ix.1930, *Milne-Redhead* 1082 (K). N: Kasama Dist., 15 km E of Kasama, seasonal dambo, 20.xii.1961, *Robinson* 4731 (K). C: Serenje Dist., Serenje, 22.xii.1964, *Robinson* 6331(K). **Malawi**. N: Nyika Plateau, 2350 m, 14.iii.1961, *Robinson* 4504 (K).

Also in Angola and Tanzania. Shallow soil over rocky outcrops, grassland and *Brachystegia* woodland; 1000–2350 m.

Conservation notes: Widespread; not threatened.

This species has been collected in burnt grassland, particularly in N Zambia, and may be fire-resistant. Robinson noted on collections 4732 and 4731, collected at the same locality on the same day, that the former had very hairy stems whilst in the latter they were glabrous. There is affinity with *Bulbostylis contexta* and *B. boeckeleriana*; in our opinion *B. igneotonsa* differs from the latter in the shape (long-acuminate) larger glumes and nutlet. *B. igneotonsa* was previoulsy considered synonymous with *B. boeckeleriana* by Gordon-Gray, but is treated here as a separate species.

16. **Bulbostylis macrostachya** (Lye) R.W. Haines in Haines & Lye, Sedges & Rushes E. Afr., app.3: 1 (1983). —Verdcourt in F.T.E.A., Cyperaceae: 71 (2010). Type: Tanzania, Mbeya Dist., Ruaha Nat. Park, Magangwe Hill, 20.xii.1972, *Bjørnstad* 2228 (O holotype, K).

    Abildgaardia macrostachya Lye in Nordic J. Bot. **1**: 749 (1982). —Haines & Lye, Sedges & Rushes E. Afr.: 96, fig.162,163 (1983).

Perennial up to 80 cm tall; rhizome woody, with compactly confluent swollen stem bases clothed in soft dull brown to black leaf bases, forming persistent tunics to aerial shoots. Leaves numerous, from quarter or half to equalling culm length; leaf sheaths straw-coloured, sometimes with dense masses of grey hairs, mouth truncate to sloping, long white bearded, ligule 0; leaf blades filiform to linear, 0.3–0.5 mm wide, scabrid. Culms 250–800 × 0.5–1.5 mm, terete, several ridged, scabrid. Inflorescence of a central sessile spikelet and 3–5 stalked spikelets on flattened stalks; bracts 2–4, with pale brown bearded basal part and a 3–8 mm long filiform, scabrid extension. Spikelets 10–40 × 3–5 mm, dark reddish brown. Glumes 6–8 mm long, lower with prolonged scabrid mucro, flanks shortly scabrid, bicoloured reddish brown with green excurrent midrib. Stamens 3. Style 3-branched, base 3-angled, persistent. Nutlet 2–2.3 × 1.5–1.8 mm, trigonous, obovoid, brown to slate grey, surface rugulose.

Zambia. N: Mbala Dist., 90 km W of Tunduma on Mbala road, 28.xi.1958, *Napper* 1157 (K); Mbala Dist., Chilongowelo, 19.ii.1954, *Richards* 2289 (K).

Also in Tanzania. *Brachystegia* woodland with granite outcrops; c. 1500 m.

Conservation notes: Localised, but probably not threatened.

Zambian collections of *Bulbostylis macrostachya* (with an anthelate inflorescence) are only recorded from the vicinity of Tunduma, around 113 km from those in Tanzania.

In the K isotype, the spikelets are closely placed so the inflorescence appears capitate; closer examination is needed to reveal the stalked spikelets which make up the anthela. Measurements for spikelet length of 10–40 mm appear to have been based on very old spikelets where glumes have been shed and the rachillae greatly elongated. Spikelets from the Zambian collections were 10–25 mm in length.

Gordon-Gray (pers. comm.) had problems accepting *B. macrostachya*, which Haines & Lye (1983) linkened to *Abildgaardia triflora* (L.) Abeyw.

Some collections at Kew previously identified by Lye as *Bulbostylis macrostachya*, such as *Robinson* 4163, have a single sessile spikelet, conforming more closely with *B. macra*. Others with a single spikelet are perhaps better allied with *B. schoenoides*. As noted previously (see note under *B. macra*) these unispicate examples may represent intermediates.

17. **Bulbostylis quaternella** (Ridl.) Goetgh. in Bull. Jard. Bot. Belg. **55**: 223 (1985). Types: Angola, Pundo Andongo, Pedras de Guinga, iii.1857, *Welwitsch* 6830b & 6827b (BM syntypes).

    Fimbristylis quaternella Ridl. in Trans. Linn. Soc. London, Bot. **2**: 152 (1884).

Perennial 18–40 cm tall, tightly tufted; rhizome short, inconspicuous, roots slender. Leaf sheaths short, c. 10 mm long, brown, membranous, mouth white-fringed; leaf blades 40–70 × 0.3–0.4 mm, glabrous, bristle-like, closely placed. Culms 0.3–0.5 mm wide, glabrous, scabridulous below inflorescence. Inflorescence anthelate, a central sessile spikelet with 3–4 erect, equally long

(10–15 mm) scabridulous branches, terminating in a spikelet; bracts inconspicuous, shorter than spikelets, with a few long white hairs. Spikelets 9–10 × 2 mm, lanceolate, dark red-brown. Glumes 2.8–3.5 mm long, lanceolate-oblong, obtuse to rounded, keel scabrid, pronounced, margins ciliate, lower glumes mucronulate. Stamens 3, anthers up to 2 mm long. Style 3-branched, base persistent on nutlet. Nutlet 0.8–1 × 0.6–0.7 mm, trigonous, obovoid, greyish-white, transversely rugose with 3 prominent angles.

Zambia. C: Chongwe Dist., Chakwenga headwaters, 100–129 km E of Lusaka on Lusaka–Rufunsa road, 1.xii.1963, *Robinson* 5879 (MRSC, P). S: Choma Dist., 10 km E of Choma, 1310 m, 18.xii.1956, *Robinson* 1986 (K). **Zimbabwe**. C: Harare Dist., Cleveland dam, 26.x.1956, *Robinson* 1823 (K, SRGH).

Also in D.R. Congo and Angola. Open sandy ground in damp forest clearings and margins of dambos; 1100–1300 m.

Conservation notes: Fairly widespread; not threatened.

Bulbostylis quaternella is distinguished by the scaberulous, ± equal, erect (scarcely spreading) inflorescence branches and dark brown to blood-red spikelets. It may be confused with *B. burchellii* – Gordon-Gray initially determined *Robinson* 1986 and 2557 as the latter species.

18. **Bulbostylis andongensis** (Ridl.) C.B. Clarke in Durand & Schinz, Consp. Fl. Afr. **5**: 611 (1894); in F.T.A. **8**: 443 (1902). Types: Angola, Pundo Andongo, between Condo and Quisonde, iii.1857, *Welwitsch* 6820 (BM syntype, LISU); Pungo Andongo, i.1857, *Welwitsch* 6823, 6827 (BM syntypes).

    Fimbristylis andongensis Ridl. in Trans. Linn. Soc. London, Bot. **2**: 153 (1884).

Tufted annual, 10–30 cm tall. Leaves numerous, half to sometimes equalling culm length; leaf sheaths membranous, rust-brown, mouth inconspicuously long white-fringed, ligule 0; leaf blades up to 150 mm, filiform, setaceous, some long hairs present, hairs stiff and bristly. Culms ridged and furrowed. Inflorescence anthelate, variable, a central spikelet and 2–5 spikelets added on equally long diverging, hispid-branches; bracts 3–4, filiform, shorter or sometimes longer than inflorescence, margins white fringed. Spikelets 6–7 mm long, ellipsoid, dark-brown. Glumes spiral, boat-shaped, 3–3.5 mm long, finely scabrid, margins ciliate, apex mucronate to shortly apiculate. Stamens 3. Style glabous, base dark brown, rounded, vertically flattened, persistent on nutlet, branches 3, usually shorter than style. Nutlet c. 1 mm long, obtusely trigonous, obovoid, transversely rugose (many papillate rugae closely placed).

Zimbabwe. W: Hwange Dist., Hwange (Wankie), 19.vi.1934, *Eyles* 8029 (BM, SRGH). C: Makonde Dist.?, Hunyani, 2.xii.1928, *M.& R. Young* 541 (BM).

Also in Central African Republic, Congo, Gabon, D.R. Congo and Angola; 500–1400 m.

Conservation notes: Not widely distributed; not threatened.

The Hwange collection does not have a collector name, but from the date this is likely to be Eyles.

It has not been possible to dissect to make accurate measurements of glumes and nutlets. It is likely that this is a perennial species, although described by Clarke as annual.

19. **Bulbostylis burchellii** (Ficalho & Hiern) C.B. Clarke in Durand & Schinz, Consp. Fl. Afr. **5**: 612 (1894). —Gordon-Gray in Strelitzia **2**: 30, fig.10C,F (1995). — Verdcourt in F.T.E.A., Cyperaceae: 91 (2010). Types: Zambia, W side of high plateau along upper reaches of R. Ninde by Zambezi, 14°46'S 20°56'E, viii.1878, *Serpa Pinto* 60, 62 (LISU syntypes).

    Fimbristylis burchellii Ficalho & Hiern in Trans. Linn. Soc. London, Bot. **2**: 28 (1881).

    Fimbristylis huillensis Ridl. in Trans. Linn. Soc. London, Bot. **2**: 154 (1884). Types: Angola, Huíla, Empalança, iv.1860, *Welwitsch* 6950, 6951 (BM syntypes, LISU).

Fimbristylis rhizomatosa Pires de Lima in Bol. Soc. Brot. ser. 2, **2**: 134 (1924). Type: Mozambique, Palma Dist., road between Palma and Quionga, 8.xi.1916, *Pires de Lima 52* (PO holotype).

    Abildgaardia burchellii (Ficalho & Hiern) Lye in Bot. Not. **127**: 495 (1974). —Haines & Lye, Sedges & Rushes E. Afr.: 113, fig.205 (1983).

Perennial, up to 50 cm tall, densely tufted, appearing annual in early growth; rhizome mostly inconspicuous, compact, woody, with closely packed stem bases. Leaf sheaths finely pubescent pilose, not woolly, mouth sloping, conspicuously long white-fringed, ligule 0; leaf blades filiform, pubescence mostly on nerves, apex acute. Culms erect, glabrous below to scabrid upwards. Inflorescence anthelate, usually compound, of (11)37(90) spikelets, occasionally much reduced; bracts 2–4, variable in length. Spikelets 4–10 × 1–2 mm; basal glumes tardily deciduous, rachilla shortly winged. Glumes 2.2–3.6 × 1.3–2.5 mm, prominent keel of 3 greenish veins terminating in apex (upper) or minutely mucronate (lower), margin fringed white hairy. Stamens 3. Style 3-angled, sparsely villous to glabrous, 3-branched, base rounded, vertically flattened, persistent on nutlet. Nutlet 0.6–1 × 0.4–0.7 mm, obovate to obcordate in outline, trigonous, surface transversely wavy, individual cells longitudinally oblong, 1–3 papillae developing per cell at maturity.

Botswana. N: Ngamiland Dist., near Shishikola Well, in fossil river valley, 18°07'S 23°00.66'E, 16.xi.1980, *P.A. Smith 3584* (K). SE: Central Dist., Orapa, 21°17.022'S 25°25.212'E, 960 m, 7.iv.2005, *P.A. Smith et al. 24* (K). **Zambia**. B: Mongu Dist., Mongu floodplain, 1.xii.1962, *Robinson 5512* (K). W: Chingola Dist., 11 km N of Chingola, 3.i.1961, *Robinson 4221* (K). C: Chibombo Dist., 30 km N of Lusaka, 2.i.1961, *Robinson 4214* (K). S: Choma Dist., Pengelley's Farm, 23 km N of Choma, fl. 3.ii.1960, *White 6744* (K). **Zimbabwe**. N: Guruve Dist., Mwanzamtanda R. area, foothills of Mt Chiruwa, 600 m, 29.i.1966, *Müller 257* (K, SRGH). W: Matobo Dist., Quaringa farm, 1460 m, iii.1958, *Miller 5161* (K, LISC). C: Kadoma Dist., Iwaba Estate, 50 km NE of town, 1200 m, 8.ii.2000, *Lye 23733* (K). E: Mutare Dist., Vumba, 19.iv.1993, *Browning 567* (K, NU, SRGH). S: Mberengwa Dist., 32 km N of West Nicholson, Mberengwa (Belingwe) road, 17.iii.1964, *Wild 6411* (K, SRGH). **Malawi**. S: Blantyre, Shire Highlands, 1887, *Last s.n.* (K). **Mozambique**. Z: Morrumbala Dist., Morrumbala (Morambala) Station, Lower Shire, 19.ii.1888, *Scott s.n.* (K). GI: Massinga Dist., Pomene, xii.1971, *Tinley 2271* (K).

Also in Tanzania, Namibia, Angola and South Africa (Eastern Cape, Free State, Gauteng, KwaZulu-Natal, Limpopo, Mpumalanga). In *Uapaca–Brachystegia* woodland, coastal grassland and savanna on sand; 10–1600 m.

Conservation notes: Widespread; not threatened.

Bulbostylis burchellii may be distinguished by the closely packed very narrow and fine (filiform) leaves that give a distinct tufted appearance. Plants are sometimes confused with both *B. contexta* and *B. densa* subsp. *afromontana*; the latter is annual and more delicate while *B. contexta* is more robust with larger spikelets than *B. burchellii*.

20. **Bulbostylis contexta** (Nees) Bodard in Ann. Fac. Sci. Univ. Dakar **9**: 77 (1963). —Gordon-Gray in Strelitzia **2**: 31, fig.10G,J (1995). —Verdcourt in F.T.E.A., Cyperaceae: 79 (2010). Types: South Africa, Uitenhage area, Addo, *Ecklon s.n.* (S? syntype); Olifantshoek to Boshmansrivier, *Zeyher & Ecklon s.n.* (B syntype, P).

    Trichelostylis contexta Nees in Linnaea **10**: 146 (1835).
    Fimbristylis contexta (Nees) Kunth, Enum. Pl. **2**: 245 (1837).
    Bulbostylis collina (Kunth) C.B. Clarke in Durand & Schinz, Consp. Fl. Afr. **5**: 613 (1894). —Mapaura & Timberlake, Checkl. Zimbabwe Pl.: 87 (2004). Type: South Africa, E Cape, Addo, *Drége 2037* (K lectotype, P), lectotypified by Gordon-Gray on sheet (1955).
    Bulbostylis zeyheri (Boeckeler) C.B. Clarke in Durand & Schinz, Consp. Fl. Afr. **5**: 616 (1894). Type: South Africa, Gauteng, Magaliesburg Mts, *Zeyher 1768* (S?).
    Abildgaardia contexta (Nees) Lye in Bot. Not. **127**: 495 (1974). —Haines & Lye, Sedges & Rushes E. Afr.: 103, figs.179,180 (1983).

Densely tufted perennial to 15–30(50) cm; rhizome compact or elongate in soft, sandy habitats; shoot bases 3–4 mm wide, leaf base tunics present; Leaves with sheaths ridged by 5–9 keelar veins, finely white pubescent when young, becoming glabrous; sheath mouth truncate, conspicuously long white-pilose, ligule 0; leaf blades c. 10 cm, linear or setaceous, sparsely pubescent to scabrid, apex acute. Culms terete to slightly angled upwards, c. 0.8 mm wide. Inflorescence anthelate, very variable, compound to simple to single heads with many or a solitary spikelet, with variants on a single plant; pedicels several to 0, flattened; bracts (2)3(4), blades variable in length, mostly not exceeding inflorescence. Spikelets 5–8(12) × 2–4 mm, ferruginous; basal glumes persistent, rachilla deeply notched, very slightly winged. Glumes closely imbricate, spirally arranged 3.2–5.3 × 2–4 mm, pubescent abaxially, margins white-fringed in upper younger ones, apex on lower glumes shortly mucronate, upper emarginate, splitting with age. Stamens 3, anther crest minute. Style glabrous, faintly 3-angled, base pyramidal 0.2–0.4 mm, persisting temporarily after style abscission, stigma 3-branched. Nutlet 1.3–1.6 (including style base) × 0.9–1 mm, broadly obovoid, trigonous, faces transversely rugulose, ridges slightly projecting, angles eventually papillate or smooth, epidermal cells central on face longitudinally oblong, remainder ± subquadrate and eventually with a central papilla, later velate leaving nutlet surface ± smooth.

Zambia. W: Luanshya Dist., 15 km S of Luanshya, 12.iii.1960, *Robinson* 3387 (K). N: Mbala, no locality, 19.vi.1956, *Robinson* 1689 (MRSC). C: Mkushi Dist., Fiwila, 3.i.1958, *Robinson* 2590 (K, MRSC). S: Choma Dist., Muckle Neuk, 27.xi.1954, *Robinson* 987 (MRSC). **Zimbabwe**. N: Guruve Dist., 8 km S of Kanyemba, 1.ii.1966, *Müller* 317 (K, SRGH). C: Chegutu Dist., Chegutu (Hartley), Poole Farm, 22.xii.1944, *Hornby* 2379 (K, SRGH). E: Mutare Dist., Souldrop Farm, 11.ii.1949, *Chase* 5411 (K, SRGH). S: Masvingo Dist., Makoholi Exp. Station, 10.iii.1965, *Senderayi* 144 (K, SRGH). **Malawi**. C: Mchinji Dist., Mchinji (Fort Manning), xii.1962, *Robinson* 5598 (K). **Mozambique**. MS: Marromeu Dist., Kongone R., i.1861, *Kirk* s.n. (K). GI: Inhassoro Dist., Bazaruto Is., Ponta Muldiza, 15.xi.1958, *Mogg* 29078 (K, PRE). M: Marracuene Dist., national forest, Marracuene, 10.iv.1946, *Gomes e Sousa* 3413 (K).

Also Tanzania, Namibia and South Africa (Eastern Cape, Free State, Gauteng, KwaZulu-Natal, Limpopo, Mpumalanga). Sandy soils on hillsides, on disturbed soils of old cultivated fields; 5–1800 m.

Conservation notes: Widespread; not threatened.

Goetghebeur & Coudijzer (1985) transferred all perennials to *Bulbostylis hensii* (C.B. Clarke) R.W. Haines, while we believe them better placed under *B. contexta*, an extremely widespread taxon in sub-Saharan Africa. *B. hensii* seems to be a slightly younger stage of *B. contexta* with shorter glumes and (slightly) immature nutlets. With species as widely distributed and frequent as *B. contexta*, there is a gradual size decrease north to south (and probably east to west). Measurements are not to be relied upon (except Goetghebeur & Coudijzer who state clearly (1985: 209) "middle part of full grown spikelet"). Despite its inclusion in F.T.E.A. (2010), it does not seem to be a good species. *B. subumbellata* (Lye) R.W. Haines may be another superfluous synonym. We include *B. hensii* within limits of *B. contexta*.

Bulbostylis contexta is a polymorphic species widely distributed in the Flora area. It closely resembles *B. burchellii* and may be easily misidentified, particularly as collections are frequently made before nutlets have been formed or reached maturity. Nutlets in *B. contexta* have a pyramidal style base; unfortunately this may be easily detached at or just before spikelet maturity is reached. It may also be confused with *B. boeckeleriana*, but the latter is a usually a more robust plant with a larger nutlet on which the style base is persistent at spikelet maturity.

21. **Bulbostylis hispidula** (Vahl) R.W. Haines in Haines & Lye, Sedges & Rushes E. Afr., app.3: 1 (1983). —Goetghebeur & Coudijzer in Bull. Jard. Bot. Belg. **55**: 217 (1985). Type: Guinea, *Thonning* 349 (C holotype, MO, P-JU). FIGURE 14.**17**.
  Scirpus hispidulus Vahl in Enum. Pl. **2**: 276 (1805).

Fimbristylis exilis (Kunth) Roem. & Schult., Syst. Veg. **2**: 98 (1817). Type: Venezuela, Cumama, 1833, *Bonpland* s.n. (P holotype).

Fimbristylis hispidula (Vahl) Kunth, Enum. Pl. **2**: 227 (1837).

Bulbostylis filiformis C.B. Clarke in F.T.A. **8**: 441 (1902). Type: Kenya, Machakos Dist., Kikumbuliyu, 1893 *Scott Elliot* 6231 (K holotype).

Fimbristylis rhodesiana Rendle in J. Linn. Soc., Bot. **40**: 222 (1911). Type: Zimbabwe, Upper Buzi, 19.iv.1907, *Swynnerton* 920 (BM).

Fimbristylis longibracteata Pires de Lima in Bol. Soc. Brot. ser.2, **2**: 134 (1924). Type: Mozambique, Palma Dist., near Palma town, 10.ii.1917, *Pires de Lima* 104 (PO holotype).

Bulbostylis exilis (Kunth) Lye in Mitt. Bot. Staatssamml. München **10**: 547 (1971).

Abildgaardia hispidula (Vahl) Lye in Bot. Not. **127**: 496 (1974).

Bulbostylis hispidula subsp. *filiformis* (C.B. Clarke) R.W. Haines in Haines & Lye, Sedges & Rushes E. Afr., app.3: 1 (1983). —Verdcourt in F.T.E.A., Cyperaceae: 83 (2010).

Bulbostylis hispidula subsp. *pyriformis* (Lye) R.W. Haines in Haines & Lye, Sedges & Rushes E. Afr., app.3: 1 (1983). —Gordon-Gray in Strelitzia **2**: 33 (1995). —Verdcourt in F.T.E.A., Cyperaceae: 84, fig.13 (2010). Type: Uganda, Karamoja Dist., Moroto, 2.vi.1967, *Haines* 4208 (MHU holotype, K).

Slender to tufted annual extremely variable in height, up to 40 cm tall; rhizome inconspicuous, seldom exceeding 1.5 mm wide, tunics mostly lacking. Leaves few to numerous, filiform, mostly to ⅓ culm length (shorter in recently germinated and depauperate examples); leaf sheath mouth sloping, long white-fringed, ligule 0; leaf blade apex acute, margins scabrid pubescent. Culms terete to slightly flattened, ridged, pilose to glabrescent, hairs mostly spreading from ridges. Inflorescence anthelate of 1–13 spikelets, compound or simple, sometimes reduced to a few clustered or one sessile spikelet in depauperate examples; pedicels hispidulous to scabrid when developed, often spreading becoming ± patent; bracts filiform, usually short (up to 20 mm long), sometimes long with a few long white hairs. Spikelets 5–9(15) × 1.8–3.5 mm, polygonal, angled by projecting glume keels and gaping from maturing nutlets, rachilla straight, deeply notched, shortly winged. Glumes 2–3.5 × 1–1.4 mm, broadly boat-shaped, often deep chestnut red to almost black, pubescent, margins shortly ciliate, apex emarginate or not, shortly mucronate. Stamens 2, anthers with minute rounded crest, soon drying off. Style 3-angled, 3-branched, base pyramidal to conical, not persistent on mature nutlet. Nutlet 0.8–1.3 × 0.7–1.1 mm, trigonous, outline obovoid to pyriform (base often narrow), angles smooth or papillate, gynophore minute, surfaces faintly to clearly 5–10 transversely ridged, epidermal cells long-oblong in transverse rows, eventually velate.

Botswana. N: Okavango, Chief's Is, 18.iii. 2010, *A. & R. Heath* 1988 (K). SW: Kgalagadi Dist., Werda, 26 km from Werda on road to Tsabong–Molopo valley, 9.iii.1976, *Ellis* 2618 (K). SE: Central Dist., Mahalapye, 24.i.1977, *Camerick* 6 (K). **Zambia**. B: Kaoma Dist., 85 km W of Kaoma (Mankoya), 5.iv.1966, *Robinson* 6922 (K). N: Mbala Dist., Ningi (Uninje) Pans, 12 km S of Mbala (Abercorn), *Robinson* 4540 (K). C: Serenje Dist., on road to Kundalila Falls near Serenje, 1600 m, 16.v.1972, *Kornaś* 1711 (K). E: Katete Dist., St Francis Hospital, 1070 m, 8.iii.1957, *Wright* 168 (K). S: Choma Dist., Mapanza, Simasunda, 3.iii.1957, *Robinson* 2143 (K, SRGH). **Zimbabwe**. N: Hurungwe Dist., Mensa Pan, 18 km ESE of Chirundu bridge, 460 m, 29.i.1958, *Drummond* 5336 (K, SRGH). W: Matobo Dist., Besna Kobila, 1460 m, i.1958, *Miller* 5036 (K, SRGH). C: Chegutu Dist., Chegutu (Hartley), Poole Farm, 1220 m, 19.ii.1945, *Hornby* 2390 (K, SRGH). E: Chimanimani Dist., 3 km S of Hot Springs, Sabi Valley, 520 m, 23.iv.1969, *Plowes* 3192 (K, SRGH). S: Masvingo Dist., Masvingo (Fort Victoria), 10.i.1949, *Fisher & Schweickerdt* 375 (K, NU). **Malawi**. N: Nkata Bay Dist., Nkhata Bay at Chikale beach, 460 m, 28.xii.1975, *Pawek* 7656 (K, MO, SRGH). S: Chikwawa Dist., Lengwe Nat. Park, 90 m, 3.ii.1970, *Hall Martin* 497 (K). **Mozambique**. N: Ilha de Moçambique Dist., Goa (Joa) Is., 15°03.08"S 40°47.08"E, 5.v.1947, *Gomes e Sousa* 3501 (K). MS: Manica Dist., Dororo, 24.i.1949, *Fisher & Schweickerdt* 503 (K, NU). M: Maputo Dist., 13 km N of Maputo near Marracuene, 27.ii.1946, *Gomes e Sousa* 3383 (K).

Also widely distributed in tropical and subtropical Africa. Grassland, sandy soils and disturbed areas such as old cultivated fields; 500–1600 m.

Conservation notes: Widespread; not threatened.

Bulbostylis hispidula is a polymorphic species widely spread in sub-Saharan Africa favouring sandy substrates with variation in water availability. As such, and because of style abscission leaving the more persistent base temporarily attached to the nutlet, it was included in *Fimbristylis*, later being transferred to *Bulbostylis*.

A number of subspecies have been recognised. In F.T.E.A. (2010), in addition to those cited above, there are subspp. *brachyphylla* (Cherm.) R.W. Haines, *capitata* Verd., *halophila* (Lye) R.W. Haines, *hispidula*, *intermedia* (Lye) R.W. Haines and two subspecies without names. These are not separated out here as further field study is required in our area to confirm their presence and distribution. Subsp. *pyriformis*

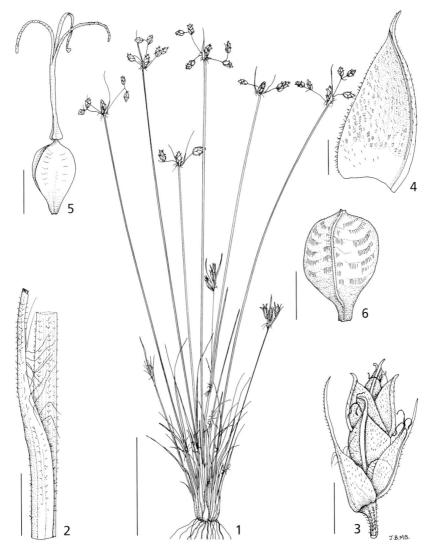

Fig. 14.**17**. BULBOSTYLIS HISPIDULA. 1, habit; 2, mouth of leaf sheath; 3, spikelet; 4, glume; 5, young nutlet with style; 6, nutlet. All from *Ward* 610. Scale bars: 1 = 40 mm; 2, 3 = 5 mm; 4–6 = 0.5 mm. Drawn by Jane Browning. Reproduced from Strelitzia (1995) by kind permission of the South African National Biodiversity Institute, Pretoria.

(Lye) R.W. Haines is illustrated, and is found in Botswana, Namibia, Swaziland and South Africa (Eastern Cape, Free State, Gauteng, KwaZulu-Natal, Limpopo, Mpumalanga).

The species is predominantly annual although it has been termed a perennial (Haines & Lye 1983, Verdcourt 2010). Here it is regarded as annual following Gordon-Gray (1959, 1995) and Goetghebeur & Coudjizer (1985), although there may be an occasional exception in highly favourable habitats.

Germination and early growth produces very small plants comprising a short leaf and a short culm bearing terminally a solitary spikelet. Further development results in great variation (depending on habitat conditions) in plant height, leaf development, inflorescence form and spikelet number per inflorescence (commonly 10–12, but 1–2 also frequent). Despite extensive morphological variation, the species is recognisable by its filiform leaf blades, some pedicels spreading to recurved below the single sessile central spikelet, glumes shortly pubescent abaxially, frequently dark coloured by numerous minute glands. This last feature requiring further study.

22. **Bulbostylis viridecarinata** (De Wild.) Goetgh. in Bull. Jard. Bot. Belg. **54**: 104 (1984). —Goetghebeur & Coudijzer in Bull. Jard. Bot. Belg. **55**: 222 (1985). —Govaets & Simpson, World Checkl. Cyperaceae: 31 (2007) as *viridicarinata*. Type: D.R. Congo, Demba region, i.1910, *Sapin* s.n. (BR lectotype), lectotypified by Goetghebeur.

 Fimbristylis hispidula (Vahl) Kunth var. *glabra* Kük. in Fries, Wiss. Erg. Schwed. Rhodesia-Kongo-Exped. 1911–1912, Erg.: 6 (1921). Type: Zambia, Bangweolo swamps, Kamindas, x.1911, *R.E. Fries* s.n. (UPS).
 Fimbristylis viridicarinata De Wild., Pl. Bequaert. **4**: 207 (1927).
 Fimbristylis tisserantii Cherm. in Arch. Bot. Mém. **4**(7): 32 (1931). Type: Central African Republic, Oubangui, 1928, *Tisserant* 2695 (P).
 Fimbristylis exilis (Kunth.) Roem. & Schult. var. *rufescens* Cherm. in Bull. Soc. Bot. France **81**: 266 (1934). Types: Togo, Sokodé, viii.1933, *Mahoux* 2160, 2161 & 2169 (P syntypes).
 Bulbostylis tisserantii (Cherm.) Lye in Mitt. Bot. Staatssamml. München **10**: 547 (1971), illegitimate name, non Cherm.
 Bulbostylis lyei Verdc. in F.T.E.A., Cyperaceae: 86 (2010). Type: Central African Republic, 25 km N of Bessou, vi.1914, *Tisserant* 157 (P).

Robust annual or short-lived perennial, 20–50(90) cm tall; stem base erect with adventitious roots and occasional branches from lower nodes, tufted or slender. Leaf sheaths glabrous or shortly pubescent, mouth with or without long white hairs; leaf blades 10–25 cm long, 0.5–1(1.5) mm wide, flat, scabrid or densely soft hairy. Culms ridged, glabrous, scabrid in upper part. Inflorescence anthelate, simple or compound of generally up to 20 spikelets variable in terms of sessile and pedicelled units, inflorescence branches glabrous or scaberulous in upper half; bracts 2–5 leaf-like, not hairy. Spikelets 8–15 × 2.5–4 mm. Glumes 3.5–6 × 2 mm, typically cinnamon-coloured, glabrous to minutely hispidulous, midnerve green when live, apex of lower ones shortly mucronate, upper emarginate. Stamens 3. Style 3-branched, style base deciduous. Nutlet 0.9–1 × 0.7–0.8 mm, obovoid, trigonous, faces conspicuously transversely rugose, 3 ribs appearing wavy due to height of rugae.

Zambia. N: Luwingu Dist., 32 km E of Luwingu, 10.ii.1962, *Robinson* 4938 (K). W: Mwinilunga Dist., Zambezi R., 15 km E of Kalene Hill, 16.xii.1963, *Robinson* 6095 (K). C: Lusaka Dist., 19 km S of Lusaka, Chilanga (Mt Makulu), 17.xii.1956, *Angus* 1469 (K). **Mozambique**. Z: Inhassunge Dist., Inhassunge (Mucupia), near Cappuccin Mission, *Amico & Bavazzano* 1970 (FT). MS: Sussundenga Dist., 3 km S of Makurupini Falls, 25.xi.1967, *Simon & Ngoni* 1309 (K, SRGH). GI: Inhambane Dist., Inhambane-Velho, 1936–8, *Gomes e Sousa* 2137 (K). M: Maputo, between Costa de Sol and Marracuene, 15.xi.1960, *Balsinhas* 269 (K).

Also in Nigeria, Guinea, Ivory Coast, Ghana, Togo, Central African Republic, Burundi,

D.R. Congo and Tanzania. Humid grassland, marsh borders, miombo woodland, often in sandy and alluvial soils; 10–1700 m.

Conservation notes: Widespread; not threatened.

Goetghebeur & Coudijzer (1985: 222) suggest that plants from Zambia and Mozambique tend to be perennial rather than annual. The noticeable variation pattern of this species seems to yield ± geographically isolated populations. They also note that West African plants (Napper 1972: 325) have smaller spikelets than those from Central Africa.

23. **Bulbostylis densiflora** (Lye) Lye in Nordic J. Bot. **28**: 516 (2010). Type: Zambia, Kawambwa Dist., Kawambwa, Timnatushi Falls, 19.iv.1957, *Richards* 9339 (K holotype, NU).

> *Abildgaardia densiflora* Lye in Lidia **1**: 31 (1985); in Nordic J. Bot. **7**: 41 (1987).

Tufted slender annual, 15–25 cm tall. Leaves ± ⅓ culm height; leaf sheath straw-coloured, mouth with few obvious fine hairs; leaf blades filiform, 20–60 × 0.3 mm, canaliculated, glabrous, apex obtuse. Inflorescence a solitary spikelet; leafy involucral bracts replaced by glume-like bracts, often persistent when glumes shed. Spikelet 6–8 × 3 mm, ovate, cylindrical, apex obtuse. Glumes spiral, 30–60 closely packed, 2–3 × 1–1.6 mm, concave, glabrous, margin sparsely ciliated, light chestnut brown, midrib (incompletely 3 nerved) ending below rounded fimbriated apex, that may appear emarginate with wear. Stamens 2, anthers c. 0.4 mm long. Style glabrous, twisted, dark brown, c. 1 mm, style base persistent on nutlet, branches 3 ± half style length. Nutlet 0.5 × 0.4 mm, obovoid, yellowish-white when young, dark grey at maturity, surface transversely wavy, papillae present on waves and nutlet angles.

Zambia. N: Kawamba Dist., Ntumbachushi Falls, 1340 m, 22.vi.1957, *Robinson* 2351 (K, NU).

Not found elsewhere. Wet, white sandy soil among rocks; 1260 m.

Conservation notes: Apparently endemic to a small area of NE Zambia; possibly Vulnerable.

Lye (1983) noted that "this species is not closely related to any other species [of *Bulbostylis*]. Its annual habit with solitary very dense-flowered spikelet is somewhat similar to *Nemum spadiceum* Lam."

24. **Bulbostylis melanocephala** (Ridl.) C.B. Clarke in Durand & Schinz, Consp. Fl. Afr. **5**: 615 (1894). Type: Angola, Huíla, Morro de Lopollo, iv.1860, *Welwitsch* 6947 (BM holotype, LISU, K drawing).

> *Fimbristylis melanocephala* Ridl. in Trans. Linn. Soc. London, Bot. **2**: 151 (1884).

Tufted annual up to 15 cm tall; rhizome and tunic of leaf bases lacking. Leaves numerous, ⅓ – ½ to occasionally equalling culm length; leaf sheaths hairy, mouth inconspicuously long white-fringed, ligule 0; leaf blades c. 60 × 0.18 mm, filiform, densely hispid to villose, hairs up to 0.5 mm long. Culm 0.2–0.2.5 mm wide, ridged and channeled, scabrid. Inflorescence a single ovate spikelet, 7–8 mm long, sometimes twinned, the second on a ray up to 10 mm long; bracts shorter than spikelet, with few white hairs, glume-like, extending into scabrid awn. Glumes c. 15, spiral, 2.5–2.8 mm long, ovate, lower minutely scabrid, upper almost glabrous, dark reddish-brown, 3-nerved, margin apically ciliate/fimbriate. Stamens 3. Style 3-branched, base persistent on nutlet. Nutlet 0.8–0.9 × 0.7–0.8 mm, obovoid, trigonous, transversely wavy, without obvious papillae on angles, light brownish yellow; pericarp fractures easily in immature white nutlets.

Zambia. W: Mwinilunga Dist., 6 km N of Kalene Hill, 12.xii.1963, *Robinson* 5905 (K, P).

Also in Angola. Gravelly soil on rocky granite outcrops; c. 1250 m.

Conservation notes: Restricted distribution, but probably not threatened.

In the Kew folders there is only one collection (*Robinson* 5905, upon which the above description is based), determined by Lye in 1984 as *Abildgaardia melanostachya* (Ridl.) Lye; this name does not appear to have been published. Our tentative identification is *Bulbostylis melanocephala*. The ovoid-globose, very dark inflorescence with 1 or occasionally 2 spikelets is distinctive. When describing the species, Clarke (1894) was unable to find any nuts on the Welwitsch collections, but these are present on the Robinson collection. It resembles *B. pusilla* and *B. moggii* Schönland & Turrill; the latter is reported from Botswana, but no specimen was located while writing this treatment.

25. **Bulbostylis wombaliensis** (De Wild.) R.W. Haines in Haines & Lye, Sedges & Rushes E. Afr., app.3:1 (1983). —Goetghebeur & Coudijzer in Bull. Jard. Bot. Belg. **55**: 220 (1985). —Verdcourt in F.T.E.A., Cyperaceae: 87 (2010). Type: D.R. Congo, Wombali, 1910, *Vanderyst* s.n. (BR holotype).

    Fimbristylis wombaliensis De Wild., Pl. Bequaert. **4**: 208 (1927).
    Abildgaardia wombaliensis (De Wild.) Lye in Bot. Not. **127**: 497 (1974); in Haines & Lye, Sedges & Rushes E. Afr.: 108, fig.193 (1983).

Small to medium sized annual, 20–30 cm tall: stem base ascending and often rooting, not easily seen, with dark scales and roots. Leaves ± half culm length; leaf sheath mouth orifice fringed with few long white hairs; leaf blades glabrous to scabridulous. Culm glabrous, except below inflorescence. Inflorescence anthelate with glabrous inflorescence branches, or reduced to 1 or 2 spikelets; bracts shorter than inflorescence, glabrous, lacking white hairs, narrowed apically to a scabrid extension. Spikelets 5–8(10) mm long, acute. Glumes spiral, 2.5–3 mm long, glabrous, pale brown with small raised darker marks, acute, midnerve green, ending at or before apex. Stamen 1. Style 3-branched, base well-developed, triangular, usually not persistent. Nutlet 1–1.2 × 0.8–0.9 mm, obovoid, trigonous with 3 pronounced pale angles (ribs), surface transversely rugose with 7–9 rugae.

Zambia. N: Kasama Dist., 95 km E of Kasama, 18.ii.1962, *Robinson* 4945 (K); Isoka Dist., 40 km SE of Nakonde, 16.iii.1961, *Robinson* 4533a (K).

Also in Uganda and D.R. Congo. Seasonally damp ground and roadside mud; 1200–1350 m.

Conservation notes: Although restricted in the Flora area, probably not threatened. The persistence of the style base may be variable and two stamens may be present.

According to Goetghebeur (Bull. Jard. Bot. Belg. **54**: 104, 1984), the nearest relative may be *Bulbostylis viridecarinata*.

26. **Bulbostylis abbreviata** (Lye) Lye in Nordic J. Bot. **28**: 516 (2010). Type: Zambia, 125 km S of Mpika, near Lusaka road, 29.iv.1961, *Robinson* 4647 (K holotype).

    Abildgaardia abbreviata Lye in Lidia **1**: 35 (1985).
    Bulbostylis pseudoperennis Goetgh. in Bull. Jard. Bot. Belg. **55**: 235 (1985). Type: D.R. Congo, Fungurume, 24.v.1974, *Malaisse* 7743 (BR holotype, GENT).

Erect annual, occasionally a perennial, up to 15 cm tall; stems clustered, lacking rhizome and tunic of leaf bases. Leaves numerous, equalling or more than culm length; leaf sheath mouth inconspicuously long white-fringed; leaf blades 40–80 × 0.2–0.3 mm, filiform, margins scabrous pubescent to glabrous. Culms ridged and furrowed, glabrous or sparsely scabrid. Inflorescence variable, capitate, sometimes with secondary heads, or anthelate with proximal branches overtopping distal ones; basal spikelets above ground may be present, but concealed among tufted stems; bracts few, longer than inflorescence, margins at base glabrous or scantily white-fringed. Spikelets 4–7 mm long, many-flowered. Glumes spiral, 2–3.5 mm long, distinctly mucronate (0.5–1.5 mm), fimbriolate with finely hairy margins, cinnamon-brown to dark brown, 3-nerved, with rounded back. Stamens 3. Style glabrous, dark brown, style base round, persistent on nutlet, branches 3, shorter than style. Nutlet 0.7–0.9 × 0.5–0.7 mm, obovoid, surface variable, usually transversely wrinkled, oblong cells with conical silica bodies; basal nutlet c. 1.6 × 1.1 mm.

Zambia. N: Mpika Dist., 148 km S of Mpika, 2.iii.1962, *Robinson* 4964 (K).

Also in D.R. Congo. Seasonally wet sand associated with quartzite rocks and metalliferous soils; c. 1400 m.

Conservation notes: Apparently with a restricted distribution. It was listed as Vulnerable B1ab(iii) in the IUCN Red List in 2013.

In 1985 Goetghebeur wrote a note on *Robinson* 4647 "*Bulbostylis* cf. *pseudoperennis* Goetgh.*" I have examined this collection and *Robinson* 4964 and accept that the collections are within the limits of *Bulbostylis pseudoperennis* as presently known. Lye's (1985) publication under *Abildgaardia*, in which he cited *A. abbreviata* with the holotype *Robinson* 4647, precedes the publication by Goetghebeur and thus Lye's specific name, more recently published under *Bulbostylis abbreviata* (Lye) Lye has priority.

Goetghebeur noted that the species was rather variable, indicating speciation in progress, particularly under unfavourable conditions such as copper-contaminated soils.

27. **Bulbostylis mucronata** C.B. Clarke in Bot. Jahrb. Syst. **38**: 135 (1906). —Podlech in Merxmüller, Prodr. Fl. SW Afr. **165**: 6 (1967). —Clarke & Mannheimer, Cyperaceae Namibia: 60 (1999). Type: Namibia, Hereroland, Otjimbingue, 1897, *Fischer* 156 (B holotype).

Tufted annual up to 8 cm (rarely to 15 cm); rhizome and tunic of leaf bases lacking. Leaves numerous, equalling culm length; leaf sheath light brown, membranous, mouth inconspicuously long white-fringed, ligule 0; leaf blades c. 80 × 0.2–0.4 mm, filiform, setaceous. Culms 50–80 × 0.25–0.3 mm, subquadrangular, ridged and furrowed, glabrous, scabrid immediately below inflorescence. Inflorescence variable, anthelate of central sessile spikelet with 1–3 stalked spikelets, or 2–3 sessile spikelets or reduced to 1 sessile terminal spikelet only; basal spikelets above ground often present but concealed among tufted stems. Bracts 1–3, 10–35 mm long, scantily white-fringed in basal sheathing part, apex narrowed, margins scabrid. Spikelets 4–5(6) × 2–3 mm (including mucros), linear-lanceolate, apex acute. Glumes 8–10, spiral, 2.3–2.5 mm long (including 0.3–0.5 mm mucro), brown to reddish brown, glabrous, 3-nerved, green midrib terminating in curved mucronate apex. Stamens 1–3. Style glabrous, dark brown basally, style base persistent on nutlet, branches 3 usually shorter than the style. Nutlet 0.7–0.8 × 0.5–0.6 mm, obovoid, trigonous, yellowish-white, darkening at maturity, transversely rugose with c. 10–13 rugae, with papillae on ridges.

Zimbabwe. W: Matobo Dist., 4.i.1948, *West* 2545 (K, SRGH).

Also in Namibia and Angola. Granite sandveld; 1200–1300 m.

Conservation notes: Widely distributed but probably undercollected; not threatened.

Clarke's description, based on a plant from Namibia, indicates that the culms are not glabrous but setaceous. Plants from W Zimbabwe which have glabrous culms were determined as this species, but a thorough examination of material throughout the range should be carried out to determine the precise morphological features. The type, *Fischer* 156, is a fragment and mounted on the same sheet is a second collection reported to be *Dinter* 33. Nambian plants are reported to have glumes 2.5–3 mm long, excluding the recurved mucro (Clarke & Mannheimer 1999).

Groenendijk, Maite & Dungo 804 (K) from Pemba, N Mozambique, may be *Bulbostylis mucronata* but the glume mucro is shorter, and the shape and nut ornamentation differs from that of Angola material. It is also quite distant geographically and ecologically from other known specimens, but further collecting may show intermediates.

Bulbostylis mucronata closely resemembles *B. abbreviata* and *B. densa*, but the glume is distinguished from the latter species by the presence of a recurved mucro up to 0.5 mm long.

28. **Bulbostylis lacunosa** (Lye) Lye in Nordic J. Bot. **28**: 517 (2010). Type: Zambia, Mbala Dist., pans 12 km S of Mbala (Abercorn), 18.iii.1962, *Robinson* 5050 (MT holotype, K).

> *Abildgaardia lacunosa* Lye in Lidia **1**: 33 (1985); in Nordic J. Bot. **7**(1): 44 (1987).

Tufted erect annual up to 20 cm tall; root system small. Leaf sheaths straw-coloured, orifice sloping with long flexuous hairs, ligule 0; leaf blades few, short, 10–30 × 0.2–0.4 mm, margins scabrous, midrib prominent, 1–2 lateral nerves on lower side. Culms 5–20 cm, ridged and furrowed, scabridulous or sparsely pubescent. Inflorescence unispicate, occasionally 1–3 spikelets; bracts 1–2, 4–10 mm long, basally glume-like, white-fringed, minutely scabrid, extending into prominent awn. Spikelet 4–8 × 2–3.5 mm, ovate to ovate-triangular, apex acute. Glumes spiral, 2.5–3 × 1–1.5 mm, ovate, brown, minutely scabrid to almost glabrous, margins shortly ciliate, apex obtuse to emarginate. Stamens 2–3. Style glabrous, dark brown basally, style base round, vertically flattened, not persistent on nutlet, branches 3. Nutlet 0.8–1 × 0.6–0.7 mm; subtrigonous, obpyriform, base pedunculate, whitish, surface foveate (lacunose).

Zambia. N: Kasama Dist., 10 km E of Kasama, 19.ii.1961, *Robinson* 4412 (K); Kasama Dist., Malole, 16.iii.1961, *Robinson* 4429 (K).

Only known from N Zambia. Damp lateritic soils; c. 1350 m.

Conservation notes: Endemic to a small area in NE Zambia; possibly Vulnerable.

Bulbostylis lacunosa with solitary spikelets may resemble some collections of *B. hispidula* and *B. taylorii* C.B. Clarke. However, the 'stalked' white nutlet is distinctive.

29. **Bulbostylis humilis** (Kunth) C.B. Clarke in Durand & Schinz, Consp. Fl. Afr. **5**: 614 (1894); in Fl. Cap. **7**: 149 (1900). —Gordon-Gray in Strelitzia **2**: 209 (1995). Type: South Africa, Cape Province, *Drège* s.n. (possibly, see note in Gordon-Gray 1995).

> *Isolepis humilis* Kunth, Enum. Pl. **2**: 207 (1837).
> *Bulbostylis breviculmis* (Kunth) C.B. Clarke in Durand & Schinz, Consp. Fl. Afr. **5**: 612 (1894); in Fl. Cap. **7**: 149 (1900). Type: South Africa, Cape of Good Hope, *Drège* s.n. (possibly *Drège* 3947 (K) from between Aliwal North & Kraai R. – Gordon-Gray, 1995).
> *Bulbostylis sphaerocarpa* (Boeckeler) C.B. Clarke in Durand & Schinz, Consp. Fl. Afr. **5**: 616 (1894). —Verdcourt in F.T.E.A., Cyperaceae: 92 (2010). Types: Ethiopia, near Matamma, 25.ix.1865, *Schweinfurth* 2046 (G lectotype, BM, BREM, K); viii.1865, *Schweinfurth* 2047 (BM syntype), lectotypified by Lye (see note in F.T.E.A.).
> *Bulbostylis striatella* C.B. Clarke in Fl. Cap. **7**: 305 (1898). Type: South Africa, KwaZulu-Natal, no locality, iv.1883, *Buchanan* 86 (K lectotype), lectotypified in Lye (1983: 233).
> *Abildgaardia humilis* (Kunth) Lye in Bot. Not. **127**: 496 (1974).
> *Abildgaardia sphaerocarpa* (Boeckeler) Lye in Bot. Not. **127**: 497 (1974). —Haines & Lye, Sedges & Rushes E. Afr.: 114, fig.212 (1983).
> *Abildgaardia striatella* (C.B. Clarke) Lye in Nordic J. Bot. **3**: 239 (1983). —Haines & Lye, Sedges & Rushes E. Afr.: 125, fig.240 (1983).

Tufted annual up to 5–12(25) cm tall, often forming dense compact cushions. Leaves numerous, equal to or overtopping culms, setaceous, sometimes microscopically scabrous-hairy; leaf sheaths light brown, sheath mouth inconspicuously long white-fringed, ligule 0; leaf blades filiform, 0.25–0.3 mm wide, margins scabrous. Culms ridged and furrowed, hispidulous. Inflorescence a solitary terminal spikelet or of 2–3 clustered sessile spikelets; basal spikelets often added at stem-base but concealed among tufted stems; bracts 1(2), filiform, c. 3 mm long or as long as spikelet, scabrid apically. Spikelets ovoid, acute, 3–8 × 1.3–2 mm. Glumes spiral, 2–4 mm long, ovate-acuminate, glabrous or scabrid, rusty-brown, 3-nerved, keel excurrent in a mucro. Stamens 2(3). Style linear, smooth, 2-branched, usually shorter than style which falls early leaving a minute button or tubercule. Nutlet 0.9–1.3 × 0.8–1 mm (excluding style base), biconvex to subtrigonous, obovoid, yellow-brown, grey-brown at maturity, slightly verrucose, transversely wrinkled, cells in longitudinal lines or with a reticulate pattern.

Botswana. N: Ngamiland Dist., Aha Hills, 19°39.7'S 21°03'E, 28.iv.1980, *P.A. Smith* 3401a (K, PRE).

Also in Ethiopia, Kenya, Tanzania, Namibia, South Africa (Eastern Cape, Free State, Gauteng, KwaZulu-Natal, Limpopo, Mpumalanga, Northern Cape, North West) and Swaziland. Moss mats over rocky outcrops, may become a weed in gardens; c. 1000 m.
Conservation notes: Widespread; not threatened.

The collection from Botswana had 3 style branches. Gordon-Gray (1995) notes that both 2 – and 3-style branches may be found within one spikelet. In Uganda, *Bulbostylis sphaerocarpa* with 3 style branches, is here considered a synonym, as is *B. striatella* from Kenya and Tanzania with 2 style branches.

30. **Bulbostylis acutispicata** (Lye) Lye in Nordic J. Bot. **28**: 516 (2010). Type: Zambia, Kabwe Dist., Muka Mwanji hills, Kalwelwe, 25 km SSW of Kabwe, 14°41'S 28°21'E, 25.ii.1973, *Kornaś* 3292 (POZ holotype, K).

   Abildgaardia acutispicata Lye in Lidia **1**: 33 (1985); in Nordic J. Bot. **7**(1): 42 (1987), as *acutespicata.*

Tufted annual up to 20 cm tall; rhizome and tunic of leaf bases lacking. Leaves numerous, 1/3–1/2 culm length, or occasionally equalling culm length; leaf sheath light brown, mouth inconspicuously long white-fringed, ligule 0; leaf blades to 10–50 × 0.2–0.4 mm, filiform, margins scabrous. Culms ridged and furrowed, glabrous or scabrid, shortly hairy immediately below inflorescence. Inflorescence variable, anthelate of central sessile spikelet with 1–3 stalked spikelets, or reduced to 1 sessile terminal spikelet; basal or sub-basal spikelets often present at ground-level but concealed among tufted stems; bracts 1–3, c. 10 mm long, scantily white-fringed in basal sheathing part, apex narrowed, margins scabrid. Spikelets 4–10 × 1–1.5 mm, linear lanceolate, apex acute. Glumes spiral, 2.5–3 mm long, slightly scabrid, brown, margins shortly ciliate to almost glabrous, 3-nerved midrib terminating at apex or excurrent into a minute mucro. Stamens 1–2, anthers 0.7–1 mm long. Style glabrous, dark brown basally, base round, vertically flattened, persistent on nutlet; branches 3, usually shorter than style. Nutlet 0.7–0.9 × 0.5–0.6 mm, obovoid, trigonous, yellowish white, darkening at maturity, transversely rugose with 10–13 rugae, with papillae on ridges.

Zambia. N: Mbala Dist., Chianga dambo, 1620 m, 23.vi.1956, *Robinson* 1722 (K). C: Kabwe Dist., Muka Mwanji Hills near Kalwelwe, 25 km SSW of Kabwe, 14°41'S 28°21'E, 25.ii.1973, *Kornaś* 3292 (K, POZ). **Malawi**. S: Mt Mulanje, path to Chambe, 1700 m, 19.vi.1962, *Robinson* 5408 (K).
Not known elsewhere. Sandy temporarily moist margins of dambos and grassy places over rocks; 1200–1700 m.
Conservation notes: Only known from Malawi and Zambia in the Flora area; but probably not threatened.

Lye (1987) noted that the plant was rare, only found in N and C Zambia. I have identified, perhaps incorrectly, *Robinson* 5408 (K) from S Malawi as this species. Lye also suggests that *Bulbostylis acutispicata* is most closely related to *B. andongensis* and *B. pusilla*.

31. **Bulbostylis pluricephala** (Lye) Lye in Nordic J. Bot. **28**: 517 (2010). Type: Kasama Dist., Chilubula, 12.ii.1961, *Robinson* 4381 (K holotype, MT).

   Abildgaardia pluricephala Lye in Lidia **1**: 35 (1985); in Nordic J. Bot. **7**(1): 49 (1987).

Annual not tufted, up to 20 cm tall,; rhizome lacking, root system not well-developed. Leaves approximately ½ culm length; leaf sheath light brown, mouth oblique, long white-fringed; leaf blades glabrous, ridged, to 60 × 0.2–03 mm, filiform, minutely scabrid near apex. Culms 0.2–0.5 mm wide, ridged and furrowed, glabrous. Inflorescence anthelate, variable, one sessile group of 2–6 spikelets with an additional 1–3 glabrous inflorescence branches with groups of spikelets, or reduced to a single group of spikelets; bracts 1–4, 3–22 mm long, shorter than inflorescence, with few long white hairs before narrowing into an arista-like extension. Spikelets 3–4 × 1–2 mm, ovate, angular, apex acute. Glumes spiral, few, 1.3–1.8 mm long, ovate-oblong, dark reddish brown, mid-nerve ending before apex, flanks glabrous to scabridulous, margins shortly hairy. Stamens 1–2. Style 3-branched, brown, flattened, base persistent on nutlet. Nutlet 0.7–0.8 × 0.5–0.6 mm, obovoid, trigonous, reddish brown at maturity with pale ribs, surface (× 40) smooth, covered with non-cellular colourless transparent layer readily detaching into fragments.

Zambia. N: Kasama Dist., 40 km ESE of Kasama, 30.iii.1961, *Robinson* 4562 (K); Mporokoso Dist., Mweru-Wantipa, floodplain below Mupundu, 1050 m, 11.iv.1957, *Richards* 9160 (K).

Not known elsewhere. Sandy ground beside roads; 1000–1400 m.

Conservation notes: Endemic to NE Zambia; probably not threatened

The widely-branched inflorescence with a central group of about 6 spikelets, and inflorescence branches (both primary and secondary in the case of *Richards* 9160) with added similar groups of spikelets, is distinctive and warrants the specific name *pluricephala*. Lye comments that it is related to *Abildgaardia wallichiana* (Roem. & Schult.) Lye, a synonym in F.T.E.A. (2010: 104) for *Bulbostylis barbata* subsp. *barbata*, which as yet is not recorded for the Flora region.

32. **Bulbostylis buchananii** C.B. Clarke in F.T.A. **8**: 437 (1902). —Goetghebeur & Coudijzer in Bull. Jard. Bot. Belg. **55**: 244 (1985). —Verdcourt in F.T.E.A., Cyperaceae: 92 (2011). Type: Malawi, no locality, 1891, *Buchanan* 1329 (K holotype).

Abildgaardia buchananii (C.B. Clarke) Lye in Bot. Not. **127**: 495 (1974); in Haines & Lye, Sedges & Rushes E. Afr.: 114, figs.209,210 (1983).

Annual or short-lived perennial, densely tufted, 10–30 cm tall; lacking rhizome, base often thickened by compacted spikelets within leaf bases. Leaves numerous, 1/2–1/3 culm length; leaf sheath mouth conspicuously long white-fringed, ligule 0; leaf blades up to 200 × 0.2–0.4 mm, filiform, margins scabrous. Culms angular, ridged and furrowed, glabrous to scabrid apically. Inflorescence a head of 5–15 spikelets; bracts 3–4(6), apically filiform, 5–30 mm long, glume-like basally, margins with pronounced white hairs. Spikelets lanceolate, 4–6 × 1–2 mm. Glumes 2.5–4 mm long, spiral, ovate, scabrid, reddish brown, midrib green, apex mucronate to shortly apiculate. Stamens 3. Style 3-branched, usually shorter than style, style base persistent on nutlet. Nutlet 0.9–1 × 0.6–0.8 mm, pale cream to light brown, trigonous, transversely rugose with 7–8 elevated bands of elongate cells, raised papillae on 3 longitudinal ribs (nutlet angles).

Zambia. W: Ndola Dist., Itawa dambo, 1300 m, 3.iv.1960, *Robinson* 3457 (K). N: Chinsali Dist., c. 60 km SW of Isoka, 1500 m, 15.iii.1975, *Hooper & Townsend* 786 (K). C: Lusaka Dist., 6 km W of Lusaka, 20.iii.1965, *Robinson* 6441 (K). E: Petauke Dist., Kachalola, by roadside, 14°51'S 30°34'E, 740 m, 1.iii.1973, *Kornaś* 3343 (K). S: Mazabuka Dist., 6.4 km from Chirundu bridge on Lusaka road, 6.ii.1958, *Drummond* 5501 (K, LISC, SRGH). **Zimbabwe**. N: Makonde Dist., Mangula, Molly South Hill, on quartzite with copper oxide, 3.ii.1965, *Wild & Simon* 6791 (K, SRGH). W: Hwange Dist., Gwayi–Lutope junction, 27.ii.1963, *Wild* 6032 (K, LISC, SRGH). C: Kwekwe Dist., 27 km along Gokwe road to NW of Skipper Mine, on copper outcrop, 16.iii.1966, *Wild* 7550 (K, SRGH). **Malawi**. S?: no locality, 1891, *Buchanan* 1329 (K). **Mozambique**. N: Malema Dist., 12 km from Mutuáli (Port Nyasa) on Malema road, fl. 25.ii.1954, *Gomes e Sousa* 4215 (K).

Also in Ethiopia, Kenya, Tanzania, Burundi and D.R. Congo. Open woodland, roadsides, rocky outcrops, sandy river and lake margins; 600–1500 m.

Conservation notes: Widespread; not threatened.

Basal flowers represented by compacted glumes were present in many collections, a few had exserted styles, but nutlets were infrequent.

The mature nutlet is distinctive with numerous papillae on the 3 angles giving a pronounced crenulated appearance when viewed from the abaxial surface.

Some collections have been recorded as being found on copper oxide lateritic soils.

33. **Bulbostylis trabeculata** C.B. Clarke in Durand & Schinz, Consp. Fl. Afr. **5**: 616 (1894). —Rendle in Hiern, Cat. Afr. Pl. **2**: 126 (1899). —Clarke in F.T.A. **8**: 437 (1902). —Haines & Lye, Sedges & Rushes E. Afr.: app.3: 1 (1983). —Verdcourt in F.T.E.A., Cyperaceae: 105 (2010). Types: Angola, Luanda, Casanga Is., 30.iv.1854, *Welwitsch* 6982 (BM syntype); Luanda, Praia do Zamba Grande & Maianga d'El Rei, 13.ii.1855, *Welwitsch* 7000 (BM syntype, LISU); Moçamedes, Cabo Negro, R. Curoca, ix.1859, *Welwitsch* 6962 (LISU syntype, BM).

    Fimbristylis barbata (Rottb.) Benth. var. *subtristachya* sensu Ridley in Trans. Linn. Soc., Bot. **2**: 152 (1884), non *Isolepis subtristachya* Boeckeler

    Abildgaardia trabeculata (C.B. Clarke) Lye in Nordic J. Bot. **1**: 758 (1982). —Haines & Lye, Sedges & Rushes E. Afr.: 123, fig.234 (1983).

Delicate annual to 15 cm tall; tunic of leaf bases and rhizome lacking. Leaves numerous, 1/3 to occasionally equalling culm length; leaf sheath pale brown, mouth conspicuously long white-fringed, ligule 0; leaf blades to 50 × 0.2–0.3 mm, filiform, margins and midribs sparsely scabrid. Culms ridged and furrowed, glabrous. Inflorescence a small head, c. 5 mm wide, chestnut coloured, of 3–4(8) sessile spikelets; bracts 2–3, filiform, longest 10–20 mm long. Spikelets 2–5 × 1.5–2 mm, narrowly elliptic. Glumes spiral, 1.5–1.7 mm long, ovate, acuminate, pubescent, margins ciliate, 3-nerved green keel, mucronate. Stamens (1)2–3. Style glabrous, dark brown basally, rounded, vertically flattened, persistent on nutlet, branches 3, usually shorter than style. Nutlet subtrigonous, obovoid, pale grey-brown at maturity, 0.5–0.6 × 0.4–0.5 mm, surface trabeculate, giving a distinctive pattern.

Botswana. N: Central Dist., Makgadikgadi pans, Mokoamoto (Ntwetwe) Pan, 19°58.2'S 25°50.2'E, 16.iv.1978, *P.A. Smith* 2388 (K, SRGH); Okavango Swamps, Moremi Wildlife Reserve, island on Gobega lagoon, 940 m, 6.iii.1972, *Biegel & Gibbs Russell* 3866 (K, LISC). **Zimbabwe**. S: Beitbridge Dist., Tuli, Shashe R. near Park Warden's house, 1.ii.1973, *Grovenor* 802 (K, SRGH).

Also Kenya, Angola and Namibia. In *Acacia* woodland and on Kalahari sand in Botswana where it can provide up to 10% of ground cover in localities, or in damp sand by rivers and other water bodies; 550–1000 m.

Conservation notes: Widespread; not threatened.

Two varieties have been recognised, of these only var. *trabeculata* is present in the Flora area; var. *microglumis* (Lye) R.W. Haines (reported to have smaller glumes) is represented only by the type in Turkana, N Kenya.

Plants resemble *Bulbostylis barbata* (Rottb.) C.B. Clarke, but the nutlet is different. They may also be confused with *B. buchananii* which has a similar inflorescence in which the involucral bracts tend to spread out on either side of the inflorescence head, while in *B. trabeculata* one bract is often longer than the others and may be held in an upright position overtopping the spikelets.

34. **Bulbostylis afromicrocephala** Lye in Nordic J. Bot. **28**: 516 (2010). Type: Zambia, Mbala Dist., pans 12 km S of Mbala, 18.iii.1962, *Robinson* 5048 (MT holotype, K).

    Abildgaardia microcephala Lye in Lidia **1**: 33 (1985); in Nordic J. Bot. **7**: 46 (1987), non *Bulbostylis microcephala* Gardner.

Erect annual, 3–12 cm tall; tunic of leaf bases and rhizome lacking. Leaves numerous, 1/4–1/3 culm height; leaf sheath mouth conspicuously long white-fringed, ligule 0; leaf blades to 30 × 0.2–0.3 mm, filiform, margins and nerves scabrid. Culms ridged and furrowed, scabrid below inflorescence. Inflorescence capitate, 3–8 × 3–7 mm, of 3–7 sessile spikelets; bracts 2–6, shorter than inflorescence, glume-like basally, with long white hairs, apically filiform. Spikelets 3–6 × 1–1.5 mm. Glumes spiral, 1.5–2 mm long, ovate, medium to dark reddish brown, margins fimbriate, flanks glabrous, pallid midrib 3-nerved, apex obtuse or subacute. Stamens 2–3. Style glabrous, dark brown, style base minute, persistent on nutlet, branches 3. Nutlet 0.4–0.5 × 0.4–0.5 mm, obtusely triangular-pyriform, greyish-white at maturity, surface appearing smooth or faintly transversely wavy with indistinct papillae.

Zambia. N: Mbala Dist., pans 12 km S of Mbala (Abercorn), 18.iii.1962, *Robinson* 5048 (K).

Not known elsewhere. Seasonally damp soil on laterite; c. 1700 m.

Conservation notes: Apparently endemic to Mbala and so far known only from the type; possibly Vulnerable.

The description is based on Lye's (1987) account.

Robinson 6619 was determined as *Abildgaardia microcephala* by Lye in 1984. Another specimen collected from the same locality on the same day, *Robinson* 6618, was determined by Lye as *A. microlaniceps* (? = *Bulbostylis fimbristyloides*). Both Robinson collections were determined by us as *B. fimbristyloides*, a species bearing some superficial resemblance to *B. afromicrocephala* but differing in the shape of the nutlet.

The species closely resembles *Bulbostylis capitata*; differences are confined to dimensional data which are not reliable when based on only on one or two collections.

The capitate inflorescence has unusual and distinctive involucral bracts, appearing almost like glumes, transparent light brown in colour, with long white hairs, 3-nerved midnerve narrowing to a green awn-like extension.

35. **Bulbostylis capitata** (Lye) Lye in Nordic J. Bot. **28**: 516 (2010). Type: Zambia, Kasama Dist., 10 km E of Kasama, 19.ii.1961, *Robinson* 4409 (K holotype).

   Abildgaardia capitata Lye in Lidia **1**: 37 (1985).

Delicate erect annual up to 17 cm tall; tunic of leaf bases and rhizome lacking. Leaves numerous, 1/4–1/3 to occasionally equalling culm length; leaf sheath mouth inconspicuously long white-fringed, ligule 0; leaf blades to 40–170 × 0.25–0.3 mm, filiform, glabrous. Culms 0.25 mm wide, ridged and furrowed, glabrous. Inflorescence capitate, c. 10 × 6 mm, of 3–7 spikelets; bracts 2–3, much shorter than inflorescence, glume-like with sheathing membranous base, margin scantily white-fringed, midnerve extending into short scabrid to glabrous narrow extension. Spikelets 6–7 × 1–2 mm, bi-coloured in dried material. Glumes spiral, 2–2.5 mm long, reddish brown with paler margins, glabrous, midrib ending before rounded or emarginate fimbriated apex. Stamens 3. Style glabrous, 3-branched, base persistent. Nutlet 0.5–0.6 × 0.4–0.5 mm, obovoid, trigonous, white to grey-white at maturity, surface transversely rugose, 3 defined ridges with small tubercles/papillae.

Zambia. N: Kasama Dist., 10 km E of Kasama, 19.ii.1961, *Robinson* 4409 (K).

Not known elsewhere. Shallow soil overlying laterite and sandstone; c. 1400 m.

Conservation notes: Known only from the type in the Kasama area; probably Vulnerable.

36. **Bulbostylis nudiuscula** (Lye) Lye in Nordic J. Bot. **28**: 517 (2010). Type: Zambia, Kasama Dist., 13 km NE of Kasama, 27.iv.1961, *Robinson* 4622 (K holotype).

   Abildgaardia nudiuscula Lye in Lidia **1**: 37, fig. p.34 (1985).

Annual or short lived perennial, erect up to 48 cm tall; rhizome and tunic of leaf bases lacking. Leaves lacking lamina or blade; sheath to 30 mm, hairs at sheath mouth few or absent, ligule 0. Culms 1–1.5 mm wide, ridged and furrowed, glabrous. Inflorescence anthelate, of central spikelet with 2–3 inflorescence branches to 30 mm long, each terminating in a spikelet; bracts 4–6, 7–9 mm long, much shorter than inflorescence, glabrous except for few marginal long hairs. Spikelets 10–12 × 2–3 mm. Glumes spiral, 5–7 mm long, glabrous to somewhat scabrid, light brown, margins paler, ciliate, midrib 3-nerved, centrally green, apex mucronate to shortly apiculate. Stamens 3. Style apically fimbriate, base triangular, vertically flattened, infrequently persistent on nutlet, branches 3, equalling style length. Nutlet 1–1.2 × 0.9–1.1 mm, mottled, dark brown with light patches, obovoid, trigonous, ribs pronounced with raised papillae, surface variable, from transversely rugose to almost smooth.

Zambia. N: Kasama Dist., 10 km NE of Kasama, 4.ii.1961, *Robinson* 4351 (K).

Not known elsewhere. Shallow lateritic pans, seasonally damp; c. 1400 m.

Conservation notes: Known only from the type locality in the Kasama area; probably Vulnerable.

Lye (1985) gave no description of this plant but it was accompanied by an illustration. He regarded it as an annual, but it may well be perennial.

At Kew *Robinson* 4622 was filed under *Bulbostylis viridecarinata*, with a pencil annotation "species 8". It was originally determined by Lye as *Abildgaardia viridecarinata*, however, according to Lye in Lidia **1**: 37 (1985) it is the type of *A. nudiuscula* (now considered a synonym of *B. nudiuscula* (Lye) Lye).

Nutlet shape and conformation, together with other features such as absence of leaf blades, is perhaps indicative of hybridisation.

37. **Bulbostylis rotundata** (Kük.) R.W. Haines in Haines & Lye, Sedges & Rushes E. Afr., app.3: 1 (1983). —Verdcourt in F.T.E.A., Cyperaceae: 89 (2010). Types: Tanzania, Dodoma Dist., Lake Chaya, *Peter* 45768 (B syntype); Tabora Dist., Goweko to Igalula, *Peter* 45934 (B syntype).

> *Fimbristylis rotundata* Kük. in Repert. Spec. Nov. Regni Veg. Beih. **40**(1): 126 (1936).
> *Abildgaardia rotundata* (Kük.) Lye in Nordic J. Bot. **1**: 758 (1982). —Haines & Lye, Sedges & Rushes E. Afr.: 110, figs.197,198 (1983).

Delicate erect annual to 30 cm tall; rhizome and tunic of leaf bases lacking. Leaves numerous, 1/3 culm length; sheath mouth inconspicuously long white fringed, ligule 0; leaf blades up to 60–80 mm, filiform, slightly hairy. Culms 0.2–0.4 mm wide, ridged and furrowed, glabrous, sometimes with scattered hairs. Inflorescence anthelate, of one sessile and 1–4 stalked spikelets on pedicels 0.2–10 mm long. Bracts 1–2, the larger 4–8 mm long, with long white hairs on clasping membranous bases, apically filiform. Spikelets 4–7 × 2–4 mm, ovoid, reddish brown. Glumes spiral, 2–3 × 1.5 mm, ovate, slighty scabrid, midrib ending before the rounded or emarginated fimbriated apex. Stamens 3. Style branches 3, base not persistent on nutlet. Nutlet 0.9–1 × 0.8–0.9 mm, white, obpyriform with cuneate base, longitudinal ridges distinct; distal part of nutlet laterally expanded, markedly trilobed, but flattened adaxially with surface irregularly transversely rugose.

Malawi. N: Chitipa Dist., 5 km SE of Chitipa (Fort Hill), 11.iii.1961, *Robinson* 4444 (K). Also in Tanzania. Seasonally flooded damp sandy soil; c. 1300 m.

Conservation notes: Not widespread but probably not threatened.

According to Haines & Lye (1983) this species is superficially similar to *Bulbostylis taylorii* C.B. Clarke, but differs in the shape and surface pattern of the nutlet. The nutlet of *B. rotundata* has an unusual and elaborate shape that sets it apart from other species of the genus.

38. **Bulbostylis johnstonii** C.B. Clarke in F.T.A. **8**: 442 (1902). —Verdcourt in F.T.E.A., Cyperaceae: 107 (2010). Type: Tanzania, Mt Kilimanjaro, v.1884, *Johnston* s.n. (K lectotype), lectotypified in Haines & Lye (1983).

> *Abildgaardia johnstonii* (C.B. Clarke) Lye in Bot. Not. **12**: 496 (1974). —Haines & Lye in Sedges & Rushes E. Afr.: 126, fig.243 (1983).

Annual or short-lived erect perennial to 30 cm tall; rhizome creeping. Leaves numerous, 1/3–1/2 to occasionally equalling culm length; leaf sheath hairy, mouth conspicuously long white-fringed, ligule 0; leaf blades c. 150 × 0.25 mm, filiform, margins scabrid. Culms ridged and furrowed, glabrous. Inflorescence anthelate, with 1–5 solitary stalked spikelets; bracts setaceous, c. 9 mm long, margins with few white hairs. Spikelets 5–9 × 2 mm, rusty brown to dark brown. Glumes spiral, 2.5–3.5 mm long, ovate, boat-shaped, pubescent, keel yellowish, minutely excurrent. Stamens ? Style glabrous, style base persistent on nutlet, branches 3, usually shorter than style. Nutlet 0.9–1 × 0.7 mm, ± globose, subtrigonous, white, turning light grey to brown when mature, 3 angles lacking papillae, surface with large quadrate cells in vertical rows.

Malawi. N: 'Tanganyika Plateau', 1070–1220 m, iv.1890, *Whyte* s.n. (K).
Also in Kenya and Tanzania. Hillsides and valley moist grasslands; 1000–1250 m.
Conservation notes: Not widespread; possibly threatened.

There is only one collection from the Flora area at Kew; this is a young specimen with many developing culms and one or two inflorescences with closely placed spikelets that do not appear to have stalks. Collections from East Africa are of larger, more mature plants and the above description is based on these.

39. **Bulbostylis congolensis** De Wild., Pl. Bequaert. **4**: 194 (1927). —Goetghebeur & Coudijzer in Bull. Jard. Bot. Belg. **55**: 244 (1985). Types: D.R. Congo, ?upper Congo, 1906, *Claessens* 1681 (BR lectotype); ?lower Congo, 1922, *Achten* s.n. (BR syntype), lectotypified by Goetghebeur & Coudijzer (1985: 245).

   Bulbostylis holotricha A. Peter in Repert. Spec. Nov. Regni Veg. Beih. **40** (Anhang); 127, tab.89 (1936). Type: Tanzania, Buha Dist., Birira to Nisusi, 27.ii.1926, *Peter* 37915 (B lectotype, GOET).
   Abildgaardia congolensis (De Wild.) Lye in Bot. Not. **127**: 495 (1974).
   Abildgaardia pusilla (A. Rich.) Lye subsp. *congolensis* (De Wild.) Lye in Nordic J. Bot. **3**: 239 (1983). —Haines & Lye, Sedges & Rushes E. Afr.: 116, fig.215 (1983).
   Bulbostylis pusilla (A. Rich.) C.B. Clarke subsp. *congolensis* (De Wild.) R.W. Haines in Haines & Lye, Sedges & Rushes E. Afr., app.3: 1 and 116 (1983). —Verdcourt in F.T.E.A., Cyperaceae: 95 (2010).

Erect annual, 35–60 cm tall; rhizome and tunic of leaf bases lacking. Leaves numerous, almost equalling culm length; leaf sheath mouth inconspicuously long white-fringed, ligule 0; leaf blades to 300 × 0.2–0.3 mm, filiform, margins and midrib hispid. Culm 0.3–0.6 mm wide, ridged and furrowed, densely hispid (hairs 0.3–0.5 mm long). Inflorescence anthelate, variable, of central spikelet or head of spikelets with additional 4–5 hispid inflorescence branches with further spikelets; bracts 3–5, 5–10 mm long, apically filiform with hispid margins, basally glume-like, ciliated, appearing almost triangular, margins without long white hairs. Spikelets 3–5 × 2 mm. Glumes 10–15, spiral, ovate to ovate-lancoelate, 1.8–2.5 mm long, dark brown, flanks scabrid, margins glabrous to scabrid, apex obtuse to emarginate, midrib 3-nerved, ending before apex. Stamens 3. Style, base persistent on nutlet, branches 3 usually shorter than style. Nutlet 0.9–1 × 0.7–0.8 mm, olive-brown, obovate to obcordate in outline, trigonous, surface transversely rugose (10–12 rugae), raised papillae present, angles finely wavy.

Zambia. W: Mwinilunga Dist., Kalenda Plain, 16.iv.1960, *Robinson* 3612 (K, LISC). N: Mbala Dist., Chilongolwelo, 17.iii.1961, *Robinson* 4536 (K).
Also in West Africa, Burundi, Central African Republic, Ethiopia, Uganda, Kenya, Tanzania, D.R. Congo and Angola. Margins of lateritic pans, and damp rock crevices within lateritic outcrops; 1300–1800 m.
Conservation notes: Widespread; not threatened.

Bulbostylis congolensis has here been accorded specific status. Other authors have included it as a subspecies or variety of *B. pusilla*, to which it is closely allied, but it differs in the shape of the spikelets (more rounded, glumes hispid) and in the densely hairy leaves and culms. There are also small differences in the nutlet.

40. **Bulbostylis flexuosa** (Ridl.) Goetgh. in Bull. Jard. Bot. Belg. **55**: 241, fig.5 (1985). Types: Angola, Pundo Andongo, near Banza do Soba de Umbilla, iii.1857, *Welwitsch* 6828 (BM syntype, LISU); between Candumba and Matamma, iii.1857, *Welwitsch* 6829 (BM syntype, LISU).

   Fimbristylis flexuosa Ridl. in Trans. Linn. Soc. London, Bot. **2**: 155 (1884).
   Abildgaardia macroanthela Lye in Lidia **1**: 35 (1985). Type: Zambia, Mongu, 14.iii.1966, *Robinson* 6877 (K holotype, M).
   Bulbostylis macroanthela (Lye) Lye in Nordic J. Bot. **28**: 517 (2010).

Erect or drooping and flexuous annual up to 50 cm tall; rhizome and tunic of leaf bases lacking. Leaves numerous, ½ to occasionally few equalling culm length; leaf sheath mouth conspicuously long white-fringed, ligule 0; leaf blades to 250 × 0.2–0.4 mm, filiform, scabrid on margins and nerves. Culms 1.5–1.8 mm wide, ridged and furrowed, glabrous to scabrid apically. Inflorescence anthelate, variable, of central spikelet or head of spikelets with 4–5 scabrid inflorescence branches, c. 60–100 mm long, sometimes spikelets on inflorescence branches very compacted; bracts 3–4, to 150 mm long, hispid, basally long white-fringed, apically narrowed. Spikelets 5–11 × 2 mm. Glumes spiral, 2.5–3.2 mm long, light brown, rounded to subacute, hispidulous, 3-nerved (ending prior to apex in lower glumes, upper ends as a small mucro with spines), margins apically fimbriolate. Stamens 3. Style glabrous, dark brown basally, style base round, vertically flattened, persistent on nutlet, branches 3, usually shorter than style. Nutlet 0.7–0.8 × 0.6–0.7 mm, white to pale brown at maturity, obovoid, trigonous, surface transversely rugulose (c. 11–12 narrow bands), 3 ridges appearing minutely papillate (× 40).

Zambia. B: Kaoma Dist., 3 km NW along Kaoma–Lukulu road from junction of Kaoma–Mongu road, 1050 m, 1.iii.1996, *Zimba et al.* 702 (K, MO). W: Mwinilunga Dist., 10 km S of Mwinilunga, 1400 m, 17.iv.1960, *Robinson* 3664 (K). C: Mkushi Dist., 35 km ENE of Kapiri Mposhi, 7.iii.1962, *Robinson* 4997 (K).

Also in Angola and D.R. Congo. *Brachystegia* and disturbed woodland on sandy soils; 1000–1400 m.

Conservation notes: Moderately widespread; not threatened.

Fimbristylis flexuosa in F.T.E.A. (2010: 97) and on the World Cyperaceae Checklist is regarded as a synonym of *Bulbostylis abortiva*. It has been reinstated as a separate species in this account as it was found not to conform with the latter; *B. macroanthela* is here placed in synonymy.

Bulbostylis flexuosa is a robust annual distinguished by its thick stem and large branching inflorescence with many spikelets on long thin stalks. The culm may be glabrous whilst in other collections it is scabridulous.

41. **Bulbostylis cupricola** Goetgh. in Bull. Jard. Bot. Belg. **54**: 92 (1984). — Goetghebeur & Coudijzer in Bull. Jard. Bot. Belg. **55**: 246 (1985). Type: D.R. Congo, Upper Katanga, Lukuni old copper mine, 1300 m, 4.i.1971, *Léonard* 5248 (BR holotype, GENT).

a) Subsp. **cupricola**

Erect, tufted annual up to 15 cm tall; rhizome and tunic of leaf bases lacking. Leaves numerous, 1/3 to occasionally equalling culm length leaf sheath mouth inconspicuously long white-fringed, ligule 0; leaf blades 30–80 × 0.2 mm, filiform, inrolled, margins scabrous, hairs when present, pilose to villous, mostly confined to nerves. Culms 0.2 mm wide, ridged and furrowed, glabrous, shortly scabrous below inflorescence. Inflorescence anthelate, of central spikelet with 0–4 scabridulous inflorescence branches bearing spikelets; bracts usually 2 visible, 3–12 mm long, margins glabrous or scantily white-fringed, apex attenuated, scabrid. Spikelets 3–6 × 1–1.5 mm, acute. Glumes spiral, 1.7–2.3 × 0.8–1 mm, ovate, margins translucent, shortly ciliate to almost glabrous; flanks veinless, glabrous with rusty-red brown patches, midrib 3-nerved, usually ending prior to apex. Stamens 3. Style glabrous, dark brown basally, style base persistent on nutlet, branches 3, usually shorter than style. Nutlet 0.7 × 0.4 mm, obovoid, to narrowly obovoid, subtrigonous, surface slightly transversely rugose, dark rusty-red.

Zambia. W: Luanshya Dist., Luanshya, 14 Shaft North Dump, 16.iv.1998, *Leteinturier & Malaisse* 52 (BR, K).

Also in D.R. Congo. Often on copper anomalies; 1200–1300 m.

Conservation notes: Endemic to the Zambian/Congo Copperbelt, very local in distribution; probably not threatened.

b) Subsp. **mucronata** Browning, subsp. nov. Differs from subsp. *cupricola* by its larger spikelets and conspicuously mucronate glumes. Type: Zimbabwe, Gokwe Dist., Copper Queen, 23.iv.1965, *Wild* 7335 (K holotype, SRGH).

Tufted annual up to 12 cm tall, resembling typical species in all respects except the larger spikelets, 8–10 × 1–2 mm, and the fewer glumes, 2.5–2.7 mm, distinctly mucronate (mucro 0.2–0.4 mm), pale with few rusty-red streaks.

Zimbabwe. N: Gokwe Dist., Copper King, 23.iv.1965, *Wild* 7342 (K, SRGH).
Not known elsewhere. On copper-rich soils; c. 850 m.
Conservation notes: Apparently endemic to the Gokwe area of C Zimbabwe and known at present from only two collections; possibly Vulnerable.
Wild noted on both known collections "On copper oxide ore outcrop with 1½% Cu, 1½% Pb, 5% Zn at highest conc."
In very mature collections, when glumes and nutlets are shed, the rachilla is slightly zig-zag. After examination of the type of *Bulbostylis cupricola*, the Zimbabwe material was considered to represent a new subspecies; unfortunately more recent collections have not been seen.

42. **Bulbostylis laxispicata** (Lye) Lye in Nordic J. Bot. **28**: 517 (2010). Type: Zambia, Mongu, 29.i.1966, *Robinson* 6820 (K holotype).
    Abildgaardia laxispicata Lye in Lidia **1**: 37 (1985) as *laxespicata*.

Delicate, tufted flexuous annual up to 40 cm tall; rhizome and tunic of leaf bases lacking. Leaves numerous, 1/3–1/2 culm length; leaf sheath light orange, mouth conspicuously long white-fringed, ligule 0; leaf blades to 150 × 0.2–0.3 mm, filiform, glabrous to slightly scabrid. Culms 0.25–0.4 mm wide, ridged and furrowed, glabrous. Inflorescence bisanthelate, of one central sessile spikelet and few to several stalked spikelets, on glabrous branches 30–35 mm long; bracts 3–4, most under 10 mm, base glume-like, with or without a few long white hairs, narrowed apically to scabrid awn. Spikelets 5–7 mm long, ovate, apex acute. Glumes few, spiral, 1.8–2 mm, brown, margins entire to shortly ciliate, flanks slightly scabrid to almost glabrous, apex shortly apiculate in lower glumes, upper not mucronate. Stamens 3. Style base persistent on nutlet, branches 3, usually shorter than style. Nutlet 0.7–0.8 × 0.5–0.6 mm, white to grey-white at maturity, obovoid, trigonous,surface smooth (velate), papillae may be visible below outer surface.

Zambia. B: Mongu Dist., Mongu, 29.i.1966, *Robinson* 6820 (K).
Not known elsewhere. Sandy grassland at edge of woodland and Kalahari sandveld; 1000–1100 m.
Conservation notes: Apparently endemic to the Mongu area of W Zambia; possibly threatened.
Lye (1985) provided a very short diagnosis and an illustration drawn by Haines in which the nutlet is shown with transverse lines. This is confusing – does this mean the nutlet is rugose? Lye mentions '*laevis*' in the Latin diagnosis. Three collections at Kew were determined by Lye as *Abildgaardia laxispicata*; one of these is the type and the other two (*Robinson* 2072 from Livingstone, SW Zambia and *Verboom* 91 from Mfuwe, Luangwa valley, SE Zambia) are doubtfully this species.
There is resemblance between *Bulbostylis laxispicata* and *B. longiradiata*, but the glumes are smaller and the culm and leaves are glabrous in the former. There is also resemblance to *B. densa*.

43. **Bulbostylis longiradiata** Goetgh. in Bull. Jard. Bot. Belg. **55**: 219, fig.1,9a-b. (1985). Type: Burundi, Rumonge, 10.iii.1976, *Reekmans* 4844 (BR holotype, GENT).

Delicate erect annual up to 35 cm tall; rhizome and tunic of leaf bases lacking. Leaves numerous, occasionally equalling culm length; sheath mouth long white fringed, ligule 0; leaf blades 60–150 × 0.2–0.3 mm, filiform, inrolled when dry, margins hispidulous. Culms c. 0.4 mm wide, ridged and furrowed, hispid. Inflorescence bisanthelate (branched to two orders), of central spikelet with 4–5 hispidulous inflorescence branches, 30–40 mm long, bearing spikelets; bracts 3–4, shorter than inflorescence branches, hispid, margins long white fringed. Spikelets 5–7(8) × 1.3–1.6 mm, acute. Glumes spiral, 2.5–2.8(3.5) × 1.4 mm, minutely scaberulous, light brown (cinnamon) with red-brown stripes, margins shortly fimbriolate, 3-nerved (central one green in life), lateral nerves prominent, apex mucronate to shortly apiculate. Stamens (1)2–3. Style glabrous, red brown basally, persistent on nutlet, branches 3, usually shorter than style. Nutlet 1 × 0.7–0.8 mm, grey-brown, obovoid, subtrigonous, angles papillate, surface transversely wavy (8–10 waves not pronounced).

Zambia. N: Mbala Dist., Kalambo R., path to Sansia Falls, 1350 m, 28 iii.1957, *Richards* 8914 (K).

Also in D.R. Congo and Burundi. *Brachystegia* woodland; c. 1350 m.

Conservation notes: Restricted distribution in the Flora area; possibly threatened.

Bulbostylis longiradiata Goetgh. is included here as it is likely to be present in Zambia, but no collections have been located at Kew. It is closely allied to *B. capitulatoradiata* and may possibly be confused with *B. laxispicata.*

44. **Bulbostylis capitulatoradiata** Browning & Goetgh., sp. nov. Affiliated to *Bulbostylis longiradiata* Goetgh. but differs as the anthela has a central group of sessile spikelets and the inflorescence branches are obliquely upright with a slight outward curve. Type: Zambia, Ndola Rural Dist., c. 33 km SW of Luanshya, Chibuli Hill, 28°14.30'E 13°24.00'S, 1250 m, 17.iv.1998, *Leteinturier, Malaisse & Matera* 69 (BR holotype, GENT, K).

Tufted erect annual up to 45 cm tall; rhizome and tunic of leaf bases lacking. Leaves numerous, shorter or occasionally equalling culm length; leaf sheath mouth long white-fringed, ligule 0; leaf blades 100–160 × 0.25–0.3 mm, filiform, inrolled when dry, margins hispidulous. Culms 0.4–0.5(0.8) mm wide, ridged and furrowed, glabrous, but scabrid below inflorescence. Inflorescence bisanthelate, a central group of 3–6(8) sessile spikelets with 4–5 scabridulous obliquely upright slightly outwardly curved branches, 30–40 mm long, bearing spikelets; bracts 3–4, shorter or equalling length of inflorescence brances, scabrid, margins with few long white hairs. Spikelets 5–7(10) × 1.5–2 mm, acute. Glumes spiral, 2.4–2.6 mm long, glabrous to slightly scabrid, light brown (cinnamon) with red-brown stripes, margins apically shortly ciliate, 3-nerved (central green in life), lateral nerves prominent, apex mucronate to shortly apiculate. Stamens 3, anther c. 1 mm long. Style glabrous, red brown basally, base persisten ton nutlet, branches 3, usually shorter than style. Nutlet 0.7–0.9 × 0.5–0.7 mm, grey-brown, obovoid, subtrigonous, angles papillate, surface transversely wavy (8–10 waves), outer layer often velate.

Zambia. W: Ndola Dist., 88 km S of Ndola, 20.iii.1960, *Robinson* 3419 (K).

Not known from elsewhere. Open forest on rocky hillslopes amongst rocks; c. 1250 m.

Conservation notes: Apparently endemic to the Ndola area of W Zambia, but possibly also in adjacent parts of D.R. Congo; probably not threatened.

The nutlet of *Leteinturier et al.* 69 (K) measures 0.82 × 0.6 mm, while the glume including mucro is 2.4 mm long and not hispidulous, but glabrous. The type and isotype are both very mature collections so that spikelets are elongated and many of the glumes have fallen.

45. **Bulbostylis tanzaniae** (Lye) R.W. Haines in Haines & Lye, Sedges & Rushes E. Afr., app.3: 1 (1983). —Goetghebeur & Coudijzer in Bull. Jard. Bot. Belg. **55**: 241 (1985). —Verdcourt in F.T.E.A., Cyperaceae: 91 (2010). Type: Tanzania, Mpanda Dist., near Mpanda, 26.v.1957, *Nye* 204 (BM holotype).

  Abildgaardia tanzaniae Lye in Nordic J. Bot. **1**: 753, fig.9 (1982); in Haines & Lye, Sedges & Rushes E. Afr.: 113, figs.207,208 (1983).

Tufted annual, 5–27(35) cm tall. Leaves numerous, ¼ culm length; leaf sheaths conspicuously long white-fringed, ligule 0; leaf blades 0.3–0.5 mm wide, filiform, inrolled, scabrid on margins. Culm erect, glabrous below to scabridulous upwards. Inflorescence anthelate of one sessile spikelet and several glabrous stalked spikelet clusters; bracts 2–5, to 3.5 cm long, occasionally longer and exceeding inflorescence, with long fine white hairs at base. Spikelets 4–5(9) × 1–2 mm, basal glumes tardily deciduous, rachilla shortly winged. Glumes 1.9–2.1 mm, glabrous,midrib terminating before margin, reddish brown, paler hyaline margins markedly fimbriolate, apex rounded. Stamens 3. Style 3-angled, sparsely villous to glabrous, base small, persistent on nutlet, branches 3. Nutlet trigonous, obovoid, off-white to pale pinkish-brown, 0.5–0.6 × 0.4–0.5 mm, faintly transversely rugose (few rugae), angles 3 with a single vertical row of papillae. In mature nutlets the pericarp readily fractures into transparent pieces.

Zambia. N: Mbala Dist., shore of Lake Chila, 14.iii.1952, *Richards* 1004 (K). W: Ndola Dist., 10 km S of Ndola, 1220 m, v.1961, *Wilberforce* 109 (K); Mwinilunga Dist., 43 km W of Mwinilunga, 17.iv.1960, *Robinson* 3662 (K). S: Choma Dist., 5 km E of Choma, 26.iii.1955, *Robinson* 1157 (K).

Also in Tanzania, Burundi and D.R. Congo. Shallow soil overlying laterite, sandy ledges and lakeshores, sometimes on copper-contaminated soils; 1200–1650 m.

Conservation notes: Widespread; not threatened.

Goetghebeur & Coudijzer (1985) noted that in collections from D.R. Congo "the inflorescence can be semicompact, i.e. with the spikelets of the partial anthelae more or less aggregated, causing a peculiar inflorescence form." *Wilberforce* 109 from W Zambia has an expanded anthela with stalked spikelets, widely spaced in comparison to *Vesey-Fitzgerald* 3372 which has a 'compact' inflorescence. The markedly fimbriolate glume margin and the nutlet are distinctive.

This species, which is well distributed in Zambia, closely resembles *Bulbostylis abortiva.*

46. **Bulbostylis abortiva** (Steud.) C.B. Clarke in Buchanan, Nyassaland: 21 (1889); in Durand & Schinz, Consp. Fl. Afr. **5**: 610 (1894). —Goetghebeur & Coudijzer in Bull. Jard. Bot. Belg. **55**: 243 (1985). —Verdcourt in F.T.E.A., Cyperaceae: 97 (2010). Type: Madagascar, Nossibé, n.d., *Boivin* 1996 (P holotype, K).

    Fimbristylis abortiva Steud., Syn. Pl. Glumac. **2**: 111 (1855).
    Scirpus schweinfurthianus Boeckeler in Linnaea **36**: 758 (1869). Type: Ethiopia, Gallabat, near Matamma, 14.ix.1865, *Schweinfurth* 2039 (B holotype, K, P).
    Bulbostylis capillaris (L.) C.B. Clarke var. *abortiva* (Steud.) H. Pfeiff. in Bot. Arch. **6**: 188 (1924).
    Bulbostylis claessensii De Wild., Pl. Bequaert. **4**: 192 (1927). Type: D.R. Congo, no locality, 1922, *Claessens* 1677 (BR syntype); Ganganyala, 1917, *Vanderyst* 6429 (BR syntype).
    Abildgaardia abortiva (Steud.) Lye in Bot. Not. **127**: 495 (1974). —Haines & Lye, Sedges & Rushes E. Afr.: 117, fig.218 (1983).

Short to medium tufted annual, occasionally up to 80 cm. Leaves of varying height; leaf sheaths scabrid or not, with long white spreading hairs added, especially from mouth, ligule 0; leaf blades 20 cm long, mostly setaceous, often curled and twisted. Culms deeply grooved, usually with short white scabrid pubescence, sometimes glabrous. Inflorescence anthelate, compound, open, of (10)20–40(60) mostly pedicelled spikelets; bracts setaceous, lengths variable, usually inconspicuous, occasionally exceeding inflorescence, margins basally long-haired, becoming scabrid upwards. Spikelets 3–7 × 1–2 mm. Glumes 1.5–3 mm, reddish brown, broadly ovate, glabrous with hyaline slightly fimbriolate margins, upper glume apex rounded, lower shortly mucronate, midrib indistinctly 3-veined. Stamens 3. Style branches 3, base persistent on nutlet. Nutlet 0.7–0.8 × 0.4–0.5 mm, obovoid, trigonous, surface punctulate with papillae in vertical rows, but often appearing smooth due to an outer covering.

Zambia. W: Mwinilunga Dist., Kalene Hill, 25.ii.1975, *Hooper & Townsend* 344 (K). S: Mazabuka Dist., Siamamba Forest Reserve, 14.i.1960, *White* 6272 (K). **Zimbabwe**. S:

Chiredzi Dist.?, Lower Save (Sabie) R., rocky outcrop, 27.i.–2.ii.1948, *Rattray* 1269 (K).
Mozambique. GI: Massinga Dist., Pomene, 24.vi.1980, *de Koning* 8260 (K).

Found throughout tropical Africa and Madagascar, Senegal, Sudan, Burundi, Uganda, Kenya, Tanzania and D.R. Congo, but not in South Africa. Sandy soils, old cultivated lands and *Brachystegia* woodland; 10–1400 m.

Conservation notes: Widespread; not threatened.

The description given mainly follows other accounts. Some of the features detailed were not found in the original account by Steudel (1889), nor could they be found on the K isotypes of *Boivin* 1996 or on *Schweinfurth* 2039.

Bulbostylis abortiva closely resembles *B. tanzaniae* – if mature nutlets are examined differences between these two species can be seen, with the former having a slightly larger nutlet with a puncticulate surface.

Collections of this species from the Flora area were difficult to locate. Several Hooper & Townsend collections from W Zambia (Mwinilunga Dist.) were determined by Hooper as "cf. *abortiva*", and Hooper was familiar with *B. abortiva* as it is present in several West African countries. It may be sympatric with *B. tanzaniae* in parts of W Zambia.

It is likely that more than one entity is included in *Bulbostylis abortiva* as presently constituted. In F.T.E.A. (2010) *Fimbristylis flexuosa* Ridley appears as a synonym of *B. abortiva*; here it is recognised as a separate species, *B. flexuosa*.

47. **Bulbostylis densa** (Wall.) Hand.-Mazz. in Karsten & Schenk, Vegetationsbilder **20**(7): 16 (1930). Type: Nepal, no locality, 1831, *Wallich* 3514c (K ?lectotype, C, P).
   Scirpus densus Wall. in Roxburgh, Fl. Ind. **1**: 231 (1820).

Subsp. **afromontana** (Lye) R.W. Haines in Haines & Lye, Sedges & Rushes E. Afr., app.3: 1 (1983). —Goetghebeur & Coudijzer in Bull. Jard. Bot. Belg. **55**: 251 (1985). —Gordon-Gray in Strelitzia **2**: 31, figs.10H,K (1995). —Verdcourt in F.T.E.A., Cyperaceae: 98, fig.15 (2010). Type: Uganda, Kigezi Dist., N slope of Mgahinga-Muhavura saddle, 24.iv.1970, *Lye* 5329 (EA holotype, K, MHU). FIGURE 14.**18**.
   Bulbostylis trifida (Nees) Nelmes var. *biegensis* Cherm. in Bull. Soc. Bot. France **82**: 341 (1935). Types: D.R. Congo, mountains W of L. Kivu, iii.1929, *Humbert* 7655 (P syntype); iii.1929, *Humbert* 7570 (P syntype); iii.1929, *Humbert* 7808 (BR syntype, P); between lakes Kivu & Edward, iv-v.1929, *Humbert* 7926 (BR syntype, P).
   Abildgaardia densa (Wall.) Lye subsp. *afromontana* Lye in Nordic J. Bot. **3**: 237, fig.9 (1983). —Haines & Lye, Sedges & Rushes E. Afr.: 120, fig.224 (1983).

Delicate, erect or drooping and flexuous annual up to 30 cm tall, rhizome and tunic of leaf bases lacking. Leaves numerous, 1/4–1/3 to occasionally equalling culm length; hairs when present pilose to villous, mostly confined to nerves; leaf sheath mouth inconspicuously long white-fringed, ligule 0; leaf blades to 160 × 0.24–0.4 mm, filiform, margins scabrous, pubescent. Culms ridged and furrowed, glabrous or sparsely pubescent or pilose apically. Inflorescence anthelate, variable, of a central spikelet or head of spikelets with 4–5 glabrous inflorescence branches added, or reduced to 2 or 1 sessile terminal spikelet(s) only; bracts 3–4, inconspicuous, filiform, margins glabrous or scantily white-fringed. Spikelets broadly ovate, 2–4 × 1.3–2 mm, angled by projecting glume keels, gaping from nutlets at full maturity. Glumes spiral, 1.4–1.8 × 0.75–1.7 mm, dark brown, glabrous or pubescent, margins shortly ciliate to almost glabrous, apex mucronate to shortly apiculate. Stamens (1)2–3. Style glabrous, dark brown basally, style base round, vertically flattened, persistent on nutlet, branches 3, usually shorter than style. Nutlet 0.6–0.9 × 0.5–0.6 mm, surface variable, smooth or transversely rugose, punctate, outermost cells longitudinally oblong to subquadrate, with or lacking a single central outgrowth (papilla), this layer of tissue eventually velate, leaving surface smooth.

Zambia. W: Mwinilunga Dist., c. 2 km E of Matonchi Farm, 1350 m, 16.ii.1975, *Hooper & Townsend* 197 (K). C: Serenje Dist., 1 km from Kanona on road to Kundalila

Falls, 1650 m, 12.iii.1975, *Hooper & Townsend* 676 (K). **Zimbabwe**. C: Harare (Salisbury), no locality, 1460 m, 20.xii.1931, *Brain* 7691 (K). **Malawi**. N: Nyika Plateau, 2250 m, 15.iii.1961, *Robinson* 4524 (K). S: Mulanje Mt, Chambe Saddle rocks, 2000 m, 13.vi.1962, *Robinson* 5337 (K). **Mozambique**. MS: Gorongosa Dist., Mt Gorongosa, Gogogo summit area, iii.1972, *Tinley* 2415 (K, LISC).

Also in Tropical West Africa, Ethiopia, Sudan, Kenya, Uganda, Tanzania, Rwanda, D.R. Congo, Angola, South Africa (Eastern Cape, Free State, KwaZulu-Natal,

Fig. 14.**18**. BULBOSTYLIS DENSA subsp. AFROMONTANA. 1, habit (× 2/3); 2, inflorescence (× 4); 3, spikelet (× 12); 4, glume (× 16); 5, floret (× 24); 5, nutlet (× 20). All from *Magogo* 20. Drawn by Juliet Williamson. Reproduced from Flora of Tropical East Africa (2010).

Limpopo, Mpumalanga) and Swaziland. Montane grassland, often amongst rocks in damp crevices; 1300–2300 m.

Conservation notes: Widespread; not threatened.

The typical subspecies, subsp. *densa*, is found only in Asia.

Goetghebeur & Coudijzer (1985) note that the difference between *Bulbostylis pusilla* and *B. densa* is not always clear. *B.densa* is very variable in size but may be recognised by the clearly pedicellate spikelets and nutlets usually "slightly transversely rugose with many rows of small cones".

48. **Bulbostylis pusilla** (A. Rich.) C.B. Clarke in Durand & Schinz, Consp. Fl. Afr. **5**: 615 (1894). —Goetghebeur & Coudijzer in Bull. Jard. Bot. Belg. **55**: 246 (1985). —Gordon-Gray in Strelitzia **2**: 35, fig.12C,F (1995). —Verdcourt in F.T.E.A., Cyperaceae: 93 (2010). Type: Ethiopia, Adua, *Quartin Dillon* s.n. (P lectotype); Ethiopia, Tigray, Gafta, 16.ix.1838, *Schimper* 796 (BR syntype, K, P); see footnote in Verdcourt (2010: 93). FIGURE 14.**19**.

> *Fimbristylis pusilla* A. Rich., Tent. Fl. Abyss. **2**: 506 (1850).
> *Fimbristylis parva* (Ridl.) C.B. Clarke in Durand & Schinz, Consp. Fl. Afr. **5**: 615 (1894). Type: Angola, Pungo Andongo, ii-iv.1857, *Welwitsch* 6823 (BM, LISU syntypes); Pungo Andongo, i.1857, *Welwitsch* 6831 (BM syntypes, LISU).
> *Bulbostylis yalingensis* Cherm. in Arch. Bot. Mém. **4**(7): 40 (1931). Type: Central African Republic, Yalinga, 21.vii.1921, *Le Testu* 2987 (P holotype, BM, BR).
> *Abildgaardia pusilla* (A. Rich.) Lye in Bot. Not. **127**: 497 (1974).
> *Abildgaardia pusilla* subsp. *yalingensis* (Cherm.) Lye in Nordic J. Bot. **3**: 239 (1983). — Haines & Lye, Sedges & Rushes E. Afr.: 116, figs.213a,b (1983).
> *Bulbostylis pusilla* subsp. *yalingensis* (Cherm.) R.W. Haines in Haines & Lye, Sedges & Rushes E. Afr., app.3: 1 (1983).

Delicate, usually erect annual, 5–40 cm tall; rhizome and tunic of leaf bases lacking. Leaves numerous, 1/4–1/2 to occasionally equalling culm length; leaf sheath mouth with long white hairs, ligule 0; leaf blades 20–70(400) × 2–3 mm, margins scaberulous. Culms ridged and furrowed, glabrous or sparsely pubescent apically. Inflorescence anthelate, variable, lax with one sessile and 1–10 stalked spikelets (stalks smooth) and up to 40 additional groups of sessile and stalked spikelets; bracts 3–4, glabrous. Spikelets 2–5 × 1–2 mm, ovoid to lanceolate, few-flowered, not gaping when nutlets mature. Glumes spiral, 1–2.5(3) mm long, glabrous to shortly scabrid, cinnamon-coloured, sometimes darker brown, with paler minutely ciliolate margins, midrib 3-veined, not reaching apex. Stamens 2–3. Style glabrous, base persistent on nutlet, branches 3, usually shorter than style. Nutlet 0.7–0.9 × 0.6–0.7 mm, trigonous, obovoid to obcordate, transversely rugose with raised papillae along tops of wrinkles, dark brown at maturity.

Zambia. W: Solwezi Dist., Mwinilunga road, Kabompa R., 1400 m, 15.ii.1975, *Hooper & Townsend* 70 (K). N: Mbala Dist., Ndundu dambo, 22.iv.1962, *Robinson* 5118 (K). C: Serenje Dist., 55 km NE of Serenje, 2.iii.1962, *Robinson* 4982 (K). S: Choma Dist., SE of Choma, 1310 m, 19.iii.1958, *Robinson* 2801 (K). **Zimbabwe**. N: Kariba Dist., Kariba Nat. Park, 25 km from Makuti on Kariba road, 1110 m, 19.ii.1981, *Philcox et al.* 8742 (K, SRGH). C: Kwekwe Dist., NE of Kwekwe town, Itwata estate, 30°06'E 18°49'S, 1200 m, 9.ii.2000, *Lye* 23772 (K, SRGH).

Also in West Africa, Central African Republic, Ethiopia, Uganda, Kenya, Tanzania, D.R. Congo, Angola, Namibia and South Africa (KwaZulu-Natal, Limpopo, Mpumalanga, North West). Moist mossy seeps over rocky outcrops, roadsides and sandy soil on riverbanks; 900–1400 m.

Conservation notes: Widespread; not threatened.

Like *Bulbostylis densa*, there is considerable variation in inflorescence form and pubescence. Goetghebeur & Coudijzer (1985) refer to the '*pusilla* group' that requires further investigation and research.

A number of subspecies have been recognised by various authors but are not accounted for here. *Bulbostylis pusilla* subsp. *congolensis* (De Wild.) R.W. Haines, as given in F.T.E.A. (2010), is here reinstated as a species following Goetghebeur & Coudijzer (Bull. Jard. Bot. Belg. **55**: 244, 1985).

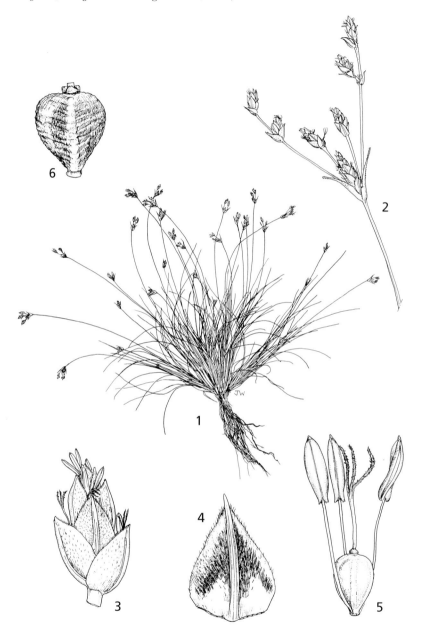

Fig. 14.**19**. BULBOSTYLIS PUSILLA. 1, habit (× 2/3); 2, inflorescence (× 3); 3, spikelet (× 12); 4, glume (× 20); 5, floret (× 24); 6, nutlet (× 30). 1, 2 from *Drummond* & *Hemsley* 2137; 3–5 from *Tanner* 5852; 6 from *Peter* 21684. Drawn by Juliet Williamson. Reproduced from Flora of Tropical East Africa (2010).

49. **Bulbostylis scrobiculata** (Lye) Lye in Nordic J. Bot. **28**: 517 (2010). Type: Zambia, Mbala Dist., Mbala (Abercorn), sandpits, 18.iii.1962, *Robinson* 5043 (K holotype, NU).

> *Abildgaardia scrobiculata* Lye in Lidia **1**: 37, fig.p.38 (1985).
> *Bulbostylis lineolata* Goetgh. in Bull. Jard. Bot. Belg. **55**: 248, fig.7 (1985). Type: Burundi, Plain of Ruzizi, Isabu, 10.ii.1980, *Caljon* 2011 (GENT holotype, BR, K).

Delicate, erect or drooping, flexuous annual up to 20 cm; rhizome and tunic of leaf bases lacking. Leaves numerous, ¼ to occasionally equalling culm length; leaf sheath mouth inconspicuously long white-fringed, ligule 0; leaf blades to 50 × 0.2–0.25 mm, filiform, margins glabrous. Culms 0.3–0.5 mm wide, ridged and furrowed, glabrous. Inflorescence bisanthelate, of central sessile spikelet with 4–5(8) glabrous primary (10–20 mm long) and secondary (c. 7 mm long) inflorescence branches bearing added spikelets; bracts 3–4, c. 5 mm long, margins lacking obvious long white hairs. Spikelets acute, 3–4 × 1–1.3 mm. Glumes spiral, ± 12, 1.3–1.6 mm long, dark brown, flanks scaberulous, margins fimbriolate, midvein 3-nerved, terminating at apex (not mucronate). Stamens 1. Style base round, dark brown, base persistent on nutlet, style branches 3. Nutlet 0.5–0.6 × 0.4–0.5 mm, creamy white to pale brown, obovoid, trigonous, surface with slightly raised, parallel transverse bands (c. 10 shallow rugae), appearing white in mature nutlets.

Zambia. N: Mbala Dist., Mbala (Abercorn) sand pits, 18.iii.1962, *Robinson* 5044 (K, NU). **Zimbabwe**. N: Hurungwe Dist., Mwami (Miami), K34 Expt. Farm, 7.iii.1947, *Wild* 1715 (BR, SRGH). C: Harare (Salisbury), 1490 m, iii.1919, *Eyles* 1563 (BM, SRGH). **Malawi**. N: Mzimba Dist., Katoto, 4.8 km W of Mzuzu, 1370 m, 13.iii.1974, *Pawek* 8213 (K, MA, SRGH).

Also in Burundi, Tanzania, D.R. Congo and Angola. Sandy damp soil and overgrazed savanna; 800–1700 m.

Conservation notes: Widespread; not threatened.

Lye (1985) gave a short Latin diagnosis, an illustration, but not a full description of this taxon, so we have described it from the type. It is not clear why the specific epithet '*scrobiculata*' (pitted) was chosen. *Bulbostylis lineolata* Goetgh., for which there was a good Latin diagnosis and an illustration, is here placed in synonymy as it was described a few months after Lye's publication of *B. scrobiculata*.

Robinson 5044, determined by Lye in 1984 as *Abildgaardia scrobiculata*, has a slightly different appearance to the type (*Robinson* 5043). Sheath hairs are more pronounced and longer, blades occasionally somewhat scabrid, bracts apically scabid, spikelets are longer (5 mm), the glumes are mucronate and the nutlet more rugose.

Bulbostylis scrobiculata is similar to *B. pusilla* and *B. densa*, but distinguished from them by the small white transverse stripes on the outer cell walls of the nutlet and by the more fimbriolate glumes.

Species doubtfully recorded.

Bulbostylis atrosanguinea (Boeckeler) C.B. Clarke in Durand & Schinz, Consp. Fl. Afr. **5**: 611 (1894). —Goetghebeur & Coudijzer in Bull. Jard. Bot. Belg. **55**: 230 (1985). —Verdcourt in F.T.E.A., Cyperaceae: 75 (2010). Type: Tanzania, Kilimanjaro, 3350 m, 1884, *Johnston* 157 (K holotype).

> *Scirpus atrosanguineus* Boeckeler in Bot. Jahrb. Syst. **7**: 276 (1886).
> *Fimbristylis atrosanguinea* (Boeckeler) K. Schum. in Engler, Pflanzenw. Ost-Afrikas **C**: 125 (1895).
> *Abildgaardia setifolia* (A. Rich.) Lye in Bot. Not. **127**: 497 (1974). —Haines & Lye in Sedges & Rushes E. Afr.: 100, fig.170 (1983). Type: Tanzania, Mt Kilimanjaro, *Johnston* s.n. (K).

This species is reported to be present in Ethiopia, Angola, Uganda, Kenya, Tanzania, Malawi and Zambia, although collections from the Flora region were not found at Kew, where many specimens named as such are referable to *Bulbostylis schoenoides sensu lato*.

Bulbostylis atrosanguinea is a high altitude species of subalpine heath and grassland at altitudes of 1800–3700 m. Goetghebeur & Coudijzer (1985: 230) describe it as "A tufted, tussocky perennial, leaves filiform, culm scabrid at the tip; inflorescence capitate, of several spikelets, rarely reduced to a single spikelet; glumes 3–4 mm long (sub)acute, often fimbriate; fruit slightly transversely rugulose to minutely papillose, style base persistent." The plants are noted to have blackish glumes with red nerves, and this may be a distinguishing character.

Species and collections of uncertain identity.

Bulbostylis malawiensis (Lye) Lye in Nordic J. Bot. **28**: 517 (2010). Type: Malawi, Mulanje Mt, Mt Chilemba, 2200 m, 7.vi.1962, *Robinson* 5307 (K holotype, herb. Lye).

> *Abildgaardia malawiensis* Lye in Lidia **1**: 35 (1985).

No detailed description of this species was given by Lye – the Latin diagnosis was for a perennial plant with a simple anthela of 1–3 spikelets, glumes 3–4.5 mm long and a small papillate nut. On examination at Kew it was difficult to place it in a genus, so it is not included here as more material is needed for study.

9. **ABILDGAARDIA** Vahl[9]

Abildgaardia Vahl, Enum. Pl. **2**: 296 (1805).

Perennial tufted herbs. Culms scapose. Leaves eligulate, well-developed or reduced to sheaths. Inflorescence of 1–5 solitary or anthelate flattened spikelets, basal glumes distichous, upper spiral; bracts not sheathing, inconspicuous. Glumes breaking at ± mid-length, basal portions persistent raggedly, clothing rachilla. Florets bisexual; bristles absent. Stamens 3. Style 3-angled, base expanded slightly, dehiscent with style. Nutlet obpyriform, stipitate; surface cells isodiametric, tuberculate or sparsely so.

Genus of c. 10 species, chiefly Australian and S American with 3 species in southern Africa.

Opinions differ regarding the independent generic status of *Abildgaardia*. It is either included within *Fimbristylis* or treated independently. Recent molecular analysis favours independent generic status.

1. Leaves reduced to sheaths, blades filiform, not longer than 3 mm, eventually deciduous . . . .**1.** *hygrophila*
– Leaves with blades well developed 2
2. Inflorescence a solitary terminal spikelet, occasionally one pedicelled or sessile spikelet added; nutlet surface prominently tuberculate . . . . **2.** *ovata*
– Inflorescence usually an anthela of one central spikelet with 1–4 pedicelled spikelets added (occasionally of one sessile spikelet only); nutlet becoming dark brown to eventually black, surface reticulate, tuberculate or not . . . . . . **3.** *triflora*

1. **Abildgaardia hygrophila** (Gordon-Gray) Lye in Mitt. Bot. Staatssamml. München **10**: 547 (1971). —Goetghebeur & Coudijzer in Bull. Jard. Bot. Belg. **54**: 87 (1984). —Gordon-Gray in Strelitzia **2**: 19, fig.2A,D (1995). —Verdcourt in F.T.E.A., Cyperaceae: 115 (2010). Type: South Africa, KwaZulu-Natal, Port Edward, i.1951, *Huntley* 701 (NU holotype, BM, BOL, K, PRE).

> *Fimbristylis hygrophila* Gordon-Gray in J. S. African Bot. **32**: 129, fig.1 (1966).

---

[9] by K.D. Gordon-Gray† and J. Browning

Perennial to 50 cm tall; rhizome woody, densely tufted. Leaf bases spongy, usually forming persistent tunics; sheaths entire, closely covering culms, mouth sloping, ligule membranous, margin of sheath mouth adaxial to leaf blade; blade obsolete, to 3 mm long, filiform, the continuation of main nerves of sheath. Culm erect to slightly drooping, terete, glabrous. Inflorescence anthelate or with one central sessile spikelet with 1–4 additional pedicelled spikelets, or reduced to a head or solitary spikelet (variable on individual plants), total size up to 23 × 35 mm; bracts 2–4, to 5 mm long, inconspicuous. Spikelets 10–22 × 3–9 mm, base compressed, becoming cylindric above, compact when young, ragged with age, rachilla twisted, wings shrivelling early, appearing absent. Glumes, lowest 1–2 sterile, the remainder fertile, 6.1–8.2 × 3.2–4.3 mm, apex mucronate. Perianth 0. Stamens 3. Style 5.3–7.1 mm long, including 3-angled to clearly winged style base 0.5–0.6 mm long, deciduous with style. Nutlet 1.2–1.4 × 0.8–1.1 mm, 3-angled, obovoid to pyriform, narrowed base (claw) c.¼ nut length, crystalline white with greenish/brownish undertones; surface faintly reticulate, markedly transversely ribbed, outermost cells subquadrate, transversely elongate over ribs.

Zambia. W: Luanshya Dist., Ibenga, 29 km S of Luanshya, 1300 m, 2.ii.1960, *Robinson* 3316 (K). C: Lusaka Dist., Nyanshishi R. near Chinkuli, 15°16'S 28°36'E, c. 1100 m, 10.xii.1972, *Kornaś* 2792 (K). **Mozambique**. Z: Maganja da Costa Dist., 14 km from Maganja da Costa to Malei, 50 m, 21.xi.1967, *Torre & Correia* 16179 (BR, LD, LISC, LUA, WAG). MS: Gorongosa Dist., Mt Gorongosa summit area, iii.1972, *Tinley* 2425 (K, SRGH). GI: Bilene Dist., S. Martinho de Bilene, 30.xi.1955, *Gonçalves-Sanches* 19 (LISC).

Also in Tanzania and South Africa (Eastern Cape, KwaZulu-Natal). In dambos and poorly aerated saline soils, organically rich turf and sandy soils. Flowering period August to March; 10–1500 m.

Conservation notes: Widespread; not threatened.

The ragged spikelets with golden brown glumes are distinctive.

2. **Abildgaardia ovata** (Burm.f.) Kral in Sida **4**: 72 (1971). —Haines & Lye, Sedges & Rushes E. Afr.: 94, fig.158 (1983). —Goetghebeur & Coudijzer in Bull. Jard. Bot. Belg. **54**: 86 (1984). —Gordon-Gray in Strelitzia **2**: 19 (1995). —Verdcourt in F.T.E.A., Cyperaceae: 113 (2010). Type: 'India Orientali', *König* s.n. in Herb. Linn. (LINN), see note.

    Carex ovata Burm. f., Fl. Indica: 194 (1768).

    Abildgaardia monostachya (L.) Vahl, Enum. Pl. **2**: 296 (1805).

    Fimbristylis monostachya (L.) Hassk., Pl. Jav. Rar.: 61 (1848). —Clarke in F.T.A. **8**: 424 (1902).

    Fimbristylis ovata (Burm. f.) J. Kern in Blumea **15**: 126 (1967).

Perennial herb up to 50 cm tall; rhizome woody, uniseriate or tufted. Leaf bases forming persistent tunic to shoots; sheaths entire, mouth truncate; ligule 0; blade to 250 × 0.5–1.7 mm, linear to almost filiform, scabrid, apex acute to acuminate. Inflorescence a solitary terminal spikelet, occasionally twinned, or with one shortly pedicelled; bracts 2–3, to 13 mm long, glume-like. Spikelets 7–19 × 3–6 mm, to 29 mm long with age, lower half compressed, upper cylindrical, green to yellowish. Glumes, basal 1–4 early deciduous, sterile, remainder fertile, 5–7.5 × 3–6 mm, midrib excurrent into a c. 1 mm mucro. Rachilla spiralled, winged by persistent glume bases. Perianth 0. Stamens 3. Style 3–5 mm long, base pyramidal, shortly 3-winged, trifid. Nutlet 2–2.8 × 1.2–2 mm, obovoid to pyriform, white to pale fawn; surface faintly to prominently tuberculate, outermost cells subquadrate, ± isodiametric, eventually velate.

Zambia. C: Lusaka Dist., Mt Makulu Pasture Res. Station, 14.5 km S of Lusaka, 1250 m, 15.iv.1956, *Robinson* 1490 (K). S: Choma Dist., Mapanza, 1067 m, 26.i.1958, *Robinson* 2755 (K). **Zimbabwe**. C: Harare Dist., Harare (Salisbury), 10.i.1945, *Weber* 101 (K). E: Chimanimani Dist., Chimanimani town, fl. 12.iii.1931, *Otterson* in SRGH 4141 (SRGH). S: Chiredzi Dist., close to Masvingo (Fort Victoria)–Birchenough Bridge road, by Zaka (Ndanga) turn off, 19.i.1960, *Goodier* 837 (K, SRGH). **Mozambique**. M: Namaacha Dist., Namaacha, 9.xii.1984, *Myre & Carvalho* 303 (LMA, NU).

In Uganda, Kenya, Tanzania, Swaziland, Lesotho, South Africa (Eastern Cape, Free State, Gauteng, KwaZulu-Natal, Limpopo, Mpumalanga, North West), Swaziland and Madagascar; also in Mauritius, India, Malaysia, S Asia, Australia, Polynesia, temperate and tropical America. A ubiquitous associate of grasses, flourishing particularly where grass cover is sparse; in sand, shale or clay, black dambo soils and in soil pockets among rocks; 10–1800 m.

Conservation notes: Widespread; not threatened.

It is not clear if the type sheet was collected by König. Gordon-Gray (1995) suggests that the type was from Java, and Verdcourt in F.T.E.A. states that the type is from "Java, collector not stated (G lectotype) (seen by Kern)".

With fruit maturation culms elongate and occasionally reach ± 100 cm tall; plants are inconspicuous in the grass cover.

3. **Abildgaardia triflora** (L.) Abeyw. in Ceylon J. Sci., Biol. Sci. **2**: 82 (1959). —Haines & Lye, Sedges & Rushes E. Afr.: 95, fig.159 (1983). —Goetghebeur & Coudijzer in Bull. Jard. Bot. Belg. **54**: 86 (1984). —Gordon-Gray in Strelitzia **2**: 20, fig.2C,E (1995). —Verdcourt in F.T.E.A., Cyperaceae: 115, fig.17 (2010). Type: India, *König* s.n. in Herb. Linn. 70.35 (LINN lectotype). FIGURE 14.**20**.

Cyperus triflorus L., Mant. Pl.: 180 (1771).
Abildgaardia tristachya Vahl, Enum. Pl. **2**: 297 (1805).
Fimbristylis tristachya (Vahl) Thwaites, Enum. Pl. Zeyl.: 434 (1864), illegitimate name. —Clarke in F.T.A. **8**: 424 (1902), non R. Br.
Fimbristylis triflora (L.) K. Schum. in Abh. Königl. Akad. Wiss. Berlin **1894**: 14 (1894). —Engler, Pflanzenw. Ost-Afrikas **C**: 124 (1895). —Napper in F.W.T.A. **3**: 324 (1972).

Perennial herb to 100 cm tall, usually 40–70 cm; rhizome woody, tufted. Leaf bases forming well-marked, persistent tunics, dark brown to black; sheath mouth truncate, occasionally sloping, oblique; blade to 400 × 1–2 mm, linear, margins scabrid. Inflorescence anthelate of one central sessile and 1–4 pedicelled spikelets (numbers variable on individual plants); bracts 2–4, to 20 mm long. Spikelets terminating pedicels to 30 mm long, 11–22 × 4–7 mm, to 32 mm with age. Glumes greenish yellow, fertile, 5.8–9.3 × 3.2–6.1 mm, strongly keeled. Perianth 0. Stamens 3. Style 4.5–5.5 mm long, base pyramidal, becoming dark brown with age, deciduous with style, faintly to prominently 3-angled. Nutlet 2.3–3.1 × 1.3–1.9 mm, obovoid to markedly pyriform, fawn to dark brown, eventually black, shining surface etuberculate to prominently tuberculate, outermost cells subquadrate, eventually velate.

Botswana. N: Chobe Dist., on Movombi–Tsimanemeha Pan track, 18°06.4'S 23°07.7'E, 27.i.1978, *P.A. Smith* 2266 (K, SRGH). **Zambia**. B: Mongu Dist., Mongu, 1.xii.1962, *Robinson* 5514 (K). S: Choma Dist., 47 km N of Choma, 1128 m, 5.i.1957, *Robinson* 2023 (K, NU, NRGH, SRGH). **Zimbabwe**. N: Binga Dist., Binga, Chikwata Hot Springs, 7.xi.1958, *Phipps* 1393 (K). **Mozambique**. N: Pemba Dist., 41 km from Pemba (Porto Amélia) to Ancuabe, 100 m, 21.xii.1963, *Torre & Paiva* 9634 (BR, LISC, LMA, LMU, MO, WAG). Z: Mopeia Dist., 3 km NE of Mopeia (Mopeia Velha), 50 m, 7.xii.1971, *Pope & Müller* 532 (K, LISC, LMA, SRGH). MS: Gorongosa Dist., Gorongosa Nat. Park (Game Reserve), W slopes of Cheringoma Plateau, 200 m, 12.xi.1971, *Ward* 7435 (K, NU). GI: Inhassoro Dist., Bazaruto Is., W coast, Ponta Este marsh, 92 m, 28.x.1958, *Mogg* 28697 (K). M: Maputo (Lourenço Marques), Costa do Sol, edge of mangroves, 28.x.1963, *Balsinhas* 635 (K, LISC).

In Ghana, Togo, Kenya, Tanzania, Angola, Namibia, South Africa (KwaZulu-Natal, Mpumalanga) and Swaziland; also in India and Sri Lanka. Halophyte growing in salt marshes, swamps and dambos; 0–1200 m.

Conservation notes: Widespread; not threatened.

The species is readily recognised by the basally flattened spikelets arranged in a simple anthela.

Fig. 14.**20**. ABILDGAARDIA TRIFLORA. 1, habit; 2, two leaf bases; 3, leaf apex; 4, spikelet; 5, rachilla; 6, 7, glume, complete and upper abscised part; 8, young floret; 9, style and branches; 10, nutlet with filaments. All from *Robinson* 2038. Scale bars: 1 = 40 mm; 2, 3, 8–10 = 2 mm; 4–7 = 5 mm. Drawn by Jane Browning.

10. **NEMUM** Desv.[10]

Nemum Desv. in Hamilton, Prodr. Pl. Ind. Occid.: 13 (1825). —Larridon, Reynders & Goetghebeur in Belg. J. Bot. **141**: 157–177 (2008).

Delicate annual herbs of temporary wet areas, or perennials of more permanent wetland grassland, tufted or rhizomatous. Leaves often short in relation to culm length, blade ± enrolled, sometimes ending in a sharp point, spiral or distichous, ligulate (most perennials) or eligulate (all annuals); long woolly hairs are present at orifice of leaf sheaths in most species. Culms and leaves glabrous, with narrow, longitudinal markings formed by silica bodies. Inflorescence terminal or ± pseudolateral, (compound) anthelate, or capitate, or reduced to 1 or a few spikelets subtended by 1 or more primary bracts. Bracts often stiffly erect, not sheathing. Spikelets (cylindrical-) ovoid or spherical. Glumes spiral, densely imbricate, reddish-brown to blackish, each subtending a bisexual flower. Bristles absent. Stamens 1–3. Style branches 2 or 3, style base not distinct, not thickened and mostly deciduous. Nutlet obovate, dorsiventrally lenticular or faintly trigonous (*N. equitans*), surface smooth, often shining black, rarely brownish to greyish.

A genus of 8 species from grasslands in tropical Sudano-Zambesian Africa, 4 species in the Flora area.

Roalson *et al.* in Taxon **67**: 642 (2018) and in Phytotaxa **395**: 199–208 (2019) have incorporated the species of *Nemum* into *Bulbostylis*, but it is kept separate here.

Species are easily recognisable due to their distinctive, reddish brown to blackish, (cylindrical-) ovoid or spherical spikelets with persistent glumes. Living plants have not yet been studied properly in their natural habitat.

1. Plant annual or perennial, with limited root system; leaves mostly spirally arranged, sometimes distichous; long hairs at orifice of leaf sheaths, if present, clearly marginal 2
– Plant clearly perennial, caespitose and/or rhizomatous; leaves distichous; long hairs at orifice of leaf sheaths distinctly submarginal 3
2. Plant annual or a short-lived perennial; leaves spirally arranged; spikelets ovoid to spherical; inflorescence anthelate; anthela with (1)2–6(11) spikelets; glumes dark reddish brown to black, 1.5–3.5 × 1–2 mm, often clearly ciliate, obtuse to somewhat acute . . . . **1**. *angolense*
– Plant annual; leaves distichous; spikelets ovoid, relatively large and thick; glumes mucronate, fimbriate and conspicuously spatulate. Inflorescence anthelate, often with only 1–2 spikelets; glumes almost black, 1–3.5 × 0.5–2.5 mm, mucronate to 1.2 mm . . . . **2**. *atracuminatum*
3. Plant with rhizome 1–2 mm thick covered in silver-grey, tightly imbricate scales; glumes 2.5–7 × 1–3 mm, ± elliptic-ovate, mucronate to 1.6 mm and only slightly ciliate; style trifid; nutlets trigonous . . . . **3**. *equitans*
– Plants with long, seriate rhizome; glumes 2.5–2.75 × 2–2.5 mm, ± ovate, obtuse; glumes and bracts strongly hirsute on margins; style bifid; nutlets lenticular. . . . . **4**. *raynalii*

1. **Nemum angolense** (C.B. Clarke) Larridon & Goetgh. in Belg. J. Bot. **141**: 158 (2009). —Beentje in F.T.E.A., Cyperaceae: 116, fig.18 (2010). Types: Angola, no locality, *Welwitsch* 6836 (BM syntype, LISU, MPU); Angola, Pungo Andongo, iv.1857, *Welwitsch* 7166 (BM syntype, COI, LISU). FIGURE 14.**21**.

   Scirpus angolensis C.B. Clarke in Durand & Schinz, Consp. Fl. Afr. **5**: 617 (1894).
   Scirpus spadiceus (Lam.) Boeckeler var. *ciliatus* Ridl. in Trans. Linn. Soc. London, Bot. **2**: 156 (1884), illegitimate name, non *Scirpus ciliatus* Rottb.

---

[10] by I. Larridon

Fig. 14.**21**. NEMUM ANGOLENSE. 1, habit (× ⅔); 2, inflorescence (× 5); 3, glume (× 10); 4, floret (× 14); 5, nutlet (× 24). All from *Milne-Redhead* & *Taylor* 8297. Drawn by Juliet Williamson. Reproduced from Flora of Tropical East Africa (2010).

Scirpus ustulatus Podlech in Mitt. Bot. Staatssamml. München **4**: 118 (1961), superfluous name. Type: Zambia, Kawambwa, 21.vi.1957, *Robinson* 2323 (K lectotype, GENT, K, MT, P, SRGH), lectotypified by Larridon *et al.* (2008).

Nemum spadiceum sensu Lye in Bot. Not. **126**: 328 (1973), non (Lam.) Ham. —Haines & Lye, Sedges & Rushes E. Afr. 130 (1983).

Small tufted annual or densely tufted perennial with many vegetative shoots, culms to 65 cm high; root system limited. Leaves to 25 cm long (often remarkably short in comparison with culms) by 0.3–1 mm, ± involute and eligulate; sheath orifice with a bundle of long hairs. Inflorescence anthelate; anthela with (1)2–6(11) spikelets. Spikelets (cylindrical-) ovoid to spherical, with many dark reddish brown to black, spirally-arranged glumes. Glumes 1.5–3.5 × 1–2 mm, ± obovate, obtuse or somewhat acute, often ciliate on margin. Stamens usually 3. Style branches 2. Nutlet lenticular, obovate to elliptical, 0.85–1 × 0.65–0.85 mm, conspicuously smooth, shiny black, often with pale mark in centre of each face.

Zambia. N: Kawambwa Dist., no locality, 21.vi.1957, *Robinson* 2323 (GENT, K, MT, P, SRGH). W: Mwinilunga Dist., Kalenda, 8.x.1937, *Milne-Redhead* 2657 (BR, K). C: Kapiri Mposhi Dist., Mpunde Mission, 53 km NW from Kabwe, Kelongwe R., 14°06'S 28°06'E, 1130 m, 20.i.1973, *Kornaś* 3030 (K). **Mozambique**. N: Marrupa Dist., Okoewangoe, 16 km along road from Marrupa to Nungo, 6.viii.1981, *Jansen, de Koning & de Wilde* 97 (K).

Also in Nigeria, Cameroon, Central African Republic, Congo, Angola, Uganda and Tanzania. In seasonally wet areas (marshes, riverbanks, depressions, wet grasslands), woodlands and open forest, on iron-rich or sandy-peaty soils, laterite or skeletal soils; 600–1500 m.

Conservation notes: Widespread; not threatened.

Specimens of *Nemum spadiceum* (Lam.) Ham. are often confused with those now included within *N. angolense*. Various published descriptions imply both species or involve only part of the variability within one of the taxa. Nonetheless, both species are easily distinguished – *N. spadiceum* always has a single cylindrical-ovoid spikelet with (pale) reddish brown, acute or obtuse, scarcely ciliate glumes, while *N. angolense* has an anthelate inflorescence with (cylindrical-)ovoid to round spikelets with dark reddish brown to black, obtuse, often ciliate glumes.

Nemum angolense is the most variable species within the genus, being exceptionally variable in its habit (annual vs. perennial), leaf width, spikelet and glume shape, nutlet width, and in the presence of a pale mark on the nutlets.

2. **Nemum atracuminatum** Larridon, Reynders & Goetgh. in Belg. J. Bot. **141**: 162 (2009). Type: Zambia, Mporokoso Dist., Kasanshi dambo, 55 km ESE of Mporokoso, 13.vi.1962, *Robinson* 5166 (K holotype, MO).

Annual herb growing in small tufts; culms to 50 cm; root system limited. Leaves to 40 cm × c. 0.5 mm, basal, distichous, ± V-shaped to involute, eligulate; sheath orifice lacking typical bundle of hairs. Inflorescence anthelate, with 1–2(5) spikelets. Spikelets ovoid, relatively large and thick, with many nearly black, spirally arranged glumes. Glumes 1–3.5 × 0.5–2.5 mm, spathulate, shortly mucronate to 1.2 mm, distinctly fimbriate at tip. Stamens usually 3. Style branches 2. Nutlet lenticular, obovate, 1 × 0.6 mm, conspicuously smooth, shiny black, often apiculate.

Zambia. W: Mwinilunga Dist., Matonchi, Kalenda Plain, 16.iv.1960, *Robinson* 3600 (K, MT). N: Kawambwa Dist., Chishinga Ranch, near Luwingu, dambo near river, 1580 m, 17.i.1961, *Astle* 648 (BRVU, K).

Also in the southern D.R. Congo (Katanga). In temporarily wet areas (marshes, riverbanks), on shallow soils from humid woodland to edge of open forest; 1400–1700 m.

Conservation notes: Widespread; not threatened.

This species might be confused with some smaller annual specimens of *Nemum angolense*, which share an anthelate inflorescence type with nearly black spikelets; both species occur in D.R. Congo and Zambia. There are several distinguishing characters – the spikelets of *N. atracuminatum* are generally thicker and the glumes show remarkable differences, while the glumes of *N. angolense* are obovate, obtuse or somewhat acute, and often ciliate on the margin. Those of *N. atracuminatum* are conspicuously spatulate, mucronate and distinctly fimbriate at the tips.

One collection (*Mutimushi* 3296) comprises several dwarf-like plants up to 10 cm high with a single pseudo-lateral spikelet. These have black, fimbriate, mucronate and spatulate glumes, which clearly identify them as *Nemum atracuminatum*.

3. **Nemum equitans** (Kük.) J. Raynal in Adansonia, n.s. **13**(2): 150 (1973). Type: D.R. Congo, Katanga, Bulelo, n.d., *Fries* 531 (B holotype). FIGURE 14.**22**.

> *Scirpus equitans* Kük. in R.E. Fries, Wiss. Ergebn. Schwed. Rhodesia-Kongo-Exped. 1911–1912, Erg: 7 (1921).
> *Bulbostylis equitans* (Kük.) Raymond in Mém. Jard. Bot. Montréal **55**: 38 (1962).

Perennial herb, culms to c. 100 cm × 1 mm; with thin stolon-like rhizomes 1–2 mm thick, covered with silver-grey, tightly imbricate scales. Leaves 2–11 per culm, to 70 cm × c. 0.6 mm, basal, distichous, ligulate, strongly rolled, ending in a sharp ± ensiform tip; sheath orifice typically with a bundle of hairs. Inflorescence compound, anthelate, with up to 20 spikelets. Spikelets ovate, with many pale reddish brown, spirally arranged glumes. Glumes 2.5–7 × 1–3 mm, ± elliptic-ovate, mucronate to 1.6 mm, slightly ciliate. Stamens usually 3. Style branches 3. Nutlet 1.4 × 0.8 mm, dark brown or greyish, trigonous, part of style base often persistent.

Zambia. W: Mwinilunga Dist., Chingabola (Sinkabola) dambo, near Matonchi, 1350 m, 16.ii.1975, *Hooper & Townsend* 97 (K). N: Kasama Dist., Mungwi, 8.x.1960, *Robinson* 3912 (LISC, MO, SRGH). C: Serenje Dist., Kundalila Falls, 13.x.1963, *Robinson* 5712 (BR, SRGH).

Also in Angola and D.R. Congo. Permanently wet places such as swampy grasslands, riverbanks, roadsides and peat marshes; 1100–1600 m.

Conservation notes: Widespread; not threatened.

4. **Nemum raynalii** Larridon & Goetgh. in Belg. J. Bot. **141**: 168 (2009). Type: Zambia, Mwinilunga, N of Matonchi Farm, 23.i.1938, *Milne-Redhead* 4290 (K holotype).

Perennial herb, culms c. 65 cm × 2.5 mm thick; rhizomes forming a series, 3–4 mm thick, covered with relatively large, slightly overlapping brown scales. Leaves 5–6 leaves per culm, to 40 cm × c. 1 mm, basal, distichous, involute, ligulate, ending in a sharp ± ensiform tip; sheath orifice with a bundle of long pale hairs. Inflorescence compound, anthelate, with 13–22 spikelets. Spikelets elliptic-ovoid with many reddish brown, spirally-arranged glumes. Glumes 2.5–2.75 × 2–2.5 mm, ± ovate, obtuse; glumes and bracts strongly hirsute on margins. Stamens usually 3. Style branches 2. Nutlet unknown.

Zambia. W: Mwinilunga Dist., Kalenda Plain, N of Matonchi Farm, 23.i.1938, *Milne-Redhead* 4290 (K).

Not known elsewhere. In boggy ground over exposed laterite; c. 1300 m.

Conservation notes: Known only from the type; Data Deficient.

This species is similar to *Nemum equitans* in having culms and leaves with silica bodies only vaguely perceptible with a stereo-microscope. In all other species of *Nemum*, silica bodies form clear linear markings on the culms and leaves visible at ×10 magnification. The ovoid spikelets with their dark reddish brown colour and obtuse glumes strongly resemble those of *N. bulbostyloides* (S.S. Hooper) J. Raynal.

Nemum raynalii can be distinguished from all other rhizomatous perennial *Nemum* species by its more developed rhizomes and its generally larger dimensions, especially the relative thickness of the culms.

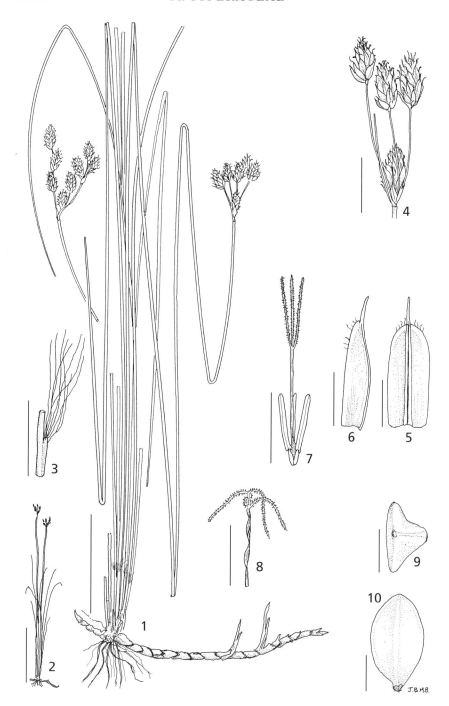

Fig. 14.**22**. NEMUM EQUITANS. 1, 2, habit; 3, leaf sheath; 4, inflorescence; 5, 6, glume, abaxial and lateral view; 7, floret; 8, style, twisted; 9, 10, nutlet, apical and adaxial view. 1–3 from *Robinson* 4065; 4 from *Robinson* 6027; 5, 6 from *Greenway* 5576; 7–10 from *Callens* 3165. Scale bars: 1 = 40 mm; 2 = 250 mm; 3, 4 = 10 mm; 5–8 =2 mm; 9, 10 = 0.5 mm. Drawn by Jane Browning.

11. **FICINIA** Schrad.[11]

Ficinia Schrad., Commentat. Soc. Regiae Sci. Gott. Recent. **7**: 143 (1832).

Perennial herbs, sparsely tufted, erect or rhizomatous, decumbent to creeping, mostly found at higher altitudes. Culms scapose. Leaves basal, grass-like, ligule variable, elongate, soon fragmenting into yellowish persistent wisps (*F. filiformis*) or cylindric, brown-streaked (*F. gracilis*), slanted or absent. Inflorescences terminal, capitate, occasionally branched, of medium-sized or extremely small spikelets. Primary bracts 3, leaflike, spreading. Glumes spiral, most subtending each a bisexual floret. Perianth 0. Stamens 3. Style 3-branched, fimbriate. Nutlet ovate to obovate in outline, trigonous, surface smooth, shining when young, becoming transversely lineolate, puncticulate. Gynophore rudimentary or cuplike, thickened, enveloping nutlet base to ⅓ nutlet length.

A genus of c. 78 species, mostly in western and southern Africa a few in eastern Africa; rare in Madagascar. Two species in the Flora area.

Inflorescence a hemispherical head, 1–20 spikelets . . . . **1.** *gracilis*
– Inflorescence not a hemispherical head, 1–8 spikelets . . . . **2.** *stolonifera*

1. **Ficinia gracilis** Schrad., Anal. Fl. Cap. **1**: 44 (1832). —Haines & Lye, Sedges & Rushes E. Afr.: 133 (1983). —Gordon-Gray in Strelitzia **2**: 85 (1995); in S. African J. Bot. **74**: 167 (2008). —Muasya in F.T.E.A., Cyperaceae: 118 (20 10). Type: South Africa, Cape Province, 1828, *Zeyher* 32 (P holotype, G).

   Isolepis commutata Nees in Linnaea **10**: 161 (1835). Type: Uncertain, possibly *Ecklon* (S?).
   Ficinia commutata (Nees) Kunth, Enum. Pl. **2**: 256 (1837).
   Melancranis commutata (Nees) Kuntze, Revis. Gen. Pl. **2**: 756 (1891).

Perennial tufted herb; rhizome abbreviated. Culms nodeless, numerous, 20–50 cm × 0.9–1.1 mm. Leaves basal, numerous; sheaths tubular, brown, ligule sloping away from lamina or appearing truncate; blade 10–30 cm × 0.6–1 mm, wiry, scabrid, channelled, shorter than culm. Involucral bracts usually horizontal, 2–3, leaf-like, main bract 2–7 cm long. Inflorescence terminal, lateral when young, head hemispherical, dark to light brown, 10–15 mm wide. Spikelets 1–20, closely packed, sometimes internally branched, 3–9 × 2–3 mm. Glumes spiral, 2.5–3 mm long, 0.8 mm wide each side of midrib, boat-shaped, elliptic in profile, dark to light brown, flanks with prominent pallid veins, keel thick, brown or pallid, running into a straight or curved apiculum (noticeable in young glumes). Stamens 3, anther crest pyramidal with microscopic projections. Style 3-branched. Nutlet 1.3–1.5 × 0.8–1.1 mm (including gynophore), trigonous, ovate to obovate in outline, apex small, projecting, dark brown overlaid greyish-white (layer peels away), surface appearing smooth. Gynophore variable, from an obpyramidal disc to a thin basal yellow cup, indistinctly 3-lobed, soon drying to leave microscopic (× 40) black lobes.

Zimbabwe. E: Chimanimani Dist., Chimanimani, Mt Peza, 16.x.1950, *Wild* 3620 (K, PRE, SRGH). **Mozambique**. MS: Sussundenga Dist., Chimanimani Mts, between Skeleton Pass and plateau, 1580 m, 27.ix.1966, *Simon* 874 (K).

Also in Kenya, Uganda and Tanzania, South Africa (Eastern Cape, Free State, KwaZulu-Natal, Limpopo, Mpumalanga), Lesotho and Swaziland. Widely scattered to clustered in montane grassland, particularly on quartzite sands; 1500–2100 m.

Conservation notes: Widespread; not threatened.

Some collections of *Ficinia gracilis* were previously erroneously recorded in the Flora area as *F. trollii* (Kük.) Muasya & D.A. Simpson.

Gordon-Gray (1995: 85) states that *Scirpus gracilis* Poir. (type: *Thouars* 17) is a separate taxon and should not be confused with *Fuirena gracilis*, as was done in F.T.E.A. (2010).

---

[11] by K.D. Gordon-Gray† and J. Browning

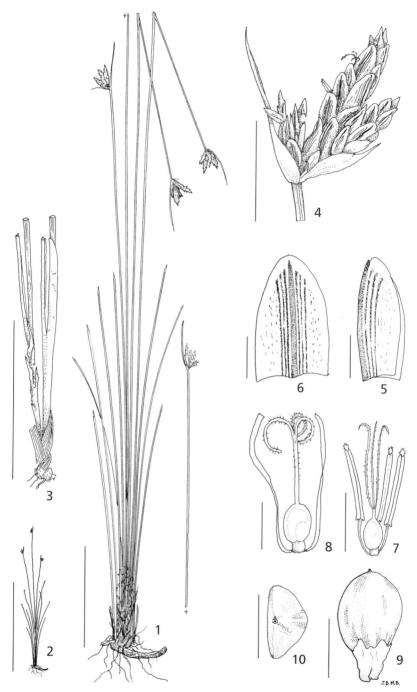

Fig. 14.**23**. FICINIA STOLONIFERA. 1, 2, habit; 3, two culms, leaf blades and ligules, intact and frayed; 4, inflorescence; 5, 6, glume, lateral and abaxial view; 7, 8, floret; 9, 10, nutlet adaxial and apical view. All from *Robinson* 6338. Scale bars: 1 = 40 mm; 2 = 250 mm; 3 = 25 mm; 4 = 5 mm; 5–10 = 1 mm. Drawn by Jane Browning.

2. **Ficinia stolonifera** Boeckeler in Linnaea **37**: 60 (1871). —Clarke in Durand & Schinz, Consp. Fl. Afric. 5: 643 (1894); in Fl. Cap. 7: 243 (1897). —Schönland in Mem. Bot. Surv. South Africa **3**: 45 **(1922)**. —Raynal in Adansonia, n.s. **14**: 211 (1974). —Gordon-Gray in Strelitzia **2**: 87 (1995). —Mapaura & Timberlake, Checkl. Zimbabwe Pl.: 88 (2004). Type: South Africa, no precise locality, *Ecklon & Zeyher* 141 (not located). FIGURE 14.**23**.

> *Ficinia filiformis* (Lam.) Schrad. f. *contorta* Nees in Linnaea **10**: 173 (1835).
> *Scirpus leucocoleus* K. Schum. in Engler, Pflanzenw. Ost-Afrikas **C**: 125 (1895). Type: Tanzania, Kilimanjaro (Kifinka Volkan), 1.ii.1894, *Volkens* 1858 (B holotype, BR, BM, G, K).
> *Ficinia filiformis* auct., non (Lam.) Schrad. —Clarke in F.T.A. **8**: 460 (1902). —Napper in J. E. Africa Nat. Hist. Soc. Natl. Mus. **26**: 1 (1965).
> *Ficinia contorta* (Nees) H. Pfeiff., Revis. *Ficinia*: 19 (1921).

Perennial tufted herb, 20–36 cm tall; stolons long, 2–2.5 mm wide. Culms 0.4–0.6 mm wide, exceeding leaves. Leaves numerous, basal; sheaths tubular, delicate, pale brown, 30 mm up from base, height free from leaf 15 mm, breaking readily into fragments; blade 0.2–0.5 mm wide, wiry, scabrid, shorter than culms, channelled. Involucral bracts 1–3, main bract 10–35 mm long, leaf-like. Inflorescence a head of 1–8 (sometimes only 1–2) radiating, sessile, easily counted spikelets. Spikelets 5–9(11) × 2–3 mm. Glumes 2–3 mm long, 0.8–1 mm wide each side of mid-nerve that terminates prior to obtuse apex, light to dark brown, flanks with several prominent veins, margin membranous. Stamens 3; anther crested with minute spikes. Ovary trigonous; style branches 3. Nutlet including gynophore 1.4–1.8 × 1–1.2 mm, elliptic to oblong, apex small and projecting, dark brown overlaying greyish white (layer peels away), surface appearing smooth. Gynophore c. ¹/₃ of nutlet length with 3 well-developed lobes.

Zimbabwe. E: Nyanga Dist., Mt Nyangani, lower ranges, 2100 m, 2.i.1991, *Laegaard* 16042 (K); Chimanimani Dist., Chimanimani Mts, 2000 m, 31.xii.1964, *Robinson* 6338 (K, SRGH). **Malawi**. S: Mulanje Dist., Mt Mulanje, Lichenya Plateau, 1850 m, 6.vi.1962, *Robinson* 5267 (K). **Mozambique**. MS: Gorongosa Dist., Mt Gorongosa, summit area, i.1972, *Tinley* 2320 (K).

Also in Tanzania, South Africa (Eastern & Western Cape, Free State, Gauteng, KwaZulu-Natal, Limpopo, Mpumalanga), Lesotho and Swaziland. Dry upland grassland, sometimes with *F. gracilis*; 1500–2150 m.

Conservation notes: Widespread; not threatened.

There are a large number of synonyms, only a few of which have been listed here.

12. **ISOLEPIS** R. Br.[12]

Isolepis R. Br., Prodr. Fl. Nov. Holland.: 221 (1810). —Muasya & Simpson in Kew Bull. **57**: 257–362 (2002).

Annuals or short-lived mat-forming or tufted perennials. Culms scapose or few noded (*I. fluitans*). Leaves eligulate, blade often reduced to a minute lobe. Inflorescences appearing pseudolateral, with 1 to several spikelets, subtended by lowermost leaf-like involucral bract, often erect, shorter or longer than spikelets. Spikelets sometimes proliferating, with few to many spiral, mostly deciduous glumes, apart from sometimes the lowest, each subtending a bisexual floret. Perianth segments absent. Stamens 1–3. Style 2–3 branched. Nutlet, elliptic to obovoid, trigonous often with 3 defined ribs, beaked, smooth, or variously ornamented.

A genus of about 60 species, mainly in the southern hemisphere.

1. Inflorescence a solitary, terminal, pedunculate spikelet, with similar inflorescences from axillary shoots so spikelets appear numerous. . . . . **1**. *fluitans*
– Inflorescence consisting of two or more spikelets closely aggregated forming a head that may or may not be overtopped by lowest involucral bract. 2

---

12 by J. Browning and K.D. Gordon-Gray†

2. Surface of mature nutlets trabeculate (longitudinally ribbed with faint transverse bars linking striations). . . . . **2.** *costata*
– Surface of mature nutlets not trabeculate, smooth, minutely reticulate, papillate or punctate. 3
3. Leaves reduced to sheaths, leaf blade a short outgrowth no longer than 5–10 mm . . . . **3.** *prolifera*
– Leaves (at least the majority) with leaf blades longer than 5–10 mm 4
4. Nutlets obovoid or rounded, slightly longer than wide . . . . **4.** *cernua*
– Nutlets elliptic, almost twice as long as wide 5
5. Involucral bract noticeably exceeding inflorescence; glumes 0.5–1 mm long . . . . **5.** *sepulcralis*
– Involucral bract shorter than or as long as inflorescence; glumes 1.3–2 mm long . . . . **6.** *natans*

1. **Isolepis fluitans** (L.) R. Br., Prod. Fl. Nov. Holland.: 221 (1810). —Haines & Lye, Sedges & Rushes E. Afr.: 138 (1983). —Gordon-Gray in Strelitzia **2**: 108 (1995). —Muasya in F.T.E.A., Cyperaceae: 121, fig.20 (2010). Type: Europe (England?), n.d., *Morison* 1699 (OXF lectotype), lectotypified by Simpson *et al.* in Kew Bull. **56**: 1011 (2001). FIGURES 14.**24**, 14.**25A**.

Scirpus fluitans L., Sp. Pl.: 48 (1753). —Clarke in F.T.A. **8**: 449 (1902).
Eleogiton fluitans (L.) Link, Hort. Berol. **1**: 285 (1827).
Isolepis inyangensis Muasya & Goetgh. in Kew Bull. **57**: 282 (2002). Type: Zimbabwe, Nyanga Dist., Gairezi Ranch, 9.6 km N of Troutbeck, 1680 m, 14.xi.1956, *Robinson* 1889 (K holotype, B, BR, LISC, NRGH, PRE, SRGH).
Isolepis fluitans var. *ludwigii* (Boeckeler) Kük. sensu Mapaura & Timberlake, Checkl. Zimbabwe Pl.: 88 (2004), uncertain name.

Aquatic or semi-aquatic, mat-forming, or dryland short-lived perennial; fine roots from branching noded rounded stems 0.2–1 mm diameter. Leaf blade glabrous, often inrolled, eligulate, 1–5 cm long; lateral branches infertile or fertile. Culms 2–40 cm tall. Inflorescence a terminal solitary, elliptic, non-proliferating spikelet, with or without overtopping involucral bract. Spikelets 4–5 × 1–2 mm. Glumes 8–10, c. 2 × 1 mm, obtuse, midrib green, with or without mucro, side with 3–4 veins, pale cream with brown patch in upper third. Stamens 2 (rarely 3). Style bifid, 0.5 mm long, branches 1.5 mm long. Nutlet 1–1.2 × 0.6–1 mm, faintly trigonous, ovoid to obovoid, apex short conical; smooth to minutely reticulate, pale brown at maturity.

Zambia. W: Mwinilunga Dist., Dobeka R., S of Dobeka bridge, 17.xii.1937, *Milne-Redhead* 3704 (K). N: Kasama Dist., Mungwi, 20.vi.1969, *Robinson* 3752 (K). **Zimbabwe**. W: Matobo Dist., Besna Kobila Farm, 1460 m, v.1957, *Miller* 4381 (K). C: Harare Dist., Harare, 1460 m, 19.ii.27, *Eyles* 4737 (K, SRGH). E: Chimanimani Dist., Chimanimani Mts, Upper Bundi, 1600 m, 1.ii.1957, *Phipps* 279 (K, SRGH). **Malawi**. N: Nkhata Bay Dist., South Vipya plateau, by Luwawa Dam, 1560 m, 8.v.1970, *Brummitt* 10493 (K). S: Mt Mulange, above Ruo Gorge, 1950 m, 8.iv.1989, *Iversen & Martinsson* 89195 (K). **Mozambique**. MS: Gorongosa Dist., Gorongosa Mt, summit area, i.1972, *Tinley* 2286 (K, LISC).

Distributed widely across the Old World but absent from the Americas; in Uganda, Kenya, Tanzania, South Africa, Lesotho and Swaziland. Mainly in freshwater streams and along margins of large water bodies, streams in montane grasslands, or as a tufted upright terrestrial plant in frequently damp situations; 1000–2000 m.

Conservation notes: Widespread; not threatened.

Isolepis fluitans, reported to be non-proliferous, is an extremely variable species collected from all countries within the Flora area except Botswana. It is particularly common in the Eastern Highlands of Zimbabwe. When submerged or floating in freshwater pools the culms, each bearing a single spikelet, rise above the water surface.

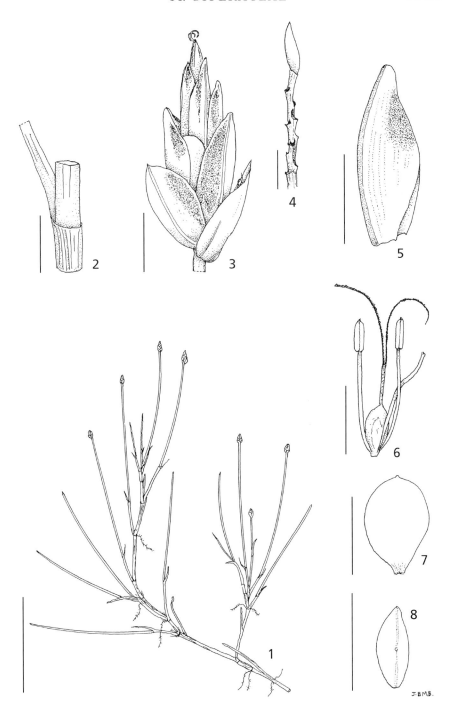

Fig. 14.**24**. ISOLEPIS FLUITANS. 1, habit; 2, leaf sheath; 3, spikelet; 4, rachilla, apical glume attached; 5, glume, lateral view; 6, floret; 7, 8, nutlet, abaxial and apical view. All from *Iversen & Martinsson* 89195. Scale bars: 1 = 40 mm; 2–8 = 1 mm. Drawn by Jane Browning.

The bright green plants frequently grow in damp soil at the margins of pools or in moss mats on rocky slopes receiving run-off, and appear different in form being tufted, with thicker bases and longer, darker spikelets with more glumes. These plants have been separated as *I. inyangensis* Muasya & Goetgh., but this is not recognised here.

2. **Isolepis costata** A. Rich., Tent. Fl. Abyss. **2**: 499 (1850). —Haines & Lye, Sedges & Rushes E. Afr.: 135 (1983). —Gordon-Gray in Strelitzia **2**: 107 (1995). —Muasya in F.T.E.A., Cyperaceae: 123 (2010). Types: Ethiopia, Ouodgerate, *Petit* s.n. (P syntype); no locality, i.1842, *Schimper* 1153 (P syntype, BM, BR, K, MO, UPS). FIGURE 14.**25B**.

> *Scirpus costatus* (A. Rich.) Boeckeler in Linnaea **36**: 511 (1870). —Clarke in F.T.A. **8**: 451 (1902).
> *Scirpus macer* Boeckeler in Bot. Jahrb. Syst. **5**: 503 (1884). Type: Madagascar, Ost Imerina, Andrangoloaka, xi.1880, *Hildebrandt* 3738 (B† holotype, BM, K, P).
> *Isolepis costata* var. *macra* (Boeckeler) B.L. Burtt in Notes Roy. Bot. Gard. Edinburgh **43**: 363 (1986).

Tufted perennial; rhizomes delicate, linking tufts. Leaf blade reduced to oblong lobe, 1–5 mm long; leaf sheath purple to brown, mouth truncate. Culms nodeless, slender, 0.3–1 mm diameter. Inflorescence a head that may proliferate, of 2–16(26) spikelets, appearing lateral, overtopped by curving involucral bract 7–8 mm long with apical gland. Spikelets sessile, ovate, 2.5–2.8 mm long. Glumes numerous, 1.2–1.3 × 0.8–0.9 mm, flanks veined, brown, with paler midrib. Stamens 3, anthers crested. Style 3(2). Nutlet 0.8–1 × 0.5–0.6 mm, trigonous, ovoid with conical apex, 6–8 conspicuous longitudinal ribs with fainter cross bars linking striations, light brown to cream.

Zimbabwe. E: Mutare Dist., Penhalonga, 1310 m, 30.x.1956, *Robinson* 1824 (K, LISC, SRGH). **Malawi**. N: Nyika Plateau, Chelunduo stream at Chelinda Camp, 2200 m, 26.x.1958, *Robson* 375 (K, LISC). C: Dedza Dist., Chongoni Forest Reserve, fl. 31.x.1968, *Salubeni* 1184 (K, LISC). S: Mt Mulanje, 1891, *Whyte* s.n. (K).

Also in Uganda, Kenya, Tanzania, Namibia, South Africa, Lesotho, Swaziland and Madagascar. By streams, in lakes and occassionally in damp woodland; frequently in mountainous areas; 1000–2200 m.

Conservation notes: Widespread; not threatened.

It is distinguished from other species of *Isolepis* in the Flora area by the nutlet with longitudinal ribs and linking cross bars, inflorescences usually with many spikelets in a somewhat rounded head, and in the filiform culms with dark bases.

3. **Isolepis prolifera** (Rottb.) R. Br., Prodr. Fl. Nov. Holland.: 223 (1810). —Gordon-Gray in Strelitzia **2**: 109, fig.43 (1995). Type: South Africa, Cape of Good Hope?, *Rottbøll* s.n. (C holotype, LINN?). FIGURE 14.**25C**.

> *Scirpus prolifer* Rottb., Descr. Icon. Rar. Pl.: 55 (1773). —Clarke in Fl. Cap. **7**: 226 (1897).
> *Cyperus punctatus* Lam., Tabl. Encycl. **1**: 144 (1791). Type: South Africa, *Sonnerat* s.n. (P-LAM holotype).

Tufted annual or short-lived perennial; rhizomes short and thin. Leaf blade reduced to remnants, seldom exceeding 5 mm, sheaths dark brown, mouths truncate or sloping. Culms 5–60 cm tall, soft, often decumbent. Inflorescence of compact heads, proliferating, appearing pseudolateral; lowest involucral bract c. 8 mm, shorter than or slightly overtopping spikelets. Spikelets numerous, 4–20, ovate to cylindric, c. 12 × 2 mm. Glumes numerous, 1–3 mm long, ovate, with rounded apex, midrib terminating in short mucro, flanks prominently veined, light brown to greenish. Stamens 3, crested. Styles 3. Nutlet triquetrous, ellipsoid, apiculate, 0.7–1.2 × 0.5–0.7 mm, surface smooth at low magnification, isodiametric cells visible at higher magnification, glistening yellow to dark brown.

Zimbabwe. C: Chegutu Dist., Chegutu, Aroudale Farm dam, fl. 25.ii.1969, *Mavi* 984 (SRGH).

Also in Australia, New Zealand, St Helena, Tristan da Cunha and South Africa (Eastern and Western Cape, KwaZulu-Natal). Grows gregariously in sandy moist situations at margins and amongst rocks in streams and seepage areas; 100–1000 m.

Conservation notes: Widespread; not threatened.

Plants of this species are robust, coarser than other species, most often proliferating and frequently decumbent over freshwater bodies. They can be recognised by the absence of leaves.

4. **Isolepis cernua** (Vahl) Roem. & Schult., Syst. Veg. **2**: 106 (1817). —Gordon-Gray in Strelitzia **2**: 107 (1995). —Muasya in F.T.E.A., Cyperaceae: 125 (2010). Type: Portugal (Lusitania), *Rathke* s.n. (C holotype). FIGURE 14.**25D**.

> *Scirpus cernuus* Vahl, Enum. Pl. **2**: 245 (1805). —Clarke in F.T.A. **8**: 45 (1902).
> *Eleogiton cernua* (Vahl) A. Dietr., Sp. Pl. **2**: 96 (1832).
> *Schoenoplectus cernuus* (Vahl) Hayek in Repert. Spec. Nov. Regni Veg. Beih. **30**(3): 153 (1932).

Tufted annual or short-lived perennial. Leaves slender, usually much shorter than culms; leaf sheaths brown or green. Culms 2–25 cm tall, slender, 0.1–0.6 mm diameter. Inflorescence generally not proliferating, appearing pseudolateral by lowest involucral bract, c. 5 mm, overtopping or shortly exceeding spikelets. Spikelets 1 (occasionally 3), ovate to cylindric, 2–8 mm long. Glumes numerous, 1–2 mm long, ovate, green midrib terminating shortly before rounded glume apex, flanks prominently 2–6 veined, straw coloured to dark brown. Stamens 3(2), crested. Style 3(2). Nutlet 0.8–1 × 0.4–0.8 mm, obovoid, trigonous, abaxial ridge slightly asymmetric in basal third (slightly longer than wide), surface smooth to minutely punctate, glistening yellow to dark brown.

Zimbabwe. E: Nyanga Dist., Nyanga, Nyangombe (Inyangambi) Falls, 1520 m, 3.xi.1950, *Chase* 3119 (BM, SRGH).

Predominantly in the northern hemisphere, it is suggested that the typical variety has been introduced into Kenya, Zimbabwe and across South Africa. Mud banks of streams and other water bodies, generally in mountainous or higher altitude situations; 1500–2000 m.

Conservation notes: Rarely collected; unlikely to be threatened.

There are many synonyms for this species; only three are given above. Muasya & Simpson (2002: 295) provide an exhaustive list.

Muasya & Simpson (2002: 295) also list five varieties including the typical, which is the only one recorded from the Flora area. It is possible that the Nyanga collection was an introduction as the area was a popular picnic site.

Plants of *Isolepis cernua*, *I. natans* and *I. sepulcralis* can be confused. Examination of nutlets is necessary to distinguish these three species – *I. cernua* (with a rounded nutlet), differs from *I. natans* and *I. sepulcralis* with elliptic nutlets that are longer than wide.

5. **Isolepis sepulcralis** Steud., Syn. Pl. Glumac. **2**: 94 (1855). —Haines & Lye, Sedges & Rushes E. Afr.: 140 (1983). —Gordon-Gray in Strelitzia **2**: 109 (1995). —Muasya in F.T.E.A., Cyperaceae: 125 (2010). Type: St Helena, near Napoleon's grave, n.d, *D'Urville* 69a (P holotype, BR). FIGURE 14.**25E**.

> *Fimbristylis exigua* Boeckeler in Bot. Jahrb. Syst. **5**: 506 (1884). Type: Madagascar, Ost-Imerina, Andrangoloaka, xi.1880, *Hildebrandt* 3739 (P holotype, K, P).
> *Scirpus griquensium* C.B. Clarke in Durand & Schinz, Consp. Fl. Afr. **5**: 623 (1894) name only; in Fl. Cap. **7**: 222 (1897) description. Type: South Africa, former Transkei (East Griqualand), streams near Clydesdale, xii.1884, *Tyson* 2861 (K holotype, PRE).
> *Scirpus chlorostachyus* auct., non Levyns.

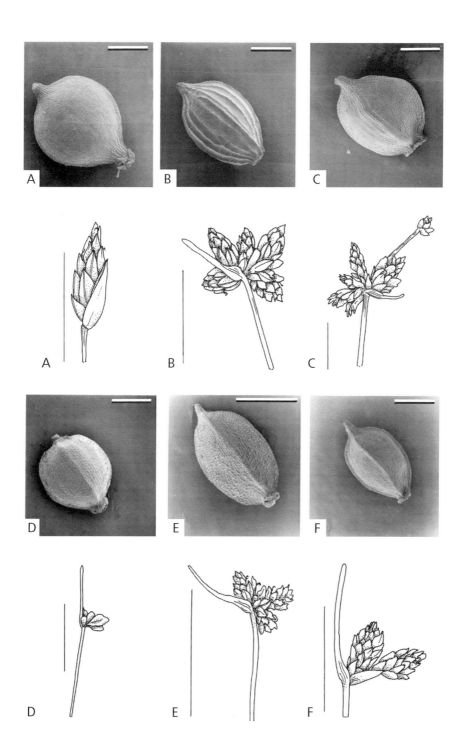

Small tufted annual; slender rhizomes linking tufts. Leaves much shorter than culms; sheaths brown. Culms 2–20 cm tall, slender, 0.1–0.5 mm diameter. Inflorescence appearing pseudolateral as overtopped 5–10 mm by lowest involucral bract. Spikelets ovate, 1–3, 1.5–3 × 1–2 mm long and wide. Glumes numerous, 0.5–1 mm long, midrib green, ending below acute apex, flanks with 3–4 lateral veins, reddish-brown. Stamens 1–2. Styles 3. Nutlets 0.5–0.8 × 0.2–0.4 mm, trigonous, elliptic to ovate elliptic, minutely apiculate, surface minutely papillate, dark brown to black.

Botswana. N: Ngamiland Dist., Chief's Is., beside airstrip, 19°18'26.7"S 22°54'16.38"E, 960 m, 28.viii.2009, *A. & R. Heath* 1672 (K, BNHBG).

Also found in Kenya, Tanzania, Angola and South Africa (Cape Provinces, Free State, KwaZulu-Natal, Limpopo, Mpumalanga); in Tristan da Cunha, Madagascar and introduced into Australia and New Zealand. On alluvium by streams and other waterbodies; 900–1100 m.

Conservation notes: Widespread outside the Flora area; not threatened.

Plants of this species may be confused with *Isolepis natans* and *I. cernua*, but are generally more slender. In common with *I. natans*, nutlets are elliptic and twice as long as wide.

The specific epithet refers to Napoleon's tomb, which is the type locality.

6. **Isolepis natans** (Thunb.) A. Dietr., Sp. Pl. **2**: 106 (1832). —Haines & Lye, Sedges & Rushes E. Afr.: 141 (1983). —Gordon-Gray in Strelitzia **2**: 108 (1995). —Muasya in F.T.E.A., Cyperaceae: 126 (2010). Type: South Africa, Cape of Good Hope (Cap. b. spei.), *Thunberg* 1633 (UPS holotype). FIGURE 14.**25F**.

Scirpus natans Thunb., Prod. Pl. Cap.: 17 (1794).
Isolepis rivularis Schrad., Anal. Fl. Cap. **1**: 19 (1832). Type: "Ex Orient" (Arabia?), *Forsskål* s.n. (BM holotype?).
Scirpus rivularis (Schrad.) Boeckeler in Linnaea **36**: 504 (1870). —Clarke in Fl. Cap. **7**: 220 (1897)

Tufted annual or short-lived perennial; rhizomes delicate, linking tufts. Leaves shorter than culms, bright green; sheaths brown. Culms nodeless, coarse, soft, spongy, 5–25 × c. 2 mm. Inflorescence may proliferate, appearing pseudolateral as ± overtopped by lowest involucral bract. Spikelets 1–5, sessile to shortly pedicelled, ovate, 2–5 × 1.5–2 mm. Glumes numerous, 1.3–2 mm long, midrib green, mucro short, flanks veined, light-brown. Stamens 2. Style 3. Nutlet trigonous, 0.8–0.9 × 0.4–0.5 mm, elliptic to ovate-elliptic, minutely apiculate, surface minutely papillate, brown.

Zimbabwe. E: Nyanga Dist., Gairesi Ranch on Mozambique border, 10 km N of Troutbeck, 1680 m, 14.xi.1956, *Robinson* 1888 (K, LISC, PRE, SRGH).

Also in Angola, Tanzania and South Africa (East and Western Cape, Mpumalanga, KwaZulu-Natal). Creeping or erect, growing amongst rocks or on mud banks of streams and waterbodies; 1000–1800 m.

Conservation notes: Infrequently collected, but probably not threatened.

Isolepis natans is recognised by its bright green colour, which changes almost to black on drying. It is slightly coarser than *I. sepulcralis*, but the mature nutlets of these two species are indistinguishable.

Fig. 14.**25**. ISOLEPIS nutlets and inflorescences. A. —ISOLEPIS FLUITANS, from *Iversen & Martinsson* 89195. B. —ISOLEPIS COSTATA, from *Sturgeon & Painter* 30670. C. —ISOLEPIS PROLIFERA, from *Ward* 10704. D. —ISOLEPIS CERNUA, from *Chase* 3119. E. —ISOLEPIS SEPULCRALIS, from *Heath* ARH1672. F. —ISOLEPIS NATANS, from *Robinson* 1888. Scale bars: SEM = 0.25 mm; inflorescences = 5 mm. Drawn by Jane Browning. SEM photographs reproduced from Strelitzia (1995) by kind permission of the South African National Biodiversity Institute, Pretoria.

13. **SCIRPOIDES** Ség.[13]

Scirpoides Ség., Pl. Veron. **3**: 73 (1754).
Holoschoenus Link., Hort. Berol. **1**: 293 (1827).

Perennial herbs, tufted or shortly rhizomatous. Culms scapose, often basally thickened. Leaves setaceous, frequently reduced to sheaths, ligulate or eligulate. Inflorescences pseudolateral, anthelate to capitate spikelets numerous; lowest primary bract overtopping inflorescence and continuing line of culm. Spikelets with spirally arranged glumes, each subtending a single bisexual floret; bristles generally absent. Stamens 2–3. Style branches 3, base undefined, persistent. Nutlets obovoid, trigonous, beaked, surface finely puncticulate.

A genus with 6 species worldwide and additions likely; two species in the Flora area. A segregate genus from *Scirpus* L., *Scirpoides* is perhaps somewhat paramorphic (based on inadequate data) while transfer of additional taxa proceeds.

Inflorescence a pseudolateral globose head, with or without other stalked heads; individual spikelets closely aggreggated appearing markedly echinate; glumes aristate . . . . **1.** *varia*
– Inflorescence a pseudolateral globose head, lacking other stalked heads; individual spikelets discrete, cylindrically oblong, not appearing echinate; glumes not aristate . . . .**2.** *dioeca*

1. **Scirpoides varia** Browning in S. African J. Bot. **77**: 507 (2011). Types: South Africa, Gauteng, Apies Poort, *Rehmann* 4036 (?); South Africa, Houtbosch, *Rehmann* 5635 (?); Lesotho (Basutoland), Leribé, n.d., *Buchanan* 225 (K000671266 lectotype), lectotypified by Browning & Gordon-Gray (2011).

 Scirpus varius C.B. Clarke in Fl. Cap. **7**: 229 (1898), invalid name. —Gordon-Gray in Strelitzia **2**: 177 (1995).

Perennial herb 80–100 cm tall, tufted or rhizomatous with woody rhizome. Culms 3-angled apically, 1.25–2 mm wide, smooth. Leaves basal, occasionally sparse, stiff, mostly shorter than or less often as long as culm length, 2–4 mm wide, glabrous. Inflorescence pseudolateral, overtopped by erect main primary bract c. 1.3 cm long continuing culm line; additional bracts 2, each reduced by half length of precursor to 3–5 small bracts invisible beyond limits of capitulum; all bracts scabrid. Flowering head solitary or in close groups of (2)3–4(5), 8–18 mm in diameter, sessile or shortly branched, each bracteate; collective heads 15–45 × 10–15(17) mm. Spikelets numerous per head, c. 4 mm long. Glumes spiral, occasionally appearing almost distichous, 2.25–2.75 mm long (including 0.6–0.85 mm arista), green keel and arista strongly developed, flanks delicate, brown, sometimes translucent with 2 prominent lateral nerves. Perianth 0. Stamens 3. Style branches 3, ± as long as style. Nutlet c. 0.75 × 0.4 mm, trigonous, pale brown, basal foot expanded, surface indistinctly patterned by faintly marked longitudinal striations, faintly puncticulate at maturity.

Botswana. SE: Southern Dist., Kanye, downstream of Mmakgoduma Dam, below Mosupa road crossing, 24°56.6'–56.9'S 25°20.5'–21.0'E, 1280 m, 5.v.1990, *P.A. Smith* 5376 (NU, SRGH).

Endemic to southern Africa with isolated records from South Africa (North West, Gauteng, KwaZulu-Natal, Limpopo, Mpumalanga), Lesotho, Swaziland and Botswana. Forming small scattered colonies in muddy habitats with semi-permanent water and sometimes rock-strewn soil; c. 1300 m.

Conservation notes: Limited distribution with only one known collection from the Flora area; possibly Vulnerable.

As *Scirpus varius*, sometimes mistakenly written as *Scirpus variabilis*, is an invalid name, Browning published the name *Scirpiodes varia* in 2011.

---

[13] by J. Browning and K.D. Gordon-Gray†

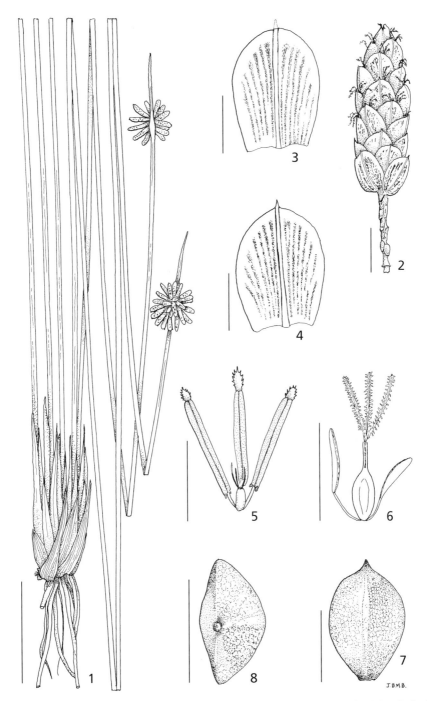

Fig. 14.**26**. SCIRPOIDES DIOECA. 1, habit; 2, female spikelet; 3, 4, glume, abaxial and adaxial view; 5, male floret; 6, female floret with two abortive stamens; 7, 8, nutlet, adaxial and apical view. 1–4, 6–8 from *de Winter & Leistner* 5799; 5 from *Ward* 10102. Scale bars: 1 = 40 mm; 2 = 2 mm; 3–6 = 1 mm; 7, 8 = 0.5 mm. Drawn by Jane Browning.

2. **Scirpoides dioeca** (Kunth) Browning in S. African J. Bot. **60**: 318 (1994). Type: South Africa, no locality, 1840?, *Drège* s.n. (K000416621 lectotype), lectotypified by Browning (1989). FIGURE 14.**26**.

> *Isolepis dioeca* Kunth, Enum. Pl. **2**: 199 (1837).
>
> *Scirpus dioecus* (Kunth) Boeckeler in Linnaea **36**: 719 (1870). —Browning in S. African J. Bot. **55**: 426 (1989).
>
> *Scirpus schinzii* Boeckeler in Verh. Bot. Vereins Prov. Brandenburg **29**: 47 (1888).

Perennial herb 25–150 cm tall; rhizome abbreviated, 2–5 mm in diameter, formed of contiguous stem bases clothed with persistent, erect sharply tapering scale-leaves. Culms scapose, somewhat terete, glabrous, 2–3 mm wide. Leaves absent, reduced to scale leaves. Inflorescence a pseudolateral head, 5–25 mm in diameter of 10–60 sessile spikelets, overtopped c. 30 mm by pointed bract appearing as continuation of culm. Spikelets imperfectly unisexual, male and female on different plants, somewhat cylindrical, oblong in outline at maturity. Glumes spiral, 2–2.3 × 1.2–1.5 mm, obovate, midrib well-defined, fawn to pale brown with darker markings, apex acute to mucronate, each subtending a floret. Male florets – stamens 3, anthers c. 1.3 mm long before anthesis, clearly crested, gynoecium rudimentary; female florets – stamens 3, poorly-developed, usually uncrested; gynoecium robust; style branches 3, strongly fimbriate. Perianth segments 0, gynophore 0 in both sexes. Nutlet 0.7–1.2 × 0.6–0.9 mm, obovoid with short stylar beak, greyish brown at maturity; surface topography reticulate at × 20 magnification, sometimes with faintly irregular transverse lineation.

Botswana. N: Ngamiland Dist., locality, 19°13.7S 21°10.25E, watercourse by Qangwa village, 29.iv.1980, *P.A. Smith* 3535 (K).

Endemic to southern Africa; also recorded from Namibia, South Africa (Northern Cape, Western Cape, Eastern Cape, Free State, Limpopo, Mpumalanga, North West) and Swaziland. In brackish water and salt-pan marshes, along streams and waterholes; c. 1100 m.

Conservation notes: Widespead outside the Flora area; not threatened.

14. **OXYCARYUM** Nees[14]

Oxycaryum Nees in Martius, Fl. Bras. **2**(1): 90 (1842).

Crepidocarpus Boeckeler in Linnaea **36**: 44 (1870). —Lye in Bot. Not. **124**: 280–286 (1971).

Rhizomatous leaf-bearing perennial. Culms triangular, smooth, nodeless. Leaves basal, often exceeding culm length, ligulate. Inflorescence anthelate to capitate, mostly of pedunculate heads each subtended by a primary foliose bract. Heads globose; secondary bracts glume-like, subtending close-packed spikelets, eprophyllate. Glumes spiral; florets bisexual. Perianth 0. Stamens 3. Style branches 2. Nutlet narrow, dorsiventrally plano-convex, beaked, base margins and top corky, surface smooth.

A monotypic genus, widespread, but not frequent in tropical to sub-tropical climates in eastern Africa. Also in Central and South America. Larridon in Pl. Ecol. Evol. **144**: 327–356 (2011) has incorporated the species of *Oxycaryum* into *Cyperus*, but it is kept separate here.

Oxycaryum cubense (Poepp. & Kunth) Palla in Denkschr. Kaiserl. Akad. Wiss., Wien. Math.-Naturwiss. Kl. **79**: 169 (1908) —Haines & Lye, Sedges & Rushes E. Afr.: 144, fig.282 (1983) as *O. cubensis*. —Gordon-Gray in Strelitzia **2**: 137, fig.57 (1995). —Beentje in F.T.E.A., Cyperaceae: 126, fig.21 (2010). Type: Cuba, 'in paludosis', *Poeppig* s.n. (P). FIGURE 14.**27**.

> *Scirpus cubensis* Poepp. & Kunth in Kunth, Enum. Pl. **2**: 172 (1837). —Clarke in F.T.A. **8**: 451 (1902).

---

[14] by J. Browning and K.D. Gordon-Gray†

Robust, hydrophytic perennial, floating or attached to mud by elongate, hairy roots from stem nodes; stolons 50–200 × 2–3 mm, with dark scales, apices often terminated by a plantlet. Leaves 40–90 cm × 4–10 mm, V-shaped, often purplish in lower section, mostly becoming elongate, midrib and margins minutely scabrid, lowest usually inflated; ligule a hairy rim. Inflorescence a sub-umbellate group of 3–10 radiating peduncled heads, or peduncles reduced; prophylls tubular with ciliate margin; heads hemispherical to globose, c. 15 mm diameter; foliose bracts

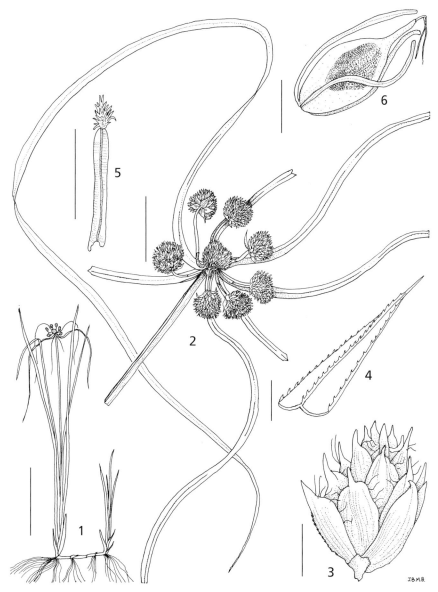

Fig. 14.**27**. OXYCARYUM CUBENSE. 1, habit; 2, inflorescence; 3, spikelet; 4, leaf apex; 5, anther; 6, nutlet with styles and filaments attached. 1, 2 from *Ward* 8044; 3–6 from *Ward* 8546. Scale bars: 1 = 250 mm; 2 = 25 mm; 3 = 2mm; 4 = 5 mm; 5, 6 = 1 mm. Drawn by Jane Browning. Reproduced from Strelitzia (1995) by kind permission of the South African National Biodiversity Institute, Pretoria.

radiating, conspicuous, secondary bracts subtending spikelets much reduced, scale-like, relatively inconspicuous. Spikelets numerous per head, close-packed, prophylls lacking. Glumes spiral, c. 3.5 mm long, stiff, apically thickened, acuminate, greenish or reddish-brown, margins long-ciliate. Florets 1 per glume, bisexual. Perianth absent. Stamens 3. Style bifid, base persistent not distinct. Nutlet to 3 mm long (beak c. 0.7 mm) × 0.6–0.8 mm, dorsiventrally flattened, texture corky, surface smooth, whitish at maturity, mainly water distributed.

Botswana. N: Ngamiland Dist., Okavango R. backwater, few km S of Sepopa, 6.v.1975, *Gibbs Russell* 2908 (K). **Zambia**. B: Mongu Dist., 20 km NW of Mongu, 10.x.1962, *Robinson* 5473 (K). E: Chipata Dist., South Luangwa Nat. Park (Game Reserve), Big Lagoon Camp, 600 m, 10.x.1960, *Richards* 13318 (K). N: Sanfya Dist., S Lake Bangweulu, swamps between Ncheta Is. and Chibambo Lagoon, 1062 m, 11.ii.1996, *Renvoize* 5584 (K). C: Kabwe Dist., Chilondo Lake, Lukanga Swamp, 19.viii.1958, *Seagrief* 3184 (K, NU). **Zimbabwe**. N: Hurungwe Dist., Mana Pools Nat. Park (Game Reserve), Tree Tops Pool, viii.1970, *Guy* 1039 (K, LISC, SRGH). **Malawi**. C: Salima Dist., Salima, shore of Lake Nyasa, 10.vii.1962, *Robinson* 5457 (K). S: Chikwawa Dist., Elephant Marsh, Shire R. near Alimenda, 60 m, 20.iv.1980, *Blackmore, Osborne & Dudley* 1303 (K).

Also widespread in tropical Africa (Kenya, Uganda and Tanzania), Namibia, South Africa (KwaZulu-Natal), and in C & S America. Floating in open water or attached to wet soil in permanent marshes or slow-flowing rivers; 60–1100 m.

Conservation notes: Widespread; not threatened.

Reported to be a very successful and important colonizer.

15. **KYLLINGIELLA** R.W. Haines & Lye[15]

Kyllingiella R.W. Haines & Lye in Bot. Not. **131**: 175–177 (1978).

Tufted small perennials, bulbous thickened bases clothed in old leaf sheaths. Leaves eligulate. Culms scapose. Inflorescence a compact head of several spikelets which may consist of smaller spikelets, with spirally arranged glumes each subtending a sessile bisexual floret. Perianth 0. Stamens 1(2). Style 3. Nutlet obovate in outline, compressed subtrigonous, with a short beak, minutely punctate.

A genus of 3 or 4 species, mostly in tropical and subtropical Africa.

Larridon in Pl. Ecol. Evol. **144**: 327–356 (2011) has incorporated the species of *Kyllingiella* into *Cyperus*, but it is kept separate here. Morphologically *Kyllingiella* is close to *Isolepis* but differs in its inflorescence and glume structure.

1. Inflorescence greenish; heads squarrose with tips of glumes projecting outwards
. . . .**1.** *polyphylla*
– Inflorescence whitish or grey; heads not particularly squarrose 2
2. Glumes 1.2–1.7 mm long, nutlets 0.6–0.7 mm long; widespread . . .**2.** *microcephala*
– Glumes 1.9–2.5 mm, nutlets 1.3–1.7 mm long; not widespread . . . . . . **3.** *simpsonii*

1. **Kyllingiella polyphylla** (A. Rich.) Lye in Haines & Lye, Sedges & Rushes E. Afr.: 143, fig.280 (1983). —Beentje in F.T.E.A., Cyperaceae: 128 (2010). Type: Ethiopia, Chiré plain, n.d., *Quartin Dillon* s.n. (P holotype).

　　Isolepis polyphylla A. Rich., Tent. Fl. Abyss. **2**: 503 (1850).
　　Scirpus steudneri Boeckeler in Linnaea **36**: 733 (1870). —Clarke in F.T.A. **8**: 458 (1902). Type: Eritrea, Keren, *Steudner* 904 (B holotype, not located).
　　Cyperus steudneri (Boeckeler) Larridon in Pl. Ecol. Evol. **146**: 138 (2013).

---

[15] by J. Browning and K.D. Gordon-Gray†

Small tufted perennial, 4–17 cm tall. Bases contiguous, swollen, covered with fibrous remains of old leaf sheaths; roots 0.25 mm wide. Leaves half to shortly exceeding culm height, eligulate, 7–25 cm × 1.5 mm, margin and midrib with spine-like hairs. Culm trigonous. Inflorescence appearing as terminal globose head, 4–6 × 6–7 mm. Bracts 3–5, longest c. 6 cm, spreading, leaf-like. Spikelets numerous, 'glumes' (scales) spirally arranged, white with green midrib extended into mucro 0.5–1 mm long, giving spiky appearance to inflorescence. Perianth 0. Stamens ?1. Style 3-branched. Nutlet pale brown, obovoid, subtrigonous, c. 0.7 × 0.5 mm, minutely papillose.

Zambia. N: Mbala Dist., Kawimbe, 1680 m, 24.i.195, *Richards* 7987 (K); Mbala Dist., Chinakila, Mpukutu Forest, 1200 m, 14.i.1965, *Richards* 19530 (K). **Zimbabwe**. N: Hurungwe Dist., 1.6 km SE of Chirundu bridge, 3.ii.1958, *Drummond* 5464 (K, LISC, SRGH).

Also in Ethiopia, Kenya and Tanzania. In mopane woodland, wet banks by waterfalls and on termitaria; 450–1700 m.

Conservation notes: Infrequently collected; could be threatened.

Dissection of the apparently capitate inflorescence of *Kyllingiella polyphylla* indicated that there was a 'telescoping' effect resulting in a terminal pseudospikelet composed of several spikelets borne in the axil of each of the inflorescence bracts.

Haines & Lye (1983) note that *Kyllingiella polyphylla* is separated from *Cyperus* by its spirally arranged glumes. Beentje (2010) notes that "Though Haines & Lye state the combination [*Kyllingiella polyphylla* (A. Rich.) K. Lye] was published in Nordic J. Bot. **3** (1983), this did not occur".

2. **Kyllingiella microcephala** (Steud.) R.W. Haines & Lye in Bot. Not. **131**: 176 (1978). —Haines & Lye, Sedges & Rushes E. Afr.: 142, fig.277 (1983). —Beentje in F.T.E.A., Cyperaceae: 129, fig.22 (2010). Type: Ethiopia, Gon Ambra, 20.vi.1840, *Schimper* 650 (P holotype, BR, G, K, S, WAG). FIGURE 14.**28**.

Kyllinga microcephala Steud. in Flora **25**: 597 (1842).
Isolepis kyllingioides A. Rich., Tent. Fl. Abyss. **2**: 502 (1850). Types: Ethiopia, Gon Ambra, 20.vi.1840, *Schimper* 650 (P syntype, BR, K); Ethiopia, Chiré, *Quartin Dillon* s.n. (P syntype, S).
Scirpus kyllingioides (A. Rich.) Boeckeler in Linnaea **36**: 733 (1870). —Clarke in F.T.A. **8**: 457 (1902).
Scirpus microcephalus (Steudel) Dandy in Andrews, Fl. Pl. Sudan **3**: 366 (1956).
Isolepis microcephala (Steud.) Lye in Bot. Not. **124**: 480 (1971).
Cyperus kyllingiella Larridon in Pl. Ecol. Evol. 144: 351 (2011), new name, non *Cyperus microcephalus* R. Br. (1810).

Small tufted perennial, 5–20(30) cm tall, bases swollen, clothed in fibrous remains of leaf-sheaths; roots 0.3–0.4 mm wide. Leaves 2–3, shorter than culm, c. 1 mm wide, scabrid apically, enrolled when dry. Culm 0.25–0.5 mm wide, terete. Inflorescence capitate, 5–10 mm wide with several 'branchlets' or pseudospikelets. Bracts 3(5), longest 6–7 cm, spreading or recurved, leaf-like. Glumes spiral, 1.2–1.4(1.7) mm long, boat-shaped, lanceolate, apex incurved, white, midrib thickened, flanked by 2–3 lateral veins, with a bisexual floret. Perianth 0. Stamen 1. Styles 3(2), style base persistent as small dark knob. Nutlet 0.6–0.7 × 0.3–0.4 mm, obovoid, subtrigonous, with short beak, minutely papillose, brown.

Caprivi: Kavango Region, 3 km E of Mashari, 1720CC, 9.ii.1976, *Vorster* 2765 (K). **Botswana**. N: Okavango swamps, Nqautsha Is., 19°15.3'S 23°09'E, 21.xi.1979, *P.A. Smith* 2893 (K). **Zambia**. W: Mwinilunga Dist., Kalenda Plain, N of Matonchi Farm, 8.xii.1937, *Milne-Redhead* 3558 (K). N: Mbala Dist., c. 2 km above Sansia Falls, Kalambo R., 1740 m, 29.xii.1958, *Richards* 10369 (K). C: Mpika Dist., South Luangwa Nat. Park, near Kapamba R., 7.i.1966, *Astle* 4262 (K). E: Chama Dist., Nyika Plateau, W side, 2130 m, 6.i.1959, *Robinson* 3084 (K). S: Mazabuka Dist., Ross' Farm, 21 km N of Choma, 1.xi.1960, *White* 6679 (K). **Zimbabwe**. N: Hurungwe Dist., near Mensa Pan, 18 km ESE of Chirundu bridge, 460 m, 4.ii.1958, *Drummond* 5474 (K, LISC, SRGH). W: Hwange

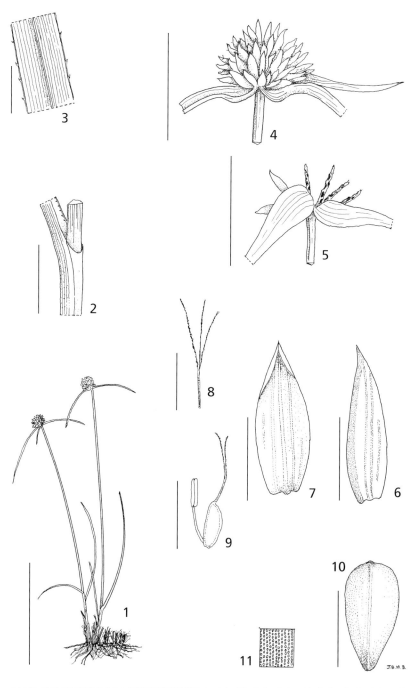

Fig. 14.**28**. KYLLINGIELLA MICROCEPHALA. 1, habit; 2, leaf sheath; 3, leaf, adaxial surface; 4, 5, inflorescence (pseudospike), mature and old with 4 pedicels; 6, 7, glume, lateral and abaxial view; 8, 9, trifid and bifid styles; 10, 11, nutlet and surface detail. 1–4, 6–11 from *Milne Redhead* 3558; 5 from *Hooper & Townsend* 471. Scale bars: 1 = 40 mm; 2, 3, 6, 7 = 1 mm; 4, 5 = 5 mm; 8–10 = 0.5 mm. Drawn by Jane Browning.

Dist., Ngamo Forest (Gwaai Reserve), Ngamo Pans, 910 m, 16.xii.1956, *Paterson* 1 (K, SRGH). C: Chegutu Dist., 45 km W of Harare, 15.ii.1946, *Weber* 207 (K). **Malawi**. N: Chitipa Dist., 18 km SSE of Chisenga, Jembya Forest Reserve, 1870 m, 20.i.1989, *Thompson & Rawlins* 6070 (K). C: Kasungu Dist., Kasungu Nat. Park, 21.xii.1970, *Hall-Martin* 1272 (K). S: Machinga Dist., Liwonde Nat. Park, 550 m, 11.xii.1988, *Dudley* 1405 (K). **Mozambique**. Z: Lugela Dist., Mocuba, Namagoa, iv.?, *Faulkner* K142 (K).

Also in West Africa, Ethiopia, D.R. Congo, Kenya, Uganda, Tanzania, Angola, Namibia and South Africa (Limpopo), and in India. In miombo and mopane woodland in damp sandy ground, on dambo margins, and open saline short grassland; 100–1900 m.

Conservation notes: Widespread; not threatened.

3. **Kyllingiella simpsonii** Muasya in Kew Bull. **57**: 997 (2002). —Beentje in F.T.E.A., Cyperaceae: 131 (2010). Type: Tanzania, Chunya Dist., Rungwa Game Reserve, 1 km W of Itigi–Mbeya road, 28.i.1969, *Sayalel* 5320 (EA holotype, K).

Cyperus simpsonii (Muasya) Larridon in Pl. Ecol. Evol. **144**: 352 (2011).

Tufted perennial; rhizome short, horizontal, 3 mm thick, base covered by fibrous remains of old leaf sheaths. Leaves several, 5–15 cm × 1.5–2.3 mm, flat or inrolled, margin and midrib with minute spine-like hairs; leaf sheath pale brown, 4.4–7.5 cm long, glabrous. Culm 30–62 cm tall, 0.7–1.5 mm thick, glabrous. Inflorescence a dense terminal off-white head, 3–7 × 5–9 mm, with many tightly packed spikelets. Bracts c. 3, leaf-like, largest 3–10 cm long. Spikelets cylindrical, c. 3 mm long, many-flowered. Glumes boat-shaped, 1.9–2.5 × 0.3–0.5 mm, apex obtuse, bearing bisexual floret. Perianth 0. Stamens 1–3. Style 3-branched. Nutlet subtrigonous, narrowly obovoid, 1.3–1.7 × 0.3–0.4 mm, with short beak, whitish turning dark brown, minutely papillose.

Zambia. W: Mwinilunga Dist., Kalenda, x.1937–ii.1938, *Milne-Redhead* 3555 (K). N: Mbala Dist., Chinakila, 11.i.1965, *Richards* 19476 (K). S: Namwala Dist., 8.i.1957, *Robinson* 2048 (K, NU, SRGH).

Also in D.R. Congo and Tanzania. Seasonally wet depressions in miombo woodland; 1000–1400 m.

This species resembles *Kyllingiella microcephala* but has larger glumes and nutlets of different size and shape.

16. CYPERUS L.[16]

Cyperus L., Sp. Pl.: 44 (1753).

Mariscus Gaertn., Fruct. Sem. Pl. **1**: 11 (1788), illegitimate name, non Scop. (1754).

Juncellus C.B. Clarke in Hooker, Fl. Brit. India **6**: 594 (1893).

Annual or perennial herbs; if perennial, then with thick or thin rhizomes. Mostly scapose in habit. Leaves usually laminate, but sometimes reduced to scaly bracts; ligule absent. Inflorescence bracts one to several, the lowest largest, the others progressively smaller. Inflorescence usually terminal but sometimes pseudolateral, capitate or anthelate; if anthelate, then the secondary branches ending in digitate clusters of spikelets, or the spikelets arranged along the axis forming a spike. Spikelets linear to ovoid, laterally flattened; axis persistent, or spikelets falling away entire when the nutlets are ripe (subgenus *Mariscus* (Vahl) C.B. Clarke); glumes few to many, two-ranked, dehiscent (or persistent in subgenus *Mariscus*), glabrous, keel prominent and obtuse to mucronate at the apex. Flowers bisexual. Perianth absent. Stamens 1–3. Stigma (2)3-branched; style base not thickened. Nutlets trigonous, dorsiventrally compressed, with a flat side facing the rachilla, sometimes almost rounded.

A genus of about 550 species in a range of habitats, particularly seasonally wet, mainly in the tropics but extending into warmer temperate zones.

---

16 by M. Lock

There is little agreement in generic delimitation. This treatment separates several genera which have been placed within *Cyperus* by other workers; these include *Pycreus*, *Kyllinga*, *Alinula*, *Oxycaryum* and *Courtoisina*. The most likely confusions are between *Cyperus* and *Kyllinga*, and between *Cyperus* and *Pycreus*. The differences are obscure and technical and require careful examination of good quality material with ripe fruits. See the generic key for more information. Cladistic analysis, although helpful in pointing up clear groupings, does not help greatly in deciding at what level these clades should be recognised, and which groups are recognized at generic level remains very much a matter of personal opinion among cyperologists.

Where the species occurs in both areas, the species descriptions in this account are taken very largely from F.T.E.A. Specimens from the Flora Zambesiaca area have been compared with the F.T.E.A. descriptions and changes made to measurements and morphological details where appropriate. The descriptions of inflorescence structure have often been changed to a standard format as used here. I should here express my gratitude to, and admiration for, Kim Hoenselaar and Henk Beentje who prepared the F.T.E.A. account against a strict and restrictive time-table. They were also hampered by the absence of some types from Kew, and this problem remains to this day, so that the types of some recently described species, which are said to be at Kew, cannot be found in the collections.

Collecting notes and specimens: It is extremely difficult and often impossible to name imperfect material. Good material should include carefully excavated and cleaned underground parts as well as flowers and ripe fruits.

The keys presented here are entirely artificial, and makes no pretence of representing true relationships. I have not followed the order used in F.T.E.A., which in some cases leads to species that look very similar appearing far from one another both in the keys and in the text. The key is based first on whether a species does or does not have green foliage leaves, then on the habit (annual or perennial) and finally on inflorescence structure.

In the keys I have attempted to use gross morphological characters that are readily visible to the naked eye and have made little use of measurements. This may have to be changed, at least in some cases.

Conservation notes are based solely on the herbarium specimens that I have seen and not on any field experience; I feel I should echo my comments on the conservation of *Xyris* species in an earlier volume of the Flora. Many *Cyperus* species normally grow in wet places which are at risk everywhere from drainage, pollution/eutrophication and general overuse. The small annual species have seeds that are probably capable of lying dormant in the soil for many years so it is hard to know their status outside particularly favourable seasons.

Key to Groups

1. Plants leafless (no green laminate leaves although brown basal leaf-sheaths may be present) . . . . **Group A**
 – Plants with obvious green laminas in addition to basal, often brown, leaf sheaths
    2
2. Plants annual (see note below) . . . . **Group B**
 – Plants perennial (see note below) 3
3. Inflorescence capitate (including compound-capitate) . . . . **Group C**
 – Inflorescence not capitate, usually anthelate with spreading branches 4
4. Terminal groups of spikelets arranged in small umbels . . . . **Group D**
 – Terminal groups of spikelets spicate, or inflorescence subunits with spicate arrangement of spikelets . . . . **Group E**

Note: Annual and perennial. Annual species usually have a rather poorly developed root system. The plant may consist of a single stem, or a group of stems arising from the same point, forming a simple tuft. Most annual species do not exceed 30 cm in height. Swollen bulbous bases, tubers, and both thick and thin rhizomes are absent from annual species. However, some perennial species such as *C. imbricatus* may flower in their first year and be a source of confusion.

Group A: Plants lacking green laminate foliage leaves.

Bear in mind that some species included here may have conspicuous whorls of green leafy inflorescence bracts. These arise in a whorl beneath the inflorescence, not from the plant base.

1. Inflorescence a dense head of > 20 terete green peduncles, often sterile; involucral bracts short, brownish 2
–   Inflorescence with fewer spikelet-bearing branches, all fertile; involucral bracts usually green and as long or longer than the inflorescence 3
2. Culms 3.5 cm in diameter towards base and often much more; basal bracts of inflorescence branches brown, dark-tipped; inflorescence units spicate; spikelets terete . . . . **A1.** *papyrus*
–   Culms less than 1 cm in diameter; basal bracts of inflorescence brown, without dark tips; inflorescence units capitate; spikelets flattened . . . . **A2.** *prolifer*
3. Involucral bracts numerous (>10), leafy, forming a dense involucre around the inflorescence; stem 3-angled . . . . **A3.** *involucratus*
–   Involucral bracts fewer, not forming a complete involucre; stems 3 – or 6-angled, or terete 4
4. Culms markedly 6-angled; inflorescence twice capitate-anthelate . . . .
    . . . .**A4.** *sexangularis*
–   Culms terete or 3-angled; inflorescence various 5
5. Stems terete, articulate (run finger down dried stem); involucral bracts <2 cm long; rhizomes long, horizontal 6
–   Stems terete or 3-angled, not articulate 7
6. Involucral bracts very short, brown; stem articulations prominent . **A5.** *articulatus*
–   Involucral bracts longer, green; stem articulations obscure . . . . **A6.** *corymbosus*
7. Basal leaf sheaths disintegrating into blackish fibres . . . . **A7.** *marginatus*
–   Basal leaf sheaths remaining entire or not persisting 8
8. Inflorescence a capitate head of flattened spikelets; aquatic plants often growing in floating mats 9
–   Inflorescence capitate and sometimes appearing lateral; not floating aquatics 10
9. Stems very strongly three-angled; spikelets very pale green, flattened . . . .
    . . . .**A8.** *colymbetes*
–   Stems terete or trigonous, slender and flexuous; spikelets brown, subterete . . . .
    . . . . **A9.** *pectinatus*
10. Inflorescence usually of 1–10 spikelets, appearing lateral; stems terete; often in saline or alkaline sites . . . . **A10.** *laevigatus*
–   Inflorescence terminal, capitate or branched 11
11. Inflorescence capitate or branched, of more than 10 spikelets; culms subterete, ridged; usually in sandy sites near sea . . . . **A11.** *natalensis*
–   Inflorescence branched, not capitate; culms trigonous or triquetrous, smooth 12
12. Culms 0.5–1.6 mm across; involucral bracts 2(3) . . . . **A12.** *denudatus*
–   Culms 3–7 mm across; involucral bracts 1(2) . . . . **A13.** *platycaulis*

A1. **Cyperus papyrus** L., Sp. Pl.: 47 (1753). —Clarke in F.T.A. **8**: 374 (1901). —Kükenthal in Engler, Pflanzenr. **4**, 20(101): 45 (1935). —Haines & Lye, Sedges & Rushes E. Afr.: 177, figs.336,337 (1983). —Gordon-Gray in Strelitzia **2**: 66 (1995). —Hoenselaar & Beentje in F.T.E.A., Cyperaceae: 209, fig.31 (2010). Type: 'Habitat in Calabria, Sicilia, Syria, Aegypto', *Herb. Linn.* Papyrus 15 (UPS lectotype), lectotypified by Simpson (2004).

   Papyrus antiquorum Willd. in Abh. Königl. Akad. Wiss. Berlin **1812–1813**: 70 (1812). Type unclear.

   Cyperus papyrus var. *antiquorum* (Willd.) C.B. Clarke in Durand & Schinz, Consp. Fl. Afr. **5**: 571 (1894).

   Cyperus nyassicus Chiov. in Lav. Reale Ist. Bot. Modena **1**: 73, 76 (1931). Type: Mozambique (or Malawi), Shamo, banks of Shire R., *Scott* s.n. (K syntype).

   Cyperus ugandensis Chiov. in Lav. Reale Ist. Bot. Modena **1**: 77 (1931). Type: Uganda, Ruwenzori Exped., *Scott-Elliot* s.n. (type unknown).

   Cyperus papyrus subsp. *ugandensis* (Chiov.) Kük. in Engler, Pflanzenr. **4**, 20(101): 47 (1935).

Perennial, very robust, up to 5.5 m high, with a creeping rhizome, 2–5 cm in diameter, with a white central pith and a lighter brown (green in life) harder outside cylinder; the outside of the rhizome densely covered by blackish scales 5–10 × 5–10 cm wide, rhizome with many roots; culms 200–550 cm long, basally 1–3.5 cm or wider, apically 0.4–1 cm wide, trigonous, sometimes almost rounded, glabrous. Leaf sheath brown to black, thick and leathery to sometimes almost woody, 4–26 cm long, glabrous; leaf blades absent. Involucral bracts pale-brown, 3–10, leaf-like, spreading, lowermost 6–18 cm long, 0.8–1.7 cm wide, glabrous. Inflorescence simple, primary branches up to 350, 7–40 cm long and 1–3 mm in diameter, triquetrous to rounded, with reddish brown tubular prophylls 2.5–6 cm long at the base of the branches; spikelets spicately arranged towards the end of primary or secondary branches, up to 40 per axis, lanceolate to cylindric, 2.7–10 mm long, 0.4–1.3 mm wide, axis straight; glumes pale brown to golden, ovate to obovate, 1.3–2.3 mm long, 0.8–1.2 mm wide, keel flattened, sometimes green, apex obtuse. Stamens 3; filaments 1.6–2.2 mm long; anthers 0.6–1.4 mm long. Style with 3 branches. Nutlet grey, ellipsoid to ovoid, trigonous, 0.9–1.4 mm long, 0.4–0.5 mm wide, ± smooth.

Caprivi. Kakumba Is., Chobe R., 17.i.1959, *Killick & Leistner* 3419 (K, PRE). **Botswana**. N: Ngamiland Dist., between Dumatau Lagoon and Kwando R. system, 28.i.2004, *A & R. Heath* 494 (K). **Zambia**. B: Mongu Dist., between Lealui and Sendaula pontoon, 12.xi.1959, *Drummond & Cookson* 6366 (K, SRGH). N: Luapula Dist., Lake Bangweulu, S of Ncheta Is., 6.ii.1996, *Renvoize* 5527 (K). C: Lusaka Dist., Lake Kafue below Kafue Bridge, 26.xii.1972, *Kornaś* 2857 (K). S: Victoria Falls, Livingstone Is., 21.xi.1949, *Wild* 3131 (K, SRGH). **Zimbabwe**. W: Hwange Dist., Matetsi Safari Area, 12.iii.1981, *Gonde* 354 (K, SRGH). **Malawi**. C: Dowa Dist., 2 miles S of Lake Nyasa Hotel, 9.viii.1951, *Chase* 3909 (BM, SRGH). S: Nsanje Dist., Lower Shire Valley, Elephant Marsh, ix.1956, *Robertson* s.n. (K). **Mozambique**. GI: Homoine Dist., Inhanombe R., vi.1936, *Gomes e Sousa* 1751 (K). M: Maputo Dist., Inhaca Is, Hlanganisa swamp, 29.ix.1957, *Mogg* 27576 (K).

Widespread in Africa; Mediterranean. Margins of lakes, rivers and swamps, generally in places of relatively high nutrient content. Can form floating mats in deeper water; 20–1200 m.

Conservation notes: Least Concern due to its wide distribution, but liable to local extinctions due to drainage works.

Cyperus papyrus is likely to be more widespread than the records suggest, firstly because its familiarity makes it unlikely to be collected, and secondly because its sheer size makes collecting and preservation difficult. Detailed observation that I have made in the Sudd Region of the Sudan show that the stature of *C. papyrus* varies greatly according to nutrient availability. Plants close to flowing rivers can be very large while those far from such a nutrient source are much shorter and less vigorous.

Chiovenda described a number of taxa from Africa, all of them apparently at specific rank although they are often cited as subspecies. Although Chiovenda's taxa are based on characters of the inflorescence and spikelets, I would be very wary of accepting them without extensive work on a much larger sample of collections than is currently available.

F.T.E.A. gives the measurements of the rhizome scales in mm; surely a misprint for cm.

A2. **Cyperus prolifer** Lam., Tabl. Encycl. **1**: 147 (1791). —Clarke in F.T.A. **8**: 339 (1901). —Kükenthal in Engler, Pflanzenr. **4**, 20(101): 256 (1936). —Haines & Lye, Sedges & Rushes E. Afr.: 171, figs.326,327 (1983). —Gordon-Gray in Strelitzia **2**: 66 (1995). —Hoenselaar in F.T.E.A., Cyperaceae: 189 (2010). Type: Mauritius [Insula Franciae], *J. Martia* s.n. (P holotype).

   Cyperus isocladus Kunth, Enum. Pl. **2**: 37 (1837). —Clarke in F.T.A. **8**: 339 (1901). Type: South Africa, East, *Drège* s.n. (B holotype).

   Cyperus prolifer var. *isocladus* (Kunth) Kük. in Engler, Pflanzenr. **4**, 20(101): 257 (1936).

Perennial, fairly robust, up to 130 cm tall, with a thick creeping rhizome and purple to blackish-brown roots; culms crowded, 55–120 cm long, 3–7 mm wide, terete to trigonous, smooth. Leaf blades absent; leaf sheath reddish-brown to dark purple, 2–32 cm long. Involucral bracts scale-like to almost leaf-like, spreading, 3–4, lowermost 1.5–3(11) cm long, 2–5 mm wide. Inflorescence once to twice anthelate, usually with no spikelets at the primary branch point, primary branches up to 50–100, all equal in length giving the inflorescence a spherical to umbel-like appearance, 3–11 cm long; spikelets in digitate clusters, at the end of primary and sometimes secondary branches, 1–5 per cluster, linear to ovate-lanceolate, 2.7–15 mm long, 0.9–1.9 mm wide, rachis straight; glumes pale reddish-brown, ovate, 1.1–1.6 mm long, 0.7–1 mm wide, keel pale brown to green, apex rounded to acute, slightly excurrent. Stamens 3; filaments 0.9–1.6 mm long; anthers 0.5–1.2 mm long, with spiny apex. Nutlet white to almost brown, obovoid, 0.4–0.5 mm long, 0.3–0.4 mm wide, almost smooth to minutely papillose.

Zimbabwe. C: Harare (cultivated), 30.vi.1976, *Biegel* 5328 (K, SRGH). **Mozambique**. N: Palma Dist., Palma to Pundanhar road, c. 11 km W of junction with road to Nhica do Rovuma, 8.xi.2009, *Goyder et. al.* 6039 (K). Z: Lugela Dist., road to Moebede, 25.i.1948, *Faulkner* 181 (K). MS: Beira Dist., 25.xii.1906, *Swynnerton* 931 (K). GI: Chibuto Dist., Campo de Ensaios, 9.vii.1948, *Myre* 59 (K). M: Maputo Dist., Inhaca Is., 38 km (23 miles) E of Maputo, 4.iii.1958, *Mogg* 31431 (K).

Also in Somalia, Kenya, Tanzania, and South Africa; also Madagascar. Swamp edges, stream-sides, seasonally flooded grasslands and in and beside permanent pools, almost always along the coast; occasionally cultivated elsewhere; 0–50m.

Conservation notes: Probably Least Concern, but more restricted in its distribution than *Cyperus papyrus*.

The inflorescence is similar in form to that of *C. papyrus*, of which this taxon resembles a dwarf version. In spite of this superficial resemblance, the arrangement and form of the spikelets is quite different in the two species and they are probably not closely related.

In *Goyder et al.* 6039 (K), some of the inflorescence branches have the tips modified into dense bunches of short shoots, resembling an artist's paintbrush. These may be galls.

Cyperus prolifer may be under-collected as it can easily be mistaken for stunted *C. papyrus*.

A3. **Cyperus involucratus** Rottb., Descr. Pl. Rar.: 22 (1772). —Haines & Lye, Sedges & Rushes E. Afr.: 154, fig.283 (1983). —Hoenselaar in F.T.E.A., Cyperaceae: 187, fig.28 (2010). Type: Ethiopia, Adua, *Schimper* 55 (P holotype, HAL, K). FIGURE 14.**29**.

   Cyperus flabelliformis Rottb., Descr. Icon. Rar. Pl.: 42 (1773). —Clarke in F.T.A. **8**: 336 (1901), illegitimate name, based on the Rottbøll name.

   Cyperus alternifolius L. subsp. *flabelliformis* Kük. in Engler, Pflanzenr. **4**, 20(101): 193 (1936).

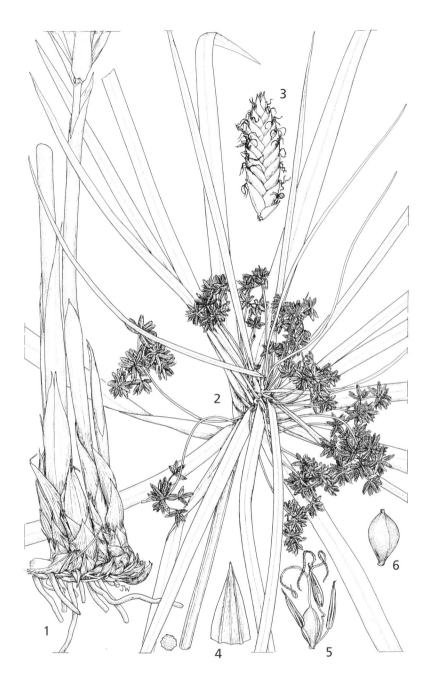

Fig. 14.**29**. CYPERUS INVOLUCRATUS. 1, habit (× ²/₃); 2, inflorescence with involucral bracts (× ²/₃); 3, spikelet (× 6); 4, glume (× 14); 5, flower (× 14); 6, nutlet (× 24). 1 from *Haines* 4012; 2–5 from *Lind* 208, 6 from *Thomas* 2184. Drawn by Juliet Williamson. Reproduced from Flora of Tropical East Africa (2010).

Perennial, robust, up to 2 m tall, with a creeping rhizome 2–10 mm in diameter and several culms usually arranged in a row, the bases touching one another; culms 66–160 cm long, 3–8 mm wide, rounded to trigonous, with longitudinal ridges, glabrous to sometimes minutely scabrid, the base covered with black fibrous remains from old leaf sheaths; sheath almost black, pale brown to green, 1–36 cm long; blade absent. Involucral bracts many, leaf-like, spreading, spirally arranged along a 1–5 cm long axis, the longest 18–37 cm long, 0.8–1.5 cm wide, linear, flat, scabrid, apex acute, the upper ones progressively narrower and shorter. Inflorescence once to twice anthelate, primary branches many, 3.5–10 cm long; spikelets in digitate clusters at the end of secondary and tertiary branches, 5–20 per cluster, lanceolate to elliptic-ovoid, much compressed, 3.5–11 mm long, 1.4–3 mm wide; glumes very pale brown, golden brown to reddish brown, elliptic-ovate, 1.4–2.2 mm long, 0.9–1.6 mm wide, 3-keeled, green, apex acute, sometimes slightly mucronate, glabrous. Stamens 3; filaments 1.6–2.2 mm long; anthers 0.8–1.4 mm long; the connective protruding into a needle-like apex. Nutlet yellow to brown, narrowly ovoid to oblong, trigonous, 0.8–1.1 mm long, 0.5–0.6 mm wide, papillose.

Botswana. SE: Central Dist., Palapye, Malata, 7.ii.1958, *de Beer* 595 (K, SRGH). **Zambia**. C: Lusaka Dist., Mt Makulu, 12 km [8 miles] S of Lusaka, 11.vi.1955, *Robinson* 1294 (K). E: Lundazi Dist., Kazembe, 80 km [50 miles] W of Lundazi, in Luangwa Valley, 3.vi.1954, *Robinson* 831 (K). S: Choma Dist., Mapanza W, 5.ix.1953, *Robinson* 305 (K). **Zimbabwe**. N: Gokwe Dist., between Zane & Gwave rivers, 5.vi.1963, *Bingham* 680 (K, SRGH). W: Hwange Dist., Victoria Falls, 10–15.iii.1932, *Brain* 8929 (K). C: Harare, Mount Pleasant (cultivated), 14.vii.1972, *Biegel* 3974 (K, SRGH). E: Mutare (Umtali) Commonage, bank of Menini R., 9.ix.1948, *Chase* 873 (K, SRGH). S: Beitbridge Dist., Chipisi Hot Spring, 25.ii.1961, *Wild* 5386 (K, SRGH). **Malawi**. N: Rumphi Dist., 5 km [3 miles] NE of Rumphi, by Chelinda R., 8.ix.1976, *Pawek* 11760 (K, MAL, MO, SRGH, UC). C: Kasungu Dist., Kasungu, 27.viii.1946, *Brass* 17441 (K, NY). S: Mulanje Dist., Mt Mulanje foot, Phalombe R., 19.vii.1986, *Chapman & Chapman* 7848 (K, MO). **Mozambique**. N: Pemba Dist., gallery forest of R. Ridi, W of Pemba [Porto Amelia], 1.x.1964, *Gomes e Sousa* 4833 (K). T: Tete, between Mutarara Velha and Sinjal, 38.5 km from Mutarara Velha, 18.vi.1949, *Barbosa & Carvalho* 3136 (K, LISC). MS: Marromeu Dist., Shupanga forest, 5.x.1887, *Scott* s.n. (K). GI: Inhambane Dist., banks of R. Inhambane, vi.1936, *Gomes e Sousa* 1752 (K).

Widespread in tropical Africa from Senegal to Somalia and south to South Africa. Streambanks, riverbanks, lakesides, by hot springs; usually in permanently wet sites; 50–1550 m.

Conservation notes: Least Concern; widespread in Africa.

The absence of collections from northern and western Zambia, and from northern Botswana, seems surprising, but perhaps collectors have just ignored this species which is common and widespread in wet places, as well as being large and relatively difficult to press.

Used for making mats; also widely planted in gardens and parks as an ornamental, and cultivated as a house plant in temperate regions.

A4. **Cyperus sexangularis** Nees in Linnaea **10**: 135 (1835). —Gordon-Gray in Strelitzia **2**: 72 (1995). Type: South Africa, Cape, circa Castellum Beaufort (Fort Beaufort), *Drège* s.n. (P, JZU, HAL, JE, S, not seen).

Perennial; culms densely spaced on a very short rhizome. Culms (28)60–120 cm. tall, green, triquetrous with a ridge along the flat faces, thus 6-angled (hence the name), lightly scabrid on the margins, otherwise glabrous. Leaves absent; basal bracts brown, 5–10 cm long, sheathing. Involucral bracts 6–11, green, leafy, the longest 8–20(26) × 0.4–1 cm, flat or plicate, scabrid on the margins. Inflorescence twice anthelate-capitate, primary branches to 5 cm long. Spikelets in capitate clusters of 3–12 at the tips of the primary and secondary branches, narrowly ovate, 4–6 × 1–1.5 mm, flattened, 18–24(32)-flowered, the glume tips projecting and giving a saw-edged appearance to the spikelet. Glumes broadly ovate, 1.4–1.6 × 1 mm, pale to dark red-brown with pale green margins. Nutlet blackish, broadly ellipsoid, 0.5–0.6 × 0.4–0.5 mm, apiculate and with a persistent base, papillose.

Zimbabwe. C: Chikomba Dist., 8 km [5 miles] N of Lalapansi on Great Dyke, 17.i.1962, *Wild* 5623 (K, SRGH). E: Chiredzi Dist., Savi Valley, Honde dip, 26.ix.1947, *Whellan* 252 (K, SRGH). Without locality, xi.1929, *Parday* 4698 (K).

Also in South Africa. Swamps, wet places; 1000–1200 m.

Conservation notes: Probably Least Concern as it is widespread and also cultivated in South Africa.

The species is both widespread and widely cultivated in South Africa and one wonders if these collections may refer to plants escaped from cultivation. However, notes on the *Whellan* collection describe the plant as 'abundant' and in a natural habitat, and there is no mention of cultivation on the *Wild* specimen. Govaerts & Simpson, World Checklist of Cyperaceae, record this species from Zambia and Botswana but I have not seen specimens.

The description has been prepared largely from South African material. The 6-angled stem and the ellipsoid papillose nutlets are distinctive.

There are two sheets at Kew apparently collected by *Drège* but neither claims to be from the Fort Beaufort type locality. A sheet at P has the locality 'Ceded Territory. Prope Fort Beaufort 1000'–2000', and was collected by Ecklon & Zeyher but states 'communicavit Drège. This may well be the type.

A5. **Cyperus articulatus** L., Sp. Pl.: 44 (1753). —Clarke in F.T.A. **8**: 356 (1901). — Kükenthal in Engler, Pflanzenr. **4**, 20(101): 77 (1935). —Haines & Lye, Sedges & Rushes E. Afr.: 183, fig.353 (1983). —Gordon-Gray in Strelitzia **2**: 54 (1995). —Hoenselaar & Beentje in F.T.E.A., Cyperaceae: 208 (2010). Type: Jamaica, "Habitat in Jamaicae rivulis", *Herb. Sloane* 2:63 (BM-SL lectotype), designated by Tucker (1983). FIGURE 14.**30**.

Cyperus articulatus var. *erythrostachys* Graebn. in Repert. Spec. Nov. Regni Veg. **16**: 25 (1919). —Kükenthal in Engler, Pflanzenr. **4**, 20(101): 80 (1935). Type: Tanzania, Ufipa Dist., Lake Kwera, *Fromm & Münzner* 140 (B holotype).

Robust perennial up to 250 cm tall, with solitary culms from horizontal rhizomes to 10 cm or more long and 2–8 mm thick, often woody, clothed with blackish or purple scales; culms basally swollen, 80–250 cm long, 0.4–1 cm wide, rounded, pith-filled with transverse partitions at 5–50 mm intervals (septate), smooth. Leaves reduced to sheaths only, 3–5 sheaths covering the lower part of the culm, straw-coloured, purple to blackish, 3–28 cm long, ending in a triangular limb; leaf blade absent. Involucral bracts scale-like, 3–5, brown, lowermost 0.7–1.2 cm long. Inflorescence once anthelate, primary branches 5–8, 1–10 cm long, spikelets in loosely digitate clusters, sessile and at the end of primary branches, 9–20 per cluster, linear, terete to flattened, 7.5–33 mm long, 0.9–2 mm wide, rachilla straight to slightly curved; glumes pale brown to reddish-brown, ovate, 2.7–4(5.1) mm long, 1.3–1.9 mm wide, keel pale brown, apex obtuse. Stamens 3; filaments 2.7–3.5 mm long; anthers 0.9–1.5 mm long. Nutlet reddish-brown to almost black, narrowly ellipsoid, 1.3–1.6 mm long, 0.4–0.6 mm wide, shortly apiculate, smooth to minutely papillose.

Caprivi. Mpilila Is., banks of Zambezi R., 13.i.1959, *Killick & Leistner* 3369 (K, PRE). **Botswana**. N: Ngamiland Dist., Matlapeneng, Thamalakane R. bridge, 8 km E of Maun, 19.iii.1965, *Wild & Drummond* 7158 (K, SRGH). **Zambia**. N: Mporokoso Dist., Mweru Wantipa, floodplain below Mpundu, 11.iv.1957, *Richards* 9166 (K). C: Binga Dist., Sable Farm, Chisamba, 11.i.1996, *Bingham* 10793 (K, MRSC). S: Namwala Dist., Maala, 40 km [25] miles E of Namwala, edge of Kafue Flats, 24.iv.1954, *Robinson* 730 (K). **Zimbabwe**. N: Mount Darwin Dist., Mkumburu R. on road to Tete, 22.i.1960, *Phipps* 2389 (K, SRGH). E: Chipinge Dist., East Savi/Rupembe, 22.i.1957, *Phipps* 82a (K, SRGH). S: Nuanetsi Dist., Nuanetsi R., gorge upstream from Buffalo Bend,

Fig. 14.**30**. CYPERUS ARTICULATUS. 1, habit; 2, portion of stem, cut open; 3, spikelet; 4, flower. Original specimen(s) and scales not stated but 1 & 2 approx. × 1. Drawn by W.E. Trevithick. Reproduced from Flora of West Tropical Africa, ed. 1 (1936).

28.iv.1971, *Drummond & Rutherford-Smith* 7577 (K, SRGH). **Malawi**. C: Nkhotakota Dist., Nkhotakota, Lake Malawi, 2.v.1963, *Verboom* 104(S). S: Zomba Dist., Lake Chirwa, near Nsenga, 28.i.1979, *Robson* 1326 (K). **Mozambique**. MS: Beira Dist., Cheringoma Section, Nhemissembe R., 14.vii.1974, *Ward* 7914 (K). GI: Guijá Dist., Sul do Save, Guija, Lagoa Chinan, 6.vii.1947, *Pedro & Pedrogão* 1327 (K). M: Maputo, Matola Road, 28.ix.1925, *Moss* 11744 (K).

Widespread throughout Africa; also in the New World. In swamps, lake-shores, river margins, wet grasslands and pools, often in standing water up to 30 cm deep, often forming large pure patches; sea-level up to 1600 m.

Conservation notes. Least Concern; widespread and common in wet habitats.

Readily recognisable because of the septate stems and the lack of leaf blades. It can, however, easily be confused with the apparently much rarer *Cyperus corymbosus*, which has a less prominently septate culm and green inflorescence bracts.

A6. **Cyperus corymbosus** Rottb., Descr. Pl. Rar.: 19 (1772); Descr. Icon. Rar. Pl.: 42, tab.7 fig.4 (1773). —Clarke in F.T.A. **8**: 357 (1901). —Kükenthal in Engler, Pflanzenr. **4**, 20(101): 80 (1935). —Haines & Lye, Sedges & Rushes E. Afr.: 183, fig.354 (1983). —Hoenselaar in F.T.E.A., Cyperaceae: 233 (2010). Type: India, *König* s.n. (C holotype).

Perennial, fairly robust, up to 160 cm tall, with rather thick scale-covered stolons, 2–4 mm in diameter; culms 86–137 cm long, 4.4–8.3 mm wide, only very slightly articulated, rounded to trigonous, smooth. Leaves up to 41 cm long; leaf sheath greyish-brown to brown, 8.5–18 cm long; leaf blade linear, flat, 10–24 cm long, 4.4–7.5 mm wide, glabrous, apex acute. Involucral bracts leaf-like, green, spreading, 3–4(7), lowermost 12.5–20 cm long, 3–7 mm wide. Inflorescence a compound anthela, with primary and secondary branches, primary branches 7–9, 4–20 cm long; spikelets in loose clusters on elongated axis, at the end of secondary branches, 4–10 per cluster, linear, 6.8–12.9 × 0.8–1.3 mm wide, rachis straight; glumes grey to dark reddish-brown, ovate, 1.9–2.1(4?) × 0.9–1.1 mm wide, keel green, apex acute. Stamens 3; filaments 1.4–2.4 mm long; anthers 1.3–1.4 mm long. Nutlet pale brown, ellipsoid-obovoid, trigonous, c. 1.1 × 0.5 mm wide, apiculate, smooth.

Botswana. N: Ngamiland Dist., Okavango Delta, Chief's Camp, Chief's Is., 14.ix.2009, *A. & R. Heath* 1741 (K). **Mozambique**. M: Moamba Dist., Ressano Garcia, 22.iv.1897, *Schlechter* 11885 (K).

Also in Ivory Coast, Togo, Angola, Mozambique, Tanzania and South Africa, as well as India. Riverbanks, swamps; 300–1000 m.

Conservation notes: Probably Least Concern as widespread, although apparently never common.

Very similar to *Cyperus articulatus* but differs in its less strongly articulated stems and green and more prominent inflorescence bracts. Apparently scarce and scattered in Africa but almost certainly overlooked and undercollected because of confusion with *C. articulatus*.

A7. **Cyperus marginatus** Thunb., Prodr. Pl. Cap.: 18 (1794). —Clarke in F.T.A. **8**: 339 (1901). —Haines & Lye, Sedges & Rushes E. Afr.: 264, figs.537,538 (1983). —Hoenselaar in F.T.E.A., Cyperaceae: 256 (2010). Type: South Africa, *Thunberg* s.n. (UPS holotype).

Perennial, forming clumps from a short thick vertical rhizome. Culms 60–80 cm long and 1–3 mm in diameter, trigonous, glabrous. Leaves absent or reduced to brownish sheaths near the base of the culms. Sheaths prominent, the oldest persistent and splitting into blackish fibres. Involucral bracts 3–5, up to 3 cm long, usually shorter than the anthela branches. Inflorescence once anthelate with a single sessile subunit and up to 5 stalked subunits on glabrous peduncles

0.5–3 cm long. Spikelets ovate, strongly flattened, 6–12 × 2–3 mm, 20–30-flowered. Glumes c. 2 mm long, brownish, with a single prominent nerve on each side of the midrib, which ends in the apex. Stamens 3. Style with 3 long branches. Nutlet c. 0.6 × 0.4 mm, ovate-elliptic, trigonous, minutely papillose, dark brown to black at maturity.

Botswana. SW: Ghanzi Dist., Olifants Kloof (Farm 168), 14 km [10 miles] NE of Mamuno, 13.ii.1970, *Brown* 27 (K).

Probably in Kenya, also in Namibia, Lesotho, Swaziland and South Africa. Seasonal watercourses in very dry country; 1000 m.

Conservation notes. Data Deficient; insufficient information.

The single specimen from the Flora area lacks basal parts. Most of the (copious) material from South Africa at K has a short creeping rhizome bearing shoots at regular intervals, like *Cyperus denudatus*. Material from Angola appears much more tussocky with lots of black fibres (the remains of leaf sheaths) at the base. More than one taxon may be involved.

A8. **Cyperus colymbetes** Kotschy & Peyr., Pl. Tinn.: 49, tab.24 (1867). —Clarke in F.T.A. **8**: 317 (1901). —Kükenthal in Engler, Pflanzenr. **4**, 20(101): 289 (1936). —Haines & Lye, Sedges & Rushes E. Afr.: 173, fig.329 (1983). —Hoenselaar & Beentje in F.T.E.A. Cyperaceae: 151 (2010). Type: Sudan, *Tinné* s.n. (W holotype).

    Anosporum colymbetes (Kotschy & Peyr.) Boeckeler in Bot. Zeitung (Berlin) **27**: 26 (1869).

Perennial, fairly robust, up to 54 cm tall, with an erect or creepy subwoody rhizome from which new culms develop at irregular intervals, often floating; culms green, 20–70 cm long, 3–5 mm wide, triquetrous to winged, smooth. Leaves up to 20 long; leaf sheath reddish-brown to purple, very wide, ending in a thin ligule and a thick triangular apex, basal sheaths short, c. 1 cm, longer sheaths up to 20 cm long; leaf blade absent. Involucral bract leaf- to bract-like, sometimes culm-like, erect, 8–12 mm long. Inflorescence capitate; spikelets 3–15(20) per head, ovoid, 6–15 mm long, 4–10 mm wide, rachilla straight; glumes pale greenish or whitish, sometimes drying reddish-brown, ovate, 4.2–6 mm long, 2.4–3.7 mm wide, with large surface cells, 3–9-veined, keel thicker, scabrid towards apex, apex acute. Stamens 3; filaments 3–5.4 mm long; anthers 0.9–1.6 mm long. Nutlet brown, surrounded by yellow sterile tissue, 4.2–5.5 mm long, 1–1.4 mm wide, nutlet ellipsoid-oblong, 1.7–2.4 mm long, 0.9–1.1 mm wide, minutely papillose.

Mozambique. MS: Chinde Dist., Expedition Is., Zambesia, ix.1888, *Kirk* s.n. (K).

Also in Sudan, Somalia, Uganda, Kenya and Tanzania. No habitat data from the Flora area; elsewhere it grows in boggy pools and in floating vegetation mats. In the Flora 0–50 m; elsewhere in Africa to 1100m.

Conservation notes: Least Concern; fairly widespread in Tropical Africa.

This species is very close to *Cyperus pectinatus*. Both species have nutlets surrounded by corky tissue which makes them long-floating. *C. colymbetes* has much fewer and thicker culms which are almost winged. It also has larger involucral bracts, more spikes in the head, and the living spikes are whitish rather than brownish as in *C. pectinatus*.

It is remarkable that this taxon has not been collected in the Flora area for well over a century, possibly because it often grows in inaccessible floating mats of vegetation. The description is taken from F.T.E.A.

The date on the specimen, in Kirk's hand, is clearly 1888. The Zambezi Expedition was 1858–1864. However, there is also a note: Rec 10/83. There must be an error somewhere.

A9. **Cyperus pectinatus** Vahl, Enum. Pl. **2**: 298 (1805). —Haines & Lye, Sedges & Rushes E. Afr.: 172 (1983). —Gordon-Gray in Strelitzia **2**: 66 (1995). —Hoenselaar in F.T.E.A., Cyperaceae: 151, fig.24 (2010). Type: Guinea, *Isert* s.n. (C holotype). FIGURE 14.**31**.

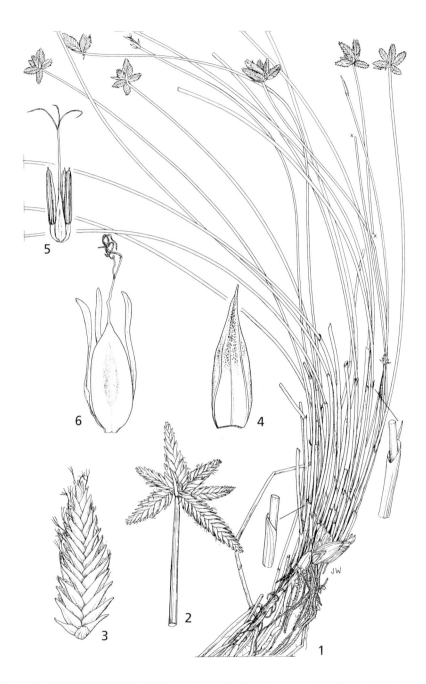

Fig. 14.**31**. CYPERUS PECTINATUS. 1, habit (× ²/₃); 2, inflorescence (× 1½); 3, spikelet (× 3); 4, glume (× 10); 5, flower (× 10); 6 nutlet (× 10). 1 from *Peter* 8794; 2–4, 6 from *Greenway & Kanuri* 12298; 5 from *Richards* 24601. Drawn by Juliet Williamson. Reproduced from Flora of Tropical East Africa (2010).

Cyperus nudicaulis Poir. in Lamarck, Encycl. **7**: 240 (1806). —Clarke in F.T.A. **8**: 316 (1901). —Kükenthal in Engler, Pflanzenr. **4**, 20(101): 284 (1936). Type: Madagascar, *du Petit Thouars* s.n. (P holotype).

Perennial, up to 122 cm tall, with a short rhizome, sometimes stoloniferous, roots pale to reddish-brown, sometimes spongy; culms arising at very short intervals along the rhizome, 25–120 cm long, 0.8–1.3 mm wide, rounded to trigonous, grooved longitudinally, when young culms erect, but mature culms often curving so that the inflorescence eventually touches the ground (or water). Leaves up to 14 cm long; leaf sheath greyish-black to purple, 1.5–14 cm long; leaf blade absent, the sheath ending in a short triangular limb. Involucral bracts culm-like, erect, 1–2, 0.5–2 × 1–1.5 mm wide. Inflorescence capitate; spikelets in a digitate crowded head, 2–11(20) per head, ovoid-lanceolate, 5.6–16(28) × 3.5–6 mm wide, rachis straight; glumes reddish-brown, ovate to boat-shaped, 3.3–5.5 × 1.7–2.4 mm wide, keel green, 3-veined, scabrid, apex obtuse to acute. Stamens 3; filaments 3.5–4.4 mm long; anthers 1.4–1.8 mm long. Nutlet surrounded by spongy yellow corky tissue, 2.7–4 × 0.8–1.3 mm wide, nutlet itself (visible part) brown, narrowly ovoid, 1.4–2 × 0.5–0.8 mm wide, smooth.

Caprivi. Diyona Camp, 3 km (2 miles) E of Nyangana Mission Station, 27.ii.1956, *De Winter & Marais* 4888 (K, PRE). **Botswana**. N: Ngamiland Dist., Okavango Delta, near Gwetsa, Monachira R. N of Gcobega Lagoon, 4.i.2010, *A. & R. Heath* 1835 (K). **Zambia**. B: Mongu Dist., Kande Plain, E of Mongu, 20.vii.1961, *Angus* 3008 (K, FHO). N: Chinsali Dist., Shiwa Ngandu, 1.vi.1956, *Robinson* 1514 (K). W: Mwinilunga Dist., by Kamweshi R., 17.xii.1937, *Milne-Redhead* 3712 (BM, K). C: Kabwe Dist., Kabwe [Broken Hill], Lukanga Swamp, 19.viii.1958, *Seagrief* 3174 (K, SRGH). **Zimbabwe**. N: Binga Dist., Lake Kariba, Masangwa R. mouth, 4.ix.1973, *Gibbs Russell* 2587 (K, SRGH). **Malawi**. S: Kasupe Dist., Domasi R. mouth, Lake Chilwa, 14.viii.1970, *Howard Williams* 210 (K, SRGH).

Widespread in western, central and eastern Africa, south to South Africa. Marshes, dambos, lake, stream and river margins and floating aquatic vegetation mats (sudd); 700–1850 m.

Conservation notes: Least Concern; a widespread species of wet places.

This species is close to *Cyperus colymbetes*; for differences see under that species.

A10. **Cyperus laevigatus** L., Mant. Pl. **2**: 179 (1771). —Kükenthal in Engler, Pflanzenr. **4**, 20(101): 321 (1936). —Haines & Lye, Sedges & Rushes E. Afr.: 264, fig.539 (1983). —Gordon-Gray in Strelitzia **2**: 59, fig.22 (1995). —Hoenselaar in F.T.E.A., Cyperaceae: 153 (2010). Type: South Africa, Cape of Good Hope, *König* s.n. (LINN 70.13 lectotype), designated by Tucker & McVaugh (1993).

 Pycreus laevigatus (L.) Nees in Linnaea **10**: 130 (1835).
 Cyperus subaphyllus Schinz in Verh. Bot. Vereins Prov. Brandenburg **30**: 139 (1888). Type: Namibia, Lüderitz, *Schinz* s.n. (B holotype).
 Juncellus laevigatus (L.) C.B. Clarke in Hooker, Fl. Brit. India **6**: 596 (1893); in F.T.A. **8**: 308 (1901).
 Cyperus laevigatus var. *subaphyllus* (Schinz) Kük. in Fries, Wiss. Ergbn. Schwed. Rhodesia-Kongo-Exped. 1911–1912, Erg.: 2 (1921).

Perennial, up to 60(96) cm tall, with a long creeping rhizome, to 30 cm or more long, 1–5 mm in diameter, pale brown to purple-black, scales quickly disappearing; culms tufted, crowded, or spaced along the rhizome, 3–60(96) cm long, 0.5–4.4 mm wide, rounded to trigonous, sometimes triquetrous, glabrous, the base covered with short scales. Leaves up to 16 cm long; leaf sheath pale to dark purple brown, 1.5–14 cm long, rather loose; leaf blade usually absent but when present linear, inrolled, almost culm-like, 2.2–6.5 cm long, 0.5–2 mm wide, scabrid on margin but appears glabrous as margins are inrolled, apex acute. Involucral bract one, leaf-like, upright and continuing in the direction of the culm, making the inflorescence appear lateral, 1.1–3.8 cm long, 1–1.5 mm wide. Inflorescence capitate; spikelets 1 to 24 per head, loosely crowded, linear to lanceolate, 5–25 × 1.5–4.1 mm wide, rachis straight to curved; glumes pale yellowish with reddish brown dots in the central part, nerve and margins usually pale, sometimes dark red to almost black on the wings, broadly elliptic, 2.5–4.1 × 1.8–2.1 mm wide, very closely overlapping, apex acute, shortly mucronate or frayed. Stamens 2; filaments 2.7–3.2 mm long; anthers 1.2–1.9 mm long. Style with 2 long linear branches. Nutlet grey to brown, often shiny, obovoid to ellipsoid, flat on one side, rounded on the other, 1.4–2.1 × 0.8–1.2 mm wide, shortly apiculate, smooth.

Botswana. N: Ngamiland Dist., Okavango Delta, Kaporota Camp, 10.iii.2010, *A. & R. Heath* 1967 (K). SE: Central Dist., Tsokotse Pan, 7.xii.1978, *P.A. Smith* 2566 (K). **Zambia**. W: Kasempa Dist., Kiambwe Salt Pan, x.1934, *Trapnell* 1643 (K). E: Chipata Dist., Luangwa, edge of alkaline pan near Nsefu, 11.v.1963, *van Rensburg* KBS 2140 (K). S: Monze Dist.,, W section of Lochinvar Ranch near Chirundu, 8.viii.1963, *van Rensburg* KBS 2403 (K). **Zimbabwe**. N: Binga Dist., Sebungwe, 6.xi.1958, *Phipps* 1365 (K, SRGH). W: Hwange Dist., Milibisi R., hot springs, x.1955, *Davies* 1580 (K). **Malawi**. S: Karupe Dist., sandbar between Lake Chilwa and Lake Chinta, 31.vii.1969, *Howard-Williams* 178 (K, SRGH). **Mozambique**. GI: Inharrime Dist., near Lagos Polela, 13.ix.1948, *Myre & Carvalho* 171 (K, LMA).

Widespread in Africa. Lake shores, usually in saline or alkaline sites; riverbanks, and around hot springs, where it often forms dense monospecific mats; sea-level to 1400 m.

Conservation notes: Least Concern; widespread in Africa in its restricted habitat.

Confused with *Pycreus* but the flat side of the nutlet is pressed against the rachilla, while in *Pycreus* one of the edges is pressed into the rachilla.

A characteristic species of saline or alkaline lake shores (but sometimes in freshwater sites), often forming dense monospecific stands around saline or alkaline hot springs. Trapnell states that the plants are hoed up and burned to make salt.

The inflorescence is distinctive in usually appearing more or less lateral.

A11. **Cyperus natalensis** C. Krauss in Flora **28**: 755 (1845). —Gordon-Gray in Strelitzia **2**: 65 (1995). Type: South Africa, KwaZulu-Natal, Umlaas R., xii.1840, *Krauss* 207 (?B† holotype, K, PRE).

Perennial from a long rhizome 3–6 mm in diameter, covered in brown scales, creeping. Culms erect, 80–130 cm tall, 3–5 mm in diameter, glabrous, subterete, ridged. Leaves usually absent, when present basal, linear, 15–30 cm long, 3–5 mm wide, glabrous or margin minutely denticulate above. Basal leaf sheaths pale brown with darker margins, 10–15 cm long. Inflorescence once anthelate, with a basal sessile umbel of spikelets and 3–6 stalked umbels of spikelets; stalks 5–15 cm long, sometimes simply capitate. Involucral bracts usually small and inconspicuous, the longest 0.5–3.5 cm long (but see below). Spikelets 15–25 per dense cluster, sessile or at the ends of the primary branches, linear, 1.5–3.5(5) cm long, 2–3 mm wide. Glumes 4 × 2 mm, ovate-elliptic, apex obtuse and/or bifurcate, brownish with paler membranous margins, longitudinally ridged. Stamens 2; filaments c. 3 mm long; anthers 2.5–3 mm long. Style 3-branched. Nutlets obovoid, trigonous, blackish, 2 × 0.7 mm, smooth.

Mozambique. GI: Xai-Xai Dist., 3 km, dunes of Chongoano, 25.i.1980, *de Koning* 8128 (LMU, K, WAG). M: Maputo Dist., Inhaca Is., 40 km (23 miles) E of Maputo, 7.iii.1958, *Mogg* 31567 (K, PRE); Marracuene Dist., near Marracuene, 33 km N of Maputo, 27.ii.1946, *Gomes e Sousa* 3397 (K).

Also in South Africa (KwaZulu-Natal). Usually close to sea either in sandy places, including dunes, or by brackish pools or estuaries; 0–50 m.

Conservation notes: Least Concern as apparently common on and near seashores in southern Africa.

Two other specimens may belong here. *Kirk* s.n. (River Kongone, mouth of Zambesi, i.1861) is leafy and has the lowest involucral bract 20 cm long. C.B. Clarke annotated this as 'C. natalensis Hochst. var. ß longibracteatus, type of var.'. *Vesey-Fitzgerald* 3007

(Malawi, Chinteche, Lake Nyasa, 21.ii.1961 (K)) has leaves to 50 cm long and a lowest involucral bract 20 cm long. The inflorescence is much laxer, with the stalks of the inflorescence subunits up to 15 cm long. It was determined as *Cyperus natalensis* by J. Raynal in 1972. The *de Koning* specimen cited above is also leafy.

Gordon-Gray points out that there appear to be two forms of this species which grow close to one another in the same dune system but in different environments. A leafy form grows in loose wind-blown sand, while the leafless form is found in more stable sandy sites in low-lying open grassland.

A12. **Cyperus denudatus** L.f., Suppl. Pl.: 102 (1782). —Clarke in F.T.A. **8**: 338 (1901). —Kükenthal in Engler, Pflanzenr. **4**, 20(101): 255 (1936). —Haines & Lye, Sedges & Rushes E. Afr.: 169, figs.319–321 (1983). —Gordon-Gray in Strelitzia **2**: 55 (1995). —Hoenselaar in F.T.E.A., Cyperaceae: 190 (2010). Type: South Africa, Cape of Good Hope, without indication of collector (type unknown).

Cyperus platycaulis Baker var. *recedens* Peter & Kük. in Engler, Pflanzenr. **4**, 20(101): 254 (1936). Types: Tanzania, many localities, *Holst* 2045, *Peter* 12528, 14506, 14506a, 23914, 24805, 35546, 36926, 39724, 40683, 46211, 46674 & *Schlieben* 2467 (B syntypes).

Cyperus kasamensis Podlech in Mitt. Bot. Staatssamml. München **4**: 108 (1961). Type: Zambia, Kasama Dist., Mungwi, 29.vi.1960, *Robinson* 4021 (M holotype, K).

Perennial, with a 3 cm thick creeping scale-covered rhizome, up to 95 cm tall; culms tufted, crowded, 19–89 cm long, 0.5–1.6 mm wide, trigonous to slightly triquetrous, smooth. Leaves with leaf sheath reddish-brown to purple, 2–18 cm long; leaf blade absent or reduced to 1 cm long, then reddish-brown. Involucral bracts leaf-like, erect to spreading, 2(3), lowermost 1–4 cm long, 1–2.3 mm wide. Inflorescence a simple to compound anthela, primary branches 5–12, 1.5–6.6(16) cm long; spikelets in digitate clusters, sessile and at the end of primary and secondary branches, 2–5 per cluster, linear-lanceolate, 2.4–8(20) × 0.9–1.7 mm wide, rachis straight, sometimes slightly curved; glumes pale brown to reddish-brown to dark brown, ovate, 1.1–1.7 × 0.6–0.8 mm wide, keel greenish, acute, apex slightly excurrent. Stamens 3; filaments 1.1–1.6 mm long; anthers 0.6–1 mm long. Nutlet whitish when young or depauperate, brown when mature, ovoid to obovoid, 0.5–0.7 × 0.3–0.6 mm wide, base cuneate, muricate when young or depauperate, tuberculate when mature.

Botswana. N: Ngamiland Dist., Kwando, 23.iv.1975, *Williamson* 40 (K, SRGH). **Zambia**. B: Senanga Dist., Siloana Plain, 10.x.1964, *Verboom* 1764 (K). N: Mbala Dist., Saisi R. by bridge on Mbala road, 23.x.1954, *Richards* 2139 (K). C: Mpika Dist., Mutinondo Wilderness area, Kapisano dambo, 6.iv.2003, *Chisala & P.A. Smith* 23 (K). S: Kazangula Dist., Katambora, banks of Zambezi R., 26.viii.1947, *Greenway & Brenan* 7989 (K). **Zimbabwe**. N: Guruve Dist., Nyamnyetsi Estate, 1.ix.1978, *Nyariri* 319 (K, SRGH). W: Matobo Dist., SW Matopos, Maleme Valley, 7.i.1963, *Wild* 5958 (K, SRGH). C: Harare Dist., Kaola Estate, 12 km [7 miles] S of Harare [Salisbury], 14.x.1955, *Drummond* 4896 (K, SRGH). E: Nyanga Dist., Inyanga, 20.x.1935, *Eyles* 8481 (K, SRGH). S: Masvingo Dist., Makoholi Expt. Station, 10.xi.1976, *Senderayi* 58 (K, SRGH). **Malawi**. N: Mzimba Dist., Mzuzu, Marymount, 7.xi.1969, *Pawek* 2952 (K). **Mozambique**. MS: Chimanimani Dist., c. 28 km W of Demba, base of SE slope of Chimanimani Mts, 25.iv.1974, *Pope & Müller* 1300 (K, SRGH). GI: Xai-Xai Dist., Xai-Xai, 7.x.1978, *de Koning* 7270 (K, LMU).

Widespread in west, central and southern tropical Africa. Riverbanks, floodplains, dambos, damp grassland; 900–1750 m.

Conservation notes: Least Concern on account of its wide distribution.

Much confused with *Cyperus haspan*, of which it is essentially a leafless form. In general, *C. denudatus* differs in having no leaf blades, or very short ones, in the short involucral bracts, and in more often having a clear elongated rhizome with regularly spaced culms.

Cyperus denudatus var. *sphaerospermus* is here treated as a good species, *C. sphaerospermus*, following Gordon-Gray (1995). *Cyperus kasamensis* is here placed in synonymy; it comprises particularly well grown and well collected specimens of *C. denudatus* showing the elongated woody rhizome bearing regularly spaced culms.

A13. **Cyperus platycaulis** Baker in J. Linn. Soc., Bot. **22**: 532 (1887). —Kükenthal in Engler, Pflanzenr. **4**, 20(101): 253 (1936). —Hoenselaar in F.T.E.A., Cyperaceae: 190 (2010). Type: Madagascar, *Baron* 4456 (K holotype).

    Cyperus lucentinigricans K. Schum. in Abh. Königl. Akad. Wiss. Berlin **39**: 59 (1894); in Engler, Pflanzenw. Ost-Afrikas **C**: 119 (1895). —Clarke in F.T.A. **8**: 339 (1901). Type: Tanzania, Lushoto Dist., Usambara Mts, *Holst* 3851 (B holotype)

    Cyperus denudatus L.f. var. *lucentinigricans* (K. Schum.) Kük. in Notizbl. Bot. Gart. Berlin-Dahlem **9**: 303 (1925); in Engler, Pflanzenr. **4**, 20(101): 254 (1936). —Haines & Lye, Sedges & Rushes E. Afr.: 170, fig.322 (1983).

Perennial, robust, up to 95 cm tall; culms tufted, 55–90 cm long, 3–7 mm wide, triquetrous to slightly winged, smooth. Leaves up to 23 long; leaf sheath reddish-brown, 5–23 cm long; leaf blade absent or 1 cm long maximum. Involucral bracts bract to leaf-like, 1(2), erect, sometimes giving the inflorescence a lateral appearance, 1–4 cm long, 2–5 mm wide. Inflorescence almost capitate or simple anthela, primary branches 4–13, 1–6 cm long; spikelets in digitate clusters, sessile or at the end of primary branches, 2–7(10) per cluster, linear-lanceolate, 4.8–12 ×1–2 mm wide, rachis straight; glumes dark reddish-brown to almost black, ovate, 1.7–2 × 0.8–1 mm wide, keel acute, apex slightly excurrent. Stamens 3; filaments 1.7–1.8 mm long; anthers 0.9–1 mm long. Nutlet only seen immature.

Botswana. N: Ngamiland Dist., Nhabe R. (Lake River), ix.1896, *Lugard* 10 (K). **Zambia**. N: Mporokoso Dist., Mporokoso-Sanga Hill road, 19.i.1960, *Richards* 12428 (K). **Zimbabwe**. C: Marondera Dist., 27.iv.1948, *Corby* 108 (K, SRGH). **Malawi**. N: Mzimba Dist., Mzuzu, Lusamgazi, Mbowe Dam, 24.ii.1961, *Vesey-Fitzgerald* 3030 (K).

Also in Uganda, Kenya, Tanzania and Madagascar. Mainly in wet areas, swamps, bordering lakes and ponds; (950)1100–2950 m.

Conservation notes: Least Concern because although there are few specimens from the Flora area, it is more common in East Africa.

Hoenselaar in F.T.E.A. regarded this as very close to *Cyperus denudatus*, with which I agree. The differences are largely in size and the stem cross-section (markedly triquetrous in *C. platycaulis*) and there is very little African material that is as large as the type, from Madagascar. Only *Richards* 12428 attains the stature of the East African material. I cannot be sure that most of the material cited here is not just exceptionally vigorous *C. denudatus*. The species is included here with some hesitation.

Group B: Annual species.

Annual species have no rhizomes or tubers, and usually have a weak root system so that specimens are almost always of whole plants, with the root system. The division between annual and perennial species is usually clear, but a few species such as *Cyperus distans*, can persist from one year to the next, and some quite large species such as *C. imbricatus* may flower in their first year. See also the notes on the genus.

Beware also of some species with capitate inflorescences, particularly *Cyperus clavinux* and *C. remotiflorus*, which can appear to be annual but in fact are small perennials forming loose clumps from which single stems may be collected; these are covered by Key C.

1.   Median nerve of glumes excurrent at apex, markedly recurved, spreading 2
–   Apices of glumes ± straight or nerve not excurrent 8

2. Inflorescence dense, almost capitate, often with a further single capitate stalked unit; glumes dark, aristate, but only slightly curved . . . . **B1.** *hamulosus*
– Inflorescence clearly either digitately or capitately one or twice anthelate. 3
3. Inflorescence once or twice spicate-anthelate; glumes yellow-brown; spikelets densely packed; plants usually <20 cm tall. 4
– Inflorescence once capitate-anthelate 5
4. Nutlets densely papillose with small papillae. . . . . **B2.** *squarrosus*
– Nutlets sparsely papillose with conical papillae. . . . . **B3.** *atriceps*
5. Inflorescence open; glumes yellowish or red-brown. 6
– Inflorescence dense; glumes pale with a dark midrib. 7
6. Glumes yellowish; glumes well separated and recurved, giving a serrate appearance to the spikelets; plants often >20 cm in height . . . . **B4.** *reduncus*
– Glumes red-brown; glumes closely packed, not recurved; plants usually <20 cm in height . . . . **B5.** *cancellatus*
7. Culms 0.2–0.5 mm in diam.; involucral bracts 0.3–0.7 mm wide; glumes 1.3–2.2 mm long; nutlet 0.5–0.8 mm long . . . . **B6.** *cuspidatus*
– Culms 0.5–0.9 mm in diam.; involucral bracts 1.5–2.3 mm wide; glumes 2.2–2.8 mm long; nutlet 1–1.1 mm long. . . . . **B7.** *maderaspatanus*
8. Inflorescence a single group of densely capitate heads; glumes greenish. . . . . **B8.** *michelianus*
– Inflorescence more open, mostly diffusely capitate-anthelate 9
9. Glumes with a conspicuous black patch on the proximal side towards the base; glumes rounded at apex 10
– Glumes without conspicuous black or dark brown centres 11
10. Spikelets 2–5 mm broad, flattened, obtuse at apex. . . . . **B9.** *pustulatus*
– Spikelets 1–1.5 mm broad, more or less terete, acute at apex . . . . **B10.** *sphacelatus*
11. Inflorescence bracts 1–2(3), the lowermost at least 4 times as long as the inflorescence; glumes ovate, blunt, with brown centres. . . . . **B11.** *submicrolepis*
– Inflorescence bracts >2; glumes usually acute to acuminate 12
12. Spikelets ovate to oblong 13
– Spikelets very narrowly ovate to linear 15
13. Inflorescence very dense with very numerous spikelets; glumes small, golden brown, concave, rounded at apex. . . . . **B12.** *difformis*
– Inflorescence more open so that the individual spikelets are clearly visible; glumes various but not as above 14
14. Inflorescence lax and more-or-less spicate; glumes 1.4–1.9 mm long . . . **B13.** *iria*
– Inflorescence diffuse; spikelets flattened; glumes dull pale green with a conspicuous excurrent median nerve, 2.7–4.7 mm long . . . . **B14.** *compressus*
15. Robust plants with very narrow (>1 mm wide) dark brown spikelets . . . . **B15.** *distans*
– Delicate plants; spikelets broader and not usually dark brown 16
16. Spikelets in dense groups of 5–25, brown 17
– Spikelets in smaller groups, yellowish or greenish 18
17. Spikelets in spherical clusters, broad; involucral bracts 3–5 mm wide. . . . . **B16.** *pyrotechnicus*
– Spikelets in fan-shaped clusters, narrower; involucral bracts 0.5–2.8 mm wide . . . . **B.17.** *amabilis*
18. Spikelets >1.5 cm long, greenish brown, flattened . . . . **B18.** *zollingeri*
– Spikelets <1 cm long, yellowish or brownish, flattened. 19
19. Involucral bract filiform, usually <3 mm wide . . . . **B19.** *tenuispica*
– Involucral bract leaf-like, at least 4 mm wide. . . . .**B20.** *foliaceus*

B1. **Cyperus hamulosus** M. Bieb., Fl. Taur.-Caucas. **1**: 35 (1808). Type: Ukraine, "ad ripas Borysthenis circa oppidulum Berislaw" (?LECB, not seen).

> *Mariscus hamulosus* (M. Bieb.) S.S. Hooper in Kew Bull. **26**: 578 (1972).
> *Scirpus lugardii* C.B. Clarke in F.T.A. **8**: 458 (1902). Type: Botswana, Okavango Valley, vi.1898, *Lugard* 290 (K holotype).

Caespitose annual; culms 5–15 cm tall, erect, trigonous. Leaves narrowly linear, 0.5–1 mm broad, usually shorter than the stems; lower leaf-sheaths reddish-purple. Bracts (2)3(4), the longest up to 6.5 cm long, exceeding the inflorescence. Inflorescence a very simple anthela, with 1–3 dense sessile broadly ovoid spikes 5–6 mm long and 4–5 mm broad, and a single elongated branch 0.5–2 cm long bearing 1–3 usually somewhat smaller spikes. Spikelets 2–2.5(4) × 1.5–2.5 mm, broadly ovoid with 6–8 flowers. Glumes c. 0.8 mm long, excluding the 0.6–1 mm-long arista, in 3 rows of irregularly distichous, dark red-brown to purplish-black with a green keel, narrowly to very narrowly ovate, acuminate, aristate. Stamen 1; stigmas 3. Nutlet 0.75–1 mm long, oblong-ovoid, trigonous, with face against rachilla, brownish.

Botswana. N: Ngamiland Dist., Okavango Swamps, junction of Karongana & Thaoge rivers, 5.x.1974, *P.A. Smith* 1138 (K, SRGH); Okavango Delta, Kundum R. area, Beacon Is., 9.ix.1975, *Hiemstra* 257 (K).

Also in NE Nigeria, Senegal, north Africa, SE Europe, west Asia. Seasonally wet sandy areas; 900–1000 m in the Flora area.

Conservation notes: Data Deficient. All annual *Cyperus* have seeds that can probably persist for many years so that collections may be a poor guide to real potential abundance.

The description is taken largely from Davis (ed.), Fl. Turkey **11**, suppl.2: 306 (2000) but has been checked against the few specimens from the Flora area.

This taxon has been placed in *Cyperus, Mariscus* and *Scirpus*; present opinion places it in *Cyperus*.

A remarkably widely distributed species; the discontinuity between its localities in Eastern Europe, northern Nigeria and Botswana is very large; as a wetland species, could it have arrived on the feet of migrating birds?

B2. **Cyperus squarrosus** L., Cent. Pl. II: 6 (1756). —Kükenthal in Engler, Pflanzenr. **4**, 20(101): 505 (1936). —Haines & Lye, Sedges & Rushes E. Afr.: 253, figs.501,502 (1983). —Hoenselaar in F.T.E.A., Cyperaceae: 215, fig.33. Type: India, (LINN lectotype). FIGURE 14.**32**.

> *Cyperus aristatus* Rottb., Descr. Icon. Rar. Pl.: 22 (1773). —Clarke in F.T.A. **8**: 348 (1901). —Kükenthal in Engler, Pflanzenr. **4**, 20(101): 502 (1936), superfluous name.
> *Mariscus squarrosus* (L.) C.B. Clarke in Hooker, Fl. Brit. India **6**: 623 (1893); in F.T.A. **8**: 400 (1902). —Gordon-Gray in Strelitzia **2**: 135 (1995).

Annual, fairly slender, up to 25 cm tall, not swollen at base, with a small root system; culms solitary or crowded and tufted, 1–26 cm long, 0.4–3 mm wide, trigonous, almost glabrous. Leaves up to 16 cm long; leaf sheath green to purple, rather wide, 0.6–4.7cm long; leaf blade linear, flat, 2–13.5 cm long, 1–4 mm wide, slightly scabrid to glabrous, apex acuminate. Involucral bracts leaf-like, erect to spreading, 2–5, lowermost 1.6–12 cm long, 1–5 mm wide. Inflorescence a simple anthela of 1–2(3) sessile and 1–5(6) stalked branches, 0.5–7 cm long; spikelets in dense spikes, sessile and at the end of primary branches, 6–41 per spike, crowded, linear, 2–7 × 1.5–3 mm wide, flattened, squarrose with recurved glume-apices, 5–15-flowered; glumes yellowish to reddish-brown, elliptic, 1.4–2.7 mm long (including 0.3–1.1 mm long mucro), 0.3–0.4 mm wide, keel green, strongly excurrent with 3–4 veins on either side, apex strongly mucronate, recurved. Stamens 1; anthers 0.2–0.3 mm long. Nutlet dark grey, narrowly oblong to obovoid, trigonous, 0.5–0.8 × 0.2–0.4 mm wide, minutely papillose, disarticulating at its base but held by persistent glume so that it falls with the spikelet.

Botswana. N. Ngamiland Dist., on side of track North to South Gate, Moremi Nat. Park, 19 ii.2008, *A. & R. Heath* 1517 (K). SW: Kgalagadi Dist., Mabuasehube Game Reserve, Mabuasehube Pan, 12.iii.1949, *Ellis* 2649 (K, PRE). **Zambia**. N: Chinsali Dist., c. 60 km W of Isoka, 15.iii.1975, *Hooper & Townsend* 787 (K). W: Solwezi Dist., 5 km E of Kabompo R. on Mwinilunga–Solwezi road, 27.ii.1975, *Hooper & Townsend* 425 (K). C: Kabwe Dist., 1 km W of Kamaila Forest Station, 36 km N of Lusaka, 9.ii.1975, *Brummitt et al.* 14313 (K). E: Chipata Dist., Chadiza Hill, 2.iii.1973, *Kornaś* 3383 (K). S: Choma

Fig. 14.**32**. CYPERUS SQUARROSUS. 1, habit (× 2/3); 2, habit of small plant (× 1); 3, spike (× 5); 4, spikelet (× 8); 5, glume (× 32); 6, flower (× 32); 7, nutlet (× 40). 1 from *Bogdan* 2992; 2 from *Gillett* 13039; 3, 5, 6 from *Leippert* 5041; 4, 7 from *Muasya et al.* NMK 305. Drawn by Juliet Williamson. Reproduced from Flora of Tropical East Africa (2010).

Dist., Mapanza W, 7.ii.1954, *Robinson* 510 (K). **Zimbabwe**. N: Gokwe Dist., 5 km N on Gokwe to Chinyenyetu road, 25.iii.1963, *Bingham* 593 (K, SRGH). W: Matobo Dist., Farm Besna Kobila, iii.1957, *Miller* 4176 (K, SRGH). C: Kwekwe Dist., c. 50 km N of Kwekwe Town, Iwaba Estate, 10 km S of camp, 12.ii.2001, *Lye* 24544 (K). E: Hurungwe Dist., Km 349.5, Harare to Chirundu, Hurungwe Nat. Park, S to Chirundu, 14.ii.1981, *Philcox et al.* 8536 (K). S: Beitbridge Dist., Beitbridge, near bridge, 17.iii.1967, *Rushworth* 380 (K, SRGH). **Malawi**. N: Rumphi Dist., Nyika Plateau, 10 km [6.5 miles] north of M1, 11.iii.1978, *Pawek* 14058 (K, MO). C: Dedza Dist., 11km NW of Dedza on Lilongwe road, 18.ii.1978, *Brummitt* 8617a (K). S: Blantyre Dist., Nyambadwe, 7.v.1980, *Townsend* 2146 (K). **Mozambique**. T: Tete (Tette), Zambesia, ii.1859, *Kirk* s.n. (K).

Widespread in Africa, India, Australia, and the Americas. Seasonally wet grassland, roadsides and temporarily wet sandy places; 500–1650 m.

Conservation notes: Least Concern; a common and widespread species.

Vorster, in his unpublished thesis on the genus *Mariscus*, placed *Cyperus squarrosus* in his unpublished genus *Monandrus*. This is not the place to discuss the validity of this genus. He also described but did not validly publish two new species, *Monandrus longicarpus* and *M. atriceps* (see below under *Cyperus atriceps*), as well as a subspecies of *Mariscus squarrosus*, subsp. *ovamboensis*. Only very limited material of most of these taxa is available at Kew and I am unable to assess their validity, although the name *Cyperus atriceps* has recently been validated by Archer and Goetgebuhr and I accept it here, although with some hesitation. Relevant specimens are (all listed by Vorster in his unpublished thesis): '*Monandrus longiscapus*': Botswana N: Ngamiland Dist., Mboma Camp, 14.ii.1974, *P.A. Smith* 836 (K); Zimbabwe N: Gokwe Dist., 8 km N of Gokwe on road to Chinyenyetu, *Bingham* 593 (K). '*Monandrus squarrosus* subsp. *ovamboensis*': Botswana N: Ngamiland Dist., 18 km S of Khardoum valley, Namibia border, 14.iii.1965, *Wild & Drummond* 7036 (K, SRGH); Zambia S/Zimbabwe W: Victoria Falls, ii.1906, *Allen* 286 (K).

B3. **Cyperus atriceps** (Kük.) C. Archer & Goetgh. in Bothalia **41**: 300 (2011). Type: Namibia, Grootfontein, 25.iv.1934, *Dinter* 7377 (B lectotype, K, PRE), lectotypified by Archer & Goetghebeur (2011).

   Cyperus aristatus Rottb. var. *atriceps* Kük. in Mitth. Thüring. Bot. Vereins **2**(50): 8 (1943).
   Mariscus aristatus (Rottb.) Cherm. var. *atriceps* (Kük.) Podlech in Mitt. Bot. Staatssamml. München **3**: 523 (1960); in in Merxmüller, Prodr. Fl. SW Afr. **165**: 36 (1967).

Small erect annual, very similar to *Cyperus squarrosus* but differing mainly in the nutlets, which are obovoid, with sparsely spaced conical papillae.

Botswana. N: Ngamiland Dist., Mompswe [Mumpswe] Pan, *Drummond & Seagrief* 5170 (K, SRGH); Ghanzi Dist., 140 km NW of Ghanzi on road to Maun via Kuki, 17.iii.1976, *Ellis* 2701 (K).

Also in Namibia and South Africa. Temporarily wet places, pan edges; 900–1000 m.

Conservation notes: Data Deficient. Difficult to assess because of potential confusion with *Cyperus squarrosus*.

The distinction between this and *Cyperus squarrosus* is fairly thin, and perhaps subspecific or varietal status might be more appropriate for both this and Vorster's unpublished *Monandrus longiscapus*.

B4. **Cyperus reduncus** Boeckeler in Linnaea **35**: 580 (1868). —Clarke in F.T.A. **8**: 329 (1901). —Haines & Lye, Sedges & Rushes E. Afr.: 160, fig.297 (1983). — Hoenselaar in F.T.E.A., Cyperaceae: 217 (2010). Type: Ethiopia, Gapdia, 19.ix.1838, *Schimper* s.n. (B holotype).

   Cyperus aristatus C.B. Clarke in J. Linn. Soc., Bot. **21**: 90 (1884), illegitimate name, non Rottb. (1772).

Annual, up to 40 cm tall; culms tufted, 3.5–35 cm long, 0.8–2.2 mm wide, trigonous, glabrous to scabrid. Leaves up to 36 cm long; leaf sheath green to pale brown, 1.5–6 cm long; leaf blade linear, flat with few longitudinal ribs, 7–30 cm long, 2–4 mm wide, scabrid on ribs and margins, apex acuminate. Involucral bracts leaf-like, erect or spreading, overtopping the inflorescence, 4–7, lowermost 7–33 cm long, 2–6 mm wide. Inflorescence simple to compound, primary branches 4–8, 2–8.5 cm long; spikelets in laxly sub-digitate clusters, sessile and at the end of primary and secondary branches, 3–10 per cluster, 4.8–12.4 × 3.2–4.9 mm wide, with spreading glumes; glumes greenish-yellow to reddish-brown, the margins uncoloured, linear-elliptic, 1.9–2.5 × 0.4–0.8 mm wide, keel green, 3-veined, apex strongly mucronate, recurved. Stamens 3; filaments 1.6–2.2 mm long. Nutlet brownish-red, linear-oblong, 1.6–2.2 × 0.3–0.5 mm wide, minutely papillose in longitudinal rows.

Zambia. B: Kabompo Dist., 65 km W of Kabompo, 7.vii.1963, *Robinson* 5566 (K). N: Mbala Dist., Namkolo (Niamkolo), 11.vi.1961, *Robinson* 4702 (K).

Widespread in tropical west and central Africa, Ethiopia, Sudan, Uganda and Tanzania. In wet, often disturbed areas; sometimes a weed of rice fields; 950–1200 m.

Conservation notes: Data Deficient. See note under *C. hamulosus*.

Like most annual *Cyperus*, this species is very variable in size. *Richards* 8977 is at least 40 cm tall, while *Robinson* 4702, described as a weed in a rice field, is 15 cm tall at most.

B5. **Cyperus cancellatus** Ridl. in Trans. Linn. Soc. London, Bot. **2**: 131 (1884). — Clarke in F.T.A. **8**: 327 (1901). Types: Angola, *Welwitsch* 6916 & 6917 (BM syntypes).

Cyperus haspan auth var. *cancellatus* (Ridl.) Kük. in Engler, Pflanzenr. **4**, 20(101): 249 (1936).

Annual forming small clumps. Stems less than 30 cm tall, triquetrous, strongly ribbed. Leaves linear, flat, glabrous, acuminate, much shorter than the culms; sheaths thin, split, faintly purple-spotted. Involucral bracts 2–3, linear, one of them about 5 cm long. Inflorescence once – (very occasionally twice-) anthelate, the secondary peduncles slender, terete, unequal. Spikelets 10–18-flowered, 2–10 mm long. Glumes ovate, obtuse, mucronate, spreading, red-brown, with a recurved short dark mucro; keel paler, punctate. Rachilla slender, flexuous, the scars narrow, oblong. Stamens 3; filaments long; anthers very short, ovate-oblong. Style slender, shortly trifid, reddish. Nutlet tiny, subglobose, trigonous, less than half the length of the glumes, whitish, irregularly ridged and papillose.

Zambia. N: Mporokoso Dist., Chibya R., tributary of Choma R., NW of Mweru Wantipa, 1 mile N of Shadrakaputa, 7.viii.1962, *Tyrer* 350 (BM). **Zimbabwe**. E: Mutare Dist., Nyumquara Valley, ii.1935, *Gilliland* K1555 (K). **Mozambique**. N: Ribáuè Dist., Ribáuè, viii.1931, *Gomes e Sousa* 773 (K).

Also in Angola. In seasonally damp places, or in shallow soil over rock; 600–1000 m.

Conservation notes: Data Deficient. See note under *Cyperus hamulosus*.

A small and probably inconspicuous annual, known from the flora area only from these two collections. Possibly overlooked. The very slender ridged stems and the red-brown spikelets are distinctive. Very different from *Cyperus haspan*.

Clarke (1901) recognised Ridley's var. *gracillimus* of which the type is *Welwitsch* 6862 (BM).

B6. **Cyperus cuspidatus** Kunth in Humboldt, Bonpland & Kunth, Nov. Gen. Sp. **1**: 204 (1816). —Kükenthal in Engler, Pflanzenr. **4**, 20(101): 261 (1936). —Haines & Lye, Sedges & Rushes E. Afr.: 252, figs.507,508 (1983). —Gordon-Gray in Strelitzia **2**: 55 (1995). —Hoenselaar in F.T.E.A., Cyperaceae: 191 (2010). Type: Venezuela, "flumilis Orinoci prope Maypures, Atures et Carichana", *Bonpland* 5049 (P holotype.).

Annual, slender, up to 25 cm tall, often much less, with a slender root system; culms tufted, 1–17 cm long, 0.2–0.5 mm wide, trigonous, glabrous. Leaves up to 13.4 cm long; leaf sheath reddish-brown to purple, 0.1–1.4 cm long; leaf blade linear, flat or inrolled, 1–12 cm long, 0.2–1.1 mm wide, slightly scabrid near the apex, apex acuminate. Involucral bracts leaf-like to filiform, spreading, 2–7, lowermost 2–13 cm long, 0.3–0.7 mm wide. Inflorescence usually capitate, sometimes a simple anthela, primary branches 1–4, 0.5–8.5 cm long; spikelets in digitate clusters, sessile and at the end of primary branches, 4–25 per spike, linear, squarrose, 4–10 × 1–2 mm wide, 8–25-flowered, rachis straight; glumes orange to reddish-brown, truncate, 1.3–2.2 mm long (including 0.4–0.8 recurved mucro) × 0.4–0.5 mm wide, keel green, excurrent, strongly 3-veined, apex mucronate, recurved. Stamens 1–3; anthers c. 0.2 mm long. Nutlet reddish-brown with darker, almost black, angles, obovoid, trigonous, (0.5)0.7–0.8 × (0.2)0.3–0.4 mm wide, densely papillose, most markedly on the angles.

Zambia. B: Mongu Dist., Mongu, 11.iv.1966, *Robinson* 6928 (K). N: Kasama Dist., Lua Lua Hotel, 6 km (4 miles) N of Kasama, 26.vi.1956, *Robinson* 1746 (K). C: Kabwe Dist., Muka Mwanje Hills near Kalwelwe, 25 km SW of Kabwe, 25.ii.1973, *Kornaś* 3289 (K). S: Livingstone Dist., Livingstone, 7.iv.1956, *Robinson* 1454 (K). **Zimbabwe**. N: Hurungwe Dist., Mwami [Miami], K34 Expt. Farm, 6.iii.1947, *Wild* 1763 (K, SRGH). W: Matobo Dist., SW Matopos, Mt Silozwe, 9.i.1963, *Wild* 5971 (K, SRGH). C: Marondera Dist., Marondera (Marandellas), 14.v.1931, *Brain* 4275 (K). E: Nyanga Dist., Mandea Range, Honde Valley, 30.iii.1969, *Plowes* 3176 (K, SRGH). S: Beitbridge Dist., Tshiturapadsi Dip Camp, c. 80 km [55 miles] E of Beitbridge, 18.iii.1967, *Rushworth* 411 (K, SRGH). **Malawi**. S: Chikwawa Dist., Lengwe Game Reserve, 5.iii.1970, *Brummitt* 8889 (K). **Mozambique**. Z: Namagoa Dist., Mocuba, 6.v.1948, *Faulkner* 258a (K). MS: Dondo Dist., Beira, 2.iv.1898, *Schlechter* 12175 (K).

Widespread in Africa, Asia and the Americas. In seasonally wet grassland, usually short, patchy and on sandy soil; also on and around rocky outcrops (usually granite); 200–1650 m.

Conservation notes: Least Concern; a widespread species.

B7. **Cyperus maderaspatanus** Willd., Sp. Pl. **1**: 278 (1797). —Haines & Lye, Sedges & Rushes E. Afr.: 253, fig.509 (1983). Type: India, no further details (B-W holotype).

> *Mariscus maderaspatanus* (Willd.) Napper in J. E. Africa Nat. Hist. Soc. Natl. Mus. **28**: 10 (1971).
>
> *Mariscus squarrosus* C.B. Clarke in F.T.A. **8**: 400 (1902), non *Cyperus squarrosus* L.

Dwarf annual, delicate, up to 20 cm tall; culms tufted, crowded, 2.5–11(20) cm long, 0.5–0.9 mm wide, trigonous, glabrous. Leaves up to 13 cm long; leaf sheath green to purple, 0.9–3.1 cm long; leaf blade linear, flat or inrolled, 1–10 cm long, 0.7–1.1 mm wide, scabrid on margin near apex, apex acuminate. Involucral bracts leaf-like, erect to spreading, 3–5, lowermost 3.5–14.5 cm long, 1.5–2.3 mm wide. Inflorescence a simple anthela, primary branches 1–4, 0.5–1.5 cm long; spikelets in digitate clusters, sessile and at the end of primary branches, 3–16 per cluster, linear-lanceolate, flattened, 7.2–9 × 1–1.3 mm wide, rachis straight, 10–14-flowered; glumes green with red streaks, 2.2–2.8 mm long (including 0.8–1.1 mm long recurved mucro) × 0.4–0.5 mm wide, imbricate at flowering, later diverging and spreading, keel 3-veined, apex mucronate. Stamens 1–2; anthers ± 0.5 mm long. Nutlet brownish, oblong, trigonous, 1–1.1 × c. 0.4 mm wide, minutely tuberculate.

Malawi. S: Chikwawa Dist., Lengwe Game Reserve, 10.iv.1970, *Hall-Martin* 603 (K). **Mozambique**. Z: Mocuba Dist., Namagoa, 6.v.1948, *Faulkner* 258b (K).

Also in Kenya, Uganda and Tanzania, and India; usually as a weed in crops or grassland; 0–500 m.

Conservation notes: Data Deficient. See note under *C. hamulosus*.

B8. **Cyperus michelianus** (L.) Delile, Descr. Egypte, Hist. Nat. **2**: 50 (1813). Type: "Cyperus italicus, omnium minimus, locustis in capi rubro collectis D. Micheli" in Tilli, Cat. Pl.: tab.20, fig.5 (1723), lectotypified on icon by Simpson in Taxon

53: 180 (2004); Epitype: Italy, Lombardia, Prov. di Mantova, juxta Padum (Po) ad Ponte di Borgoforte, 17 m, 19.ix.1904, *Fiori* s.n., Fl. Italica Exsicc. no. 745 (BM epitype), designated by Simpson (2004).

> *Scirpus michelianus* L., Sp. Pl.: 52 (1753).

Annual, tussocky, to 15(23) cm tall; culms crowded, 1–14(22) cm long, 0.5–1.5 mm wide, trigonous, glabrous. Leaves to 12(18) cm long; leaf sheath reddish to purple, 0.7–3 cm long; leaf blade linear, flat but often folded and twisted when dried, 1–12 cm long, 1–2 mm wide, scabrid on margin and primary vein near apex. Involucral bracts leaf-like, spreading, 4–6, lowermost 3–9 cm long, 1.5–2.5 mm wide. Inflorescence capitate, made up out of several spikes, spikelets crowded, many per spike, oblong-lanceolate, 2.5–4.5 × 1–1.8 mm wide; glumes uncoloured below, pale greenish or reddish brown above, narrowly ovate, 1.3–1.8 × 0.7–1 mm wide, keel green, rather thick, apex slightly mucronate but blunt. Stamens 1–2; filaments 1.8–2 mm long; anthers 0.3–0.7 mm long. Style 2-branched. Nutlet yellow to apricot, oblong, lenticular with one flat and one rounded side, 1–1.2 × 0.3–0.4 mm wide, minutely papillose.

Subsp. **pygmaeus** (Rottb.) Asch. & Graebn., Syn. Mitteleur. Fl. **2**(2): 273 (1904). — Kükenthal in Engler, Pflanzenr. **4**, 20(101): 312 (1936). —Haines & Lye, Sedges & Rushes E. Afr.: 262, figs.532,533 (1983). —Hoenselaar in F.T.E.A., Cyperaceae: 156 (2010). Type: India, *König* s.n. (C holotype).

> *Cyperus pygmaeus* Rottb., Descr. Icon. Rar. Pl.: 20 (1773).
> *Pycreus pygmaeus* (Rottb.) Nees in Linnaea **9**: 283 (1834).
> *Juncellus pygmaeus* (Rottb.) C.B. Clarke in Hooker, Fl. Brit. India **6**: 596 (1893); in F.T.A. **8**: 308 (1901).

Zambia. C: Mpika Dist., Luangwa Game Reserve, Mpika Mushilashi/Luangwa confluence, 4.v.1965, *Mitchell* 2794 (K). **Zimbabwe**. N: Kariba Dist., Zambezi R., c. 3 km [2 miles] upstream of Chirundu Bridge, 31.iii.1961, *Drummond & Rutherford-Smith* 7509 (K, SRGH). S: Nuanetsi Dist., Lundi R. near Fishans, 28.iv.1962, *Drummond* 7778 (K, SRGH). **Mozambique**. MS: Gorongosa Dist., Gorongosa Nat. Park (Game Reserve), banks of Urema R., 25.viii.1958, *Chase* 6983A (K, SRGH).

Also in Ghana, Nigeria, Sudan, Ethiopia, Somalia, Kenya, Tanzania, Namibia; Mediterranean, S and E Asia, Australia. In seasonally wet habitats, often on riverside sand or mud after water levels have fallen; 30–1200 m.

Conservation notes: Least Concern from its wide distribution.

A small and rather inconspicuous species growing in unattractive habitats which is probably much overlooked. The papillae on the nutlet are very small indeed, and probably represent individual projecting cells.

B9. **Cyperus pustulatus** Vahl, Enum. Pl. **2**: 341 (1805). —Kükenthal in Engler, Pflanzenr. **4**, 20(101): 161 (1936). —Haines & Lye, Sedges & Rushes E. Afr.: 265, figs.540,541 (1983). —Hoenselaar in F.T.E.A., Cyperaceae: 157 (2010). Type: Guinea, *Thonning* s.n. (C holotype).

> *Juncellus pustulatus* (Vahl) C.B. Clarke in Durand & Schinz, Consp. Fl. Afric. **5**: 546 (1894); in F.T.A. **8**: 307 (1901).

Annual, slender to robust, up to 60 cm tall; culms 20–45 cm long, 0.7–1.5 mm wide, trigonous, glabrous. Leaves up to 45 cm long; leaf sheath brown-grey to purplish red, 2–10 cm long; leaf blade linear, flat to canaliculate, glabrous, 13–33 cm long, 1.7–4 mm wide, apex acuminate, asymmetric, blunt, glabrous. Involucral bracts 2–5, leaf-like, erect to spreading, glabrous, the lowermost 13–40 cm long, 1.2–3.5 mm wide. Inflorescence simple, sometimes capitate; when simple primary branches 2–7, 1.8–20 cm long; spikelets in digitate clusters, sessile and at the end of primary branches, 3–21 per cluster, linear-lanceolate to elliptic, slightly compressed, 7–12 mm long, elongating to 25 mm long in fruit, 2–5 mm wide; glumes greyish green to pale brown, usually with dark red to purple spot (blackish when dry) on the lower part of the margin, giving the spikelet a red to purple band across the centre, ovate, 2.7–3.2 × 2.5–3 mm wide, with prominent lateral veins on either site of the keel, keel rounded, apex rounded. Stamens 2; filaments 1.9–2.5 mm long; anthers 0.6–0.8 mm long. Style with 2 stigma branches. Nutlet brown to greyish-black, sometimes glaucous, trigonous, or more rarely, dorso-ventrally compressed, ellipsoid to rounded to obovoid, 1.4–2 × 1–1.7 mm wide, minutely papillose.

Zambia. N: Kasama Dist., 95 km E of Kasama, 2.ii.1961, *Robinson* 4363 (K). W: Mwinilunga Dist., 6 km N of Kalene Mission, 12.xii.1963, *Robinson* 5926 (K). C: Serenje Dist., Kasanka Nat. Park, 8.iii.2005, *Bingham* 12884 (K). **Mozambique**. MS: Chimoio Dist., Bandula, 6.iv.1952, *Chase* 4607 (BM, K, PRE, SRGH).

Also in Uganda and Tanzania, and widespread in West Africa. Usually in seasonally wet places on shallow soil over laterite or granite; 700–1350 m.

Conservation notes: Widely distributed in Africa, therefore Least Concern.

Often grows with *Pycreus melanacme*; *Bingham* 12884 includes both species, and *Robinson* 5926 notes "Growing with *Pycreus melanachme*".

In the F.T.E.A. area it seems that trigonous fruits are rarer than compressed ones; I have seen one specimen (*Richards* 4663) with compressed fruits in the material at Kew; all others have trigonous fruits. Also, the maximum height in F.T.E.A. is given as 80 cm but none of the F.Z. material exceeds 55 cm. The species is very variable in stature, probably reflecting soil moisture at the time of growth. *Milne-Redhead* 3600, for instance, is very slender with simple inflorescences.

B10. **Cyperus sphacelatus** Rottb., Descr. Icon. Rar. Pl.: 26 (1773). —Clarke in F.T.A. **8**: 346 (1901). —Kükenthal in Engler, Pflanzenr. **4**, 20(101): 129 (1935). —Haines & Lye, Sedges & Rushes E. Afr.: 195, figs.383,384 (1983). —Hoenselaar in F.T.E.A., Cyperaceae 218 (2010). Type: Surinam, *Rolander* s.n. (C holotype).

Annual up to 61 cm tall, slender to medium-sized, with a slightly swollen stem base and many slender roots; culms 27–50 cm long, 1.1–1.8 mm wide, trigonous, smooth. Leaves up to 29 cm long long; leaf sheath green to reddish brown, 2.5–5 cm long; leaf blade linear, flat or W-shaped, 10–24 cm long, 1.6–3.1 mm wide, scabrid on margins and major veins, apex acute to acuminate. Involucral bracts leaflike, spreading, 3–4, lowermost 7–20 cm long, 2.5–3.3 mm wide. Inflorescence simple, primary branches 3–5, 2.5–12.5 cm long; spikelets in loose clusters on an elongated axis, sessile and at the end of primary (and sometimes secondary) branches, 7–12(20) spikelets per cluster, linear, 11–23 × 1.0–1.5 mm wide, axis slightly zig-zag; glumes light brown with dark reddish-brown patch on the margin, ovate, 2.2–2.9 × 1.3–1.9 mm wide, keel slightly green, excurrent, apex acute to acuminate. Stamens 3; filaments 2–2.9 mm long. Nutlet brown, obovoid, trigonous, 1.1–1.4 × 0.5–0.8 mm wide, smooth.

Zambia. N: Nchelenge Dist., E of Lake Mweru, vi.1961, *Chongo* 9 (K); Mbala Dist., Sumbu Bay, Lake Tanganyika, 29.xii.1963, *Richards* 18715 (K).

Widespread throughout tropical Africa and the Americas. Edge of village at margin of small stream; sandy ground at lake edge; 750–1000 m.

Conservation notes: Data Deficient. See note under *Cyperus hamulosus*. There are only two collections from the Flora area.

B11. **Cyperus submicrolepis** Kük. in Engler, Pflanzenr. **4**, 20(101): 241 (1936). — Haines & Lye, Sedges & Rushes E. Afr.: 164, fig.306 (1983). —Hoenselaar in F.T.E.A., Cyperaceae: 192 (2010). Types: Ivory Coast: Man, *Portères* s.n. (P syntype); Nigeria, R. Niger, near Jebba, *Barter* s.n. (K, syntype); Central African Republic, upper Ubangi, *Tisserant* 121, 1559, 1978, 2224 (P syntypes); Sudan, Kulikoro, *Chevalier* 2469 (P syntype); Sudan, Djur Ghattas, *Schweinfurth* 2328 (K syntype, PRE); Bongo, Gir, *Schweinfurth* 5295 (?, syntype); Uganda, District unclear, Zumbua, *Dummer* 2811 (K syntype); Angola, between Chibia and Quihita, *Pearson* 2638 (K syntype).

Cyperus microlepis Boeckeler in Flora **62**: 551 (1879). —Clarke in F.T.A. **8**: 330 (1901), illegitimate name, non Baker (1877).

Annual, slender, with slightly purplish roots, up to 27 cm tall; culms 11–23.5 cm long, 0.8–1.3 mm wide, trigonous, smooth. Leaves up to 30 cm long; leaf sheath green to purplish with a wide transparent margin near the throat, 1–6 cm long; leaf blade linear, flat, 7–26 cm long, 1.1–3 mm wide, with strong longitudinal ribs, glabrous, apex acute to acuminate. Involucral bracts leaf-like, erect or spreading, 2–3, the uppermost much longer than the rest and suberect so that the inflorescence can appear lateral, the lowermost 7.5–26.5 cm long, 1.6–2.7 mm wide. Inflorescence simple, primary branches 3–8, 0.5–2(5) cm long; spikelets in dense digitate clusters, sessile or at the end of primary branches, 7–20 per cluster, ovoid, glumes spreading and showing nutlet when mature, 2.4–6.3 × 1.1–2.4 mm wide; glumes whitish-green, elliptic-ovate, 0.8–1.3 × 0.3–0.6 mm wide, keel with 3 greenish-brown veins, apex acute. Stamens 2. Nutlet grey to brown, ellipsoid-obovoid, 0.9–1.3 × 0.3–0.6 mm long, apiculate, smooth to sometimes minutely papillose.

Zambia. W: Mufumbwe Dist., 7 km E of Mufumbwe (Chizera), 27.iii.1961, *Drummond & Rutherford-Smith* 7459 (K, SRGH).

Also in Senegal, Mali, Guinea, Ivory Coast, Ghana, Nigeria, Central African Republic, D.R. Congo, Uganda, Sudan, and Angola. Edge of shallow laterite pan; 1200 m.

Conservation notes: Least Concern based on its wide distribution in West Africa.

Easy to recognize as its nutlet is larger than the glumes, and is visible at maturity. Very close to *Cyperus difformis* but differs slightly in size, leaf blade and culm width, and the keel is less winged. There is only one collection from the Flora area and the description is taken largely from F.T.E.A.

B12. **Cyperus difformis** L., Cent. Pl. II: 6 (1756). —Clarke in F.T.A. **8**: 330 (1901). — Haines & Lye, Sedges & Rushes E. Afr.: 165, figs.310,311 (1983). —Gordon-Gray in Strelitzia **2**: 56 (1995). —Hoenselaar in F.T.E.A., Cyperaceae: 173 (2010). Type: India, Herb. Linn. No. 70.10 (LINN lectotype), lectotypified by Tucker (Syst. Bot. Monogr. **43**: 50, 1994).

   Cyperus difformis var. *subdecompositus* Kük. in Engler, Pflanzenr. **4**, 20(101): 240 (1936). Types: Nigeria, Sokoto, *Dalziel* 460 (P syntype); Lagos, ix.1905, *Dawodun* 37 (BR syntype, P); Cameroon: near Yaounde, 1.i.1897, *Zenker* 1513 (BR syntype, P); Tanzania, Lushoto Dist., Handei, 24.i.1915, *Peter* 8248 (K syntype, GOET, WAG); 1.v.1915, *Peter* 10114 (K syntype, GOET, WAG); Pangani Dist. Hale, 15.v.1926, *Peter* 40272 (GOET syntype); Madagascar, Bemarivo, *Perrier de la Bathie* 2388; Mauritius, *Sieber* 137 (B syntype).

Annual or sometimes short-lived perennial to 55(67) cm tall; culms 2.5–55 cm long, 1.4–3.2 mm wide, trigonous, smooth. Basal leaves without blades. Leaves up to 46 cm long; leaf sheath green to reddish-brown, 2–10.5 cm long; leaf blade linear, flat, 9–38 cm long, 2.4–8.3 mm wide, glabrous to scabrid on primary veins and margins, apex acuminate. Involucral bracts leaf-like, spreading, 2–4, (3)8–35 cm long, 3–6.4 mm wide. Inflorescence simple to compound, sometimes almost capitate, primary branches 0–14, 0–7 cm long; spikelets in dense digitate clusters, sessile or at the end of primary and secondary branches, 10 to many per cluster, ovoid, 2.2–8.7 × 0.8–1.4 mm wide, glumes spreading and showing nutlet when matures, axis straight, elongating when fruit matures, red with white scars where the glumes were attached; glumes pale yellowish-brown with dark reddish-brown centres, obovate to rounded, 0.5–1 × 0.3–0.8 mm wide, keel green, winged, apex shortly mucronate, rounded. Stamens 2; filaments 0.3–0.63 mm long. Nutlet pale yellowish-brown, darker when fully ripe, ellipsoid-obovoid, 0.3–0.5(0.8) × 0.3–0.5 mm wide, slightly exceeding the glume, smooth to minutely papillose.

Botswana. N: Ngamiland Dist., Roger's Ridge, Transit Road, CNA/NG18, Selinda Reserve, 3.iii.2006, *A. & R. Heath* 1201 (K). SW: Ghanzi Dist., Old Winkel, Farm 39, Ghanzi, 25.i.1970, *Brown* 7975 (K). SE: Central Dist., Orapa Game Park, 12.iv.2005, *P.A. Smith et al.* 137 (K). **Zambia**. N: Mpika Dist., 15 km [10 miles] NE of Mfuwe, 26.iii.1060, *Astle* 5661 (K). W: Solwezi Dist., c. 5 km E of Kabompo R. on Mwinilunga–

Solwezi Road, 27.ii.1975, *Hooper & Townsend* 424 (K). S: Monze Dist., Namanansa R., 3 km [2 miles] E of Becugwa Is., 18.iii.1964, *van Rensburg* KBS2845 (K). **Zimbabwe**. N. Mwami [Miami] Dist., K34 Experimental Farm, 6.iii. 1947, *Wild* 1758 (K, SRGH). W: Bulalima-Mangwe Dist., Manandu Dam, 50 km [30 miles] N of Marula, 20.iv.1972, *Gibbs-Russell* 1723 (K). C: Harare Dist., Harare, University College site, 29.iii.1963, *Loveridge* 639 (K, SRGH). E: Chipinge Dist., Chipinga, 7.iii.1968, *Plowes* 2881 (K, SRGH). S: Masvingo Dist., Kyle Nat. Park, Chenyati Dam, 21.v.1971, *Mavi* 1239 (K, SRGH). **Malawi**. N: Rumphi Dist., Nyika Plateau, ii-iii.1903, *McClounie* 110 (K). C: Nkota-Kota Distr., Benga, W shore of Lake Nyasa, 2.ix.1946, *Brass* 17492 (K). S: Kasupe Dist., Milinje Village, 14.ii.1965, *Salubeni* 245 (K, SRGH). **Mozambique**. Z: Lugela Dist., Namagoa, n.d., *Faulkner* 139 (K). MS: Beira Dist., Cheringoma Section, Lower Chinizua R. 1 km from Gano, 13.vii.1972, *Ward* 7885 (K, NH). GI: Xai-Xai city, pool in city centre, 7.x.1978, *de Koning* 7271 (K, LMU).

Widespread in Africa; SE Asia and the Pacific. In swamps, watersides, in and by temporary pools, seasonally wet grasslands, and roadside ditches, a weed of rice fields; sea-level up to 2300 m.

Conservation notes: Least Concern; a widely distributed and common species.

According to F.T.E.A. this is close to *Cyperus submicrolepis* but much coarser with broader leaves and culm. Hoenselaar gives the length of the nutlets as up to 0.8 mm but I have seen nothing longer than 0.5 mm from the Flora area. *Robinson* 574, from Zambia S: Muckle Neuk, 18 km N of Choma, 27.ii.1954 (K), shows the size range at a single site and date, with flowering individuals between 2.5 and 55 cm tall. In the field the densely packed golden-brown inflorescences are distinctive.

B13. **Cyperus iria** L., Sp. Pl.: 45 (1753). —Clarke in F.T.A. **8**: 346 (1901). —Kükenthal in Engler, Pflanzenr. **4**, 20(101): 150 (1935). —Haines & Lye, Sedges & Rushes E. Afr.: 198, figs.391,392 (1983). —Gordon-Gray in Strelitzia **2**: 59 (1995). —Hoenselaar in F.T.E.A., Cyperaceae: 214 (2010). Type: India, *Osbeck* 70.16 (LINN lectotype), lectotypified by Tucker in Syst. Bot. Monogr. **43**: 91 (1994). FIGURE 14.**33**.

Chlorocyperus iria (L.) Rikli in Pringsh. Jahrb. **27**: 564 (1895).

Annual, up to 60 cm tall; culms tufted, 21–36 cm long, 1.4–2.2 mm wide, trigonous, glabrous. Leaves up to 37 cm long; leaf sheath green to reddish-brown, 3.5–8 cm long; leaf blade linear, flat to folded, 17–29 cm long, 3.2–5 mm wide, scabrid on margins and veins, apex acuminate. Involucral bracts leaf-like, spreading, 2–4, lowermost 14–30 cm long, 2.1–5.0 mm wide. Inflorescence anthelate-paniculate, primary branches 3–8, 1–11 cm long, umbellately or near-umbellately branched at apex; spikelets in irregular clusters, sessile and at the end of primary branches, few to many per cluster, 4–14.3 mm long, 1.6–3.3 mm wide, the rachis straight to zigzag; glumes golden brown with an uncoloured margin, obovate to rounded, 1.4–1.9 mm long, 1.4–1.7 mm wide, keel greenish, with a slightly excurrent midrib, apex rounded. Stamens 2–3; filaments ± 1.3 mm long. Nutlet dark brown to almost black, ellipsoid-obovoid, strongly trigonous, 1.3–1.6 mm long, 0.6–0.8 mm wide, apiculate, smooth on the ridges, minutely papillose to almost smooth elsewhere.

Botswana. N: Ngamiland Dist., Chadun Valley, Namibia (SWA) border, 16 km W of Knau Knau, 14.iii.1965, *Wild & Drummond* 7013 (K, SRGH). **Zambia**. C: Mambwe Dist., Luangwa Valley Game Reserve, Mushilashi Exclus. Plot, 20.iv.1967, *Prince* 515 (K). **Zimbabwe**. N: Lomagundi Dist., Raffingora, Yomba Estate, ii.1974, *Dolloway* GHS231731 (K, SRGH). E: Chipinge Dist., Chibuwe Irrigation Scheme, 7.iii.1968, *Plowes* 2878 (K, SRGH). S: Chiredzi Dist., Hippo Valley Estate, 28.ii.1973, *Lonsdale* 302 (K, SRGH). **Malawi**. S: Zomba Dist., Lake Chilwa road 0.5 km after Govalo Market, 13.i.1978, *Banda* 1257 (K, SRGH). **Mozambique**. GI: Chokwe Dist., Massavasse, 22.i.1987, *Compton* 1596 (K).

Fig. 14.**33**. CYPERUS IRIA. 1, habit; 2, inflorescence detail; 3, spikelet; 4, nutlet. All from *Plowes* 2898. Scale bars: 1 = 3 cm; 2 = 1 cm; 3 = 3 mm; 4 = 1 mm. Drawn by Juliet Williamson.

Widespread throughout Africa and Asia, Australia. Swamp grassland, edge of seasonal pools, streambanks in dry zones, a weed of rice fields; 90–1000 m.

Conservation notes: Least Concern; a widespread species.

Probably more widespread than the records suggest; it is often small and easily overlooked, although material from the flora area is often larger than stated in the description in F.T.E.A.

B14. **Cyperus compressus** L., Sp. Pl.: 46 (1753). —Clarke in F.T.A. **8**: 347 (1901). — Kükenthal in Engler, Pflanzenr. **4**, 20(101): 156 (1935). —Haines & Lye, Sedges & Rushes E. Afr.: 266, fig.542 (1983). —Gordon-Gray in Strelitzia **2**: 54 (1995). — Hoenselaar in F.T.E.A., Cyperaceae: 195 (2010). Type: America (LINN lectotype).

Annual up to 60 cm tall, slender to robust; culms (5)8–60 cm long, 1–3 mm wide, trigonous, glabrous. Leaves up to 49 cm long; leaf sheath pale brown to red to purplish, 0.5–6 cm long; leaf blade linear, flat, 10–43 cm long, 2–8 mm wide, apex acute to acuminate, glabrous to minutely scabrid on the margins. Involucral bracts 2–6, leaf-like, spreading, the lowermost 12–32 cm long, 2–5.5 mm wide. Inflorescence a simple anthela, occasionally capitate, primary branches 2–7, (0)0.5–13 cm long; spikelets in sub-digitate clusters, sessile and at the end of primary branches, 4–12 per cluster, linear-oblong, 10–29 mm long, 2–4 mm wide; glumes (pale) green to (pale) brown, ovate-elliptic, 2.7–4.7 × 1.8–2.9 mm wide, with lateral veins and sometimes a chestnut patch on either side of the keel, keel green, acute, apex mucronate, up to 1 mm long. Stamens 3; filaments 2.9–3.5 mm long; anthers 0.6–1.1 mm long. Nutlet reddish brown to almost black, shiny, ellipsoid to obovoid, 1.4–2 × 0.9–1.3 mm wide, smooth.

Botswana. N: Ngamiland Dist., Okavango R. between Mohembo & Shakawe, 28.iv.1975, *Gibbs Russell* 2849 (K, SRGH). SE: Kweneng Dist., W side of Takatokwane Pan, next to Takatokwane village. 14.ii.2008, *Farrington et al.* MSB474 (K). **Zambia**. B: Mongu Dist., Bulozi Plain, near Malebo gauging station, 12.ii.1999, *Bingham & Luwiika* 11831 (K). N: Mpika Dist., North Luangwa Nat. Park, 7.i.1995, *P.A. Smith* 1024 (K). W: Sesheke Dist., Mwandi, *Bingham* 7389 (K, NU). **Zimbabwe**. C: Gweru Dist., Mlesu School Farm, 13.v.1976, *Biegel* 5298 (K, SRGH). S: Masvingo Dist., Makaholi Expt. Farm, 10.iii.1978, *Senderayi* 155 (K, SRGH). **Malawi**. S: Zomba Dist., W of Lake Chilwa, 21.vi.1962, *Robinson* 5403 (K). **Mozambique**. N: Pemba Dist., Pemba, 27.i.1984, *Groenendijk et al.* 785 (K, WAG). Z: Lugela Dist., Mocuba, Posto Agrícolo, 6.vi.1949, *Barbosa & Carvalho* 2982 (K, LISC). MS: Chimoio Dist., Chimoio (Vila Pery), 15.x.1925, *Surcouf* 94 (K, P). M: Maputo Dist., Delagoa Bay, 5.i.1898, *Schlechter* 11993 (K).

Widespread in tropical Africa from to Somalia, and south to South Africa; Asia, Americas. In roadside ditches, permanent and seasonal pools, on both sandy and clay soil; 0–1500 m.

Conservation notes: Laest Concern; common and widely distributed.

B15. **Cyperus distans** L.f., Suppl. Pl.: 103 (1782). —Clarke in F.T.A. **8**: 349 (1901). — Kükenthal in Engler, Pflanzenr. **4**, 20(101): 137 (1935). —Haines & Lye, Sedges & Rushes E. Afr.: 200, fig.398 (1983). —Gordon-Gray in Strelitzia **2**: 56 (1995). —Hoenselaar in F.T.E.A., Cyperaceae: 249 (2010). Type: India, Herb. Linn. 70.42 (LINN lectotype).

Cyperus distans var. *niger* C.B. Clarke in F.T.A. **8**: 350 (1901). —Kükenthal in Engler, Pflanzenr. **4**, 20(101): 140 (1935). Types: Malawi, Fwambo, *Carson* 14 (K syntype); Ethiopia, *Schimper* 1255 (B syntype).

Cyperus keniensis Kük. in Notizbl. Bot. Gart. Berlin-Dahlem **9**: 306 (1925) as *keniaeensis*. Type: Kenya, N Nyeri Dist., Liki R., *Fries & Fries* 1476 (B holotype, K).

Cyperus distans var. *pseudonutans* Kük. in Engler, Pflanzenr. **4**, 20(101): 140 (1935). Types: many cited, including Tanzania, Lushoto Dist., Usambara, *Holst* 2764, *Peter* 23684 (B syntypes); Tabora Dist., Ngulu, Malongwe, *Peter* 34632, 45882 (B syntypes); Bukoba, *Stuhlmann* 3728 (B syntype).

Cyperus longibracteatus (Cherm.) Kük. var. *subdistans* Kük. in Repert. Spec. Nov. Regni Veg. **26**: 250 (1929). Type: Tanzania, Lushoto Dist., Amani, *Warnecke* 384 (B not found, K).

Mariscus keniensis (Kük.) S.S. Hooper in J. E. Africa Nat. Hist. Soc. Natl. Mus. **28**(124): 12 (1971); in Kew Bull. **26**: 579 (1972).

Annual or perennial, (15)30–150 cm tall, with a short thick rhizome; stems tufted, usually set in a row, or solitary, trigonous to triangular, green and shiny, 1.5–5 mm in diameter, glabrous, the basal part covered with leaf sheaths. Leaves with leaf sheath grey to dark purple, black on old culms; leaf blade green above, linear, slightly channelled, 5–45 × 0.2–1 cm, scabrid on margin and primary vein, attenuate. Involucral bracts 3–5, leaf-like, erect or spreading, longest to 33 cm long. Inflorescence twice anthelate-spciate with the spikelets tending to be aggregated towards tips of branches, to 25 cm in diameter, with 5–15 primary branches to 15 cm long; secondary and tertiary branches very slender, a few cm long or spikelets sessile; spikelets rather laxly set, often at right angles to axis, brown to pale brown, sometimes tinged with green, 6–20 × 0.5–2 mm; rachilla straight to zig-zag when glumes are spreading, with wide transparent wing on two sides, the spikelet often breaking at base with glumes and nutlets persistent on the rachilla; glumes laxly placed, red-brown with green keel, oblong-elliptic, 1.7–2.6 mm long, 3–5-veined, apex obtuse. Stamens 3. Style with 3 branches, white. Nutlet yellowish when young, grey with metallic shine when mature, narrowly ellipsoid, 1.4–1.7 × 0.4–0.5 mm, minutely papillose in longitudinal rows.

Botswana. N: Ngamiland Dist., Tsodilo Hills, 2.v.1975, *Gibbs Russell & Biegel* 2885 (K, PRE). **Zambia**. N: Mansa Dist., Chibongo, Mapula Valley, 10.iv.1961, *Angus* 2800 (FHO, K). W: Mwinilunga Dist., by Matonchi R., 6.xi.1937, *Milne-Redhead* 3116 (K). C: Lusaka Dist., 26 km SE of Lusaka, Lazy J Botanical Reserve, 21.iii.1999, *Bingham* 11994 (K). E: Katete Dist., St Francis' Hospital, 25.v.1963, *Wright* 3 (K). S: Monze Dist., Lochinvar Ranch, 3.i.1964, *Van Rensburg* 2729 (K). **Zimbabwe**. N: Kariba Dist., Chipepo Cleared Area, Lake Kariba, 6.i.1965, *Mitchell* 1002 (K, SRGH). W: Hwange Dist., Victoria Falls, 10–15.iii.1932, *Brain* 8683 (K). E: Hurungwe Dist., Mansa Pan, 16 km [11 miles] ESE of Chirundu Bridge, *Drummond* 5373 (K, SRGH). S: Nuanetsi Dist., Malangwe R., SW Mateke Hills, 5.v.1958, *Drummond* 5575 (K, SRGH). **Malawi**. N: Nkhata Bay Dist.,, Chombe State, Limpasa Drainage, 21.ii.1961, *Vesey-Fitzgerald* 3016 (K). C: Lilongwe Dist., Lilongwe Nature Sanctuary, Lingadzi R., 1.ii.1985, *Patel & Banda* 2033 (K, MAL). S: Blantyre Dist., Maone Estate, 2 km N of Limbe, 9.ii.1970, *Brummitt* 8497 (K). **Mozambique**. Z: Namagoa Dist., Lugela–Mocuba, 1944, Faulkner P20 (K, PRE). MS: ? Dist., Dombe, near Cherene (Chuaca). 26.x.1953, *Gomes Pedro* 4419 (K). M: Maputo (Lourenco Marques), v.1893, *Quintas* 203 (K).

Widespread in Africa, Asia and the Americas. Streamsides, permanently or seasonally swampy or moist sites often in shade, moist sites in cultivated land, forest margins; 200–2300 m.

Conservation notes: Least Concern; common and widely distributed.

B16. **Cyperus pyrotechnicus** Lock, sp. nov. Similar to *Cyperus amabilis* but all its parts bigger; differs by its involucral bracts 3–5 mm wide (not 1–3 mm) and nutlets 1.5 mm long (not 0.4–0.7). Type: Zimbabwe, North Region, Gokwe Dist., 8 km [5 miles] N of Gokwe on road to Chinyenyetu Nhongo, 12.iii.1963, *Bingham* 505 (K holotype, SRGH). FIGURE 14.**34**.

Annual with 1–5 stems from a single base. Culms 10–15(27) cm long and 1–2 mm in diameter, trigonous, glabrous. Leaves all basal, 4–10(17) cm long, 1.5–4 mm broad, flat, glabrous except for a few small teeth on the margins and midrib beneath towards the apex. Involucral bracts 3–5, 9–20 cm long and 3–5 mm wide, glabrous but margins denticulate, particularly towards the apex. Inflorescence once-anthelate with 5–8 long rays and some shorter; peduncles of long rays 6–12 cm long so that the inflorescence is larger than the rest of the plant. Spikelets 1–1.5(2) cm long, very narrowly elliptic, serrate in outline, 8–22-flowered; glumes 2–2.5 × 0.8–1 mm, elliptic, curved, apiculate, somewhat squarrose. Anthers 3, about 0.3 mm long. Style 3-branched. Nutlets narrowly obovoid, 1.5 × 0.5 mm, trigonous, black, minutely papillose.

Zambia. S: Choma Dist., Kabanga, 8 km [5 miles] E of Mapanza, 1.iii.1956, *Robinson* 1356 (K). **Zimbabwe**. W: Hwange Dist., Victoria Falls, 10.iii.1932, *Brain* 8695 (K, SRGH). C: Harare Dist., Harare [Salisbury] 9.v.1931, *Brain* 4143 (K, SRGH).

Not known elsewhere. Margins of temporary pools, sometimes among other vegetation; 900–1500 m.

Conservation notes: Data Deficient; known only from four collections, all in the Flora area.

A distinctive small annual with the inflorescence dwarfing the rest of the plant. The extreme measurements come from *Robinson* 1356 which was noted by the collector as growing among other vegetation and therefore perhaps somewhat etiolated.

The specific name comes from a fancied resemblance between the spherical inflorescence units and a bursting firework.

Fig. 14.**34**. CYPERUS PYROTECHNICUS. 1, habit; 2, spikelet; 3, flower; 4, nutlet; 5, detail of nutlet surface. 1–3 from *Bingham* 505; 4, 5 from *Brain* 4143. Scale bars: 1 = 2 cm; 2 = 2 mm; 3 = 1 mm; 4 = 2 mm; 5 = 0.2 mm. Drawn by Margaret Tebbs.

B17. **Cyperus amabilis** Vahl, Enum. Pl. **2**: 318 (1805). —Clarke in F.T.A. **8**: 327 (1901). —Kükenthal in Engler, Pflanzenr. **4**, 20(101): 265 (1936). —Haines & Lye, Sedges & Rushes E. Afr.: 266, figs. 544,545 (1983). —Hoenselaar in F.T.E.A., Cyperaceae: 158, fig.26. Type: Ghana, *Thonning* s.n. (C holotype).

    Cyperus muelleri Boeckeler in Flora **42**: 434 (1859). —Clarke in F.T.A. **8**: 376 (1901). Type: Mozambique, no specimen indicated.

    Cyperus castaneus Willd. subsp. *amabilis* (Vahl) Lye in Hedberg & Edwards, Fl. Ethiopia Eritrea **6**: 460 (1997).

Slender annual 7–40 cm tall; culms solitary or more often tufted, 4.7–27 cm long, 0.5–1.8 mm wide, trigonous, glabrous. Leaves up to 20 cm long; leaf sheath purplish red, 0.5–3 cm long; leaf blade linear, glabrous, flat or inrolled, 1.5–17 cm long, 1–2.5 mm wide, apex acuminate, glabrous. Involucral bracts leaf-like, 2–7, spreading, 1.2–14.5 cm long, 0.5–2.8 mm wide. Inflorescence occasionally capitate, more often simply anthelate but occasionally compound-anthelate, primary branches 3–10, 1.5–9 cm long; spikelets in digitate, ovoid clusters, sessile and at the end of primary and (when present) secondary branches, 5–25 per cluster, linear, 5–12 × 0.9–1.6 mm wide; glumes pale orange brown, reddish brown or golden brown, linear-elliptic, glabrous, 0.9–1.9 × 0.6–1 mm wide, keel green in life, darker when dry, acute to excurrent, with a single strong vein on each side of the keel, apex shortly mucronate. Stamens 1; filament 0.8–1.6 mm long; anthers 0.3–0.4 mm long. Nutlet trigonous, ellipsoid-obovoid, pale brown, often with darker angles, 0.4–0.7 × 0.3–0.5 mm wide, minutely papillose in longitudinal rows.

Botswana. N: Ngamiland Dist., Okavango, Vei Xlaba, 27.iv.1967, *Geiss* 10053 (K, SWA). SW: Ghanzi Dist., Farm 46, 5.vi.1969, *R.C. Brown* s.n. (K). **Zambia**. B: Sesheke Dist., Barotseland, 31.i.1963, *Angus* 3560 (FHO, K). N: Kaputa Dist., L. Mweru Wantipa, 20 km N of Nsama, *Goyder et al.* 3052 (K). W: Solwezi Dist., 5 km E of Kabompo R. on Mwinilunga–Solwezi road, 27.ii.1975, *Hooper & Townsend* 420 (K). C: Lusaka Dist., Chalimbana Agric. Station, 40 km (25 miles) E of Lusaka, 8.iii.1961, *Angus* 2501 (FHO, K). E: Katete Dist., St Francis' Hospital, 8.iii.1957, *Wright* 169 (K). S: Choma Dist., Dundwa, 10 km (6 miles) S of Mapanza, 6.iv.1953, *Robinson* 167 (K). **Zimbabwe**. Mwami Dist., K34 Experimental Farm, 7.iii.1947, *Wild* 1716 (K, SRGH). W: Hwange Dist., Victoria Falls, 12.iii.1932, *Brain* 8800 (K). C: Marondera Dist., (Marandellas), 20.iii.1948, *Corby* 45 (K, SRGH). E: Mutare Dist., 50 km (35 miles) S of Umtali, 23.iv.1969, *Plowes* 3200 (K, SRGH). **Malawi**. N: Nkhata Bay Dist., Chinteche, Luweya R., 21.ii.1961, *Vesey-Fitzgerald* 3010 (K). S: Blantyre Dist., near Matope Mission, N of Shire R., 12.ii.1970, *Brummitt & Banda* 8534 (K). **Mozambique**. N: Mandimba Dist., Dist., 14 km (10 miles) N of Mandimba Border post, 3.v.1960, *Leach* 9911 (K, SRGH). Z: Mocuba Dist., Mocuba to Namagoa, 1944, *Faulkner* G44 (K). MS: Beira, iv.1921, *Dummer* 4654 (K). GI: Xai-Xai Dist., between Manjacaze and Chongoene, Missão de S. Benedito dos Muchopes, Mangunza, 2.iv.1959, *Barbosa & Lemos* 8472 (K).

Widespread in West Africa, Sudan, Somalia, Uganda, Kenya, Tanzania, South Africa; Asia, Americas. Seasonally wet habitats, often on sandy soil near roads, lakes and swamps, sandy hollows on rocky soil; 20–1700 m.

Conservation notes. Least Concern; common and widely distributed.

The inflorescence is very variable in form but is usually large in proportion to the size of the plant. The papillae on the nutlet are extremely small and barely visible at x40; they are probably at the cellular level.

B18. **Cyperus zollingeri** Steud., Syn. Pl. Glum. **2**: 17 (1854). —Clarke in F.T.A. **8**: 360 (1901). —Kükenthal in Engler, Pflanzenr. **4**, 20(101): 133 (1935). —Haines & Lye, Sedges & Rushes E. Afr.: 196, fig.387 (1983). —Hoenselaar in F.T.E.A., Cyperaceae, 194 (2010). Type: Indonesia, Java, *Zollinger* 2689 (GH).

    Cyperus rubroviridis Cherm. in Bull. Soc. Bot. France **66**: 350 (1920). —Kükenthal in Engler, Pflanzenr. **4**, 20(101): 135 (1935). Types: Madagascar, Berorona, *Perrier* 2395; Ankarafantsika, *Perrier* 2433; Lake Kinkony, *Perrier* 2458 (P syntypes).

Cyperus ramosii Kük. in Repert. Spec. Nov. Regni Veg. **21**: 326 (1925). Type: Philippines, Luzon, Ilocos, *Ramos* 7672 (B holotype).

Annual, up to 53 cm tall; culms solitary or tufted, 12–34 cm long, 0.7–1.9 mm wide, trigonous, smooth. Leaves up to 22.5 cm long; leaf sheath greyish-brown to purple, 2–7 cm long; leaf blade linear, plicate to w-shaped, 8–17 cm long, 1.6–4.3 mm wide, apex acuminate, inrolled, serrulate. Involucral bracts leaf-like, spreading, 5–9, lowermost 10–28 cm long, 1.8–4.8 mm wide. Inflorescence simply anthelate (sometimes capitate), primary branches 5–9, 2–17 cm long; spikelets in digitate clusters, sessile and at the end of long slender primary branches, 1–7 per cluster, linear, subquadrangular in cross-section, 12–53 × 1.6–3 mm wide, rachis zig-zag when glumes shed; glumes light brown, ridged, margins uncoloured, ovate, 2.7–3.8 × 1.4–2.9 mm wide, keel green, slightly excurrent, apex acute. Stamens 3; filaments 2.4–2.9 mm long. Nutlet grey to blackish, obovoid, trigonous, 1.5–1.9 × 0.9–1.2 mm long, almost smooth.

Botswana. N. Ngamiland Dist., Chief's Island Camp, Chief's Is., Okavango Delta, 21.i.2010, *A. & R. Heath* 1901 (K). **Zambia**. N: Mbala Dist., Mpulungu, Cassawa sand dunes, 14.iv.1957, *Richards* 9239 (K). C: Luangwa Valley, i.1971, *Abel* 305, (K, SRGH). S: Livingstone Dist., 3.iv.1956, *Robinson* 1385 (K). **Zimbabwe**. N: Hurungwe Dist., Chirundu Sugar Estate, 6km [4 miles] NE of Chirundu Bridge, 1.ii.1958, *Drummond* 5405 (K, SRGH). W: Hwange Dist., Victoria Falls, 10–15.iii.1932, *Brain* 8767 (K). E. Mutare Dist., Maranki Reserve, 27.ii.1953, *Chase* 4795 (K, SRGH). **Malawi**. N: Nkhata Bay Dist., Chintaca road, 12 km [8 miles] S of N8 junction, 4.iv.1971, *Pawek* 4572 (K, MAL). S: Chikwawa Dist., Lengwe Game Reserve, 10.xi.1970, *Hall-Martin* 602 (K, PRE). **Mozambique**. N: Nampula Dist., Campo Experimental do CICA, 8.iv.1961, *Balsinhas & Marrime* 362 (K, LISC). T: Tette, 1859, *Kirk* s.n. (K). M: Maputo Dist., Delagoa Bay, 27.i.1898, *Schlechter* 12083 (K).

Widespread throughout West Africa and south to South Africa; tropical Asia. In seasonally wet habitats; sea-level up to 1100 m.

Conservation notes: Least Concern; common and widely distributed.

The cited specimens span the size range of this annual species. *Chase* 4795 is barely 10 cm tall and single-stemmed, while *Richards* 9239 is robust and multistemmed with culms up to 50 cm long.

The download from the GH specimen database lists the specimen *Zollinger* 2689 as the type of *Cyperus zollingeri* Steudel, but gives the correct name as *C. ramosii* Kük. Further checking is needed. I am grateful to Christine Bartram (CGE) for directing me to this database.

B19. **Cyperus tenuispica** Steud., Syn. Pl. Glum. **2**: 11 (1854). —Kükenthal in Engler, Pflanzenr. **4**, 20(101): 245 (1936). —Haines & Lye, Sedges & Rushes E. Afr.: 167, figs.313,314 (1983). —Gordon-Gray in Strelitzia **2**: 75 (1995). —Hoenselaar in F.T.E.A., Cyperaceae: 193 (2010). Type: India, Mangalore, *Hohenacker* 1607 (P holotype, K, M).

Annual, up to 30 cm tall, with a small root system; culms few or several, 10–24 cm long, 0.2–1 mm wide, trigonous to 6-angular, glabrous. Leaves up to 18.5 cm long; leaf sheath pale reddish brown to dark brown, 1–4.5 cm long; leaf blade linear, flat, 6.5–14 cm long, 1.3–8 mm wide, glabrous, apex acute to acuminate. Involucral bracts filiform, spreading and often inconspicuous, 3–5, lowermost 5–14 cm long, 1–3(8) mm wide. Inflorescence 2–3-times anthelate, primary branches 1–8, 1–9.5 cm long; spikelets in digitate clusters at the end of primary branches, 2–8 per cluster, linear-lanceolate, 5.3–9.5 × 1.3–1.9 mm wide; glumes red-brown, sometimes with a paler margin, ovate-truncate, 1.1–1.4 × 0.6–1 mm wide, keel excurrent, apex (shortly) mucronate and slightly recurved. Stamens 2–3; filaments 0.7–1.2 mm long. Nutlet whitish, pale brown at maturity, rounded to obovoid, weakly trigonous, 0.5–0.6 × 0.3–0.5 mm wide, reticulate-foveolate.

Botswana. N: Ngamiland Dist., Maun–Mababe road at 19°17.7'S 23°55.5'E, 16.v.1977, *P.A. Smith* 2011 (K, SRGH). **Zambia**. N: Mporokoso Dist., Mweru Wantipa, Kangiri, road through Mawe swamp, 8.iv.1957, *Richards* 9095 (K). W: Solwezi Dist., 100 km W of Solwezi, 15.iv.1960, *Robinson* 3550 (K). C: Chisenga area, Lusaka–Kabwe (Broken Hill) road, 22.iii.1963, *van Rensburg* KBS1788 (K). E: Lundazi Dist., Nsefu, 10.v.1963, *Verboom* 131 S (K). S: Choma Dist., 5 km [3 miles] SW Mapanza, 17.v.1953, *Robinson* 239 (K). **Zimbabwe**. N: Gokwe Dist., Sengwa Research Station, 20.v.1974, *Guy* 2141 (K, SRGH). W: Hwange Dist., Hwange Nat. Park, Makwa Pan, 17.iv.1972, *Gibbs Russell* 1614 (K, SRGH). E: Chimanimani Dist., Vumba Mts, Eastern Beacon slope, 17.iv.1993, *Browning* 555 (K, NU). **Malawi**. N: Rumphi Dist., Kondowe to Karonga, vii.1896, *Whyte* s.n. (K). C: Nkhota Kota Dist., 6 km N of Nkhota Kota, 16.vi.1970, *Brummitt* 11447 (K). **Mozambique**. MS: Beira Dist., Cheringoma Section, Lower Chiniziua Sawmill, 13.vii.1972, *Ward* 7898 (K, NU).

Widespread in tropical Africa and South Africa, also India. Seasonally wet habitats, swamps and rice fields; 0–1700 m.

Conservation notes: Least Concern; common and widely distributed.

Often confused with *Cyperus haspan* and *C. foliaceus*; from the first it is distinct by being very short-lived ('annual') and the lack of rhizomes; from the second it has traditionally been distinguished by number of stamens (2 rather than 3) and nutlet (smooth rather than tuberculate), but Beentje in his F.T.E.A. account found these characters too variable, believing that the most satisfactory difference lies in the more slender habit, with the involucral bract being filiform, whereas in *C. foliaceus* it is consistently leaf-like and more than 4 mm wide.

B20. **Cyperus foliaceus** C.B. Clarke in Bot. Jahrb. Syst. **38**: 134 (1906). —Kükenthal in Engler, Pflanzenr. **4**, 20(101): 247 (1936). —Haines & Lye, Sedges & Rushes E. Afr.: 167, figs.315,316 (1983). —Hoenselaar in F.T.E.A., Cyperaceae: 193 (2010). Type: Tanzania, Lushoto Dist., Amani, *Warnecke* 388 (B holotype).

Annual, slender to robust, up to 78 cm tall, with a minute root system; culms 18–59 cm long, 1.6–4 mm wide, trigonous, glabrous, with longitudinal grooves. Leaves up to 52 cm long; leaf sheath green to greenish-brown, 1–7.5 cm long; leaf blade linear, flat, 18–44 cm long, 2–10 mm wide, often with distinct transverse bars and prominent veins, apex acute to acuminate. Involucral bracts leaf-like, spreading, 3–4(7), lowermost 20–31 cm long, 3–9 mm wide. Inflorescence 1–2-capitate-anthelate, primary branches 7–15, 2–13 cm long, with a green to pale brown tubular prophyll at base; spikelets in digitate clusters, sessile and at the end of primary, secondary and tertiary branches, 2–5 per cluster, linear-lanceolate, 2.9–11 × 1.2–1.9 mm wide, wider during maturation due to spreading of glumes, rachis straight; glumes pale green with reddish-brown bases, margin translucent, truncate, 1.1–1.6 × 0.5–0.7 mm wide, keel green, excurrent, apex mucronate, slightly recurved. Stamens 2–3; filaments 1.1–1.5 mm long; anthers 0.3–0.7 mm long. Nutlet shiny greyish-white, obovoid to almost orbicular, 0.4–0.7 × 0.3–0.5 mm wide, base cuneate, with isodiametric usually tuberculate surface-cells.

Zambia. N: Kawambwa Dist., Luapula Leper Settlement, 2.xii.1961, *Richards* 15487 (K). W: Mufulira Dist., 28.v.1934, *Eyles* 8209 (K, SRGH). **Malawi**. S: Mangochi Dist., Cape Maclear, Chembe village, 13.viii.1987, *Salubeni & Patel* 5103 (K, MAL). **Mozambique**. MS: Dondo Dist., 25 Mile Station (Dondo), 11.iv.1898, *Schlieben* 12266 (K).

Also in Togo, Ethiopia, Uganda, Kenya and Tanzania. Seasonally wet places, including roadsides, sometimes beneath a grass canopy and often on sandy soil; 500–1200 m.

Conservation notes: Data Deficient, but quite widely distributed so perhaps Least Concern.

Group C: Perennial species with capitate inflorescences.

In this group the spikelets all arise from (more or less) on point at the tip of the peduncle, with the inflorescence bracts forming a whorl below them. A few species have more than one capitate inflorescence making up the whole. A number of the species have white or yellow spikelets so that the capitate inflorescence is whitish or yellowish. Be aware that other genera of Cyperaceae also produce spherical whitish inflorescences (*Kyllinga*, *Ascolepis*, *Rhynchospora*, *Sphaerocyperus*, *Kyllingiella*).

Cyperus atractocarpus, which appears first in the key, may look more like a grass than a sedge; it has only 1–3 elongate spikelets but they arise from a single point and it is included in this part of the key.

Note that is essential to have complete specimens with basal parts and inflorescences.

A number of species with either spicate or umbellate terminal inflorescence parts can have condensed inflorescences that may appear capitate at first sight. These include *Cyperus tenax* with almost black spikelets, *C. kipasensis* with numerous small yellowish brown spikelets, and *C. nyererei* with dark spikelets. *Cyperus hemisphaericus* has a dense mass of whitish spicate inflorescence units that may appear capitate. Check if in doubt.

Cyperus cuspidatus and *C. michelianus* have capitate inflorescences but are annual. See the notes on and key to annual species, and check if in doubt.

Two leafless species of wet swampy ground or floating vegetation mats, *C. pectinatus* and *C. colymbetes*, are dealt with in the Group A key.

Cyperus neoschimperi and *C. amauropus* are not yet separated in the key, although this will be possible.

1. Inflorescence of more than one capitate unit, each on a separate elongated pedicel 2
–   Inflorescence of a single capitate unit 3
2. Inflorescence once-anthelate; spikelet clusters 2–3 cm across . . . . . . **C1.** *congestus*
–   Inflorescence twice anthelate; spikelet clusters 1–1.5 cm across . . . . . . **C2.** *solidus*
3. Spikelets 1–3(4), 1.5–4.5 cm long, glumes dark reddish-brown but lowest glumes longer and acuminate, with pale green margins . . . . **C3.** *atractocarpus*
–   Spikelets >3 or, if 3 or fewer then <1.5 cm long; glumes variously coloured 4
4. Inflorescence capitate to subcapitate; spikelets forming a loose head, acute to acuminate, brownish; stems arising from corms on lateral rhizomes 5
–   Inflorescence capitate; spikelets variously coloured, usually white, yellow, orange or green, if brown then stems not arising from corms on lateral rhizomes 6
5. Spikelets darker, red-brown, 6–12 mm long . . . . **C4.** *usitatus*
–   Spikelets straw-yellow, 13–26 mm long . . . . **C5.** *palmatus*
6. Glumes and spikelets bright yellow or orange 7
–   Glumes and spikelets whitish, greenish or brownish 10
7. Rhizome c. 1 cm in diameter, horizontal; shoots single . . . . **C6.** *altochrysocephalus*
–   Rhizome generally <5 mm in diameter, often upright; shoots in groups 8
8. Spikelets flattened, 4–17 in each inflorescence . . . . **C7.** *flavissimus*
–   Spikelets not flattened, >17 in each inflorescence 9
9. Base of plant a mass of old fibrous leaf bases; slender rhizomes absent . . . . **C8.** *chrysocephalus*
–   No basal mass of old leaf sheaths; slender brown-scaly rhizomes present . . . . **C9.** *rhynchosporoides*
10. Large rhizomatous perennials of sandy seashores; leaves leathery, 9–14 mm wide at base; spikelets c. 10–30 mm long . . . . **C10.** *crassipes*
–   Not as above 11

11. Shoots clearly separated from each other by rhizomes. 12
–　Shoots forming dense clumps, or solitary without connecting rhizomes; bases cormose or bulbose. 14
12. Spikelets 7–20 per inflorescence; involucral bracts 4–8, 6–13 mm broad . . . .
. . . .**C11.** *mapanioides*
–　Spikelets >20 per inflorescence; involucral bracts narrower 13
13. Rhizome robust, 3–6 mm diameter, covered by stiff brown scales; leaves 4–8 mm broad . . . . **C12.** *angolensis*
–　Rhizome slender, 1–2 mm diameter, with delicate pale brown scales, leaves <4 mm broad. . . . . **C13.** *diurensis*
14. Spikelets pale to dark brown or reddish brown. 15
–　Spikelets white or whitish 17
15. Base a dense cluster of elongated slender bulbs . . . .
. . . . **C14.** *amauropus*/**C.15.** *neoschimperi*
–　Base a fibrous mass of leaf sheaths and stem bases 16
16. Inflorescences 4–5.5 cm broad; spikelets to 2.5 cm long . . . . **C16.** *longispicula*
–　Inflorescences 1–2(3) cm broad; spikelets to 1.5 cm long . . . . . . **C17.** *semitrifidus*
17. Culms minutely but densely hairy, especially immediately below inflorescence; leaves short, cauline. . . . . **C18.** *albopilosus*
–　Culms glabrous (although the ridges may be scabrid) 18
18. Inflorescence of more than 10 spikelets 19
–　Inflorescence of fewer than 10 spikelets 25
19. Lowermost involucral bracts usually up to 1.5 cm long or less, hardly exceeding inflorescence . . . . **C19.** *nduru*
–　Involucral bracts longer 20
20. Spikelets with numerous very small flowers; stems usually solitary without marked basal bulb or corm (though base may be slightly swollen). . . . . . . **C20.** *pulchellus*
–　Spikelets fewer-flowered; stems usually clumped, often with basal bulb or corm 21
21. Spikelets less than 6 mm long; basal bulb covered with delicate chestnut-brown scales. . . . . **C21.** *dubius*
–　Spikelets larger, the smallest more than 6 mm long; base with hard scales or fibrous. 22
22. Culm bases joined side by side, forming a short segmented rhizome . . . .
. . . . **C22.** *margaritaceus*
–　Culm bases either solitary or in a cluster, without a rhizome 23
23. Plant more than 30 cm tall; base not fibrous. . . . . **C23.** *niveus*
–　Plant less than 30 cm tall; base fibrous. 24
24. Glumes broad, without prominent longitudinal veins, apart from green excurrent midrib. . . . . **C24.** *clavinux*
–　Glumes ovate, with several prominent longitudinal veins; midrib not green . . . .
. . . . **C25.** *remotiflorus*
25. Culm base swollen, bearing several 1–10 cm long slender, horizontal rhizomes .
. . . . **C13.** *diurensis*
–　Culm base swollen, lacking slender horizontal rhizomes 26
26. Spikelets very small, more than 10, not clearly distinct; plant usually more than 30 cm tall from a dense fibrous base. . . . . **C26.** *mollipes*
–　Spikelets larger, usually <10, flattened; basal old leaf sheaths papery, not splitting into fibres. . . . . **C23a.** *niveus* var. *leucocephalus*

C1. **Cyperus congestus** Vahl, Enum. Pl. **2**: 358 (1805). Type: South Africa, Cape of Good Hope "Cap. b. spei." *Stadtmann* s.n. (C, not seen).

> *Mariscus congestus* (Vahl) C.B. Clarke in J. Bot. **35**: 72 (1897), illegitimate name, non (Boeckeler) Kuntze (1891). —Gordon-Gray in Strelitzia **2**: 128 (1995).

Perennial; culms forming a small clump. Leaves numerous, bases often purplish; laminas up to 140 cm long and 4–6 mm wide in middle, tapering gradually to the acute apex, minutely scabrid on the margins and midrib. Culms trigonous, glabrous, 17–90 cm tall and 1.5–3.5 mm in diameter when dry. Inflorescence bracts 3–8, up to 22 cm long and 3–6 mm wide, green. Inflorescence either simply capitate (rarely), or capitate-anthelate with up to 8 rays each up to 11 cm long; inflorescence units subspherical. Spikelets 6–13-flowered, 5–40 × 1–2.5 mm wide, very narrowly ovate to linear, crowded. Glumes ovate, 2.5–4.8 mm long, acute, with a single non-excurrent median nerve and 6–8 smaller lateral veins; keel greenish, sides reddish brown. Stamens 3; anthers 0.5–0.7 mm long. Style 3-branched. Nutlets ellipsoid to narrowly ellipsoid, 3-angled, 1.2–2 × 0.3–0.9 mm, sometimes slightly curved, smooth to somewhat papillose.

Zimbabwe. N: Mazowe Dist., Henderson Research Station, 15.iv.1973, *Gibbs Russell* 2565 (K, SRGH). W: Umzingwane Dist., c. 10 km from Esigodini (Essexvale) on Bulawayo road, 4.ii.1973, *Mavi* 1532 (K, SRGH). C: Harare Dist., University College of Rhodesia and Nyasaland site, Harare [Salisbury], 29.iii.1963, *Loveridge* 638 (K, SRGH).

Also widespread in South Africa. Wet places; sometimes weedy in temporarily wet sites; 1300–1500 m.

Conservation notes: Probably Least Concern as widespread in South Africa.

It appears that this taxon can flower in its first year (as in *Loveridge* 638 above), and may then have a simple or almost simply capitate inflorescence; plants that grow in more permanently wet sites (such as *Mavi* 1532) can have a much larger and more complex inflorescence. However, none of the four specimens from the Flora area at K has simply capitate inflorescences, and there are very few such among the extensive South African collections at K.

A specimen at C has the locality 'Cap.b. spei' and the name 'Stadtmann'. It is labelled as the holotype and this was confirmed by Vorster in 1979. Jean Frederic Stadtmann (1762–1807) was a collector in South Africa.

The type of the var. *pseudonatalensis* (*Marloth* 1028) appears to be labelled 'Kachau'. This may be Kachun Wells, in South Africa; *Marloth* 1027, the type of *Cyperus betschuanus*, is clearly labelled with this locality, which is not in Botswana as I first believed.

I thank Dr M.A.Garcia for these two pieces of information.

C2. **Cyperus solidus** Kunth, Enum. Pl. **2**: 76 (1837). Type: South Africa, Cape, "Ora orientalis Africae australis, Coloniam inter et Port Natal", *Drège* 4410 (B lectotype), lectotypified by Vorster (1986).

> *Mariscus solidus* (Kunth) Vorster in S. African J. Bot. **52**: 265 (1986). —Gordon-Gray in Strelitzia **2**: 134 (1995).

Robust perennial from a thick horizontal rhizome c. 1.5 cm in diameter, covered in old leaf bases. Leaves up to 75 cm long, 3.8–14.5 mm wide, acuminate, V-shaped in transverse section, scabrid on the margins and midrib beneath, otherwise glabrous. Inflorescence bracts 5–11, to 110 cm long, 4–11 mm wide, acuminate, spreading. Inflorescence once or twice spicate-anthelate, the spikes short and dense so as to appear almost spherical; peduncle up to 150 cm long and 2.4–7.3 mm in diameter, obtusely trigonous, glabrous; primary branches up to 22 cm long and 0.7–2.3 mm in diameter; spikes 13–34 × 11–30 mm, subspherical or broadly ovoid. Spikelets 4.6–17 × 1–3 mm, slightly imbricate, 0.6–1.5 mm apart, ovate, acute, concave, midrib sometimes excurrent into a short apiculus, with 6–10 lateral nerves, dark red-brown, midrib greenish. Rachilla flattened, broadly winged. Stamens 3; filaments flattened; anthers 1.5 3.2 × 0.2 mm, linear. Style 3-branched. Nutlets 1.4–2.5 × 0.4–0.7 mm, narrowly obovoid or narrowly ellipsoid, trigonous, papillose, dark brown.

Zimbabwe. E: Nyanga Dist., 10 km [6 miles] N of Troutbeck, Gairesi Ranch, 13.xi.1956, *Robinson* 1876 (K, SRGH). C: Harare Dist., [Salisbury], *Brain* 7707 (fide Vorster). **Mozambique**: GI: Inharrime Dist., Ponta Barra Falsa, 21.xi.1958, *Mogg* 28950 (K). M: Maputo (Lourenço Marques), Maotas (Mahotas), between Mahotas and Costa do Sol, 25.ix.1963, *Macuácua* 111 (K, LMA).

Also in Swaziland and South Africa. Boggy ground, swamps; 0–1400 m.

Conservation notes: Probably Least Concern as widespread in South Africa.

On some leaves the teeth are black, at least towards the leaf apex.

C3. **Cyperus atractocarpus** Ridl. in Trans. Linn. Soc. London, Bot. **2**: 141 (1884). Type: Angola, Empalanca, ii-iv.1850, *Welwitsch* 6863 (BM holotype, LISU). FIGURE 14.**35**.

Perennial from a dense bulbous basal mass of old leaf bases. Leaves linear, 0.2–0.4 mm wide, 2–6 cm long. Sheaths persistent, brown, striate, strongly longitudinally ribbed and 3–4 mm wide at base, disintegrating to leave ribs as fibres. Culms 6–25 cm long, 0.5 mm in diameter. Involucral bracts (1)2–3, linear, 1–3 cm long, shorter than the spikelets. Inflorescence capitate, of 1–3(4) spikelets, erect, linear, each 1.5–4.5 cm long, 10–30-flowered. Glumes reddish brown, narrowly elliptic, 5–7 mm long, strongly longitudinally ribbed; apex acute; keel greenish, ribbed. Stamens probably 2, 4 mm long including 2 mm anthers. Styles flexuous, deeply trifid, branches to 3 mm long. Nutlet [from original description] fusiform-subcylindric, with two pale lines.

Zambia. N: Mbala Dist., Old Katwe road, 25.xi.1954, *Richards* 2355 (K). W: Mwinilunga Dist., 5–6 km (3–4 miles) SE of Angola border, 2–6 km SW of Mujileshi R., 7.xi.1962, *Richards* 16937 (K).

Also in Angola. In dry places on deep sand, sometimes among rocks; 1250–1400 m.

Conservation notes: Data Deficient; known from only a few collections but is small and probably easily overlooked.

This small species is distinctive in its dense fibrous basal mass, inflorescences of 1–3 long spikelets, narrow deeply furrowed red-brown glumes, and long flexuous styles. Dr Sylvia Phillips remarks that it could easily be mistaken for a small species of *Andropogon* (Poaceae).

C4. **Cyperus usitatus** Burch. in Schultes, Mant. **2**: 477 (1824). —Clarke in F.T.A. **8**: 353 (1901). —Kükenthal in Engler, Pflanzenr. **4**, 20(101): 122 (1935). —Haines & Lye, Sedges & Rushes E. Afr.: 192, fig.376 (1983). —Gordon-Gray in Strelitzia **2**: 76 (1995). —Hoenselaar in F.T.E.A., Cyperaceae: 164 (2010). Type: South Africa, Cape of Good Hope, Vyentjes, *Burchell* s.n. (P, not seen).

Cyperus stuhlmannii K. Schum. in Engler, Pflanzenw. Ost-Afrikas **C**: 118 (1895). —Clarke in F.T.A. **8**: 354 (1901). —Kükenthal in Engler, Pflanzenr. **4**, 20(101): 125 (1935). —Hoenselaar in F.T.E.A., Cyperaceae: 165 (2010). Type: Tanzania, Bukoba Dist., Karagwe, Kafuro, *Stuhlmann* 1826 (K isotype).

Cyperus usitatus var. *stuhlmannii* (K. Schum.) Lye in Nordic J. Bot. **3**: 231 (1983). —Haines & Lye, Sedges & Rushes E. Afr.: 193 (1983).

Perennial, slender, up to 44 cm tall, producing thin rhizomes often ending in small bulbs; rhizomes 0.5–10 cm long, 0.2–1 mm in diameter, covered in light reddish-brown scales, sometimes fibrous; bulbs 5–10(20) mm in diameter; culms solitary, 5–42 cm long, 1–3.8 mm wide, triquetrous to trigonous, smooth. Leaves up 30 cm long; leaf sheath grey, straw-coloured to pale brown, 1–5.5 cm long; leaf blade linear, flat, rather thick and sometimes semi-fleshy, shrivelling when dry, 3.5–25 cm long, 1–3.8 mm wide, scabrid on margin at least above, apex acuminate, trigonous. Involucral bracts leaf-like, spreading, 2–4, lowermost 3–10 cm long, 0.6–2.5 mm wide, shape and surface as leaves. Inflorescence more often (loosely) capitate, sometimes anthelate, when simple, primary branches 0–2; spikelets in crowded digitate clusters, 7–25 to many more per cluster, linear, rachis straight, 8–21 × 1.3–3.2 mm wide; glumes golden brown to almost black,

Fig. 14.**35**. CYPERUS ATRACTOCARPUS. 1, habit of small plant; 2, habit of large plant; 3, inflorescence; 4, spikelet. 1 from *Richards* 2355; 2–4 from *Richards* 16937. Scale bars: 1, 2 = 3 cm; 3, 4 = 1 cm. Drawn by Juliet Williamson.

ovate-lanceolate to ovate, 3–6.4 × 1.1–2.5 mm wide, keel sharp and slightly excurrent, with 3–9 ribs on either side, apex acuminate. Stamens 3; filaments 2–5.1 mm long; anthers 1.6–2.9 mm long. Nutlet grey, reddish-brown to dark brown, ellipsoid-oblong to obovoid, trigonous, 1.3–1.6 × 0.6–1 mm wide, apiculate, minutely tuberculate to minutely papillose in longitudinal rows.

Bulbs 5–10 mm in diameter . . . . a) var. *usitatus*
Bulbs 10–20 mm in diameter . . . . b) var. *macrobulbus*

a) Var. **usitatus**

Bulbs 5–10 mm in diameter; glumes dark reddish-brown to almost black. Nutlet ellipsoid-oblong, 1.3–1.6 × 0.6–0.8 mm.

Botswana. N: Central Dist., stream below Mosu Village, Sua Pan, *Ngoni* 348 (K). **Zambia**. N: Mporokoso Dist., Kabwe Plain, Mweru Wantipa, 18.xii.1960, *Richards* 13750 (K). C: Luangwa Dist., Feira road, 24 km S from Kakondwe Mission (Kapoche), 1.i.1973, *Kornaś* 2904 (K). S: Choma Dist., Muckle Neuk, 17 km [12 miles] N of Choma, 11.i.1954, *Robinson* 444 (K).

Also in Ethiopia, Uganda, Kenya, Tanzania, and South Africa. Seasonally wet dambos and grasslands; rough short grassland; 360–1250 m.

Conservation notes: Probably Least Concern as it is widely distributed, although there is no evidence that it is ever common.

Subsp. *palmatus* of F.T.E.A. is now treated as a full species, *Cyperus palmatus* (see C5 below).

b) Var. **macrobulbus** Kük. in Engler, Pflanzenr. **4**, 20(101): 124 (1935). —Haines & Lye, Sedges & Rushes E. Afr.: 192 (1983). —Hoenselaar in F.T.E.A., Cyperaceae: 164 (2010). Types: Namibia, Rehoboth-Aub, 13.iv.1911, *Dinter* 2247 (B syntype); Windhoek, *Foermer* 4 (?B† syntype); Botswana, Ntochokuta, date, *Seiner* 130 (?B† syntype); Tanzania, Turu, near Itigi, 30.xii.1925, *Peter* 33738 (B syntype).

Bulbs 10–20 mm in diameter.

Botswana. SW: Ghanzi Dist., 4 km N of Dondong borehole, 1.ii.1976, *Skarpe* 5-25 (K). Also in Tanzania, Namibia, South Africa. Deep Kalahari sand in *Acacia reficiens* wooded grassland; 1300 m.

Conservation notes: Probably Least Concern as it is widely distributed, although there is no evidence that it is ever common.

C5. **Cyperus palmatus** (Lye) C. Archer & Goetgh. in Bothalia **41**: 300 (2011). Type: Tanzania, Ufipa Dist., Ndago, Milepa-Zimba, 22.i.1951, *Bullock* 3625 (K holotype).

Cyperus usitatus Burch. subsp. *palmatus* Lye in Nordic J. Bot. **3**: 228, fig.22 (1983). —Haines & Lye, Sedges & Rushes E. Afr.: 193, fig.377 (1983). —Hoenselaar in F.T.E.A., Cyperaceae: 165 (2010).

Cyperus fulgens C.B. Clarke var. *contractus* Kük. in Engler, Pflanzenr. **4**, 20(101): 122 (1935), non *Cyperus contractus* Steud.

Perennial. Culms solitary, at the end of a short vertical rhizome that itself arises from a scaly bulb 6–9 mm in diameter, (10)16–33 cm long, trigonous, glabrous, 0.5–1 mm in diameter when dry. Leaves all basal, the lowest reduced to pale brown sheaths, 7–15 cm long and 0.5–1 mm broad, inrolled but flat and with narrow scarious margins at apex. Involucral bracts 2–3, the lowest 5–7 cm long and 2–3 mm wide at base, much exceeding inflorescence. Upper bracts much shorter, usually not exceeding inflorescence. Inflorescence capitate, occasionally with a single short branch bearing another capitate unit. Spikelets 3–12, very narrowly ovate to linear, 1.3–2.6 cm long, acute or acuminate at apex, pale yellowish brown, 15–20-flowered. Glumes narrowly ovate, 0.4–0.6 × 0.2–0.25 mm, pale yellowish brown, often with dark dots, strongly 7–9-nerved. Anthers 1.5 mm long. Styles 3-branched. Nutlet obovoid, 1.3–1.4 × 0.7–1 mm wide.

Botswana. N: Ngamiland Dist., Okavango Delta, 0.8 km S of Chief's Camp, Chief's Is., 2.i.2012, *A. & R. Heath* 2330 (K). SE; Central Dist., stream below Mosu Village, Soa Pan, 16.i.1974, *Ngoni* 348 (K, SRGH). **Zambia**. S: Namwala Dist., Namwala, 9.i.1957, *Robinson* 2088 (IRLCS, K, NRGH, NU, SRGH).

Also in Tanzania. Sandy soils, fringe of seasonal pan, 900–1000 m.

Conservation notes: Data Deficient; no evidence of status and few collections.

All material from the Flora area lacks mature nutlets and few specimens have the bulbs, but is generally identical to East African material.

C6. **Cyperus altochrysocephalus** Lye in Candollea **43**: 508, fig.4 (1988). Type: Zambia, Western Province, 20 miles from Mwinilunga along road to Kalene Hill, 21.xi.1972, *Strid* 2576 (NLH holotype).

Perennial herb. Rhizome 3–7 cm long and up to 13 mm in diameter when dry, fleshy in life, horizontal, covered with sticky brownish or purplish scales; roots reddish. Culms 8–90 cm tall, 1.5–5 mm in diameter, scarcely thickened at the base, trigonous, scabrid on the angles. Leaf sheaths minutely villous. Leaves about equalling the culms, to 50 cm long and 3–9 mm wide, somewhat glaucous; sheaths pale brown, the lowest lacking a lamina. Inflorescence bracts 3–4, reflexed, the lowest up to 10 cm long and 6 mm wide at base, with ciliate margins. Inflorescence capitate, subglobose, 9–18 mm in diameter, of numerous small spikelets. Spikelets linear, 7–8 mm long, 1-flowered. Glumes oblong-elliptic, 6–7 mm long, obtuse, margins inrolled, yellow to orange, brown when old, many-nerved. Stamens 3, anthers linear, c. 1.5 mm long. Styles long, stiff, stigmas 3. Nutlet oblong, trigonous, apiculate, smooth (but cells domed at x40).

Zambia. B: Mongu Dist., 80 km on Mankoya Road, 25.xi.1964, *Verboom* 1782 (K). W: Mwinilunga Dist., Cha Mwana (Chibara's) Plain, 14.x.1937, *Milne-Redhead* 2784 (K).

Also in Angola. In *Cryptosepalum, Guibourtia–Brachystegia, Guibourtia–Baikiaea* or other woodlands on Kalahari sands; 1200–1400 m.

Conservation notes: Probably Near Threatened as it occurs only in a fairly small area on Kalahari sand.

The specimens (of which there are several at Kew, all from Zambia B and W) suggest that each section of rhizome, terminated by a leafy shoot, represents a single season's growth; the burned base of the previous year's shoot is present in a few specimens.

There is substantial variation in leaf width and length – *Verboom* 1782 has very narrow leaves while *Brummitt et al.* 14163 has broader ones.

Browning & Gordon-Gray in Nordic J. Bot. **13**: 507–510 (1993) have reinterpreted the structure of the spikelets which they consider to be consistently one-flowered.

C7. **Cyperus flavissimus** Schrad. in Gött. Gel. Anz. **3**: 2067 (1821). —Hoenselaar in F.T.E.A. Cyperaceae: 149 (2010). Type: South Africa, *Hesse* s.n. (LE holotype).

    Cyperus obtusiflorus Vahl var. *flavissimus* (Schrad.) Boeckeler in Linnaea **35**: 529 (1868). —Kükenthal in Engler, Pflanzenr. **4**, 20(101): 286 (1936).

    Cyperus compactus Lam. var. *flavissimus* (Schrad.) C.B. Clarke in Durand & Schinz, Consp. Fl. Afric. **5**: 552 (1894); in F.T.A. **8**: 320 (1901).

    Cyperus niveus Retz. var. *flavissimus* (Schrad.) Lye in Haines & Lye, Sedges & Rushes E. Afr.: 257 (1983).

Perennial, up to 58 cm tall; culms crowded, bases swollen and fused into a horizontal rhizome, 14–56 cm long, 1.1–1.9 mm wide, trigonous to rounded, with longitudinal grooves, glabrous. Leaves up to 42 cm long; leaf sheath almost black at the base, brown on the culm, 2.5–5 cm long, leaf sheaths at base breaking up into thin fibres; leaf blade linear, flat, 14–37 cm long, (1.9)2.9–4.3 mm wide (see note below), scabrid on margins and primary vein, apex acuminate. Involucral bracts leaf-like, spreading, 3–5, lowermost 4–11.5 cm long, 2.5–4.4 mm wide. Inflorescence capitate, spikelets in a dense head, 4–17, ovoid, 9–19 × 4.8–10 mm wide, rachis straight; glumes bright yellow-orange, ovate to boat-shaped, 6.4–10.3 × 2.7–4.6 mm wide, keel acute, with 6–8 conspicuous striations on either side of keel, apex acute. Stamens 3; filaments 5.4–8.7 mm long; anthers 3.6–4.3 mm long. Nutlet brown to black, obovoid, trigonous, 2.2–3.3 × 1.6–2.5 mm wide, smooth, shortly apiculate.

Zambia. B: Mongu Dist., Barotse floodplain, 5.xii.1964, *Verboom* 1792 (K). N: Mbala Dist., Chilongowelo, 22.xii.1954, *Richards* 3741 (K). W: Mufulira Dist., Nkana to Kitwe, 16.ix.1959, *Shepard* 92 (K). C: Serenje Dist., Kasanka Nat. Park, c. 11 km E of Musande Tent Camp, 4 km W of pontoon over Kasanka R., 19.xi.1993, *Harder et al.* 1987 (K, MO). **Zimbabwe**. E: Mutare Dist.,(Umtali), 16.xii.1954, *Chase* 5352 (K, SRGH). **Malawi**. N: Chitipa Dist., Chitipa [Fort Hill], 1.i.1959, *Robinson* 2981 (K, PRE). C: Mchinji Dist., near Tamanda Mission, 8.i.1959, *Robson* 1099 (K). **Mozambique**. MS: Manica, Chua Valley, 24.xi.1906, *Johnson* 40 (K). [The locality could be read as 'Lua Valley' but Lua and Lua R are in Z according to the Gazetteer].

Also in Tanzania, Swaziland and South Africa. Woodland and grassland, on sandy soil; 1100–1900 m.

Conservation notes: Probably Least Concern as it is widely distributed.

Treated by Haines and Lye as a variety of *Cyperus niveus*, but Hoenselaar (2010) gave it full specific rank because of the striking inflorescence colour and much larger nutlets.

Thompson & Rawlings 609 (CM, K) from Malawi, Chitipa Dist., has very narrow leaves (<0.5 mm) and only 1–2 spikelets in the inflorescence. It may be a high altitude form, or just young. *Pawek* 3117 (K), *Robinson* 2981 (K) and *Chapama et al* 584 (K) from the same area are similar but with 3–4 spikelets in the inflorescence and leaves c. 1 mm wide. *Harder et al.* 1987 (K, MO) from Zambia (C), Kasanka Nat. Park, c. 11 km E of Musande Tent Camp and 4 km W of pontoon over Kasanka R., 19.xi.1993, is also similar, as is *Richards* 2290 from Zambia N. Further study may show that these are recognisable at varietal level. It is also possible that they may belong to Lye's *C. austrochrysocephalus* but I have not seen any authentic material of this taxon.

Milne-Redhead 984 and 2709 from Zambia, Mwinilunga Dist. appear very similar to the narrow-leaved form, if not identical, and have been determined as *Cyperus remotus* (C.B. Clarke) Kük, mentioned by Hoenselaar in F.T.E.A., Cyperaceae: 150 (2010). The description in F.T.A. **8**: 382 (1901), as *Mariscus remotus* C.B. Clarke, is not really adequate and one needs to see the type (*Deschamps* s.n., from D.R. Congo). Hoenselaar did not see this for her F.T.E.A account.

C8. **Cyperus chrysocephalus** (K. Schum.) Kük. in Fries, Wiss. Ergebn. Schwed. Rhodesia-Kongo-Exped. 1911–1912, Erg.: 5 (1921). —Haines & Lye, Sedges & Rushes E. Afr.: 220, fig.446 (1983). —Hoenselaar in F.T.E.A., Cyperaceae: 150 (2010). Types: Angola, Kinebe R., Malumgue, 23.xi.1899, *Baum* 311 (K syntype, M); Mapalauna, 31.xii.1899, *Baum* 311a (K syntype).

Mariscus chrysocephalus K. Schum. in Warburg, Kunene-Sambesi Exped.: 178 (1903).

Perennial, up to 78 cm tall, often showing signs of burning, with a somewhat swollen tussocky base, covered by black and dark brown fibrous remains of leaf sheaths; culms tufted, 18–77 cm long, 0.6–1.6 mm wide, trigonous, longitudinally ridged, glabrous. Leaves up to 42 cm long; leaf sheath black, breaken up into fibres when older, 4–8 cm long; leaf blade linear, folded to canaliculate, 13–34 cm long, 1.3–1.9 mm wide, slightly scabrid on margins. Involucral bracts leaf-like, spreading to recurved, 1–2, lowermost (1.2)3–5(17) cm long, 0.9–1.5(2) mm wide, the upper one much shorter, 0.5–1.5 cm long. Inflorescence capitate, a dense globose head, 7–12 × 7–11 mm wide; spikelets many per head, linear-lanceolate, 5.2–8.9 mm long, 1.1–1.3 mm wide, producing one nutlet only; glumes yellow, linear-lanceolate, 3.8–6 × 1.4–2 mm wide, keel flat with many veins on either side, apex obtuse. Stamens 3; filaments 3.7–5.8 mm long; anthers 1.9–2.4 mm long. Nutlet greyish to blackish, linear-oblong, 2.8–3.2 × 0.6–0.8 mm wide, minutely papillose.

Zambia. N: Mbala Dist., Lunzua Swamp near Mbala (Abercorn), 17.i.1962, *Richards* 15922 (K). W: Solwezi Dist., W of Solwezi on Solwezi–Chingola road, 1.x.1947, *Greenway & Brenan* 8149 (K). C: Serenje Dist., 8 km [5 miles] E of Serenje, 14.x.1967, *Simon & Williamson* 992 (K, SRGH). **Malawi**. N: Mzimba Dist., 5 km W of Mzuzu, Katote, grounds of Bishop's house, 23.v.1970, *Brummitt & Pawek* 11077 (K).

Also in Tanzania, D.R. Congo, Burundi and Angola. Seasonally wet grassland, or thin soil over rock; 1300–1700 m.

Conservation notes: Probably Least Concern because of its wide distribution.

Cyperus boreochrysocephalus Lye (Nordic J. Bot. **3**: 216, 1983) is known from Uganda, Kenya, and N Tanzania but has not yet been found in the Flora area; it could perhaps occur in N Malawi or N Zambia. The base of *C. boreochrysocephalus* is a cluster of slender bulbs, with brown leaf sheaths that do not break up into fibres, recurved flattened leaves, and usually three, not two, involucral bracts.

C9. **Cyperus rhynchosporoides** Kük. in Engler, Pflanzenr. **4**, 20(101): 537 (1936), new name for *C. ochrocephalus*. —Haines & Lye, Sedges & Rushes E. Afr.: 220, fig.447 (1983). Type: Angola, Lunda, Kimbundo, viii.1876, *Pogge* 412 (B holotype). FIGURE 14.**36**.

> *Rhynchospora ochrocephala* Boeckeler in Flora **62**: 568 (1879). Type as for *C. rhynchosporioides*.
> *Cyperus ochrocephalus* (Boeckeler) C.B. Clarke in Durand & Schinz, Consp. Fl. Afr. **5**: 571 (1894); in Trans. Linn. Soc. London, Bot. **4**: 53 (1894); in F.T.A. **8**: 322 (1901); in Illustr. Cyp.: t.8, figs.1–3 (1909), illegitimate name, non Steud (1842). Type as for *C. rhynchosporoides*.

Perennial; shoots solitary or 2–3 together, bearing from base 1–2 slender, 2–3 mm diameter rhizomes, covered in brown scales with purple spots and patches. Culms solitary, erect, 10–35 cm long, trigonous, glabrous but scabrid on angles above. Leaves 15–40 cm long by 3 mm wide, glabrous but scabrid on margins. Inflorescence bracts 3, lowest up to 6 cm long. Inflorescence capitate, yellow, 8–14 mm in diameter, of c. 50 spikelets. Spikelets 1(3)-flowered, 3–4 mm long; glumes bright yellow. Stamens 3. Style 3-branched. Nutlet (probably immature) narrowly ellipsoid, 2 × 0.6 mm, black, slightly rough at the cellular level (x40).

Zambia. W: Mwinilunga Dist., source of Zambezi R., 213.xii.1963, *Robinson* 6000 (K); by Mwinilunga road, between Kanyama turn-off and Samuteba, 25.ii.1975, *Hooper & Townsend* 402 (K).

Also in Angola and D.R. Congo. Dry sandy soil at edge of pan; sandy woodland; c. 1400 m.

Conservation notes: Data Deficient. There are few specimens and they come from a fairly restricted area.

Clarke's description in F.T.A. shows that he believed the rhizomes spread above ground, but neither of the specimens from the flora area shows this and it is probably an error.

A poorly known taxon, which has been confused with *Cyperus altochrysocephalus*; the rhizomes are completely different. I have not seen the Congo material from which Haines' illustration was prepared.

C10. **Cyperus crassipes** Vahl, Enum. Pl. **2**: 299 (1805). —Haines & Lye, Sedges & Rushes E. Afr.: 262, fig.534 (1983). —Gordon-Gray in Strelitzia **2**: 55 (1995). — Hoenselaar & Beentje in F.T.E.A., Cyperaceae: 162 (2010). Type: Guinea, *Isert* s.n. (C holotype).

> *Cyperus maritimus* Poir. in Lamarck, Encycl. **7**: 240 (1806). —Clarke in F.T.A. **8**: 326 (1901). —Kükenthal in Engler, Pflanzenr. **4**, 20(101): 269 (1936). Type: Madagascar, *du Petit Thouars* s.n. (P holotype).
> *Cyperus maritimus* Poir. var. *crassipes* (Vahl) C.B. Clarke in Durand & Schinz, Consp. Fl. Afric. **5**: 569 (1894); in F.T.A. **8**: 326 (1901), superfluous name.

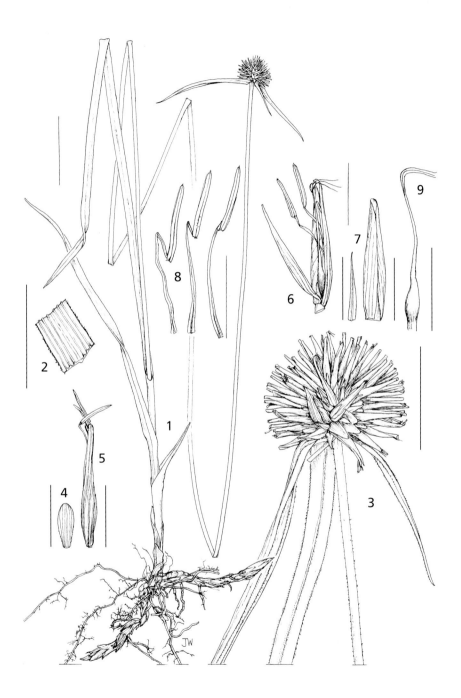

Fig. 14.**36**. CYPERUS RHYNCHOSPOROIDES. 1, habit; 2, leaf detail; 3, inflorescence; 4, bract; 5, single floret; 6, single floret, slightly opened; 7, floral bracts; 8, stamens; 9, style and ovary. All from *Robinson* 6000. Scale bars: 1 = 3 cm; 2, 3 = 1 cm; 4–9 = 3 mm. Drawn by Juliet Williamson.

Cyperus frerei C.B. Clarke in F.T.A. **8**: 327 (1901). —Kükenthal in Engler, Pflanzenr. **4**, 20(101): 283 (1936). —Haines & Lye, Sedges & Rushes E. Afr.: 256, fig.516 (1983). Type: Kenya, Frere Town and Rabai Hills, near Mombasa, 2.xii.1885, *Taylor* s.n. (BM holotype).

Perennial, leafy shoots in small groups, up to 35(70) cm tall, with long horizontal creeping rhizomes 3–5 mm in diameter covered with brown acute scales; culms few, 15–35(62) cm long, 2–4.5 mm in diameter, trigonous to almost terete, glabrous. Leaves up to 40(85) cm long; leaf sheath green to reddish-brown, at the base almost black, 4.5–10 cm long; leaf blade linear, flat or folded, rather thick, 25–40(80) cm long, 3–14 mm wide, apex acute, scabrid on margin. Involucral bracts leaf-like, spreading to reflexed, (2)4–8, lowermost 15–38 cm long, 4–8 mm wide. Inflorescence almost capitate to a simple anthela, primary branches 0–8, 0–7 cm long; spikelets in digitate clusters, sessile and at end of primary branches, 8 to many per cluster, linear-lanceolate, 9–27 × 2.4–5.2 mm wide, slightly compressed to almost terete, rachilla straight to sometimes slightly curved; glumes pale reddish-brown to pale brown, concave, elliptic-ovate, 5.1–8.1 × 1.8–4.8 mm wide, keel slightly excurrent, sometimes slightly green, many veins on both sides, apex very shortly mucronate. Stamens 3; filaments 2.5–5(8) mm long; anthers 1.9–3.7 mm long. Nutlet dark brown to black, obovoid, flattened on one side, this side pressed against the rachilla, 1.9–3 × 1–1.4 mm wide, smooth or minutely papillose.

Mozambique. Z: Chinde Dist., Chinde, 6.i.1970, *Amico & Bavazzano* s.n. (FI, K). MS: Beira Dist., Macuti, 23.iii.1960, *Wild & Leach* 5232 (K, SRGH). GI: Inhambane Prov., Inhanombe, vi.1936, *Gomes e Sousa* 1750 (K). M: Maputo Bay (Lourenço Marques), vii.1931, *Muir* 4972 (K).

Also on tropical African coasts from Senegal to D.R. Congo in the west and Somalia to Tanzania in the east. Seashores and dunes just above high tide mark, often with *Ipomoea pes-caprae, Sophora imhambanensis, Scaevola plumieri* etc.; 0–25m.

Conservation notes: Least Concern because of its wide distribution.

This taxon seems smaller in the Flora area than in F.T.E.A., where the bracketed measurements apply.

C11. **Cyperus mapanioides** C.B. Clarke in F.T.A. **8**: 340 (1901). —Haines & Lye, Sedges & Rushes E. Afr.: 160, figs.295,296 (1983). —Hoenselaar in F.T.E.A., Cyperaceae: 167 (2010). Types: D.R. Congo, Stanley Pool, 12.viii.1888, *Hens* B7 (BR syntypes), 69 (BR syntype, K) & 389 (BR syntype). FIGURE 14.**37**.

　　Cyperus dichromeniformis Kunth var. *major* Boeckeler in Flora **62**: 549 (1879). Types: Sudan, Niamniam and Monbuttu, *Schweinfurth* 3461 and 3886 (B syntypes).
　　Cyperus mapanioides var. *major* (Boeckeler) Kükenthal in Engler, Pflanzenr. **4**, 20(101): 230 (1936).

Perennial, slender, with a short thick creeping rhizome covered with brown strongly-veined scales, up to 56 cm tall; culms 17–54 cm long, 1.4–3.9 mm wide, triquetrous, glabrous. Leaves up to 38 cm long; leaf sheath reddish-brown to deep purple, 1.5–7 cm long; leaf blade linear, flat, 10–31 cm long, 0.4–0.7(1.2) cm wide, with 2 main veins next to primary vein, scabrid on margins and veins, apex acute to acuminate. Involucral bracts 4–7, leaf-like, spreading, 10–15(34) × 0.6–0.8(1.3) cm wide. Inflorescence capitate, sometimes loosely so; spikelets 7–20 per head, linear-lanceolate to ovoid, 7–18 × 2.4–4 mm wide, glumes spreading when mature, whitish grey, lanceolate-ovate, boat-shaped, many-veined, 2.7–4.7 × 1.3–3.5 mm wide, scabrid on margins, keel greenish-white, apex acute. Stamens 3; filaments 1.3–3 mm long; anthers 0.6–1.3 mm long. Nutlet shiny brown, ellipsoid-obovoid, trigonous, angles prominent or ridged, 1.4–1.9 × 0.9–1.3 mm wide, smooth, sometimes slightly minutely papillose.

Zambia. N: Kawambwa Dist., Nchelenge, Lake Mweru, 22.iv.1957, *Richards* 9404 (K). W: Luanshya Dist., 28.ii.1960, *Robinson* 3355 (K). **Mozambique**. Z: Tacuane Dist., Mabu Mountain, 24.x.2008, *Harris* 646 (K); Mt Mabu, near summit, 14.x.2008, *Patel* s.n. (K).

Widespread in tropical West and central Africa, south to Tanzania and Angola. In forest or moist woodland (mushitu), often along paths, in clearings and by streams; 1200–1700 m.

Fig. 14.**37**. CYPERUS MAPANIOIDES. 1, habit; 2, rhizome detail; 3, inflorescence; 4, spikelet; 5, nutlet. 1, 2, 5 from *Robinson* 3355; 3, 4 from *Milne-Redhead* 3721. Scale bars: 1 = 3 cm; 2–4 = 1 cm; 5 = 1 mm. Drawn by Juliet Williamson.

Conservation notes: Least Concern because of its wide distribution.

Richards 9404 bears the note 'Roots with tiny tubers'. These are not visible on this specimen but *Robinson* 3355 has short bulbil-like branches on the rhizome which may be what Richards was referring to.

Hoenselaar gives the leaf width as up to 1.3 cm but none of the F.Z. area material at K has leaves wider than 7 mm.

C12. **Cyperus angolensis** Boeckeler in Flora **63**: 435 (1880); in F.T.A. **8**: 321 (1901). — Kükenthal in Engler, Pflanzenr. **4**, 20(101): 281 (1936). —Haines & Lye, Sedges & Rushes E. Afr.: 255, figs.514,515 (1983). —Hoenselaar in F.T.E.A., Cyperaceae: 167 (2010). Type: Angola, Malange, vii.1879, *von Mechow* 182 (BR holotype, B, BR, M).

    Cyperus angolensis var. *amplibulbus* Peter & Kük. in Engler, Pflanzenr. **4**, 20(101): 282 (1936). Type: Tanzania, Kigoma Dist., Ujiji, E of Kigoma, 13.ii.1926, *Peter* 36846 (B holotype).

Perennial, up to 100 cm tall, from a horizontal rhizome 1–14(24) cm long, 3–6 mm in diameter, densely covered by brown multi-veined scales which sometimes split into fibres; culms solitary, culm base swollen and covered in leaf sheaths, 14–98 cm long, 1.4–2.7 mm wide, trigonous, smooth. Leaves up to 64 cm long; leaf sheath at base brown, higher up green, 2.5–10 cm long; leaf blade linear, flat or folded, rather thick, 7–56 cm long, 3.5–8 mm wide, scabrid on margin and primary vein, apex acuminate. Involucral bracts leaf-like, spreading or reflexed, 3(4), lowermost 2.5–13 cm long, 3.1–5 mm wide. Inflorescence capitate, 12–30 mm in diameter; spikelets crowded, c. 6–15-flowered, many per head, ovoid, 6–11.2 × 4–5.7 mm wide, rachis straight; glumes greyish-white to sometimes pinkish-white, lanceolate, 4.2–7 × 1.5–2.7 mm wide, keeled, apex acute. Stamens 3; filaments 5–6 mm long; anthers 1.6–3 mm long. Nutlet yellowish brown to olive green to black, obovoid, trigonous, 2.5–3.2 × 1.3–1.6 mm wide, minutely papillose.

Zambia. B: Mankoye Dist., Mankoye, near Luene R., 20.xii.1959, *Drummond & Cookson* 6687 (K). N: Mbala Dist., top of source of Inono, 14.xi.1954, *Richards* 2221 (K). W: Mwinilunga Dist., by Mwinilunga road between Kanyama turnoff and Samuteba, 25.ii.1975, *Hooper & Townsend* 403 (K). C: Lusaka Dist., 6 km [4 miles] E of Lusaka on Great East Road, 30.xi.1952, *Best* 35 (K). E: Isoka Dist., Nyika Plateau, by main road c. 3 km [2 miles] SW of resthouse, 21.x.1958, *Robson* 213 (K). S: Choma Dist., Muckle Neuk, 16 km [12 miles] N of Choma, 28.xi.1954, *Robinson* 1000 (K). **Zimbabwe**. N: Gokwe Dist., 5 km [3 miles] N of Gokwe, 13.xii.1963, *Bingham* 1022 (K, SRGH). C: Chegutu Dist., Poole Farm, 22.xii.1944, *Hornby* 2382 (K, SRGH). E: Nyanga Dist., Honde Valley, hillsides, 16.xi.1950, *Chase* 3064 (BM, SRGH). **Malawi**. N: Chitipa Dist., Nyika Plateau, above upper Mondwe Valley, 6.ix.1972, *Synge* WC396 (K). C: Dedza Dist., Dedza Mt, 6.ix.1950, *Jackson* 166 (K). S: Zomba Dist., Zomba Plateau, 23.x.1941, *Greenway* 6355 (K). **Mozambique**. T: Angónia Dist., Ulongue, near Catsanha Village, 3.xii.1980, *Macuácua* 1398 (K, LMA).

Also in Ghana, Nigeria, Cameroon, Congo, Gabon, Rwanda, Uganda, Tanzania, D.R. Congo and South Africa. *Brachystegia/Julbernardia* woodlands, and in open grasslands, extending into upland *Exotheca/Loudetia* grasslands, often in places where burning is frequent; (450)1400–2500 m.

Conservation notes: Least Concern because of its wide distribution.

Kassner 2154 from "Michangu Spruit, N.W. Rhodesia" has a compound inflorescence of one sessile and two stalked capitula.

The varietal name is misspelled as "*ampibulbus*" in both F.T.E.A. and World Checklist of Cyperaceae.

C13. **Cyperus diurensis** Boeckeler in Flora **62**: 556 (1879). —Clarke in F.T.A. **8**: 381 (1901). —Kükenthal in Engler, Pflanzenr. **4**, 20(101): 560 (1936). —Haines & Lye, Sedges & Rushes E. Afr.: 223, fig.454 (1983). —Hoenselaar in F.T.E.A., Cyperaceae: 166 (2010). Type: Sudan, Djur, Seriba Ghattas, 20.viii.1869, *Schweinfurth* 198, series III (B holotype, K).

   Cyperus gondanus Boeckeler in Beitr. Cyper. **1**: 3 (1888). Type: 'Africa orient. tropica', Gonda, date, *Boehm* s.n. (B holotype).
   Mariscus diurensis (Boeckeler) C.B. Clarke in Durand & Schinz, Consp. Fl. Afr. **5**: 586 (1894); in F.T.A. **8**: 381 (1901).
   Cyperus diurensis var. *laetevirens* Peter & Kük. in Engler, Pflanzenr. **4**, 20(101): 560 (1936). Type: Tanzania, Tanga Dist., Ukereni Hill near Amboni, *Peter* 39476 (B holotype).
   Cyperus diurensis var. *longistolon* Kük. in Wiss. Ergebn. Schwed. Rhodesia-Kongo-Exped. 1911–1912, Erg.: 4 (1921); in Engler, Pflanzenr. **4**, 20(101): 560 (1936). Types: Tanzania, Dist. unclear, Niakagunda, *Fries* 1472 (? syntype); Tabora, Unyanyembe, *Peter* 35342 (? syntype); Tabora Dist., Ngulu near Malongwe, *Peter* 34761(? syntype); Dodoma Dist., Uyansi near Chaya [Tschaya], *Peter* 45830b & 45831 (B syntypes).
   Cyperus diurensis var. *gondanus* (Boeckeler) Kük. in Engler, Pflanzenr. **4**, 20(101): 560 (1936).
   Cyperus diurensis var. *acuminatosquamatus* Kük. in Engler, Pflanzenr. **4**, 20(101): 561 (1936). Types: Tanzania, Tanga Dist., W Usambaras, Tonya near Nashera, 9.ix.1915, *Peter* 13879 (K syntype, WAG, P); Mirele, Gombero, 30.ix.1918, *Peter* 25094 (K syntype, P).

Perennial, up to 80 cm tall, with a slightly swollen culm base bearing several 1–10 cm long slender horizontal brown-scaly rhizomes about 2 mm in diameter; culms few, 25–80 cm long, 0.7–2.5 mm wide, trigonous, glabrous. Leaves up to 45 cm long; leaf sheath 2.5–9 cm long; leaf blade linear, flat or sometimes folded, 22–39 cm long, 2–3.8 mm wide, scabrid on margin and primary vein, apex acuminate. Involucral bracts leaf-like, spreading to reflexed, 3–6, lowermost 5.5–30 cm long, 1.9–3.2 mm wide. Inflorescence capitate, a solitary globose head, 10–20 × 11–23 mm wide; spikelets many per inflorescence, ovate-lanceolate, 6.5–14.1 × 2.1–4 mm wide, falling off entire when mature; glumes whitish with a reddish-brown tinge, especially near apex, ovate to boat-shaped, 4–5.1 ×1.6–2 mm wide, keel, apex obtuse, frayed. Stamens 2–3; filaments c. 5.4 mm long; anthers 2.2–2.7 mm long. Nutlet grey-brown, ellipsoid-oblong, trigonous, 1.7–2 × 0.7–0.8 mm wide, minutely papillose.

Zambia. N: Kasama Dist., 95 km E of Kasama, 2.ii.1961, *Robinson* 4343 (K). **Zimbabwe**. C: Gweru Dist., Whitewaters Dam, 20.i.1965, *Loveridge* 595 (K, SRGH). **Malawi**. S: Zomba Dist., Zomba, Lambulira village near Chikewi, 16.v.1958, *Jackson* 2139, in part (K).

Also in Ethiopia, Sudan, Uganda, Kenya, Tanzania, Rwanda and D.R. Congo. In grassland and woodland, often in shallow sandy soil over rocky outcrops and inselbergs; (5)1000–1400(1800) m.

Conservation notes: Least Concern because of its wide distribution.

Although a mixed collection, *Jackson* 2139 clearly contrasts specimens of *Cyperus dubius* with bulbous bases and *C. diurensis* with stoloniferous bases.

The characteristic, relatively slender horizontal rhizomes with bud-like apices from which new shoots arise are, unfortunately, often lacking in collections from the Flora area.

Notes on *Robinson* 4343 state "Stems and leaves more or less succulent; took three months to dry".

C14. **Cyperus amauropus** Steud., Syn. Pl. Glum. **2**: 33 (1854). —Haines & Lye, Sedges & Rushes E. Afr.: 213, figs.428,429 (1983). —Hoenselaar in F.T.E.A., Cyperaceae: 237, fig.36 (2010). Type: Ethiopia, Mt Schoata, 12.vii.1838, *Schimper* 1391 (P holotype, K).

   Mariscus leptophyllus C.B. Clarke in F.T.A. **8**: 385 (1902).
   Cyperus leptophyllus (C.B. Clarke) Kük. in Engler, Pflanzenr. **4**: 20(101): 548 (1936).
   Mariscus amauropus (Steud.) Cufod. in Bull. Jard. Bot. Natl. Belg. **40**(3, suppl.): 1448 (1970).

Perennial, fairly robust, succulent, up to 70 cm tall; base a cluster of slightly swollen pseudobulbs 4–5 cm long and 0.4–0.6 cm in diameter, adjacent to one another and forming a row; sometimes with 1–5 cm long stolons; culms tufted, 15–60 cm long, 1–3 mm wide, trigonous, glabrous. Leaves many, basal, up to 40 cm long; leaf sheath uncoloured or pale brown, sometimes partly purplish, 3–8 cm long; leaf blade linear, often inrolled or folded, 16–34 cm long, 1.3–2.5 mm wide, scabrid on margin, apex acuminate. Involucral bracts leaf-like, erect to spreading, 3–4, lowermost 3–15 cm long, 1.2–3 mm wide. Inflorescence a simple anthela, sometimes very loosely capitate, primary branches 0–4, 0–3.7 cm long; spikelets in loose clusters, sessile and at the end of primary branches, 3–10 per cluster, spreading or reflexed, very narrowly ovate, 5.8–24 × 1.6–3.7 mm wide, rachilla straight; glumes pale to dark reddish-brown, ovate-oblong, 3.1–4.2 × 1.4–1.9 mm wide, keel green to reddish-brown, with 4–8 slender veins on either side, apex rounded to acuminate. Stamens 3; filaments 2.8–4.3 mm long; anthers 1.7–2.5 mm long. Nutlet brown, oblong-ellipsoid, trigonous, 1.6–2.1 × 0.5–0.8 mm wide, densely papillose.

Zambia. N: Mbala Dist., Kawimbe Rocks, 26.xii.1967, *Simon et al.* 1576 (K, SRGH). C: Mkushi Dist., Fiwila Rocks, 3.i.1958, *Robinson* 2678 (K). S: Choma Dist., c. 40 km [23 miles] NNE of Choma, 3.i.1957, *Robinson* 2009 (K). **Zimbabwe**. W: Bulawayo Dist., Hillside Dams, 4.xii.1967, *Best* 672 (K, SRGH). C: Mazoe Dist., hills above Mazoe Dam, 24.xii.1956, *Robinson* 1980 (K, NDO, NU, SRGH).

Also in Sudan, Eritrea, Ethiopia, Somalia, Uganda, Kenya, Tanzania and Rwanda. Almost always in soil pockets and crevices of rocky (usually granite) hills, and rocky woodland; 850–1700 m.

Conservation notes: Least Concern because of its wide distribution.

Harris 134 (K) from Mt Chiperone, Zambezia, Mozambique (14.xii.2006), resembles this species but differs in its pale greenish spikelets. It is probably an abnormal form, and would be the only record of the species from Mozambique.

The species appears to be much less common in the Flora Zambesiaca area than in the F.T.E.A. area; why, is not clear, as there is plenty of the usual habitat available. It is perhaps undercollected because of its habitat. The F.T.E.A. account includes a much fuller synonymy.

C15. **Cyperus neoschimperi** Kük. in Engler, Pflanzenr. **4**, 20(101): 552 (1936), new name for *C. variegatus* Boeckeler —Hoenselaar in F.T.E.A., Cyperaceae: 172 (2010). Types: Ethiopia, Scholoda Mt, 20.vi.1837, *Schimper* I.173 (B syntype); Schoata Mts, 25.vii.1839, *Schimper* II.578 (B syntype), 588 (not found), 1363 (B syntype).

 Mariscus schimperi A. Rich., Tent. Fl. Abyss. **2**: 491 (1850). —Clarke in F.T.A. **8**: 383 (1901). Type as for *C. neoschimperi*.

 Cyperus variegatus Boeckeler in Linnaea **36**: 337 (1870), illegitimate name, non *C. variegatus* Kunth (1816). Type as for *C. neoschimperi*.

 Cyperus neoschimperi var. *subvirescens* Kük. in Engler, Pflanzenr. **4**, 20(101): 552 (1936). Type: Tanzania, Morogoro Dist., Ukami area, *Peter* 39072 (B holotype, B).

 Cyperus vexillatus Kük. in Engler, Pflanzenr. **4**, 20(101): 547 (1936). Types: Tanzania, Kilosa Dist., Usagara, km 32 of railroad, km 327 to Kidete, 4.xii.1925, *Peter* 32783 (B syntype, K); SW of Kidete, 15.xii.1925, *Peter* 32802 (B syntype), syn. nov.

 Cyperus pseudovestitus sensu Haines & Lye, Sedges & Rushes E. Afr.: 212, fig.427 (1983), non (C.B. Clarke) Kük.

Perennial, with short creeping rhizome; culms several, tufted, 10–60 cm high, 0.5–2 mm across, trigonous, glabrous, at base slightly swollen. Leaves with leaf sheaths reddish brown with wide translucent margin, darker near very base, to 12 cm long; blade 10–66 cm long, 1.5–4 mm wide, scabrid on margin and midrib. Involucral bracts 3–5, hanging or reflexed, wider at base, the longest 7–30 cm long, 3 mm wide. Inflorescence a simple anthela with 3–4 spikes to capitate and ± ovoid, mid – to dark brown, sessile or briefly stalked; spikes ovoid to narrowly ovoid,

10–25 × 10–20 mm; spikelets 10–20, closely set along and at the end of primary branches, 5–10 × 1–2.5 mm, 2–6-flowered, compressed, falling off entire when mature; rachilla slightly winged; glumes densely imbricate, reddish brown or yellowish with green keel, ovate-oblong, 3.2–5 mm long, several-veined, apex acute. Stamens 3; filaments 3.5–4 mm long; anthers 1.9–2 mm long. Style 3-branched. Nutlet reddish brown, oblong-ellipsoid, 1.8–2 × 0.6–0.7 mm, trigonous, apiculate, densely papillose.

Zambia. B: Mongu Dist., 80 km [55 miles] from Mongu on Mankoya road, near turn-off to Ndanda Dispensary, 19.xi.1959, *Drummond & Cookson* 6617 (K, SRGH). C: Mkushi Dist., Fiwila. 9.i.1958, *Robinson* 2709 (K). S: Choma Dist., Choma, 18.xii.1954, *Robinson* 1021 (K). **Zimbabwe**. W: Hwange Dist., 15 km [10.6 miles] SSW of Main Camp along Dopi Pan road, Hwange (Wankie) Nat. Park, 20.xii.1968, *Rushworth* 1372 (K, SRGH). C: Harare Dist., [Salisbury], 19.i.1927, *Eyles* 4672 (K, SRGH). E: Mutare Dist., Burma Valey, Bomponi, 6.xii.1961, *Wild & Chase* 5561 (K, SRGH). **Malawi**. N: Rumphi Dist., 8 km [5 miles] N of Livingstonia, 9.i.1959, *Robinson* 3126 (K). S: Zomba Dist., Namasi, Zomba, xi-xii.1899, *Cameron* 93 (K). **Mozambique**. Z: Nampula Dist., Moebede road, 15.xi.1948, *Faulkner* 353 (K).

Also in Sudan, Ethiopia, Somalia, Kenya, Uganda and Tanzania. Rocky slopes and hilltops, in cracks or on flat rocks; 500–2000 m.

Conservation notes: Least Concern because of its wide distribution.

According to F.T.E.A. it is close to *C. obsoletenervosus* but lacks stolons.

Of *Cyperus vexillatus* Kük. says 'in between *vestitus* and *pseudovestitus*' but distinct from the first in hard keeled leaf sheaths and the blunt glumes, from the second in dark red-purple sheaths; in F.T.E.A. Beentje placed this with the more capitate forms of *C. neoschimperi*.

Vorster and Kew Herb use the name *pseudovestitus*.

C16. **Cyperus longispicula** Muasya & D.A. Simpson in Kew Bull. **59**: 595, fig.3 (2004).
    Type: Zimbabwe, Lower Sabi, 27.i-2.ii.1948, *Rattray* 1273 (K holotype, SRGH).

Shortly rhizomatous perennial. Culms moderately to densely tufted, 18–32 cm by 0.7–0.9 mm, terete to trigonous, smooth, green, densely covered at base by mid- to dark brown leaf sheath remains. Leaves: blade narrowly linear, 6–29 cm by 0.4–0.7 mm, very gradually tapering to an acute apex, flat in cross-section, margins smooth, green; sheath 2.5–3.5 cm long, mid-brown. Involucral bracts 2–3, leaf-like, the longest 6–12 cm long, margins smooth. Inflorescence lax capitate, 2.5–5 × 4–5.5 cm. Spikelets 14–18 per inflorescence, linear, (1)1.8–2.5 cm by 1.8–2.4 mm, flattened, reddish brown, rachilla straight. Glumes 18–44 per spikelet, ovate-lanceolate, 2.4–2.7 × 1–1.6 mm, ± acute, sides membranous, 2–3-nerved on each side, mid-reddish brown, keel green to pale brown. Stamens 3; anthers 1.2 mm long. Stigmas three. Nutlets obovate, trigonous with convex sides, 0.7–0.8 × 0.4 mm, maturing dark brown, very minutely papillose.

Zimbabwe. W: Matobo Dist., 12.i.1948, *West* 2595 (K, SRGH). E: Chiredzi Dist. (Lower Sabi), 27.i-2.ii.1948, *Rattray* 1273 (K, SRGH). **Mozambique**. MS: Sussundenga Dist., 15 km [10miles] N of Mavita, 1.ii.1962, *Wild* 5633 (K, SRGH).

Not known elsewhere; Peaty soil in rock fissures; shallow soil over granite and Permian rocks; dambo grassland on sand; 400–500 m.

Conservation notes: Known from only four collections, all in much the same area straddling the Mozambique–Zimbabwe border. Data Deficient as few specimens and restricted distribution.

Some larger material of *Cyperus semitrifidus* approaches this taxon, but there is little if any overlap. Superficially this species resembles *C. palmatus*, but the latter has paler acute spikelets and a basal subspherical bulb.

C17. **Cyperus semitrifidus** Schrad., Anal. Fl. Cap. **1**: 6 (1832). Type: *Zeyher* 89 (?herb).

    Cyperus rupestris Kunth, Enum. Pl. **2**: 52 (1837). —Gordon-Gray in Strelitzia **2**: 69, fig.25 (1995). Type: South Africa, Cape of Good Hope (Cap. b. Spei), *Drège* s.n. (P, not seen).

    Cyperus kirkii C.B. Clarke in F.T.A. **8**: 318 (1901). —Haines & Lye, Sedges & Rushes E.Afr.: 259, fig.524 (1983). —Hoenselaar in F.T.E.A., Cyperaceae: 175 (210). Types: Mozambique, Lower Zambesi, near Lupata, *Kirk* s.n. (K syntype); Malawi, Manganja Hills, *Meller* s.n. (K syntype).

    Cyperus tanganyicanus (Kük.) Lye in Nordic J. Bot. **3**: 231 (1983). —Haines & Lye, Sedges & Rushes E. Afr.: 260 fig.527 (1983). —Hoenselaar in F.T.E.A., Cyperaceae: 174 (2010). Type: Tanzania, Iringa Dist., Lupembe, on rocks near Mpaponzi, iii.1931, *Schlieben* 436 (B lectotype, BM, K), lectotypified by Lye (1983).

    Cyperus bellus Kunth. var. *tanganyicanus* Kük. in Engler, Pflanzenr. **4**, 20(101): 304 (1936).

Densely clumped perennial to 25 cm tall. Base of culms swollen(?), covered by fibrous remains of leaf sheaths. Culms erect, trigonous, smooth, glabrous, to 20 cm tall and 0.4–0.9 mm wide. Leaves up to 21 cm long; sheaths brown, 1–3 cm long, breaking into fibres; blade linear, flat but strongly inrolled when dry, 0.3–1.0 mm wide, 4–9.5 cm long, glabrous, apex acuminate. Involucral bracts 2, leaf-like, spreading, the lowermost 2–6.5 cm long and 0.6–1.0 mm wide. Inflorescence capitate, 1–2(3) cm in diameter. Spikelets 3–15 per head, linear-lanceolate, 4.6–15 mm long, 1.4–2.4 mm wide, straight or slightly curved, with few to 40 glumes per spikelet. Glumes reddish brown with uncoloured margin, ovate or elliptic-lanceolate, with 3–5 prominent but thin uncoloured veins on each side of the keel or with 3–4 prominent ribs on each side of keel. Keel excurrent (±greenish, apex mucronate, recurved). Stamens 1–3. Nutlet purplish black or grey to almost black, 0.8–1.1 × 0.4–0.8 mm, papillose in longitudinal rows, apiculate.

Botswana. SE: South East Dist., Content Farm, northern area, Gaborone, 3.xi.1977, *Hansen* 3267 (C, K). **Zambia**. W: Mwinilunga Dist., Zambezi Rapids, 6 km [4 miles] from Kalene Mission, 9.xi.1962, *Richards* 17108 (K). S: Choma Dist., Sinachirundu, near mouth of Zongwe R., 29.xii.1958, *Robson & Angus* 1005 (BM, K). **Zimbabwe**. W: Matobo Dist., Matopos Nat. Park, foot of World's View, 9.iii.1932, *Brain* 8718 (K, SRGH). C: Harare Dist., 35 km NE of Harare [Salisbury], 26.xii.1960, *Robinson* 4212 (K). E: Mutare Dist., 50 km [35 miles] from Mutare (Umtali) Bridge, 12.i.1967, *Biegel* 2065 (K, SRGH). S: Masvingo Dist., Zimbabwe Nat. Park, 27.iii.1973, *Chiparawasha* 607 (K, SRGH). **Malawi**. N: Rumphi Dist., Nyika Plateau, Chosi Hill, *Pawek* 2129 (K). S: Zomba Dist., Lambulina village, 10.i.1958, *Jackson* 2137 (K). **Mozambique**. N: Marrupa Dist., estrade para Nungo 20 km, R. Messabo, 20.ii.1981, *Nuvunga* 657, (K, LMU). Z: Gurue Dist., Mt Namuli Mt, Muretha Plateau, 25.v.2007, *Harris et al.* 168 (K). M: Namaacha Dist., Namaacha, 20.i.1958, *Barbosa & Lemos* 8237 (K, LISC).

Also in Tanzania, Angola and South Africa. Crevices of usually granite rocks; near sea-level to 2300 m.

Conservation notes: Least Concern because of its wide distribution.

The differences between *Cyperus kirkii* and *C. tanganyicanus* given in F.T.E.A. are very small; these small perennial sedges, growing in shallow soil, are strongly influenced by moisture conditions and I suspect that *C. tanganyicanus* is just a stunted form of *C. semitrifidus* in a dry year. Gordon-Gray remarks on the variability of *C. rupestris* and recognises three varieties, but also states that much more work is needed on the group. The type of *C. kirkii* is certainly rather pale throughout but it is my impression that a number of specimens from the Zambezi Expedition collections are pale; possibly some unorthodox method was used to dry and/or preserve them. The oldest name for the complex appears to be *C. semitrifidus*.

As Gordon-Gray states, the complex is in need of thorough revision but at present I feel it is better to place all these small rock-dwelling plants under a single name. When the revision is done, the selection of neotypes or interpretive epitypes for the recognised taxa would be very helpful to future workers on the group.

The distinction between this taxon and the previous one, *Cyperus longispicula*, is not always clear. Some specimens (e.g. *Gibbs Russell* 2709 (K) from Chipinge, Zimbabwe E and *Wild* 5407 from Tshitshurapadzi, near Beitbridge, Zimbabwe S) could either be interpreted as vigorous *C. semitrifidus* or small *C. longispicula*. When the thorough revision of the group is carried out, *C. longispicula* should be included.

C18. **Cyperus albopilosus** (C.B. Clarke) Kük. in Bot. Not. **1934**: 69 (1934). —Haines & Lye, Sedges & Rushes E. Afr.: 217, figs.438,439 (1983). —Hoenselaar in F.T.E.A., Cyperaceae: 155 (2010). Type: Malawi, Zomba, xii.1896, *Whyte* s.n. (K holotype).

> *Mariscus albopilosus* C.B. Clarke in F.T.A. **8**: 394 (1902).

Perennial, up to 55 cm tall, with a short thick horizontal somewhat moniliform rhizome; culms few, 15–47 cm long, 0.7–1.3 mm wide, trigonous, minutely but densely hairy, particularly on the upper parts. Leaves up to 30 cm long, to halfway up the culm; leaf sheath pale green with dark dots, 2.5–10.5 cm long; leaf blade linear, flat, rather stiff, 1–11 cm long, 3.1–4 mm wide, usually densely hairy on upper surface and margins, apex acute. Involucral bracts leaf-like, spreading to reflexed, 2–3, densely hairy on midrib and margins, lowermost 1.5–6 cm long, 3–4.2 mm wide. Inflorescence a single capitate, globose spike 7–9 × 8–10 mm wide; spikelets many per inflorescence, lanceolate, 3.2–5.5 × 1–1.3 mm wide; glumes dirty white to yellowish, lanceolate, 3.1–4.2 × 1.7–2 mm wide, keel with many veins on either side, apex (long) acuminate. Stamens 3; filaments 2.2–4.1 mm long; anthers 0.9–1.1 mm long. Nutlet almost black, obovoid, trigonous to triquetrous, 1.9–2.1 × 0.9–1 mm wide, minutely papillose.

Zambia. W: Mwinilunga Dist., 1 km (0.5 mile) S of Matonchi Farm, 7.xi.1937, *Milne-Redhead* 3121 (K). N: Kasama Dist., 80 km [50 miles] N of Kasama, 30.xii.1958, *Robinson* 2965 (K). E: Nyika Plateau, 2.i.1959, *Robinson* 2990 (K). S: Choma Dist., Beckett's Farm, 18.i.1959, *Robinson* 3238 (K). **Zimbabwe**. E: Nyanga Dist., Nyanga Nat. Park, Mare R., 25.i.1960, *Wild* 4922 (K, SRGH). **Malawi**. S: Zomba Dist., Mulugushi, off Zomba Plateau, 19.xii.1907, *Kassner* 2062 (K).

Also in Ethiopia, Kenya and Tanzania. Grassland, wooded grassland, *Brachystegia* woodland, and roadsides; one collection from a laterite outcrop; 1300–2150 m.

Conservation notes: Least Concern because of its wide distribution.

Superficially rather similar to *Cyperus mapanioides*, but the latter grows in moister, shadier places and has multiflowered spikelets, longer and broader involucral bracts and longer leaves.

According to Robinson (R.3268) the rhizome is brittle and orange or yellow inside. He also comments that the mature inflorescence looks grey becasue the black nutlets show through the translucent whitish glumes.

C19. **Cyperus nduru** Cherm. in Arch. Bot. Mém. **4**(7): 18 (1931). —Hoenselaar in F.T.E.A., Cyperaceae: 179 (2010). Type: Central African Republic, Bambari, *Tisserant* 332 (P holotype).

> *Cyperus margaritaceus* Vahl var. *nduru* (Cherm.) Kük. in Bot. Not. **1934**: 67 (1934); in Engler, Pflanzenr. **4**, 20(101): 285 (1936). —Haines & Lye, Sedges & Rushes E. Afr.: 257 (1983).

Perennial, fairly robust, up to 32(80) cm tall, swollen bulb-like base; culms tufted, 7–80 cm long, 0.7–1.4 mm wide, trigonous to almost terete, smooth. Leaves few, up to 12 cm long; leaf sheath black to brown at base, often burnt at apex, brown on culm, 1–4.5 cm long; leaf blade linear, flat to inrolled, 2–7.5 cm long, 0.7–1.6 mm wide, glabrous to scabrid on margins, apex acute. Involucral bracts bract-like, spreading, 1–3, lowermost 0.5–1.5(5) cm long, 0.8–1.6 mm wide, not or only slightly exceeding the inflorescence. Inflorescence capitate; spikelets in a dense head, sometimes single, 1–6 per head, ovoid, 7–11 × 4–5.6 mm wide, rachis straight; glumes white to brownish, ovate to boat-shaped, 4.7–5.6 × 2.9–4.7 mm wide, keel acute, many veins on either side of keel, apex obtuse. Stamens 3; filaments 4.3–6.4 mm long; anthers 1.8–2.5 mm long. Nutlet pale brown to olive, trigonous, 2.4–2.9 × c. 1.6 mm wide, smooth.

Zambia. N: Mbala Dist., pans near Mbala (Abercorn), 14.xii.1954, *Richards* 3642 (K). W: Ndola Dist., Forest Reserve, Coppice Plot 2, 26.xi.1947, *Brenan* 8371 (FHO, K). C: Lusaka Dist., Mt Makulu, 17 km (12 miles) S of Lusaka, 3.xii.1956, *Angus* 1450 (FHO, K). E: Isoka Dist., Nyika Plateau, c. 2 miles SW of Rest House, 21.x.1958, *Robson & Angus* 214 (K). S: Kalomo Dist., 1 km (0.75 mile) S of Zimba Township, 22.x.1958, *Bainbridge* 942 (K). **Zimbabwe**. C: Harare Dist., Spelonken Farm, near University College Farm, 1.xii.1968, *Grosvenor* 459 (K, SRGH). E: Nyanga Dist., Honde Valley, 1935, *Gilliland* K1137 (K). **Malawi**. N: Rumphi Dist., Nyika Plateau, 10 km [6 miles] SW of Rest House, 25.x.1958, *Robson & Angus* 360 (K). S: Blantyre Dist., Manganja Hills, ix–xi.1861, *Meller* s.n. (K). **Mozambique**. N: road to Chelinda, 16.x.2005, *Chapama et al.* 332 (K). T: Angónia Dist., arredores de Dómuè, 20.xi.1980, *Macuácua* LM1276 (K, LMU). M: Maputo, Vila Luiza, 1.x.1957, *Barbosa & Lemos* 7878 (K, LMA).

Also in Sierra Leone, Guinea, Ghana, Nigeria, Cameroon, Congo, D.R. Congo, Kenya, Tanzania. Open regularly burned grassland, wooded grassland, open *Brachystegia* woodland, or upland grassland with *Protea*; 1100–2150 m.

Conservation notes: Least Concern because of its wide distribution.

F.T.E.A. notes that this is very similar to *Cyperus margaritaceus* but slightly smaller, with shorter involucral bracts, spikelets and glumes, and a smooth, narrower nutlet. Also similar to *C. niveus* var. *tisserantii* but with fewer leaves, shorter involucral bracts and fewer spikelets per head.

The F.T.E.A. key (but not the description) distinguishes this from *C. margaritaceus* by, *inter alia*, the smooth, not minutely papillose nutlet. As far as I can see, any papillae are at the cellular level so that the nutlets of both species appear smooth except at high magnification (x40).

C20. **Cyperus pulchellus** R. Br. in Prodr. Fl. Nov. Holland.: 213 (1810). —Haines & Lye, Sedges & Rushes E. Afr.: 173, figs.330,331 (1983). —Simpson in Kew Bull. **45**: 489–492. —Hoenselaar in F.T.E.A., Cyperaceae: 163 (2010). Type: Australia, Arnhem Bay, *Brown* 5915 (BM holotype, K).

Cyperus zambesiensis C.B. Clarke in Trans. Linn. Soc. London, Bot. **4**: 53 (1894); in F.T.A. **8**: 344 (1901). Type: Malawi, Mlanje, *Buchanan* 647 (K holotype).

Sorostachys pulchellus (R. Br.) Lye in Nordic J. Bot. **1**: 189 (1981).

Slender perennial up to 45 cm high, with swollen stem-bases, forming small tufts; rhizome more or less absent; culms 11–36 cm long, 0.9–1.4 mm wide, trigonous, longitudinally grooved, smooth to slightly scabrid, the base covered with fibrous remains of old leaf sheaths. Leaves up to 18.5 cm long; leaf sheath pale green to pale reddish-brown, basal sheaths darker and split into fibres, 1–3.5 cm long; leaf blade linear, flat to v-shaped, 2–15 cm long, 1.8–2.9(4.2) mm wide, apex acute to acuminate, scabrid on margins and primary vein towards apex. Involucral bracts linear, spreading to reflexed, 2–3, lowermost 5–11 cm long, 1.8–2.9 mm wide. Inflorescence capitate; spikelets 15–60 in a very dense globose head, each with up to 20 flowers, lanceolate-ovate, 3.3–7.2 × 1.5–2.5 mm wide, rachis straight; glumes greyish-white, with cinnamon tinge, lanceolate, 1.4–1.8 × 0.5–1 mm wide, margin often curved inwards, keel indistinct, apex acute to rounded. Stamen 1. Nutlet blackish to grey to (pale) brown, flattened-trigonous, (narrowly) obovoid to ellipsoid, 0.8–1 × 0.2–0.4 mm wide, papillose in longitudinal rows.

Zambia. C: Mazabuka Dist., c. 0.8 km [0.5 mile] from Chirundu Bridge, near Lusaka Road, 5.ii.1958, *Drummond* 5482 (K, SRGH). S: Sesheke Dist., Katombora, 3.iv.1956, *Robinson* 1411 (K). **Zimbabwe**. N: Kariba Dist., Charare Fish Camp, 12.xi.1964, *Jarman* 72 (K, SRGH). **Malawi**. N: Chitipa Dist., 5 km SE of Chitipa (Fort Hill), 13.iii.1961, *Robinson* 4528 (K).

Widespread through West Africa, Chad, Ethopia, Somalia, Uganda, Kenya, Tanzania, South Africa; SE Asia, Australia. Seasonally wet places including seasonal bogs, damp flush on sandstone and pool margins; 400–1300 m.

Conservation notes: Least Concern because of its wide distribution.

The white subspherical heads make this species look superficially like a *Kyllinga*, but a lens immediately shows that the small spikelets each have up to 20 flowers.

C21. **Cyperus dubius** Rottb. in Descr. Icon. Rar. Pl.: 20, t.4 fig.5 (1773). —Kükenthal in Engler, Pflanzenr. **4**, 20(101): 563 (1936). —Haines & Lye, Sedges & Rushes E. Afr.: 221, figs.449,450 (1983). —Hoenselaar in F.T.E.A., Cyperaceae: 186 (2010). Type: India, *König* s.n. (C holotype).

Perennial, perhaps sometimes annual, with bulbous-based tufted culms up to 45 cm tall, the base covered in translucent brown scales, without rhizomes; culms many, crowded, sometimes semi-succulent, 8–40 cm long, 0.5–2 mm wide, bluntly to sharply triangular, glabrous. Leaves many, up to 33 cm long, often recurved; leaf sheath pale brown, thin and membranous, the lower somewhat thicker, brown, to 4 cm long; leaf blade bright green or glaucous in very dry situations, linear, flat or slightly v-shaped, 5–33 cm long, 1–4(5) mm wide, scabrid on at least margin and primary vein, apex attenuate. Involucral bracts leaf-like, erect to spreading, 3–6, lowermost 4–23 cm long, 0.5–3.5 mm wide. Inflorescence capitate, green to almost white, hemispherical to ovoid, 5–15 mm in diameter, often of a single spike but sometimes of 3–6 congested spikes and then appearing lobed; spikelets narrowly ovoid, 2–6 mm long, 1–2 mm wide, 3–9(18)-flowered but with only few maturing nutlets; glumes greenish with uncoloured margin, ovate, 2–3 mm long, keel narrow, with 5–8 slender veins on either side, apex concave. Stamens 2–3, with white filaments and yellow anthers; style white. Nutlet black to brown with dark brown angles, obovoid, trigonous, 1.2–1.4 mm long, (including 0.1–0.2 mm long apiculus), 0.8–0.9 mm wide, papillose.

Var. **dubius**

> *Cyperus coloratus* Vahl, Enum. Pl. **2**: 312 (1805). Type: 'Guinea', *Thonning* 396 (C holotype).
> *Cyperus capitatus* Poir. in Lam., Encycl. **7**: 246 (1806), illegitimate name, non Vand. (1771). Type: Madagascar, *du Petit Thouars* s.n. (P holotype).
> *Mariscus coloratus* (Vahl) Nees in Linnaea **9**: 286 (1834). —Clarke in F.T.A. **8**: 381 (1901).
> *Isolepis boeckeleri* Oliv. in Trans. Linn. Soc. London **29**: 167 (1875). Type: no locality, *Grant* s.n. (K holotype).
> *Mariscus dubius* (Rottb.) C.E.C. Fisch. in Gamble, Fl. Madras: 1644 (1931).
> *Cyperus dubius* var. *capitatus* (Cherm.) Kük. in Engler, Pflanzenr. **4**, 20(101): 564 (1936).
> *Cyperus dubius* var. *coloratus* (Vahl) Kük. in Engler, Pflanzenr. **4**, 20(101): 565 (1936). —Haines & Lye, Sedges & Rushes E. Afr.: 223, fig.453 (1983).
> *Cyperus dubius* var. *polyactis* Kük. in Engler, Pflanzenr. **4**, 20(101): 565 (1936). Types: Tanzania, Kigoma Dist., Uvinza, N of Malagarasi, *Peter* 35923 (B holotype, B).
> *Cyperus dubius* var. *stenactis* Kük. in Engler, Pflanzenr. **4**, 20(101): 565 (1936). Types: Tanzania, various localities, *Peter* 4442b (B syntype), 19555 (B syntype, K), 11306 (B syntype), 24546 (B syntype).
> *Cyperus dubius* subsp. *coloratus* (Vahl) Lye in Nordic J. Bot. **3**: 231 (1983); in Haines & Lye, Sedges & Rushes E. Afr.: 223, fig.453 (1983).

Leaves 1–5 mm wide. Head 5–15 mm in diameter; glumes 2–3 mm long. Nutlets 1.2–1.4 mm long.

Botswana. N: Ngamiland Dist., Chobe R. at Kubu Lodge, 20.i.2006, *A. & R. Heath* 1113 (K). **Zambia**. B: Mongu Dist., Mongu, 22.xii.1965, *Robinson* 6741 (K). N: Mbala Dist., Chinakila, Loye Flats, 11.i.1965, *Richards* 19479 (K). S: Choma Dist., 48 km SE of Choma, 17.xii.1956, *Robinson* 1784 (K). **Zimbabwe**. N: Gokwe Dist., Sengwa Research Station, 21.i.1977, *Guy* 2466 (K, SRGH). W: Hwange Dist., Victoria Falls, 12.iii.1932, *Brain* 8841 (K). E: Mutare Dist., Vumba Mts., E Beacon to Maonza trees, 22.i.1990, *Browning* 278 (K, NU, SRGH). S: Ndanga Dist., Chionja Plateau, 29.i.1957, *Phipps* 219 (K, SRGH). **Malawi**. N: Rumphi Dist., Njakwa Gorge, 3 km [2 miles] E of Rumphi, 30.xii.1973, *Pawek* 7665 (K, MO, SRGH). C: Nkhata Bay Dist., Chinteche, Luweya R., 21.ii.1961, *Vesey-Fitzgerald* 3009 (K). S: Blantyre Dist., Mpatamanga Gorge, 5.ii.1979, *Blackmore* 285 (K). **Mozambique**. N: Marrupa Dist., 40 km on road Marrupa to Mukwajaja, *Jansen & Boane* 7986, 21.ii.1982 (K). Z: Lugela Dist., Namagoa, Mocuba, n.d., *Faulkner* P32 (K). MS: Dombe Dist., E bank of Makuripini R., 5 km above confluence with Haroni, 6.i.1969, *Bisset* 4 (K, SRGH). GI: Xai-Xai Dist., Xai-Xai, 6.ix.1993, *Hooper* s.n. (K). M: Matutuine Dist., between Zitundo and Manhoca, 29.xi.1979, *de Koning* 7709 (K, LMU).

Widespread in the drier parts of tropical Africa; also in India. In various types of woodland, bushland or grassland, often in shallow soil over rock outcrops; occasionally ruderal; 0–1600 m.

Conservation notes: Least Concern because of its wide distribution.

According to F.T.E.A., the bulbous base is eaten by rodents, francolin and guinea fowl; the whole plant is grazed by cattle, sheep, goats and hares. F.T.E.A. notes that the roots are fragrant but this probably refers to the bulbous stem-base.

Var. **macrocephalus** Boeckeler described from Sudan, occurs in the F.T.E.A. area in similar habitats. It has broader leaves (4–8 mm, not 1–5 mm), larger heads (13–20 mm, not 5–15 mm), longer glumes (3–4 mm not 2–3 mm) and nutlets (1.6–1.7 mm not 1.2–1.4 mm). I have not seen any material from the Flora area that can unequivocally be attributed to this taxon, but it may well occur. As there appears to be a continuous gradation in inflorescence size and leaf width, I question whether this variety is worth recognition anyway.

C22. **Cyperus margaritaceus** Vahl, Enum. Pl. **2**: 307 (1805). —Clarke in F.T.A. **8**: 321 (1901). —Kükenthal in Engler, Pflanzenr. **4**, 20(101): 284 (1936). —Haines & Lye, Sedges & Rushes E. Afr.: 257, fig.519 (1983). —Hoenselaar in F.T.E.A., Cyperaceae: 179 (2010). Type: Guinea, *Thonning* s.n. (C holotype).

  Cyperus pseudoniveus Boeckeler in Verh. Bot. Vereins Prov. Brandenburg **29**: 45 (1888). Type: Namibia, Olukonda, *Schinz* s.n. (B holotype).

  Cyperus margaritaceus var. *pseudoniveus* (Boeckeler) C.B. Clarke in F.T.A. **8**: 322 (1901). —Kükenthal in Engler, Pflanzenr. **4**, 20(101): 284 (1936).

Perennial, robust, up to 85 cm tall, with swollen bulb-like base; rhizome absent, or very short so that adjacent stem bases touch one another. Culms 30–85 cm long, 0.8–2 mm wide, trigonous, with longitudinal ribs, glabrous. Leaves up to 45 cm long; basal leaf sheaths reddish-brown to blackish, covering the base, leaf base on culm pale brown, 2–8 cm long; leaf blade linear, flat to inrolled, 5–37 cm long, 1.6–4 mm wide, scabrid on margins, apex acuminate. Involucral bracts leaf-like, spreading or reflexed, 2–4, lowermost 2–8 cm long, 1–2.9 mm wide. Inflorescence capitate; spikelets 1–10 per head, (broadly) ovate, 6–22 mm long, 5–10 mm wide, rachis straight; glumes dirty white, yellowish to pale reddish-brown, boat-shaped to elliptic-lanceolate, 5.7–11 mm long, 3.2–6.4 mm wide, keel prominent, many veins on either side of keel, apex acute, sometimes slightly excurrent. Stamens 3; filaments 4.8–8.8 mm long; anthers 1.9–4.1 mm long. Nutlet pale brown to dark olive, ovoid to orbicular, trigonous, 2.2–3 mm long, 2–2.1 mm wide, smooth, apiculate.

Botswana. N. Ngamiland Dist., Okavango Delta, Chief's Island, 1.8 km SE of Chief's Camp, 6.i.2012, A. *& R. Heath* 2345 (K). SE. Kgatleng Dist., railway/road crossing c. 5 km N of Gaborone, 6.xi.1977, *Hansen* 3272 (K). SW: Kgalagadi Dist., Kang, 300 km W of Gaborone, 4.iii.1977, *Mott* 1122 (K, SRGH, UCBG). **Zambia**. B: Mongu Dist., Looma, Nanganda, 21.xii.1993, *Bingham* 9883 (K). N: Mporokoso Dist., Nsama, 3.iv.1957, *Richards* 8970 (K). W: Mwinilunga Dist., Kafweka Forest Reserve near Lisombo R., 18 km SW of Kalene Hill, 22.ii.1975, *Hooper & Townsend* 309 (K). C: Lusaka Dist., Mt Makulu, 18 km [12 miles] S of Lusaka, 10.i.1956, *Angus* 1460 (FHO, K). S: Namwala Dist., Namwala, 9.i.1957, *Robinson* 2065 (K). **Zimbabwe**. N: Hurungwe Dist., Mwami (Miami), K34 Expt. Farm, 6.iii.1947, *Wild* 1683 (K, SRGH). W: Matobo Dist., Farm Besna Kobila, i.1957, *Miller* 3230 (K, SRGH). C: Marondera Dist., [Marandellas], Cave Farm, ii-iii.1946, *Weber* 228 (K). **Malawi**. N: Rumphi Dist., Nyika Plateau, Dembo

Bridge, 24.xii.1975, *Phillips* 718 (K, MO). C: Kasungu Dist., Kasungu Game Reserve, by Katete Dambo, 22.vi.1970, *Brummitt* 11626 (K). **Mozambique**. N: Marrupa Dist., Marrupa, about 20 km on road to Lichinga, 17.ii.1981, *Nuvunga* 552 (K, LMU).

Widespread in west tropical Africa, central Africa and southern Africa. Open *Brachystegia*, *Isoberlinia* or other woodland, or grassland, usually on sandy or rocky soils; 600–2100 m.

Conservation notes: Least Concern because of its wide distribution.

This species is very variable in size and it is possible that more than one taxon may be involved.

C23. **Cyperus niveus** Retz. in Observ. Bot. **5**: 12 (1788). —Haines & Lye, Sedges & Rushes E. Afr.: 256 (1983). —Hoenselaar in F.T.E.A., Cyperaceae: 168 (2010). Type: India, Midnapur, Tschandrancone, *König* s.n. (LD holotype).

Perennial, up to 118 cm tall; culms crowded, bases swollen and fused into a horizontal rhizome, 5.5–116 cm tall, 0.8–3.7 mm wide, trigonous to rounded, smooth. Leaves up to 118 cm long; leaf sheath nearly black at the base, brown up to the culm, 1–13 cm long; leaf blade linear, canaliculate or flat, 5–56 cm long, 1.4–8 mm wide, glabrous or scabrid on margin and primary vein, apex acute to acuminate. Involucral bracts leaf-like, spreading, sometimes reflexed, 2–5, lowermost 2.5–20 cm long, 1.2–6 mm wide. Inflorescence capitate, spikelets 5 to many per head, ovoid-lanceolate, 7.2–17 × 4–8.7 mm wide, rachis straight; glumes dirty white, sometimes with a pinkish or pale brown tinge, elliptic-lanceolate, ovate to boat-shaped, 4.3–8.9 × 2.4–3.8 mm wide, keel flat to acute, mostly with 6–8 conspicuous striations on either side of keel, apex acute to sometimes obtuse. Stamens 3; filaments 4.2–7.2 mm long; anthers 1.6–3.7 mm long. Nutlet (dark) brown to black, shiny, obovoid, trigonous, 1.6–2.5 × 1.3–1.8 mm wide, almost smooth to minutely papillose.

Cyperus niveus is quite variable and several infraspecific taxa have been described. Of the three varieties Haines and Lye only recognize two, var. *leucocephalus* and var. *tisserantii*. Var. *tisserantii* has previously been described as a variety of *C. margaritaceus*, along with *C. nduru*.

Leaf blade less then 1.2 mm wide; often in burnt areas, plant showing signs of burning; spikelets up to 12 per head . . . . a) var. *tisserantii*
Leaf blade 1.4–8 mm wide; spikelets many per head . . . . b) var. *leucocephalus*

a) Var. **tisserantii** (Cherm.) Lye in Nordic J. Bot. **3**: 231 (1983). —Hoenselaar in F.T.E.A., Cyperaceae: 168 (2010). Type: Central African Republic, Ouaka region, near Ippy, *Tisserant* 1878 (P holotype, P).

   Cyperus tisserantii Cherm. in Arch. Bot. Mém. **4**(7): 18 (1931).
   Cyperus margaritaceus Vahl var. *tisserantii* (Cherm.) Kük. in Engler, Pflanzenr. **4**, 20(101): 285 (1936). —Haines & Lye, Sedges & Rushes E. Afr.: 257 (1983).

Culms 5.5–20 cm long, 0.6–1.3 mm wide. Leaf blade canaliculate, 5–11 cm long, 0.6–1.2 mm wide. Involucral bracts 2–3, lowermost 2.5–7 cm long, 0.6–1.7 mm wide; spikelets 5–12 per head, ovoid, 7.2–13.5 mm long, 4–7.1 mm wide; glumes dirty white, ovate-lanceolate, 4.4–6.4 mm long, 2.5–3.8 mm. Nutlet ± 2.5 mm long, ± 1.8 mm wide.

Not so far collected from the Flora area, but may well occur. Senegal, Burkina Faso, Ivory Coast, Ghana, Togo, Benin, Nigeria, Cameroon, Gabon, Central African Republic, Ethiopia, Uganda, Kenya, Tanzania. In dry grassland and wooded grassland, where recently burnt; (0)100–2200 m.

This variety often occurs in areas which are recently burnt. It shows similarities with *Cyperus nduru*, which has far fewer leaves and fewer spikelets per inflorescence. It is debatable whether if *tisserantii* is treated as a variety, *C. nduru* should described as a subspecific taxon as well.

b) Var. **leucocephalus** (Kunth) Fosberg in Kew Bull. **31**: 835 (1977). —Haines & Lye, Sedges & Rushes E. Afr.: 256, figs.517,518 (1983). —Hoenselaar in F.T.E.A., Cyperaceae: 170 (2010). Type: South Africa, Cape of Good Hope, *?Drège* s.n. (B holotype).

   Cyperus compactus Lam., Tabl. Encycl. **1**: 144 (1791), illegitimate name, non Retz. (1788). — Clarke in F.T.A. **8**: 319 (1901).

   Cyperus obtusiflorus Vahl, Enum. Pl. **2**: 308 (1805). —Kükenthal in Engler, Pflanzenr. **4**, 20(101): 285 (1936). Type: Madagascar, no collector (P-LAM holotype).

   Cyperus sphaerocephalus Vahl var. *leucocephalus* Kunth, Enum. Pl. **2**: 45 (1837).

   Cyperus obtusiflorus var. *ledermannii* Kük. in Engler, Pflanzenr. **4**, 20(101): 287 (1936). Type: Cameroon, *Ledermann* 5257 (B holotype).

   Cyperus ledermannii (Kük.) S.S. Hooper in Kew Bull. **26**: 578 (1972).

   Cyperus niveus var. *ledermannii* (Kük.) Lye in Nordic J. Bot. **3**: 231 (1983); —Haines & Lye, Sedges & Rushes E. Afr.: 257 (1983).

Culms 16–116 cm long, 1–3.7 mm wide, trigonous (to rounded). Leaf blade flat, 10–56 cm long, 1.4–8 mm wide. Involucral bracts 2–4, lowermost 3.5–23 cm long, 1.2–6 mm wide; spikelets 5 to many per head, 9–17 × 4–9 mm wide; glumes (dirty) white, with a pinkish or pale brown tinge, elliptic-lanceolate to ovate, 4.3–8.9(11.6) × 2.4–3.8(5.6) mm wide. Nutlet brown to black, obovoid, trigonous, 1.6–2.9 × 1.3–1.8 mm wide.

Botswana. N: Chobe Dist., 43 km N of Nata on road to Kazangulu via Pandamatenga, 23.iii.1976, *Ellis* 2764 (K, PRE). SW: Kgalagadi Dist., 13 km N of Hukuntsi on Nojane road, 14.iii.1976, *Ellis* 2674 (K, PRE). **Zambia**. N: Mbala Dist., Nakatali, near Mbala, 4.ii.1951, *Bullock* 3656 (K). C: Mkushi Dist., Fiwila, 4.i.1958, *Robinson* 2606 (K). E: Petauke Dist., Nyimba to Luenbe road, 15.xii.1958, *Robson* 935 (K). S: Mazabuka Dist., Tara Protected Forest Area, 26.i.1960, *White* 6395 (FHO, K). **Zimbabwe**. N: Mutoko Dist., 160 km [100 miles] ENE of Harare, 8.xii.1946, *Weber* 292 (K). C: Gweru Dist., Gweru (Gwelo) Teachers' College, 23.xi.1966, *Biegel* 1459 (K, SRGH). E: Chimanimani Dist., Chimanimani Mts, 1900 m., 31.xii.1964, *Robinson* 6340 (K). S: Ndanga Dist., N side of Lundi River, Chipinda Pools area, 19.xii.1959, *Goodier* 743 (K). **Malawi**. N. Mzimba Dist., 5 km [3 miles] W of Mzuzu, Katoto, 13.iii.1974, *Pawek* 8210 (K, MO). C: Dedza Dist., Mwalawankhondo, 13.xi.1950, *Jackson* 261 (K). **Mozambique**. N: Angoche Dist., (Antonio Enes), Matangula littoral, 24.x.1965, *Mogg* 32505 (K). MS: Gorongosa Dist., Cheringoma section, 4 km W of Safrique Hunting Camp, 12.vii.1972, *Ward* 7849 (K, DURB). GI: Inhassoro Dist., Bazaruto Is., Ponta Estoue, 28.x.1958, *Mogg* 28693 (K). M: Inhaca Is., Shilepuzambeti, 28.xii.1956, *Mogg* 27022 (K).

Also in Benin, Nigeria, Cameroon, Gabon, D.R. Congo, Rwanda, Burundi, Ethiopia, Somalia, Kenya, Uganda, Tanzania, Namibia, South Africa. Miombo woodland, dry grassland, on stony slopes, on shallow soil over rocks, in dried up riverbeds, swampy areas and in cultivated areas; sea-level to 2000 m.

Conservation notes: Least Concern because of its wide distribution.

Beentje in F.T.E.A. commented as follows: Haines & Lye already remarked on the few differences between *Cyperus margaritaceus* and *C. niveus* (in key separated only by presence/absence of rhizomes, but also separated them by the slightly larger stem base in *margaritaceus* and less compressed spikelets) with the taxa *tisserantii* and *nduru* intermediate, and possibly linked to fire regimes. It is quite possible this is all one species; *niveus* is the oldest name.

C24. **Cyperus clavinux** C.B. Clarke in F.T.A. **8**: 319 (1901). —Kükenthal in Engler, Pflanzenr. **4**, 20(101): 304 (1936). —Haines & Lye, Sedges & Rushes E. Afr.: 261, figs.530,531 (1983). —Hoenselaar in F.T.E.A., Cyperaceae: 175 (2010). Types: Nigeria, Bornu, *Vogel* 64 & 65 (K syntypes).

   Cyperus monostigma C.B. Clarke in Bull. Soc. Bot. France **54**, Mém. 8: 26 (1907). Types: Congo, *Chevalier* s.n.; ?Chad, Lac Fiottri & Baguirmi N, Moula, *Chevalier* 9609, 9610 (P syntypes).

Perennial, slender, up to 26 cm tall, with a slightly swollen culm-base covered by the fibrous remains of old leaf sheaths; culms 3–23 cm long, 0.5–0.9 mm wide, trigonous, smooth. Leaves up to 15.5 cm long; leaf sheath pale brown, 1–2.5 cm long; leaf blade linear, folded, sometimes canaliculate, 7–13 cm long, 0.8–0.9 mm wide, glabrous, apex pale, solid, acuminate. Involucral bracts 2–3, leaf-like, spreading, lowermost 7–12 cm long, 0.8–1 mm wide. Inflorescence capitate, spikelets up to 25 per inflorescence, elliptic-lanceolate, 7.5–10.5 × 2.2–2.5 mm wide; glumes very pale brown to greyish-white, sometimes with slight reddish dots on the wings, ovate, 2.1–2.2 × 1.6–1.9 mm wide, keel greenish, 3-veined, apex acuminate, slightly recurved. Style usually unbranched. Nutlet dark grey to black, pear-shaped, trigonous, c. 0.8 × 0.6 mm wide, minutely papillose.

Botswana. N:Ngamiland Dist., Near Kwando Hunters Camp, 18°06.6'S, 23°21.1'E, 23.i.1978, *P.A. Smith* 2215 (K, SRGH). **Malawi**. N: Rumphi Dist., Chitipa (Fort Hill), 11.iii.1961, *Robinson* 4443 (K). **Zambia**. C: Mpika Dist., Lubi R., Mfuwe, Luangwa Nat. Park, 5.iii.1967, *Astle* 5048 (K). E: Chipata Dist., Rukuzi R. on Chipata–Lundazi road, 24.i.1964, *Wilson* 47 (K, SRGH). S: Livingstone Dist., 34.8 km W of intersection of Lusaka and Livingstone-Sesheke roads, 0.5 km from Mandia School, 20.ii.1997, *Harder et al.* 3887 (K, MO). **Zimbabwe**. N: Hurungwe Dist., 1.5 km [1 mile] N of R. Mauora, 28.ii.1958, *Phipps* 964 (K, SRGH). W: Hwange Dist., Victoria Falls, 12.iii.1932, *Brain* 8820 (K, SRGH). C: Chegutu Dist., Poole Farm, 27.ii.1945, *Hornby* 2386 (K, SRGH).

Also in Nigeria, Chad, and Tanzania. Seasonally wet ground in mopane and miombo woodland; sandy ground; 550–1450 m.

Conservation notes: Least Concern because of its wide distribution.

A small perennial, but easily recorded as annual because the base is small; very similar to *Cyperus remotiflorus* but the broad flattened glumes with a prominent green excurrent midrib are distinctive.

C25. **Cyperus remotiflorus** Kük. in Repert. Spec. Nov. Regni Veg. **18**: 345 (1922); in Engler, Pflanzenr. **4**, 20(101): 303 (1936). Type: Namibia, *Range* 270 (presumably B, not seen).

Small densely tufted perennial. Culms 4–8 cm long, slender, trigonous, glabrous, somewhat thickened to bulbous at the base, with brown leaf sheaths which break up into fibres. Leaves few, basal equalling or shorter than the culms, setaceous, inrolled, with enlarged membranous sheaths. Spikelets 5–25, narrowly oblong, 6–10 mm long and 1.5–2 mm wide, stellately arranged in a capitate terminal inflorescence. Rhachilla straight, stiff, not winged. Inflorescence bracts 2, much exceeding the inflorescence. Glumes well spaced, later spreading, membranous, narrowly ovate, acute, conspicuously 7-nerved, reddish, margins whitish-hyaline. Stamens (2)3. Style long, three-branched, the branches much shorter than the style. Nutlet about half the length of the glume, small, broadly obovate, trigonous, grey, apiculate, densely puncticulate.

Botswana. N: Ngamiland Dist., Quangwa [Kangwa] (Xanwe), 27 km N of Aha Hills, 12.iii.1965, *Wild & Drummond* 6932 (K, SRGH). SW: Ghanzi Dist., Ghanzi Pan, 14 km [9 miles] E of Ghanzi, 30.i.1970, *Brown* 8275 (K). SE: Central Dist., Thalamabele-Mosu area, near Soa Pan, 13.i.1974, *Ngoni* 312 (K, SRGH). **Zambia**. W: Ndola Dist., Ndola, 27.ii.1960, *Robinson* 3349 (K). C: Lusaka Dist., Liempe, 12 km E of Lusaka, 5.i.1972, *Kornaś* 0796 (K).

Also in Namibia and South Africa (Mpumalanga, Northern Cape). Usually in shallow soil over rock, at least sometimes on limestone or dolomite, 900–1200 m.

Conservation notes: Least Concern because of its wide distribution.

Superficially rather similar to *Cyperus clavinux*, with a similar delicate perennial habit which could easily lead to it being recorded as an annual, but the strongly nerved ovate glumes and the absence of a strong green excurrent midrib on the glumes should distinguish it.

C26. **Cyperus mollipes** (C.B. Clarke) K. Schum. in Engler, Pflanzenw. Ost-Afrikas **C**: 122 (1895). —Kükenthal in Engler, Pflanzenr. **4**: 20(101): 557 (1936). — Hoenselaar in F.T.E.A., Cyperaceae: 180 (2010). Type: Sudan, Djur, Majob, *Schweinfurth* 1547 (B holotype, K).

Cyperus macropus Boeckeler in Flora **62**: 550 (1879), illegitimate name, non Miq. (1861). Type: Sudan, Djur, Seriba Ghattas, *Schweinfurth* 1917 (B holotype, B).

Rhynchospora bulbocaulis Boeckeler in Flora **62**: 567 (1879). Type: Sudan, 26.iv.1859, *Schweinfurth* 1547 (B, P, S isotypes).

Mariscus macropus C.B. Clarke in Durand & Schinz, Consp. Fl. Afric. **5**: 590 (1894), based on *Cyperus macropus* Boeckeler, non Miq. (1861).

Mariscus mollipes C.B. Clarke in Durand & Schinz, Consp. Fl. Afric. **5**: 590 (1894); in F.T.A. **8**: 387 (1902), new name for *Rhynchospora bulbocaulis* Boeckeler.

Mariscus oblonginux C.B. Clarke in F.T.A. **8**: 526 (1902). Type: Malawi, Zomba, *Cameron* 90 (K holotype).

Cyperus mollipes var. *amomodorus* (K. Schum.) Kük. in Engler, Pflanzenr. **4**, 20(101): 557 (1936).

Cyperus globifer (C.B. Clarke) Lye in Nordic J. Bot. **3**: 232 (1983); —Haines & Lye, Sedges & Rushes E. Afr.: 219, fig.443 (1983).

?*Cyperus firmipes* (C.B. Clarke) Kük. in Engler, Pflanzenr. **4**, 20(101): 562 (1936). Type: Malawi, Zomba, *Whyte* s.n. See note below.

Perennial, tufted, with a bulbous or tuberous culm-base covered by rather thick brown or blackish old fibres from leaf sheaths, intact leaf sheaths strongly pale-veined; culms few to many and crowded, 10–50 cm tall and 0.5–2.0 mm wide, trigonous to triquetrous, glabrous. Leaves with leaf sheath pale to dark brown, 3–7 cm long; leaf blade linear, flat or slightly channelled, 5–30 × 0.1–0.7 cm, scabrid on at least margin and primary vein, particularly towards apex, attenuate; apex rounded. Involucral bracts leaf-like, scabrid like the leaves, often conspicuously dilated at the base, erect, spreading or reflexed, 2–5, lowermost 2–6(20) × 0.1–0.6 cm. Inflorescence capitate, hemispherical to spherical, whitish to straw-yellow, 8–21 mm in diameter; spikelets many per head, very narrowly ovate, 4–8 × 0.7–2 mm wide, 2–4-flowered but often only perfecting 1 nutlet; glumes whitish or yellowish, concave, 3–5.9 × 1.1–1.4 mm wide, keel obscure, with 4–10 veins on either side, apex acuminate. Stamens 3, filaments 3.7–7 mm long, anthers yellow, 1.6–2.3 mm long. Nutlet dark brown to blackish, oblong to broadly obovoid and slightly trigonous to cylindrical, 1.4–3.5 × 0.4–1.2 mm wide, conspicuously apiculate, minutely papillose, completely enveloped by the glume when mature, dark grey to blackish.

Zambia. N: Mbala Dist., Mbala, above Ndundu, 7.xii.1958, *Richards* 10247 (K). C: Petauke Dist., Mushilashi to Lubi R., Luangwa Valley, 13.xii.1966, *Prince* 25 (K). S: Choma Dist., 10 km [6 miles] E of Choma, 18.xii.1956, *Robinson* 1796 (K). **Zimbabwe**. N: Guruve Dist., 0.5 km N of Station, Nyamunyeche Estate, 4.xii.1978, *Nyariri* 530 (K, SRGH). **Malawi**. S: Zomba Dist., Namasi, xi-xii.1899, *Cameron* 90 (K, type of *M. oblonginux* C.B. Clarke). **Mozambique**. N: Malema Dist., Malema, near river, 1.xii.1931, *Gomes e Sousa* 802 (K).

Also in Ethiopia, Sudan, Somalia, Kenya, Uganda, Tanzania, Rwanda, D.R. Congo. In mopane and miombo woodland on sandy or rocky soils, sandveld.

Conservation notes: Least Concern because of its wide distribution.

F.T.E.A. has a much fuller synonymy for this species, with extensive notes.

Cyperus firmipes (C.B. Clarke) Kük. (*Mariscus firmipes* C.B. Clarke) may well belong here, but the type (from Malawi) lacks basal parts and it is not really possible to assign it to a known taxon without better material, although the strong veining of the tops of the basal bracts suggests this species. See Hoenselaar in F.T.E.A., Cyperaceae: 257 (2010).

Group D: Species with umbellate terminal inflorescence units.

1. Inflorescences with numerous very small usually dark spikelet clusters; leaves broad; plants of forest or forest edge 2
– Inflorescences with less numerous spikelets, usually pale; leaves narrow; other habitats 8
2. Glumes rounded at apex . . . . **D1.** *derreilema*
– Glumes acute, acuminate or mucronate 3
3. Glumes 1.2–1.8 mm long 4
– Glumes 1.8–3 mm long 5
4. Spikelets a few together; nutlet 1.1–1.4 mm long . . . . **D2.** *laxus*
– Spikelets crowded; nutlet 0.7–1.1 mm long . . . . **D3.** *renschii*
5. Plants with stolons; stamens 2; glumes with conspicuous black patch; lowest involucral bract much longer than rest, held upright . . . . **D4.** *dichrostachyus*
– Plants with rhizomes; stamens 3 6
6. Leaves up to 80 cm long and 12 mm wide; largest involucral bract 10–23 cm long . . . . **D5.** *glaucophyllus*
– Leaves 54–220 cm long and 9–37 mm wide; largest involucral bract 30–125 cm long 7
7. Culms 0.8–2 m long; leaf blade 14–37 mm wide; spikelets 3.2–6.7 mm long . . . . **D6.** *ajax*
– Culms 0.5–1.2 m long; leaf blade 9–16 mm wide; spikelets 4–13 mm long . . . . **D7.** *fischerianus*
8. Spikelets green, few, flattened, up to three times as long as broad . . **D8.** *chinsaliensis*
– Spikelets brownish, yellowish, reddish or blackish; more than three times as long as broad 9
9. Inflorescence of few spikelets, surrounded by a conspicuous involucre of overlapping green bracts . . . . **D9.** *albostriatus*
– Inflorescence many-flowered or, if few-flowered, without an involucre of overlapping green bracts 10
10. Plant base arising from a short vertical rhizome from a bulb; spikelets flattened, dark red-brown . . . . **C3.** *usitatus* var. *macrobulbus*
– Plant base and spikelets not as above 11
11. Plant without a distinct thin elongated rhizome(s) 12
– Plant with distinct thin (thick in *C. latifolius*) elongated rhizome(s) 22
12. Inflorescence consistently once anthelate 13
– At least some inflorescences more than once anthelate 17
13. Many spikelets in clusters of 5 or more, 2.5 mm broad or less 16
– Spikelets in clusters of 2–4, 2–3 mm broad, dark red-brown 14
14. Base without fibrous leaf sheath remnants; slender rhizomes and tubers often present . . . . **D10.** *mwinilungensis*
– Base with fibrous leaf sheath remnants; no slender rhizomes or tubers 15
15. Inflorescence small and subspherical, with 3–5 short branches; glumes blackish . . . . **D11.** *nyererei*
– Inflorescence larger and more open; glumes dark red-brown . **D12.** *matagoroensis*
16. Spikelets usually less than 1.5 mm broad, usually very dark brown to black; base tussocky . . . . **D13.** *tenax*
– Spikelets larger, usually brown; base a short knotty rhizome . . . . **D14.** *haspan*
17. Spikelets straw-yellow; culm base swollen/bulbous . . . . **D15.** *hensii*
– Spikelets darker, usually shades of reddish brown 18
18. Involucral bracts 5–8, much longer than the inflorescence branches . . . **D2.** *laxus*
– Involucral bracts 1–4, equalling, shorter than or slightly longer than the inflorescence branches 19

19. Involucral bracts 2–4, exceeding the inflorescence branches; inflorescence 2–3 times anthelate. . . . . **D16.** *sphaerospermus*
–   Involucral bracts 2–3, equalling or shorter than the inflorescence branches; inflorescence 1–3 times anthelate 20
20. Rhizomes short, horizontal; shoots almost contiguous; spikelets initially solitary at branch apices, later proliferating to produce new spikelets or plantlets. . . . . **D17.** *deciduus*
–   Not as above. 21
21. Leaves up to 4.5 mm broad; no elongated rhizomes . . . . **D14.** *haspan*
–   Leaves 5–9 mm broad; short thick rhizomes usually present . .**D18.** *aureobrunneus*
22. Roots or rhizomes with tubers 23
–   No tubers on roots or rhizomes 24
23. Plants bearing small tubers at the ends of slender rhizomes; involucral bracts 2–4; spikelet clusters often with more than 20 spikelets. . . . . C3. *usitatus*
–   Plants with ovoid or ellipsoid root-bearing tubers at intervals on slender rhizomes; spikelet clusters usually with six spikelets or fewer . . . . **D19.** *rotundus*
24. Spikelets small and narrow, yellow-brown, up to 1 mm broad; leaves <3.5 mm broad . . . . **D20.** *kipasensis*
–   Spikelets larger, >1 mm broad, brown or red-brown; leaves usually broader . . . 25
25. Spikelets terete, with smooth outline; glumes 20 or more per spikelet. 26
–   Spikelets flattened, often with serrate outline; glumes 8–12 per spikelet 27
26. Spikelets dark red-brown; 1–1.2 cm long. . . . . **D10.** *mwinilungensis*
–   Spikelets pale greenish or yellowish, 0.7–2 cm long . . . . **D21.** *maculatus*
27. Basal bulb or tuber present; glumes pale-edged . . . . **D21.** *maculatus*
–   No basal bulb or tuber; spikelets 8.5 mm or more long 28
28. Spikelets densely clustered at the ends of the branchlets, >15 mm long. . . . . .A11. *natalensis*
–   Spikelets well spaced so that the bases are clearly visible 29
29. Spikelets usually dark red-brown, well spaced; some at least 20 mm long. . . . .**D22.** *procerus*
–   Spikelets usually yellowish brown, lowest sometimes well-spaced, upper forming a cluster, mostly less than 16 mm long . . . . **D23.** *esculentus*
30. Spikelets with serrate outline; glumes 8–12 per spikelet. . . . . .E22. *turrillii*
–   Spikelets terete with smooth outlline; glumes 20 or more per spikelet. . . . . **D21.** *maculatus*

D1. **Cyperus derreilema** Steud. in Flora **25**: 585 (1842). —Clarke in F.T.A. **8**: 343 (1901). —Kükenthal in Engler, Pflanzenr. **4**, 20(101): 199 (1935), as *dereilema*. —Haines & Lye, Sedges & Rushes E. Afr.: 155, fig.285 (1983). —Hoenselaar in F.T.E.A., Cyperaceae: 197 (2010). Type: Ethiopia, near Mt Silke, *Schimper* 659 (P holotype, BR, HAL, K).

> *Cyperus deckenii* Boeckeler in Linnaea **38**: 361 (1874). —Clarke in F.T.A. **8**: 342 (1901). Type: Tanzania, Kilimanjaro, 6500–8500', *Kersten s.n.* (B holotype).
>
> *Cyperus derreilema* subsp. *deckenii* (Boeckeler) Kük. in Engler, Pflanzenr. **4**, 20(101): 199 (1936).
>
> *Cyperus derreilema* var. *brevispiculosus* Kük. in Engler, Pflanzenr. **4**, 20(101): 199 (1936). Types: Kenya, Mt Kenya, *Fries & Fries* 1215 & 772; Mt Kenya, Coles Mill, *Fries & Fries* 1112; Aberdares, *Fries & Fries* 2497; Mt Elgon, *Granvik* 54. Tanzania, Kilimanjaro, 25.vi.1926, *Peter* 41961 (K, WAG syntype)

Perennial, robust, up to 2.25 m tall, with a thick woody rhizome; culms tufted, 135–200 cm long, 3.5–8 mm wide, trigonous to triquetrous, smooth to sometimes slightly scabrid. Leaves up to 100 cm long; leaf sheath not prominent, only seen at the very base of the culm, brown; leaf blade linear, w-shaped or flat, 58–100 cm long, 1.1–2.2 cm wide, scabrid on veins and margins, apex acute to acuminate. Involucral bracts leaf-like, erect to spreading, 5–9, lowermost 24–85 cm long, 0.7–2.2 cm wide. Inflorescence a compound anthela, primary branches 6–15, 5–17 cm long; spikelets in digitate clusters, at the end of primary, secondary and tertiary branches, (1)2–6 per cluster, ovoid, 4.3–8 mm long, 1.9–2.5 mm wide, rachilla straight to slightly curved; glumes reddish-brown, ovate, 2.2–2.7 mm long, 1.1–1.4 mm wide with several veins on either side, keel green, not excurrent, apex rounded. Stamens 3; filaments 1.6–2.7 mm long; anthers 1.3–1.9 mm long. Nutlet reddish-brown, ellipsoid-ovoid, 1 mm long, 0.6–0.7 mm wide, almost smooth.

Zambia. N: Mbala [Abercorn] Dist., Sumbawanga road, 5 km [3 miles] from Kawimbe, 25.ii.1959, *Richards* 10987 (K). **Malawi**. N: Rumphi Dist., Nyika Plateau, 10km [6 miles] SW of Rest House, 25.x.1958, *Robson* 363 (BM, K). S: Mulanje Dist., Mt. Mulanje, Lichenya River below forestry hut, 17.ii.1982, *Hepper* 7359 (K).

Also in Ethiopia, Uganda, Kenya, Tanzania, D.R. Congoand Rwanda. Moist places in upland forest; valley bottom by permanent stream; clearing in *Widdringtonia* forest; 1800–2200 m.

Conservation notes: Data Deficient because of confusion with the other large species in the group.

This species is related to *Cyperus ajax* but can be distinguished by its rounded (not mucronate) glume apex.

Often spelled *dereilema* but the protologue uses *derreilema* (F.T.E.A.).

Cyperus derreilema, together with *C. ajax, C. dichrostachyus, C .glaucophyllus, C. fischerianus* and *C. renschii*, form a distinct group of very similar large species, mostly of upland forest and forest margins, which are extremely hard to distinguish. *Cyperus derreilema* is supposed to have the glumes rounded at the apex, but there are specimens in which the glume apex is rounded but also mucronate. The character 'plants with stolons', used as one of the distinguishing characters for *C. dichrostachyus* is somewhat ambiguous as the definition of a stolon is itself variable, and the basal parts are often missing from collections. The characters of leaf length, leaf width and culm length, also used in the F.T.E.A. key, overlap wherever they are used in the key. The illustration of *C. renschii* in F.T.E.A. shows the glumes to have toothed margins but this is not mentioned in the text and certainly not all specimens show it. *Cyperus renschii* is also placed in a completely different part of the key in F.T.E.A. although the form of the inflorescence is very similar to that of the other species mentioned here.

A final difficulty is that, for purely practical reasons, these large sedges are rarely collected as whole plants and partial specimens are much less easy (if not impossible) to identify. Much of the material from the Flora area is incomplete and/or poor and I have used the descriptions from F.T.E.A. virtually unchanged.

While I have retained these species as distinct, it is likely that further studies in the field using whole plants and populations may well show that these are just forms of a single variable entity.

D2. **Cyperus laxus** Lam., Tabl. Encycl. **1**: 146 (1791). —Haines & Lye, Sedges & Rushes E. Afr.: 163 (1983). —Hoenselaar in F.T.E.A., Cyperaceae: 201 (2010). Type: Brazil, E Cayenne, no collector indicated (BM-Sloane holotype)

Perennial up to 124 cm tall, with short woody rhizome; culms tufted, 28–113 cm long, 1.1–4 mm wide, trigonous to triquetrous, smooth. Leaves up to 60 cm long; leaf sheath reddish-brown to purple, 4–9 cm long; leaf blade linear, flat, 25–53 cm long, 0.3–1.3 cm wide, scabrid on margins and veins, apex acuminate to acute, with 2 main veins next to primary vein. Involucral bracts leaf-like, erect or spreading, much overtopping the inflorescence, 5–8, lowermost 15.5–31 cm long. Inflorescence simple to compound, primary branches 6–10, 2–7 cm long; spikelets in small digitate clusters at the end of primary, secondary and sometimes tertiairy branches, 3–5 per cluster, 4–6 mm long, 1.6–2.4 mm wide; glumes pale reddish-brown, ovate, 1.1–3 mm long, 0.6–1.6 mm wide, keel green, excurrent, apex mucronate, slightly recurved. Stamens 3; filaments

1.3–1.9 mm long. Nutlet reddish brown to black, ovoid to obovoid, 1.1–1.4 mm long, 0.6–1.1 mm wide, almost smooth to minutely papillose to minutely irregularly pitted.

Culms up to 50 cm long; leaves up to 35 cm long; glumes glabrous. . . . .
 . . . . a) subsp. *buchholzii*
Culms more than 50 cm long; leaves more than 35 cm long; glumes scabrid at apex
 . . . . b) subsp. *laxus*

a) Subsp. **buchholzii** (Boeckeler) Lye in Nordic J. Bot. **3**: 232 (1983). —Haines & Lye, Sedges & Rushes E. Afr.: 163, figs.303,304 (1983). —Hoenselaar in F.T.E.A., Cyperaceae: 201 (2010). Type: Cameroon, Bonjongo, Victoria, *Buchholz* s.n. (B holotype).

> *Cyperus buchholzii* Boeckeler, Beitr. Cyper. **1**: 3 (1888).
> *Cyperus diffusus* Vahl subsp. *buchholzii* (Boeckeler) Kük. in Engler, Pflanzenr. **4**, 20(101): 210 (1936).

Culms 28–49 cm long, 1.1–2.4 mm wide. Leaves 23–31 cm long, 3.2–8 mm wide; glumes 1.1–1.4 mm long, 0.6–0.9 mm wide, glabrous.

Zambia. B: Mongu Dist., Looma, Nanganda, 21.xii.1993, *Bingham* 9910 (K). N: Kasama Dist., Chibutubutu, Lukulu River (50 km [30 miles] S of Kasama) on Mpika road, 23.ii.1960, *Richards* 12551 (K). W: Mwinilunga Dist., Mwinilunga–Solwezi road at Lumwana River, 27.ii.1975, *Hooper & Townsend* 430 (K). C: Muchinga Dist., Mutinondo Wilderness area, Lodge Rock, 27.iii.2013, *Merrett* 1359 (K). **Zimbabwe**. N: Kariba Dist., Miami, Experimental Farm, 5.iii.1947, *Wild* 1723 (K, SRGH). W: Matobo Dist., Farm Besna Kobila, i.1958, *Miller* 5031 (K, SRGH).

Widespread in west and central Africa. In forest, secondary vegetation, on stream banks and in glades; 1200–1600 m.

Conservation notes: Least Concern because of its wide distribution.

b) Subsp. **sylvestris** (Ridl.) Lye in Nordic J. Bot. **3**: 232 (1983). —Haines & Lye, Sedges & Rushes E. Afr.: 163 (1983); Hoenselaar in F.T.E.A., Cyperaceae: 201 (2010). Type: Angola, *Welwitsch* 6898 (LISU holotype, BM).

> *Cyperus sylvestris* Ridl. in Trans. Linn. Soc. London, Bot. **2**: 134 (1884).
> *Cyperus diffusus* Vahl subsp. *sylvestris* (Ridl.) Kük. in Engler, Pflanzenr. **4**, 20(101): 210 (1936).

Culms 50–113 cm long, 2.8–4 mm wide. Leaves 36–53 cm long, 9–13 mm wide; glumes 1.9–3 mm long, 1.3–1.6 mm wide, scabrid near the apex.

Zambia. N: Chinsali Dist., Shiwa Ngandu, 1.vi.1956, *Robinson* 1515 (K). W. Mufulira Dist., 2.iv.1948, *Cruse* 312 (K). C: Kabwe Dist., Kamaila Forest Station, 36 km N of Lusaka, 9.ii.1975, *Brummitt* 14287 (K). S: Choma Dist., Mapanza E., 29.i.1954, *Robinson* 496 (K). **Zimbabwe**. N: Binga Dist., Lusulu Veterinary Ranch, 28.ii.1965, *Bingham* 1398 (K, SRGH). W: Matobo Dist., Farm Besna Kobila, xii.1957, *Miller* 4871 (K, SRGH). **Malawi**. N: Chitipa Dist., Chisenga, Mizimero area, 11.iii.2007, *Chapama et al.* 597 (K, MAL). C: Lilongwe Dist., Nature Sanctuary Zone A, along S trail, 13.i.1985, *Patel & Banda* 1967 (K, MAL).

Also in Tanzania and Angola. In miombo woodland and riverine woodland, sometimes around termite mounds; 1000–1700 m.

Conservation notes: Least Concern because of its wide distribution.

Intermediates between the two subspecies exist, such as **Zambia**. B. Kaoma Dist., c. 5 km along road to Luampa Hospital from Lusaka-Mongu road, 2.iii.1996, *Harder et al.* 3636 (K, MO), which has leaves 8 mm wide and one or two teeth on the glumes.

D3. **Cyperus renschii** Boeckeler in Flora **65**: 11 (1882). —Clarke in F.T.A. **8**: 345 (1901). —Kükenthal in Engler, Pflanzenr. **4**, 20(101): 206 (1936). —Haines & Lye, Sedges & Rushes E. Afr.: 161, figs.300,301 (1983). —Hoenselaar in F.T.E.A. Cyperaceae: 247, fig.37 (2010). Type: Comoro Islands, Anjouan [Johanna], *Hildebrandt* 1740 ('Herb. Resch', not found at B).

> *Cyperus deremensis* Engl., Pflanzenw. Ost-Afrikas **C**: 119 (1895). Type: Tanzania, Lushoto Dist.: Derema [Nderema], *Holst* 2257 (B holotype).
> *Cyperus ochrocarpus* K. Schum. in Engler, Pflanzenw. Ost-Afrikas **C**: 122 (1895). Type: Tanzania, Moshi Dist., Marangu, *Volkens* 903 (B holotype).
> *Cyperus renschii* var. *scabridus* Lye in Nordic J. Bot. **3**: 229 (1983). —Haines & Lye, Sedges & Rushes E. Afr.: 162, fig.302 (1983). Type: Uganda, Toro Dist., Ntandi, *Haines* 4227 (MHU holotype, K).

Perennial, robust, up to 1 m tall, with thick woody rhizome, 1–1.5 cm in diameter; culms tufted, 70–83 cm long, 4–8 mm wide, trigonous to slightly triquetrous, smooth. Leaves many, up to 140 cm long; leaf sheath reddish-purple near the base, 3–13 cm long; leaf blade linear, w-shaped, 68–130 cm long, 1–1.9 cm wide, scabrid on margins and major veins, apex acute. Involucral bracts leaf-like, spreading, 7–9, lowermost 40–90 cm long, 1.1–3 cm wide. Inflorescence compound, with primary, secondary and tertiairy branching, primary branches few to many, 3.5–18 cm long; spikelets in small, crowded clusters, at the end of on secondary and tertiairy branches, 3–9 per cluster, ovoid, 1.4–3.5 mm long, 0.8–1.9 mm wide; glumes reddish-brown, ovate-lanceolate, 1.3–1.6 mm long, 0.8–1 mm wide, keel green, apex strongly mucronate, recurved. Stamens 3; filaments 1–1.1 mm long; anthers 0.3–0.7 mm long. Nutlet brown, ellipsoid, 0.7–1.1 mm long, 0.5–0.6 mm wide, glabrous to sometimes minutely papillose.

Zambia. B: Mongu Dist., Looma, Nanganda, 21.xii.1993, *Bingham* 9910 (K). W: Solwezi Dist., Solwezi, 7.iv.1960, *Robinson* 3500 (K). **Zimbabwe**. E: Chimanimani [Melsetter] Dist., Melsetter-Chipinga road, near 'Skyline', 28.ii.1956, *Drummond* 5105 (K, SRGH). **Malawi**. N: Nkhata Bay Dist., 8 km [5 miles] S of Mzuzu, Roseveare's, 21.iv.1973, *Pawek* 6555 (K, MO, MAL). Nkhotakota Dist., Nchisi Mt., 29.vii.1946, *Brass* 17020 (K, NY). S: Zomba Dist., Zomba Rock, 1896, *Whyte* s.n. (K).

Also widespread in west, east and central Africa; Comoro Is. In and around dry evergreen, riverine and montane evergreen forests and *Pinus* plantations; 1000–1900 m.

Conservation notes: Least Concern because of its wide distribution.

D4. **Cyperus dichrostachyus** A. Rich., Tent. Fl. Abyss. **2**: 481 (1850). —Clarke in F.T.A. **8**: 331 (1901), as *dichroostachyus*. —Kükenthal in Engler, Pflanzenr. **4**, 20(101): 233 (1936), as *dichroostachyus*. —Haines & Lye, Sedges & Rushes E. Afr.: 165, figs.308,309 (1983). —Hoenselaar in F.T.E.A., Cyperaceae: 198 (2010). Type: Ethiopia, Mt Scholoda 3 km W of Adua, *Schimper* 391 (P lectotype, HAL, K). FIGURE 14.**38**.

Perennial, robust, up to 100 cm tall, with slender horizontal reddish brown to almost black rhizomes up to 12 cm long, 2–5 mm in diameter; culms usually solitary, connected by rhizomes, 33–90 cm long, 3–8 mm wide, triquetrous, glabrous. Leaves up to 95 cm long; leaf sheath light to dark brown, sometimes reddish to purplish brown, 2–11 cm long; leaf blade linear, flat, longitudinal veins sometimes clearly visible, 22–87 cm long, 0.7–1.8 cm wide, apex acuminate, scabrid towards the apex beneath on midrib and margins. Involucral bracts 2–4, leaf-like, spreading, the lowermost much longer than the rest and sometimes erect, 10–48 cm long, 0.5–1.4 cm wide. Inflorescence usually twice capitate-anthelate, primary branches 5–12, 1–9 cm long; spikelets in digitate clusters, sessile and at the end of primary, secondary and tertiary branches, 3–20 per cluster, ovoid-lanceolate, 2.4–5 mm long, 0.9–2 mm wide; glumes dark brown to almost black in the centre, margins sharply defined, pale brown to greenish, ovate to obovate, 1.3–1.8 mm long, 0.6–1 mm wide, keel pale brown to greyish-green, 3-veined, apex acuminate to shortly mucronate. Stamens 2; filaments 1.2–1.4 mm long; anthers 0.4–0.8 mm long. Style with 3 stigma

Fig. 14.**38**. CYPERUS DICHROSTACHYUS. 1, habit (× 2/3); 2, primary inflorescence branch (× 2); 3, spikelet (× 10); 4, glume (× 20); 5, flower (× 20); 6, nutlet (× 24). 1–5 from *Verdcourt* 1020; 6 from *Napier* 5837. Drawn by Juliet Williamson. Reproduced from Flora of Tropical East Africa (2010).

branches. Nutlet pale greyish brown, lanceolate to ellipsoid, 1–1.3 mm long, 0.5–0.6 mm wide, short-apiculate, minutely wrinkled to irregularly papillose.

Zambia. N: Mbala [Abercorn] Dist., Kawimbe, Lumi River, 29.xii.1959, *Richards* 12042 (K). W: Mwinilunga Dist., Mundeji River on Kanyama road, 25.ii.1975, *Hooper & Townsend* 396 (K). **Zimbabwe**. E: Inyanga Dist., Gaireshi Ranch, on PEA border 10 km [6 miles] N of Troutbeck, 15.xi.1956, *Robinson* 1944 (K). **Malawi**. N: Rumphi Dist., Nyika Plateau, Western valleys, 6.i.1959, *Robinson* 3148 (K). C: Dedza Dist., Dedza Mt., 23.x.1956, *Banda* 305 (K). S: Zomba Dist., Between Zomba and Ncheu, Domasi River, 20.viii.1950, *Jackson* 120 (K).

Also in Cameroon, D.R. Congo, Uganda, Kenya, Tanzania, Rwanda, Burundi, Angola, Sudan, Ethiopia, and South Africa. River banks, streamsides, margins of pools and dams, and in swamps; 1000–2000 m.

Conservation notes: Least Concern; because of its wide distribution.

There are no mature nutlets in the material at Kew from the Flora area; the description is taken from F.T.E.A. Haines & Lye say this is somewhat similar to *C. difformis*, but differs in stoloniferous perennial habit, less congested inflorescence, usually larger leaves and bracts. Very variable in size; some flowering specimens are about 30 cm tall.

D5. **Cyperus glaucophyllus** Boeckeler, Beitr. Cyper. **1**: 4 (1888). —Clarke in F.T.A. **8**: 345 (1901). —Kükenthal in Engler, Pflanzenr. **4**, 20(101): 202 (1936). —Haines & Lye, Sedges & Rushes E. Afr.: 157, fig.290 (1983). Type: Malawi, presumably from the Shire Highlands, *Buchanan* 24 (B holotype, K).

    Cyperus zambesiensis C.B. Clarke in Trans. Linn. Soc. London, Bot. **4**:53 (1894); in Durand & Schiz, Consp. Fl. Afr. **5**: 581 (1894); in F.T.A. **8**: 345 (1901). Type: Malawi, Shire Highlands, *Buchanan* 47 (K holotype).

    Cyperus pseudoleptocladus Kük. in Repert. Spec. Nov. Regni Veg. **29**: 196 (1931); in Engler, Pflanzenr. **4**, 20(101): 201 (1936). —Haines & Lye, Sedges & Rushes E. Afr.: 156, fig.288, 289 (1983). Types: Tanzania, Kilosa Dist.: Marangu, vii.1893, *Volkens* 650 & 652 (B syntypes, K); Lushoto Dist.: Usambara, *Holst* s.n. & *Engler* 1267 (B syntypes); Lushoto Dist.: Lutindi, *Holst* 3385 (B syntype); Rungwe Dist.: Kyimbila, *Stolz* 1146 (B syntype). Malawi, Mt. Malosa, *Whyte* s.n. (B syntypes, K).

    Cyperus pseudoleptocladus var. *polycarpus* Kük. in Repert. Spec. Nov. Regni Veg. **29**: 196 (1931); in Engler, Pflanzenr. **4**, 20(101): 201 (1936). Types: Kenya, Naivasha Dist., Masai Highlands, Mau Plateau, *Herb. For. Dep. Nairobi* 153. Tanzania, Kilosa Dist., Marangu, *Volkens* 704 (B syntype). Malawi, Mt Zomba, *Whyte* s.n. South Africa, Transvaal, Pietersburg, Drift Helpmekaar, *Pott* 4748 (PRE syntype).

    Cyperus glaucophyllus Boeckeler var. *longispiculosus* Kük. in Engler, Pflanzenr. **4**, 20(101): 203 (1936). Type: Tanzania, Lushoto Dist., West Usambara, track between Lushoto and Mombo, *Peter* 40902 (B holotype, B, K).

    Cyperus glaucophyllus var. *zambesiensis* (C.B. Clarke) Kük. in Engler, Pflanzenr. **4**, 20(101): 203 (1936).

Perennial, up to 122 cm tall, with a creeping woody rhizome; culms few to many, 28–112 cm long, 1.1–5 mm wide, trigonous to triquetrous, smooth. Leaves up to 80 cm long; leaf sheath purplish at the base, brown, 2–11 cm long; leaf blade linear, flat, 22–75 cm long, 3.7–12 mm wide, scabrid on major veins and margin, apex acute to acuminate. Involucral bracts leaf-like, erect to spreading, 2–5, lowermost 10–23 cm long, 2.7–10 mm wide. Inflorescence a compound anthela, primary branches 5–11, 1–11.5 cm long; spikelets in digitate clusters, at the end of primary and secondary branches, 2–8 per cluster, linear-lanceolate, 4–12.5 mm long, 1–2.5(4.4) mm wide, rachis straight, 10–12-flowered; glumes reddish-brown, ovate-lanceolate, 2–2.7 mm long, 0.6–1.3 mm wide, keel green, slightly excurrent, apex acuminate to mucronate. Stamens 3; filaments 1.4–2.4 mm long; anthers 1.2–2 mm long. Nutlet reddish-brown to dark grey, (narrowly) ellipsoid-oblong, 1.3–1.8 mm long, 0.4–0.6 mm wide, smooth to minutely papillose.

Zambia. C: Serenje Dist., Kundalila Falls, S of Kanona, 13.iii.1975, *Hooper &
Townsend* 701 (K). **Malawi**. N: Rumphi Dist., Nyika Plateau, Businande, 28.xii.1975,
Phillips 766 (K). C: Ntchisi Dist., Ntchisisi Mt., forest patch SE of main peak, 3.v.1980,
Blackmore et al. 1393 (K, MAL). S: Zomba Dist., Zomba Plateau, 2.vi.1946, *Brass* 16148
(K, NY). **Mozambique**. N: Lichinga Dist., Massangulo, xii.1932, *Gomes e Sousa* 1202
(K). Z: Gurué Dist., Serra do Gurué, 18.x.1949, *Grandvaux Barbosa* 4514 (K, LISC).

Also in D.R. Congo, Uganda, Kenya, Tanzania, Rwanda, Burundi, Swaziland and
South Africa. Forest edges, streamsides, wet roadsides, 1250–1800 m.

Conservation notes: Least Concern; because of its wide distribution.

D6. **Cyperus ajax** C.B. Clarke in F.T.A. **8**: 343 (1901). —Kükenthal in Engler, Pflanzenr.
4, 20(101): 198 (1936). —Haines & Lye, Sedges & Rushes E. Afr.: 155, fig.284
(1983). —Hoenselaar in F.T.E.A., Cyperaceae: 202 (2010). Type: Malawi, Mt
Malosa, *Whyte* s.n. & Mt Zomba, *Whyte* s.n. (K syntypes).

  Cyperus derreilema Steud. var. *ajax* (C.B. Clarke) Kük. in Notizbl. Bot. Gart. Berlin-Dahlem
  9: 302 (1925)

Perennial, robust, up to 210 cm tall, with thick woody rhizome, up to 1.2 cm in diameter;
culms 85–200 cm long, 4–7 mm wide, trigonous to triquetrous, smooth to somewhat scabrid.
Leaves many, crowded, up to 215 cm long; leaf sheath reddish-brown, 4–14 cm long; leaf blade
linear, flat, 64–200 cm long, 1.4–3.7 cm wide, with several prominent veins, scabrid on veins and
margins, apex acute. Involucral bracts leaf-like, spreading to sometimes erect, 3–many, lowermost
34–100 cm long, 1.1–2.9 cm wide. Inflorescence compound, primary branches 6–many, 3–20 cm
long; spikelets in digitate clusters, sessial and at the end of primary, secondary and tertiary
branches, 3–7 per cluster, lanceolate-ovoid, 3.2–6.7 mm long, 0.8–2.2 mm wide, rachilla straight;
glumes reddish-brown, sometimes reddish-green, ovate, 1.8–2.7 mm long, 1.2–1.8 mm wide, keel
green, excurrent, apex mucronate and sometimes recurved. Stamens 3; filaments 1.3–1.9(3.1)
mm long; anthers 0.6–1.1 mm long. Nutlet reddish-brown, ellipsoid (obovoid), 0.9–1.3 mm long,
0.4–0.63 mm wide, almost smooth to minutely papillose.

Zambia. N: Mbala [Abercorn] Dist., Saisi River Marsh, 27.ii.1957, *Richards* 8359 (K).
C: Kabwe Dist., Chitakata River 24 km NW of Kabwe, 21.i.1973, *Kornaś* 3067 (K). E:
Chama Dist., Makutu Hills, 27.x.1972, *Fanshawe* 11579 (K). **Malawi**. N: Rumphi Dist.,
Nyika Plateau, Lake Kaulime, 4.1.1959, *Robinson* 3046 (K). S: Mlanje Dist., Mlanje Mt.,
Litchenya Plateau, 7.vii.1946, *Brass* 16709 (K, NY). **Mozambique**. Z: Milange Dist.,
Chiperone Mountain, 30.xi.2006, *Harris et al* 96 (K).

Also in D.R. Congo, Burundi, Rwanda, Uganda, Kenya, Tanzania. Within and at
margins of upland evergreen forest, and at edges of riverine forest; 1500–2350 m.

Conservation notes: Least Concern because of its wide distribution.

This species is similar to *Cyperus derreilema* but differs in the mucronate (not
rounded) apex to the glume.

D7. **Cyperus fischerianus** A. Rich., Tent. Fl. Abyss. **2**: 488 (1850). —Clarke in F.T.A. **8**:
342 (1901). —Kükenthal in Engler, Pflanzenr. **4**, 20(101): 203 (1936). —Haines
& Lye, Sedges & Rushes E. Afr.: 155, fig.286 (1983). Type: Ethiopia, Mt Scholoda,
3 km W of Adua, *Schimper* 348 (P lectotype, BR, HAL, K, UPS), lectotypified by
Hoenselaar in F.T.E.A. (2010).

  Cyperus fischerianus var. *ugandensis* Lye in Nordic J. Bot. **3**: 230 (1983). —Haines & Lye,
  Sedges & Rushes E. Afr.: 156, fig.287 (1983). Type: Uganda, Teso Dist., Soroti, *Haines* 4288
  (MHU holotype, K).

Perennial to 133 cm tall, robust, with short thick woody rhizomes forming dense tussocks;
culms densely tufted, 55–120 cm long, 2.4–5.7 mm wide, trigonous to triquetrous, smooth.
Leaves up to 145 cm long; leaf sheath dark purple and glossy and the base, rather thick,
5–15(27) cm long; leaf blade linear, with several major veins, flat, 54–130 cm long, 0.9–1.6 cm

wide, scabrid on margins and major veins, apex acuminate. Involucral bracts leaf-like, spreading, 5–10 to many, lowermost 30–125 cm long, 0.8–1.6 cm wide. Inflorescence compound, often proliferating, primary branches 7–17, 3–10 cm long; spikelets in digitate clusters, at the end of primary, secondary and tertiary branches, 1–5 per cluster, linear-lanceolate, 4.3–12.7 mm long, 1.6–2.2 mm wide, rachis straight to slighty curved; glumes rusty reddish-brown, ovate-lanceolate, 2.1–2.4 mm long, 0.9–1.6 mm wide, keel shortly excurrent, apex acuminate. Stamens 3; filaments 2.1–2.7 mm long; anthers 1.1–1.6 mm long. Nutlet reddish-brown, obovoid-ellipsoid, 0.9–1.3 mm long, 0.6–0.7 mm wide, minutely papillose.

Malawi. N: Rumphi Dist., Nyika Plateau, Chelinda Camp, 3.ii.1978, *Pawek* 13735 (K, MAL, MO, SRGH). C: Dedza Dist., Chongoni Forest, 3.iii.1961, *Chapman* 1162 (K, SRGH). **Mozambique**. Z: Gurué Dist., Namuli Mt., 27.v.2007, *Patel et al.* 7351 (K). MS: Manica Dist., S tip of Chimanimani Mts, 28.v.1969, *Muller* 1120 (K, SRGH).

Also in Eritrea, Ethiopia, D.R. Congo, Uganda, Kenya, and Tanzania. In or at margins of montane and riverine forest, often near streams; 1000–2300 m.

Conservation notes: Least Concern because of its wide distribution.

Humbert 117078 from Mlanje Mt. (Malawi S), ix.1937, (K, P) is probably this species but lacks basal parts.

D8. **Cyperus chinsaliensis** Podlech in Mitt. Bot. Staatssamml. München **4**: 107 (1961). —Haines & Lye, Sedges & Rushes E. Afr.: 159, fig.294 (1983). —Hoenselaar in F.T.E.A., Cyperaceae: 171 (2010) (all as *chinsalensis*). Type: Zambia, 42 km S of Chinsali, 14.i.1959, *Robinson* 3207 (M holotype, K).

Perennial, with a short thick rhizome and persistent swollen stem-bases, up to 95 cm tall; culms triquetrous, 41–92 cm long, 1.8–2 mm wide, scabrid on the margins. Leaves up to 55 cm long; leaf sheaths pale brown to greenish-brown, 2–7 cm long; leaf blade linear, plicate, 20–48 cm long, 5–6 mm wide, scabrid on the margins and veins, apex acute to acuminate. Involucral bracts leaf-like, spreading, 2–4(10), 3.5–10 cm long, 3–4 mm wide. Inflorescence loosely capitate, primary branches 0–3, 0–1.5 cm long; spikelets congested in loose clusters, sessile or at the end of short primary branches, 2–5 per cluster, broadly ovoid, 8–10 mm long, 4–6 mm wide, glumes spreading when mature; glumes whitish to greenish to very pale brown, elliptic, glabrous, 3.5–4 mm long, 1.3–1.8 mm wide, keel not prominent, with many distinct veins on either side, apex obtuse. Stamens 3: filaments 1.9–3 mm long; anthers ± 1.6 mm long. Stigma 3-branched. Nutlet brown, broadly obovoid, trigonous, 1.4–1.7 mm long, 1–1.1 mm wide, smooth.

Zambia. N: Mbala Dist., Mbala [Abercorn], 5.xii.1958, *Vesey-Fitzgerald* 1998 (K); Mbala Dist., Kawimbe Rocks, 17.i.1964, *Richards* 18810A (K).

Also in Tanzania. *Brachystegia* woodland, roadsides, and amongst granite rocks; 1300–2000 m.

Conservation notes: Data Deficient. There are not many collections and the distribution area is quite small.

The nutlet is identical to that of *Cyperus angolensis. Cyperus chinsaliensis* also resembles *C. mapanioides* but the stem base is more swollen, the leaves are narrower, there are fewer and smaller involucral bracts, and the spikelets are larger.

I have treated *chinsalensis* as an orthographic error to be corrected. The name of the place whence it was described is Chinsali.

Richards 18810 was a mixed gathering with *Cyperus albostriatus*; her 18810B is cited here under that species.

D9. **Cyperus albostriatus** Schrad., Anal. Fl. Cap. **1**: 7 (1832). —Clarke in Fl. Cap. **7**: 176 (1897). —Schonland in Mem. Bot. Surv. S. Africa **3**: 27, pl.15 (1922). —Gordon-Gray in Strelitzia **2**: 51 (1995). Type: South Africa, Uitenhagen, *Zeyher* s.n. (SAM); a sheet at B numbered 213 but without a collector's name, determined by Kükenthal, may also be type material.

Rhizomatous perennial to 40 cm tall. Rhizomes slender, horizontal, 2–4 mm in diameter, covered by red-brown scales. Culms 30–50 cm tall, 1–2 mm diam., trigonous, grooved when dry, glabrous. Leaf sheaths up to 3 cm long, reddish; lamina 15–30 cm long, 3–8 mm wide, glabrous but minutely serrate near the acuminate apex. Involucral bracts leaf-like, 6–10, the lowest 13–21 cm long and 3–12 mm wide, thin, spreading. Inflorescences once or twice-anthelate, longest primary branch 5–15 cm long, each branch with 2–5 apical spikelets, a few branches sometimes apically branched, the branchlets each bearing 2–3 spikelets. Spikelets 4–9(14) × 1–1.5 mm, very narrowly ovate to very narrowly oblong. Glumes 1.7–1.9 × 0.5–0.7 mm, dull brown with a darker midrib; midrib shortly excurrent, slightly reflexed. Nutlets blackish, ellipsoid, 1.0–1.2 × 0.6 mm, smooth.

Zambia. N: Mbala [Abercorn] Dist., Kawimbe Rocks, 17.i.1964, *Richards* 18810B (K). **Zimbabwe**. W: Matobo Dist., Farm Besna Kobila, iii.1958, *Miller* 5163 (K, SRGH). E: Umtali Dist., Vumba Mts., Charleswood, 26.i.1964, *Chase* 6114 (K, SRGH). S: Mberengwa Dist., Mberengwa Mt., [Mt. Belingwe], north slope, 1.v.1973, *Biegel et al.* 4232 (K, SRGH). **Mozambique**. MS: Manica Dist., Southern tip of Chimanimani Mts., on slopes above Haroni-Makurupini Forest, 29.v.1969, *Muller* 1116 (K, SRGH). M: Maputo, 15.iv.1947, *Hornby* 2682 (K, SRGH).

Also in Lesotho, Swaziland and South Africa. Moist woodland, usually in shade, sometimes on rocks; 50–1700 m.

Conservation note: Least Concern because of its wide distribution.

A distinctive plant with its few-spikeletted widely spreading inflorescence, and involucral bracts that are as prominent and abundant as the leaves.

Note on typification: The type description does not refer to a specimen but only to "*Cyperus pulcher* Zeyh., Catal. Pl. Capens. in Flora 1839: 5, non Thunb.". The reference in this paper is to a specimen, number 213, presumably collected by *Ecklon & Zeyher*.

D10. **Cyperus mwinilungensis** Podlech in Mitt. Bot. Staatssamml. München **4**: 109 (1961). Type: Zambia, 40 km ESE of Kasama, *Robinson* 4561 (M holotype, BR, K).

   Cyperus mwinilungensis var. *maior* Podlech in Mitt. Bot. Staatssamml. München **4**: 110 (1961). Type: Zambia, Kasama Dist., 103 km E of Kasama, *Robinson* 4431 (M holotype, K).

Perennial, fairly slender, up to 40 cm tall, stoloniferous, with small round tubers at the base of the culm (often lost in herbarium material); culms few, 20–50 cm long, 5–14 mm wide, trigonous to triquetrous, smooth (sometimes slightly scabrid). Leaves up to 33 cm long; leaf sheath reddish-brown, greenish-brown to brown, 1.5–5 cm long; leaf blade linear, flat, 5–30 cm long, 1.1–2.9 mm wide, glabrous to scabrid on margins, apex acute to acuminate. Involucral bracts leaf-like, erect, at least the lowermost, 2(3), lowermost 1.5–7 cm long, 1.2–1.9 mm wide, usually hardly equalling the inflorescence. Inflorescence a simple anthela, primary branches 1–4, 0.5–3 cm long; spikelets in digitate clusters, sessile and at the end of primary branches, 2–11 per cluster, very narrowly ovate to linear, 5–13 mm long, 1.6–2.5 mm wide, rachis straight; glumes dark reddish-brown to black, ovate-elliptic, 1.2–2.1 mm long, 1.1–1.4 mm wide, with 5–9 veins on either side of the keel, keel inconspicuous, apex rounded to slightly excurrent. Stamens 3; filaments 0.8–2.1 mm long; anthers 0.9–1.3 mm long. Nutlet trigonous, obovoid-ellipsoid, 0.6 × 0.3 mm, smooth, grey-black.

Botswana. N: Ngamiland Dist., Okavango Swamps, at edge of Mboma-Gadikwe Channel, 15.viii.1979, *P.A. Smith* 2800 (K, SRGH). **Zambia**. B: Mongu Dist., 10 km. E of Mongu, 31.x.1965, *Robinson* 6698 (K). N: Kasama Dist., Misamfu, 19.ii.1961, *Robinson* 4402 (K). W: Mwinilunga Dist., source of Matonchi Dambo, 26.x.1937, *Milne-Redhead* 2958 (K). C: Lusaka Dist., Chakwenga Headwaters, 100–129 km E of Lusaka, 1.xii.1963, *Robinson* 5878 (K). E: Isoka Dist., 5 km E of Chitipa [Fort Hill], 11.iii.1961, *Robinson* 4435 (K). S. Mapanza Dist., Choma, 31.1.1959, *Robinson* 3242 (K). **Zimbabwe**. E: Chimanimani [Melsetter] Dist., Chimanimani, Stonehenge, 1.ii.1957, *Phipps* 357 (K, SRGH).

Also in Tanzania. Dambos, mostly seasonal but occasionally permanently wet, often on sandy soil; 1000–1400 m.

Conservation note: Least Concern because although it was described fairly recently, it is a rather nondescript species, which has probably been overlooked, and of which there are now numerous collections.

Now that there is a greater range of material available than Podlech saw when he described the species and the variety, it appears that there is a continuity of size from small to large and that it is no longer right to recognize the variety.

D11. **Cyperus nyererei** Lye in Nordic J. Bot. **3**: 225 (1983). —Haines & Lye, Sedges & Rushes E. Afr.: 158, fig.293 (1983). —Hoenselaar in F.T.E.A., Cyperaceae: 176 (2010). Type: Tanzania, Mbeya Dist., Kitulo Plateau, Igoma–Kitulo road 5 km beyond Kikondo, *Wingfield* 529 (DSM holotype, K).

Perennial up to 42 cm tall, densely tufted, producing many black roots, culm base hard, covered with black fibres from old leaf sheaths; culms tufted, 15–40 cm long, 0.5–1 mm wide, trigonous, almost smooth to slightly scabrid. Leaves up to 21 cm long; leaf sheath purple to black, 1–3 cm long; leaf blade linear, flat, rather stiff, 5–18 cm long, 1.3–3 mm wide, scabrid on major veins and margins, apex rounded to acuminate. Involucral bracts leaf-like, spreading or erect, 2–3, lowermost 2.3–10 cm long, 1–2.2 mm wide. Inflorescence loosely capitate to simple, primary branches 0–3, 0–3 cm long; spikelets in digitate clusters, sessile and at the end of primary branches, 3–12 per cluster, lanceolate to ovoid, 6.8–10 mm long, 1.9–4 mm wide, the glumes spreading through maturity, rachis straight; glumes dark reddish-brown with blackish margins, as if burned, lanceolate-ovate, 2.7–4 mm long, 0.6–1.4 mm wide, keel greenish brown, apex acute to slightly acuminate. Stamens 3; filaments ± 1.6 mm long; anthers 1–1.3 mm long. Nutlet greyish to reddish-brown, ellipsoid, 1.3–1.5 mm long, 0.6–0.7 mm wide, slightly apiculate, minutely papillose.

Malawi. N. Rumphi Dist., Nyika Plateau, eastern side, 5.i.1959, *Robinson* 3074 (K).

Also in Tanzania. Dry hillside; in F.T.E.A. area, thin soil over rocks; (1950)2500 (2750) m.

Conservation note: Probably Near-Threatened because its distribution area is small and its habitat restricted.

This is the only specimen from the Flora area. The description has been only slightly modified from that in F.T.E.A.

D12. **Cyperus matagoroensis** Muasya & D.A. Simpson in Kew. Bull. **59**: 593 (2004). —Hoenselaar in F.T.E.A., Cyperaceae: 205 (2010). Type: Tanzania, Songea Dist.: Matagoro Hills, 3.ii.1956, *Milne-Redhead & Taylor* 8595 (K holotype, BM, BR).

Perennial up to 55 cm tall, with short rhizome; culms moderately to densely tufted, 30–50 cm long, 0.7–0.9 mm wide, trigonous, smooth, densely covered at base by fibrous, dark reddish-brown to blackish leaf sheath remains. Leaves up to 32 cm long; leaf sheath brownish, 1.5–2 cm long; leaf blade narrowly linear, v-shaped in cross-section, 17–30 cm long, 0.5–0.7 mm wide, scabrid on the margins, apex acute. Involucral bracts leaf– like, spreading, 3–4, lowermost 1.5–2.2 cm long. Inflorescence simple, primary branches 3–4, 0.5–3 cm long; spikelets in digitate clusters, at the end of primary branches, 2–4 per cluster, elliptic-lanceolate to lanceolate, 8–10 mm long, 2–3 mm wide, rachilla straight; glumes dark reddish-brown with pale brown margins, ovate, 2.3–2.7 mm long, 0.5–0.7 mm wide, sides membraneous, 1-veined on each side, keel green, apex acute. Stamens 3; anthers 1.3–1.5 mm long. Nutlet dark reddish, ellipsoid, trigonous, 1.3–1.6 mm long, 0.6–0.8 mm wide, minutely papillose.

Zambia. N: Kasama Dist., 8 km. E of Kasama, 4.xii.1961, *Robinson* 4721 (K). E: Isoka Dist., Nyika Plateau, 2.i.1959, *Robinson* 2991 (K).

Also in Tanzania; should be looked for on the Malawi side of the Nyika Plateau. Dry open grassland; dry bank of stream; on and between large rocks on sandy soil; 1400–2150 m.

Conservation note: Data Deficient; very few specimens and not well known.

Cyperus matagoroensis was regarded by its describers as being close to *C. nyererei*, from which it differs in its narrower leaves, longer inflorescence branches, reddish brown spikelets and shorter glumes.

D13. **Cyperus tenax** Boeckeler in Linnaea **35**: 504 (1868). —Clarke in F.T.A. **8**: 334 (1901). —Kükenthal in Engler, Pflanzenr. **4**, 20(101): 259 (1936). —Haines & Lye, Sedges & Rushes E. Afr.: 267, figs.546,547 (1983). —Gordon-Gray in Strelitzia **2**: 74 (1995). —Hoenselaar in F.T.E.A., Cyperaceae: 203 (2010). Type: South Africa, Cape, *Zeyher* 13 (B holotype).

   Cyperus grantii Boeckeler in Flora **58**: 260 (1875). Type: 'Afr. orient. tropica, 3800'', *Grant* s.n. (K holotype).

   Cyperus boehmii Boeckeler in Bot. Jahrb. Syst. **5**: 498 (1884). —Clarke in F.T.A. **8**: 335 (1901). Type: Tanzania, Lake Tanganyika, Mpanda Dist., Ugalla R., iii.1882, *Boehm* s.n. (B holotype).

   Cyperus amabilis Vahl var. *pseudocastaneus* Kük. in Fries, Wiss. Ergebn. Schwed. Rhodesia-Kongo-Exped. 1911-1912, Erg: 2 (1921). Type: Zambia, Kali, *Fries* 637 (B holotype?).

   Cyperus tenax var. *pseudocastaneus* (Kük.) Kük. in Engler, Pflanzenr. **4**, 20(101): 260 (1936).

Perennial, densely tufted, up to 73 cm tall, with an erect rhizome covered by often fibrous blackish old leaf bases; culms tufted, 5–68 cm long, 0.6–1.8 mm wide, rounded, striate, glabrous. Leaves up to 33 cm long; leaf sheath straw-coloured to purple, 1.5–7 cm long; leaf blade linear, flat or folded, 5–30 cm long, 1.1–4 mm wide, scabrid on margins and primary vein at least on young leaves, apex acuminate. Involucral bracts leaf-like, spreading, the lowermost (1)3–14 cm long, 1.2–2 mm wide. Inflorescence simply capitate-anthelate, sometimes condensed so as to appear virtually capitate, the primary branches sometimes with a few apical branches, primary branches 3–12, 0.5–5 cm long; spikelets in digitate clusters, sessile and at the end of primary and, occasionally, secondary branches, 5–14 per cluster, linear, 3–19.1 mm long, 0.7–1.3 mm wide, up to 2.5 mm wide with glumes spreading, rachilla straight; glumes glossy, pale reddish-brown to almost black with pale longitudinal lines, ovate, 1.2–2.3 mm long, 0.8–1.3 mm wide, 3-veined, apex obtuse to slightly mucronate; glumes falling off with nutlet. Stamens 3; filaments 2–2.1 mm long; anthers 1.2–1.9 mm long. Nutlet yellowish-brown when young, blackish with metallic shine when mature, narrowly obovoid to oblong or ellipsoid, 0.7–1.1 mm long, 0.3–0.4 mm wide, minutely tuberclate in longitudinal rows.

Zambia. B: Mongu Dist., 17 km SE of Mongu Town, 6 km of the Mongu-Kaoma road, 17.iii.1996, *Zimba et al.* 783 (K, MO). N: Chinsali Dist., Shiwa Ngandu, 2.vi.1956, *Robinson* 1553 (K). W: Mwinilunga Dist., base of Kalene Hill, 22.ii.1975, *Hooper & Townsend* 312 (K). C: Serenje Dist., Kundalila Falls, S of Kanona, 14.iii.1975, *Hooper & Townsend* 756 (K). S: Mapanza Dist., Muckle Neuk, 18 km [12 miles] N of Choma, 29.xi.1954, *Robinson* 995 (K). **Zimbabwe**. N: Lomagundi Dist., Mangula, Old Hat Claims, 14.ii.1968, *Wild* 7682 (K, SRGH). W: Hwange [Wankie] Dist., 15 km [10.6 miles] SSE of Main Camp along Dopi Poan road, Wankie National Park, 20.xii.1968, *Rushworth* 1369 (K, SRGH). C: Harare [Salisbury], 19.i.1927, *Eyles* 4671 (K, SRGH). E: Chimanimani [Melsetter] Dist., Chimanimani Mts, Stonehenge area, 14.ii.1960, *Goodier* 898 (K, SRGH). **Malawi**. N: Mzimba Dist., 5 km [3 miles] S of Mzuzu, Katoto, 1.xii.1973, *Pawek* 7544 (K, MAL, MO, SRGH). C: Kasungu Dist., Kasungu National Park, 2.xii.1970, *Hall-Martin* 1316 (K, SRGH). S: Thyolo Dist., Shire Highlands, 6.iv.1906, *Adamson* 192 (K). **Mozambique**. N: Marrupa Dist., arredores do viveiro do projeto Florestal do Niassa, 20.i.1981, *Nuvunga* 433 (K, LMU). MS: Manica Dist., c. 3 km [2 miles] S of Makurupini Falls, Chimanimani Mts., 25.xi.1967, *Simon & Ngoni* 1310 (K, SRGH). M: Maputo Dist., Bela Vista para Porto Henrique, 30.x.1980, *de Koning & Nuvunga* 8553 (K, LMU).

Also in Sierra Leone, Liberia, Ivory Coast, Benin, Nigeria, Equatorial Guinea, Gabon, D.R. Congo, Uganda, Kenya, Tanzania, Angola and South Africa. *Isoberlinia* and *Brachystegia* woodland; vlei grasslands; usually on sand (often Kalahari Sand); one collection (*Wild* 7682) from site with 'highest copper values'; sea-level to 1700 m.

Conservation notes: Least Concern because of its wide distribution.

F.T.E.A. notes that this species has a very leafy culm base; this, with the long narrow spikelets, make it easy to recognize. The colour of the glumes is highly variable. In the past two varieties have been described based on colour, and glumes in the darker specimens are more open and spreading, and occur more inland. Too many intermediate specimens exist and the varieties are not upheld here. I follow this.

D14. **Cyperus haspan** L., Sp. Pl.: 45 (1753). —Clarke in F.T.A. **8**: 332 (1901). — Kükenthal in Engler, Pflanzenr. **4**, 20(101): 247 (1936). —Haines & Lye, Sedges & Rushes E. Afr.: 168, figs.317,318 (1983), as *halpan.* —Hoenselaar in F.T.E.A., Cyperaceae: 205, fig.30 (2010). Type: Sri Lanka [Ceylon], *Hermann* 2: 43, No. 37 (BM lectotype), lectotypified by McGivney (1938).

> *Cyperus phaeorhizus* K. Schum. in Engler, Pflanzenw. Ost-Afrikas **C**: 119 (1895). —Clarke in F.T.A. **8**: 331 (1901). —Kükenthal in Engler, Pflanzenr. **4**, 20(101): 252 (1936). Type: Tanzania, Tanga Dist., Marungu, *Volkens* 2275 (B holotype, K).
> *Cyperus princeae* C.B. Clarke in Bot. Jahrb. Syst. **38**: 133 (1906). Type: Tanzania, Uhehe, Udzungwa Mts at 1600 m, *Mrs Prince* s.n. (B holotype).
> *Cyperus phaeorhizus* var. *princeae* (C.B. Clarke) Kük. in Engler, Pflanzenr. **4**, 20(101): 253 (1936).
> *Cyperus kipasensis* sensu Haines & Lye, Sedges & Rushes E. Afr.: 170, figs.323,324 (1983), non Cherm.

Perennial 15–60(90) cm tall, with short sometimes creeping rhizome; roots reddish; culms pale to bright green, crowded, 0.8–2.5 mm wide, trigonous or triangular, smooth, glabrous. Leaves with leaf sheath reddish-brown or purple, 1–11 cm long; leaf blade present at least at some shoots, pale to bright green, linear, slightly v-shaped, 5–22(33) cm long, 1.5–4.5 mm wide, attenuate, glabrous. Involucral bracts leaf-like, spreading or semi-erect or erect, 1–3, lowermost up to 7(12) cm long. Inflorescence a simple to compound anthela, primary branches 1–10, 1–7 cm long; spikelets in sessile digitate clusters, at the end of primary and secondary branches, 2–9 per cluster, narrowly ovoid to cylindrical, 7–15(30)-flowered, 3–12 × 1–3 mm, with straight rachis; glumes light to dark brown or reddish-brown to almost black, ovate, 1.3–2.8 × 0.9–1.3 mm, keel pale green, slightly excurrent, margin sometimes pale green. Stamens 3; anthers sulphur-yellow, 0.7–1.2 mm long; filaments white, 1.5–1.9 mm long. Styles white. Nutlet greyish-brown, ellipsoid, obovoid or almost orbicular, 0.5–1 mm long, 0.3–0.8 mm wide, irregularly tuberculate when mature.

Botswana. N: Ngamiland Dist., Motswiri camp waterside, 15.ii.2003, *A. & R. Heath* 316 (K). SE: Southern Dist.. 3 km [2 miles] S of Lobatsi, E of railway, 17.i.1960, *Leach & Noel* 164 (K, SRGH). **Zambia**. W: Mwinilunga Dist., Kalenda Plain, 18.x.1937, *Milne-Redhead* 2834 (K). N: Kasama Dist., Chibutubutu, Lukulu River, 50 km [30 miles] S of Kasama, on Mpika road, 23.ii.1960, *Richards* 12553 (K). C: Lusaka Dist., Chongwe River, near Constantia (N of Kasisi), 29.xi.1972, *Kornaś* 2184 (K). E: Chipata Dist., Katete, c. 80 km [50 miles] W of Chipata [Fort Jameson], 14.vi.1954, *Robinson* 863 (K). **Zimbabwe**. C: Marondera Dist., Marandellas, Grasslands, 9.i.1948, *Colville* 36 (K, SRGH). E: Chimanimani Dist., Chimanimani National Park, Perry's Cave area, 26.iv.1993, *Browning* 599 (K, NU). S: Nuanetsi Dist., Malangwe River, SW Matoke Hills, 6.v.1958, *Drummond* 5664 (K, SRGH). **Malawi**. S: Thyolo Dist., Shire Highlands, 8.iv.1906, *Adamson* 103 (K).

Widespread in Africa, Asia and the Americas. Swampy or marshy sites, often by streams or rivers; in seasonally or more usually permanently wet grassland, thin wet soil over rock, ditches; 550–1700 m.

Conservation notes: Least Concern because of its wide distribution.

See Kartesz & Gandhi in Phytologia **72**: 19 (1992) for a discussion on the spelling of *haspan* vs. *halpan*. They conclude it should be *haspan*, as does Wilson in Telopea **5**: 598 (1994).

Plants without the rhizome and a more short-lived look are referred to *Cyperus foliaceus*, which might only be a form of *C. haspan*.

D15. **Cyperus hensii** C.B. Clarke in Mém. Couronnés Autres Mém. Acad. Roy. Sci. Belgique **53**: 289 (1896); in F.T.A. **8**: 334 (1901). —De Wildeman & Schinz in Ann. Mus. Congo Belge, Bot. sér. 1 **1**(1): 15, pt.8 (1898). Type: D.R. Congo, Lisha, *Hens* C364 (BR). FIGURE 14.**39**.

Clump-forming perennial, base somewhat bulbous, covered by old leaf bases. Culms 40 cm long and 2 mm in diam., subterete. Leaves 15–18 cm long, sheaths split to the base, margins scarious. Lamina very narrow, 2–3 mm wide, inrolled when dry. Involucral bracts 4–5, similar to the leaves, up to 7 cm long, broadened at the base. Inflorescence simply or twice anthelate-capitate, usually congested but sometimes more lax, with 1–6 primary branches 0–15 mm long. Spikelets 12–20 in digitate groups, linear, 16–24-flowered, 10–12 mm long, 1–1.5 mm wide, compressed with parallel sides, straw-coloured; rachilla somewhat zigzag, with narrow hyaline wings. Glumes obtuse, sometimes shortly mucronate, nerved only on the keel, closely crowded, loosely imbricate, straw-coloured, often somewhat darker towards the tip. Styles red, short, exserted a little from the glumes. Nutlet broadly oblong, trigonous, half the length of the glumes, reddish brown, smooth, or puncticulate by virtue of the outer quadrate shining cells remaining turgid.

Zambia. N: Mbala [Abercorn] Dist., Lufubu River, Iyendwe Valley, 8.xii.1959, *Richards* 11915 (K); Mbala Dist., Kambole, 19.ii.1957, *Richards* 8266 (K).

Also in D.R. Congo. Woodland, or grassland among rocks, on sandy soil; 750–1500 m.

Conservation notes: Probably Least Concern; although its distribution is relatively restricted there are now a fair number of collections.

Differs from *Cyperus tenax* in forming a dense clump, and also in the colour of the glumes which are straw-coloured, not blackish. In the material at K, the inflorescences are usually congested, whereas those of *C. tenax* may be either congested or expanded.

D16. **Cyperus sphaerospermus** Schrad., Anal. Fl. Cap. **1**: 8 (1832). —Clarke in F.T.A. **8**: 336 (1901). —Gordon-Gray in Strelitzia **2**: 73 (1995). Type: South Africa, Cape, Zeekoe Valley, *Ecklon* 82 (TUB, M, PRE, not seen).

   Cyperus denudatus L.f. var. *sphaerospermus* (Schrad.) Kuk. in Engler, Pflanzenr. **4**, 20(101): 255 (1936).
   Cyperus huillensis Ridl. in Trans. Linn. Soc. London, Bot. **2**: 139 (1884). Type: Angola, *Welwitsch* 6865 (BM syntype), 6867 (BM syntype, LISU), 6868 (BM syntype, LISU), 6869 (BM syntype), 6889 (BM syntype, LISU).

Tussock-forming perennial to 50 cm tall. Rhizomes very short, thick, scale-covered, usually barely visible. Culms 18–65 cm long, 1–1.5 mm wide, trigonous, smooth. Leaves up to 30 cm long, 1.5–4 mm wide, slightly scabrid on margins and midrib, tapering to the smooth acuminate apex. Involucral bracts 2–4, the longest up to 15 cm long and 3 mm wide, scabrid on the margins, exceeding ther inflorescence branches. Inflorescence twice or three times anthelate, the secondary branches 0–14, to 6 cm long. Spikelets in digitate clusters, sessile and at the ends of the branches, 2–6 per cluster, narrowly ovate to linear, 3–9 × 1–2 mm; glumes greenish brown with a broad chestnut-brown margin (the spikelets thus appearing reddish brown), c. 1.5 mm long, narrowly ovate, apex blunt, midrib slightly excurrent. Stamens 3, anthers c. 1 mm long. Styles 3. Nutlet subspherical, weakly trigonous, whitish, tuberculate.

Caprivi. Kake Camp, between Andara and Bagani, 21.i.1956, *de Winter & Wiss* 4376 (K, PRE). **Botswana**. N: Ngamiland Dist., Khwai River floodplain, Moremi Game Reserve, 27.xi.2007, *A. & R. Heath* 1463 (K). SE: Kweneng Dist., Aedume Pt, Gaborone

Fig. 14.**39**. CYPERUS HENSII. 1, habit; 2, inflorescence; 3, spikelet detail; 4, nutlet. 1 from *Richards* 16904; 2, 3 from *Richards* 3895; 4 from *Richards* 8266. Scale bars: 1, 2 = 3 cm; 3 = 5 mm; 4 = 1 mm. Drawn by Juliet Williamson.

Dam, 22.ix.1977, *Hansen* 3245 (C, K). SW: Kgalagadi Dist., Okwa Valley, 1 km NE of Namibia border, 13.xii.1976, *Skarpe* S105 (K). **Zambia**. C: Lusaka Dist., Chongwe River near Kanubwe, 3.xii.1972, *Kornaś* 2754 (K). S: Mapanza Dist., Simansunda, 3 km [2 miles] E of Mapanza, 5.i.1954, *Robinson* 430 (K). **Zimbabwe**. N: Sebungwe Dist., headwaters of Masumo River below Chizarira Hill, 11.xi.1958, *Phipps* 1448 (K, SRGH). W: Matobo Dist., Matopos, ii.1931, *Rattray* 264 (K, SRGH). C: Harare Dist., [Salisbury], Mazoe Dam, 6.ix.1956, *Robinson* 1800 (K). E: Chipinga Dist., near Chibuwe, 1.vi.1972, *Gibbs Russell* 2079 (K, SRGH). S: Nuanetsi Dist., SW Mateke Hills, Malangwe River, 5.v.1958, *Drummond* 5572 (K, SRGH). **Mozambique**. MS: Manica Dist., Moribane, 3 km from the Muxamba crossroads, 5.x.1953, *Gomes Pedro* 4215 (K, LISC). M: Maputo Dist., Inhaca Island, Ponta Prasa, 17.vii.1959, *Mogg* 2940 (K, PRE).

Also in South Africa. Riverbanks, lake margins, floodplains; 0–1400 m.

Conservation note: Least Concern because of its wide distribution.

Formerly placed as a variety of *Cyperus denudatus*, but Gordon-Gray has pointed out that it can readily be distinguished by the presence of well-developed leaves, the very short rhizomes, the culms which are trigonous, not triquetrous, the longer bracts, and the dull, not shiny, spikelets. The whitish nutlets may not be fully mature; Gordon-Gray states that they are often not fully developed.

D17. **Cyperus deciduus** Boeckeler in Flora **62**: 547 (1879). —Haines & Lye, Sedges & Rushes E. Afr.: 166, fig.312 (1983). Type: Angola, Kimbundo, *Pogge* 466 (B holotype).

   Mariscus deciduus (Boeckeler) C.B. Clarke in Fl. Cap. **7**: 191 (1897); in F.T.A. **8**: 394 (1902).

Perennial, rather slender, up to 70 cm tall, with a brown or red woody rhizome up to 10 cm long and 5 mm in diameter, roots brown or purplish; culms usually more or less contiguous from the horizontal rhizome, 21–52 cm long, 0.6–1 mm wide, trigonous, smooth. Leaves up to 23.5 cm long; leaf sheath grey to reddish brown, 2–5.5 cm long; leaf blade linear, 8–18 cm long, 1.6–2.1 mm wide, scabrid on margins and veins, apex acute. Involucral bracts bract-like, erect, 1–2, lowermost 1–2.5 cm long, 1.1–1.7 mm wide. Inflorescence a simple anthela, primary branches 2–4, 1–3.5 cm long; spikelets in digitate clusters, sessile and on primary branches, disarticulate, 2–8 per cluster, linear, 6.3–13.7 mm long, 1.2–1.6 mm wide, rachis straight; glumes greenish-brown, ovate, 1.4–1.6 mm long, 0.9–1.1 mm wide, keel flat and indistinct, apex obtuse. Stamens 3; filaments 1.2–1.4 mm long; anthers 0.6–0.8 mm long. Nutlet white when dry and young, later dark grey, broadly ellipsoid, 0.6–0.7 mm long, ± 0.4 mm wide, wrinkly to roughly papillose in longitudinal rows.

Zambia. B: Kalabo Dist., 5 km [3 miles] W of Kalabo, 16.xi.1959, *Drummond & Cookson* 6528 (K, SRGH). N: Kasama Dist., Chambeshi Flats via Mbwayalola Village, 10.xii.1964, *Richards* 19341 (K). W: Mwinilunga Dist., Chinabola Dambo near Matonchi, 16.ii.1975, *Hooper & Townsend* 91 (K). C: Serenje Dist., Kundalila Falls, S of Kanone, 13.iii.1975, *Hooper & Townsend* 705 (K). S: Mapanza Dist., 8 km [5 miles] S of Choma, 26.iii.1965, *Robinson* 1166 (K). **Zimbabwe**. W: Matobo Dist., Farm Besna Kobila, x.1956, *Miller* 3674 (K, SRGH). C: Kwekwe Dist., c. 50 km N of Kwekwe town, Iwaba Estate, c. 8 km S of camp, 10.ii.2001, *Lye* 24524 (K). E: Umtali Dist., Lake Alexander, 7.i.1972, *Gibbs Russell* 1249 (K, SRGH). **Malawi**. N: Rumphi Dist., Nyika Plateau, Lake Kaulime, 16.v.1970, *Brummitt* 10796 (K).

Also in Tanzania, D.R. Congo, Angola, Namibia, Swaziland, and South Africa. Permanently or seasonally wet peaty or sandy ground, usually with grasses or other Cyperaceae; 900–2350 m.

Conservation notes: Least Concern because of its wide distribution.

In *Richards* 735 the inflorescences are producing plantlets (viviparous).

D18. **Cyperus aureobrunneus** C.B.Clarke in F.T.A. **8**: 346 (1901). Type: Malawi, Chitipa, 'British Central Africa, Nyasaland, Tanganyika Plateau at Fort Hill', vii.1896, *Whyte* s.n. (K).

> *Cyperus denudatus* L.f. var. *aureobrunneus* (C.B. Clarke) Kük. in Engler, Pflanzenr. **4**, 20(101): 256 (1936).

Perennial up to 75 cm tall; rhizomes, when present, short, up to 3 mm in diameter with initially pale, later dark, brown scales, and sometimes with 2–4 ovoid tubers at the stem base; roots dark reddish. Culms few, 35–75 cm long, 2–4 mm wide, trigonous, glabrous or slightly scabrid on the angles above. Leaves 30–50 cm long; leaf sheath to 5 cm long, pale brown; leaf blade linear, flat, 20–45 cm long, 5–9 mm wide, slightly scabrid on the margins, otherwise glabrous; apex acuminate. Involucral bracts 2–3, shorter than or equalling the inflorescence, the largest to 10 cm long and 7 mm wide, but mostly much smaller. Inflorescences 1–3 times capitate-anthelate, primary branches 7–10, 3–10 cm long; spikelets in capitate clusters of 3–6, sessile and at the ends of primary and secondary branches, narrowly elliptic, 8–10 × 1–2 mm, 18–24-flowered; rachilla weakly zigzag, ridged at the nodes; glumes reddish brown with pale keel and margins, 1.5–2 × 0.5–0.9 mm, boat-shaped; keel slightly excurrent as a short mucro. Stamens 2, filaments 1.2–1.5 mm long, anthers 4 mm long. Style 3-branched. Nutlet pale grey, ellipsoid to obovoid, trigonous, 0.5–0.6 × 0.2–0.3 mm, pale brown.

Zambia. N: Chinsali Dist., Shiwa Ngandu, 7.vi.1956, *Robinson* 1623 (K). C: Serenje Dist., Kundalila Falls, S of Kanona, 14.iii.1975, *Hooper & Townsend* 753 (K). E: Mpika Dist., Katete Hills, 10.vi.1963, *Verboom* 139 (K). **Malawi**. N: Mzimba Dist., 5 km [3 miles] W of Mzuzu, Katoto, 10.ii.1974, *Pawek* 8077 (K, MAL, MO, SRGH).

Not known elsewhere. Marshes, wet dambos.

Conservation note: Data Deficient; not many specimens and poorly known.

Very similar to *Cyperus sphaerospermus* which, like *C. aureobrunneus*, differs from *C. denudatus* in being leafy. *Cyperus aureobrunneus* is much larger than *C. sphaerospermus* in all its parts.

D19. **Cyperus rotundus** L., Sp. Pl.: 45 (1753). —Clarke in F.T.A. **8**: 364 (1901). — Kükenthal in Engler, Pflanzenr. **4**, 20(101): 107 (1935). —Haines & Lye, Sedges & Rushes E. Afr.: 186, figs.362,363 (1983). —Gordon-Gray in Strelitzia **2**: 65 (1995). —Hoenselaar in F.T.E.A., Cyperaceae: 211, fig.32 (2010). Type: India, *Herb. Hermann* 1, 3: 36 (BM lectotype), lectotypified by Tucker (1994).

> *Cyperus tuberosus* Rottb., Descr. Icon. Rar. Pl.: 28 (1773). Type: India, Malabar, *König* s.n. (?C holotype)
> *Cyperus rotundus* var. *platystachys* C.B. Clarke in Fl. Cap. **7**: 182 (1897); in F.T.A. **8**: 365 (1901), in part. —Kükenthal in Engler, Pflanzenr. **4**, 20(101): 114 (1935). Type: many specimens mentioned by Clarke in Durand & Schinz, Consp. Fl. Afric. **5**: 575 (1894) but none in F.T.A.
> *Cyperus taylorii* C.B. Clarke in F.T.A. **8**: 367 (1901). Type: Kenya, Kilifi Dist., Rabai Hills, near Mombasa, *Taylor* s.n. (BM holotype).
> *Cyperus merkeri* C.B. Clarke in Bot. Jahrb. Syst. **38**: 134 (1906). Type: Tanzania, Mbulu Dist.: Mbugwe [Umbugwe] and Iraku, *Merker* 114 (B holotype).
> *Cyperus rotundus* subsp. *tuberosus* (Rottb.) Kük. in Engler, Pflanzenr. **4**, 20(101): 113 (1935); Haines & Lye, Sedges & Rushes E. Afr.: 188, fig.36 (1983).
> *Cyperus rotundus* var. *taylorii* (C.B. Clarke) Kük. in Engler, Pflanzenr. **4**, 20(101): 114 (1935). —Haines & Lye, Sedges & Rushes E. Afr.: 188 (1983).
> *Cyperus rotundus* subsp. *merkeri* (C.B. Clarke) Kük. in Engler, Pflanzenr. **4**, 20(101): 115 (1935). —Haines & Lye, Sedges & Rushes E. Afr.: 187, fig.364 (1983).

Perennial or sometimes seemingly annual, 10–100 cm tall, with a somewhat swollen culm-base arising from rather thick scale-covered stolons; nodules on roots white turning brown; gregarious, but not clump-forming; culms few, green, 1–3 mm wide, triangular, glabrous. Leaves glossy green; leaf sheath green to reddish-brown; leaf blade linear, 10–40 × 0.2–0.8 cm wide, slightly M-shaped in cross-section, scabrid on margin and major veins, attenuate. Involucral bracts leaf-like, 1–5, erect or spreading, lowermost 3–26 cm long, 2–9 mm wide. Inflorescence a simple to compound anthela, primary branches 1–8, 0.5–12 cm long; spikelets in rather dense clusters, 3–15 per cluster, bright to golden to dark brown, linear-lanceolate and slightly flattened, 6–70 mm long, 1–2.5 mm wide, rachilla straight, remaining attached to rachis while lower glumes and nutlets are shed; glumes pale to dark reddish brown, ovate, 2.7–4.3 mm long, keel green, glabrous or slightly scabrid, with 1–2 veins on either side, apex obtuse. Stamens 3, yellow, 1.6–2.2 mm long. Style white, 3-branched. Nutlet greyish to brown, obovoid, trigonous, 1.3–1.7 mm long, 0.5–1 mm wide, minutely papillose.

Botswana. N: Ngamiland Dist., Selinda Spillway, Selinda Reserve, Motswiri Camp, 17.iii.2005, *A. & R. Heath* 921 (K). SE: Southern Dist., 4.9 km from Kanye on Mosopa-Kanye road, 8.iii.1991, *Cook et al.* 50 (K). **Zambia**. C: Serenje Dist., Mfuwe, 21.i.1969, *Astle* 5424 (K, NDO). S. Mazabuka Dist., Mazabuka, 27.xii.1962, *van Rensburg* KBS 1123 (K). **Zimbabwe**. W: Hwange Dist., Victoria Falls, 10–15.iii.1932, *Brain* 8666 K, SRGH). C: Gweru Dist., Gwelo, Mlezu Government Agricultural School Farm, 30 km [18 miles] SSE of Que Que, 7.iii.1966, *Biegel* 960 (K, SRGH). S: Beitbridge Dist., Tuli River at Tuli Breeding Station, 6.i.1961, *Wild* 5319 (K, SRGH). **Malawi**. C: Liliongwe Dist., 1–5 km S of Lilongwe, 2.ii.1989, *Thompson* 6319 (CM, K). S: Nsanje Dist., Port Herald, between Muona and Shire Rivers, 20.iii.1960, *Phipps* 2570 (K, SRGH). **Mozambique**. N: Mossuril Dist., Lumbo, near airport, 5.v.1948, *Pedro & Pedrogão* 3107 (K). Z: Gurué Dist., Namuli Mt., plateau N of two peaks, 20.xi.2007, *Harris* 401 (K). MS: Marínguè Dist., Shupanga, i.1863, *Kirk* s.n. (K). GI: Chibuto Dist., Maniquenique, 12.i.1970, *Sousa de Almeida* M8 (K). M: Maputo [Lourenco Marques], Jardim Vasco da Gama, 11.xi.1971, *Balsinhas* 2261 (K).

Widespread in Africa; India. Swamps, damp sites, riverbanks, drainage lines in coastal bush or forest glades, common weed in rice and maize fields, seasonally wet grassland; 0–1950 m.

Conservation notes: Least Concern; one of the most widespread tropical species and often included among 'the World's worst weeds'.

Beentje, in F.T.E.A., noted: Haines & Lye kept up four subspecies and varieties, based on Kükenthal's taxa; Kükenthal had combined these taxa into *Cyperus rotundus*. Differences between them were only expressed in short sentences (rather than a key) and these have given me problems. Colour and degree of compression of the spikelet, plus glume length and glume apex configuration, were the characters used to distinguish these four taxa.

Regarding the taxon *Cyperus merkeri*, C.B. Clarke, when describing his new species (based on a now destroyed type) only compared it to an unknown taxon, *C. neuerensis*; he gave the length of the culm as 30 cm. Kükenthal in Engler, Pflanzenr. **4**, 20(101): 115 (1935) combined *C. merkeri* into a subspecies of *rotundus*. He did not give a key, but from his brief descriptions we can see he thought this subspecies differed from the typical one by its longer culms (45–60 cm rather than 15–30 cm), and darker-coloured spikelets (dark dull red, rather than brown or dull red). Haines & Lye distinguished these two subspecies (again, without giving a key) by 'its shorter glumes with a usually much darker colour': glumes 2.7–3.2, rather than 3.3–4.3 mm; description of glume colour in the text overlaps for the two subspecies.

D20. **Cyperus kipasensis** Cherm. in Rev. Zool. Bot Africaines **24**: 297 (1934); in Bull. Jard. Bot. État. Bruxelles **13**: 281 (1935). Types: D.R. Congo, *Vanderyst* 30869, 31934, 33334, 33337, 33353 & 34013 (B syntypes).

Cyperus platycaulis var. *kipasensis* (Cherm.) Kük. in Engler, Pflanzenr. **4**, 20(101): 636 (1936).

Perennial forming clumps of culms joined by slender creeping rhizomes 1–2 mm in diameter covered by brown scales that become black with age. Culms 15–60 cm long and 1.5 mm in diameter, subtriquetrous, not bulbous at the base. Leaves equalling or shorter than the culms, 2–3.5 mm wide, flat, or plicate, slender, margins smooth but scabrid towards the triquetrous apex; sheaths dark purplish. Involucral bracts 2–3, the lowest erect, 3–12 cm long. Inflorescence usually a simple dense anthela, but sometimes twice-anthelate and more spreading, 1.5–4 cm in diameter, 3–8-radiate, primary branches unequal, 0–2 cm long, suberect, smooth; ochreae pale yellowish; secondary branches spreading, 0–0.5 cm long. Spikelets in bunches of 7–12, 16–20-flowered; rachilla straight, unwinged. Glumes loosely imbricate, 1–1.25 mm long, ovate, acute, not or scarcely mucronate, margins nerveless, straw or chestnut, keel green, smooth, 3-nerved. Stamens 3; anthers linear. Styles 3-fid, branches exserted. Achenes narrowly oblong, apiculate, trigonous, 1 mm long, brown to black, smooth.

Zambia. W: Solwezi Dist., Chingola, 6.v.1960, *Robinson* 3712 (K). N: Mporokoso Dist., 50 km [30 miles] S of Mporokoso, 23.xii.1967, *Simon & Williamson* 1528 (K, SRGH).

Also in D.R. Congo and possibly W Tanzania. In wet often peaty and sometimes acid (pH 5) bogs and dambos; sometimes in wet places between rock outcrops; 1250–1550 m.

Conservation notes: Probably Least Concern; although confined in our area to N and W Zambia, whence there are about 20 collections.

The largest involucral bract is generally help upright making the inflorescence appear somewhat lateral. Certainly distinct from *Cyperus platycaulis* which has short thick rhizomes with culms at short intervals, larger spikelets, and no leaves.

D21. **Cyperus maculatus** Boeckeler in Peters, Naturw. Reise Mossambique **6**(2): 539 (1864). —Clarke in F.T.A. **8**: 363 (1901). —Kükenthal in Engler, Pflanzenr. **4**, 20(101): 103 (1935). —Haines & Lye, Sedges & Rushes E. Afr.: 189, figs.367,368 (1983). —Hoenselaar in F.T.E.A., Cyperaceae: 230 (2010). Type: Mozambique, *Peters* s.n. (B holotype).

  Cyperus longus L. var. *maculatus* (Boeckeler) Boeckeler in Linnaea **36**: 282 (1870).

Perennial up to 92 cm tall, slender to robust, with up to 15 cm long brown-scaly horizontal rhizomes, but when growing in narrow rock-cracks the rhizomes are reduced and the basal part of the plant mainly consist of many densely crowded swollen woody culm-bases; culms few, 21–80 cm long, 1.3–3.3 mm in diameter, trigonous to subterete, glabrous. Leaves up to 44 cm long; leaf sheath green to pale reddish-brown, 1.5–15 cm long; leaf blade somewhat bluish-green, linear, flat, 13–35 cm long, 3–5.5 mm wide, scabrid on margin and primary vein, particularly towards the acuminate apex. Involucral bracts leaf-like, erect to spreading, 3–5, lowermost 12–43 cm long, 2.2–4.2 mm wide. Inflorescence a simple to compound anthela, primary branches 4–6, 1.5–10 cm long; spikelets in lax almost digitate spikes, sessile and on the end of primary (and occasionally secondary) branches, 7–15 per cluster, linear-lanceolate, 6.5–19 mm long, 1.2–1.9 mm wide, rachilla straight or curved; glumes reddish-brown with a wide uncoloured margin, ovate-elliptic, 2.4–3.5 mm long, 1.3–1.7 mm wide, closely overlapping, keel greenish, apex rounded to acute. Stamens 3; filaments 1.7–2.6 mm long; anthers 0.9–1.7 mm long. Nutlet brown, obovoid, trigonous, 1–1.3 mm long, 0.5–0.7 mm wide, minutely papillose.

Botswana. N: Ngamiland Dist., Kwemotahaa stream bridge, 22.i.1977, *P.A. Smith* 1882 (K, SRGH). **Zambia**. B: Mongu Dist., Mongu, 6.i.1966, *Robinson* 6783 (K). N: Luapula Dist., Mbereshi-Luapula River Swamp, 14.i.1960, *Richards* 12355 (K). **Zimbabwe**. N: Mutoko Dist., Mkota Reserve, by Mazoe River, 1.x.1948, *Wild* 2693 (K, SRGH). C: Harare Dist., Hunyani River at junction of Gwebi River, 5.i.1973, *Gibbs Russell* 2501 (K, SRGH). S: Masvingo Dist., Sabi-Lundi River junction, Chitsa's Kraal, 4.vi.1950, *Wild* 3349 (K, SRGH). **Malawi**. N: Karonga Dist., 29 km [18 miles] N of Chilumba, 6.i.1978, *Pawek* 13552 (K, MAL, MO, SRGH). S: Chikwawa Dist., Lower Mwanza River, 4.x.1946, *Brass* 17968 (K, NY).

Widespread in tropical and southern Africa. In sandy habitats near lakes and rivers, in rock-crevices; 180–1400 m.

Conservation notes: Least Concern because of its wide distribution.

This taxon has pale yellowish spikelets and is almost always found in sandy habitats – lake, shores, river banks, etc.

D22. **Cyperus procerus** Rottb., Descr. Icon. Rar. Pl.: 29 (1773). —Kükenthal in Engler, Pflanzenr. **4**, 20(101): 91 (1935). —Haines & Lye, Sedges & Rushes E. Afr.: 182, fig.352 (1983). —Hoenselaar & Beentje in F.T.E.A., Cyperaceae: 234 (2010). Type: Egypt, *Forskåhl* s.n. (C holotype)

> *Mariscus procerus* A. Rich., Tent. Fl. Abyss. **2**: 489 (1850), illegitimate name. —Clarke in F.T.A. **8**: 395 (1902), non Nees (1842).
>
> *Cyperus procerus* var. *stenanthus* Kük. in Engler, Pflanzenr. **4**, 20(101): 92 (1935). Types: Sudan, Djur, 5.vii.1869, *Schweinfurth* 2017 (P syntype, PRE, S); Tanzania, Uzaramo Dist., Dar es Salaam, Magagoni Lake, *Peter* 44989 (?B syntype); Tanzania, Singida/Dodoma Dist., Turu, Itigi–Bangayega, *Peter* 33742b (?B syntype); Zimbabwe, Mamlova, *Pocock* 107 (B syntype).

Perennial up to 135 cm tall, robust, stoloniferous, thin horizontal rhizomes 1–3 mm in diameter covered by distantly spaced black scales; culms 42–119 cm long, 2–5 mm wide, trigonous, smooth. Leaves up to 90 cm long; leaf sheath brown, sometimes slightly fibrous and blackish at base, 1–11 cm long; leaf blade linear, flat, 30–86 cm long, 3–10 mm wide, glabrous, apex acute to acuminate. Involucral bracts 2–4, spreading, leaf-like, lowermost 8–30 cm long, 3–8 mm wide. Inflorescence simply to compound-anthelate, primary branches 3–7, of different lengths, 0.5–12 cm long; spikelets loosely clustered, sessile and at the end of primary branches, spreading at right angles or almost so to the axis, 7–20 per cluster, linear, 8.5–28 mm long, 1.9–2.9 mm wide, glumes spreading with age, straight to slightly curved; glumes reddish-brown, with an uncoloured margin, ovate, 2.2–3.2 mm long, 1.6–1.9 mm wide, keel pale brown to brown, flattish, apex rounded to emarginate. Stamens 3; filaments 1.6–4 mm long; anthers 1–1.9 mm long. Nutlet brown, obovoid, trigonous, 1–1.6 mm long, 0.9–1.6 mm wide, smooth to slightly papillose in longitudinal rows.

Zambia. N: Luapula Dist., Mbereshi, close to R. Luapula Swamp, 11.i.1960, *Richards* 12323 (K). S: Mapanza Dist., Kabulamwanda, 110 km [70 miles] N of Choma, 13.ii.1955, *Robinson* 1106 (K). **Zimbabwe**. W: Hwange Dist., Victoria Falls, 4.iv.1956, *Robinson* 1425 (K). **Malawi**. C: Nkhotakota Dist., Senga Hill Forest, 16.ii.1959, *Robson* 1635 (K).

Widespread in west and tropical Africa, extending into South Africa; Egypt; in seasonally wet grasslands and swamps, growing in water; 30–1500 m.

Conservation notes: Least Concern because of its wide distribution.

Robinson, in his collecting notes, mentions 'bulbs' (R.1325) and 'small tubers' (R.1350). These structures are not apparent on the sheets, but Robinson was an excellent observer. East African material (*Bidgood et al.* 5053) shows thin rhizomes ending in a tuber c. 5–8 mm in diameter from which a new upright shoot arises.

D23. **Cyperus esculentus** L., Sp. Pl.: 45 (1753). —Clarke in F.T.A. **8**: 355 (1901). —Kükenthal in Engler, Pflanzenr. **4**, 20(101): 116 (1935). —Haines & Lye, Sedges & Rushes E. Afr.: 190, figs.372,373 (1983). —Gordon-Gray in Strelitzia **2**: 57(1995). —Hoenselaar in F.T.E.A., Cyperaceae: 227 (2010). Type: 'Habitat Monspelii, inque Italia, Oriente' ('Cyperus rotundus esculentus angustifolius' in Bauhin, Theatr. Bot.: 222, (1658) lectotype), lectotypified on icon by Simpson (1993).

> *Cyperus esculentus* var. *cyclolepis* Kük. in Engler, Pflanzenr. **4**, 20(101): 119 (1935). Types: South Africa, Pretoria, *Rehmann* 4776 (K syntype); Kenya, Nairobi, *Thomas* 113 p.p. (?B syntype).

Perennial, stoloniferous, up to 1 m high; stolons to ± 15 cm long, 0.5–1.5 cm thick, covered with brown scales and ending in a blackish tuber 3–8 mm in diameter (only rarely present in herbarium material); culms 18.5–74 cm long, 1.6–3.5 mm wide, trigonous to triquetrous, glabrous. Leaves up to 38 cm long, crowded near the base; leaf sheath pale brownish-green to green, 2–9 cm long; leaf blade linear, flat, 11.5–29 cm long, 2.3–12 mm wide, with 2 main veins next to primary vein, apex acuminate, glabrous to minutely scabrid on margins and veins. Involucral bracts 2–6, leaf-like, spreading, the lowermost 4–19 cm long, 2.1–9 mm wide. Inflorescence usually simply anthelate but a few of the primary branches are sometimes themselves branched at the apex, primary branches (3)5–10, 1–18 cm long; spikelets in loose clusters on elongated axis, sessile and at the end of primary and secondary branches, 9–20 per cluster, very narrowly ovate to very narrowly oblong, 5.5–16 mm long, 1.2–2.5 mm wide, the apex of the spikelet obtuse; glumes yellowish-brown to reddish-brown, elliptic-ovate to obovate, with 3–4 distinct veins on each side of the keel, glabrous, 2.4–3.5 mm long, 1.4–2 mm wide, keel green to reddish-brown, apex obtuse, slightly reflexed so that the spikelet outline is serrulate. Stamens 3; filaments 2.5–4 mm long; anthers 1–2 mm long. Nutlet shiny grey, ellipsoid to obovoid, trigonous, 1.3–2.0 mm long, 0.6–0.9 mm wide, smooth.

Botswana. N: Ngamiland Dist., 0.79 km S of Chief's Camp, Chief's Island, Okavango Delta, 3.i.2012, *A. & R. Heath* 2333 (K). **Zambia**. B: Mongu Dist., Kande Pan, 26.xi. 1964, *Verboom* 1783 (K). N: Mbala [Abercorn] Dist., about 2.5 km [1.5 miles] above Sansia Falls, Kalambo River, 29.xii.1958, *Richards* 10363 (K). W: Solwezi Dist., c. 5 km E of Kabompo River on Mwinilunga-Solwezi road, 27.ii.1975, *Hooper & Townsend* 419 (K). C: Mpika Dist., Luangwa N.P., Mfuwe, 28.i.1969, *Mitchell* 5428 (K). S: Mazabuka Dist., Ross's Farm, 20 km [13 miles] N of Choma, 1.xi.1960, *White* 6695 (FHO, K). **Zimbabwe**. W: Bulawayo Dist., Hillside Dams, xii.1958, *Miller* 5687 (K, SRGH). C: Marondera Dist., [Marandellas], Grasslands Research Station, 3.i.1949, *Corby* 328 (K, SRGH). S: Beitbridge Dist., Tuli Research Station, 6.i.1961, *Wild* 5320 (K, SRGH). **Malawi**. N: Rumphi Dist., 8 km [5 miles] N of Rumphi, 14.i.1976, *Phillips* 968 (K, MO). C: Lilongwe Dist., near Lilongwe, 26.ii.1961, *Vesey-Fitzgerald* 3055 (IRLCS, K). S: ?Nsanje Dist., Namasi, v.1899, *Cameron* 14 (K). **Mozambique**. T: Angónia Dist., Ulongue, entre Tsangano e Chitambe, 17.xii.1980, *Macuácua* 1471 (K, LMA). MS: Gorongosa Dist., Gorongoza National Park, Road 5, 12.iii.1969, *Tinley* 1857 (K, SRGH).

Widespread in Africa; also in S Europe; marshes, bogs, lake margins, seasonally damp grasslands and swamps, shallow boggy soil over laterite; sometimes a weed of cultivation (rice fields); 50–1750 m.

Conservation notes: Least Concern because of its wide distribution.

Similar to *Cyperus fulgens* but bulbs are absent, the spikelets are paler (golden brown rather than reddish brown) and form an angle of less than 90° to the axis.

Group E: Species with spicate terminal inflorescence units.

1. Inflorescence compound (2–3 times anthelate) 2
– Inflorescence simple (once anthelate) 9
2. Flowers arranged all round the spikelet axis; spikelets relatively broad, >1.8 mm wide. . . . . **E1.** *alopecuroides*
– Flowers distichously arranged; spikelets narrower, 1–1.7 mm wide 3
3. Spikelets >1 cm long and c. 1 mm broad, brown, well separated. . . . . B13. *distans*
– Spikelets usually < 1 cm long and > 1 mm broad, tightly arranged 4
4. Glumes not mucronate 5
– Glumes mucronate 8
5. Leaves and involucral bracts less than 5 mm wide 6
– Leaves and involucral bracts >5 mm wide and often >10 mm wide 7
6. Inflorescence lax and untidy; glumes yellow-brown with median red-brown band . . . . D2. *laxus*
– Inflorescence neater and compact; spikelets parallel-sided; glumes mostly dark red-brown. . . . . **E2.** *longus*

7. Glumes blunt, obtuse, strongly keeled . . . .**E3.** *latifolius*
– Glumes acute or obtuse but not mucronate, tightly appressed . . . . . .**E4.** *digitatus*
8. Glumes mucronate with short appressed tips; spikelets often >6 mm long . . . . . .
   . . . . **E5.** *exaltatus*
– Glumes mucronate with longer spreading tips; spikelets up to 6 mm long . . . . .
   . . . . **E6.** *imbricatus*
9. Some inflorescence subunit stalks > 5 cm long; inflorescence subunits not densely
   spicate with the spikelets at right angles to the axis; spikelets > 1 cm long . . . . . .
   . . . . A11. *natalensis* (if spikelets blackish, check *C. rigidifolius*)
– Stalks of inflorescence subunits shorter; inflorescence units densely spicate or
   not; most spikelets smaller 10
10. Leaves and upper part of culm pubescent (×10) . . . . **E7.** *pubens*
– Leaves and culms glabrous 11
11. Spikelets all sessile on the rachis of the subunits, crowded 12
– Spikelets well separated from one another 25
12. Inflorescence dense, congested, of very narrow whitish to pale cream spikelets . . .
   . . . . **E13.** *hemisphaericus*
– Inflorescence less dense, of relatively short and broad brwnish or greenish
   spikelets 13
13. Large plants; culms 3–8 mm in diameter; lamina 3–13 mm wide 14
– Smaller plants; culms 2 mm or less in diameter; laminas <4 mm wide 15
14. Rhizome thick, elongated, covered with old fibrous leaf bases; inflorescence
   dense . . . . **E8.** *tomaiophyllus*
– Rhizome short; inflorescence lax; spikelets with long curved tips . . . . . .**E9.** *luteus*
15. Some inflorescence units with clear spikelet-free peduncles 16
– Inflorescence units covered by spikelets to the base 19
16. Spikelets well separated so that the axis is clearly visible 17
– Spikelets densely arranged so that the axis is hidden 18
17. Base bulbous, covered by thin whitish leaf bases, connected to other shoots by
   slender rhizomes; spikelets often orange-tinged . . . . **E10.** *trigonellus*
– Spikelets in dense clusters at ends of peduncles, >10-flowered; centres of glumes
   blackish . . . . **E11.** *rigidifolius*
18. Glumes green; 10–12(14) secondary longitudinal veins on glume . . **E12.** *sublimis*
– Glumes yellow-green; 6–8 secondary veins on glume . . . . **E13.** *corynephorus*
19. Plant base a distinct pseudobulb, usually covered by thin brown scales 20
– Plant base nodular or tuberous, or neither, without a scaly covering 22
20. Pseudobulbs elongated, 5–12 cm long, covered by pale brown membranous leaf
   sheaths . . . . **E15.** *chersinus*
– Pseudobulbs not as above 21
21. Pseudobulbs shorter and more nearly spherical, hard, covered by brown or
   purplish leaf sheaths . . . . **E16.** *pseudoflavus*
– Relatively small; spikelets shorter and blunt . . . . **E17.** *nyasensis*
22. Tips of glumes long-acuminate, recurved . . . . **E18.** *decurvatus*
– Tips of glumes apiculate or rounded, straight 23
23. Inflorescence very dense, subunits touching one another; thin rhizomes present
   . . . . **E19.** *myrmecias*
– Inflorescence units not touching one another 24
24. Robust plants of coastal sand dunes; spikelets 5–6 mm long . . . .**E20.** *macrocarpus*
– Relatively robust; spikelets long and narrow . . . . **E21.** *cyperoides*
25. Spikelets sessile and towards the ends of primary branches; glumes yellow-brown,
   separate, giving a serrate outline; base bulbous . . . .**E22.** *turrillii*
– Not as above 26

26. Spikelets dark purple-brown; glumes with distinct whitish membranous margins
. . . . D22. *procerus*
– Spikelets and glumes not as above 27
27. Glumes dark with pale keel; lowermost involucral bracts to 30 cm long; rhizome
short and knotty. . . . . **E23.** *schimperianus*
– Without the above combination of characaters. 28
28. Culm base arising at the top of a slender vertical rhizome from a bulb c. 1 cm in
diam.; glumes dark reddish brown, apiculate, striate . . . . **E24.** *callistus*
– Culms not arising in this way; spikelets long and narrow, up to 2 mm wide and
8–20 mm long, numerous, well separated 29
29. Spikelets clustered at the ends of the branches, >15 mm long. 30
– Spikelets well spaced so that the axis is clearly visible. 31
30. Stem base swollen, adjacent to the next stem, forming a horizontal nodular
rhizome; slender rhizomes usually also present . . . . **E25.** *tenuiculmis*
– Stem bases well separated, linked by a horizontal rhizome . . . . . . .A11. *natalensis*
31. Spikelets golden yellow; clump-forming perennial. . . . . **E26.** *aureospicatus*
– Spikelets brownish or dark with pale margins. 32
32. Perennial up to at least 100cm tall; spikelets almost black; upland forest,
uncommon. . . . . **E27.** *aterrimus*
– Annual or perennial; spikelets brownish or dark with pale margins 33
33. Annual or short-lived perennial; spikelets brownish. . . . . .B13. *distans*
– Perennial with slender horizontal rhizomes; glumes dark with pale margins. . . .
. . . .**E28.** *ferrugineoviridis*

E1. Cyperus alopecuroides Rottb., Descr. Icon. Rar. Pl.: 38 (1773). —Kükenthal in
Engler, Pflanzenr. **4**, 20(101): 71 (1935). —Haines & Lye, Sedges & Rushes E.
Afr.: 181, fig.348 (1983). —Gordon-Gray in Strelitzia **2**: 53 (1995). —Hoenselaar
in F.T.E.A., Cyperaceae: 219 (2010). Type: Arabia, *Forskåhl* s.n. (C holotype).

Juncellus alopecuroides (Rottb.) C.B. Clarke in Hooker, Fl. Brit. India **6**: 595 (1893); in
F.T.A. **8**: 307 (1901).

Cyperus alopecuroides f. *pallidiflorus* (Peter) Kük. in Engler, Pflanzenr. **4**, 20(101): 72 (1935).
Types: Tanzania, Pare Dist., Pare, *Peter* 8400 (B syntype); Pangani near Hale, *Peter* 8372 (K
syntype, WAG).

Perennial, fairly robust, up to 170 cm tall; rhizomes very short or absent; culms few, 22–140
cm long, 3.5–8.4 mm wide, trigonous, glabrous. Leaves basally crowded, up to 100 cm long; leaf
sheath reddish-brown to blackish, 7–31 cm long; leaf blade linear, flat or pleated, 37–75 cm long,
4–15 mm wide, strongly scabrid to smooth on margin, apex acuminate. Involucral bracts leaf-
like, spreading, 5–8, lowermost 44–73 cm long, 5–16 mm wide. Inflorescence twice anthelate-
spicate, primary branches 6–10, 3–22 cm long; spikelets in crowded clusters, sessile and at the
end of primary and secondary, sometimes tertiary branches, many per cluster, ovoid, 2.5–6.7
mm long, 1.8–2.7 mm wide, rachilla flattened, straight; glumes golden to reddish-brown, ovate,
the margins inrolled, 1.2–1.7 mm long, 0.8–1.2 mm wide, keel green, rounded, apex excurrent.
Stamens 3; filaments 1.4–2 mm long; anthers 0.6–0.8 mm long. Style with 2 stigma branches.
Nutlet brown, flattened, pressed against the rachilla with the flat side, 0.7–1 mm long, 0.5–0.75
mm wide, smooth or minutely reticulate-striate.

Botswana. N: Ngamiland Dist., Okavango Delta, Chief's Camp, beside airstrip,
16.ix.2009, *A. & R. Heath* 1756 (K). SE: Central Dist., Lake Xau [Dow], 1.xii.1968,
Child 129.162 (K, SRGH). **Zambia**. N: Mbala Dist., Mpulungu, Lake Tanganyika,
22.x.1967, *Simon & Williamson* 1159 (K, SRGH). E: Lundazi Dist., Kazembe, 80 km [50
miles] W of Lundazi, in the Luangwa Valley, 3.vi.1954, *Robinson* 832 (K). **Zimbabwe**.
C: Gwelo Dist., Mlezu Agricultural School Farm, 10.vi.1976, *Biegel* 5310 (K, SRGH). S:

Chiredzi Dist., Gona-re-zhou between Chitsa's Store and Sabi-Lundi Rivers junction, in Tamboharta Pan, 31.v.1971, *Grosvenor* 584 (K, SRGH). **Malawi**. S: Chikwawa Dist., Shire River near Sucoma Sugar Estate, Nchalo, 21.iv.1980, *Blackmore & Dudley* 1320 (K, MAL). **Mozambique**. GI: Gaza Dist., Massangir, margen do rio dos Elefantes, 19.xi.1957, *Grandvaux Barbosa* 8190 (K).

Widespread in tropical Africa from Egypt to South Africa; Madagascar, Arabia, S Asia. Riverbanks, seasonally dry riverbeds, lake shores, sometimes with *Ludwigia* sp.; 50–1300 m.

Conservation notes: Least Concern because of its wide distribution.

E2. **Cyperus longus** L., Sp. Pl.: 45 (1753). —Clarke in F.T.A. **8**: 366 (1901). —Kükenthal in Engler, Pflanzenr. **4**, 20(101): 97 (1935). —Haines & Lye, Sedges & Rushes E. Afr.: 189, fig.369 (1983). —Gordon-Gray in Strelitzia **2**: 61 (1995). —Hoenselaar in F.T.E.A., Cyperaceae: 230 (2010). Type: Southern Europe, "habitat in Italiae, Galliae paludibus", *Herb. A. van Royen* 909.89–686 (L lectotype), lectotypified by Kukkonen (2004).

    Cyperus tenuiflorus Rottb., Descr. Icon. Rar. Pl.: 30 (1773). Type: grown in Hort. Hagensis by Kaesemaker (Tab. 14.1 iconotype).

    Cyperus longus var. *tenuiflorus* (Rottb.) Boeckeler in Linnaea **36**: 281 (1870). —Kükenthal in Engler, Pflanzenr. **4**, 20(101): 102 (1935).

Perennial, fairly robust, up to 100 cm tall, with 1–3 mm thick horizontal, dark brown, scale-covered horizontal rhizomes and only slightly swollen culm bases; culms few, 25–95 cm long, 1.9–4 mm wide, trigonous above, terete below, glabrous. Leaves few, withering early, up to 50 cm long; leaf sheath pale to dark reddish-brown, 3–10 cm long; leaf blade flat, 16–40 cm long, 3–5 mm wide, slightly scabrid on margin and primary vein, apex acuminate. Involucral bracts leaf-like, erect to spreading, 3–5, lowermost 6–30 cm long, 2–6 mm wide, margins scabrid towards the base. Inflorescence a simple to compound anthela, primary branches 4–8, 0.5–10 cm long; spikelets in almost digitate spikes, sessile and at the end of primary branches, 3–15 spikelets per spike, linear to very narrowly ovate, 8–25 mm long, 1.2–2 mm wide, rachilla straight; glumes reddish brown with narrow uncoloured margin, ovate, 2.7–3.5 mm long, 1.4–1.7 mm wide, keel green, apex obtuse. Stamens 3; filaments 2.9–3.8 mm long; anthers 1.4–2.2 mm long. Nutlet brown, ellipsoid, trigonous, 1.0–1.6 mm long, 0.5–0.75 mm wide, almost smooth.

Caprivi. Okavango River, 19 km N of Shakawe, on Botswana [Bechuanaland] border, 16.iii.1965, *Wild & Drummond* 7094 (K, SRGH). **Botswana**. N: Ngamiland Dist., Kaporota Camp, Okavango Delta, 9.iii.2010, *A. & R. Heath* 1959 (K). **Zambia**. S: Mapanza Dist., Choma–Namwala road, 35 km [22 miles] N of Choma, 20.ii.1956, *Robinson* 1354 (K); Livingstone Dist., Victoria Falls, banks of R. Zambezi, 5.i.1955, *Robinson* 1063 (K). **Zimbabwe**. N: Darwin Dist., Mzarabani Tribal Trust land, Muzingwa River, 30.iv.1972, *Mavi* 1349 (K, SRGH). W: Hwange Dist., Victoria Falls, banks of R. Zambezi, 10–15.iii.1952, *Brain* 8794 (K). C: Gweru Dist., Gwelo, 11.xii.1931, *Brain* 7452 (K). S: Umzingwane Dist., Umzingwane River 8 km [5 miles] E of Balla Balla, 28.xii.1965, *Simon* 560 (K, SRGH).

Widespread in Africa; S Europe. Lake edges, black cotton soils, ditches, periodically flooded depressions in grassland or bushland, in swamps and temporary pools; sea-level to 1600 m.

Conservation notes: Least Concern because of its wide distribution.

This species has deep reddish brown glumes with pale margins, and the spikelets tend to cluster at the ends of the branchlets of the inflorescence.

F.T.E.A. note: Haines & Lye state this differs from *C. rotundus* only in the faintly swollen stem base and somewhat shorter glumes. Very polymorphic.

E3. **Cyperus latifolius** Poir. in Lamarck, Encycl. **7**: 268 (1806). —Clarke in F.T.A. **8**: 351 (1901). —Kükenthal in Engler, Pflanzenr. **4**, 20(101): 87 (1935). —Haines & Lye, Sedges & Rushes E. Afr.: 182, figs.350,351 (1983). —Gordon-Gray in Strelitzia **2**: 61 (1995). —Hoenselaar in F.T.E.A., Cyperaceae: 233 (2010). Type: Madagascar, *Du Petit-Thouars* s.n. (P holotype).

Perennial up to 200 cm tall, robust, with a hardened base producing 1–3 mm thick stolons covered with brownish or blackish scales; culms few, 46–160 cm long, 4–8.2 mm wide, triquetrous, usually slightly scabrid below the inflorescence. Leaves up to 2.7 m long; leaf sheath green to reddish-brown, slightly fleshy below, the lowest leafless sheaths almost black, 8–20 cm long; leaf blade linear, flat or V- or W-shaped in section, 53–256 cm long, 9–28 mm wide, scabrid on margin and primary vein, apex acute to acuminate. Involucral bracts leaf-like, usually spreading, 3–6, lowermost 29–53 cm long, 9–21 mm wide. Inflorescence a compound anthela, primary branches 6–9, 2.5–23 cm long; spikelets in loose clusters, spreading and rather distantly placed, sessile and at the end of primary and secondary branches, 5–20 per cluster, linear, 7–30 mm long, 1.4–2.2 mm wide, rachilla straight; glumes straw-coloured to pale or dark reddish brown, with an indistinct uncoloured margin, oblong-elliptic, 2.4–3.2 mm long, 1.3–1.8 mm wide, keel green, apex rounded. Stamens 3; filaments 1.4–3 mm long; anthers 1.4–1.9 mm long. Nutlet pale brown when young, turning dark brown or grey to black when mature, obovoid, sometimes almost obcordate, trigonous, 1.3–1.6 mm long, 0.6–1 mm wide, minutely papillose.

Botswana. N: Ngamiland Dist., Banks of Okavango River at Sepona, 18°44.25'S, 22°11.75'E, 26.i.1976, *P.A. Smith* 1346 (K). **Zambia**. N: Mbala Dist., bog about 2.5 km [5 miles] above Sansia Falls, Kalambo River, 29.xii.1958, *Richards* 10362 (K). W: Mwinilunga Dist., Mundeji River on Kanyama road, 25.ii.1975, *Hooper & Townsend* 397 (K). **Malawi**. N: Rumphi Dist., Misuku Hills, 12.i.1959, *Robinson* 3176 (K). **Mozambique**. GI: Chibuto Dist., road to Manjacaze, 12.x.1957, *Grandvaux et al.* 8013 (K). M: Maputo, 26.x.1919, *Borle* 97 (K).

Also in Benin, Cameroon, D.R. Congo, Rwanda, Burundi, Ethiopia, Uganda, Kenya, Tanzania, Angola, Swaziland, and South Africa; Madagascar. In swamps, marshes, boggy grasslands, in roadside ditches and along streams; sea level up to 2100 m.

Conservation notes: Least Concern because of its wide distribution.

The large size of the plant means that most collections are incomplete, often consisting of the inflorescence only. Probably more widespread than the records suggest.

E4. **Cyperus digitatus** Roxb., Fl. Ind. **1**: 209 (1820). —Kükenthal in Engler, Pflanzenr. **4**, 20(101): 55 (1935). —Hoenselaar in F.T.E.A., Cyperaceae: 244 (2010). Type: India, *Roxburgh* s.n. (BM holotype?).

Subsp. **auricomus** (Spreng.) Kük. in Bot. Not. **1934**: 65 (1934). —Kükenthal in Engler, Pflanzenr. **4**, 20(101): 57 (1935). —Haines & Lye, Sedges & Rushes E. Afr.: 178 (1983). —Hoenselaar in F.T.E.A., Cyperaceae: 244 (2010). Type: Egypt, 'Aegypt. infer.', *Sieber* s.n. (G syntype, LD, P).

Cyperus auricomus Spreng., Syst Veg. **1**: 230 (1824). —Clarke in F.T.A. **8**: 373 (1901).

Perennial, robust, up to 1.75(2) m high, with a thick woody creeping rhizome covered with dark red-brown to blackish scales, 1–1.5 cm in diameter; culms closely spaced on the horizontal rhizome, 77–153 cm long, 0.5–1 cm wide, triquetrous, sometimes ± winged, smooth to scabrid on the margins. Leaves few, up to 100 cm long; leaf sheath reddish to yellowish-brown, 10–30 cm long; leaf blade 18–85 cm long, 0.7–1.9 cm wide, pleated, linear 2 main veins next to primary vein, flat, scabrid on the veins and margins, apex acuminate. Involucral bracts 2–4, leaf-like, spreading, lowermost 15.5–67 cm long, 0.8–1.9 cm wide. Inflorescence usually twice spicate-anthelate, primary branches 3–8, 2.7–12 cm long, prophylls 1.5–2.5 cm long; spikelets in crowded clusters on elongated axis, sessile and at the end of primary, and secondary branches, 20–many per cluster, 4.5–13.5 mm long, 0.6–0.9 mm wide, linear, terete or angular; glumes reddish brown to golden, elliptic to obovate, 1.9–2.4 mm long, 0.7–1.3 mm wide, keel green, acute, 3–4-veined, apex (shortly) mucronate. Stamens 3: filaments 2–2.6 mm long; anthers 0.9–1.6 mm long. Nutlet dark grey, narrowly ellipsoid-oblong, trigonous-triquetrous, 0.8–1.2 mm long, 0.2–0.3 mm wide, minutely papillate in longitudinal rows.

Caprivi. Calcareous banks of Okavango River below Diyona Camp near Nyangana Mission Station, 9.i.1956, *De Winter* 4157 (K, PRE). **Botswana**. N: Ngamiland Dist., Bototi River, Samedupe Bridge, 13.xii.1976, *P.A. Smith* 1839 (K, SRGH). **Zambia**. B: Mongu Dist., Barotse floodplain on Lizulu, 27.v.1964, *Verboom* 1746 (K). N: Luapula Dist., Lake Bangweulu, S of Ncheta Island, 6.ii.1996, *Renvoize* 5529A (K). W: Mwinilunga Dist., Lunga River, 29.xi.1937, *Milne-Redhead* 3442 (K). S: Mazabuka Dist., Whitbread's Farm near Kanchombe, 18.xi.1960, *White* 7158 (FHO, K). **Zimbabwe**. N: Binga Dist., Chizarira Game Reserve, Bimba Vlei, 16.i.1968, *Thomson* 17 (K, SRGH). W: Matobo Dist., Sandy Spruit Dam, 20.ii.1965, *Simon* 146 (K, SRGH). C: Makoni Dist., Tsungwesi River, by bridge on Harare–Umtali road, 29.xi.1955, *Drummond* 5058 (K, SRGH). S: Mberengwa Dist., Doddieburn Ranch, Doddieburn Dam, on Tsibizini River, 10.v.1972, *Pope* 729 (K). **Malawi**. S: Chikwawa Dist., Lengwe National Park, 15.xii.1970, *Hall-Martin* 1174 (K, SRGH). **Mozambique**. N: Marrupa Dist., 6 km on road to Nungo, 20.ii.1982, *Jansen* 7948 (K). Z: Morrumbala Dist., Megaza, Shire River, 26.x.1971, *Babrick* R35B (K, SRGH). MS: Gorongosa Dist., Urema Floodplain, Gorongosa National Park, ii.1971, *Tinley* 2037 (K, SRGH). GI: Massingir Dist., lake edge, 27.vi.1980, *Schäfer* 7160 (K).

Widespread in west, central, and southern Africa. In swamps, by rivers, dams and lake shores, often with the base submerged; usually in permanently wet sites but also sometimes in seasonally wet grasslands with *Echinochloa* sp., sometimes forming extensive stands; 50–1600 m.

Conservation notes: Least Concern because of its wide distribution.

E5. **Cyperus exaltatus** Retz., Observ. Bot. **5**: 11 (1788). —Clarke in F.T.A. **8**: 370 (1901). —Kükenthal in Engler, Pflanzenr. **4**, 20(101): 64 (1935). —Haines & Lye, Sedges & Rushes E. Afr.: 179, fig.342 (1983). —Gordon-Gray in Strelitzia **2**: 56 (1995). —Hoenselaar in F.T.E.A., Cyperaceae: 244. Type: India, Tranquebar, *König* s.n. (LD holotype).

Perennial, very robust, up to 180 cm tall, with crowded culms on a short woody rhizome, 1 cm in diameter, the scales of the rhizome breaking up in fibrous remains; culms crowded, 40–150 cm long, 3–15 mm wide, trigonous, glabrous, the base slightly swollen. Basal leaves many; leaf sheath green to purple, 7–15 cm long; leaf blade linear, flat, up to 64–140 cm long, 8–35 mm wide, strongly scabrid on margins and primary vein, apex acuminate. Involucral bracts leaf-like, spreading, 5–9, lowermost 20–75 cm long, 8–28 mm wide; secondary anthelae also bracteate but bracts much smaller. Inflorescence usually twice anthelate-spicate, primary branches 7–11, 5–29 cm long; spikelets in dense, elongate clusters, sessile and at the end of primary and secondary branches, 15–120 per cluster, 2–12 mm long, 1–1.7 mm wide, rachilla straight; glumes reddish-brown to golden with darker reddish-brown margin, ovate-elliptic, 1.2–2.9 mm long, 1.1–1.5 mm wide, keel green, with 2–3 veins on either side, apex acuminate to mucronate. Stamens 3; filaments 1.3–2.7 mm long; anthers 0.7–0.8 mm long. Nutlet greyish, ellipsoid, trigonous, 0.6–1 mm long, 0.4–0.6 mm wide, almost smooth.

a) Var. **exaltatus**

Culms 3–10 mm wide; leaf blade 8–12 mm wide. Involucral bracts 8–12 mm wide; spikelets 6–12 mm long. Glumes 1.8–2.9 mm long.

Zimbabwe. E: Chimanimani [Melsetter] Dist., Lusitu River, 12.i.1969, *Mavi* 903 (K, SRGH). **Mozambique**. N: Palma Dist., Nhica do Rovuma, 40 km W of Palma. Nhica do Rovuma Lagoa, c. 3 km N of village, 12.xi.2009, *Goyder et al.* 6063a (K).

Widespread in tropical west and central Africa, down into Angola; S and SE Asia, Australia, Central and S America. At margins of lakes and rivers, and in swamps and in open water; sea-level to 1500 m.

Conservation notes: Least Concern because of its wide distribution.

b) Var. **dives** (Delile) C.B. Clarke in J. Linn. Soc., Bot. **21**: 187 (1884); in F.T.A. **8**: 370 (1901). —Hoenselaar in F.T.E.A., Cyperaceae: 245 (2010). Type: Egypt, *Delile* s.n. (MPU holotype).

> *Cyperus dives* Delile, Descr. Egypte, Hist. Nat.: 149, fig.3 (1813). —Kükenthal in Engler, Pflanzenr. **4**, 20(101): 68 (1935). —Haines & Lye, Sedges & Rushes E. Afr.: 180, figs.344,345 (1983). —Gordon-Gray in Strelitzia **2**: 56 (1995).
>
> *Cyperus immensus* C.B. Clarke in J. Linn. Soc., Bot. **20**: 294 (1883); in F.T.A. **8**: 371 (1901). —Kükenthal in Engler, Pflanzenr. **4**, 20(101): 67 (1935). Type: Madagascar NE, *Pervillé* 483 (P holotype).
>
> *Cyperus immensus* var. *taylori* C.B. Clarke in F.T.A. **8**: 372 (1901). Types: Kenya, Rabai Hills, *Taylor* s.n. (BM syntype); Tanzania, Zanzibar, *Taylor* s.n. (BM syntype).

Culms 5–15 mm wide; leaf blade 13–35 mm wide. Involucral bracts 14–28 mm wide; spikelets 3.4–7 mm long. Glumes 1.2–1.7 mm long.

Botswana. N: Ngamiland Dist., Sides of Thaoge River, 22.vii.1975, *P.A. Smith* 1445 (K, SRGH). **Zambia**. N: Mbala Dist., Vomo Gap, Fwambo side, 27.ix.1960, *Richards* 13296 (K). C: Lusaka Dist., Mount Makulu, 12 km [8 miles] S of Lusaka, 11.vi.1955, *Robinson* 1293 (K). E: Chipata Dist., Chikowa Mission, 13.x.1958, *Robson & Angus* in *Robson* 75 (K). S: Mumbwa Dist., Big Concession, Hot Springs, Piamadzi River, 18.viii.1970, *Verboom* 3101 (K). **Zimbabwe**. N: Kariba Dist., Lake Kariba, S shore, near islands, 23–24.xi.1964, *Mitchell* 1072 (K). E: Chimanimani Dist., Melsetter, Lusitu River, 12.i.1969, *Mavi* 903 (K, SRGH). S: Chipinge Dist., Changadzi River, c. 8 km [5 miles] E of Birchenough Bridge, i.1938, *Obermeyer* 2413a (K). **Malawi**. N: Chitipa Dist., Chisenga, Mbalizi river bank, 13.x.2007, *Chapama et al* 766 (K, FRIM). C: Kasungu Dist., Dowa, Lingadzi River, 31.x.1950, *Jackson* 209 (K). S: ?Nsanje Dist., River Shire, Elephant Marsh, ii.1863, *Kirk* s.n. (K). **Mozambique**. N: Palma Dist., Nhica do Rovuma, 40 km W of Palma. Nhica do Rovuma Lagoa, c. 3 km N of village, 12.xi.2009, *Goyder et al.* 6063b (K). MS: Manica Dist., Dombe, Mevumoze River, 26.x.1953, *Gomes Pedro* 4410 (K). M: Inhaca Island, Mdalandala, 30.xii.1956, *Mogg* 27050A (K).

Also in Egypt, Senegal, Ivory Coast, Nigeria, D.R. Congo, Rwanda, Burundi, Sudan, Eritrea, Ethiopia, Somalia, Uganda, Kenya and Tanzania. In swamps, on river-banks and in open water; sea-level up to 1550 m

Conservation notes: Least Concern; because of its wide distribution.

Recorded as being grazed by stock.

The two parts of the collection *Goyder et al.* 6063 have been placed in the two subspecies. This perhaps suggests that the two varieties are no more than forms induced by different environmental conditions.

Hoenselaar, in F.T.E.A., regarded this as close to *Cyperus alopecuroides*.

E6. **Cyperus imbricatus** Retz., Observ. Bot. **5**: 12 (1788). —Kükenthal in Engler, Pflanzenr. **4**, 20(101): 69 (1935). —Haines & Lye, Sedges & Rushes E. Afr.: 180, figs.346,347 (1983). —Gordon-Gray in Strelitzia **2**: 58, fig.21H,K (1995). — Hoenselaar in F.T.E.A. Cyperaceae: 251 (2010). Type: India, Tranquebar, *König* s.n. (LD holotype).

> *Cyperus radiatus* Vahl, Enum. Pl. **2**: 369 (1805). —Clarke in F.T.A. **8**: 369 (1901), illegitimate name, based on the same type.
>
> *Cyperus flexifolius* Boeckeler in Flora **62**: 549 (1879). —Clarke in F.T.A. **8**: 375 (1901). Type: D.R. Congo, island off Ponte da Lenha, *Naumann* 143, 150 (B syntypes).

Perennial (but probably often behaving as an annual) up to 135 cm tall but usually less, with a short woody rhizome; culms few, 26–60 cm long, 2.9–4.3 mm wide, trigonous, smooth. Leaves few, up to 50 cm long; leaf sheath pale brown and purple, 3.5–17 cm long; leaf blade linear, flat, 22–33 cm long, 2–7 mm wide, scabrid on margin and primary vein, apex acuminate. Involucral bracts leaf-like, spreading, 5–8, lowermost 27–48 cm long, 3–8 mm wide. Inflorescence twice spicate-anthelate, primary branches 4–8, 2.5–6 cm long; spikelets in very dense spikes, spikes 2–3.5 cm long, 0.3–0.8 cm wide, spikelets 30–80 per spike, narrowly ovoid to ovoid, 3–6.2 mm long, 1–1.5 mm wide, rachilla straight; glumes pale brown or golden with or without purplish streaks, ovate, 0.9–1.4 mm long, 0.8–1 mm wide, keel green, excurrent, apex usually shortly mucronate. Stamens 3; filaments 1.1–1.4 mm long; anthers 0.3–0.4 mm long. Nutlet reddish-brown, ellipsoid to obovoid, trigonous, 0.5–0.7 mm long, 0.4–0.5 mm wide, smooth or irregularly pitted.

Botswana. N: Ngamiland Dist., Toteng, 18.iii.1965, *Wild & Drummond* 7142 (K, SRGH). **Zambia**. B: Kalabo Dist., Barotse Plain, margins of R. Zambesi, 8.xii.1964, *Verboom* 1793 (K). N: Luapula Dist., Mbereshi–Luapula River swamp, 14.i.1960, *Richards* 12357 (K). C: Lusaka Dist., Kafue Bridge, 20.i.1964, *Van Rensburg* KBS2803 (K). S: Livingstone Dist., Victoria Falls, banks of Zambezi above falls, 4.i.1955, *Robinson* 1054 (K). **Zimbabwe**. N: Mazowe Dist., Henderson Research Station Fisheries, 15.iv.1973, *Gibbs-Russell* 2564 (K, SRGH). W: Hwange Dist., 16 km [11 miles] W of Binga, 8.xi.1958, *Phipps* 1405 (K, SRGH). S: Nuanetsi Dist., Gona-re-zhou, S bank of Lundi River just N of Chitove Gorge, 29.v.1971, *Grosvenor* 564 (K, SRGH). **Malawi**. N: Nkhata Bay Dist., Chinteche, Luweya River, 21.ii.1961, *Vesey-Fitzgerald* 3013 (K). S: Nsanje [Port Herald] Dist., James's Lagoon, 10 km [6 miles] N of Chiromo, 22.iii.1960, *Phipps* 2618 (K, SRGH). **Mozambique**. Z: Mopeia Dist., loc. Vicente, 27.xii.1969, *Amico & Bavazzano* s.n. (FI, K). T: Moatize Dist., Ilha Micune, 19.vi.1949, *Grandvaux et al.* 3175 (K). MS: Expedition Island, lower Zambezi, vi.1858, *Kirk* s.n. (K). GI: Xai-Xai Dist., city centre, 7.x.1978, *de Koning* 7275 (K).

Widespread in tropical west Africa, central Africa, Chad, Sudan, Ethiopia, Uganda, Tanzania and south to South Africa; S and SE Asia and South America; river and stream banks, swamps and marshes; sea-level to 1000 m.

Conservation notes: Least Concern due to its wide distribution.

C.B. Clarke distinguished plants with mucronate glumes as forma *mucronata*.

Gordon-Gray states that this species is relished by hippo which may explain why it is less common in eastern and south-central Africa where hippos are (or were) more common than in Natal.

Can be confused with small forms of *Cyperus alopecuroides* but the latter has much smaller spikelets.

E7. **Cyperus pubens** Kük. in Repert. Spec. Nov. Regni Veg. **29**: 200 (1931). —Haines & Lye, Sedges & Rushes E. Afr.: 203, fig.405 (1983). —Hoenselaar in F.T.E.A., Cyperaceae: 226 (2010). Type: Zimbabwe, Malangusti R., 19.xii.1907, *Kassner* 2060 (B holotype, K).

   Mariscus pubens (Kük.) Podlech in Mitt. Bot. Staatssaamml. München **4**: 115 (1961).

Perennial, fairly robust, up to 68 cm tall, with a slightly swollen stem-base emitting thick scale-covered stolons 3–5 mm in diameter; culms few, 31–62 cm long, 1.4–2 mm wide, trigonous, with short white densely set hairs below the inflorescence, almost glabrous near the base. Leaves up to 44 cm long; leaf sheath straw-coloured above, purplish-brown below, 3–11 cm long, densely white-hairy at least on the margins; leaf blade linear, flat, 24–33 cm long, 4.4–7.3 mm wide, densely white-hairy, especially beneath; apex acute. Involucral bracts leaf-like, spreading, 3–4, very sparely hairy except at the extreme base, lowermost 12–17 cm long, 3.2–4.1 mm wide. Inflorescence a simple anthela, primary branches 4–6, 3–8 cm long, bearing spikelets in the apical third; rachis of spike densely set with white hairs; spikelets spicately arranged, closely spaced, 7–25 per spike, linear-lanceolate, 7–12 mm long, 0.9–1.1 mm wide, falling off entirely when mature; glumes pale brown with darker margins, lanceolate, 4.7–6.6 mm long, 1.2–1.9 mm wide, keel flat, apex acute. Stamens 3; filaments 5.2–6.9 mm long; anthers 2.8–3.1 mm long. Nutlet grey, narrowly obovoid, trigonous, 3.2–3.4 mm long, 0.6–1 mm wide, minutely papillose.

Zambia. B: Kaoma Dist., 3 km NW along Kaoma-Lukulu road from junction of Kaoma-Mongu road, 1.iii.1996, *Zimba et al.* 711 (K, MO). N: Kaputa Dist., Kundabwika Falls on the Kalungwishi River, 10.xii.2002, *Bingham* 12627 (K). W: Mwinilunga Dist., just W of R. Dobeka, S of Dobeka Bridge, 17.xi.1937, *Milne-Redhead* 3294 (K). C: Kabwe Dist., Kamaila Forest Station, 36 km N of Lusaka, 9.ii.1975, *Brummitt* 14297 (K). E: Lundazi Dist., 3 km [2 miles] E of Lundazi Boma on Mzimba road, 27.i.1961, *Feely* 103 (K, SRGH). S: Mazabuka Dist., Nachibanga stream, Choma to Pemba road Km 40 [mile 26], 9.ii.1960, *White* 6896 (FHO, K). **Zimbabwe**. N: Mwami Dist., K34 Experimental Farm, Mfuti area, 6.iii.1947, *Wild* 1689 (K, SRGH). C: Goromonzi Dist., Ruwa Tanglewood Farm, i.1961, *Miller* 8137 (K, SRGH). **Malawi**. N: Mzimba Dist., Kasitu Valley, 27.i.1938, *Fenner* 222 (K).

Also in Tanzania. In *Brachystegia* woodland, on sandy soil; 1000–1700 m.

Conservation notes: Least Concern because of its wide distribution.

F.T.E.A. note: Easily recognized due to its hairy leaf sheaths and rachis.

E8. **Cyperus tomaiophyllus** K. Schum. in Engler, Pflanzenw. Ost-Afrikas **C**: 122 (1895). —Clarke in F.T.A. **8**: 392 (1902). —Kükenthal in Engler, Pflanzenr. **4**, 20(101): 429 (1936). —Haines & Lye, Sedges & Rushes E. Afr.: 207, fig.414 (1983). —Hoenselaar in F.T.E.A., Cyperaceae: 239 (2010). Type: Tanzania, Kilimanjaro, Rua Stream, *Meyer* 272 (B lectotype).

   Cyperus alpestris K. Schum. in Engler, Pflanzenw. Ost-Afrikas **C**: 122 (1895). Types: Tanzania, Kilimanjaro, Mawenzi, Ruassi stream, *Volkens* 872 (B syntype); Tanzania, Moshi Dist., Useri, *Haarer* 1702 (B syntype).
   Mariscus tomaiophyllus (K. Schum.) C.B. Clarke in F.T.A. **8**: 392 (1902).
   Mariscus alpestris (K. Schum.) C.B. Clarke in F.T.A. **8**: 401 (1902).
   Mariscus magnus C.B. Clarke in Bot. Jahrb. Syst. **38**: 134 (1906). Type: Tanzania, Lushoto Dist., Usambara, Kwai, *Albers* 145 (B holotype).
   Cyperus tomaiophyllus var. *magnus* (C.B. Clarke) Kük. in Engler, Pflanzenr. **4**, 20(101): 429 (1936).
   Cyperus tomaiophyllus var. *alpestris* (K. Schum.) Kük. in Engler, Pflanzenr. **4**, 20(101): 430 (1936).

Perennial, very robust, up to 150 cm tall, with a branching, scale covered woody rhizome up to 2 cm in diameter; culms few, the base covered with brown scales and old leaf bases split by the new culms arising in their axil, culms 45–140 cm long, 4–8 mm wide, trigonous, glabrous. Leaves up to 100 cm long; leaf sheath brown, 5–14 cm long; leaf blade linear, flat or v-shaped, 35–90 cm long, 6–13 mm wide, scabrid on margin and primary vein, apex acuminate. Involucral bracts leaf-like, erect to spreading, 6–12, lowermost 26–50 cm long, 9–19 mm wide. Inflorescence spicate-anthelate, usually simply so, but the primary branches sometimes themselves branched, primary branches 7–15, 1.5–9 cm long; spikelets in long, crowded clusters, sessile and at the end of primary branches, many per cluster, linear-oblong, 5.6–13.5 mm long, 1.2–2.5 mm wide, rachilla straight, falling off entire when mature; glumes pale brownish with uncoloured margin, oblong-lanceolate, 4–6.7 mm long, 1.8–2.1 mm wide, keel with 5–7 slender veins on each side, apex acute. Stamens 3; filaments 5.2–7 mm long; anthers 2.8–3.4 mm long. Nutlet brown, oblong, trigonous, 2.2–3.5 mm long, 0.6–0.7 mm wide, minutely papillose.

Malawi. N: Rumphi Dist., Nyika Plateau, Chelinda Camp, 3.ii.1978, *Pawek* 13736 (K, MAL, MO, SRGH); Rumphi Dist., Nyika Plateau, Lake Kaulime, 24.x.1958, *Robson* 328 (K).

Also in Nigeria, Cameroon, Ethiopia, D.R. Congo, Rwanda, Uganda, Kenya and Tanzania. Montane bogs, wet places at forest edges, lake edges; 2150–2350 m.

Conservation Notes: Least Concern on a world scale but in the flora area known only from the Malawi portion of the Nyika Plateau, whence there are several collections at K, of which two are cited above (one for the inflorescence, the other for the rhizome).

None of the F.Z. specimens at K has leaves or involucral bracts as wide as in the F.T.E.A. description but as there are so few F.Z. collections at K, I retain the dimensions given in F.T.E.A.

E9. **Cyperus luteus** Boeckeler in Linnaea **38**: 370 (1874). —Kükenthal in Engler, Pflanzenr. **4**, 20(101): 414 (1936). —Haines & Lye, Sedges & Rushes E. Afr.: 203, figs.406,407 (1983). —Hoenselaar in F.T.E.A., Cyperaceae: 171 (2010). Type: Madagascar, Nosy Be [Nossi-Bé], 29.i.1841, *Pervillé* 516 (P holotype, K, P).

> *Mariscus luteus* (Boeckeler) C.B. Clarke in Durand & Schinz, Consp. Fl. Afric. **5**: 589 (1894).
> *Mariscus foliosus* C.B. Clarke in F.T.A. **8**: 399 (1902). Types: Uganda, Ruwenzori, *Scott Elliot* 7674 (K syntype); Kenya, Kilifi Dist., Rabai Hills, *Taylor* s.n. (BM syntype); Malawi, Mt Zomba, *Whyte* s.n. (K syntype); Malawi, Nyika Plateau, *Whyte* s.n. (K syntype).

Robust perennial with a swollen stem base and a short creeping rhizome, up to 100 cm tall; culms few, 30–80 cm long, 1.6–4.4 mm wide, trigonous, glabrous. Leaf sheath greenish or pale purple above, dark purple below, 3.5–11 cm long; leaf blade linear, flat, 21–49 cm long, 2.5–7.3 mm wide, scabrid on margin and primary vein, apex acute to acuminate, trigonous, scrabrid. Involucral bracts leaf-like, erect to spreading, 5–9, lowermost 13–46 cm long, 3.7–7 mm wide. Inflorescence a simple anthela, primary branches 4–8, 6–15 cm long, spikelets single, spicately arranged along the terminal third to quarter of the primary branches, 15 to many per branch, very narrowly ovate to linear, 11–13 mm long, 1–1.4 mm wide, falling off entire when mature, leaving the lowest bract attached to the short spikelet stalk; glumes green with or without a golden or reddish-brown tinge, lanceolate-elliptic, 4.4–6 mm long, 1.3–1.5 mm wide, keel with several veins on either side, apex acute to acuminate. Stamens 3; filaments 4.5–5.6 mm long; anthers 1–2.9 mm long. Nutlet grey to black, linear-ellipsoid, trigonous, 2.5–3.7 mm long, 0.6–0.9 mm wide, rather smooth to minutely papillose.

Malawi. N: Rumphi Dist., Nyika Plateau, SW side, 7.i.1959, *Robinson* 3100 (K). S: Zomba Dist., Zomba Mountain, Chawe Plateau, 2.iv.1934, *Lawrence* 161 (K).

Also in Cameroon, D.R. Congo, Rwanda, Uganda, Kenya, Tanzania, and Madagascar. Mainly at upland forest margins; 1200–2000 m.

Conservation notes: Least Concern because of its wide distribution.

For the present I am restricting this name to plants from upland forest margins with large spicate anthelae. Plants with much more condensed inflorescences (e.g. *Abel Saimon* 40 (K) from Zambia. N: Mbala, 6.xi.1950) have been placed here both in F.T.E.A. and in the flora folders at Kew. The spikelets are smaller than in F.T.E.A. material, as are the nutlets.

E10. **Cyperus trigonellus** Suess. in Trans. Rhodesia Sci. Assoc. **43**: 80 (1951). Type: Zimbabwe, Marandellas, 8.xi.1943, *Dehn* 803 (M holotype).

> *Mariscus psilostachys* C.B. Clarke in J. Bot. **34**: 225 (1896); in F.T.A. **8**: 384 (1901). Type: Kenya, Njoro, *Gregory* s.n. (K holotype).
> *Cyperus psilostachys* (C.B. Clarke) Kük. in Bot. Not. **1934**: 69 (1934); in Engler, Pflanzenr. **4**: 20(101): 546 (1936). —Haines & Lye, Sedges & Rushes E. Afr.: 211, fig.423 (1983), illegitimate name, non Steud. (1854).
> *Cyperus psilostachys* var. *pluribracteatus* Kük. in Engler, Pflanzenr. **4**, 20(101): 546 (1936). Type: Tanzania, Dodoma Dist., Saranda, 24.xii.1925, *Peter* 33483 (B syntype); Tanzania, Makutupora, 27.xii.1925, *Peter* 33699a (B syntype).
> *Cyperus psilostachys* var. *subrufus* Kük. in Engler, Pflanzenr. **4**, 20(101): 546 (1936). Types: Tanzania, between the coast and Uyui, *Taylor* s.n. (B syntype); Tanzania Lushoto Dist., Mt Gomba at Makuyuni, *Peter* 15366, 15486 (B syntypes).
> *Cyperus pluribracteatus* (Kük.) Govaerts in Govaerts & Simpson, World Checkl. Cyperaceae: 350 (2007). —Hoenselaar in F.T.E.A., Cyperaceae: 223 (2010).

Perennial, robust, up to 75 cm tall, with swollen fleshy culm-bases 0.8–1.3 cm in diameter connected by a thin rhizome up to 10 cm long and 1–2 mm in diameter, covered by reddish brown scales to 12 mm long when yoiung; culms densely crowded in groups of 2–20, rarely solitary at the end of a stolon, 35–74 cm long, 1–3.1 mm wide, trigonous, hairy at least above. Leaves up to 40 cm long; leaf sheath pale brown to greyish, fleshy, 3–10.5 cm long; leaf blade linear, flat, 14–32 cm long, 2.5–5.5 mm wide, hairy, apex acuminate. Involucral bracts leaf-like, erect to spreading, hairy, 2–3, lowermost 2.5–10 cm long, 2–5 mm wide. Inflorescence a simple anthela, often without any sessile spikelets, primary branches 3–8, 1.2–9.3 cm long; spikes with 20 to many spikelets, sessile and at the end of primary branches, 1.3–2.5 cm long, 1.2–1.6 cm wide; spikelets linear-lanceolate, 5.9–9 mm long, 0.7–1.1 mm wide, falling off entirely when matured, rachis straight; glumes pale reddish brown with paler margins, narrowly ovate, 2.9–3.8 mm long, 1–1.3 mm wide, densely hairy, keel green, apex somewhat recurved. Stamens 3; filaments 3–4.5 mm long; anthers 1.5–2.1 mm long. Style branches 3. Nutlet reddish-brown, oblong-obovoid, trigonous, ± 1.8 mm long, ± 0.6 mm wide, minutely papillose.

Zambia. C: Mkushi Dist., Kalomo, 1.i.1958, *Robinson* 2549 (K, SRGH). S: Mapanza Dist., Muckle Neuk, 16 km [12 miles] N of Choma, 28.xi.1954, *Robinson* 993 (K). **Zimbabwe**. N: Harare Dist., Hills above Mazoe Dam, Harare [Salisbury], 24.xii.1956, *Robinson* 1981 (K). C: Harare [Salisbury] Dist., Prince Edward Dam, 20.xii.1931, *Brain* 7695 (K, SRGH). E: Mutare Dist., 1.2 km beyond Stapleford turn on Penhalonga-Watsomba road, 19.xii.1994, *Wilkin* 723 (K). **Malawi**. N: Rumphi Dist., Njakwa Gorge, 3km [2 miles] S of Rumphi, 30.xii.1973, *Pawek* 7666 (K, MAL, MO, SRGH).

Also in Uganda, Kenya, Tanzania, Rwanda, and Burundi; usually on rocks, in cracks on in leaf litter; occasionally in *Brachystegia* woodland at dambo margins or by termite mounds, in dryish grassland and on rocky outcrops; 1000–1600 m.

Conservation notes: Least Concern because of its wide distribution.

Simpson and Govaerts made a new name for this taxon because the name *Cyperus psilostachys* is a later homonym. It would appear, however, that they overlooked *C. trigonellus* of which I have seen images of the type.

E11. **Cyperus rigidifolius** Steud. in Flora **25**: 593 (1842). —Clarke in F.T.A. **8**: 367 (1901). —Kükenthal in Engler, Pflanzenr. **4**, 20(101): 104 (1935). —Hedberg, Afroalpine Vasc. Pl.: 55 (1957). —Haines & Lye, Sedges & Rushes E. Afr.: 185, figs.359,360 (1983). —Gordon-Gray in Strelitzia **2**: 67 (1995). —Hoenselaar in F.T.E.A., Cyperaceae: 165. Type: Ethiopia, Enchadcap, *Schimper* 991 (P holotype, HAL, K, MUN, P, UPS, WAG).

    Cyperus adoensis A. Rich., Tent. Fl. Abyss. **2**: 484 (1850). Type: Ethiopia, Adua, *Schimper* I:186 (P holotype, BM, K, S).

    Cyperus longus L. var. *adoensis* (A. Rich.) Boeckeler in Linnaea **36**: 281 (1870).

    Cyperus rigidifolius var. *intercedens* Kük. in Notizbl. Bot. Gart. Berlin-Dahlem **9**: 304 (1925); in Engler, Pflanzenr. **4**, 20(101): 75 (1935). Types: Kenya, Nyeri, *Fries & Fries* 101 (UPS); Tanzania, Kilimanjaro, Moshi, *Merker* 401 (?B); Tanzania, Ngaruka, *Merker* 407 (B syntype).

Perennial, slender to fairly robust, up to 72 cm tall, with a woody base and curving thin horizontal rhizomes up to 15 cm long, 1.5–3 mm in diameter; culms few, 15–70 cm long, (0.7)2–3(4.4) mm in diameter, trigonous, sometimes slightly winged on the angles, glabrous. Leaves up to 35 cm long; leaf sheath green to pale brown, 3–10 cm long; leaf blade linear, flat, rather stiff, 7–30 cm long, (1.5)2–3.5(5) mm wide, scabrid on margin and primary vein, apex acuminate. Involucral bracts leaf-like, erect to spreading, 3–5, lowermost 4.5–22 cm long, 2–4.8 mm wide. Inflorescence capitate or a simple and compact anthela, primary branches 0–6, 0–11 cm long; spikelets in crowded digitate spikes, erect, sessile and at the end of primary branches, 5–10 spikelets per spike, lanceolate, somewhat compressed, 7–18 mm long, 2–2.8 mm wide, rachilla straight; glumes dark reddish-brown to almost black, ovate, 2.7–4 mm long, 1.8–2.2 mm wide, keel green, apex rounded. Stamens 3; filaments 3.4–4.2 mm long; anthers 1.5–2.7 mm long. Nutlet greyish brown to olive green, obovoid, trigonous, 1.6–1.9 mm long, 0.9–1.1 mm wide, with minute isodiametric surface cells.

Zambia. N: Mbala Dist., Left-Hand side Pans, Mbala, 11.ii.1957, *Richards* 8161 (K).
Zimbabwe. E: Chimanimani Dist., Gairesi Ranch, on PEA border 10 km [6 miles] N of
Troutbeck, 18.xi.1956, *Robinson* 1942 (K). **Malawi**. N: Rumphi Dist., Malawi National
Park, Nyika Plateau, Chosi Hill, 12.iv.1969, *Pawek* 2133 (K).

D.R. Congo, Rwanda, Burundi, Ethiopia, Uganda, Kenya, Tanzania, South Africa,
Swaziland. In upland seasonally wet grassland and swamps; 1500–1750(2800) m.

Conservation notes: Least Concern because of its wide distribution.

This is a species with a somewhat montane distribution and is much commoner in
the F.T.E.A. area.

The following eight taxa form a complex and confusing group. Hoenselaar and
Beentje, in F.T.E.A., took a very much broader view, and, included within *Cyperus
cyperoides* the taxa here treated separately as *C. sublimis*, *C. corynephorus*, *C. pseudoflavus*,
C. macrocarpus and *C. decurvatus*. These taxa mainly occur in natural or semi-natural
habitats and seem to have distinct habitat preferences. It is possible that the disruption
of these natural habitats may have brought normally well separated taxa together
in sites such as roadsides, a hypothesis put forward by R.M. Polhill to explain the
taxonomic difficulties in *Chamaecrista mimosoides* (Leguminosae, Caesalpinioideae),
and hybridisation is possible although one hesitates to use this as a catch-all excuse for
difficulties in distinguishing closely related taxa. The species were well described and
illustrated (as *Mariscus*) in P. Vorster's thesis, which was unfortunately never published,
as a letter from Vorster to Linder inside the Kew copy makes clear. However, most of
Vorster's new taxa have been published subsequently elsewhere.

E12. **Cyperus sublimis** (C.B. Clarke) Dandy in Exell, Cat. Vasc. Pl. S.Tomé: 362
(1944), in part, excluding *Schweinfurth* 1508. Type: Sudan, Jur Ghattas, 2.vi.1869,
Schweinfurth 1842 (K lectotype, B, PRE, S), lectotypified by Vorster.
  Mariscus sublimis C.B. Clarke in F.T.A. **8**: 390 (1902).
  Cyperus subumbellatus Kük. var. *sublimis* (C.B. Clarke) Kük. In Pflanzenr. **4**, 20(101): 525
(1936).

Perennial from a corm-like base 8–12 mm in diameter, forming a new corm laterally after
flowering so that a chain of corms is eventually formed. Leaves basal, up to 25 cm long and
3.5–5 mm broad, tapering gradually to the apex, green but purplish towards the base, glabrous
but minutely scabrid on the margins and midrib beneath. Culms 23–77 cm long and 1.2–2.5
mm in diameter, weakly trigonous, glabrous. Inflorescence bracts 6–11, up to 22 cm long and
3.0–4.5 mm broad, similar to the foliage leaves. Inflorescence a simple umbel of 7–11 rays,
the secondary peduncles up to 10 cm long. Spikes 10–27 mm long and 6 mm wide, cylindrical
with a rounded apex. Spikelets 2.1–3.6 mm long and 0.7–1.0 mm wide, 1-flowered, spreading,
narrowly ovate, acute, trigonous to more-or-less terete. Glumes 2.3–3.7 mm long, third glume
reaching more-or-less to apex of spikelet, concave, ovate when flattened, with a single median
dark green non-excurrent vein and 10–12 lateral veins. Rhachilla widely winged. Stamens 3.
Style 3-branched. Nutlets 1.5–1.9 mm long, 0.6–0.9 mm broad, obovoid to ellipsoid, obscurely
triquetrous, papillose

Zambia. N: Mporokoso Dist., E of Lake Mweru, vi.1961, *Chongo* 15 (K). W:
Mwinilunga Dist., 1 km (0.5 mile) S of Matonchi Farm, 6.xii.1937, *Milne-Redhead*
3514 (K). C: Mkushi Dist., junction of Mtuza and Great North Roads, 13.i.1958,
Robinson 2744 (K, SRGH). E: Chipata Dist., Chipata [Fort Jameson], ii.1954, *Grout*
122 (K, SRGH). S: Mapanza Dist., Simansunda, 2 km E of Mapanza, 5.i.1954, *Robinson*
429 (K). **Zimbabwe**. N: Mazowe Dist., Mazoe, N of Jumbo, 13.i.1958, *Phipps* 847 (K,
SRGH). W: Matobo Dist., Matobo, Farm Chesterfield, xii.1957, *Miller* 4815 (K, SRGH).
C: Ruwa Dist., Tanglewood Farm, xii.1960, *Miller* 7549 (K, SRGH). E: Mutare Dist.,
Nyamkwarara Valley, naturelands, ii.1935, *Gilliland* K1502 (K). **Malawi**. N: Mzimba
Dist., Mzuzu, Marymount, 25.i.1974, *Pawek* 2978 (K, MA, MO, SRGH). C: Dedza Dist.,

Mua-Livulezi Forest Reserve, 6.i.1965, *Banda* 606 (K, SRGH). S: Blantyre Dist., Maperi (?Maperere Mission), 5 km from Blantyre, 17.i.1938, *Lawrence* 521 (K). **Mozambique**. N: Nampula, Dist. Mossuril, Reserva Florestal de Necrusse de Matibane, 17.ii.1984, Groenendijk et al 1169 (LMU, K). Z: Mocuba Dist., Namagoa, *Faulkner* G29 (K).

Also in Sudan, Angola, Tanzania and South Africa. Often recorded from *Brachystegia* and other woodlands, but also river banks, old fields, and cultivated ground; 900–1500 m.

Conservation notes: Least Concern because of its wide distribution.

Vorster describes the nutlets as covered with flat-topped papillae. Although, to me, the surface is papillose at the cellular level at x40, I cannot make out the supposed 'flat-topped' papillae; I suspect that SEM images are needed to make out this feature.

E13. **Cyperus corynephorus** Lock, new name (see note below).

> *Mariscus cylindristachyus* Steud., Syn. Pl. Glumac. **2**: 65 (1855). —Raynal in Adansonia, n.s. **17**: 274 (1978). Type: Gabon [Côtes de Guinée], without locality, 1846, *Jardin* s.n. (P holotype, P).

Perennial, forming small tufts, stem base somewhat swollen, 6–7 mm in diameter, often attached laterally to the bases of former shoots. Basal sheaths purplish. Leaves 15–35 cm long, 3–6 mm wide, green, glabrous but sometimes minutely scabrid on the midrib, margins and principal veins beneath. Culms up to 41(60) cm long, 1.2–2.5 mm in diameter, trigonous, glabrous. Inflorescence bracts 7–9, the lowest up to 31 cm long and 3–4 mm wide at the base, glabrous but minutely scabrid on the margins. Inflorescence once anthelate-spicate with 8–15 rays. Secondary peduncles up to 5 cm long and 0.5 mm in diameter. Spikes at the apex of the ray peduncles, 8–15(20) × 4–6 mm in diameter, cylindrical. Spikelets 2.0–3.2 × 0.5–0.8 mm, densely crowded, spreading, narrowly ovoid, acute at base and apex, terete to somewhat trigonous. Third glume 2.0–2.6 mm long, reaching more or less to the apex of the spikelet, concave, ovate when flattened, acute, with a single central non-excurrent vein and 6(8) secondary longitudinal veins. Rhachilla 1.3–1.5 mm long, flattened. Stamens 3. Style 3-branched. Nutlet 1.4–1.8 × 0.5–0.8 mm, ellipsoid, straight, trigonous, papillose.

Zambia. N: Mbala Dist., Mbala [Abercorn], lake shore, 4.iii.1950, *Bullock* 2604 (K). W: Mwinilunga Dist., by Mwinilunga road, between Kanyama turn-off and Samuteba, 25.ii.1975, *Hooper & Townsend* 404 (K). C: Mkushi Dist., Fiwila, 7.i.1958, *Robinson* 2665 (K). S: Mazabuka Dist., 40 km [25 miles] N of Pemba, near Kanchale Village, 11.ii.1960, *White* 6949 (FHO, K). **Zimbabwe**. E: Chimanimani [Melsetter] Dist., N bank Lushitu River, c. 2 km W of Haroni confluence, 12.i.1969, *Biegel* 2808A (see note). **Malawi**. N: Nkhata Bay Dist., 5 km [3 miles] S of junction, Nkhata Bay Secondary School, 31.iii.1974, *Pawek* 8274 (K, MO). S: Zomba Dist., near Salisbury Bridge, 5.iv.1984, *Banda & Salubeni* 2139 (K, MAL).

Widespread in tropical Africa from Liberia to Uganda and Tanzania, Gabon, D.R. Congoand Madagascar. Forest margins, woodland on sandy soil, *Brachystegia boehmii* woodland on rocky slope, lake shores, roadsides; 400–1650 m.

Conservation notes: Least Concern because of its wide distribution.

Vorster notes that the duplicate of *Biegel* 2808 at SRGH is *Mariscus sublimis*.

Cyperus cylindrostachys Boeckeler in Linnaea **36**: 383 (1870) refers to a different taxon. While this is not identical in spelling to *C. cylindristachys* I feel that the similarity is so great that confusion is very likely and that therefore it is right to propose a new name for this taxon. 'Corynephorus' (club-bearing) refers to the shape of the inflorescence subunits.

As for *Cyperus sublimis*, I have problems with the supposed 'flat-topped' papillae on the nutlet, reported by Vorster. The inflorescence subunits are usually shorter that in *C. corynephorus*. The two species may occur together and there are mixed collections. As for *C. sublimis*, many of the collections are from natural of semi-natural habitats.

E14. **Cyperus hemisphaericus** Boeckeler in Flora **42**: 436 (1859). —Kükenthal in
Engler, Pflanzenr. **4**, 20(101): 406 (1936). —Haines & Lye, Sedges & Rushes E.
Afr.: 206, fig.413 (1983). —Hoenselaar in F.T.E.A., Cyperaceae: 173 (2010). Type:
Mozambique, Tete, *Peters* s.n. (B holotype).

　　Mariscus hemisphaericus (Boeckeler) C.B. Clarke in Durand & Schinz, Consp. Fl. Afric. **5**:
589 (1894); in F.T.A. **8**: 400 (1902).

　　Mariscus gregorii C.B. Clarke in J. Bot. **34**: 225 (1896); in F.T.A. **8**: 401 (1902). Type: Kenya,
River Tana, Kiroruma, 15.vii.1893, *Gregory* 93 (K holotype, BM).

　　Cyperus hemisphaericus var. *gregorii* (C.B. Clarke) Kük. in Engler, Pflanzenr. **4**, 20(101): 407
(1936)

　　Cyperus hemisphaericus var. *longibracteus* Kük. in Engler, Pflanzenr. **4**, 20(101): 407 (1936).
Types: Tanzania, Tanga Dist.: East Usambaras, *Peter* 39855 (B syntype, K); Tanzania Useguha,
Mnyussi, *Peter* 10329 (B syntype, K); Tanzania, Uzaramo Dist., Dar es Salaam, *Holtz* 601, *Peter*
39341, 39389 & 39440 (B syntypes); Mozambique, between Mapinga and Kondutschi, *Peter*
14816 (B syntype); Mozambique, Beira, *Peter* 31116 (B syntype).

Robust tussocky perennial herb to 130 cm tall; culms 15–115 cm long, 2.2–9 mm wide,
trigonous, sometimes almost rounded, with longitudinal grooves, glabrous, arising at short
intervals from short horizontal creeping rhizomes. Leaves numerous, crowded at the base,
up to 1.3 m long; leaf sheath dark purple at base, (pale) brown higher up, 3.5–10 cm long;
leaf blade linear, flat or folded, 15–125 cm long, 6–13 mm wide, scabrid on primary vein and
margin, apex acuminate. Involucral bracts leaf-like, spreading, 6–10, lowermost 13–56 cm long,
4.5–12 mm wide. Inflorescence a simple anthela, sometimes very congested to almost capitate,
primary branches (0)2–8, (0)1–9 cm long; spikelets sessile and at the end of primary branches,
12 to many per spike, linear-lanceolate, 9–16 mm long, 1.3–2.2 mm wide, falling off entire
when mature; glumes striate, yellowish-white with a paler smooth membranous margin, ovate-
lanceolate, 3.8–6 mm long, 1.6–2.4 mm wide, keel flat with many veins on either side, apex acute.
Stamens 2; filaments 4.6–6.2 mm long; anthers 2.5–2.7 mm long. Nutlet dark reddish-brown to
black, narrowly oblong to obovoid, trigonous, 2.4–2.7 mm long, 0.6 mm wide, minutely papillose
in longitudinal rows.

　　Zimbabwe. W: Bulawayo Dist., 1930, *Cheesman* 18 (BM). C: Harare Dist., [Salisbury],
30.xii.1924, *Eyles* 4442 (K, SRGH). E: Chimanimani [Melsetter] Dist., Tarka Forst
Reserve, 27.xi.1967, *Simon & Ngoni* 1360 (K, SRGH). S: Ndanga Dist., N of Chipundu
pools, on hills, 17.i.1960, *Goodier* 817 (K, SRGH). **Malawi**. N: Nkhata Bay Dist., 21 km
SW of Nkhata Bay on road to Chinteche, 11.v.1970, *Brummitt* 10624 (K). C: Dedza
Dist., Chiwao Hill slopes, 12.i.1967, *Jeke* 51 (K, SRGH). S: Blantyre Dist., Upper Hynde
Dam, 2 km W of Limbe, 14.ii.1970, *Brummitt* 8570 (K). **Mozambique**. N: Marrupa Dist.,
Jansen & Boane 7811, 18.ii.1982 (K). Z: Zambezia Province, Chiperone Mt., 3.xii.2006,
Harris et al. 121 (K). T: Tete, *Peters* s.n. (B). MS: Dondo Dist., 40 km N of Dondo, on
road to Inhaminga, 3.xii.1971, *Pope & Muller* 502 (K, SRGH).

　　Also in Somalia, Kenya and Tanzania. In open grassland, *Brachystegia* woodland and
wooded grassland; sea-level up to 1850 m.

　　Conservation notes: Least Concern because of its wide distribution.

　　There are no records from Zambia; is this real?

　　The pagination for the protologue appears to be a typographic error; page 436 was
used three times in volume 42 of Flora and the third is the relevant one.

E15. **Cyperus chersinus** (N.E. Br.) Kük. in Engler, Pflanzenr. **4**, 20(101): 525 (1936).
Types: Botswana, Kwebe Hills, 1.ii.1898, *Lugard* 142 (GRA syntype, K); Botswana,
Mochudi, 1.iii.1914, *Rogers* 6310 (BOL syntype, K).

　　Mariscus chersinus N.E. Br. in Bull. Misc. Inform. Kew **1921**: 300 (1921). —Gordon-
Gray in Strelitzia **2**: 128 (1995).

　　Cyperus capensis (Steud.) Endl. var. *polyanthemus* Kük. in Engler, Pflanzenr. **4**, 20(101):
540 (1936). Type: Mozambique, Maputo, Ressano Garcia, 9.xii.1897, *Schlechter* 11683 (G
syntype, K).

Clump-forming perennial from a narrowly bulbous base 5–12 cm long and 0.6–1.4 cm wide when dry, enclosed by thin pale brown leaf sheaths; rhizomes absent. Leaves mainly basal, linear, 25–45 cm long, 3.7–5.9 mm wide in the middle, glabrous but minutely scabrid on margins and midrib and on the triquetrous tip. Inflorescence bracts 7–10, to 29 cm long and 3.5–4.2 mm wide; similar to the foliage leaves. Inflorescence simply spicate-anthelate with 7–11 primary branches; peduncle 21–65 cm long and 1.3mm in diameter, weakly trigonous. Primary branches up to 9 cm long but usually less. Spikes cylindrical, 17–25 mm long and 5–6 mm wide. Spikelets 1(2)-flowered, 2.6–3.5 × 0.8–1.9 mm, densely crowded, spreading to slightly ascending. Glumes ovate, yellowish or pale brown; midrib not excurrent, dark green; rachilla flattened and widely winged. Stamens 3; filaments flattened; anthers 1–1.3 × 0.1 mm. Style 3-branched. Nutlet 1.6–2 × 0.8–1.2 mm, ellipsoid to obovoid, trigonous, papillose.

Botswana. N: Ngamiland Dist., near the Selinda-Kwando road, 10.1 km N of the Selinda Bridge, 22.i.1978, *P.A. Smith* 2206 (K, SRGH). SW: Kgaligadi/Ghanzi Dist., 80 km [50 miles] N of Kang, 18.ii.1960, *Wild* 5071 (K, SRGH). SW: Kgaligadi Dist., Central Kalahari Game Reserve, W part, 91 km E of Lonetree in Okwa Valley, 12.i.1977, *Verhagen & Barnard* 37 (K, PRE). **Zimbabwe**. E: Chipinge Dist., Lower Sabi River, 27.i.1948, *Rattray* 1271 (K, SRGH). **Mozambique**. M: Maputo, Costa do Sol, 26.ii.1985, *Groenendijk* 1609 (K, LMU).

Also in South Africa. Dry *Acacia* or *Colophospermum* or *Terminalia* woodland on deep Kalahari sand; 5–1000 m.

Conservation notes: Least Concern because of its wide distribution.

Within its group, distinct in its long bulbous bases covered by pale brown sheaths. This species is extremely similar to *Cyperus phillipsiae* (C.B. Clarke) Kük. from Somalia and coastal N Kenya. The latter would be the older name, based on *Mariscus phillipsiae* C.B. Clarke in F.T.A. **8**: 391 (1902).

E16. **Cyperus pseudoflavus** (Kük.) Lock comb. & stat. nov. Type: Tanzania, Kilimanjaro, x.1884, *Johnston* 393 (B lectotype), lectotypified here.

   Mariscus macer Kunth, Enum. Pl. **2**: 121 (1837). —Clarke in Durand & Schinz, Consp. Fl. Afr. **5**: 589 (1894); in Fl. Cap. **7**: 190 (1897); in F.T.A. **8**: 329 (1901), for the most part. —Podlech in Mitt. Bot. Staatssamml. München **4**: 113 (1961). Type: South Africa, "Cap b. spei.", *Drège* 4449 (B holotype).
   Mariscus pseudoflavus C.B. Clarke in Durand & Schinz, Consp. Fl. Afr. **5**: 591 (1894), invalid name; in Ill. Cyperaceae, tab.23 figs.8,9 (1909), nomen.
   Cyperus macer (Kunth) K. Schum. in Engler, Pflanzenw. Ost-Afrikas **C**: 122 (1895), illegitimate name, non C.B. Clarke (1884).
   Cyperus pseudoflavus (C.B. Clarke) K. Schum. in Engler, Pflanzenw. Ost-Afrikas **C**: 123 (1895), invalid name.
   Cyperus macrocarpus (Kunth) Boeckeler var. *pseudoflavus* Kük. in Engler, Pflanzenr. **4**, 20(101): 529 (1936).

Perennial forming small clumps, or shoots single. Base swollen, 4–9 mm in diameter, covered by brown and purple sheaths. Basal leaves 1.7–4.4 mm wide in the middle, tapering gradually to the apex, glabrous except for minute teeth on the margins and on the midrib beneath; sheaths brown to purple. Inflorescence bracts 4–9, up to 31 cm long and 1.7–4.5 mm wide, spreading. Inflorescence once anthelate-spicate, contracted, or 3–9 rays; culm 11–43 cm long, trigonous, glabrous; secondary peduncles short or absent. Spikes 7–15 long and 5–7 mm wide, simple, cylindrical. Spikelets 1(3)-flowered, crowded, ascending, ellipsoid to narrowly so, acute; 1-flowered spikelets 2.3–3 × 0.7–1 mm; 3-flowered up to 4.7 mm long. Glumes 2.3–2.9 mm long, pale green, concave, ovate, median nerve not excurrent, broad, dark green, 6–8 additional longitudinal veins. Stamens 3; anthers 0.5–0.9 mm long. Style 3-branched. Nutlets ovoid, 1.5–2 × 0.7–0.9 mm, reddish brown.

Botswana. N: Central Dist., Near main Maun-Francistown road, 63 km from Francistown, 6.ii.1980, *P.A. Smith* 3023 (K, SRGH). **Zimbabwe**. N: Kariba Dist., Mwami

[Miami], K34 Experimental Farm, 4.iii.1947, *Wild* 1775 (K, SRGH). W: Matobo Dist., Farm Besna Kobila, i.1958, *Miller* 4975 (K, SRGH).

Also in Uganda, Kenya, Tanzania, D.R. Congo and South Africa (fide Vorster). Dry grasslands; by termite mound, between rocks, in crack in rock; 1350–1500 m.

Conservation notes: Least Concern because of its wide distribution.

Similar to *Cyperus chersinus* in its swollen base, but in *C. pseudoflavus* the base is small, rounded and hard, while in *C. chersinus* the base is much longer and more bulb-like, and covered with long delicate pale sheaths. The inflorescence of *C. pseudoflavus* is smaller and more compact.

E17. **Cyperus nyasensis** (Podlech) Lye in Haines & Lye, Sedges & Rushes E. Afr.: 211, fig.424 (1983), as *nyassensis*. —Hoenselaar in F.T.E.A., Cyperaceae: 155 (2010), as '*nyassensis*'. Type: Malawi, Nyika Plateau, W side, 6.i.1959, *Robinson* 3083 (M holotype, K, SRGH).

> *Mariscus nyasensis* Podlech in Mitt. Bot. Staatssamml. München **4**: 114 (1961).

Perennial, tussocky, up to 36 cm tall, with a succulent culm base; culms tufted, 14–34 cm long, 1–1.3 mm wide, trigonous to almost terete, hairy, sometimes only in the upper part. Leaves up to 23.5 cm long; leaf sheath greyish to pale brown, 3.5–8 cm long, hairy, covering the culm base; leaf blade linear, 8–20 cm long, 1.3–3 mm wide, hairy on lower surface, apex acute, hairy. Involucral bracts leaf-like, spreading to reflexed, 3–4, lowermost 2.5–12 cm long, 1.3–2 mm wide, hairy on lower surface. Inflorescence capitate; spikelets in dense spikes, spikes sessile, 3–6 per head, 20–44 spikelets per spike, spikelets linear-lanceolate, 2.2–3.1 mm long, 0.5–9 mm wide, falling off entirely when mature; glumes golden to reddish-brown, lanceolate-obovate, 2.3–2.9 mm long, 1–1.2 mm wide, hairy to almost glabrous, keel slightly excurrent, apex slightly mucronate. Stamens 3; filaments 2.5–3.1 mm long. Nutlet grey to reddish-brown to blackish, oblong to ellipsoid, 1.6–2 mm long, 0.6–0.7 mm wide, minutely papillose.

Zambia. E: Lundazi Dist., Nyika Plateau, 18.ii.1961, *Vesey-Fitzgerald* 2984 (K). **Malawi**. N: Chipata Dist., Nyika Plateau, falls about 6 km (4 miles) downstream of L. Kaulime, 11.ii.1963, *Simon et al.* 1731 (K, SRGH).

Also in Tanzania. In rock crevices and on shallow soil over rocks; 1650–2100 m.

Conservation notes: Data Deficient as it has a fairly restricted range and there are taxonomic uncertainties.

The original spelling of the name was '*nyasensis*' and there seems no logical reason to change this; Lye does not give an explanation that I can find.

The isotype at K does not appear to have any mature nutlets. Those on *Vesey-Fitzgerald* 2984 are narrowly ellipsoid, 2 × 0.3 mm, blackish and more-or-less parallel-sided.

E18. **Cyperus decurvatus** (C.B. Clarke) C. Archer & Goetgh. in Bothalia **41**: 300 (2011). Type: Mozambique, Ressano Garcia, *Schlechter* 11952 (K lectotype, B, BOL, BR, G, GRA, PRE, Z), lectotypified by Vorster and confirmed here.

> *Mariscus rehmannianus* C.B. Clarke in Durand & Schinz, Consp. Fl. Afr. **5**: 591 (1894); in Fl. Cap. **7**: 196 (1898). Types: South Africa, viii.1880, *Nelson* 13 (K syntype); South Africa, Natal, 1875, *Rehmann* 4479 (K syntype), non *Cyperus rehmannianus* (C.B. Clarke) Kuntze.
> *Mariscus vestitus* (C. Krauss) C.B. Clarke var. *decurvatus* C.B. Clarke in Bot. Jahrb. Syst. **38**: 134 (1906).
> *Cyperus indecorus* Kunth var. *decurvatus* (C.B. Clarke) Kük. in Engler, Pflanzenr. **4**, 20(101): 545 (1936).

Small clump-forming perennial; base a narrowly or very narrowly ovoid pseudobulb, 2.5–6 × 1–1.5 cm, covered in brown membranous leaf sheaths. Basal leaves up to 30 cm long and 2–5 mm wide in the middle, linear, glabrous except for minute teeth on the margin and on the midrib beneath. Inflorescence bracts 4–7, up to 29 cm long and 1.2–4.4 mm broad in the middle, similar to foliage leaves. Culms up to 54 cm long but usually less. Inflorescence once spicate anthelate, contracted, of 5–9 rays. Secondary peduncles up to 4.5 cm long; spikes 10–20 × 11–28 mm, shortly cylindrical. Spikelets 6–12 × 0.6–3 mm, 2–5-flowered, narrowly ellipsoid, fairly densely crowded; glumes 3–6.5 mm long, very narrowly ovate with long-acuminate recurved tips; median vein not excurrent; 10–16 lateral veins also present. Stamens 3; anthers 0.8–1 mm long. Style 3-branched. Nutlet narrowly ellipsoid, straight or curved, trigonous, 1.5–2.4 × 0.4–0.7 mm, minutely papillose.

Botswana. SW: Ghanzi Dist., Farm 56, Tr. 93, 2.ii.1970, *Brown* 8402 (K, SRGH). ?Dist., Andalucia, Bechuanaland, 1941-42, *Weber* 4 (K). **Zimbabwe**. W: Matobo Dist., Farm Besna Kobila, i.1957, *Miller* 4032 (K, SRGH). E: Chipinga Dist., 1 km W of confluence of Musirizwi & Bwazi Rivers, 30.i.1975, *Gibbs Russell* 2679 (K, SRGH). S: Beitbridge Dist., between Chiturupazi and Chikwerekwera, 24.ii.1961, *Wild* 5376 (K, SRGH). **Mozambique**. M: Maniça Dist., Ressano Garcia, 23.xii.1897, *Schlechter* 11952 (B, BOL, BR, G, GRA, K, PRE, Z).

Also in Namibia and South Africa. Open dry grasslands and shrublands on deep sand; 450–700 m.

Conservation notes: Least Concern because of its wide distribution.

E19. **Cyperus myrmecias** Ridl. in Trans. Linn. Soc. London, Bot. **2**: 144 (1884). Type: Angola, between Lopollo and Monino, iii-v.1860, *Welwitsch* 7060 (BM holotype, LISU).

Perennial. Rhizome horizontal, 3–5 mm in diameter, covered in dark red-brown scales. Culms up to 3 cm apart on the rhizome, usually more closely spaced, 25–50 cm long, 1–2 mm in diameter, glabrous, longitudinally furrowed. Leaves linear, 8–30 cm × 3–5 mm, margins serulate above, otherwise glabrous. Involucral bracts 4–, the lowest 6–15 cm × 3–5 mm, serrulate on the margins, oherwise glabrous. Inflorescence 3–6-branched, the units sometimes sessile forming a lobed capitate head but more usually with some units with stalks to 3.5 cm long, but at least some units sessile. Spikelets 2.5–2.7 mm long, the glumes pale greenish with strong longitudinal veins. Nutlet weakly trigonous, ellipsoid-ovoid, apiculate, 1.8–2 mm long, blackish.

Zambia. N: Mbala Dist., track by Lake Chila, 10.iv.1959, *McCallum Webster* 621 (K). W: Solwezi Dist., Meheba River on road from Solwezi to Mwinilunga, 15.ii.1975, *Hooper & Townsend* 68 (K). C: Mkushi Dist., junction to Mkuza road on Great North Road, 13.i.1958, *Robinson* 2745 (K). S: Mazabuka Dist., Tsiknaki's Farm, 18 km [13.4 miles] N of Choma, 30.i.1960, *White* 6643 (FHO, K). **Zimbabwe**. C: Ruwa Dist., Tanglewood Farm, xii.1960, *Miller* 7591 (K, SRGH).

Also in Angola. *Brachystegia* woodland and *Acacia–Combretum* scrub, sometimes at roadsides and around the bases of termite mounds (whence the specific epithet); 1200–1650 m.

Conservation notes: Data Deficient because of its relatively restricted distribution and because there are taxonomic uncertainties.

E20. **Cyperus macrocarpus** (Kunth) Boeckeler, Linnaea **36**: 380 (1870). —Kükenthal in Engler, Pflanzenr. **4**, 20(101): 528 (1936). Type: South Africa, 1834, *Drège* 4421 (B holotype).

Mariscus macrocarpus Kunth, Enum. Pl. **2**: 120 (1837). —Gordon-Gray in Strelitzia **2**: 132 (1995).

Mariscus angularis Turrill in Bull. Misc. Inform. Kew **1925**: 69 (1925). Type: South Africa, Cape, *Schonland* 3843 (GRA holotype).

Mariscus radiatus Hochst. in Flora **28**: 757 (1845), non *Cyperus radiatus* Vahl (1806). Type: South Africa, Natal, xii.1839, *Krauss* 35 (B syntype, G, MO, TUB).

Perennial, probably short-lived, forming small clumps. Culm bases adjacent, without elongated rhizomes. Culms 15–30 cm long, covered in brownish or purplish sheaths at the base, weakly trigonous, glabrous. Leaves linear, 15–30 cm long and 3–4 mm wide, tapering gradually to the apex, glabrous but minutely scabrid on margins and back of midrib, especially towards the apex. Inflorescence bracts 5–7, similar to foliage leaves, the lowest up to 20 cm long and 4–5 mm wide at the base. Inflorescence once anthelate-spicate with 6–10 spicate subunits borne on peduncles up to 6 cm long, with one or two sessile subunits. Spikelets 1–3-flowered, loosely arranged, very narrowly ovoid, 5–6 × 1–1.2 mm, greenish. Glumes 2.6–4.6 mm long, median vein extending into a very short mucro or non-excurrent. Stmens 3. Style three-branched. Nutlet ellipsoid, trigonous, c. 2 × 0.8 mm, smooth, black.

Mozambique. GI: Xai-xai Dist., Xai-Xai, 5.ix.1993, *Hooper* s.n. (K). M: Inhaca Island, W coast ridge, 15.vii.1959, *Mogg* 29352 (J, K, SRGH).

Also in South Africa (Natal, Cape). Coastal and near-coastal grasslands and sand dunes; 0–50 m.

Conservation notes: Probably Least Concern because of its common and relatively undisturbed habitat.

Apart from a single doubtful record from Swaziland, all of Vorster's records come from coastal South Africa. Material from Zambia has been identified as this, but differs in its much more compact inflorescence with short secondary peduncles, and narrower spikelets.

E21. **Cyperus cyperoides** (L.) Kuntze, Revis. Gen. Pl. **3**(2): 333 (1898). —Clarke in F.T.A. **8**: 404 (1902). —Kükenthal in Engler, Pflanzenr. **4**, 20(101): 514 (1936). —Haines & Lye, Sedges & Rushes E. Afr.: 204 (1983). —Hoenselaar in F.T.E.A., Cyperaceae: 223 (2010), in part. Type: India, *König* s.n. (LINN 71.42 lectotype), lectotypified by Gordon-Gray (1995).

    Scirpus cyperoides L., Mant. Pl. **2**: 181 (1771).

    Kyllinga sumatrensis Retz., Obs. Bot. 4: 13 (1786). Type: Sumatra, *Wennerberg* s.n (LD fide Raynal).

    Mariscus sieberianus C.B. Clarke in Hooker, Fl. Brit. India 6: 622 (1893). —Clarke in F.T.A. **8**: 388 (1902). Type: as for *Cyperus cyperoides*.

    Mariscus sumatrensis (Retz.) J. Raynal in Adansonia, n.s. **15**: 110 (1975).

Perennials, sometimes flowering in the first year, forming small clumps; base globular, hard, 7–10 mm in diameter, covered by brownish or purplish leaf bases, forming new shoots laterally after flowering. Leaves 10–40 cm long and 3–7.5 mm wide in the middle, glabrous except for minute teeth on the margins and on the midrib below, and sometimes on the main veins below. Inflorescence bracts 6–9, up to 23 cm long and 3.5–7 mm wide, similar to the foliage leaves. Culms up to 40(70) cm long, trigonous. Inflorescence once spicate-anthelate, with 11–14 rays; peduncles of sub-units up to 5 cm long; spikes 18–35 × 6–12 mm, cylindrical, often ascending. Spikelets 3–6 × 4–9 mm, 1(2)-flowered, appearing laxly spaced due to their linear shape, spreading or slightly ascending; glumes 3.3–3.9 mm long, with a single non-excurrent dark green median vein and 8–12 lateral veins. Stamens 3; anthers 0.5–0.8 mm long. Style 3-branched. Nutlets narrowly ellipsoid, 1.6–2.3 × 0.4–0.7 mm, straight or slightly curved, minutely papillose, pale yellowish brown.

Zambia. W: Ndola Dist., Sacred Lake near St Anthony's Mission, c. 48 km SW of Luanshya, 14.ii.1975, *Hooper & Townsend* 45 (K). C: Lusaka Dist., 8 km [5 miles] N of Lusaka, 13.xii.1953, *Best* 47 (K). S: Mapanza Dist., Mapanza Mission, 24.xii.1952, *Robinson* 25 (K). **Zimbabwe**. N: Gokwe Dist., 5 km [3 miles] S of Gokwe, 13.xii.1953, *Bingham* 1019 (K, SRGH). C: Marandellas Dist., Grasslands, 6.i.1949, *Corby* 348 (K, SRGH). E: Inyanga Dist., Pungwe River, *Fries et al.* 3749, (K, PRE). **Malawi**. N: Rumphi Dist., Nyika Plateau, 22 km [14 miles] N of M1, 23.xii.1977, *Pawek* 13328 (K, MAL, MO). S: Zomba Dist., Neighbourhood of Zomba, 1936, *Cormack* 311 (K).

Also in India and southeast Asia; introduced and widespread in tropical Africa and tropical South America. Weed of cultivation and roadsides, sometimes extending into semi-natural habitats; 1200–1800 m.

Conservation notes: Least Concern because of its wide distribution.

E22. **Cyperus turrillii** Kük. in Repert. Spec. Nov. Regni Veg. **29**: 199 (1931). —Haines & Lye, Sedges & Rushes E. Afr.: 201, fig.402 (1983). —Hoenselaar in F.T.E.A., Cyperaceae: 241 (2010). Type: Angola, Benguella, country of the Ganguellas and Ambuellas, 31.iii.1906, *Gossweiler* 3723 (B syntype, BM, K); Zimbabwe, Wilde 81, 83 (PRE); Botswana, *Kaessner* 2061 (B syntype).

Mariscus laxiflorus Turrill in Bull. Misc. Inform. Kew **1914**: 171 (1914). Type as for *Cyperus turrillii.*

Perennial (but some collectors record as annual), fairly robust, up to 70 cm tall with a short creeping rhizome; culms rather crowded, 30–50 cm long, 1–2.1 mm wide, trigonous, glabrous, basal part bulbous, 6–8 mm in diameter. Leaves up to 38 cm long; leaf sheath grey to pale reddish-brown, thin, 4–7.5 cm long, old basal sheaths disintegrating into fibres; leaf blade linear, flat, 15–31 cm long, 2.0–5.1 mm wide, scabrid at least along the margin, apex acuminate. Involucral bracts leaf-like, erect to spreading, 3–4, lowermost 3–35 cm long, 2–6 mm wide. Inflorescence a simple anthela with one sessile spikelet cluster and 1–11 pedunculate clusters on 1.5–8 cm long peduncles; spikelets in loose clusters, sessile and at the end of primary branches, 4–15 per cluster, sometimes reflexed, linear to narrowly ovate, 4.7–9.2 mm long, 2.4–3.3 mm wide, spreading when nutlets mature, spikelet falling off as a unit, rachis straight; glumes green, golden or reddish-brown, ovate, spreading, 3.2–4.6 mm long, 1.7–2 mm wide, keel green when young, with 3–4 strong veins on either side, apex acute to acuminate. Stamens 3; filaments 4–5 mm long, anthers 1.3–1.5 mm long. Nutlet brown to blackish, ellipsoid to obovoid, strongly trigonous, apiculate, 2.0–2.8 mm long, 1.1–1.4 mm wide, minutely papillose.

Botswana. N: Ngamiland Dist., 3.66 km S of Chief's Camp, Chief's Island, Okavango Delta, 4.i.2012, *A. & R. Heath* 2336 (K). SE: Kagatlang Dist., Masama Ranch, 10.xi.1978, *Hansen* 3550 (C, GAB, K, PRE, SRGH). **Zambia**. B: Mongu Dist., Mongu, 31.xii.1965, *Robinson* 6758 (K). N: Luapula Dist., Mbereshi Settlement, 15.i.1960, *Richards* 12364 (K). C: Mkushi Dist., Fiwila, 3.i.1958, *Robinson* 2593 (K). S: Mapanza Dist., 10 km [6 miles] E of Choma, 18.xii.1956, *Robinson* 1794 (K). **Zimbabwe**. W: Hwange Dist., Victoria Falls, 10–15.iii.1932, *Brain* 8805 (K, SRGH). C: Marondera [Marandellas] Dist., Grasslands Research Station, 19.xi.1948, *Corby* 241 (K, SRGH). S: Masvingo [Victoria] Dist., Makaholi Experimental Station, 10.iii.1978, *Senderayu* 154 (K, SRGH). **Malawi**. C: Kasungu Dist., Kasungu-Bua road, 13.i.1959, *Robson* 1131 (K).

Also in Tanzania, D.R. Congo, and Angola. Mainly in deciduous woodlands, often of *Brachystegia* and/or *Julbernardia*, but also at roadsides, usually on sand; 950–1650 m

Conservation notes: Least Concern because of its wide distribution.

Very vigorous specimens (e.g. *Robinson* 6758) may be twice-anthelate.

E23. **Cyperus schimperianus** Steud., Syn. Pl. Glumac. **2**: 34 (1854). —Clarke in F.T.A. **8**: 358 (1901). —Kükenthal in Engler, Pflanzenr. **4**, 20(101): 84 (1935). —Haines & Lye, Sedges & Rushes E. Afr.: 184, fig.355 (1983). —Hoenselaar in F.T.E.A. Cyperaceae: 241 (2010). Type: Ethiopia, near Adua, *Schimper* 57 (P holotype, K).

Perennial, robust, up to 105 cm tall, with a woody rhizome covered in brown to blackish scales; culms 56–94 cm long, 2–4 mm wide, trigonous, sometimes almost rounded near the apex, slightly longitudinally ridged, smooth. Leaves up to 30 cm long; leaf sheath grey to reddish-brown, 10–21 cm long, fairly wide and loosely surrounding the culm; leaf blade linear, flat, glabrous to slightly scabrid, 4–10 cm long, 1.5–3 mm wide, apex acute to acuminate, slightly scabrid. Involucral bracts leaf-like, spreading, 4–6, lowermost 19–30 cm long, 2–6 mm wide. Inflorescence simple, primary branches 4–9, 2–9 cm long; spikelets in loose clusters at the end of primary branches, 6–14 per cluster, linear-lanceolate, 9.5–22 mm long, 1–2.5 mm wide; glumes reddish-brown, sometimes pale, ovate-lanceolate, glabrous, 1.3–2.5 mm long, 0.6–1.6 mm wide, keel flat, apex rounded. Stamens (2)3; filaments 1.3–2.4 mm long; anthers 0.8–1.8 mm long. Nutlet grey-brown, (narrowly) ellipsoid-obovoid, 0.9–1.6 mm long, 0.3–0.6 mm wide, smooth to slightly papillose in longitudinal rows.

Malawi. C: Ntchisi Dist., E bank of Bua River on Kasungu–Nkhota-kota road, 20.vi.1970, *Brummitt* 11587 (K).

Also in Cameroon, D.R. Congo, Sudan, Ethiopia, Uganda, Kenya, Tanzania. Sandy or stony river banks, near or in water; 450–1600 m.

Hoenselaar, in F.T.E.A., note that this taxon is easily recognized because of its wide leaf sheaths and short leaf blades.

Only one specimen seen from the Flora area.

E24. **Cyperus callistus** Ridl. in Trans. Linn. Soc. London, **2**: 143 (1884). Type: Angola, Loanda Dist., nr. Quicuxe, iii.1854, *Welwitsch* 7079 (BM holotype, LISU).

> *Cyperus fulgens* C.B. Clarke in F.T.A. **8**: 355 (1901). —Haines & Lye, Sedges & Rushes E. Afr.: 192, fig.375 (1983). —Hoenselaar in F.T.E.A., Cyperaceae: 229 (2010). Type: Namibia, Hereroland, *Fleck* 642 (B syntype); Botswana, Koobie to N Shaw Valley, 1.iii.1863, *Baines* s.n. (K syntype); Botswana, Kwebe Hills, Nyamiland, 13.i.898, *Lugard* 104 (K syntype).

Perennial up to 75 cm tall, slender to robust, with ± 10–15 mm thick bulbs and slender stolons; culms few to tufted, 16–68 cm long, 1.1–4 mm wide, triquetrous to trigonous, smooth. Leaves many at the base, up to 52 cm long; leaf sheath brown, 2–7 cm long; leaf blade linear, flat to folded, 19–28 cm long, 1.6–5.7 mm wide, glabrous, apex acuminate. Involucral bracts leaf-like, spreading, 2–4, lowermost 9–33 cm long, 1.1–6 mm wide. Inflorescence simply anthelate, primary branches (3)5–10, 1.5–5.5 cm long; spikelets in lax clusters on an elongated axis, at the end of primary branches up to 18 cm long, 10–32 per cluster, very narrowly elliptic to linear, 8–18 mm long, 1–2 mm wide, rachis straight; glumes reddish brown, very narrowly ovate, 3.2–5.0 mm long, 1.3–1.8 mm wide, with prominent lateral veins, keel greenish-brown, greyish when dry, apex rounded to acute. Stamens 3; filaments 4–5 mm long; anthers 1.7–2.7 mm long. Nutlet brown to greyish-black, ellipsoid, trigonous, 1.4–2.2 mm long, 0.8–1 mm wide, smooth to minutely papillose.

Botswana. N: Ghanzi Dist., Bed of Groot Laagte fossil river, 17.iii.1980, *P.A. Smith* 3222 (K, SRGH). SW: 80 km [50 miles] N of Ghanzi, 17.iv.1969, *Brown* s.n. (K). SE: Central Dist., Thlalamabele, Mosu area, near Soa Pan, 8.i.1974, *Ngoni* 273 (K, SRGH). **Zambia**. S: Mapanza Dist., 60 km [45 miles] SE of Choma, 17.xii.1956, *Robinson* 1780 (K). **Zimbabwe**. N: Urungwe Dist., Mansa Pan, 17 km [11 miles] ESE of Chirundu Bridge, 29.i.1958, *Drummond* 5351 (K, SRGH). W: Hwange [Wankie] Dist., Victoria Falls, 11.xii.1978, *Mshasha* 131 (K, SRGH); Chipinge [Chipinga] Dist., 8 km [5 miles] E of Hippo Mine, 8.iii.1968, *Plowes* 2882 (K, SRGH). S: Beitbridge Dist., Tuli R at Tuli Breeding Station, 6.i.1961, *Wild* 5318 (K, SRGH).

Also in Kenya, Tanzania, Namibia, and South Africa. In woodland, often *Colophospermum mopane* but also *Combretum/Terminalia*, on sandy soil or, in two Botswana collections, on secondary limestone (calcrete); also in open grasslands on sandy soil, sometimes a weed of cultivation; 400–1350 m.

Conservation notes: Least Concern because of its wide distribution.

Hooper (notes on K material) suggested that *Cyperus callistus* might be the correct (earlier) name. Having seen the type at BM I can now confirm this.

Well-collected specimens (including *Lugard* 104, one of the syntypes of *Cyperus fulgens*) show the scaly subterranean bulb attached either to the base of the leafy stem, or producing a slender vertical rhizome from which the leafy stem arises. In the absence of bulbs, the dark reddish brown spikelets, spreading at 90° or slightly more from the axis, are distinctive. Plants from the region are generally larger and more vigorous than the sparse material from the F.T.E.A. area.

E25. **Cyperus tenuiculmis** Boeckeler in Linnaea **36**: 286 (1870). —Haines & Lye, Sedges & Rushes E. Afr.: 196, figs.388,389 (1983). —Hoenselaar in F.T.E.A., Cyperaceae: 242 (2010). Types: Sierra Leone, *Afzelius* s.n. (BM); Nigeria, Nupe, *Barter* 1573 (K); Nepal, *Wallich* 3321 (BR syntype, K, P); E India, Khasia Hills,

Hooker & Thomson s.n. (K); Indonesia, Batavia, *Junghuhn* s.n. (?LD); Sri Lanka, *Thwaites* 807 (?BM); Philippines, Luzon, *Haenke* s.n., *Meyen* s.n. (B syntype).

Cyperus majungensis Cherm. in Bull. Soc. Bot. France **67**: 329 (1921), new synonym. Type: Madagascar, ii.1920, *Perrier de la Bâthie* 13033 (K isotype, P).

Cyperus zollingeri Steud. var. *longiramulosus* Kük. in Engler, Pflanzenr. **4**, 20(101): 135 (1935). Types: Sierra Leone, *Thomas* 2059 (?B syntype); Cameroon, Boruru, *Tessman* 2558 & 2676 (?B, syntypes); D.R. Congo, Mukenge, *Pogge* 1583 & 1610 (?B syntypes), Lunda, *Pogge* 1578a (?B syntype), Nlenfu, *Butaye* s.n. (?B syntype); Sudan, Land der Djur, *Schweinfurth* 2280 (?B syntype); Tanzania, Kigoma Dist., Machaso near Kigoma, 18.ii.1926, *Peter* 37039 (B syntype).

Cyperus robinsonii Podlech in Mitt. Bot. Staatssamml. München **4**: 111 (1961). Type: Zambia, Choma, *Robinson* 3243 (M holotype, K).

Perennial, medium-sized to robust, up to 150 cm tall, with a rather thick creeping rhizome and swollen stem-bases; culms 34–116 cm long, 0.8–5 mm wide, trigonous to triquetrous, smooth to slightly scabrid, sometimes only scabrid just below the inflorescence. Leaves up to 65 cm long; leaf sheath green to brown, 2.5–10 cm long; leaf blade sometimes rather stiff, linear, flat, 12–55 cm long, 2.5–11 mm wide, with multiple major veins, scabrid on margins and major veins, apex acute to acuminate. Involucral bracts leaf-like, erect to spreading, 2–5, lowermost 6.5–28 cm long, 2–6.4 mm wide. Inflorescence a simple anthela, occasionally twice-anthelate, primary branches 3–10, 2.5–25 cm long; spikelets spicately arranged in loose clusters, at the end of primary (sometimes secondary) branches, 2–11 per cluster, linear-lanceolate, 15–46 mm long, 1.6–2.2 mm wide, rachilla strongly zig-zag when glumes are shed; glumes pale brown to dark reddish brown, ovate, 2.7–4.1 mm long, 1.6–2.2 mm wide, keel green, sometimes excurrent, apex rounded, acute to acuminate. Stamens 3; filaments 1.9–4.1 mm long; anthers 0.6–1.4 mm long. Nutlet dark reddish-brown to almost black, obovoid-ellipsoid, trigonous, 1.6–2 mm long, 0.8–1.1 mm wide, minutely papillose.

Glumes pale or yellowish-brown with paler margins; midrib not excurrent; culms
 smooth . . . . a) var. *tenuiculmis*
Glumes reddish brown with pale margins; midrib of glumes excurrent; culms scabrid,
 at least towards the apex . . . . .b) var. *schweinfurthianus*

a) Var. **tenuiculmis**

Culm 0.8–2.2 mm wide, smooth. Spikelets 10–20 × 2–2.5 mm, acute; glumes with yellowish-brown band along the ridged midrib which is not or only slighty excurrent.

Botswana. N. Ngamiland Dist., Transit road A, N of tunring to Selinda Camp, 7.ii.2004, *A. & R. Heath* 539 (K). **Zambia**. N: Luapula Dist., Kalasa Mukoso Flats, 19.ii.1996, *Renvoize* 5628 (K). C: Kabwe Dist., Mfukushi R. near Chipepo, 51 km NW of Kabwe, 20.i.1973, *Kornaś* 3053 (K). S: Mazabuka Dustrict, Choma to Pemba, km 15 [mile 10], 30.i.1960, *White* 6609 (FHO, K). **Zimbabwe**. E: Mutare Dist., iNywmquara Valley, ii.193, *Gilliland* Q1633 (K). **Malawi**. C: Lilongwe Dist., Lilongwe to Dzalanyama road, near Singala market, 6.ii.1959, *Robson* 1485 (BM, K). S: Nsanje Dist., Nchenge, c. 1.5 km [1 mile] from shore of Lake Chilwa, 10.i.1963, *Brown* 385K (K).

Widespread through west tropical and central Africa, S to Angola; S and SE Asia. Seasonally wet grassland, seepage areas, road margins and drainage ditches, usually on sandy soil; 650–1250 m.

Conservation notes: Least Concern because of its wide distribution.

It is most unfortunate that the species named by Podlech for the most prolific and oustanding collector of Cyperaceae in the Flora area should have to fall into synonymy, but rules are rules.

b) Var. **schweinfurthianus** (Boeckeler) S.S. Hooper in Kew Bull. **26**: 578 (1972). —Haines & Lye, Sedges & Rushes E. Afr.: 197, fig.390 (1983). —Hoenselaar in F.T.E.A., Cyperaceae: 242 (2010). Type: Sudan, Seriba Ghattas, 27.viii.1869, *Schweinfurth* 2318 (K, P).

Cyperus schweinfurthianus Boeckeler in Flora **62**: 553 (1879). —Clarke in F.T.A. **8**: 361 (1901).

Cyperus zollingeri Steud. var. *schweinfurthianus* (Boeckeler) Kük. in Engler, Pflanzenr. **4**, 20(101): 134 (1935).

Culm 1.3–5 mm wide, scabrid, at least towards the apex. Glumes reddish brown with pale margins; midrib excurrent.

Zambia. N: Kasama Dist., Kasama, Chambeshi Pontoon, 27.ii.1960, *Vesey-Fitzgerald* 2687 (K). W: Mwinilunga Dist., Zambezi River, 15 km E of Kalene Hill, 16.xii.1963, *Robinson* 6093 (K). C: Kabwe Dist., Kamaila Forest Station, 36 km N of Lusaka, 9.ii.1975, *Brummitt et al.* 14294 (K). **Mozambique**. N: Pemba Dist., 27.i.1984, *Groenendijk et al.* 786 (K, LMU). MS: Beira, xii.1906, *Swynnerton* 941 (BM, K). M: Manhiça Dist., near Manhiça, 29.ii.1952, *Barbosa & Balsinhas* 4851 (K).

Widespread through west tropical and central Africa. Dry or damp grasslands, ditches, marshy ground and swamp; 1050–1450 m.

Conservation notes: Least Concern because of its wide distribution.

F.T.E.A. note: This variety is distinguishable from var. *tenuiculmis* by its coarse habit and its excurrent keel on the glumes. It has been accepted with some doubt, as there are some specimens with a coarse habit, glabrous culms, and slightly excurrent glumes, and therefore intermediates between var. *tenuiculmis* and var. *schweinfurthianus*.

Kuntze 202 from Mozambique is a good match for the type of *Cyperus majungensis* (*Perrier de la Bâthie* 13033) and was so named by Kükenthal, but seems identical with *C. tenuiculmis* var. *schweinfurthianus*. A final decision on whether these two are synonymous is left for a future monographer of Madagascar Cyperaceae.

E26. **Cyperus aureospicatus** Lock, sp. nov. Perennial, similar to *Cyperus tenax* but differs by its spikelets arranged in spikes, not in umbels, and golden in color, not blackish. Type: Zambia, Eastern Region, between Chadiza turn-off and Nsadzu River, 27.xi.1958, *Robson* 739 (K holotype, BM). FIGURE 14.**40**.

Perennial forming loose clumps from a short knotty rhizome bearing thick roots. Culms 12–37 cm tall and 1–1.5 mm in diameter, terete to weakly trigonous, glabrous. Leaves mainly basal, 10–17 cm long, 2–3 mm wide, folded, glabrous but sparsely scabrid on the margins towards the apex. Inflorescence bracts 3, the longest up to 8 cm long, just exceeding the mature inflorescence, glabrous but scabrid on the margins. Inflorescence a simple anthela with 5–8 rays of various lengths, the longest 2–4.5 cm to the lowest spikelet. Spikelets 8–16 per ray, spicately arranged, spreading at right angles to the axis when mature, very narrowly ovate, 0.8–1.3 cm long, golden, 16–22-flowered. Glumes elliptic, folded, 1.7–2 mm long, 1–1.2 mm wide; keel narrow, green. Main body of glumes thin, membranous, golden yellow. Anthers 0.6–0.8 mm long. Style 3-branched. Mature nutlet not seen.

Zambia. C: Lusaka Dist., Chongwe River near Constantia (N. of Kasisi), 24.x.1972, *Kornaś* 2431 (K), and same locality, 3.xii.1972, *Kornaś* 2749 (K). E: Mapanza Dist., Lundazi –Choma, 50 km [32 miles], 19.x.1958, *Robson* 162 (BM, K).

Not known elsewhere. Dambo margins; sandy grassland; 900–1150 m.

Conservation notes: Data Deficient. Not enough collections to assess.

A distinctive plant represented by just these four collections. The yellow spikelets, short stature and its habit of flowering early in the wet season must make it quite conspicuous in the field and it is surprising that there are not more collections; perhaps it is genuinely scarce.

The specific epithet refers to the golden yellow spikelets.

Fig. 14.**40**. CYPERUS AUREOSPICATUS. 1, habit; 2, spikelet; 3, flower. All from *Robson* 739.
Scale bars: 1 = 2 cm; 2 = 2 mm; 3 = 1 mm. Drawn by Margaret Tebbs.

E27. **Cyperus aterrimus** Steud., Syn. Pl. Glumac. **2**: 31 (1854). —Clarke in F.T.A. **8**: 358 (1901). —Kükenthal in Engler, Pflanzenr. **4**, 20(101): 141 (1935). —Hoenselaar in F.T.E.A., Cyperaceae: 232 (2010). Type: Ethiopia, Debra Eski, 22.x.1850, *Schimper* 233 (P holotype, P)

    Cyperus atroviridis C.B. Clarke in F.T.A. **8**: 359 (1901). —Haines & Lye, Sedges & Rushes E. Afr.: 199, figs.395,396(1983). Type: Bioko [Fernando Poo], iv.1862, *Mann* 1466 (K holotype, B).

    Cyperus aterrimus var. *agglomeratus* Kük. in Notizbl. Bot. Gart. Berlin-Dahlem **9**: 304 (1925); in Engler, Pflanzenr. **4**, 20(101): 142 (1935). Type: Kenya, Mt Kenya, Coles Mill, *Fries & Fries* 1089 (B holotype).

    Cyperus aterrimus var. *atroviridis* (C.B. Clarke) Kük. in Engler, Pflanzenr. **4**, 20(101): 142 (1935).

Perennial, robust, up to 200 cm tall although usually less, with a short 3–6 mm thick creeping rhizome and many crowded roots, occasionally with more slender curving scale-covered stolons; culms few, 26–82 cm long, 1.9–7 mm wide, trigonous, sometimes almost triquetrous, glabrous. Leaves up to 60 cm long; leaf sheath green to reddish-brown, 4–20 cm long; leaf blade linear, flat, 20–40 cm long, 5–12 mm wide, scabrid on margins and major veins, apex acute. Involucral bracts leaf-like, spreading, 4–7, lowermost 21–47 cm long, 4–14 mm wide. Inflorescence anthelate, primary branches compound-spicate, 4–10, 1.5–12 cm long; spikelets in pedunculate crowded spikes, giving the inflorescence a brush-like appearance, sessile and at the end of primary and secondary branches, 82 spikelets per spike, very narrowly ovate, 8–15 mm long, 2–3 mm wide, rachilla straight; glumes dark reddish-brown, sometimes almost black, narrowly ovate, 3.2–3.5 mm long, 1.4–1.6 mm wide, keel usually green, apex acuminate to slightly mucronate. Stamens 3; filaments 2.7–3.7 mm long; anthers 0.5–0.6 mm long. Nutlet brown, narrowly ellipsoid-ovoid, trigonous, 2–2.4 mm long, 0.6–0.8 mm wide, almost smooth.

Malawi. N: Rumphi Dist., Nyika Plateau, Businande, 28.xii.1975, *Phillips* 765 (K, MO). S: Zomba Dist., Mt Zomba, xii.1896, *Whyte* s.n. (K).

Also in Bioko, D.R. Congo, Rwanda, Burundi, Ethiopia, Kenya, Uganda and Tanzania. Rocky outcrops by streams in upland forest; 1500–2150 m.

Conservation notes: Least Concern because of its wide distribution.

Much less frequent in the flora area than in East Africa because of its restriction to upland forest; either absent from, or not yet collected from, the Eastern Highlands of Zimbabwe and the mountains of Mozambique.

None of the flora area specimens has complete basal parts or complete culms, and there are no mature nutlets; these details are taken from the F.T.E.A. account.

E28. **Cyperus ferrugineoviridis** (C.B. Clarke) Kük. in Engler, Pflanzenr. **4**, 20(101): 412 (1936). —Haines & Lye, Sedges & Rushes E. Afr.: 202, figs.403,404 (1983). —Hoenselaar in F.T.E.A., Cyperaceae: 231 (2010). Types: Uganda, Ruwenzori, *Scott-Elliot* 7590 (K syntype); Tanzania, Kilimanjaro, *Volkens* 1620 (B syntype).

    Cyperus maranguensis K. Schum. var. *ferrugineoviridis* C.B. Clarke in F.T.A. **8**: 359 (1901).

    Cyperus ferrugineoviridis var. *distantiformis* Kük. in Engler, Pflanzenr. **4**, 20(101): 413 (1936). Types: Tanzania, Songea Dist., Lupembe, Ugololo, *Schlieben* 307 pro parte (BR syntype); Njombe Dist., Mpoponzi, v.1931, *Schlieben* 791 (BR syntypes, S) & 873 pro parte (?BR syntype); Morogoro Dist., Uluguru Mts, 6.ii.1933, *Schlieben* 3393 (EA syntype, BR); W Kilimanjaro, Sanya, *Petzholtz* 92 (B syntype).

    Cyperus ferrugineoviridis var. *luteiformis* Kük. in Engler, Pflanzenr. **4**, 20(101): 412 (1936). Types: Tanzania, Bukoba Dist., Karagwe, Kaforu, *Stuhlmann* 1839 (?B†); Bukoba Dist., Kagera R. at Kavingo, *Stuhlmann* 1949 (?B†); Kondoa, *Burtt Davy* 1100 (?PRE); Morogoro Dist., Uluguru Mts, Lukwangule plateau, 22.ii.1933, *Schlieben* 3547 (B syntype, BR); South Africa, Pretoria saltpan, *Leemann* 27585 (B syntype).

    Mariscus ferrugineoviridis (C.B. Clarke) Cherm. in Bull. Jard. Bot. État Bruxelles **14**: 330 (1937).

    Mariscus bequaertii Cherm. in Bull. Jard. Bot. État Bruxelles **14**: 329 (1937). Type: D.R. Congo, Rutshuru, *Bequaert* 5605, 6239 (BR syntypes); Mokoto Lakes, *Claessens* 36 (BR syntype); Mulungu, *de Craene* 202, 202b (BR syntypes).

Cyperus bequaertii (Cherm.) Robyns & Tournay in Robyns, Fl. Spermatophyt. Parc Nat. Albert **3**: 246 (1955).

Perennial up to 120 cm tall, robust, usually with a hard swollen cormose stem base bearing thin brown horizontal rhizomes 2–4 mm in diameter with dark brown spaced scales; culms few, 47–102 cm long, 2–4.2 mm wide, trigonous, sometimes almost triquetrous, glabrous. Leaves on lower half of culm, up to 58 cm long; leaf sheath rather conspicuous, green to brownish above, dark brown to purple near culm-base, 3–12.5 cm long; leaf blade linear, flat or W-shaped in cross-section, 15–45 cm long, 4.9–12 mm wide, scabrid on margins and primary vein, apex acute. Involucral bracts leaf-like, erect to spreading, 4–8, lowermost 9–40 cm long, 4–23.5 cm long. Inflorescence a lax anthela with 5–10 main branches; spikelets in loose clusters, sessile and at the end of primary (and sometimes secondary) branches, 10–30 per cluster, linear-lanceolate, 7.2–24 mm long, 0.9–1.6 mm wide, falling off entirely when mature; glumes greenish, golden or reddish-brown, with a translucent border, ovate-lanceolate, 4–5.2 mm long, 1.6–1.9 mm wide, keel green, sometimes slightly excurrent, apex rounded to sometimes acute. Stamens 3; filaments 4.4–5.5 m long; anthers 1.8–3.2 mm long. Nutlet grey, obovoid, trigonous, 2.2–2.5 mm long, 0.8–0.9 mm wide, slightly apiculate, minutely papillose.

Botswana. N: Ngamiland Dist., Thamalakane River, by the bridge, 14.iii.2004, *Kabelo et al.* MSB-52 (K). **Zambia**. N: Mbala Dist., Chilongwelo, 24.xii.1951, *Richards* 137 (K). W: Ndola Dist., Ndola, 10.i.1960, *Robinson* 3292 (K). **Zimbabwe**. E: Chimanimani [Melsetter], 1.i.1965, *Robinson* 6354 (K). **Malawi**. N: Mzimba Dist., Mzuzu, Marymount, 19.vii.1973, *Pawek* 7205 (K, MAL, MO, SRGH). C: Nkhota-kota Dist., Ntchisi Mt Forest reserve, 11.v.1984, *Banda & Kaunda* 2169 (K, MAL). S: Zomba Dist., Zomba, xii.1896, *Whyte* s.n. (K).

Also in Somalia, Uganda, Kenya, Tanzania, D.R. Congo, Rwanda, Burundi, and South Africa. In grassland and cleared forest, also found as a weed in cultivated land; 900–2450 m.

Conservation notes: Least Concern because of its wide distribution.

17. **COURTOISINA** Soják[17]

Courtoisina Soják in Cas. Nár. Mus., Odd Prír. **148**: 193 (1980).
Courtoisia Nees in Linnaea **9**: 286 (1834), non Rchb. (1829). —Clarke in F.T.A. **8**: 403 (1902), illegitimate name.
Cyperus L. subgen. *Courtoisia* (Nees) Lye in Nordic J. Bot. **3**: 230 (1983); in Haines & Lye, Sedges & Rushes E. Afr.: 174 (1983).
Cyperus L. subgen. *Courtoisina* (Soják) Lye in Lidia **3**(2): 52 (1992).

Larridon in Pl. Ecol. Evol. **144**: 327–356 (2011) has incorporated the species of *Courtoisina* into *Cyperus*, but it is kept separate here.

Medium-sized annual herbs, forming small tufts. Basal leaves linear, V-shaped in cross-section, ± straight, glabrous except for few small, sparely-spaced teeth along margins near apices. Bracts subtending inflorescence up to 9, spreading-erect, morphologically similar to basal leaves. Inflorescence a long-peduncled, 2-compound umbel of spheroid spikes; spikes c. 10 mm in diameter. Spikelets to 8 mm long, 1–9-flowered, densely spaced, spreading, deciduous between glumes 2–3, narrowly elliptic with acute apex, conspicuously compressed laterally. Glumes distichously arranged, narrowly ovate when flattened, folded with prominent abaxial wing, shortly apiculate, median longitudinal vein shortly excurrent, 1 pair of secondary longitudinal veins spaced close to midrib to leave greater part of glume veinless, yellow-green, eventually maturing into warm yellow-brown. Flowers bisexual, but 2 lowest glumes empty. Stamens 3; pollen grains obtusely obpyramidal with 4 lateral facets. Style 3-branched. Nutlets narrowly spindle-shaped, tapering to both ends, straight, obtusely trigonous, not compressed, covered with acute-conical, flat-topped papillae.

---

[17] by P. Vorster

A genus of 2 species, along waterbodies in dry, hot parts of southern and eastern Africa, Madagascar and India.

In Larridon *et al.* in Pl. Ecol. Evol. **144**: 327–356 (2011) *Courtoisina* is regarded as a section of *Cyperus*. However, here the genus is retained as being separate.

Fresh material, and dried specimens even more, have a characteristic and strong curry-like odour.

Spikelet 5–9-flowered, 4.7–7.5 mm long; fruit 1.5–2.3 mm long . . . . **1.** *assimilis*
Spikelet 1-flowered, 3.5–4.5 mm long; fruit 2.2–3.4 mm long . . . . **2.** *cyperoides*

1. **Courtoisina assimilis** (Steud.) Maquet in Bull. Jard. Bot. Belg. **58**: 265 (1988). —
Gordon-Gray in Strelitzia **2**: 209 (1995). —Verdcourt in F.T.E.A., Cyperaceae: 257
(2010). —Larridon *et al.* in Pl. Ecol. Evol. **144**: 349–350 (2011). Types: Ethiopia,
Gapdia, *Schimper* 1208 (P lectotype, HAL, K, UPS); Gaffa, 19.ix.1838, *Schimper*
1252 (K syntype, G, P, S), lectotypified by Lye (Nordic J. Bot. **3**: 230, 1984).
FIGURE 14.**41**.

Cyperus assimilis Steud. in Flora **25**: 584 (1842). —Boeckeler in Linnaea **35**: 579 (1868). —
Engler, Hochgebirgsfl. Afrika: 140 (1892). —Kükenthal in Engler, Pflanzenr. **4**, 20(101):
499 (1936). —Haines & Lye, Sedges & Rushes E. Afr.: 174, fig.332 (1983).

Courtoisia assimilis (Steud.) C.B. Clarke in Durand & Schinz, Consp. Fl. Afr. **5**: 596 (1894); in
F.T.A. **8**: 404 (1902); in Illustr. Cyp.: t.32 (1909).

Mariscus assimilis (Steud.) Podlech in Mitt. Bot. Staatssamml. München **3**: 523 (1960); in
Merxmüller, Prodr. Fl. SW Afr. **165**: 36 (1967). —Napper in J. E. Africa Nat. Hist. Soc. Natl.
Mus. **28**: 10 (1971).

Annual herb forming small tufts. Basal leaves numerous, longer than inflorescence, 1.1–5 mm wide in middle, erect, ± straight, broadly V-shaped in section, very soft in texture, bright green, glabrous except for few minute, sparsely-spaced teeth along margins near apex. Involucral bracts 2–8, to 280 × 1.3–5.2 mm wide in middle, spreading-erect, otherwise like basal leaves. Inflorescence a terminal, long-peduncled simple or compound umbel of 4–9 rays with 3–60 terminal spikes; primary peduncle to 240 × 0.7–3 mm in dried specimens, usually trigonous, glabrous; secondary peduncles to 65 × 0.3–0.8 mm; spikes 8–17 × 7–18 mm wide when dry, roughly spherical. Spikelets 4.7–7.5 × 2–3.3 mm, 5–9-flowered, densely packed, ovate-lanceolate or narrowly ovate with acute apex, strongly compressed laterally. Glumes 2.3–3.3 mm long including apical, slightly recurved 0.1–0.5 mm mucro, distichously arranged, tightly imbricate, 0.4–0.7 mm apart, ovate, folded lengthwise with median longitudinal 'keel' of sponge-like tissue closely flanked by secondary vein on each side leaving wide flanks of glume veinless, greenish yellow; rachilla of spikelet winged. Stamens 3; pollen grains obtusely pyramidal; filaments ribbon-shaped; anthers 0.25–0.5 × 0.1–0.15 mm, narrowly ellipsoid. Style 3-branched. Nutlet 1.5–2.3 × 0.4–0.5 mm, narrowly spindle-shaped, tapering at both ends, straight, obtusely trigonous; surface dense with acute-conical, warm brown, flat-topped papillae.

Botswana. N: Central Dist., Motsetso (Marsupe) R., 121 km WNW of Francistown on Maun road, 2.v.1957, *Drummond* 5301b (GRA, K, PRE, SRGH). **Zambia**. N: Mbala Dist., Lumi R. marsh, 1680 m, 30.iii.1957, *Richards* 8933 (K). W: Solwezi Dist., Mulenga Protected Forest Area, 20 km NW of Kansanshi, 19.iii.1961, *Drummond & Rutherford-Smith* 7072a (K, PRE). C: Lusaka Dist., Kabulonga, 8 km E of Lusaka, 17.vi.1955, *Robinson* 1306 (K). S: Choma Dist., 5 km SE of Mapanza, 1070 m, 18.v.1955, *Robinson* 1260 (K). **Zimbabwe**. W: Bulilimamangwe Dist., Dombodena Mission Station, 1 km E of secondary school, 1300 m, 17.v.1972, *Norgrann* 164 (J, K, SRGH). C: Harare (Salisbury), 1460 m, 9.v.1934, *Brain* 4126 (K, PRE).

Also in D.R. Congo, Rwanda, Eritrea, Ethiopia, Kenya, Uganda, Tanzania, South Africa (Free State, Limpopo, North West) and Namibia, but not known from Malawi, Mozambique or Angola. Riversides, seasonal pools, ditches and moist clay areas; 1000–1700 m.

Conservation notes: Widespread; not threatened.

Fig. 14.**41**. COURTOISINA ASSIMILIS. 1, habit; 2, spikelet; 3, glume; 4, 5, glume and part of rachilla, respectively mature nutlet and floret; 6, nutlet. 1 from *Robinson* 1306; 1–6 from *Ngoni* 386. Scale bars: 1 = 40 mm; 2–6 = 2 mm. Drawn by Jane Browning.

2. **Courtoisina cyperoides** (Roxb.) Soják in Cas. Nár. Mus., Odd. Prír. **148**: 193 (1980).
—Maquet in Bull. Jard. Bot. Belg. **58**: 265 (1988). —Gordon-Gray in Strelitzia
2: 209 (1995). —Verdcourt in F.T.E.A., Cyperaceae: 259 (2010). Type: India, no
locality or collector given.

> *Kyllinga cyperoides* Roxb., Fl. Ind. **1**: 182 (1820).
> *Mariscus cyperoides* (Roxb.) A. Dietr., Sp. Pl. **2**: 348 (1833). —Gordon-Gray in Strelitzia **2**:
> 129 (1995).
> *Courtoisia cyperoides* (Roxb.) Nees in Linnaea **9**: 286 (1834).
> *Cyperus pseudokyllingioides* Kük. in Engler, Pflanzenr. **4**, 20(101): 14 (1935), new name,
> non *Cyperus cyperoides* (L.) Kuntze (1898). —Haines & Lye, Sedges & Rushes E. Afr.: 175,
> fig.335 (1983).

Subsp. **africana** (Kük.) Vorster, comb. nov. Type: Tanzania, Ulanga Dist., Mahenge,
14.vi.1932, *Schlieben* 2389 (B lectotype, G, K), lectotypified by Vorster (1978, DSc
thesis, University of Pretoria) from 15 syntypes.

> *Courtoisia cyperoides* Roxb. var. *africana* C.B. Clarke in Durand & Schinz, Consp. Fl. Afr. **5**:
> 596 (1894), invalid name.
> *Cyperus pseudokyllingioides* Kük. var. *africanus* Kük. in Engler, Pflanzenr. **4**, 20(101): 501
> (1936).
> *Mariscus cyperoides* (Roxb.) A. Dietr. subsp. *africanus* (Kük.) Podlech in Mitt. Bot.
> Staatssamml. München **3**: 523 (1960). —Gordon-Gray in Strelitzia **2**: 129 (1995).

Annual herb, forming small tufts. Basal leaves numerous, longer than inflorescence, 2–8.6 mm
wide in middle, tapering gradually to apex, erect, ± straight, V-shaped in section, soft in texture,
bright green, glabrous except for few minute, sparsely-spaced teeth along margins near apex.
Involucral bracts 3–9(5), to 420 × 2.1–7.6 mm wide in middle, spreading-erect, otherwise like
basal leaves. Inflorescence a terminal, long-peduncled, simple or often compound umbel of 3–7
primary rays with 6–40(112) terminal spikes; primary peduncle to 380 (rarely to 500) mm long
by 1.4–3.9 mm thick in dry specimens, obtusely trigonous, glabrous; secondary peduncles to 130
× 0.4–1 mm; spikes 9–14 × 9–12 mm when dry, roughly spherical. Spikelets 3.5–4.5 × 1.5–2.3 mm
wide, 1-flowered, densely packed, ovate, strongly compressed laterally. Glumes 3.6–4.1 mm long
including apical mucro 0.4–0.7 mm long, slightly recurved, distichously arranged, imbricate,
ovate, folded lengthwise with median longitudinal 'keel' of sponge-like tissue closely flanked by
secondary vein on each side leaving wide flanks of glume veinless, greenish yellow maturing to
yellow-brown; rachilla of spikelet winged. Stamens 3; pollen grains obtusely pyramidal; filaments
ribbon-shaped; anthers 0.4–0.65 × 0.1–0.15 mm wide, narrowly ellipsoid. Style 3-branched.
Nutlet 2.2–3.4 × 0.5–0.6 mm, narrowly spindle-shaped, tapering at both ends, straight, obtusely
trigonous; surfaces dense with warm brown, acute-conical, flat-topped papillae.

Botswana. N: Central Dist., Motsetso (Marsupe) R., 120 km WNW of Francistown
on Maun road, 960 m, 2.v.1957, *Drummond* 5301a (GRA, K, LISC, PRE, SRGH). SE:
Kweneng Dist., Thamaga, 25.iii.1977, *Camerik* 144 (PRE). **Zambia**. B: Kaoma Dist.,
Luena R., 13.v.1993, *Bingham* (NU). W: Ndola Dist., Itawa dambo, 1300 m, 29.iii.1960,
Robinson 3437 (K). C: Lusaka Dist., Chinyunyu Hot Spring, 27.ii.2000, *Bingham* 12141
(K). E: Lundazi Dist., South Luangwa Nat. Park (Nsefu Game Reserve), Chichele salt
pans, 13.vi.1991, *Bingham & Lewis* 7754 (NU). S: Gwembe Dist., Chipepo, Lake Kariba,
485 m, 6.i 1965, *Mitchell* 1013 (K). **Zimbabwe**. N: Hurungwe Dist., Mwami (Miami),
Expt. Farm, Mfuti area, 1370 m, 7.iii.1947, *Wild* 1738 (K, SRGH). W: Hwange Dist.,
Hwange Nat. Park, 14 km S of Maun Camp, 18°20'S 27°03'E, 17.iv.1972, *Gibbs Russell*
1600 (K, SRGH). C: Harare (Salisbury), University College, 29.iii.1963, *Loveridge*
636 (K, PRE, SRGH). S: Beitbridge Dist., Shashi R., 5 km from Tuli Police Camp,
3.v.1959, *Drummond* (K, SRGH). **Malawi**. N: Nyika Plateau, 2.iii.1903, *McClounie* 67
(B). **Mozambique**. MS: Tambara Dist., Nhacolo (Tambara), 15.v.1971, *Torre & Correia*
18456 (COI, LISC, LMA, LMU, PRE).

Also in Angola, Tanzania, South Africa (Gauteng, KwaZulu-Natal, Limpopo,
Mpumalanga), Swaziland, Namibia and Madagascar. Cultivated lands, seasonally wet

areas, roadsides, lakes and rivers; 400–1400 m.

Conservation notes: Widespread; not threatened.

Subsp. *cyperoides* is widespread on the Indian subcontinent and is geographically well-separated from the African material. It is similar but the glumes have an apical mucro 0.1–0.4(5) mm long, while in subsp. *africana* these are 0.4–1 mm long. Although I do not consider these differences sufficient to justify separate specific ranking, they are sufficient to warrant subspecific status rather than the varietal status previously recognised.

18. **SPHAEROCYPERUS** Lye[18]

Sphaerocyperus Lye in Bot. Not. **125**: 214 (1972).

Strongly stoloniferous leaf-bearing perennial. Culms scapose, slender. Leaves without ligule, lower reduced to sheaths. Inflorescence capitate of closely-packed greenish-white spikelets; primary bracts foliose, spreading, becoming reflexed. Spikelets lanceolate, laterally flattened, shed as units leaving a basal reduced bract and persistent tiny prophyll. Glumes distichous, sterile, except uppermost enfolding a bisexual floret (and occasionally a continuation of rachilla). Perianth 0. Stamens (2)3. Style branches 3, base not distinct, not or slightly thickened, persistent. Nutlets narrowly obovoid to ellipsoid, subtrigonous, beaked, surface smooth, becoming minutely punctate.

A monotypic genus, endemic to South Tropical Africa.

The affinity of the genus is with *Cyperus*, but with considerable specialization. Larridon *et al.* in Bot. J. Linn. Soc **172**: 106–126 (2013) placed the species of *Sphaerocyperus* in *Cyperus*, but it is kept separate here.

Sphaerocyperus erinaceus (Ridl.) Lye in Bot. Not. **125**: 214 (1972). —Haines & Lye, Sedges & Rushes E. Afr.: 293, fig.609 (1983). —Browning & Gordon-Gray in Nordic J. Bot. **13**: 507 (1993). —Verdcourt in F.T.E.A., Cyperaceae: 261, fig.39 (2010). Type: Angola, Huíla, near Monino and Mupanda streams, iv.1860, *Welwitsch* 6788 (BM holotype, LISU). FIGURE 14.**42**.

Schoenus erinaceus Ridl. in Trans. Linn. Soc. London, Bot. **2**: 165 (1884).

Rhynchospora erinacea (Ridl.) C.B. Clarke in Durand & Schinz, Consp. Fl. Afr. **5**: 654 (1894); in F.T.A. **8**: 479 (1902). —Robinson in Kirkia **1**: 41 (1961). —Napper in J. E. Africa Nat. Hist. Soc. Coryndon Mus. **24**(5): 42 (1964).

Cyperus erinaceus (Ridl.) Kük. in Boissiera **7**: 103 (1943).

Actinoschoenus erinaceus (Ridl.) Raymond in Mitt. Bot. Staatssamml. München **10**: 588 (1971).

Robust perennial, 70–100 cm tall; stolons stout, creeping, clothed in leaf bases with reduced or fully developed blades. Culms erect, slender, base slightly swollen, remainder nodeless. Leaves few, 3–4 mm wide, lowermost reduced to a sheath, upper shorter than inflorescence; blade stiff, 30–50 cm long, apex acuminate. Inflorescence a solitary, terminal, greenish-white head c. 20 mm wide, subtended by 2–3 leaf-like foliose bracts, 20–120 mm long, reflexed at maturity. Spikelets numerous, 10–12 × 1–2 mm, closely-packed in indistinct groups, each spikelet narrowly lanceolate, comprising small bract and smaller propyhll with (5)7 glumes that increase in size upwards (lower 8–9 mm, upper 10–11 mm), sterile except uppermost (largest) subtending one bisexual floret and not always a continuation, to 5 mm long, of spikelet rachilla; spikelets deciduous as units, bract and prophyll persistent. Perianth 0. Stamens (2)3, anthers extruded from enveloping glume at anthesis; gynoecium developing late. Style branches 3. Nutlet 3–4 × 1.15 mm, linear 2-angled, one face flat, other rounded, sometimes with detectable faintly marked third angle; surface smooth to faintly cellular, eventually minutely punctate.

---

[18] by J. Browning and K.D. Gordon-Gray†

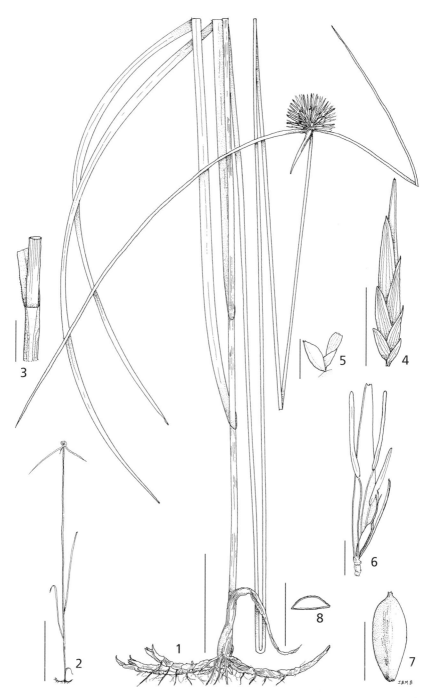

Fig. 14.**42**. SPHAEROCYPERUS ERINACEUS. 1, 2, habit; 3, leaf sheath; 4, spikelet without bract and prophyll; 5, bract and prophyll; 6, floret and rachilla within apical glume; 7, 8, nutlet immature, surface view and section. 1–6 from *Robinson* 3553; 7, 8 from *Lisowski* 972. Scale bars: 1 = 40 mm; 2 = 250 mm; 3, 4 = 5 mm; 5 = 1 mm; 6–8 = 2 mm. Drawn by Jane Browning.

Zambia. N: Mbala Dist., by Lunzua waterfall, 23.ii.1955, *Richards* 4650 (K). W: Mwinilunga Dist., 95 km E of Mwinilunga, 1350 m, 15.iv.1960, *Robinson* 3553 (K). E: Lundazi Dist., Nyika foothills, 1600 m, 13.iii.1961, *Robinson* 4472 (K).

Also in Tanzania, D.R. Congo and Angola. Undergrowth in miombo woodland; 1000–1600 m.

Conservation notes: Widespread and locally common; not threatened.

Plants are strongly rhizomatous, leafy perennials.

19. **ALINULA** J. Raynal[19]

Alinula J. Raynal in Adansonia, n.s. **17**: 43 (1977). —Goetghebeur & Vorster in Bull. Jard. Bot. Belg. **58**: 457–465 (1988).

> *Aliniella* J. Raynal in Adansonia, n.s. **13**: 157 (1973).
> *Marisculus* Goetgh. in Bull. Jard. Bot. Belg. **47**: 444 (1977).
> *Cyperus* L. subgen. *Alinula* (J. Raynal) Lye in Nordic J. Bot. **3**: 230 (1938).
> —Haines & Lye, Sedges & Rushes E. Afr.: 250 (1983).

Small annual herbs 3–25 cm high. Culm scapose, (flattened) rounded to trigonous, glabrous and smooth. Leaves all basal, to 1.3 mm wide, leaf sheaths brown to reddish-brown, eligulate. Inflorescence (bis)anthelate or congested, occasionally pseudolateral, 2 to many pedicelled to semi-sessile spikes. Spikes 1–6 mm, ovoid to cylindrical, bright yellow-green to dark reddish-brown, with tubular prophyll near base; pedicels absent or to 30 mm long; spikes with numerous densely spirally-set spikelet bracts, 0.6–1 mm long, each subtending a highly reduced, lateral 1-flowered spikelet with an adaxial empty prophyll and a flower-bearing abaxial glume. The single bisexual flower has a trimerous ovary with 3 stigmatic branches, 1–2 lateral stamens, and no perianth. Fruits narrow, linear to obovate, straight to slightly curved, rounded trigonous, shortly beaked, (pale)brownish to reddish-brown, rarely with a white, slightly lobed collar.

A genus of 4 species from seasonally wet open grasslands or disturbed wet alluvial soils; rarely collected and probably often overlooked.

Larridon *et al.* in Bot. J. Linn. Soc **172**: 106–126 (2013) placed the species of *Alinula* in *Cyperus*, but it is kept separate here.

1. Inflorescence open, conspicuously branched in 2 orders; spikelet rachilla present . . . . **3.** *paradoxa*
 – Inflorescence condensed or with 1–2 pedicelled spikes; spikelet rachilla absent 2
2. Inflorescence pale, yellowish; glume utriculiform, adaxially closed . . . . . **4.** *peteri*
 – Inflorescence dark, brownish; glume not utriculiform. 3
3. Inflorescence with 3–6 pedicelled spikes often present; fruit 1.4–1.6 mm long, without hypogynous disc; stamens 1. . . . . **2.** *malawica*
 – Inflorescence congested; fruit 1–1.2 mm long, with hypogynous disc; stamens 2 . . . . **1.** *lipocarphioides*

1. **Alinula lipocarphioides** (Kük.) J. Raynal in Adansonia, n.s. **17**: 43 (1977). —Beentje in F.T.E.A., Cyperaceae: 266 (2010). Type: Tanzania, Mangati, Mdumgari, 11.viii.1926, *Peter* 43922b (B holotype).

> *Ficinia lipocarphioides* Kük. in Repert. Spec. Nov. Regni Veg. Beih. **40**(1), Anh.: 125, t.87 (1936).
> *Aliniella lipocarphioides* (Kük.) J. Raynal in Adansonia, n.s. **13**: 157 (1973).
> *Cyperus lipocarphioides* (Kük.) Lye in Nordic J. Bot. **3**: 230 (1983). —Haines & Lye, Sedges & Rushes E. Afr.: 251, fig.504 (1983).

---

[19] by K. Bauters

Small annual herb with few basal leaves; stems 4–15 cm long, trigonous, glabrous and smooth. Leaves narrow, to 6 cm long by 1.3 mm wide. Inflorescence congested, with 2–6 spikes, rarely with a single pedicelled spike. Spikes 2–5 mm long, ovoid, dark red-brown, with numerous densely spirally set spikelet bracts, each c. 0.8 mm long, subtending a reduced lateral, 1-flowered spikelet; spikelets with a small prophyll c. 0.9 mm long and first glume c. 1.8 mm long. Stamens 2, lateral. Ovary with 3 stigmatic branches. Fruit c. 1.2 mm long, narrowly ellipsoid, almost straight, brownish, enclosed at base by a white, slightly lobed collar.

Zambia. N: Mbala Dist., Sansia Falls on Kalambo road, 8.v.1961, *Richards* 15131 (K).

Also in D.R. Congo, Ethiopia, Uganda, Kenya and Tanzania. In seasonally wet grassland or in damp hollows in open ground: c. 1500 m.

Conservation notes: Restricted distribution in the Flora area, but probably not threatened globally (although Haines & Lye state it is very rare in Uganda).

The illustration in Haines & Lye (1983: fig.504) is partly wrong – a spikelet is given with a bract, a prophyll, a first glume and a mysterious second glume; this second glume has not been observed in other specimens.

2. **Alinula malawica** (J. Raynal) Goetgh. & Vorster in Bull. Jard. Bot. Belg. **58**: 462 (1988). Type: Malawi, 5 km SE of Chitipa (Fort Hill), 11.iii.61, *Robinson* 4440 (NY holotype, M).

   Mariscus malawicus J. Raynal in Adansonia, n.s. **13**: 159 (1973).
   Cyperus malawicus (J. Raynal) Lye in Nordic J. Bot. **3**: 231 (1983). —Haines & Lye, Sedges & Rushes E. Afr.: 251, fig.505 (1983).

Small annual herb with few basal leaves; stems 5–15 cm long, trigonous, glabrous and smooth. Leaves narrow, to 5 cm long by 1.5 mm wide. Inflorescence anthelate or completely congested, with 3–6 spikes, rays to 15 mm long, each with tubular prophyll at base. Spikes 4–6 mm long, ovoid to cylindrical, dark grayish brown, with numerous densely spirally set spikelet bracts, each c. 0.9 mm long, subtending a reduced lateral, 1-flowered spikelet; spikelet with a small prophyll c. 1 mm long and first glume c. 2 mm long. Stamen 1, lateral. Ovary with 3 stigmatic branches. Fruit c. 1.6 mm long, narrowly linear, slightly curved, brownish.

Zambia. N: Isoka Dist., 40 km SE of Nakonde, 23.iii.1961, *Robinson* 4533B (M). **Malawi**. N: Chitipa Dist., 5 km SE of Chitipa (Fort Hill), 11.iii.1961, *Robinson* 4440 (M, PRE).

Known only from Malawi and Zambia, but likely to be found in S Tanzania. In seasonally wet sandy soil; 1200–1700 m.

Conservation notes: Said to be very rare (Haines & Lye 1983); possibly Vulnerable.

3. **Alinula paradoxa** (Cherm.) Goetgh. & Vorster in Bull. Jard. Bot. Belg. **58**: 461 (1988). —Gordon-Gray in Strelitzia **2**: 21, fig.4 (1995). —Beentje in F.T.E.A., Cyperaceae: 265 (2010). Type: Madagascar, Stampika, 1913, *Perrier de la Bâthie* 2423b (P holotype). FIGURE 14.**43**.

   Lipocarpha paradoxa Cherm. in Bull. Soc. Bot. France **68**: 425 (1922).
   Mariscus paradoxus (Cherm.) Cherm. in Bull. Soc. Bot. France **72**: 169 (1925). —Haines & Lye in Bot. Not. **124**: 477 (1971).
   Cyperus subparadoxus Kük. in Engler, Pflanzenr. **4**, 20(101): 525 (1936). —Haines & Lye, Sedges & Rushes E. Afr.: 251, fig.506 (1983).
   Cyperus fimbristyloides T. Koyama in Bot. Mag. Tokyo **73**: 438 (1960), superfluous name, non *C. paradoxus* Steud.
   Pseudolipocarpha paradoxa (Cherm.) Vorster in Rev. Mariscus S. Afr.: 319 (1978).

Small annual herb with few basal leaves; stems 5–25 cm long, (flattened) trigonous, glabrous and smooth. Leaves narrow, to 10 cm long by 1.2 mm wide. Inflorescence (bis)anthelate, with 5 to many spikes, rays to 30 mm long, each with a tubular prophyll. Spikes 1–4 mm long, spherical to ovoid, dark reddish brown, with numerous densely spirally set spikelet bracts, each c. 0.6 mm long, subtending a reduced lateral, 1-flowered, spikelet. Spikelets with a small prophyll c. 0.5 mm long, first glume c. 1 mm long and rachilla c. 1.1 mm long. Stamens 2, lateral. Ovary with 3 stigmatic branches. Fruit c. 1 mm long, narrowly ellipsoid, slightly curved, reddish brown.

Mozambique. MS: Sussundenga Dist., Moribane Forest, roadside from Chimoio (Vila Pery) to Forozi R., 8.iv.1952, *Chase* 4534 (BR, BRLU, LISC). M: Maputo Dist., Inhaca Is., 4.iii.1958, *Mogg* 30373 (J, SRGH).

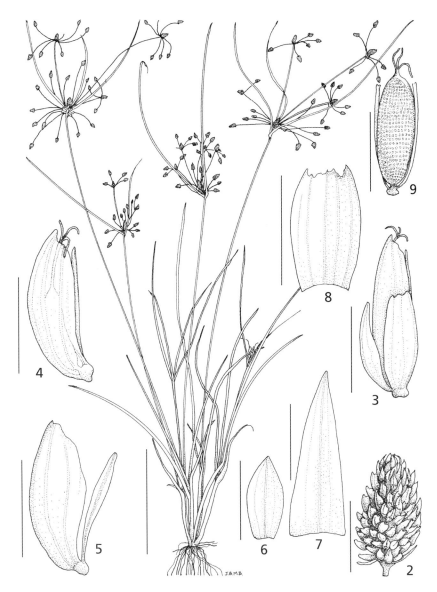

Fig. 14.**43**. ALINULA PARADOXA. 1, habit; 2, spike; 3, spikelet; 4, glume and floret; 5, glume and axis of spikelet; 6, 7, upper and lower spike bracts; 8, prophyll; 9, nutlet. All from *Reid* 1207. Scale bars: 1 = 40 mm; 2 = 2 mm; 3–9 = 0.5 mm. Drawn by Jane Browning. Reproduced from Strelitzia (1995) by kind permission of South African National Biodiversity Institute.

Also in Kenya, Tanzania, Namibia, South Africa and Madagascar. In and near ricefields near sea-level and mangrove/coconut plantation fringe, on sandy soil; 0–20 m.

Conservation notes: Apparently localised (Haines & Lye 1983 state it is very rare); possibly Near Threatened.

4. **Alinula peteri** (Kük.) Goetgh. & Vorster in Bull. Jard. Bot. Belg. **58**: 464 (1988). —Beentje in F.T.E.A., Cyperaceae: 265 (2010). Type: Tanzania, Dodoma Dist., Uyansi, Chaya towards Kazikazi, km 675.5, 1250 m, 6.i.1926, *Peter* 34327a (B holotype).

Ascolepis peteri Kük. in Repert. Spec. Nov. Regni Veg. Beih. **40**: 386 (1936).
Mariscus peteri (Kük.) Goetgh. in Bull. Jard. Bot. Belg. **47**: 444 (1977). —Haines & Lye, Sedges & Rushes E. Afr.: 311, fig.642 (1983).
Cyperus microaureus Lye in Lidia **3**: 132 (1994).

Small annual herb with few basal leaves; stems 3–20 cm long, rounded, glabrous and smooth. Leaves narrow, to 5 cm long by 1.2 mm wide. Inflorescence congested, with 2–8 spikes. Spikes 2–6 mm long, ovoid to spherical, bright yellow-green, with numerous densely spirally set spikelet bracts, each c. 1 mm long, subtending a reduced lateral, 1-flowered, spikelet. Spikelets with a small prophyll c. 0.6 mm long, first glume c. 1.8 mm long, utriculiform with an apical adaxial slit. Stamens 2, lateral. Ovary with 3 stigmatic branches. Fruit c. 1 mm long, narrowly obovate, straight, (pale)brownish.

Zambia. B: Sesheke Dist., Machili, 20.ii.1961, *Fanshawe* 6290 (SRGH). N: Mbala Dist., Kawimbe, ii.1957, *Richards* 8323 (B, K). S: Namwala Dist., 160 km NNW of Choma, 8.i.1957, *Robinson* 2076 (K, LISC, SRGH). **Malawi**. N: Chitipa Dist., 5 km SE of Chitipa (Fort Hill), 11.iii.1961, *Robinson* 4438 (B, K, M, MO, SRGH).

Also in Ethiopia, Kenya and Tanzania. On seasonally damp sand in rough grassland, sometimes in dambos; 1000–1500 m.

Conservation notes: Widespread; not threatened.

20. ASCOLEPIS Steud.[20]

Ascolepis Steud., Syn. Pl. Glumac. **2**: 105 (1855), conserved name. —Goetghebeur in Adansonia, n.s. **19**: 269–305 (1980).

Annuals or tufted perennials, rarely rhizomatous or stoloniferous. Culms scapose, often bulbously thickened at base. Leaves without a ligule. Inflorescences terminally capitate with 1 to few spikes. Primary bracts leaflike, spreading to reflexed. Spikes often vividly colored, with many densely spirally arranged spikelet bracts, much smaller than spikelet; spikelets reduced to 1 glume, without prophyll, accompanied by rachilla remnant in some species, glume polymorphic, rarely utriculiform, often with elongated or swollen upper parts. Flowers bisexual. Bristles absent. Stamens 1–3(5). Style 2–3(5)-fid, style base not distinct, deciduous. Achenes obovate to oblong, rounded trigonous to lenticular, rarely stipitate, surface finely puncticulate.

A genus of about 20 species, pantropical, mainly in Africa. The perennials usually grow in marshes while the annuals prefer temporarily wet soils.

Larridon *et al.* in Bot. J. Linn. Soc **172**: 106–126 (2013) and in Phytotaxa **166**: 33–48 (2014) incorporated the species of *Ascolepis* into *Cyperus*, but it is kept separate here.

1. Glume margins adaxially joined for at least half their length 2
– Glume margins free to base 4
2. Glumes conspicuously dorsiventrally flattened, laterally winged, basal part

---

[20] by K. Bauters

obovate to rhombic, gradually narrowing into a rather broad apical part, tip
rounded; spikes 1–4(6), 6–10 mm long . . . . **2.** *capensis*
– Glumes not dorsoventrally flattened, trumpet-shaped; spikes (1)2–6, very small 3
3. Inflorescence pale brown or greyish; glume tips rounded or minutely spinulose;
rudimentary rachilla present in spikelet . . . . **10.** *pusilla*
– Inflorescence reddish; glume tips reflexed at maturity; rudimentary rachilla
absent in spikelet . . . .**4.** *erythrocephala*
4. Spikelet 3-scaled, with bract, glume and rachilla. 5
– Spikelet 2-scaled, tip of bract and glume ± rounded. 8
5. Very slender annual species; inflorescence less than 10 mm wide, composed
of 2–3 easily recognizable spikes; spikes with whitish appearance due to hollow,
swollen glume tips . . . .**1.** *ampullacea*
– ± Robust perennial species, inflorescence 15–60 mm wide, in a head or head-
like 6
6. Inflorescence head-like, of 3–7 tightly packed spikes, recognizable as bundles of
long and narrow radiating elongated glumes . . . . **6.** *majestuosa*
– Inflorescence with 1 spike, spikelets sessile on a broadly conical axis, no separate
bundles of elongated glumes present 7
7. Glumes bent, basal and apical part almost at right angles; involucral bracts,
spikelet bracts and basal part of glumes conspicuously reddish-nerved . . **7.** *pinguis*
– Glumes not bent, reddish brown with the apical part yellowish white, middle part
inflated, not shining; all nerves prominent . . . .**5.** *fibrillosa*
8. Heads very dense, glume apical parts tightly packed; glume 1.8–3.5 mm long,
head 5–9 mm wide; slender non-stoloniferous perennials 9
– Heads less dense, glume apical parts ± spreading 10
9. Apical part of glume very slender, narrowly triangular, tip subacute . . . . . **3.** *densa*
– Apical part of glume dorsiventrally flattened, plumb triangular, surface cells
shiny, ± inflated, tip rounded . . . .**11.** *trigona*
10. Spike very small, 4–6 mm wide; glumes yellow to orange-brown with shiny ±
inflated surface cells, crescent-shaped in cross-section, rather narrow . . . .
. . . . **9.** *pseudopeteri*
– Not as above . . . .**8.** *protea*

1. **Ascolepis ampullacea** J. Raynal in Adansonia, n.s. **13**: 159 (1973). —Goetghebeur
in Adansonia, n.s. **19**: 296 (1980). —Haines & Lye, Sedges & Rushes E. Afr.: 310,
fig.639 (1983). Type: Zambia, Mporokoso, 8 km N of Muzombwe, W side of Mweru
Wantipa, 16.iv.1961, *Phipps & Vesey-FitzGerald* 3233 (NY holotype, K, P, SRGH).
Cyperus ampullaceus (J. Raynal) Bauters in Phytotaxa **166**: 36 (2014).

Slender, loosely tufted annual; stem 5–10 cm high, 0.2–0.5 mm across. Inflorescence of 2–3
spikes, whitish, marginal glumes not elongated, apical spike c. 4 mm long, ovoidal, lateral ones
c. 3 mm, more spheroidal; 2–3 larger involucral bracts to 2 cm long. Spikelets densely spirally
imbricate on slender cylindrical axes. Spikelet bract 1–1.5 mm long, narrowly triangular, hyaline,
central nerve pale brown, thickened at tip. Glume c. 2 mm long, whitish hyaline, basal part very
concave, tightly packed round fruit, hyaline, prominently ribbed, apical part inflated, bladder-
like, rounded at tip, conspicuously clear whitish. Stamen 1, lateral; filament 1.5 mm long. Style
c. 1 mm long, deeply 3-cleft. Fruit c. 1 mm long, oblong, subtrigonous, red-brown. Rachilla c. 1.5
mm long, linear hyaline, with a narrow central nerve, enclosed by hyaline first glume wings.

Zambia. N: Mporokoso Dist., 8 km N of Muzombwe, W side of Mweru Wantipa,
16.iv.1961, *Phipps & Vesey-Fitzgerald* 3233 (K, NY, P, SRGH).
Not found elsewhere. On laterite, grassy areas, hard-pans in mateshi thickets; c.
1000 m.
Conservation notes: Known only from the type locality; probably Vulnerable.

2. **Ascolepis capensis** (Kunth) Ridl. in Trans. Linn. Soc. London, Bot. **2**: 164 (1884). —
Goetghebeur in Adansonia, n.s. **19**: 291 (1980). —Haines & Lye, Sedges & Rushes
E. Afr.: 310, fig.641 (1983). —Gordon-Gray in Strelitzia **2**: 23 (1995). —Beentje
in F.T.E.A., Cyperaceae: 273 (2010). Type: South Africa, Cape, Omsamcaba,
12.v.1832, *Drège* 4389 (B† holotype, K, P). FIGURE 14.**44**.

  Platylepis capensis Kunth, Enum. Pl. **2**: 269 (1837).

  Platylepis dioica Steud., Syn. Pl. Glumac. **2**: 131 (1855). Type: South Africa, Cape, Katberg,
  xi.1832, *Drège* 3953 (P holotype).

  Cyperus ascocapensis Bauters in Phytotaxa **166**: 36 (2014), new name, non *Cyperus capensis*
  (Steud.) Endl. (1842).

Tufted perennial herb on an ascending rhizome, often with slender underground runners;
stem 20–60 cm high, 0.3–1.3 mm across; stem base slightly bulbously thickened, covered in dark
brown to blackish fibrous leaf sheaths; yellowish to reddish underground runners break through
the leaf sheath mass, internodes 0.5–1.5 cm long, nodes with small or sometimes well-developed
cataphylls. Inflorescence of 1–4(6) creamish white spikes, ± spheroidal, 6–10 mm long, marginal
glumes not or only slightly elongated; larger involucral bracts 2, to 6 cm long. Spikelets densely
spirally imbricate on broadly conical axis, sometimes on a slender cylindrical pale brown axis.
Spikelet bract 2–3.5 mm long, ± spathulate, hyaline and red-dotted, 1-nerved. Glume 3.5–5.5(7)
mm long, dorsiventrally compressed, margins adaxially joined, an adaxial 2-cleft valve at top
of basal part, basal part 2.5–3 mm long, obovate to rhombic, thicker part often reddish brown,
1–3-nerved; thin lateral wings whitish hyaline, 0.25–0.5 mm broad, apical part 0.5–3(5) mm long,
not sharply differentiated from basal part, whitish, slightly swollen, ± broadly triangular, obtuse
to rounded at tip. Stamens 2–3, lateral and anterior; filament 2.5–3 mm; anther 1–2 mm long.
Style 1–2 mm long, deeply 2-cleft. Fruit 1–1.5 mm long, narrowly obovate, ± lenticular, dark red
brown on whitish 0.2–0.6 mm long stipe.

Zambia. B: Senanga Dist., 16 km N of Senanga, 1040 m, 31.vii.1952, *Codd* 7311
(K, PRE). N: Mbala Dist., L. Chila, 19.vi.1956, *Robinson* 1685 (K). W: Solwezi Dist.,
W of Solwezi on Chingola road, 1.x.1947, *Greenway & Brenan* 8147 (K). C: Serenje
Dist., Kanona, Kundalila Falls, 13.x.1963, *Robinson* 5733 (K). S: Choma Dist., 8 km E
of Choma, 26.iii.1955, *Robinson* 1177 (K). **Zimbabwe**. N: Guruve Dist., Nyamunyeche
Estate, W of farm, 16.x.1918, *Nyariri* 414 (GENT, SRGH). C: Marondera Dist.,
Marondera, Digglefold Farm, 1525 m, ii.1948, *Corby* 14 (K, SRGH). E: Nyanga Dist.,
Nyanga Nat. Park, Mare dam, 1830 m, 4.iii.1969, *Jacobsen* 3721 (K, SRGH). **Malawi**. N:
Mzimba Dist., Mzuzu, Katoto, 1370 m, 24.x.1969, *Pawek* 2909 (K). S: Zomba Dist., Mt
Zomba, xii.1896, *Whyte* s.n. (K). **Mozambique**. N: Marrupula Dist., between Marrupula
and Chefe Ancoreta, 24.i.1961, *Carvalho* 459 (K, LMA). Z: Gurué Dist., Mt Namuli,
Muretha plateau, 1860 m, 17.xi.2007, *Mphamba* 1 (K, LMA, MAL). T: Angónia Dist.,
Ulónguè, veterinary station, 11.xii.1980, *Macuácua* 1425 (K, LISC, LMA).

From Mali and Sierra Leone east to Ethiopia and south to South Africa. In wet or
seasonally wet grassland, at swamp edges and on tussocks; 1000–1900 m.

Conservation notes: Widespread; not threatened.

3. **Ascolepis densa** Goetgh. in Adansonia, n.s. **19**: 285 (1980). —Haines & Lye,
Sedges & Rushes E. Afr.: 307, fig.634 (1983). —Beentje in F.T.E.A., Cyperaceae:
273 (2010). Type: Zambia, Siamambo, Choma, 1300 m, 20.iii.1958, *Robinson* 2814
(SRGH holotype, K, P).

  Cyperus ascodensus Goetgh. in Phytotaxa **166**: 36 (2014), new name, non *Cyperus densus*
  Link (1820).

Tufted, small and slender perennial herb; stem 10–40 cm high, 0.3–0.7 mm across; stem base
bulbously thickened, covered by a few withering leaf sheaths. Inflorescence capitate, solitary spike
5–8 mm wide, whitish, ± spheroidal; marginal glumes not elongated; larger involucral bracts 2(3),
to 6 cm long. Spikelets very densely spirally imbricate on a broadly conical axis; spikelet bract
1.5–2.3 mm long, narrowly triangular, whitish-hyaline, upper part terete, tip subacute. Glume

Fig. 14.**44**. ASCOLEPIS CAPENSIS. 1, habit (× ⅔); 2, inflorescence (× 3); 3, 4, front view of 1-flowered spikelet (the second, 4, higher up in inflorescence) (× 12, 8); 5, back view of 1-flowered spikelet (× 8); 6, enclosed nutlet (× 8). 1 from *Richards* 7681; 2 from *Vesey-Fitzgerald* 2863; 3–6 from *Goyder et al.* 3769. Drawn by Juliet Williamson. Reproduced from Flora of Tropical East Africa (2010).

1.8–3.4 mm long, basal part c. 1 mm long, concave, hyaline, nerves poorly developed, floral parts adaxially enclosed by glume wings, apical part 1–2 mm long, whitish, narrowly triangular, tip subacute. Stamens 2, lateral; filament to 2 mm long; anthers 0.6–1 mm long. Style 1.5 mm long, deeply 3(4)-cleft. Fruit 0.5 mm long, obovate, subrigonous, dark red-brown.

Zambia. N: Mbala Dist., Ngundu, 1740 m, 8.v.1959, *Richards* 11385 (K, P). C: Kabwe Dist., 1 km W of Kamaila Forest Station, 36 km N of Lusaka, 9.ii.1975, *Brummitt, Hoper & Townsend* 14304 (K). S: Choma Dist., 8 km E of Choma, 26.iii.1955, *Robinson* 1172 (K).

Also in Tanzania, D.R. Congo and Angola. In dambo or seasonally wet habitats in grassland; 1300–1740 m.

Conservation notes: Widely distributed, but rare.

4. **Ascolepis erythrocephala** S.S. Hooper in Kew Bull. **37**: 605 (1983). —Haines & Lye, Sedges & Rushes E. Afr.: 309, fig.638 (1983). —Beentje in F.T.E.A., Cyperaceae: 269 (2010). Type: Tanzania, Songea Dist., 6.5 km W of Songea, 28.iv.1956, *Milne-Redhead & Taylor* 9940 (K holotype, BM, NY, P).

  Cyperus erythrocephalus (S.S. Hooper) Bauters in Phytotaxa **166**: 37 (2014).

Small, slender, erect ephermeral with thin fibrous roots; stem 4–10 cm high, 0.25 mm across. Inflorescence capitate, of 1–3 spikes, reddish at maturity, main spike hemispherical to oblong, 3–4 mm long, with or without 1–2 small rounded lateral spikes; involucral bracts 3, leaf-like, very unequal, to 3 cm, becoming sharply deflexed from a widened membranous base. Spikelet bract 1.5 mm long, lanceolate, membranous with faint median nerve. Glume closed, 1.25 mm long, trumpet-shaped, enclosing floral parts, conical, membranous, dark-ribbed basal portion much expanded, curved into succulent, dark orange-red collar with small orifice and reflexed, pale or green, sharp apical part, c. 0.4 mm long. Rachilla none. Stamens 1(2). Style 3-cleft. Fruit narrowly obovate, 0.75 mm long, very shortly stipitate, brown.

Zimbabwe. N: Harare South, Old Charter Road, 1370 m, 27.iii.1952, *Wild* 3792 (K). Also in S Tanzania. In secondary grasslands; c. 1350 m.

Conservation notes: Only known from 3 specimens, and within the Flora area known from only one; possibly Vulnerable but overlooked.

5. **Ascolepis fibrillosa** Goetgh. in Bull. Jard. Bot. Belg. **47**: 439 (1977). Type: D.R. Congo, Kwango, Manzala plain, 19.ii.1954, *Devred* 1517 (BR holotype, BRVU, K, P).

  Cyperus ascofibrillosus Goetgh. in Phytotaxa **166**: 36 (2014), new name, non *Cyperus fibrillosus* Kük. (1921).

Robust, tufted perennial herb; stem 40–70 cm high, 1–1.5 mm across; stem base bulbously thickened, covered in dense fibrous coat of withered leaf sheaths. Inflorescence capitate, solitary spike 15–20(25) mm wide, spheroidal, yellowish white, all glumes ± elongated; larger involucral bracts 3–4, to 12 cm long. Spikelets densely spirally imbricate on a broadly conical axis; spikelet bract 3–4.5 mm long, narrowly triangular, whitish hyaline, with pale brown central nerve, tip subterete. Glume 6–10 mm long, middle part inflated, often laterally compressed, basal part 3–4 mm long, very concave, hyaline, all nerves very prominent, reddish brown, floral parts adaxially enclosed by glume wings, apical part 3–7 mm long, yellowish white, subtrigonous to subrhombic, rounded at tip. Stamens 3–5, lateral and anterior; filament 3–4 mm long. Style c. 2 mm long, deeply 3(5)-cleft. Fruit 2 mm long, obovate, subtrigonous, dark red brown, basal epidermal cells inflated; rachilla 2.5–3 mm long, whitish hyaline, reddish brown at base, subterete, slightly thickened to narrowly winged at tip, completely enclosed by glume wings, ± persistent.

Zambia. W: Mwinilunga Dist., 10 km on road from Matonchi Farm, 17.xi.1962, *Richards* 17260 (BR, K).

Also in D.R. Congo and Angola. In grass savanna and degraded open *Brachystegia* woodland; c. 1200 m.

Conservation notes: Known from only one locality in the Flora area; possibly Vulnerable.

6. **Ascolepis majestuosa** P.A. Duvign. & G. Léonard in Bull. Soc. Roy. Bot. Belgique **90**: 188 (1958). —Goetghebeur in Adansonia, n.s. **19**: 299 (1977). Type: D.R. Congo, between Kasji & Dilolo, 14.viii.1956, *Duvigneaud & Timperman* 2316 A1 (BRVU holotype).

> *Cyperus majestuosus* (P.A. Duvign. & G. Léonard) Bauters in Phytotaxa **166**: 37 (2014).

Robust, tufted perennial herb; stem 20–40 cm high, 1–1.5 mm across; stem base bulbously thickened, covered in dense coat of dark brown to almost blackish leaf sheaths, becoming fibrous. Inflorescence condensed, (10)20–40(50) mm wide, hemispherical, yellowish white, brownish at base, of 3–7 tightly packed spike-like structures; upper glumes of each spike extremely elongated, to 10–20(25) mm; larger involucral bracts 2–3, to 10 cm long. Spikelets densely spirally imbricate on a few slender axes; spikelet bract 2.5–3.5 mm, narrowly ovate-triangular, central nerve poorly developed, wings often red-dotted, tip subterete, subacute. Glumes ± dimorphic, lower glumes 3.5–4.5(6) mm long, basal part c. 2.5 mm long, very concave, reddish hyaline, nerves whitish, floral parts enclosed by glume wings, apical part 1–4 mm long, yellowish, subtriangular, straight, tip subacute; upper glumes 10–20(25) mm long, mostly empty, basal part 1–1.5 mm long, yellowish, subtriangular, tip subacute. Stamens 3, lateral and anterior; filament 2.75–3 mm long; anthers c. 1.5 mm long. Style 2.5 mm long, deeply 3-cleft. Fruit 1.25–1.5 mm long, obovate, subtrigonous, pale brownish; rachilla 2–3 mm long, subterete but sometimes narrowly winged, slightly broadened at tip, whitish hyaline, completely enclosed by glume wings, ± persistent.

Zambia. B: Mongu Dist., 61 km from Mongu on road to Kaoma, 30.i.1975, *Brummitt, Chisumpa & Polhill* 14187 (K, SRGH); Kalabo Dist., between Lukona & Angolan border, 9.v.1925, *Pocock* 223 (PRE). N: Kawambwa Dist., Kawambwa Nchelengi road, 1350 m, 26.xi.1961, *Richards* 15371 (K).

Also in D.R. Congo. In dambos and seasonally wet grassland; 1000–1400 m.

Conservation notes: Apparently rare (Goetghebeur 1980).

7. **Ascolepis pinguis** C.B. Clarke in Ann. Mus. Congo Belge, Bot., sér.2 **1**: 69 (1900). —Goetghebeur in Adansonia, n.s. **19**: 299 (1980). —Haines & Lye, Sedges & Rushes E. Afr.: 303, fig.625 (1983). —Beentje in F.T.E.A., Cyperaceae: 267 (2010). Types: D.R. Congo, Kitope, iii.1896, *Descamps* s.n. (BR syntype); Kalemie (Albertville), vi.1894, *Descamps* s.n. (BR lectotype), lectotypified by Goetghebeur (1980).

> *Cyperus ascopinguis* Goetgh. in Phytotaxa **166**: 36 (2014), new name, non *Cyperus pinguis* (C.B. Clarke) Mattf. & Kük. (1936).

Very robust, tufted perennial herb; stem 20–60(80) cm high, 0.8–2 mm across; stem base bulbously thickened, covered in a dense coat of broad reddish brown to dark brown leaf sheaths, becoming ± fibrous. Inflorescence capitate, solitary spike 30–50(60) mm wide, hemispherical, yellowish white to pale yellowish brown, all glumes extremely elongated; larger involucral bracts 2–4, to 15 cm long, conspicuously reddish nerved at base. Spikelets densely spirally imbricate on a broadly conical axis; spikelet bract 3–3.5 mm long, narrowly ovate-triangular, 3–5 nerved, nerves red-brown, wings hyaline, apical part subterete, subacute at tip. Glume 15–25(30) mm long, ± laterally compressed, basal part 2–3 mm long, very concave, reddish-hyaline, nerves red brown, floral parts adaxially enclosed by glume wings, apical part 13–22(27) mm long, almost perpendicular on basal part, subrhombic in cross section, yellowish white to yellowish brown, subacute at tip. Stamens 2–3, lateral and anterior; filament 2.5–3 mm; anthers 1–1.25 mm long. Style 1.5–2 mm long, deeply 3-cleft. Fruit 1–1.25 mm long, obovate, subtrigonous, pale brownish; rachilla 2–3 mm long, linear, swollen near top, hyaline with red brown central nerve, completely enclosed by first glume wings, ± persistent.

Zambia. N: Mbala Dist., Chinakila, Kaniyka Flats, 1200 m, 15.i.1965, *Richards* 19545 (K); Chinakila, iv.1960, *Fanshawe* 5617 (K).

Also in Tanzania, Burundi and D.R. Congo. In swamps and wet grasslands; c. 1200 m.

Conservation notes: Very localised in the Flora area; possibly Vulnerable.

Both syntypes bear a label with *Ascolepis pinguis* sp. nov. in Clarke's handwriting, but on the lectotype this label is completed by the short diagnosis, as published. There can be no doubt that both sheets, although badly collected specimens, represent the same taxon.

The thickened upper part of the rachilla is a compound structure of a small but clearly differentiated second glume surrounding the minute rachilla tip.

8. **Ascolepis protea** Welw. in Trans. Linn. Soc. London **27**: 75 (1869). —Goetghebeur in Adansonia, n.s. **19**: 275 (1980). Type: Angola, Pungo Andongo, Fundo Quilombo, xii.1856, *Welwitsch* 1667 (BM holotype, K, LISU). FIGURE 14.**45**.

Ascolepis protea var. *protea* in Trans. Linn. Soc. London **27**: 75 (1869). —Haines & Lye, Sedges & Rushes E. Afr.: 304 (1983).

Ascolepis protea var. *santolinoides* Welw. in Trans. Linn. Soc. London **27**: 77 (1869). —Haines & Lye, Sedges & Rushes E. Afr.: 305, fig.627 (1983).

Ascolepis protea var. *kyllingioides* Welw. in Trans. Linn. Soc. London **27**: 76 (1869), invalid name.

Ascolepis protea var. *bellidiflora* Welw. in Trans. Linn. Soc. London **27**: 76 (1869). —Mapaura & Timberlake, Checkl. Zimbabwe Pl.: 87 (2004). —Beentje in F.T.E.A., Cyperaceae: 270 (2010). Type: Angola, Pungo Andongo, Barrancos de Catete, ii.1857, *Welwitsch* 1668 (BM lectotype, K, LISU).

Ascolepis elata Welw. in Trans. Linn. Soc. London **27**: 79 (1869). —Haines & Lye, Sedges & Rushes E. Afr.: 304 (1983). Type: Angola, Pungo Andongo, iii.1857, *Welwitsch* 1670 (BM holotype, K, LISU).

Ascolepis protea var. *splendida* K. Schum. in Warburg, Kunene-Sambesi Exped.: 177 (1903). —Mapaura & Timberlake, Checkl. Zimbabwe Pl.: 87 (2004). Type: Angola, Okachitanda R., 26.ix.1899, *Baum* 158 (B holotype, BM, E, K, M, Z).

Ascolepis protea var. *protea*. —Goetghebeur in Adansonia, n.s. **19**: 275 (1980). —Beentje in F.T.E.A., Cyperaceae: 270 (2010).

Ascolepis protea var. *anthemiflora* (Welw.) Goetgh. in Adansonia, n.s. **19**: 277 (1980). —Beentje in F.T.E.A., Cyperaceae: 270 (2010). Type: Angola, Pungo Andongo, iii.1857, *Welwitsch* 1669 (BM holotype, K, LISU).

Ascolepis protea var. *floribunda* Goetgh. in Adansonia, n.s. **19**: 277 (1980). Type: D.R. Congo, Tukpwo, 13.vi.1958, *Gérard* 3873 (BR holotype).

Ascolepis protea var. *ochracea* (Meneses) Goetgh. in Adansonia, n.s. **19**: 277 (1980). —Beentje in F.T.E.A., Cyperaceae: 271 (2010). Type: Angola, Benguella, n.d., *Gossweiler* 3469 (LISU holotype, BR, K).

Ascolepis protea var. *stellata* Goetgh. in Adansonia, n.s. **19**: 279 (1980). —Mapaura & Timberlake, Checkl. Zimbabwe Pl.: 87 (2004). Type: Zimbabwe, Makonde Dist. Mhangura, Whindale Farm, 14.ii.1968, *Wild* 7684 (SRGH holotype, K, L, P).

Ascolepis lineariglumis Lye in Nordic. J. Bot. **2**: 561 (1983). —Beentje in F.T.E.A., Cyperaceae: 267 (2010). Type: Zambia, 13 km NW of Kabwe-Bonanza, 28°23'E 14°22'S, 1200 m, 8.iv.1972, *Kornaś* 1559 (KRA holotype, K).

Ascolepis lineariglumis var. *pulcherrima* Lye in Nordic J. Bot. **2**: 562 (1983). Type: Zambia, Mbala Dist., Saisi R. marsh, 1500 m, 27.ii.1957, *Richards* 1559 (K holotype).

Ascolepis protea subsp. *rhizomatosa* Lye in Nordic J. Bot. **2**: 562 (1983). —Haines & Lye, Sedges & Rushes E. Afr.: 305, fig.630 (1983). Type: Tanzania, 25 km S of Sumbawanga, 3.i.1962, *Robinson* 4890 (K holotype, M).

Ascolepis protea subsp. *atropurpurea* Lye in Nordic J. Bot. **2**: 563 (1983). —Haines & Lye, Sedges & Rushes E. Afr.: 306, fig.631 (1983). Type: Zambia, Mbala Dist., Saisi R. marsh, 27.ii.1957, *Richards* 8350 (K holotype).

Ascolepis protea subsp. *chrysocephala* Lye in Nordic J. Bot. **2**: 564 (1983). —Haines & Lye, Sedges & Rushes E. Afr.: 307 (1983). Type: Tanzania, 28 km S of Sumbawanga, 3.i.1962, *Robinson* 4893 (K holotype).

Ascolepis protea subsp. *anthemiflora* (Welw.) Lye in Nordic J. Bot. **2**: 566 (1983). —Haines & Lye, Sedges & Rushes E. Afr.: 305, fig.629 (1983).

Ascolepis protea subsp. *bellidiflora* (Welw.) Lye in Nordic J. Bot. **2**: 566 (1983). —Haines & Lye, Sedges & Rushes E. Afr.: 305, fig.628 (1983).

Cyperus proteus (Welw.) Bauters in Phytotaxa **166**: 38 (2014).

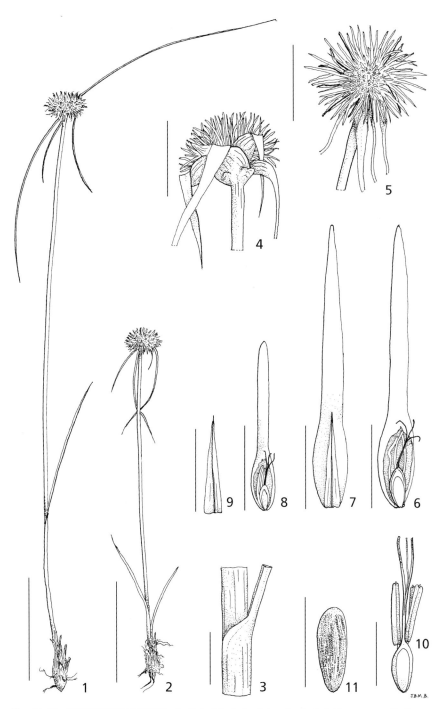

Fig. 14.**45**. ASCOLEPIS PROTEA. 1, 2, habit; 3, leaf sheath; 4, 5, inflorescence; 6, 7, outer squamella, adaxial and abaxial view; 8, inner squamella; 9, bract; 10, young floret; 11, nutlet. All from *Richards* 8072, except 11 from *Richards* 540. Scale bars: 1, 2 = 40 mm; 4, 5 = 10 mm; 3, 6–9 = 2 mm; 10, 11 = 1 mm. Drawn by Jane Browning.

Rather small, slender, ± tufted perennial herb, without runners. Stem 5–50(60) high, 0.5–2.3 mm across; stem base sometimes bulbously thickened, covered in a fibrous remains of leaf sheaths. Inflorescence capitate, solitary spike 5–30 mm wide, globose or flattened, white to yellowish orange, glumes short to elongated, sometimes only marginal glumes elongated; involucral bracts 5–10, to 8 cm long. Fruit 0.6–1.2 mm long, dark brown to black.

Zambia. N: Mbala Dist., L. Chila, 19.vi.1956, *Robinson* 1676 (K?). W: Mwinilunga Dist., source of Matonchi dambo, 26.x.1937, *Milne-Redhead* 2960 (GENT, K). C: Chongwe Dist., Chakwenga headwaters, 100–129 km W of Lusaka, 27.iii.1965, *Robinson* 6539 (K). E: Lundazi Dist., 72 km from Chipata (Fort Jameson) on road to Lundazi, 4.iv.1951, *Hodge* 8559 (K). S: Choma Dist., Muckle Neuk, 19.3 km N of Choma, 12.i.1954, *Robinson* 451 (K). **Zimbabwe**. N: Hurungwe Dist., Karoi, Fiddlers Green Farm, Happy Valley, i.1976, *Nicholas* 477 (GENT, SRGH). W: Lupane Dist., along river, 4.iv.1972, *Chiparawasha* 434 (K). C: Kwekwe Dist., 50 km NE of Kwekwe, Iwaba Estate, 30°07'E 18°49'S, 1200 m, 10.ii.2001, *Lye* 24528 (K). E: Nyanga Dist., near van Niekerk ruins, 17.v.1943, *Allen in SRGH* 13242 (K, SRGH). **Malawi**. N: Mzimba Dist., c. 11 km NW of Marymount towards Lupaso, 1370 m, 31.i.1976, *Pawek* 10737 (K, MAL, MO, SRGH, UC). S: Dedza Dist., Chongoni Forest, 22.v.1963, *Banda* 503 (K, SRGH). **Mozambique**. N: Marrupa Dist., Marrupa, beyond airport on road to Nungo, 800 m, 18.ii.1982, *Jansen & Boane* 7827 (K, LISC, LMA). MS: Sussundenga Dist. Moribane, x-xi.1901, *Dawe* 494 (K).

Also in Guinea, D.R. Congo, Tanzania and Angola. In swamps and wet grasslands; 800–1500 m.

Conservation notes: Widespread; not threatened.

Both *Ascolepis elata* and *A. lineariglumis* were previously treated as separate species, although they are clearly part of the *A. protea* species complex. Goetghebeur (1980) had already merged *A. elata* with *A. protea*.

Ascolepis lineariglumis is described as an annual, this character being its biggest difference with *A. protea*. However, material identified by Lye as *A. lineariglumis* proved to be perennial (e.g. *Kornaś* 397).

Ascolepis protea is a common species that has caused much trouble in the past, and will continue to do so into the future. In many localities there exist ± differentiated populations, indicating an active speciation process; many may deserve taxonomic recognition at varietal or even specific level. Autecological and reproductive studies along with accurate observations on glume shape and texture at different developmental stages would be very useful. Delineating varieties, subspecies and species is at this moment almost impossible as intermediates are frequent, although many authors have tried. The type specimens of described varieties and subspecies always seem very recognizable, but intermediates always appear, hence here we regard it as one variable taxon. Detailed local field studies are necessary.

9. **Ascolepis pseudopeteri** Goetgh. in Adansonia, n.s. **19**: 286. —Haines & Lye, Sedges & Rushes E. Afr.: 311 (1983). Type: Zambia, Mwinilunga Dist., Zambezi Rapids, c. 4.8 km N of Kalene Hill, 1310 m, 24.xii.1969, *Simon & Williamson* 1991 (SRGH holotype, K, P).

   Cyperus pseudopeteri (Goetgh.) Bauters in Phytotaxa **166**: 37 (2014).

Loosely tufted, small, slender perennial herb; stem 5–15 cm high, 0.2–0.5 mm across; stem base slightly thickened, surrounded by few ± withering leaf sheaths. Leaves relatively abundant. Inflorescence capitate, solitary spike 4–6 mm wide, spheroidal, yellow to orange brown, marginal glumes not elongated; larger involucral bracts 2–3, to 5 cm long. Spikelets densely spirally imbricate on a broadly conical axis; spikelet bract 1.5–2 mm long, lanceolate, yellow to orange-brown, tip subterete. Glume 1.7–2.3 mm long, basal part 1–1.5 mm long, ± concave, middle part swollen, floral parts adaxially enclosed by glume wings, apical part 0.5–0.75 mm long, yellow to orange-brown, with shiny ± inflated surface cells, crescent-shaped in section. Stamens (2)3, lateral and anterior; filament c. 1.7 mm; anthers c. 1 mm long. Style c. 1 mm long, deeply 3(4)-cleft. Fruit 0.75 mm long, obovate, obscurely subtrigonous, dark red brown.

Zambia. W: Mwinilunga Dist., 6 km N of Kalene Hill, granite outcrops, 12.vii.1963, *Robinson* 5896 (GENT, K).

Not known elsewhere. In seasonally wet grasslands; c. 1250 m.

Conservation notes: Apparently endemic to NW Zambia; possibly Near Threatened.

10. **Ascolepis pusilla** Ridl. in Trans. Linn. Soc. London, Bot. **2**: 164 (1884). —Goetghebeur in Adansonia, n.s. **19**: 297 (1980). —Haines & Lye, Sedges & Rushes E. Afr.: 308, fig.636 (1983) as var. *pusilla.* —Beentje in F.T.E.A., Cyperaceae: 268 (2010). Types: Angola, Huíla, v.1860, *Welwitsch* 1678 (BM syntype, BR, LISU); Huíla, v.1860, *Welwitsch* 6773 (BM syntype).

Ascolepis pusilla var. *echinata* S.S. Hooper in Kew Bull. **37**: 607 (1983). —Haines & Lye, Sedges & Rushes E. Afr.: 309 (1983). Type: Tanzania, Ufipa Dist., 8 km N of Sumbawanga, 10.vi.1980, *Hooper & Townsend* 1927 (K holotype, DAR).

Ascolepis pusilla var. *cylindrica* S.S. Hooper in Kew Bull. **37**: 608 (1982). —Haines & Lye, Sedges & Rushes E. Afr.: 309 (1983). Type: Tanzania, Iringa Dist., just N of Iringa, 10.vii.1956, *Milne-Redhead & Taylor* 11202 (K holotype).

Ascolepis pusilla Ridl. var. *microcuspis* Lye in Nordic J. Bot. **2**: 564 (1983). —Haines & Lye, Sedges & Rushes E. Afr.: 308, fig.637 (1983). Type: Zambia, Choma Dist., Siamambo, 20.iii.1958, *Robinson* 2815 (MHU holotype, K, PRE).

Cyperus ascopusillus Goetgh. in Phytotaxa **166**: 36 (2014), new name, non *Cyperus pusillus* Vahl (1805).

Slender, loosely clustered annual herb; stem 1–20 cm high, 0.2–0.6 mm across. Inflorescence of 1–5 spikes, yellowish brown, marginal glumes not elongated; apical spike 3–5 mm long, ovoidal, lateral ones 2–3 mm long, more spheroid; larger involucral bracts 2–4, to 7 cm long. Spikelets densely spirally imbricate on slender cylindrical axes. Spikelet bract 1–2 mm long, narrowly triangular, hyaline, central nerve reddish brown, wings often red-dotted, tip subacute, sometimes minutely spinulose. Glume 1–2.3 mm long, ± trumpet-shaped, reddish brown, enclosing floral parts and rachilla; basal part ± tubular, widening near upper margin, epidermis cells of upper third inflated, abaxially 3–5 nerved, nerves yellowish, prominent, wings adaxially joined but with a shallow incision, apical part to 0.5 mm long, subterete, rounded or minutely spinulose at tip. Stamen 1, lateral; filament 1–1.5 mm long; anther c. 0.4 mm long. Style 0.5–0.75 mm long, deeply 3-cleft. Fruit 0.75–1 mm long, obovate, subtrigonous, dark red brown, rachilla 1–1.5 mm long, club-shaped, apical part with inflated cells, ± winged, sometimes with red brown central nerve.

Zambia. N: Isoka Dist., 45 km NE of Isoka, 23.v.1962, *Robinson* 5234 (K). W: Solwezi Dist., 100 km W of Solwezi, 15.iv.1960, *Robinson* 3551 (K, SRGH). C: Kabwe Dist., 40 km N of Kabwe (Broken Hill), n.d., *Robinson* 4633 (K). S: Choma Dist., Simasunda, 8.iii.1953, *Robinson* 652 (K, SRGH). **Zimbabwe**. C: Gutu Dist., Makwiro, 1340 m, 16.iv.1932, *Brain* 8904 (K, SRGH). E: Mutare Dist., Mutare, Darlington, 1100 m, 28.iii.1953, *Chase* 4885 (K, SRGH). **Malawi**. C: Kasungu Dist., 48 km N of Kasungu, 3.vi.1962, *Robinson* 5254 (K). S: Machinga Dist., Liwonde Nat. Park, Mvuu Camp, 17.iv.1985, unknown collector 1527 (K). **Mozambique**. N: Montepuez Dist., 43 km from Nantulo to Mueda, 10.iv.1964, *Torre & Paiva* 11859 (LISC).

From Senegal to Tanzania south to Namibia; also known from Madagascar and Vietnam. In seasonally wet grassland, particularly dambos; 500–1400 m.

Conservation notes: Widespread; not threatened.

11. **Ascolepis trigona** Goetgh. in Adansonia, n.s. **19**: 286 (1980). Type: Zambia, 10 km E of Kasama, 12.i.1961, *Robinson* 4253 (SRGH holotype, BR, K, P).

> *Cyperus ascotrigonus* Goetgh. in Phytotaxa **166**: 37 (2014), new name, non *Cyperus trigonus* Boeckeler (1888).

Loosely tufted, small and slender perennial herb; stem 10–25 cm high, 0.3–0.6 mm across, stem base slightly bulbously thickened, surrounded by few withering leaf sheaths. Inflorescence capitate, solitary spike 5–7 mm wide, spheroidal, whitish or yellowish, marginal glumes not elongated; larger involucral bracts 2–4, to 6 cm long, often reflexed. Spikelets densely spirally imbricate on a broadly conical axis; spikelet bract c. 2 mm long, narrowly triangular to linear, whitish or yellowish, tip subacute to rounded. Glume 2–3 mm long, basal part 1.3–1.6 mm long, very concave, ± hyaline, nerves 3, floral parts adaxially ± enclosed by glume wings; apical part 0.5–1 mm long, whitish or yellowish with shiny, inflated surface cells, dorsiventrally flattened, plump triangular, rounded at tip. Stamens 3, lateral and anterior; filament c. 1.75 mm; anthers c. 1 mm long. Style 1–1.7 mm long, deeply 3(4) cleft. Fruit not seen.

Zambia. N: Kasama Dist., 10 km E of Kasama, 12 i 1961, *Robinson* 4253 (BR, K, P, SRGH); Mbala Dist., Kambole–Mbala road, 1500 m, 31.i.1964, *Richards* 18918 (K).

Also in the D.R. Congo. In seasonally wet grasslands; 1300–1500 m.

Conservation notes: Known from only few specimens in the Flora area; rare and probably Near Threatened.

21. **PYCREUS** P. Beauv.[21]

Pycreus P. Beauv. in Fl. Oware **2**: 48, t.86 (1816).
> *Cyperus* L. subgen. *Pycreus* (P. Beauv.) Miq. in Fl. Ind. Batav. **3**: 254 (1861).
> —Haines & Lye, Sedges & Rushes E. Afr.: 268 (1983).
> *Cyperus* L. sect. *Pycreus* (P. Beauv.) Boeckeler in Linnaea **35**: 437 (1868).

Annual or perennial herbs; root system slender and fibrous in annuals, tussocky or rhizomatous in perennials. Culms usually scapose, triangular. Leaves usually all basal, filiform or flat, scabrid on margins, leaf-sheaths glabrous, ligule absent. Inflorescence terminal, anthelate, of few to many sessile or subsessile spikelets and then head-like, or of one sessile and few to many pedunculate clusters of spikelets and then appearing subumbellate. Involucral bracts usually several and leafy, erect or spreading; prophylls at base of inflorescence branches conspicuous, tubular, dark brown. Spikelets white, yellow to light or dark brown or almost black, linear to oblong, ellipsoid or obovate, laterally flattened; axis persistent; glumes distichous, usually ovate to obovate, often concave, apex obtuse to acuminate, sometimes mucronate, dorsally with a usually prominent rib-like keel. Flowers bisexual; perianth absent; stamens (1)3; style continuous with ovary, with 2 (very rarely 3) stigma branches. Nutlets biconvex, with one margin facing the axis, sometimes almost rounded, ovate to obovate or almost sphaerical or oblong to cylindrical, brown to dark grey or almost black, smooth to (usually) finely papillose or transversally ridged to muricate.

A pantropical genus of about 100 species. Particularly numerous in central, eastern and southern Africa.

Larridon *et al.* in Bot. J. Linn. Soc **172**: 106–126 (2013) and Phytotaxa **166**: 33–48 (2014) incorporated the species of *Pycreus* into *Cyperus*, but it is kept separate here.

Note: It is sometimes difficult to decide whether a specimen is from an annual or perennial plant, so a number of species have been keyed out in both parts of the key.

----

[21] by K. Vollesen

1. Acaulous (stemless) annual herb . . . . **20.** *acaulis*
  – Plant with clearly defined stems 2
2. Plant wholly or partially submerged, neither culms nor leaves able to self-support out of water 3
  – Plant self-supporting, partly submerged or not 4
3. Plant entirely submerged except for spikelets; leaf blade flat; inflorescence branches inflated and spongy; nutlet weakly transversely ridged . . . . **4.** *waillyi*
  – Upper part of stems and inflorescence above water; leaf blade filiform; inflorescence branches not inflated and spongy; nutlet densely papillose . . . . **35.** *demangei*
4. Glumes pale green to green apart from almost translucent hyaline veins; glumes c. 50% overlapping; mature nutlet smooth . . . . **25.** *okavangensis*
  – Glumes varying from white through yellow and brown to black, not hyaline or translucent; glumes usually c. 75% overlapping; mature nutlet variously ornamented, never smooth 5
5. Glumes greenish white or white, sometimes with a faint brown tinge or faint brown lines; perennial herb; inflorescences capitate or some with a single branch . . . 6
  – Glumes yellowish green or yellow to golden yellow, or pale brown to reddish brown or dark brown to black; annual or perennial herb; inflorescences capitate or open 8
6. Plant in dense compact tussocks; leaves filiform, 0.5–1 mm wide, surrounded at base by dense old fibrous leaf bases; glumes 3–3.5 mm long . . . .**23.** *scaettae*
  – Plant with short creeping rhizome or stoloniferous; leaves flat, 1.5–4 mm wide, not with old fibrous leaf bases; glumes 1.5–3 mm long. 7
7. Spikelets 2–3 mm wide; inflorescences all strictly capitate, rachis curved in fruiting stage; glumes 1.5–2 mm long . . . . **22.** *smithianus*
  – Spikelets 3–4 mm wide; some inflorescences with a single stalked cluster, rachis straight in fruiting stage; glumes 2–3 mm long . . . . **24.** *richardsiae*
8. Spikelets 1–2 times as long as wide; glumes ± as long as wide . . . .**29.** *sanguineosquamatus*
  – Spikelets more than twice as long as wide; glumes distinctly longer than wide . . 9
9. Ephemeral or annual herb with fibrous root systems 10
  – Tussocky, rhizomatous or stoloniferous perennial 35
10. Leaves filiform 11
  – Leaves with a distinct flat lamina 21
11. Glumes with a conspicuous dark chocolate brown to almost black tip; nutlet with evenly distributed papillae . . . . **36.** *melanacme*
  – Glumes not with a conspicuous dark chocolate brown to almost black tip. . . . 12
12. Nutlet with sculpturing of transverse ridge . . . . **26.** *pauper*
  – Nutlet papillose 13
13. Glumes 3–4 mm long; nutlet 1–1.5 mm long . . . .**12.** *xantholepis*
  – Glumes 1–2.5 mm long; nutlet 0.5–1(1.5) mm long 14
14. Some spikelets over 20 mm long . . . .**27.** *sp. B*
  – Spikelets 4–20 mm long 15
15. Nutlets with papillae evenly distributed over surface 16
  – Nutlets with papillae in distinct longitudinal rows 17
16. Spikelets 1.5–3 mm wide; glumes 1(1.5) mm wide, without a hyaline margin in apical part. . . . . **35.** *demangei*
  – Spikelets 1–1.5 mm wide; glumes c. 0.5 mm wide, with a narrow hyaline margin in apical part. . . . . **37.** *melas*
17. Nutlets c. 0.5 mm long 18
  – Nutlets 1–1.5 mm long 19

18. Spikelets 0.5–1 mm wide; glumes c. 1 × 0.5 mm, about 33% overlapping . . . . . .
    . . . . **38.** *micromelas*
 – Spikelets 2.5–3 mm wide; glumes c. 2 × 1.5 mm, about 75% overlapping . . . . . .
    . . . .**34.** *atrorubidus*
19. Spikelets 2–3.5 mm wide, some over 1 cm long; glumes 1.5–2 mm wide . . . .
    . . . . **33.** *poikilostachys*
 – Spikelets 1–2 mm wide, all 5–10 mm long; glumes 0.5–1 mm wide 20
20. Glumes golden yellow; nutlet oblong. . . . .**31.** *sp. C*
 – Glumes pale brown to dark brown or almost black; nutlet ellipsoid to obovoid or
    obtrigonous . . . .**32.** *capillifolius*
21. Inflorescence capitate, of a cluster of sessile or subsessile spikelets. 22
 – Inflorescence compound, at least with some stalked clusters. 25
22. Glumes acute, midrib extended into an often 2–3-fid mucro; nutlet with numerous
    minute papillae in longitudinal rows . . . . **21.** *polystachyos*
 – Glumes obtuse to acute, midrib not extended into a mucro; nutlet smooth to
    transversely or reticulate ridged 23
23. Nutlet c. 0.5 mm long, obtrigonous with retuse apex . . . .**30.** *subtrigonus*
 – Nutlet 0.5–1.5 mm long, ellipsoid to obovoid or subglobose with subacute to
    truncate apex 24
24. Nutlet ellipsoid to obovoid, smooth to finely transversely or reticulately ridged .
    . . . . **7.** *flavescens*
 – Nutlet subglobose, strongly reticulately ridged . . . .**28.** *zonatissimus*
25. Spikelets with a broad upcurved dark brown margin making central area appear
    as a large groove . . . . **2.** *sanguinolentus*
 – Spikelets not with a broad upcurved margin, central area not appearing as a large
    groove 26
26. Glumes black, shiny, with a black membranaceous margin towards apex . . . .
    . . . . **3.** *elegantulus*
 – Glumes greenish yellow to golden yellow or pale to dark brown, with or without a
    translucent hyaline margin 27
27. Nutlets with papillae in distinct longitudinal rows, or smooth to transversely
    ridged 28
 – Nutlets with papillae evenly distributed over surface . . . . **35.** *demangei*
28. Nutlets almost smooth to transversely or reticulately ridged 29
 – Nutlets papillate with papillae in distinct longitudinal rows. 30
29. Spikelets linear to lanceolate, 1.5–3(4) mm wide; glumes 1–1.5 mm wide, without
    hyaline margin; anthers c. 0.5 mm long; nutlet 0.5–1 mm long . . . . . . **7.** *flavescens*
 – Spikelets ovate or narrowly so, 3–4.5(6) mm wide; glumes 1.5–2.5 mm wide, with
    narrow hyaline margin; anthers 1–1.5 mm long; nutlet c. 1 mm long . . . .
    . . . .**6.** *unioloides*
30. Stalked clusters with 1–6 spikelets; spikelets 4–5 mm wide in flower; glumes about
    66% overlapping . . . . **12.** *xantholepis*
 – Some or all stalked clusters with more than 6 spikelets; spikelets 1–3(4) mm wide
    in flower; glumes usually less than 50% overlapping 31
31. Glumes acute to acuminate, with a broad hyaline margin forming two conspicuous
    apical lobes, with a recurved mucro c. 0.5 mm long . . . . **16.** *pumilus*
 – Glumes obtuse to acute, hyaline margin if present not forming apical lobes, with
    or without a straight mucro 32
32. Spikelets 0.8–1.8 mm wide; nutlet 0.5–1 mm long . . . . **21.** *polystachyos*
 – Spikelets 1–3(3.5) mm wide; nutlet 1–2 mm long. 33
33. Glumes without hyaline margin . . . . **11.** *intactus*
 – Glumes with hyaline margin in upper half 34

34. Slender herb; stems rounded, 0.5–1.5 mm wide; nutlet c. 1 mm long, with truncate apex . . . . **15.** *pelophilus*
– Coarse herb; stems triangular, (1)1.5–6 mm wide; nutlet 1.5–2 mm long, with subacute apex. . . . . **14.** *macrostachyos*
35. Leaves immature at time of flowering; glumes acuminate . . . .**13.** *sp.* A
– Leaves fully developed at time of flowering; glumes obtuse to acute. 36
36. Leaves filiform, rolled (appearing cylindrical) or folded 37
– Leaves with a distinctly flat lamina 39
37. Spikelets lanceolate to ovate, 2–5(6) mm wide; glumes 2.5–4.5 mm long, about 75% overlapping, without a hyaline margin towards apex; nutlet broadly ellipsoid to obovoid, transversely to reticulately ridged to strongly muricate (rarely papillate) . . . . **19.** *nigricans*
– Spikelets linear to lanceolate, 1–2 mm wide; glumes 1.5–3 mm long, about 50% overlapping, with a narrow hyaline margin towards apex; nutlet oblong, with numerous papillae in distinct longitudinal rows. 38
38. Culms 1–1.5 mm wide; leaf blade 0.5–1(1.5) mm wide; glumes dark brown to dark purple or black. . . . . **17.** *aethiops*
– Culms 1.5–4 mm wide; leaf blade (1.5)2–4 mm wide; glumes brown to dark brown . . . . **18.** *nuerensis*
39. Glumes with a broad upcurved margin making the central area appear as a large groove. 40
– Glumes without a broad upcurved margin, central area not appearing as a large groove. 41
40. Plant with elongated leafy stems; nutlet 0.5–1 mm long. . . . .**1.** *mundii*
– Leaves only at base of stem; nutlet c. 1.5 mm long . . . . **2.** *sanguinolentus*
41. Nutlet transversely rugose to transversely ridged or with a reticulum of raised ridges 42
– Nutlet with numerous minute papillae in longitudinal rows 43
42. Culms at base surrounded by old dark brown to black fibrous leaf bases; glumes 1–1.5 mm wide, without hyaline margin . . . . **9.** *robinsonii*
– Culms not with old dark brown to black fibrous leaf bases at base; glumes 1.5–2.5 mm wide, with a narrow hyaline margin . . . . **6.** *unioloides*
43. Glumes black, shiny, with a black membranaceous margin towards apex. . . . .
      . . . . **3.** *elegantulus*
– Glumes pale yellowish green to yellowish or golden brown, brown or dark brown, with or without a translucent hyaline margin 44
44. Inflorescence capitate, of a cluster of sessile or subsessile spikelets. . **8.** *lanceolatus*
– Inflorescence compound, at least with some stalked clusters. 45
45. Mature leaves c. 1 mm wide; spikelets 4–5 mm wide in flower. . . . .**12.** *xantholepis*
– Mature leaves 1–8 mm wide; spikelets 1–4 mm wide in flower. 46
46. Glumes with a conspicuous wide hyaline margin; nutlet 1.5–2 mm long . . . .
      . . . . **14.** *macrostachyos*
– Glumes without or with narrow unconspicuous hyaline margin; nutlet 0.5–1.5 mm long 47
47. Spikelets linear-lanceolate, 1–2 mm wide; glumes 25–50% overlapping; anthers 0.5–1 mm long. 48
– Spikelets lanceolate to narrowly ovate, 2–4 mm wide; glumes 50–75% overlapping; anthers 1–3 mm long. 49
48. Glumes with a narrow hyaline margin; anthers c. 0.5 mm long; nutlets 0.5–1 mm long. . . . . **21.** *polystachyos*
– Glumes without a narrow hyaline margin; anthers c. 1 mm long; nutlets 1–1.5 mm long. . . . . **11.** *intactus*

49. Glumes acute, about 50% overlapping; anthers 2–3 mm long. . . . **10.** *chrysanthus*
–   Glumes obtuse, about 75% overlapping; anthers 1–2 mm long . . . . **5.** *nitidus*

1. **Pycreus mundii** Nees in Linnaea **10**: 131 (1836). —Clarke in Fl. Cap. **7**: 157 (1897);
    in F.T.A. **8**: 294 (1901). —Gordon-Gray in Strelitzia **2**: 143 (1995). —Glen in
    Taxon **52**: 601 (2003). —Hoenselaar in F.T.E.A., Cyperaceae: 284, fig.43 (2010),
    as *P. mundtii.* Type: South Africa, W Cape, Zwellendam, George, n.d., *Mund* s.n.
    (B holotype, S), see note.
      Cyperus mundii (Nees) Kunth, Enum. Pl. **2**: 17 (1837). —Kükenthal in Engler, Pflanzenr.
    4, 20(101): 380 (1936). —Haines & Lye, Sedges & Rushes E. Afr.: 270, fig.548 (1983).
      Cyperus distichophyllus Steud. in Flora **15**: 582 (1842). Type: Ethiopia, Tigray, Adua,
    20.xii.1838, *Schimper* II.745 (P holotype, BM, BR, HAL, K, LG, TUB).
      Cyperus sanguinolentus Vahl var. *uniceps* C.B. Clarke in Bot. Jahrb. **38**: 132 (1906). Type:
    Tanzania, Usambara Mts, Kwai, xi.1899, *Albers* 290 (B holotype).
      Cyperus mundii f. *distichophyllus* (Steud.) Kük. in Fries, Wiss. Ergebn. Schwed. Rhodesia-
    Kongo Exp. 1911–1912 **1**, Erg.: 2 (1921).
      Cyperus mundii var. *uniceps* (C.B. Clarke) Kük. in Bot. Not. **1934**: 69 (1934); in Engler,
    Pflanzenr. **4**, 20(101): 382 (1936).
      Cyperus mundii var. *distichophyllus* (Steud.) Kük. in Engler, Pflanzenr. **4**, 20(101): 381
    (1936).
      Pycreus mundii var. *uniceps* (C.B. Clarke) Napper in J. E. Africa Nat. Hist. Soc. Natl. Mus.
    28(124): 3 (1971). —Hoenselaar in F.T.E.A., Cyperaceae: 287 (2010).

Stoloniferous perennial; stolons to 1 m long, 2–5 mm thick, rooting at nodes, each ending
in a single ascending to erect culm to 75 cm long, basal part of culm covered in inflated leaf
sheaths, to 8 mm wide; stems above leaves 0.5–2 mm wide, triangular, grooved. Leaves spread
along culm, stiff to almost prickly, sheaths to 4.5 cm long, greenish to yellowish brown, opposite
blade with a brown to red or purple hyaline ochre which eventually splits into two flanges; blade
flat, to 17(25) cm long and 1.5–7(8) mm wide, minutely scabrid near tip. Inflorescence 1–6(8) ×
(1)2–10(12) cm, a compound umbel-like anthela with 1–3 subsessile and (1)2–9 stalked clusters,
rarely all clusters subsessile; peduncles of stalked clusters 0.5–5(7) cm; spikelets in dense digitate
ovoid clusters, each cluster with 2–10(15) spikelets; involucral bracts leaf like, 2–5, largest 2–10
cm long, 1–6 mm wide, scabrid near apex. Spikelets 4–12 × 2–4 mm, linear-lanceolate to ovate;
rachis straight in fruiting stage; glumes appressed at first, only slightly spreading from flowering
stage, tardily dehiscent at maturity, pale brown to reddish brown or dark brown, broadly ovate,
with a broad upcurved margin making the central area appear as a large groove, 1.5–3 × 1–2
mm, obtuse, keel brown. Stamens 2–3; filaments 2–3 mm long; anthers 1–2 mm long. Stigma
branches 2, 2–6 mm long. Nutlet oblong to obovoid, tapering to truncate, apiculate, grey to
greyish brown, 0.5–1 mm long, almost smooth to minutely papillose or finely muricate.

Botswana. N: Okavango Delta, Matsaudi aga mmatalela Is., lediba, fl. 19.iii.1973,
P.A. Smith 464 (K, SRGH). **Zambia**. B: Mongu Dist., 15 km E of Mongu, fl. 12.ix.1962,
Robinson 5466 (K). N: Mbala Dist., Kamba Bay Game Camp, shore of Lake Tanganyika,
775 m, fl. 14.iv.1957, *Richards* 9228 (K). W: Mwinilunga Dist., Mwinilunga, Lunga R.,
fl. & fr. 1.xii.1937, *Milne-Redhead* 3467 (K). C: Chibombo Dist., 22 km E of Chisamba,
Chipembi, 1225 m, fl. 17.vi.1955, *Robinson* 1305 (K). E: Chipata Dist., Chipata (Fort
Jameson), 1075 m, fl. 6.vi.1954, *Robinson* 843 (K). S: Mazabuka Dist., near Mazabuka,
Nega Nega, 1000 m, fl. 25.vii.1937, *Trapnell* 390 (K). **Zimbabwe**. N: Mazoe Dist.,
Normandale Farm, 1225 m, fl. 24.x.1944, *Mossop* in SRGH 12842 (K, SRGH); Binga
(Sebungwe) Dist., Sitonka Valley, sulphur springs, 1075 m, fl. x.1955, *Davies* 1595
(K, SRGH). C: Harare North, 1425 m, fl. 18.ix.1956, *Robinson* 1804 (K). E: Chipinge
Dist., Chirinda, road to Espungabera, 1150 m, fl. 23.x.1947, *Rattray* 1221 (K, SRGH).
S: Masvingo Dist., Kyle Dam, 1425 m, fl. 9.xi.1969, *Kelly* 124 (K, SRGH). **Malawi**. N:
Mzimba Dist., 1375 m, fl. 20.xi.1973, *Pawek* 7512 (K, MAL, MO, SRGH). C: Nkhotakota
Dist., Nkhotakota (Nkota Kota), shore of Lake Malawi (Nyasa), fl. 2.v.1963, *Verboom*
100S (K). S: Zomba Dist., slopes of Zomba Plateau above Malemia Hospital, 1100 m,

fl. 19.iv.1980, *Brummitt* 15508 (K, MAL). **Mozambique**. MS: Chimoio Dist., Chimoio (Vila Pery), fl. 15.x.1925, *Surcouf* 105 (K, P). GI: Inhambane Dist., between Inhambane and Miramar, fl. 14.ix.1948, *Myre & Carvalho* 184 (K). M: Maputo Dist., Delagoa Bay, fl. viii.1887, *L. Scott* s.n. (K).

Widespread in Tropical and South Africa and Madagascar; also in Egypt, Spain and Brazil. Swamps, lakeshores, wet grassland, frequently forming floating mats, also on seashores; 0–1900 m.

Conservation notes: Widespread distribution; Least Concern. In common and widespread habitats.

The Stockholm sheet bears the number 91, but it is not clear whether this is Mundt's collection number or a later added herbarium accession number.

Hoenselaar (2010) recognises var. *uniceps*, but I agree with Haines & Lye (1983) that it is "only an edaphic form of no taxonomic value".

Often forming thick floating swards around waterholes and along streams. The notes on *Richards* 14832 (Botswana, Okavango, Toakhe R. swamp) says "...thick floating mass... so thick one can walk on it".

2. **Pycreus sanguinolentus** (Vahl) Nees in Linnaea **9**: 283 (1835). —Clarke in F.T.A. **8**: 293 (1901). —Hoenselaar in F.T.E.A., Cyperaceae: 287 (2010). Type: 'India orientalis', n.d., *Herb. Lamarck* (P-LA holotype).

   Cyperus sanguinolentus Vahl, Enum. Pl. **2**: 351 (1805), conserved name. —Kükenthal in Engler, Pflanzenr. **4**, 20(101): 385 (1936). —Haines & Lye, Sedges & Rushes E. Afr.: 270, figs.549,550 (1983).

Tufted annual or short-lived perennial; culms 3–6, erect, to 50 cm long, 0.5–1 mm wide, triangular, longitudinally striate, glabrous, smooth. Leaves basal, leaf sheaths to 10 cm long, yellowish brown to reddish; blade flat, to 20 cm long and 1–3 mm wide, scabrid on margins and midrib near tip. Inflorescence a congested head-like anthela, or more open with several sessile and 2–35 stalked clusters, 1–1.5 × 1–2 cm if capitate, to 4 × 3 cm if open, clusters each with 3–20 loosely arranged spikelets; peduncles of stalked clusters 1–3 cm long; involucral bracts leaf-like, 2–3, largest 5–16 cm long and 1.5–3 mm wide, scabrid near tip or along whole length. Spikelets 5–10 × 2–3 mm, narrowly oblong-ovoid; rachis straight in fruit; glumes appressed at flowering, later slightly spreading, only tardily falling at maturity, straw-coloured to pale brown, with a broad upcurved dark brown margin making the central area appear as a large groove, broadly ovate, 1.5–2 × 1.5 mm, subacute or obtuse, keel pale green, prominent. Stamens 3; filaments 1.5–2.5 mm long; anthers 0.5–1 mm long. Stigma branches 2, 1.5 mm long. Nutlet obovoid with truncate apex, shortly apiculate, greyish black, 1.5 mm long, minutely papillose-punctate.

Zambia. S: Choma Dist., 18 km N of Choma, Muckle Neuk, 1250 m, fl., 27.ii.1954, *Robinson* 446 (K); same locality, 26.x.1958, *Robinson* 2914 (K).

Also in Eritrea, Ethiopia, Kenya and Tanzania; also widespread in Asia from Turkey to Japan and south to Australia. Swampy dambo grassland; 1250 m.

Conservation notes: Possibly threatened in the Flora area, where it has only been collected from one locality. Widespread outside our area. In common and widespread habitats.

The grooved glumes are very similar to the ones in *Pycreus mundii*, but the two species have very different growth forms.

3. **Pycreus elegantulus** (Steud.) C.B. Clarke in Durand & Schinz, Consp. Fl. Afr. **5**: 536 (1894); in F.T.A. **8**: 302 (1901). —Hoenselaar in F.T.E.A., Cyperaceae: 303 (2010). Type: Ethiopia, Semien, Demerki, 9.viii.1838, *Schimper* II.574 (GOET holotype, BM, HAL, K, P, UPS).

   Cyperus elegantulus Steud. in Flora **25**: 583 (1842). —Kükenthal in Engler, Pflanzenr. **4**, 20(101): 342 (1936).

Cyperus elegantulus var. *submelanostachyus* Kük. in Engler, Pflanzenr. **4**, 20(101): 343 (1936). Type: Tanzania, Bukoba Dist., Ihangiro, n.d., *Stuhlmann* 3342 (B holotype).
Pycreus niger (Ruiz & Pav.) T. Koyama subsp. *elegantulus* (Steud.) Lye in Nord. J. Bot. **1**: 622 (1982). —Gordon-Gray in Strelitzia **2**: 145 (1995).
Cyperus niger Ruiz & Pav. subsp. *elegantulus* (Steud.) Lye in Haines & Lye, Sedges & Rushes E. Afr., app.3: 2 (1983). —Haines & Lye, Sedges & Rushes E. Afr.: 271, fig.551 (1983).

Annual or few-stemmed rhizomatous perennial, when annual sometimes tufted; culms 1–3(10 in annuals), erect, to 50 cm long and 1–2.5 mm wide, rounded to triangular, smooth. Leaves basal, sheaths to 8 cm long, brown, with 0.5–1 mm wide mebranaceous ochrea-like flanges; blade flat, to 30 cm long and 2–5 mm wide, scabrid in apical part. Inflorescence a simple umbel-like anthela or condensed to become almost capitate, 2–5 × 2–5.5 cm, open inflorescences with one to several sessile racemose clusters and 3–6 stalked clusters, each cluster with 10–20 spikelets; peduncles of stalked clusters 0.5–3.5 cm long; involucral bracts leaf-like, 3–4, largest 5–20 cm long and 1.5–5 mm wide, scabrid near tip or in upper half. Spikelets 4–10 × 1–2.5 mm, lanceolate; rachis straight in fruiting stage; glumes about 50% overlapping, appressed at first, soon spreading, shiny black, ovate with a black membranaceous margin towards apex, 1.5–2 × 1–1.5 mm, obtuse, sometimes apiculate, keel conspicuous, pale brown. Stamens (2)3; filaments 1 mm long; anthers 0.5 mm long. Stigma branches 2, 1 mm long. Nutlet ovoid to oblong with tapering apex, apiculate, steel-grey, 1 mm long, with numerous minute papillae in longitudinal rows.

Zambia. E: Nyika Plateau, 5 km SW of Resthouse, 2100 m, fl. & fr. 22.x.1958, *Robson & Angus* 264 (K). **Zimbabwe**. E: Nyanga Dist., no locality, iv.1941, *Ferrar* s.n. (SRGH). **Malawi**. N: Rumphi Dist., Nyika Plateau, NE of Nganda, 2000 m, fl. & fr. 10.iv.1997, *Patel et al.* 5120 (K, MAL). S: Zomba Dist., Zomba Plateau, Chagwa Dam, 1700 m, fl. & fr. 19.iv.1970, *Brummitt* 9977 (K).

Also in Nigeria, Cameroon, Bioko, Sudan, Eritrea, Ethiopia, D.R. Congo, Rwanda, Burundi, Uganda, Kenya, Tanzania, Angola and South Africa. Marshy grassland on lakeshores, riverbanks and dams; 1650–2300 m.

Conservation notes: Widespread distribution; Least Concern. In common and widespread habitats.

This species seems to be rare in our area but is widespread and common in East Africa. It grows mostly in higher altitude grassland which restricts its distribution here, but it might also be expected to occur in Mozambique.

4. **Pycreus waillyi** Cherm. in Bull. Soc. Bot. France **85**: 366 (1938). —Hooper in F.W.T.A., ed.2 **3**: 300 (1968). —Hoenselaar in F.T.E.A., Cyperaceae: 277 (2010). Type: Mali, Gao to Berra, Marigot, 19.ii.1937, *de Wailly* 5340 (P holotype, IFAN, K).
Cyperus waillyi (Cherm.) Lye in Haines & Lye, Sedges & Rushes E. Afr., app3: 2, fig.604 (1983), as *wailleyi*.

Annual, submerged in flowing streams with only inflorescence held above water; culms 1–10, to 80 cm long and to 1.5 mm wide, rounded, striate, smooth. Leaf sheaths brown, to 2 cm long; blade linear, flat, to 20 cm long and to 1.5 mm wide, scabrid near tip. Inflorescence to 5 × 4 cm, consisting of 1–3 subsessile spikelets with 5–10 stalked clusters, each with 1–3 spikelets with an inflated peduncle 1–4.5 cm long; involucral bracts 2–5, leaf-like, thin and flaccid, largest 10–22 cm long and 1–2 mm wide, scabrid near apex. Spikelets 5–13 × 2–3 mm, linear-lanceolate; glumes appressed at flowering, spreading later, very tardily dehiscing at maturity, densely imbricate (about ⅔ overlapping), reddish brown or to dark brown, broadly ovate to elliptic, without hyaline margin, 1.5–2 × 1.5 mm, obtuse with midrib reaching the tip, keel prominent, rib-like, pale green. Nutlet black, ellipsoid to subglobose, apex subacute, apiculate, 1 mm long, weakly transversely ridged.

Botswana. N: Ngamiland Dist., Okavango R., 8 km downstream from Samocima, 1025 m, fl. 29.iv.1975, *Gibbs Russell* 2860 (K, SRGH). **Zambia**. B: Kalabo Dist., Kalabo, banks of Zambezi R., fl. 7.v.1964, *Verboom* 1742 (K, L, SRGH).

Also in Mali, D.R. Congo and Tanzania. Submerged in streams or creeping on drying mud; 950–1100 m.

Conservation notes: Data Deficient. Only a few collections spread through a vast area. Almost certainly very undercollected.

5. **Pycreus nitidus** (Lam.) J. Raynal in Kew Bull. **23**: 314 (1969). —Gordon-Gray in Strelitzia **2**: 145 (1995). —Hoenselaar in F.T.E.A., Cyperaceae: 307 (2010). Type: India, no locality, *Herb. Lamark* s.n. (P-LA lectotype), lectotypified by Raynal (1969).

    Cyperus nitidus Lam., Tabl. Encycl. **1**: 145 (1791). —Haines & Lye, Sedges & Rushes E. Afr.: 272, figs.552,553 (1983).

    Cyperus lanceus Thunb., Prodr. Fl. Cap.: 18 (1794). —Kükenthal in Engler, Pflanzenr. **4**, 20(101): 333 (1936). Type not indicated.

    Pycreus umbrosus Nees in Linnaea **10**: 130 (1836). —Clarke in Fl. Cap. **7**: 158 (1897); in F.T.A. **8**: 303 (1901). Types: South Africa, Olifantsriver and Brackfontein, n.d., *Ecklon* s.n. (B syntypes).

    Cyperus fulvus Ridl. in Trans. Linn. Soc. London, Bot. **2**: 126 (1884), non R. Br. Type: Angola, Huíla, between Lopollo and Eme, iv.1860, *Welwitsch* 6872 (BM lectotype, LISU), lectotypified here.

    Cyperus lanceus var. *grantii* C.B. Clarke in J. Linn. Soc., Bot. **21**: 66 (1884). —Kükenthal in Engler, Pflanzenr. **4**, 20(101): 335 (1936). Type: Tanzania, Dodoma Dist., Mgunda-Mkali, Jiwa la Mkoa, 11.i.1861, *Speke & Grant* 605 (K holotype).

    Pycreus umbrosus var. *grantii* (C.B. Clarke) C.B. Clarke in F.T.A. **8**: 304 (1901).

    Pycreus lanceus (Thunb.) Turrill in Bull. Misc. Inform. Kew **1925**: 67 (1925).

Slender to robust stoloniferous perennial, often in water with long inflated floating stolons; culms single, erect, to 60(85) cm long and 1–4 mm wide, rounded to triangular, smooth, sometimes basal part spongy. Leaves basal, sheaths to 10 cm long, pale brown to brown or reddish, with or without 0.5–1 mm wide mebranaceous ochrea-like flanges; blade flat, to 50 cm long and 2–7 mm wide, scabrid near tip. Inflorescence a simple open or condensed umbel-like anthela, 1.5–9 × 2.5–9 cm, branches held erect or spreading, open inflorescence with 1–2 subsessile racemose clusters and 2–6 stalked clusters, each cluster with (3)5–20(25) spikelets; peduncles of stalked clusters 0.5–6 cm long; involucral bracts leaf-like, 1–2(3) large and 1–2 small, largest 5–27 cm long and 1–5.5(7.5) mm wide, scabrid near tip. Spikelets 5–17 × 2–4 mm (to 25 mm long in fruit), lanceolate to narrowly ovate; rachis straight in fruiting stage; glumes about 75% overlapping, appressed or spreading, brown to dark brown, ovate-elliptic, with or without a narrow hyaline margin towards apex, 2.5–4 × 1–2 mm, obtuse, keel pale brown, conspicuous. Stamens (2)3; filaments 1–2 mm long; anthers 1–2 mm long. Stigma branches 2, 3–5 mm long. Nutlet ellipsoid to obovoid, with slightly tapering apex, shortly apiculate, dark grey, c. 1 mm long, with numerous minute papillae in longitudinal rows.

Botswana. N: Okavango Swamps, Mboma, fl. vi.1971, *P.A. Smith* 109 (K, SRGH). **Zambia**. B: Mongu Dist., Mongu, fl. 29.iv.1964, *Verboom* 1725 (K). N: Mwense Dist., Chipili, 1225 m, fl. & fr. 28.vi.1956, *Robinson* 1763 (K). C: Mumbwa Dist., Shibuyanji, fl. 27.xii.1963, van Rensburg 2642 (K). S: Choma Dist., Simasunda, 3 km E of Mapanza, 1075 m, fl. 5.i.1954, *Robinson* 431 (K). **Zimbabwe**. W: Matobo Dist., Besna Kobila Farm, 1450 m, i.1960, *Miller* 7104 (K, SRGH). E: Nyanga Dist., 10 km N of Troutbeck, Gairesi Ranch, 1525 m, fl. 12.xi.1956, *Robinson* 1868 (K). **Malawi**. N: Mzimba Dist., Kasitu Valley, 1050 m, fl. 5.xii.1937, *Fenner* 19 (K). **Mozambique**. M: Maputo Dist., Inhaca Is., Mamahaji swamp, 0–200 m, fl. & fr. 23.ix.1957, *Mogg* 27504 (K).

Also in Cameroon, Chad, Sudan, Ethiopia, D.R. Congo, Rwanda, Burundi, Uganda, Kenya, Tanzania, Angola, Lesotho, Swaziland and South Africa; also Madagascar and India. Swampy grassland and swamps, often in water to 1.5 m deep, sometimes forming large floating mats; (0)900–1700 m.

Conservation notes: Widespread distribution; Least Concern. In common and widespread habitats.

This species almost certainly also occurs in Mozambique MS. The label on *Robinson* 1868 says: "on P.E.A. border" and Gordon-Gray (1995) says it occurs on the KwaZulu-Natal coast.

6. **Pycreus unioloides** (R. Br.) Urb., Symb. Antill. **2**: 164 (1900). —Podlech in Prodr. Fl. SW Afr. **165**: 42 (1967). —Gordon-Gray in Strelitzia **2**: 148 (1995). —Hoenselaar in F.T.E.A., Cyperaceae: 305 (2010). Type: Australia, Queensland, Shoalwater Bay, 1802-05, *R. Brown* 5900 (BM holotype, K).

> *Cyperus unioloides* R. Br. in Prodr. Fl. Nov. Holland.: 216 (1810). —Kükenthal in Engler, Pflanzenr. 4, 20(101): 338 (1936). —Haines & Lye, Sedges & Rushes E. Afr.: 273, fig.555 (1983).
> *Cyperus angulatus* Nees in Wight & Arnott, Contr. Bot. India: 73 (1834). Types: India/Nepal, 'Napalia et Ava', 1821 & 1826, *Wallich* 3324a & b (K syntypes, P).
> *Pycreus angulatus* (Nees) Nees in Linnaea **9**: 283 (1835). —Clarke in F.T.A. **8**: 305 (1901); in Fl. Cap. **7**: 156 (1897).
> *Pycreus spissiflorus* C.B. Clarke in Trans. Linn. Soc. London, Bot. **4**: 53 (1894); in Consp. Fl. Afr. **5**: 542 (1894); in F.T.A. **8**: 304 (1901). Type: Malawi, Mt Mulanje, ix.1891, *Whyte* s.n. (BM holotype, K).
> *Cyperus spissiflorus* (C.B. Clarke) K. Schum. in Engler, Pflanzenw. Ost.-Afrikas **C**: 117 (1895). —Kükenthal in Engler, Pflanzenr. 4, 20(101): 337 (1936). —Strugnell, Checkl. Mt Mulanje: 78 (2006).
> *Pycreus overlaetii* S.S. Hooper & J. Raynal in Kew Bull. **23**: 314 (1969). Type: D.R. Congo, Kapanga, 1932, *Overlaet* 246 (K holotype, BR).
> *Cyperus overlaetii* (S.S. Hooper & J. Raynal) Lye in Nord. J. Bot. **3**: 231 (1983). —Haines & Lye, Sedges & Rushes E. Afr.: 283, figs.582,583 (1983).

Slender to robust annual or perennial, loosely tufted or stoloniferous; culms 1–5, erect, to 1 m long and (0.5)1–3.5 mm wide, rounded to triangular, smooth. Leaves basal, sheaths to 10(20) cm long, pale brown to brown or reddish, with 0.5–1 mm wide mebranaceous ochrea-like flanges; blade flat, to 55 cm long and (1)1.5–4(5) mm wide, scabrid near tip or in apical third. Inflorescence a simple umbel-like anthela, more rarely condensed to almost capitate, 1.5–13(20) × 2–9 cm, branches held erect (inflorescence longer than wide), open inflorescences with one to several subsessile racemose clusters and 2–7(10) stalked clusters, each cluster with (1)3–25(40) spikelets; peduncles of stalked clusters 0.5–8(16) cm long; involucral bracts leaf-like, 2–3, largest 5–30 cm long and 1–4.5 mm wide, scabrid near tip or in upper half. Spikelets (5)7–20 × 3–4.5(6) mm (to 25 mm long in fruit), ovate or narrowly so; rachis straight in fruiting stage; glumes about 75% overlapping, appressed throughout, golden yellow to orange-brown or bronze or dark brown, ovate with a narrow hyaline margin, 2–4.5 × 1.5–2.5 mm, acute to obtuse, keel pale brown or pale green. Stamens (2)3; filaments 1–2 mm long; anthers 1–1.5 mm long. Stigma branches 2, 1–2 mm long. Nutlet broadly ellipsoid or slightly obovoid with tapering apex, shortly apiculate, dark grey to black, c. 1 mm long, surface with fine reticulum of raised ridges or weakly transversely ridged.

Botswana. N: Okavango Delta, Mboma–Xhamu channel, fl. & fr. 14.i.1980, *P.A. Smith* 2978 (K, SRGH). **Zambia**. N: Mbala Dist., Lake Chila, 1625 m, fl. & fr. 20.vi.1956, *Robinson* 1697 (K). W: Mwinilunga Dist., 1400 m, fl. 21.xii.1969, *Simon & Williamson* 1932 (K, SRGH). C: Serenje Dist., S of Kanona, Kundalila Falls, 1400 m, fl. 14.iii.1975, *Hooper & Townsend* 750 (K). S: Choma Dist., 30 km N of Choma on Namwala road, 1200 m, fl. 20.ii.1956, *Robinson* 1352 (K). **Zimbabwe**. N: Gokwe Dist., 40 km from Gokwe, fl. 18.iii.1963, *Bingham* 545 (K, SRGH). W: Hwange Dist., Kazuma Range, Katsatetsi R. at Permanent Pool, 10.v.1972, *Russell* 1951 (SRGH). C: Marondera Dist., no locality, 1525 m, fl. i.1948, *Colville* 43 (K, SRGH). E: Mutare Dist., Vumba Mts, Sambenyara, fl. & fr. 5.iii.1991, *Browning* 360 (K, SRGH). **Malawi**. N: Rumphi Dist., Livingstonia, Kaziweziwe R., 1450 m, fl. & fr. 8.i.1959, *Robinson* 3115 (K). C: Lilongwe Dist., Dzalanyama Forest Reserve, Chaulongwe Falls, 1300 m, fl. & fr. 26.iv.1970, *Brummitt* 10177 (K, MAL). S: Zomba Dist., Namasi, fl. iii.1899, *Cameron* 22 (K).

Mozambique. MS: Sussundenga Dist., Dombe to Mapira, Mucutuco R., fl. 27.viii.1953, *Gomes e Pedro* 4450 (K).

Pantropical. Wet to swampy grassland, riverbanks, lakeshores, ditches; 150–1900(?2450) m.

Conservation notes: Widespread distribution; Least Concern. In common and widespread habitats.

7. **Pycreus flavescens** (L.) Rchb., Fl. Germ. Excurs.: 72 (1830). —Clarke in F.T.A. **8**: 290 (1901). —Podlech in Prodr. Fl. SW Afr. **165**: 41 (1967). —Gordon-Gray in Strelitzia **2**: 142 (1995). —Hoenselaar in F.T.E.A., Cyperaceae: 292 (2010). Type: 'In Germaniae, Helvetiae, Galliae', *Herb. Burser* I: 81 (UPS lectotype), lectotypified by Kukkonen (Taxon 53: 178, 2004). FIGURE 14.**46**.

Cyperus flavescens L., Sp. Pl.: 46 (1753). —Kükenthal in Engler, Pflanzenr. **4**, 20(101): 398 (1936). —Haines & Lye, Sedges & Rushes E. Afr.: 281, figs.576,577 (1983).

Cyperus intermedius Steud. in Flora **25**: 581 (1842), illegitimate name, non Guss. Type: Ethiopia, Tigray, Dschomara, 5.ix.1838, *Schimper* II.1267 (GOET holotype, BM, BR, K, M, P, TUB, WAG).

Pycreus rehmannianus C.B. Clarke in Fl. Cap. **7**: 156 (1897); in F.T.A. **8**: 291 (1901). Type: South Africa, Limpopo, Houtbosch, 1875–1880, *Rehmann* 5651 (K holotype).

Cyperus rehmannianus (C.B. Clarke) Kuntze, Rev. Gen. Pl. **3**(2): 334 (1898). —Kükenthal in Engler, Pflanzenr. **4**, 20(101): 397 (1936).

Cyperus tanaensis Kük. in Engler, Pflanzenr. **4**, 20(101): 397 (1936). Type: Kenya, Tana Dist., Tana R., n.d., *Gregory* 87 (BM holotype, K). Name referred to in F.T.E.A. for F.Z. material.

Pycreus intermedius (Rikli) C.B. Clarke in F.T.A. **8**: 290 (1901). —Hoenselaar in F.T.E.A., Cyperaceae: 295 (2010). Name used for F.Z. material.

Cyperus subintermedius Kük. in Engler, Pflanzenr. **4**, 20(101): 390 (1936). —Haines & Lye, Sedges & Rushes E. Afr.: 283 (1983), new name for *Cyperus intermedius* Steud. (1842).

Cyperus rehmannianus var. *rigidiculmis* Kük. in Engler, Pflanzenr. **4**, 20(101): 398 (1936). Types: Tanzania, Kigoma Dist., Musosi, Mkuti R., *Peter* 37204 (B† syntype); Msosi, Mchaji, *Peter* 46228 (B† syntype), Njombe Dist., Lupembe, Msima, n.d., *Schlieben* 1044 (BR lectotype), lectotypified here.

Pycreus fallaciosus Cherm. in Arch. Bot. Mém. **7**(4): 7 (1936). Types: Senegal, N of Dienoudiella, Massadella, 29.v.1934, *Trochain* 3538 (P syntype, BR); same locality, v.1934, *Trochain* 3545 (P syntype, BR). Name used for F.Z. material.

Pycreus flavescens subsp. *laevinux* Lye in Nord. J. Bot. **1**: 620, fig.5 (1981). Type: Tanzania, Iringa Dist., Ruaha Nat. Park, 1 km W of Magangwe Ranger Post, 11.iii.1973, *Björnstad* 2600 (O holotype, ?K); see notes.

Pycreus flavescens subsp. *microglumis* Lye in Nord. J. Bot. **1**: 621, fig.6 (1981). —Hoenselaar in F.T.E.A., Cyperaceae: 294 (2010). Type: Uganda, Masaki Dist., 2–3 km S of West Mengo border on Kampala–Mbarara road, 11.vii.1971, *Lye & Katende* 6452b (O holotype, UPS). Name referred to in F.T.E.A. for F.Z. material.

Pycreus flavescens var. *castaneus* Lye in Nord. J. Bot. **1**: 621 (1981). —Hoenselaar in F.T.E.A., Cyperaceae: 294 (2010). Type: Kenya, Muranga Dist., Chanya R., 1.vii.1971, *Lye et al.* 6369 (EA holotype).

Pycreus flavescens subsp. *tanaensis* (Kük.) Lye in Nord. J. Bot. **1**: 622 (1981). —Hoenselaar in F.T.E.A., Cyperaceae: 295 (2010).

Pycreus flavescens subsp. *fallaciosus* (Cherm.) Lye in Nord. J. Bot. **1**: 622 (1981).

Pycreus flavescens subsp. *intermedius* (Rikli) Lye in Nord. J. Bot. **1**: 622 (1981).

Cyperus flavescens var. *castaneus* (Lye) Lye in Haynes & Lye, Sedges & Rushes E. Afr., app.3: 2 (1983). —Haines & Lye, Sedges & Rushes E. Afr.: 282, fig.578 (1983).

Cyperus flavescens subsp. *microglumis* (Lye) Lye in Haynes & Lye, Sedges & Rushes E. Afr., app.3: 2 (1983). —Haines & Lye, Sedges & Rushes E. Afr.: 283, fig.581 (1983).

Cyperus flavescens subsp. *tanaensis* (Kük.) Lye in Haynes & Lye, Sedges & Rushes E. Afr., app.3: 2 (1983). —Haines & Lye, Sedges & Rushes E. Afr.: 283, figs.579,580 (1983).

Cyperus flavescens subsp. *fallaciosus* (Cherm.) Lye in Haynes & Lye, Sedges & Rushes E. Afr., app.3: 2 (1983). —Haines & Lye, Sedges & Rushes E. Afr.: 282 (1983).

Fig. 14.**46**. PYCREUS FLAVESCENS. 1, habit; 2, branch from plant with condensed inflorescence; 3, branch from plant with open inflorescence; 4, tip of involucral bract; 5, spikelet; 6, detail of rachis; 7, glume, side view; 8, flower; 9, nutlet. 1, 4, 5 from *Robinson* 3440; 2 from *Robinson* 2356; 3, 8 from *Robinson* 5062; 6, 7, 9 from *Richards* 8105. Scale bars: 1–3 = 3 cm; 4–6 = 3 mm; 7–9 = 1 mm. Drawn by Juliet Williamson.

Cyperus flavescens subsp. *intermedius* (Steud.) Lye in Lidia **3**: 132 (1994).

Pycreus flavescens var. *rehmannianus* (C.B. Clarke) Govaerts in Govaerts & Simpson, World Checkl. Cyperaceae 604 (2007).

Single stemmed or tufted annual; culms 1 to over 20, erect, to 65 cm long and 1.5(2) mm wide, rounded, striate, smooth. Leaves basal, sheaths brown, to 6 cm long; blade flat (often wilting at time of flowering and appearing almost linear), to 20(25) cm long and 0.5–2.5 mm wide, minutely scabrid near tip. Inflorescence of 1–3 sessile clusters of 2–20 spikelets with 1–6 stalked clusters with clearly spaced spikelets, or congested and head-like, 1–3 × 1–4 cm if head-like to 9 × 13 cm if branched; peduncles of stalked clusters to 7(8) cm long; involucral bracts leaf-like, 1–3 large and 0(2) small, largest to 18(25) cm long and 0.5–3 mm wide, scabrid near tip. Spikelets 5–15 × 1.5–3(4) mm (to 25 mm long in fruit), linear to lanceolate; rachis straight or slightly zig-zag in fruiting stage; glumes appressed at flowering, slightly spreading later, falling at maturity, densely imbricate (about 50% overlapping), greenish yellow through yellow, golden yellow and pale brown to dark brown, ovate, without hyaline margin, (1)1.5–2.5(3) × 1–1.5 mm, subacute to obtuse with midrib not reaching the tip, keel prominent or not, pale brown to green. Stamens 3; filaments 1–2 mm long; anthers 0.2–0.5 mm long. Stigma branches 2, 1–1.5 mm long, easily detached. Nutlet ellipsoid to obovoid, apex subacute to truncate, apiculate, dark brown to black, 0.5–1 mm long, smooth to finely transversely or reticulately ridged.

Botswana. N: Ngamiland Dist., Xakanaxa Lediba, Moanachira R., fl. & fr. 20.viii.1976, *P.A. Smith* 1752 (K, SRGH). **Zambia**. B: Zambezi Dist., Zambezi (Balovale), fl. & fr. 8.vii.1963, *Robinson* 5580 (K). N: Mbala Dist., 1675 m, fl. & fr. 4.v.1957, *Richards* 9555 (K). W: Ndola Dist., Itawa dambo, 1300 m, fl. & fr. 29.iii.1960, *Robinson* 3441 (K). C: Kabwe Dist., 35 km N of Lusaka, Kamaila Forest Station, fl. 9.ii.1975, *Brummitt et al.* 14309 (K). S: Choma Dist., Mapanza, 1075 m, fl. & fr. 21.ii.1959, *Robinson* 3249 (K). **Zimbabwe**. N: Mazoe Dist., Henderson Res. Station, fl. 15.iv.1973, *Gibbs Russell* 2566 K, SRGH). W: Matobo Dist., Besha Kobila Farm, 1475 m, fl. iv.1957, *Miller* 4338 (K, SRGH). C: Harare Dist., University of Zimbabwe (Harare University College), fl. & fr. 29.iii.1963, *Loveridge* 641 (K, SRGH). E: Chipinge Dist., Tanganda Tea Estate, 1225 m, fl. & fr. iv.1934, *Brain* 10604 (K). **Malawi**. N: Chitipa Dist., Chisenga, 1550 m, fr. 10.vii.1970, *Brummitt* 11964 (K). C: Kasungu Dist., Mtunthama, Kamuzu Academy, 1075 m, fl. 20.ii.1979, *Blackmore et al.* 542 (K). S: Zomba Dist., Zomba to Ncheu, Domasi R., fl. 20.viii.1950, *Jackson* 117 (K). **Mozambique**. MS: Manica Dist., Bandula, 700 m, fl. & fr. 6.iv.1952, *Chase* 4536 (BM K, SRGH).

Widespread in all tropical, subtropical and temperate regions of the world. Wet and swampy grassland, often in shallow water, riverbanks, waterholes, old paddy fields, usually on clayey soils; 500–1750 m.

Conservation notes: Widespread distribution; Least Concern. In common and widespread habitats.

Lye (1982) cites an isotype of *Pycreus flavescens* subsp. *laevinux* as being at Kew but I have been unable to locate it.

As indicated by the lengthy synonymy, this extremely widespread and common species shows a bewildering variation. Several attempts (Lye, Hoenselaar) have been made to subdivide the African material but, in my opinion, none have really succeeded. Many field studies carefully recording the variation within populations and differences between populations in varying habitats are needed. *Gibbs Russell* 634, 635, 637 and 641, all collected on the same day from the same locality on the University Campus in Harare, clearly show the variation possible within a single population. For the moment it seems best just to recognise one very variable species.

Of the taxa recognised in F.T.E.A. (Hoenselaar 2010: 292), subsp. *microglumis* characterised by having glumes 1.6–1.7 mm long represents no more than one end of a continuous spectrum of glume length which actually varies from 1 mm in Botswana to 2.5(3) mm.

Specimens with dark reddish brown glumes (var. *castaneus* sensu stricto, *Pycreus rehmannianus*) look superficially very distinct. They occur scattered throughout the Flora area, but most commonly in Zimbabwe, and are connected to typical yellow- to brown-glumed specimens through all shades of intermediates. Luxuriant specimens of this taxon have often been named as *P. overlaetii*. They are linked to normal forms via all types of intermediates. However, the type collection of *P. overlaetii* is clearly perennial and here considered a synonym of *P. unioloides*.

Subsp. *tanaensis* is no more than a slender plant the like of which are also found throughout the Flora area.

8. **Pycreus lanceolatus** (Poir.) C.B. Clarke in Durand & Schinz, Consp. Fl. Afr. **5**: 538 (1894). —Hoenselaar in F.T.E.A., Cyperaceae: 291 (2010). Type: Madagascar, no locality, n.d., *du Petit-Thouars* s.n. (P-LAM holotype).

    Cyperus lanceolatus Poir. in Lam., Encycl. **7**: 245 (1806). —Kükenthal in Engler, Pflanzenr. **4**, 20(101): 349 (1936). —Haines & Lye, Sedges & Rushes E. Afr.: 276, figs.563,564 (1983).
    Pycreus propinquus Nees in Martius, Fl. Bras. **2**(1): 7 (1842). —Clarke in F.T.A. **8**: 300 (1901). Type: Brazil, Vila Rica, iv.1838, *Gardner* 714 (K holotype, E).

Single stemmed stoloniferous or loosely tufted perennial; culms 1–15, erect, to 60(80) cm long and 0.5–1.5 mm wide, rounded to triangular, smooth. Leaves basal, sheaths to 12 cm long, brown, with narrow, less than 0.5 mm wide mebranaceous ochrea-like flanges; blade flat, to 35 cm long and 1–2 mm wide, scabrid near tip. Inflorescence capitate, 1–3 × 1–5 cm, with 5–40 spikelets; involucral bracts leaf-like, 1–2, one often much larger, largest 5–20 cm long and 0.5–2 mm wide, scabrid near tip. Spikelets 7–20 × 2.5–3.5 mm (to 30 mm long in fruit), lanceolate; rachis straight in fruiting stage; glumes about 75% overlapping, appressed throughout, yellowish brown to brown, ovate, without hyaline margin, 2–2.5 × 1–1.5 mm, obtuse, keel pale brown or pale green. Stamens (2)3; filaments 1–2 mm long; anthers 1–1.5 mm long. Stigma branches 2, 1–2 mm long. Nutlet ellipsoid, with tapering apex, distinctly apiculate, dark grey, 1 mm long, with numerous minute papillae in longitudinal rows.

Zambia. B: Mongu, fl. 10.vii.1962, *Robinson* 5416 (K). N: Chinsali Dist., Shiwa Ngandu, 1525 m, fl. 10.vi.1956, *Robinson* 1632 (K, SRGH). W: Mwinilunga Dist., Matonchi, Dobeka Bridge, 1350 m, fl. 17.ii.1975, *Hooper & Townsend* 152 (K). **Zimbabwe**. E: Mutare Dist., Nyachawa Falls, fl. 19.iii.1950, *Chase* 4171 (K, SRGH). **Malawi**. N: Rumphi Dist., Livingstonia, Kaziweziwe R., 1450 m, fl. 8.i.1959, *Robinson* 3106 (K). S: Zomba Dist., Namasi, fl. iii.1899, *Cameron* 24 (K).

Widespread in tropical Africa and South Africa; also in Madagascar and South America. Swampy grassland, dambos, roadside ditches; 800–1600(1850) m.

Conservation notes: Widespread distribution; Least Concern. In common and widespread habitats.

9. **Pycreus robinsonii** Vollesen, sp. nov. Related to *Pycreus unioloides* and *P. chrysanthus*. It differs from both in the old fibrous leaf bases surrounding the culms. It differs from *P. unioloides* in the narrower glumes which have a narrow hyaline margin. *P. chrysanthus* has superficially very similar spikelets but quite differently sculptured nutlets. Type: Zambia, Mporokoso Dist., Mweru Wantipa, Kabwe Plain, 18.xii.1960, *Richards* 13751 (K holotype, K).

Robust perennial, few stemmed or forming large loose clumps; culms erect, to 60 cm long and 1.5–2 mm wide, triangular, smooth, at base surrounded by old dark brown to black fibrous leaf bases to 6 cm long. Leaves basal, sheaths above fibrous bases to 8 cm long, pale brown, with 1–1.5 mm wide mebranaceous ochrea-like flanges; blade flat, to 20 cm long and 3–4 mm wide, scabrid near tip. Inflorescence a simple umbel-like anthela or capitate, 2–7 × 2.5–6 cm, open inflorescences with 1–2 subsessile racemose clusters and 2–3 stalked clusters, each cluster with 7–25 spikelets; involucral bracts leaf-like, 3–52, largest 3–12 cm long and 2–5 mm wide, scabrid near tip. Spikelets 9–20 × 2.5–4 mm, lanceolate; glumes about 75% overlapping, appressed at first. Slightly spreading later, yellow brown to brown, ovate to elliptic, without hyaline margin, 2.5–3 × 1–1.5 mm, acute, keel pale brown. Stamens (2)3; filaments 1–2 mm long; anthers 1–1.5 mm long. Stigma branches 2, 1–2 mm long. Nutlet (immature) ellipsoid-obovoid, slightly tapering at apex, shortly apiculate, brown, c. 1 mm long, transversely rugose.

Zambia. N: Mbala Dist., Yendwe Valley, Lufubu R., 775 m, fl. 9.xii.1959, *Richards* 11950 (K); Mporokoso Dist., Mweru Wantipa, Kabwe Plain, 15.xii.1960, *Vesey-Fitzgerald* 2821 (K).

Not known elsewhere. Alluvial grassland; 750–1000 m.

Conservation notes: Possibly threatened but Data Deficient. Known from only three collections but from widespread and non-threatened habitats.

10. **Pycreus chrysanthus** (Boeckeler) C.B. Clarke in Durand & Schinz, Consp. Fl. Afr. **5**: 535 (1894); in Fl. Cap. **7**: 161 (1897). —Podlech in Prodr. Fl. SW Afr. **165**: 41 (1967). —Gordon-Gray in Strelitzia **2**: 141 (1995). Type: South Africa, KwaZulu-Natal, Umtsikaba R., vi.1840, *Drège* 4409 (B holotype, K).

    Cyperus chrysanthus Boeckeler in Linnaea **35**: 476 (1868). —Kükenthal in Engler, Pflanzenr. **4**, 20(101): 337 (1936).

    Cyperus chrysanthus var. *occidentalis* Kük. in Engler, Pflanzenr. **4**, 20(101): 337 (1936). Type: Namibia, no locality, *Dinter* s.n. (B holotype).

    Cyperus longistolon Peter & Kük. in Engler, Pflanzenr. **4**, 20(101): 333, fig.39 (1936). —Haines & Lye, Sedges & Rushes E. Afr.: 274 (1983). Type: Tanzania, near Itigi, Turu, 1.i.1926, *Peter* 33906 (B lectotype, EA, GOET), lectotypified here. Eight other syntypes from Tanzania.

    Pycreus longistolon (Peter & Kük.) Napper in J. E. Africa Nat. Hist. Soc. Natl. Mus. **28**(124): 6 (1971). —Hoenselaar in F.T.E.A., Cyperaceae: 304 (2010).

    Pycreus longistolon subsp. *atrofuscus* Lye in Nord. J. Bot. **1**: 618, fig.3 (1981). Type: Tanzania, Masai Dist., Malanyo, *Newbould* 6057 (EA holotype).

    Cyperus longistolon subsp. *atrofuscus* (Lye) Lye in Sedges & Rushes E. Afr., app.3: 2 (1982). —Haines & Lye, Sedges & Rushes E. Afr.: 274, figs.556,557 (1983).

Stoloniferous perennial; culms 1–2, erect, to 80 cm long, 1–4 mm wide, rounded to triangular, smooth. Leaves basal, sheaths to 10 cm long, brown, with 0.5–1 mm wide mebranaceous ochrea-like flanges; blade flat, to 70 cm long and (1)2–5 mm wide, scabrid near tip. Inflorescence a simple umbel-like anthela, 4–17 × 3–10 cm, with 1–2 sessile racemose clusters, sometimes with short sidebranches near base, and 2–5(7) stalked clusters, each cluster with (4)7–20 spikelets; peduncles of stalked clusters 0.5–9(12) cm long; involucral bracts leaf-like, 2 large and 1–2 small, largest 7–35(50) cm long and 2–4 mm wide, scabrid near tip. Spikelets 10–25(30) × 2–4 mm, lanceolate; rachis straight in fruiting stage; glumes about 50% overlapping, appressed or spreading, pale brown to golden brown or brown, ovate with a narrow hyaline margin towards apex, 3–4 × 2–3 mm, acute sometimes apiculate, distinctly 3–5-veined, keel pale brown. Stamens (2)3; filaments 2–3 mm long; anthers 2–3 mm long. Stigma branches 2, 2–4 mm long. Nutlet oblong to obovate with truncate apex, shortly apiculate, dark brown, 1–1.5 mm long, with numerous minute papillae in longitudinal rows.

Botswana. N: Okavango Delta, Xigera road, fl. 22.i.1977, *P.A. Smith* 1887 (K, SRGH). **Zambia**. C: Kabwe Rural Dist., Chisamba, Wardy Farm, 1100 m, fl. 1.ii.1995, *Bingham* 10338 (K). S: Namwala Dist., Namwala, Kafue R., 900 m, fl. & fr. 9.i.1957, *Robinson* 2197 (K). **Zimbabwe**. N: Gokwe Dist., Sengwa Wildlife Institute, 20.i.1982, *Mahlangu* 570 (SRGH). W: Matobo Dist., SW Matopos, Maleme Village, fl. 6.i.1963, *Wild* 5926 (K, SRGH). C: Harare Dist., no locality, 1425 m, fl. 12.ii.1939, *Brain* 8538 (BM).

Also in Kenya, Tanzania, Namibia and South Africa. Usually on sandy soil in drying pans and dambos, but occasionally in wet grassland; 900–1450 m.

Conservation notes: Widespread distribution; Least Concern. In common and widespread habitats.

11. **Pycreus intactus** (Vahl) J. Raynal in Adansonia, n.s. **17**: 46 (1977). —Gordon-Gray in Strelitzia **2**: 142 (1995). —Govaerts in Govaerts & Simpson, World Checkl. Cyperaceae: 605 (2007). Type: Senegal, Ngalam (Galam), n.d., *unknown collector no.17* (C-Vahl holotype, P-JU); see note.

Cyperus intactus Vahl, Enum. Pl. **2**: 332 (1805).

Cyperus laxespicatus Kük. var. *brunneo-tinctus* Kük. in Repert. Spec. Nov. Regni Veg. **21**: 325 (1925). Type: Zambia, Malangushi R., 18.xii.1907, *Kässner* 2047 (B holotype, K).

Cyperus atribulbus Kük. in Engler, Pflanzenr. **4**, 20(101): 363 (1936). —Haines & Lye, Sedges & Rushes E. Afr.: 279 (1983). Types: Tanzania, no locality, n.d., *Busse* 749 (B syntype); Mozambique, near Beira, Dondo (25 Miles Station), 11.iv.1898, *Schlechter* 12254 (BR lectotype, B, K, PRE, SAM), lectotypified here.

Pycreus atribulbus (Kük.) Napper in J. E. Africa Nat. Hist. Soc. Natl. Mus. **28**(124): 5 (1971). —Hoenselaar in F.T.E.A., Cyperaceae: 301 (2010).

Pycreus laxespicatus sensu Hoenselaar in F.T.E.A., Cyperaceae: 303 (2010) for *Milne-Redhead & Taylor* 7821; see note.

A single stemmed or tufted annual or perennial, sometimes with swollen culm bases; culms 1–10, erect, to 60(90) cm long and 0.5–2.5(3) mm wide, rounded to triangular, striate, smooth, older leaf bases not fibrous. Leaves basal, sheaths to 6(10) cm long, brown; blade flat (often wilting at time of flowering and appearing rolled or filiform), to 40(50) cm long and 1.5–5 mm wide, scabrid in apical part. Inflorescence a simple umbel-like anthela (rarely a compound anthela), sometimes appearing subcapitate, 2–12 × 2–15 cm, of 1–5 sessile and 2–10 stalked racemose clusters (sessile clusters sometimes with short lateral branches), each cluster with 10–20 loosely arranged spikelets; peduncles of stalked clusters 0.2–6(10) cm long; involucral bracts leaf like, (1)2–5, largest 5–25(30) cm long and 1–3(4) mm wide, scabrid in apical part. Spikelets 5–20 × 1–2 mm, linear-lanceolate; rachis becoming zig-zag in fruiting stage; glumes appressed to spreading, falling at maturity, only slightly (25–50 %) overlapping, pale yellowish green to greyish green or olive green to yellowish brown or brown, elliptic to slightly obovate, without hyaline margin, narrowing in basal part, 1.5–3 × 1–1.5 mm, acute to obtuse with hyaline tip, keel brown, 3-veined. Stamens (2)3; filaments 1 mm long; anthers 1 mm long. Stigma branches 2, 1.5 mm long. Nutlet ovate to oblong, obtuse at apex, shortly apiculate, dark steel grey to black, 1–1.5 mm long, with numerous minute papillae in longitudinal rows.

Zambia. C: Mumbwa Dist., 62 km W of Lusaka on Mumbwa road, fl. 20.iii.1965, *Robinson* 6458 (K). E: Petuake Dist., Petuake Old Boma, fl. 30.x.1962, *Robinson* 5600 (K). S: Monze Dist., Simasunda, Mapanza, 1075 m, fl. 3.iii.1957, *Robinson* 2147 (K). **Zimbabwe**. N: Kariba Dist., Chirundu Hot Springs, fl. 19.i.1967, *Wild* 7591 (K, SRGH). W: Matobo Dist., Besna Kobila Farm, 1475 m, fl. ii.1958, *Miller* 5105 (K, SRGH). C: Harare, 1475 m, fl. 20.iii.1932, *Brain* 8847 (K). E: Nyanga Dist., Mandea Range, Honde Valley, 900 m, fl. & fr. 30.iii.1969, *Plowes* 3175 (K, SRGH). **Malawi**. N: Mzimba Dist., 20 km NNW of Chikangawa, 1675 m, fl. 8.x.1978, *Phillips* 4058 (K, MO). C: Lilongwe Dist., Dzalanyama, Naminyanga dambo, fl. 4.xii.1951, *Jackson* 703 (K). **Mozambique**. N: Marrupa Dist., Naboina, 20 km on road to Lichinga from Marrupa, fl. 17.ii.1981, *Nuvunga* 558 (K, LMA). Z: Namacurra Dist., 83 km from Mocuba to Quelimane, fl. 28.v.1949, *Barbosa & Carvalho* 2908 (K). MS: Manica Dist., Bandula, 700 m, fl. 6.iv.1952, *Chase* 4608 (BM, K, SRGH).

Also in Guinée, Senegal, Tanzania, South Africa and Madagascar. Grassland on sandy soils, dambo grassland, roadsides, pans on rocky outcrops, hot springs, montane grassland, a weed in ricefields; 50–1700(2000) m.

Conservation notes: Widespread distribution; Least Concern. In common and widespread habitats; tolerates a certain amount of habitat disturbance.

Very few botanists had been to Senegal prior to 1805. The most likely 'suspect' is Adanson. This is also supported by the fact that most of his collections are in the Jussieu herbarium in Paris. Vahl (1805) states that he got his specimen from Jussieu but I have not seen the Paris sheet.

Hoenselaar (2010) makes the combination *Pycreus laxespicatus* (Kük.) Hoenselaar based on *Cyperus laxespicatus* with the type *Fries* 1052 (UPS). She then cites *Milne-Redhead & Taylor* 7821 (K) from Tanzania under this species. I have seen the type specimen from Uppsala; it is clearly a different species and is treated here as a synonym of *Pycreus aethiops* (Ridl.) C.B. Clarke.

Annual or perennial forms of this species seem to occur throughout its distribution, although perennial forms seem to be more common in the northern parts. No differences have been found in the spikelets, glumes and nutlets. It is superficially also quite similar to *Pycreus macrostachyos* but without the conspicuous white hyaline bract margin of that species.

12. **Pycreus xantholepis** Nelmes in Kew Bull. **6**: 319 (1952). —Hoenselaar in F.T.E.A., Cyperaceae: 291 (2010). Type: Angola, Moxico Dist., Zambesi R., Ikula Hot Springs, 17.i.1938, *Milne-Redhead* 4213 (K holotype, BR).

> *Pycreus dewildeorum* J. Raynal in Kew. Bull. **23**: 313 (1969). Type: Cameroon, 15 km NE of Maroua on road to Waza, 12.ix.1964, *de Wilde* 3223 (K holotype, WAG). Name has been used for material from the Flora area.
>
> *Cyperus xantholepis* (Nelmes) Lye in Lidia **7**: 97 (2011).

Slender annual or short-lived perennial; culms 1 to 10, erect, to 40(60) cm long and 1 mm wide, rounded, striate, smooth. Leaf sheaths brown to reddish brown, to 6.5 cm long; blade linear, flat or folded, to 12 cm long and 1 mm wide, scabrid near tip. Inflorescence to 5 × 5 cm, consisting of a single subsessile cluster of (1)2–10 spikelets or also with 1–4 stalked clusters each with 1–6 spikelets and peduncle to 6 cm long; involucral bracts 1–3, large, leaf-like, with or without 1–2 small ones, when single held as an extension of the culm making inflorescence appear lateral, largest 5–20 cm long and 1–2 mm wide, scabrid near apex. Spikelets 10–25 × 4–5 mm (to 35 mm long in fruit), lanceolate; rachis zig-zag in fruiting stage; glumes appressed at flowering and fruiting stages, falling at maturity, densely imbricate (about 2/3 overlapping), olive brown to dark brown, ovate to elliptic, with or without a very narrow hyaline margin, 3–4 × 2 mm, acute and often apiculate with midrib reaching the tip, keel prominent, rib-like, green. Stamens 2; filaments 1–1.5 mm long; anthers 1 mm long. Stigma branches 2, 2–3 mm long. Nutlet black, ellipsoid to subglobose or obovoid, apex truncate, apiculate or not, 1–1.5 mm long, with numerous indistinct papillae in longitudinal rows.

Zambia. W: Ndola Dist., Itawa Dambo, 1300 m, fl. 3.iv.1960, *Robinson* 3453 (K). C: Mazabuka Dist., Sala Reserve, 55 km SW of Lusaka, fl. & fr. 4.iii.1962, *Robinson* 4989 (K). S: Choma Dist., Mapanza, 1050 m, fl. & fr. 8.iii.1958, *Robinson* 2786 (K). **Malawi**. N: Chitipa Dist., 5 km SE of Chitipa (Fort Hill), fl. 11.iii.1961, *Robinson 4445 (K)*.

Also in Cameroon, D.R. Congo, Tanzania and Angola. On mud at edges of waterholes, wet grassland and old rice fields; 1000–1400 m.

Conservation notes: Least Concern. Widespread inside and outside our area in common non-threatened habitats. Has a tolerance of 'weedy' habitats.

13. **Pycreus sp. A** of F.Z. (*Robinson* 1299).

Short-lived tufted perennial; culms to 10, erect, to 80 cm long and 1.5 mm wide, rounded, striate, smooth. Leaf sheaths reddish brown, to 7 cm long; blade linear (immature at time of flowering), folded, to 10 cm long and 0.5 mm wide, scabrid in apical half. Inflorescence to 4 × 3.5 cm, consisting of a single subsessile cluster of 3–10 spikelets; involucral bracts 1 large and leaf-like, 1 small, when single held as an extension of culm and making inflorescence appear lateral, largest to 11 cm long and c. 1 mm wide, scabrid in upper half. Spikelets 10–25 × 4–5 mm (to 35 mm long in fruit), lanceolate; rachis zig-zag in fruiting stage; glumes appressed in flower, slightly spreading in fruit, falling at maturity, densely imbricate (about 75% overlapping), pale brown to reddish brown, ovate, without hyaline margin, 3.5–4 × 2 mm, acuminate, midrib reaching tip, keel prominent, rib-like, pale. Stamens 2; filaments 1–1.5 mm long; anthers 1–1.5 mm long.

Stigma branches 2, 2 × 3 mm long. Nutlet black, ellipsoid to slightly obovoid, apex tapering, apiculate, c. 1.5 mm long, with numerous indistinct papillae in longitudinal rows.

Zambia. C: Lusaka Dist., Munali, 8 km E of Lusaka, 1250 m, fl. & fr. 14.vi.1955, *Robinson* 1299 (K).

Not known elsewhere. Edge of stream and wetter parts of dambo grassland; 1250 m.

Conservation notes: Possibly threatened but Data Deficient. Known only from this collection from a widespread habitat in a fairly well-collected area. Locally abundant according to Robinson.

14. **Pycreus macrostachyos** (Lam.) J. Raynal in Kew Bull. **23**: 314 (1969). —Gordon-Gray in Strelitzia **2**: 143 (1995). —Hoenselaar in F.T.E.A., Cyperaceae: 288 (2010). Type: 'Ex Africa', *Herb. Lamarck* s.n. (P-LA holotype). FIGURE 14.**47**.

    Cyperus macrostachyos Lam., Tab. Encycl. **1**: 147 (1791). —Haines & Lye, Sedges & Rushes E. Afr.: 288, figs.596,597 (1983).

    Cyperus tremulus Poir. in Lamarck, Encycl. **7**: 264 (1806). —Kükenthal in Engler, Pflanzenr. **4**, 20(101): 361 (1936). Type: Madagascar, no locality, n.d., *du Petit-Thouars* s.n. (P holotype).

    Pycreus albomarginatus Nees in Martius, Fl. Bras. **2**(1): 9 (1842). —Clarke in F.T.A. **8**: 305 (1901). Type: Brazil, Bahia, Joazeiro, *Martius* s.n. (M holotype).

    Pycreus tremulus (Poir.) C.B. Clarke in Durand & Schinz, Consp. Fl. Afr. **5**: 542 (1894); in F.T.A. **8**: 306 (1901).

    Pycreus macrostachyos subsp. *tremulus* (Poir.) Lye in Nord. J. Bot. **1**: 622 (1982).

    Cyperus macrostachyos subsp. *tremulus* (Poir.) Lye in Haines & Lye, Sedges & Rushes E. Afr., app.3: 2 (1983). —Haines & Lye, Sedges & Rushes E. Afr.: 289, fig.598 (1983).

A usually robust annual or short-lived perennial; culms usually solitary but sometimes 2–3(4), erect, to 1.25 m long and (1)1.5–6 mm wide, triangular, smooth, lower part covered by broad membranaceous leaf sheaths. Leaves basal, sheaths to 20(30) cm long, pale brown to reddish; blade flat, to 50 cm long and 2–8 mm wide, scabrid near tip. Inflorescence a solitary or compound umbel-like anthela, 5–25 cm long, each of one to several sessile and 2–10 stalked elongated primary racemose clusters, these again often with sessile and stalked clusters, each cluster with 15–30 loosely arranged spikelets; peduncles of stalked clusters 1–16 cm long; involucral bracts leaf-like, 2–5, largest (10)15–45 cm long and (1)2–6(10) mm wide, scabrid near tip or along whole length. Spikelets 7–25 × 1.5–3 mm (to 35 mm in fruit), linear-lanceolate; rachis becoming slightly zig-zag in fruiting stage; glumes appressed at first, spreading from flowering stage, falling at maturity, yellowish to golden brown or dark brown, obovate with a distinct white hyaline margin, narrowing in basal part, 1.5–3(3.5) × 1–2 mm, obtuse to acute, keel green. Stamens (2)3; filaments 2–2.5 mm long; anthers 0.5 mm long. Stigma branches 2, 1–2 mm long. Nutlet oblong to obovoid with very slightly tapering apex, shortly apiculate, greyish black to black, 1.5–2 mm long, with numerous minute papillae in longitudinal rows.

Botswana. N: Chobe Dist., old Mababe–Joverega road, Mababe Marsh, fl. & fr. 10.vi.1978, *P.A. Smith* 2421 (K, SRGH). SE: South East Dist., Content Farm, 1050 m, fl. 27.1.1978, *Hansen* 3344 (C, GAB, K, PRE, SRGH). **Zambia**. N: Mbala Dist., track to Casacalawe, 750 m, fl. & fr. 18.iii.1955, *Richards* 5008 (K). W: Kasempa Dist., 7 km E of Chizera, fl. 27.iii.1961, *Drummond & Rutherford-Smith* 7440 (K, SRGH). C: Kabwe Dist., 36 km N of Lusaka, Kamaila Forest Station, fl. 9.ii.1975, *Brummitt et al.* 14310 (K). E: Mambwe Dist., Chinzombo to Nyamaluma, 550 m, fl. 8.iii.1988, *Phiri* 2090 (K). S: Itezhi-Tezhi Dist., 25 km N of Mapanza, Mbeza, 1000 m, fl. 27.i.1954, *Robinson* 491 (K). **Zimbabwe**. N: Hurungwe Dist., north bank of Mauora R., 600 m, fl. & fr. 1.iii.1958, *Phipps* 1014 (K, SRGH). W: Matobo Dist., Besna Kobila Farm, 1475 m, fl. & fr. iv.1961, *Miller* 7885 (K, SRGH). C: Chegutu Dist., Kufara Farm dam, fl. 24.ii.1969, Mavi 973 (K, SRGH). E: Chiredzi Dist., 'Lower Sabi', 450 m, fl. 27.i–2.ii.1948, *Rattray* 1236 (K, SRGH). S: Mwenezi Dist., 625 m, fr. 6.v.1958, *Drummond* 5653 (K, SRGH). **Malawi**. N: Karonga Dist., Khondowe (Kondowa) to Karonga, fl. vii.1896, *Johnston* s.n. (K). C: Dedza Dist., Chongoni Forest, fl. 24.iv.1971, *Salubeni* 1548 (K, SRGH). S: Kasupe

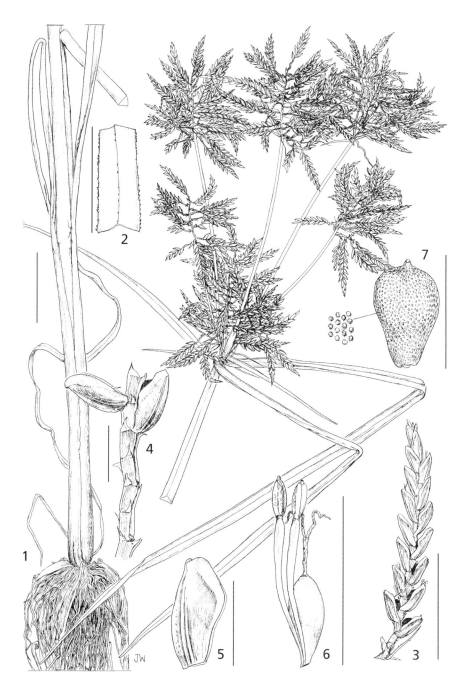

Fig. 14.**47**. PYCREUS MACROSTACHYOS. 1, habit; 2, detail of involucral bract; 3, spikelet; 4, detail of rachis; 5, glume, side view; 6, flower; 7, nutlet with detail of the surface. 1–3 from *Drummond* 7440; 4, 5, 7 from *Richards* 5008; 6 from *Kornaś* 3405. Scale bars: 1 = 3 cm; 2–3 = 1 cm; 4–7 = 1 mm. Drawn by Juliet Williamson.

Dist., Liwonde Nat. Park, 500 m, fl. & fr. 4.v.1985, *Dudley* 1626 (K). **Mozambique**. MS: Manica Dist., Bandula, fl. 6.iv.1952, *Chase* 4545 (BM, SRGH).

Widespread in tropical and South Africa; also in Madagascar and tropical America. Wet and swampy grassland around lakes and waterholes, waterholes in woodland, riverbanks, often in water; 300–1450 m.

Conservation notes: Widespread distribution; Least Concern. In common and widespread habitats.

15. **Pycreus pelophilus** (Ridl.) C.B. Clarke in Durand & Schinz, Consp. Afr. **5**: 540 (1894); in F.T.A. **8**: 298 (1901). —Podlech in Consp. Fl. SW Afr. **165**: 42 (1967). —Gordon-Gray in Strelitzia **2**: 147 (1995). —Hoenselaar in F.T.E.A., Cyperaceae: 280 (2010). Type: Angola, Bemposta, vii.1858, *Welwitsch* 7025 (BM lectotype); Luanda, Conceição, 12.vii.1854, *Welwitsch* 7082 (BM syntype); Moçamedes (Mossamedes), Giraul, vii.1859, *Welwitsch* 6887 (BM syntype), lectotypified here.

    Cyperus pelophilus Ridl. in Trans. Linn. Soc. London, Bot. **2**: 129 (1884). —Kükenthal in Engler, Pflanzenr. **4**, 20(101): 364 (1936). —Haines & Lye, Sedges & Rushes E. Afr.: 289, figs.599,600 (1983).

    Cyperus sulcinux C.B. Clarke in J. Linn. Soc., Bot. **21**: 56 (1886). Types: India. Clarke cites 14 specimens, but does not indicate a type. These are all considered syntypes.

    Pycreus sulcinux (C.B. Clarke) C.B. Clarke in Hooker fil., Fl. Brit. India **6**: 593 (1894) for African specimens, and in F.T.A. **8**: 298 (1901) for Malawi specimens. —Mapura & Timberlake, Checkl. Zimbabwe Pl.: 89 (1994).

Slender annual; culms 1–15, erect, to 30 cm long and 0.5–1.5 mm wide, rounded, longitudinally striate, smooth. Leaves basal, flat, to 20(25) cm long and (0.5)1–3 mm wide, scabrid on margins and midrib near tip; leaf sheaths brown to purple. Inflorescence 1–13 cm long, of one sessile and (0)1–10 stalked head-like or digitate clusters each with 3–20 loosely arranged spikelets; peduncles of stalked clusters 0.5–10(13) cm long; involucral bracts leaf-like, (2)3–6, 3–25 cm long and 1–4 mm wide, scabrid near tip or along whole length. Spikelets 7.5–20(25) × 1.5–4 mm (to 35 × 5 mm in fruit), linear-lanceolate; rachis becoming distinctly zig-zag in fruiting stage; glumes appressed at flowering, spreading later, falling at maturity, golden to golden brown, broadly ovate-elliptic with a white hyaline margin in apical half, narrowing in basal part, 1.5–2.5 × 1.5–2 mm, acute or with a short mucro to 0.5 mm long, keel green. Stamens 2; filaments 1–1.5 mm long; anthers 0.2 mm long. Stigma branches 2, 0.5–1 mm long. Nutlet obovoid to obcordate with flat sides and truncate apex, shortly apiculate, dark grey to greyish black, c. 1 mm long, with numerous minute papillae in longitudinal rows.

Botswana. N: Central Dist., Nata area, 6 km upstream from Nata R. Delta, fl. & fr. 14.iv.1976, *Ngoni* 494 (K, SRGH). SE: Kgatleng Dist., Mochudi, fl. iii.1974, *Mitchison* s.n. (K). **Zambia**. B: Mongu Dist., 5 km NW of Mongu, Bulozi Plain, 1050 m, fl. 19.iii.1996, *Harder et al.* 3699 (K, MO). N: Kasama Dist., Chambese R., 1200 m, fl. & fr. 7.iv.1961, *Richards* 15011 (K). C: Kapiri Mposhi Dist., 13 km NW of Kabwe, Bonanza, 1200 m, fl. & fr. 8.iv.1972, *Kornaś* 1555 (K). E: Petauke Dist., Great East Road, 17 km E of Kachalola, 850 m, fl. 1.iii.1973, *Kornaś* 3345 (K). S: Choma Dist., 18 km N of Choma, Muckle Neuk, 1275 m, fl. & fr. 27.ii.1954, *Robinson* 573 (K). **Zimbabwe**. N: Hurungwe Dist., Zambezi Valley, Menswa Pan, 500 m, fl. 27.ii.1953, *Wild* 4024 (K, SRGH). W: Bulilimamangwe Dist., Simukwe R., 4 km downstream from Mount Jim, fl. & fr. 12.iv.1974, *Ngoni* 382 (K, SRGH). C: Chegutu Dist., Nkuti Farm, fl. & fr. 24.ii.1969, *Mavi* 952 (K, SRGH). E: Chipinge Dist., Chibuwe, 700 m, fl. & fr. 1.vi.1972, *Gibbs-Russell* 2076 (K, SRGH). S: Beitbridge Dist., Tshiturapadzi (Chiturupadzi) Dip Camp, fl. & fr. 19.iii.1967, *Mavi* 252 (K, SRGH). **Malawi**. N: 'Nyasa' (probably just over border in SE Tanzania), Umbaka R., fr. 29.xii.1887, *Scott* s.n. (K). S: Blantyre Dist., near Blantyre, Nyambadwe Hill, 1125 m, fl. & fr. 6.v.1980, *Townsend* 2138 (K). **Mozambique**. N: Malema Dist., Mutuali, CICA Expt. Station, fr. 20.iv.1961, *Balsinhas & Marrime* 416 (K). MS: Manica Dist., Bandula, 700 m, fl. 6.iv.1952, *Chase* 4542 (BM

K, SRGH). GI: Chicualacuala Dist., Dumela, Limpopo R., fl. 30.iv.1961, *Drummond &*
Rutherford-Smith 7629 (K, SRGH). M: Maputo Dist., Inhaca Is., W coast, sea level, fl.
4.iii.1958, *Mogg* 31694 (K).

Also in Sudan, Somalia, D.R. Congo, Burundi, Uganda, Kenya, Tanzania, Angola,
Namibia and South Africa. Edges of drying waterholes, swampy grassland, damp places
in mopane woodland, riverbanks, fallows, a weed of maize and rice fields, usually on
clay but occasionally on sandy soils; 5–1500 m.

Conservation notes: Widespread distribution; Least Concern. Grows in common
and widespread habitats; also has a tolerance of 'weedy' habitats.

Brain 7570 (K) from Matopos, W Zimbabwe is a dwarf form with almost filiform
leaves and a single large and one small involucral bract. It has a solitary sessile spikelet
5 mm long or a single solitary cluster of 2–4 spikelets. It has not been included in the
description here.

16. **Pycreus pumilus** (L.) Nees in Linnaea **9**: 283 (1834). —Clarke in F.T.A. **8**: 296
(1901). —Podlech in Prodr. Fl. SW Afr. **165**: 42 (1967). —Hoenselaar in F.T.E.A.,
Cyperaceae: 283 (2010). Type: 'India', *Herb. Linn.* No. 70.34, R-hand specimen
(LINN lectotype), lectotypified by Kukkonen (Taxon 53: 179, 2004).

Cyperus pumilus L., Cent. Pl. II: 6 (1756).

Cyperus nitens Retz., Observ. Bot. **5**: 13 (1788). Type: India, Tranquebar, *König* s.n. (LD
holotype).

Cyperus patens Vahl, Enum. Pl. **2**: 334 (1805). Type: 'Guinea', *Thonning* s.n. (C holotype).

Pycreus nitens (Retz.) Nees in Nov. Actorum Acad. Caes. Leop.-Carol. Nat. Cur. **19**, suppl.1:
53 (1843). —Clarke in F.T.A. **8**: 295 (1901).

Cyperus pumilus var. *patens* (Vahl) Kük. in Engler, Pflanzenr. **4**, 20(101): 378 (1936). —
Haines & Lye, Sedges & Rushes E. Afr.: 290, figs.602,603 (1983).

Pycreus pumilus var. *patens* (Vahl) Kük. in Engler, Pflanzenr. **4**, 20(101): 378 (1936). —
Hoenselaar in F.T.E.A., Cyperaceae: 283 (2010).

Pycreus pumilus subsp. *patens* (Vahl) Podlech in Mitt. Bot. Staatssamml. München **3**: 523
(1960). —Gordon-Gray in Strelitzia **2**: 147 (1995).

Slender annual; culms 1–20, erect, to 20(30) cm long and 0.5–1 mm wide, rounded,
longitudinally striate, smooth. Leaves basal, sheaths pale brown, to 3 cm long; blade flat, to
15(20) cm long and 0.5–1.5(2) mm wide, scabrid on margins and midrib near tip. Inflorescence
of one sessile and 1–5(8) stalked digitate clusters each with 6–15 loosely arranged spikelets, (1)2–
6(9) × 1–5(7) cm; peduncles of stalked clusters 0.5–5(7) cm long; involucral bracts 2–4, largest
leaf-like, 2–15(20) cm long, 0.5–2 mm wide, scabrid near tip or along whole length. Spikelets
3–12 × 1–2 mm (to 20 × 3 mm in fruit), linear-lanceolate; rachis straight in fruiting stage; glumes
spreading at flowering, falling at maturity, yellow to dark brown, broadly ovate-elliptic with a
broad white to brown hyaline margin forming two conspicuous apical lobes, narrowing in basal
part, 1–2 mm long, acute to acuminate with a recurved mucro to 0.5 mm long, keel green or
brown, prominent. Stamens 1–2; filaments c. 1 mm long; anthers 0.2 mm long. Stigma branches
2, 0.5–1 mm long. Nutlet obovoid, apex truncate to slightly retuse, shortly apiculate, metallic
grey, c. 0.5 mm long, with numerous minute papillae in longitudinal rows.

Botswana. N: Okavango Delta, Mboma Camp, fl. 14.ii.1974, *P.A. Smith* 837 (K,
SRGH). **Zambia**. N: Mporokoso Dist., Mweru Wantipa, Kangiri, Mawe swamp, 1050 m,
fl. & fr. 8.iv.1957, *Richards* 9096 (K). W: Ndola Dist., Itawa Dambo, 1300 m, 1.v.1960,
Robinson 3724 (K). C: Mpika Dist., Mutinondo, 18.iii.2013, *Merrett* 1292 (K). E:
Lundazi Dist., Tumbuka, fl. & fr. iii.1962, *Robinson* 5084 (K). S: Choma Dist., 18 km
N of Choma, Muckle Neuk, 1275 m, fl. & fr. 27.ii.1954, *Robinson* 570 (K). **Zimbabwe**.
N: Kariba Dist., Lake Kariba, Sengwa West, 475 m, fl. & fr. 19.xii.1964, *Mitchell* 919
(K, SRGH). W: Matobo Dist., Besna Kobila Farm, 1475 m, fl. iv.1957, *Miller* 4326 (K,
SRGH). C: Chegutu Dist., Makwiro, 1300 m, fl. & fr. 16.iv.1932, *Brain* 8912 (K). E:
Chipinge Dist., 5 km S of Rusongo, Remai Border road, 400 m, fl. & fr. 1.ii.1975,

Gibbs-Russell 2756 (K, SRGH). S: Mwenezi Dist., Mateke Hills, Malangwe R., 625 m, fl. & fr. 6.v.1958, *Drummond* 5594 (K, SRGH). **Malawi**. C: Mchinji Dist., 7 km W of Namitete, fr. 29.iv.1989, *Radcliffe-Smith et al.* 5788 (K). S: Kasupe Dist., Liwonde Nat. Park, Chiunguni Hill, 500 m, fl. & fr. 4.v.1985, *Dudley* 1635 (K). **Mozambique**. Z: Lugela Dist., Mocuba, Namagoa, fl. & fr. iii.1943, *Faulkner* K141 (K). T: Moatize Dist., Boruma, c. 1890, *Menyhart* 1051a (not seen). MS: Manica Dist., Bandula, 700 m, fl. 6.iv.1952, *Chase* 4604 (BM, SRGH).

Pantropical. Drying mud or sand at edges of waterholes and riverbanks, dambos, various types of wet to dry grassland, roadsides, weed in fields; 400–1600 m.

Conservation notes: Widespread distribution; Least Concern. In common and widespread habitats.

17. **Pycreus aethiops** (Ridl.) C.B. Clarke in Durand & Schinz, Consp. Fl. Afr. **5**: 534 (1894); in F.T.A. **8**: 297 (1901). —Hoenselaar in F.T.E.A., Cyperaceae: 301 (2010). Type: Angola, Huíla Dist., between Ferrao da Sola and Jau, iv.1860, *Welwitsch* 6875 (BM holotype).

    Cyperus aethiops Ridl. in Trans. Linn. Soc. London, Bot. **2**: 129 (1884). —Haines & Lye, Sedges & Rushes E. Afr.: 276, fig.562 (1983).

    Cyperus betschuanus Boeckeler in Bot. Jahrb. Syst. **11**: 406 (1889). —Kükenthal in Engler, Pflanzenr. **4**, 20(101): 332 (1936). Type: South Africa, Kachun Wells, ii.1886, *Marloth* 1027 (B holotype, PRE).

    Pycreus betschuanus (Boeckler) C.B. Clarke in Durand & Schinz, Consp. Fl. Afr. **5**: 535 (1894); in Fl. Cap. **7**: 159 (1897); in F.T.A. **8**: 304 (1901). —Podlech in Prodr. Fl. SW Afr. **165**: 41 (1967). —Gordon-Gray in Strelitzia **2**: 139 (1995).

    Pycreus cooperi C.B. Clarke in Durand & Schinz, Consp. Fl. Afr. **5**: 535 (1894); in Fl. Cap. **7**: 160 (1897). —Gordon-Gray in Strelitzia **2**: 141 (1995). Type: South Africa, Orange Free State, 1861, *Cooper* 912 (K lectotype, NH); Transvaal, Houbosch, 1875–1880, *Rehmann* 5652 (K, NH syntype), lectotypified here.

    Pycreus globosus (All.) Rchb. var. *nilagiricus* sensu C.B. Clarke in F.T.A. **8**: 299 (1901) for specimens from "Mozambique Distr.".

    Cyperus laxespicatus Kük. in Fries, Wiss. Ergebn. Schwed. Rhodesia-Kongo Exped. 1911–1912, Erg.: 3 (1921); in Engler, Pflanzenr. **4**, 20(101): 332 (1936). Type: Zambia, N of Lake Bangwelu, Msombo, 20.x.1911, *R.E. Fries* 1052 (UPS holotype).

    Cyperus cooperi (C.B. Clarke) Kük. in Bot. Not. **1934**: 68 (1934), illegitimate name; in Engler, Pflanzenr. **4**, 20(101): 331 (1936).

    Pycreus laxespicatus (Kük.) Hoenselaar in F.T.E.A., Cyperaceae: 303 (2010) for type only; see note after *P. intactus*.

    Cyperus neocooperi Reynders in Phytotaxa **166**: 41 (2014), new name, non *Cyperus cooperi* (C.B. Clarke) K. Schum. (1900).

Robust tufted perennial, sometimes forming large clumps; culms 3–10, erect, to 70 cm long and 1–1.5 mm wide, rounded to triangular, smooth. Leaves basal, sheaths to 10 cm long, brown to almost black, with 0.5–2 mm wide mebranaceous ochrea-like flanges; blade rolled (appearing cylindric when dry), to 45 cm long and 0.5–1(1.5) mm wide, scabrid near tip. Inflorescence a simple umbel-like anthela, often condensed to almost capitate, 1.5–9 × 2–10 cm, open inflorescences with one to several sessile racemose clusters and 2–5(7) stalked clusters, each cluster with 10–25 spikelets; peduncles of stalked clusters 0.5–6.5 cm long; involucral bracts leaf-like (flat at base), 2–3, largest 5–17(25) cm long, 0.5–3 mm wide, scabrid near tip or in upper half. Spikelets 5–15 × 1–2 mm, linear to lanceolate; rachis straight in fruiting stage; glumes about 50% overlapping, appressed or spreading, dark brown to dark purple or shiny black, ovate with very narrow hyaline margin towards apex, 2–3 × 1.5–2 mm, acute sometimes apiculate, keel pale brown. Stamens (2)3; filaments 1–2 mm long; anthers 1–1.5 mm long. Stigma branches 2, 1–2 mm long. Nutlet oblong, with truncate to very slightly tapering apex, shortly apiculate, steel grey, 1–1.5 mm long, with numerous minute papillae in longitudinal rows.

Botswana. N: Okavango Delta, Gope R., fl. 9.i.1978, *P.A. Smith* 2180 (K, SRGH).

Zambia. B: Kalabo Dist., 5 km W of Kalabo, fl. 16.xi.1959, *Drummond & Cookson* 6551 (K, SRGH). N: Mbala Dist., Lunzua Swamp, 1500 m, fl. 17.i.1962, *Richards* 15921 (K). W: Mwinilunga Dist., Kalenda Dambo, fl. 8.x.1937, *Milne-Redhead* 2658 (BM K). C: Lusaka Dist., N of Kasisi, Constantia, Chongwe R., 1125 m, fl. 3.xii.1972, *Kornaś* 2750 (K). S: Choma Dist., 10 km E of Choma, 1325 m, fl. 18.xii.1956, *Robinson* 1988 (K). **Zimbabwe**. W: Binga Dist., Kariangwe Tsetse Camp, 850 m, fl. 15.xi.1958, *Phipps* 1468 (K, SRGH). C: Marondera Dist., Marondera (Marandellas), 1525 m, fl. i.1948, *Colville* 37 (K, SRGH). E: Nyanga Dist., Nyanga, fl. x.1946, *Wild* 1438 (K, SRGH). **Malawi**. N: Nkhata Bay Dist., Vipya Plateau, Luwawa Dam, 1625 m, fl. 11.xi.1973, *Pawek* 7502 (K, MAL, MO, SRGH). S: Thyolo Dist., Mperere Mission, Chankalama dambo, fl. 2.xi.1950, *Jackson* 251 (K). **Mozambique**. T: Angónia Dist., Ulónguè, fl. & fr. 2.xii.1980, *Macuácua* 1372 (K).

Also in Cameroon, Central African Republic, Sudan, Ethiopia, D.R. Congo, Kenya, Tanzania, Angola, Namibia and South Africa. Swampy grassland, dambos, seepages, often in places liable to burning; 850–2400 m.

Conservation notes: Widespread distribution; Least Concern. In common and widespread habitats.

18. **Pycreus nuerensis** (Boeckeler) S.S. Hooper in J. E. Africa Nat. Hist. Soc. Natl. Mus. **28**(124): 5 (1971); in Kew Bull. **26**: 579 (1972). —Hoenselaar in F.T.E.A., Cyperaceae: 302 (2010). Type: Sudan, Bahr el Ghazal, Nuer, 15.i.1869, *Schweinfurth* 1172 (B holotype).

    Cyperus nuerensis Boeckeler in Flora **62**: 555 (1879). —Haines & Lye, Sedges & Rushes E. Afr.: 275, figs.558,559 (1983).

    Cyperus globosus All. var. *nuerensis* (Boeckeler) Kük. in Engler, Pflanzenr. **4**, 20(101): 356 (1936).

    Pycreus globosus (All.) Reichb. var. *nuerensis* (Boeckeler) Troupin in Expl. Parc Natl. Garamba **4**: 126 (1956).

Robust tufted perennial; culms solitary or 2–3, erect, to 1.25 m long (over 50 cm when in flower) and 1.5–4 mm wide, rounded to triangular, sometimes inflated and hollow, smooth. Leaves basal, sheaths to 20(25) cm long, pale brown to dark brown, in aquatic specimens often inflated and spongy, with 1–3 mm wide mebranaceous flanges; blade crescent-shaped or rolled (appearing cylindric when dry), to 65 cm long and (1.5)2–4 mm wide, scabrid near tip. Inflorescence a simple or compound umbel-like anthela (rarely condensed and almost capitate), 5–12 × 5–12 cm, with one to several sessile racemose clusters and 2–8 stalked clusters, often again with sessile and stalked clusters, each cluster with 15–30 loosely arranged spikelets; peduncles of stalked clusters 0.5–8(11) cm long; involucral bracts leaf-like, 2–4, largest (10)15–40(60) cm long and 1–6 mm wide, scabrid near tip or in upper half. Spikelets 6–16 × 1–2 mm, linear-lanceolate; rachis straight in fruiting stage; glumes about 50% overlapping, appressed, only slightly spreading at maturity, brown to dark brown, ovate with a narrow hyaline margin towards apex, 1.5–2.5 × 1–1.5 mm, obtuse to acute, keel brown. Stamens (2)3; filaments 1–2 mm long; anthers 1–1.5 mm long. Stigma branches 2, 1–2 mm long. Nutlet oblong with very slightly tapering apex, shortly apiculate, steel-grey, 1–1.5 mm long, with numerous minute papillae in longitudinal rows.

    Zambia. B: Kalabo Dist., 5 km W of Kalabo, fl. 16.xi.1959, *Drummond & Cookson* 6534 (K, SRGH). N: Mbala Dist., Mbala, 1600 m, fl. & fr. 21.vi.1956, *Robinson* 1711 (K). W: Mwinilunga Dist., Lisombo R., fl. 11.vi.1963, *Edwards* 738 (K, SRGH). C: Mkushi Dist., Mwendashi Hills, 1400 m, fl. 19.xi.1993, *Bingham & Nkhoma* 9677 (K). **Zimbabwe**. C: Harare, 1450 m, fl. 30.i.1927, *Eyles* 4636 (K). **Malawi**. N: Mzimba Dist., Mzuzu, Marymount, 1375 m, fl. 7.xi.1969, *Pawek* 2953 (K).

Also in Sierra Leone, Nigeria, Cameroon, Central African Republic, Sudan, D.R. Congo, Uganda, Kenya, Tanzania and Angola. Swampy grassland including papyrus swamps; 1150–1650 m.

Conservation notes. Least concern. Widespread and in non-threatened habitats. Closely related to *Pycreus aethiops*. Differs mainly in being a taller, coarser plant with wider leaves and larger inflorescences. The glumes are generally paler and slightly smaller.

19. **Pycreus nigricans** (Steud.) C.B. Clarke in Trans. Linn. Soc. London, Bot. **4**: 53 (1894); in F.T.A. **8**: 292 (1901). —Hoenselaar in F.T.E.A., Cyperaceae: 305 (2010). Type: Ethiopia, Gonder, Semien Mts., Enschedcap, 31.vii.1838, *Schimper* II1373 (B holotype, BM, BR, GOET, HAL, HOH, K, LG, M, P, S, STU, TUB, UPS, US, WAG). FIGURE 14.**48**.

Cyperus lanceus Thunb. var. *mucronatus* Kunth, Enum. Pl. **2**: 8 (1837). Type not indicated.
Cyperus nigricans Steud. in Flora **25**: 584 (1842). —Kükenthal in Engler, Pflanzenr. **4**, 20(101): 336 (1936). —Haines & Lye, Sedges & Rushes E. Afr.: 272, fig.553 (1983).
Cyperus macranthus Boeckeler in Linnaea **35**: 462 (1868). —Kükenthal in Engler, Pflanzenr. **4**, 20(101): 388 (1936). —Haines & Lye, Sedges & Rushes E. Afr.: 280, figs.574,575 (1983). Type: South Africa, Zandplaat to Komga, 17.i.1832, *Drège* 4394 (B holotype, P).
Cyperus permutatus Boeckeler in Linnaea **35**: 477 (1868). —Haines & Lye, Sedges & Rushes E. Afr.: 275, figs.560,561 (1983). Types: South Africa, Transkei, *Drège* 4398 & 4399 (B syntypes).
Cyperus lanceus Thunb. var. *angustifolius* Ridl. in Trans. Linn. Soc. London, Bot. **7**: 126 (1884). Type: Angola, Pungo Andongo, Candumba, iii.1857, *Welwitsch* 6930 (BM holotype).
Pycreus macranthus (Boeckeler) C.B. Clarke in Durand & Schinz, Consp. Fl. Afr. **5**: 538 (1894); in Fl. Cap. **7**: 156 (1897); in F.T.A. **8**: 293 (1901). —Mapura & Timberlake, Checkl. Zimbabwe Pl.: 89: (1994). —Gordon-Gray in Strelitzia **2**: 142 (1995). —Hoenselaar in F.T.E.A., Cyperaceae: 299 (2010).
Pycreus macranthus var. *angustifolius* (Ridl.) Rendle in Hiern, Cat. Afr. Pl. **2**: 107 (1899). —Clarke in F.T.A. **8**: 293 (1901).
Pycreus nyasensis C.B. Clarke in F.T.A. **8**: 304 (1901). Type: Malawi, Mt Zomba, xii.1896, *Whyte* s.n. (K holotype).
Pycreus segmentatus C.B. Clarke in Bull. Misc. Inform. Kew, add. ser. **8**: 1 (1908). Types: Malawi, no locality, 1891, *Buchanan* 1454 (K lectotype, BM); Mt Zomba, xii.1896, *Whyte* s.n. (K syntype), lectotypified here.
Cyperus muricatus Kük. in Repert. Spec. Nov. Regni Veg. **12**: 92 (1913); in Engler, Pflanzenr. **4**, 20(101): 395 (1936). —Haines & Lye, Sedges & Rushes E. Afr.: 284, figs.585,586 (1983). Type: Malawi, Mt Zomba, xii.1896, *Whyte* s.n. (K lectotype); Tanzania, Ulanga/Songea Dist., Mampyui Ridge, 26–30.xii.1900, *Busse* 709 (B† syntype, EA), lectotypified here.
Cyperus nigricans var. *firmior* Kük. in Repert. Spec. Nov. Regni Veg. **12**: 94 (1913); in Engler, Pflanzenr. **4**, 20(101): 337 (1936). Kükenthal cites specimens from Zimbabwe under this name. Type: Tanzania, Moshi Dist., Kilimanjaro, Noholu, iii.1894, *Volkens* 2014 (K lectotype, BM), lectotypified here. Three other syntypes from Ethiopia, Congo and Tanzania; see notes.
Cyperus fibrillosus Kük. in Fries, Wiss. Ergebn. Schwed. Rhodesia-Kongo Exped. 1911–1912, Erg.: 1 (1921); in Engler, Pflanzenr. **4**, 20(101): 347 (1936). —Haines & Lye, Sedges & Rushes E. Afr.: 278, fig.569 (1983). Type: Zambia, Kulungwisi R., 28.x.1911, *R.E. Fries* 1142 (UPS holotype).
Pycreus fibrillosus (Kük.) Cherm. in Rev. Zool. Bot. Africaines **22**: 63 (1932). —Mapura & Timberlake, Checkl. Zimbabwe Pl.: 89 (2004). —Hoenselaar in F.T.E.A., Cyperaceae: 296 (2010).
Pycreus nigricans var. *firmior* (Kük.) Cherm. in Bull. Soc. Bot. France **82**: 337 (1935).
Cyperus diloloensis (Cherm.) Kük. in Engler, Pflanzenr. **4**, 20(101): 340 (1936). Type: D.R. Congo, Katanga, Dilolo, ix.1931, *de Witte* 618 (P holotype, K, PRE).
Cyperus macranthus var. *angustifolius* (Ridl.) Kük. in Engler, Pflanzenr. **4**, 20(101): 389 (1936).
Cyperus macranthus var. *mucronatus* (Kunth) Kük. in Engler, Pflanzenr. **4**, 20(101): 389 (1936). Kükenthal cites specimens from the F.Z. area under this name.

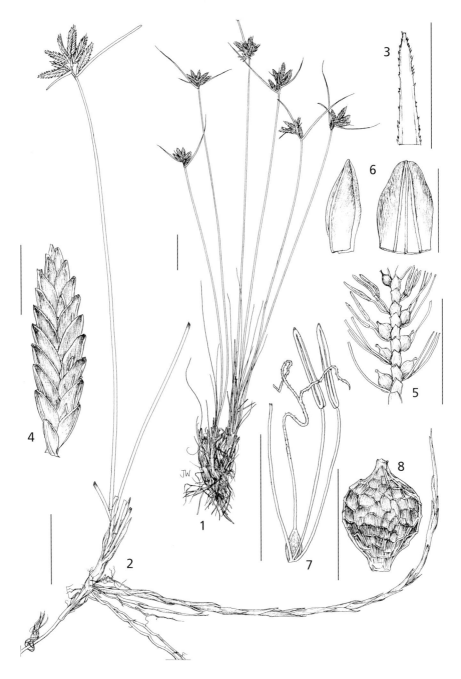

Fig. 14.**48**. PYCREUS NIGRICANS. 1, habit, tussocky plant; 2, habit, rhizomatous plant; 3, tip of involucral bract; 4, spikelet; 5, detail of rachis; 6, glume, dorsal (right) and side view; 7, flower; 8, nutlet. 1, 4 from *Milne-Redhead* 2957; 2, 3 from *Robinson* 4182; 5, 6 from *Robinson* 4139; 7 from *Angus* 351; 8 from *Hooper & Townsend* 8. Scale bars: 1, 2 = 3 cm; 4, 5 = 5 mm: 3, 6, 7 = 3 mm; 8 = 1 mm. Drawn by Juliet Williamson.

Pycreus muricatus (Kük.) Napper in J. E. Africa Nat. Hist. Soc. Natl. Mus. **28**(124): 6 (1971). —Mapura & Timberlake, Checkl. Zimbabwe Pl.: 89: (1994). —Hoenselaar in F.T.E.A., Cyperaceae: 298 (2010).

Pycreus permutatus (Boeckeler) Napper in J. E. Africa Nat. Hist. Soc. Natl. Mus. **28**(124): 6 (1971). —Hoenselaar in F.T.E.A., Cyperaceae: 300 (2010).

Pycreus sumbawangensis Hoenselaar in F.T.E.A., Cyperaceae: 297 (2010). Type: Tanzania, Ufipa Dist., Sumbawanga, xi.1954, *Richards* 3453a (K holotype); see notes.

Slender to robust perennial, stoloniferous or loosely to densely tufted (forming tussocks to 20 cm wide); culms 1–10, erect, to 75 cm long and 0.5–2 mm wide, rounded to triangular, smooth. Leaves basal, often with old fibrous leaf bases, sheaths to 7 cm long, brown to reddish brown to blackish, with 0.5–1 mm wide mebranaceous ochrea-like flanges; blade cylindric or folded, to 30(40) cm long and 0.5–1.5(2) mm wide, scabrid near tip. Inflorescence capitate to racemose or (more rarely) a simple umbel-like anthela, 1–3.5 × 1–5 cm if capitate, to 6 × 6 cm if branched, open inflorescences with one subsessile racemose cluster and 1–3(7) stalked clusters, clusters with 2–20(30) spikelets; peduncles of stalked clusters 0.5–4(5) cm long; involucral bracts leaf-like, 1 large and 1–3 small, largest 2–10(14) cm long and 0.5–1.5 mm wide, scabrid near tip. Spikelets 8–25(30) × 2–5(6) mm, lanceolate to ovate; rachis straight in fruiting stage; glumes about 75% overlapping, appressed or spreading, brown to dark brown or blackish (rarely pale brown), ovate, without hyaline margin, 2.5–4.5 × 1.5–2.5 mm, acute to obtuse, keel pale brown, distinct. Stamens (2)3; filaments 1–3 mm long; anthers 1.5–2.5 mm long. Stigma branches 2, 2–4 mm long. Nutlet broadly ellipsoid or slightly obovoid, with slightly tapering to truncate apex, apiculate, dark grey to black, c. 1 mm long, surface transversely to reticulately ridged to strongly muricate, more rarely longitudinally ribbed or with minute papillae in longitudinal rows.

Botswana. N: Okavango Delta, Moremi Wildlife Reserve, Mboroga R., fl. 9.xii.1979, *P.A. Smith* 2913 (K, SRGH). **Zambia**. B: Kalabo Dist., 5 km W of Kalabo, fl. 16.xi.1959, *Drummond & Cookson* 6529 (K, SRGH). N: Mbala Dist., Nkali Dambo, 1675 m, fl. 26.i.1967, *Simon et al.* 1571 (K, SRGH). W: Mwinilunga Dist., Matonchi dambo, fl. 26.x.1937, *Milne-Redhead* 2957 (K). C: Mkushi Dist., Fiwila, 1250 m, 5.i.1958, *Robinson* 2631 (K, SRGH). S: Mazabuka Dist., Mazabuka, fl. 14.i.1960, *White* 6260 (FHO, K). E: Nyika Plateau, 2125 m, fl. 3.i.1959, *Robinson* 3025 (K). **Zimbabwe**. N: Binga Dist., Kariangwe Tsetse Camp, 850 m, fl. 15.xi.1958, *Phipps* 1469 (K, SRGH). W: Matobo Dist., SW Matopos, Maleme Valley, fl. 6.i.1963, *Wild* 5931 (K, SRGH). C: Chegutu Dist., Poole Farm, 1225 m, fr. 11.iii.1946, *Hornby* 2438 (K, SRGH). E: Mutare Dist., Penhalonga, 1300 m, fl. & fr. 31.x.1956, *Robinson* 1830 (K, SRGH). **Malawi**. N: Nyika Plateau, Lake Kaulime, 2150 m, fl. 24.x.1958, *Robson & Angus* 333 (K). C: Dedza Dist., Chongoni Forest Reserve, fl. 4.xii.1968, *Salubeni* 1240 (K, MAL, SRGH). S: Zomba Plateau, Chingwe's Hole, fl. 20.xi.1981, *Chapman & Tawakali* 5994 (K, MAL). **Mozambique**. N: Lichinga Dist., Lichinga Plateau, 1300 m, fl. xii.1932, *Sousa* 1086 (K). Z: Gurué Dist., Mt Namuli, Ntapata Valley, 1850 m, fl. 16.xi.2007, *Timberlake et al.* 5186 (K, LMA). MS: Gorongosa Dist., Mt Gorongosa, Gogogo summit, fl. i.1972, *Tinley* 2301 (K, SRGH).

Also in Sudan, Ethiopia, Rwanda, Burundi, Uganda, Kenya, Tanzania, D.R. Congo, Angola, Swaziland, South Africa and Madagascar. Sandy to peaty seasonally wet dambo grassland, montane grassland (usually in seepages on slopes), sandy riverbeds, sandy soils in *Brachystegia* woodland; (600)850–2300 m (see notes).

Conservation notes: Widespread distribution; Least Concern. In common and widespread habitats.

Hoenselaar (2010: 307) cites *Ellenbeck* 1861 (P) as the holotype of *Cyperus nigricans* var. *firmior*, but in the original description Kükentahl cites four syntypes. I have not been able to trace this collection at P and have therefore lectotypified with the – as far as I can ascertain – only extant material.

Hoenselaar (2010) cites the type of *Pycreus sumbawangensis* as *Richards* 3452a. The correct number (accordimng to Mary Richards' collecting book) is *Richards* 3453a

(*Richards* 3452a is *Hibiscus rhodanthus*). I have not been able to check the type as the Kew material has been mislaid, but all other collections named by Hoenselaar as *P. sumbawangensis* are *P. nigricans*.

The collections normally determined as *Pycreus diloloensis* Cherm. tend to have slightly paler bracts and longer more numerous spikes. But I have not been able to find any clear differences between this and typical material. They also tend to occur at the lower end of the altitude spectrum in the Flora area. There is a general tendency for plants from the northern part of the distribution area (Ethiopia, Kenya, Tanzania) to occur at high altitudes (up to 3500 m), while plants from South Africa occur down to 1100 m.

There are several records of this species from *Brachystegia* woodland but I assume it really occurs in seasonally inundated clearings within the woodland.

20. **Pycreus acaulis** Nelmes in Kew Bull. **10**: 91 (1955). Type: Malawi, Nyika Plateau, Kaulime Pond, 27.vi.1952, *Jackson* 870 (K holotype, BR).

   Cyperus acaulescens Reynders in Phytotaxa **166**: 41 (2014), new name, non *Cyperus acaulis* Steud. (1842).

Tufted acaulous annual, several plants growing closely together forming small clumps; culms absent or extremely short and hidden under inflorescences. Leaf sheaths straw-coloured, to 2 mm long; blade filiform, to 4 cm long and 0.5 mm wide, not scabrid. Inflorescence of sessile spikelets situated amongst leaves; involucral bracts leaf-like, largest to 1 cm long and 0.5 mm wide, scabrid near apex. Spikelets 3–5 × 2–3.5 mm, broadly ovate; glumes appressed at flowering, slightly spreading later, densely imbricate (about 75% overlapping), brownish red with pale dorsal part, broadly ovate, without hyaline margin, 2–3 × 1.5–2 mm, acute, with midrib reaching tip, keel prominent, rib-like, pale green. Stamens 1; filaments 1 mm long; anthers 0.5 mm long. Stigma branches 2, 2 mm long. Nutlet dark brown, oblong, apex truncate, not apiculate, 1 mm long, with numerous papillae in longitudinal rows.

Malawi. N: Nyika Plateau, Kaulime Pond, 2200 m, fl. & fr. 27.vi.1952, *Jackson* 870 (BR, K).

Not known elsewhere. Montane grassland, 'in pure peat'; 2200 m.

Conservation notes. Possibly Endangered. Known only from the type, but it is a very inconspicuous plant when growing amongst other vegetation. Montane grasslands on the Nyika Plateau are generally not threatened and the locality is within the National Park.

The affinities of this extraordinary species are not at all clear. The relatively wide spikelets might indicate an affinity with the *Pycreus flavescens–P. unioloides* complex. The unique growth form is completely at odds with anything else in the genus.

21. **Pycreus polystachyos** (Rottb.) P. Beauv. in Fl. Oware **2**: 48, t.86, fig.2 (1816). — Clarke in F.T.A. **8**: 296 (1901). —Podlech in Prodr. Fl. SW Afr. **165**: 42 (1967). —Gordon-Gray in Strelitzia **2**: 147 (1995). —Hoenselaar in F.T.E.A., Cyperaceae: 289 (2010). Type: India, 'In regione Malabarica', *König* s.n. (C holotype).

   Cyperus polystachyos Rottb., Descr. Pl. Rar.: 21 (1772); Descr. Icon. Rar. Pl.: 39, t.11, fig.1 (1773). —Kükenthal in Engler, Pflanzenr. **4**, 20(101): 367 (1936). —Haines & Lye, Sedges & Rushes E. Afr.: 279, figs.571,572 (1983).

   Cyperus chlorostachys Boeckeler in Peters, Naturw. Reise Mossambique **2**: 540 (1864); in Linnaea **36**: 293 (1870). Type: Mozambique, no locality, n.d., *Peters* 22 (B† holotype).

   Cyperus polystachyos var. *laxiflorus* Benth., Fl. Austral. **7**: 261 (1878). Type: Australia, Arnhem Land, *F. Mueller* s.n. (BM holotype).

   Pycreus polystachyos var. *laxiflorus* (Benth.) C.B. Clarke in Hooker, Fl. Brit. India **6**: 592 (1893); in F.T.A. **8**: 297 (1901). —Kükenthal in Engler, Pflanzenr. **4**, 20(101): 370 (1936). —Mapura & Timberlake, Checkl. Zimbabwe Pl.: 89: (1994). —Hoenselaar in F.T.E.A., Cyperaceae: 290 (2010).

Cyperus polystachyos var. *chlorostachys* (Boeckeler) Kük. in Engler, Pflanzenr. **4**, 20(101): 371 (1936).

Cyperus polystachyos subsp. *laxiflorus* (Benth.) Lye in Haines & Lye, Sedges & Rushes E. Afr., app.3: 2 (1983). —Haines & Lye, Sedges & Rushes E. Afr.: 279, fig.573 (1983).

Annual or short-lived perennial, without stolons; culms tufted, more rarely solitary, erect, to 85(100) cm long and 1–3(4) mm wide, triangular, lower part slightly swollen, covered by brown to purplish leaf bases. Leaves basal, sheaths to 15 cm long, pale brown to reddish purple; blade flat, to (25)40 cm long and 1–4 mm wide, scabrid near tip. Inflorescence 1.5–3 × 2–5.5 cm if head-like, to 7 × 10 cm if with stalked clusters, a congested head-like anthela or more open with one sessile and 2–5 stalked clusters; peduncles of stalked clusters to 4.5(8.5) cm long; spikelets in dense digitate clusters, each cluster with 5–30 spikelets; involucral bracts leaf-like, 2–3 large and 2–4 smaller ones, largest 5–30(45) cm long and 1.5–4 mm wide, scabrid. Spikelets 5–15 × 0.8–1.8 mm, linear-lanceolate; rachis straight or zig-zag in fruiting stage; glumes appressed at first, slightly spreading from flowering stage, falling at maturity, pale yellow to brown, ovate, with a narrow white hyaline margin, 1.5–2 × 1–1.5 mm, acute with midrib extended into an often 2–3-fid mucro, keel yellowish brown to green. Stamens 2–3; filaments 0.5–2 mm long; anthers c. 0.5 mm long. Stigma branches 2, c. 1 mm long. Nutlet oblong to cylindrical, with very slightly tapering apex, shortly apiculate, shiny metallic grey when mature, 0.5–1 mm long, with numerous minute papillae in longitudinal rows.

Botswana. N: Ngamiland Dist., Mojeye area, fl. & fr. 26.iii.1970, *P.A. Smith* 1296 (K, SRGH). SE: Central Dist., N of Lake Xau (Dow), Boteti (Botletle) R., fl. & fr. 22.iii.1965, *Wild & Drummond* 7226 (K, SRGH). **Zambia**. B: Mongu Dist., Mongu, fl. 21.xi.1965, *Robinson* 6717 (K). N: Mbala Dist., Lake Tanganyika, Sumbu Bay, 775 m, fl., 29.xii.1963, *Richards* 18716 (K). W: Ndola Dist., Chichele Forest Reserve, 1300 m, fl. 17.i.1960, *Robinson* 3300 (K). C: Mumbwa Dist., Piamadzi R., fl. 18.viii.1970, *Verboom* 3102 (K). E: Katete Dist., 80 km W of Chipata (Fort Jameson), Luangwa Valley, Msoro, 725 m, fl. & fr. 9.vi.1954, *Robinson* 849 (K). S: Itezhi-Tezhi Dist., Mapanza North, 1075 m, fl. & fr. 10.iv.1955, *Robinson* 1224 (K). **Zimbabwe**. N: Kariba Dist., S bank of Lake Kariba, 475 m, fl. xi.1964, *Mitchell* 1070 (K, SRGH). W: Matobo Dist., 1450 m, fl. i.1962, *Miller* 8167 (K, SRGH). C: Gweru Dist., Whitewaters Dam, fl., 20.i.1963, *Loveridge* 590 (K, SRGH). E: Chimanimani Dist., Rusitu (Lusitu) R., fl. 12.i.1969, *Mavi* 901 (K, SRGH). S: Chiredzi Dist., Triangle, Mutirikwe R., fl. 9.xi.1971, *Gibbs Russell* 1175 (K, SRGH). **Malawi**. N: 'N. Nyasaland', no locality, n.d., *Whyte* s.n. (K). S: Chikwawa Dist., Lower Mwanza R., 175 m, fl. 3.x.1946, *Brass* 17934 (K, NY). **Mozambique**. N: Marrupa Dist., 23 km on Marrupa–Nungo road, Messalo R., 500 m, fl. & fr. 7.viii.1981, *Jansen et al.* 121 (K). MS: Manica Dist., Bandula, 650 m, fl. & fr. 6.iv.1952, *Chase* 4543 (BM, K, SRGH). GI: Mandlakazi Dist., Manjacaze, road to Chidenguele, fl. & fr. 2.iv.1959, *Barbosa & de Lemos* 8459 (K). M: Matutuine Dist., between Zitundo and Manhoca, fl. 29.xi.1979, *de Koning* 7717 (K, LMA).

Widespread in all tropical and subtropical regions. Riverbeds and riverbanks, lakeshores, hot springs, sandy beaches, grassland, cultivated areas, usually on drying sandy soils; 0–1450 m.

Conservation notes: Widespread distribution; Least Concern. In common and widespread habitats.

22. **Pycreus smithianus** (Ridl.) C.B. Clarke in Durand & Schinz, Consp. Fl. Afr. **5**: 542 (1894); in F.T.A. **8**: 301 (1901). Types: Congo, mouth of Congo R., n.d., *C. Smith* 47 (BM lectotype); same locality, *C. Smith* 67 (BM syntype), lectotypified here.

Cyperus smithianus Ridl. in J. Bot. **22**: 15 (1884). —Kükenthal in Engler, Pflanzenr. **4**, 20(101): 349 (1936). —Haines & Lye, Sedges & Rushes E. Afr.: 278 (1983).

Cyperus fluminalis Ridl. in Trans. Linn. Soc. London, Bot. **2**: 127 (1884). —Haines & Lye, Sedges & Rushes E. Afr.: 278, figs.567,568 (1983). Type: Angola, Pungo Andonga, Cuanza R., near Candumba, 29.i.1857, *Welwitsch* 6897 (BM holotype).

Cyperus cuanzensis Ridl. in Trans. Linn. Soc. London, Bot. **2**: 128 (1884). Type: Angola, Pungo Andongo, Cuanza R., Nbilla, iii.1857, *Welwitsch* 6899 (BM holotype).
  Pycreus cuanzensis (Ridl.) C.B. Clarke in Durand & Schinz, Consp. Fl. Afr. **5**: 536 (1894); in F.T.A. **8**: 301 (1901).
  Pycreus fluminalis (Ridl.) Rendle in Hiern, Cat. Afr. Pl. **2**: 106 (1899). —Troupin in Fl. Sperm. Parc. Nat. Garamba **1**: 126 (1956). —Hoenselaar in F.T.E.A., Cyperaceae: 298 (2010).

Perennial with short stout rhizome; culms solitary or 2(5) per plant, erect, to 50(75) cm long and 1–3 mm wide, triangular, lower part slightly swollen, often covered by dead leaves. Leaves basal, sheaths to 9 cm long, pale brown; blade flat, to 20 cm long and 1.5–4 mm wide, scabrid near tip. Inflorescence a compact globose head-like anthela, 1.5–3 × 1.5–4 cm; spikelets in dense sessile digitate clusters, each cluster with numerous spikelets; involucral bracts leaf like, 2–5, largest 5–15 cm long and 2–4 mm wide, scabrid. Spikelets 8–15 × 2–3 mm (to 25 mm in fruit), linear-lanceolate; rachis straight, becoming curved in fruiting stage, with brown dots and brown transverse bands; glumes appressed at first, only spreading from fruiting stage, falling at maturity, dull white or greenish white with brown lines on dorsal part, ovate, without white hyaline margin, 1.5–2 × 1–1.5 mm, acute, keel pale brown. Stamens 2; filaments 1–1.5 mm long; anthers c. 0.5 mm long. Stigma branches 2, c. 1 mm long. Nutlet obovoid, subacute, apiculate, pale to reddish brown when young, dark grey to blackish when mature, 0.5–1 mm long, distinctly transversely ridged or papillose.

Zambia. N: Chinsali Dist., Shiwa Ngandu, 1525 m, fl. & fr. 2.vi.1956, *Robinson* 1556 (K). W: Mwinilunga Dist., 24 km N of Mwinilunga, fl. & fr. 11.xii.1963, *Robinson* 5892 (K). C: Mpika Dist., Mutinondo Wilderness Area, 1400 m, 26.ii.2015, *Merrett* 1830 (K). S: Kazungula Dist., Katambora, fl. & fr. 14.iv.1949, *West* 2903 (K, SRGH).

Also in Guinea, Sierra Leone, Ivory Coast, Burkina Faso, Nigeria, Cameroon, Gabon, Equatorial Guinea, Congo-Brazzaville, D.R. Congo, Burundi, Uganda, Tanzania and Angola. Wet or drying grassland, riverbanks, on sandy to clayey soils, rheophytic in rocky riverbeds; 900–1550 m.

Conservation notes: Restricted distribution inside the Flora area but widespread outside; Least Concern. In common non-threatened habitats.

Viviparous specimens, where the inflorescences produce plantlets and eventually bend over to reach the ground, occur throughout the distribution, e.g. *Robinson* 1556 from Zambia.

23. **Pycreus scaettae** Cherm. in Rev. Zool. Bot. Africaines **24**: 295 (1934); in Bull. Jard. Bot. État. Bruxelles **13**: 278 (1935). —Hoenselaar in F.T.E.A., Cyperaceae: 296 (2010). Types: D.R. Congo, Mubeza, 1930, *Scaetta* 58M (BR lectotype, FHO, K), lectotypified here. Four other syntypes from Congo-Brazzaville, D.R. Congo and Gabon.
  Pycreus scaettae var. *katangensis* Cherm. in Bull. Jard. Bot. État. Bruxelles **13**: 279 (1935). —Hoenselaar in F.T.E.A., Cyperaceae: 296 (2010). Type: D.R. Congo, Katanga (Shaba), Maniema, between Kindu and Kotokokombe, viii.1932, *Lebrun* 6052 (P lectotype), lectotypified here. Three other syntypes from D.R. Congo.
  Cyperus scaettae (Cherm.) Reynders in Phytotaxa **166**: 42 (2014).

Perennial forming dense compact tussocks; culms to 10 per tussock, erect, to 35 cm long and 0.5–1 mm wide, rounded, smooth, surrounded at base by dense old fibrous leaf bases. Leaves basal, sheaths to 5 cm long, brown, with 0.5 mm wide mebranaceous ochrea-like flanges; blade filiform, to 25(40) cm long and 0.5–1 mm wide, scabrid near tip. Inflorescence of a single sessile cluster of spikelets, 1–2.5 × 1–2.5 cm, with 4–7(15) spikelets; involucral bracts leaf like, 2 (1 large and 1 small), largest 2–8 cm long and 0.5–1 mm wide, scabrid near tip. Spikelets 8–15 × 3–4 mm, lanceolate to ovate; rachis straight in fruiting stage; glumes about 75% overlapping, appressed at first, spreading from flowering onwards, greenish white to shiny white, ovate without hyaline margin, 3–3.5 × 1.5–2 mm, obtuse and apiculate, keel pale green. Stamens (2)3; filaments 1–2 mm long; anthers 1 mm long. Stigma branches 2, 1–2 mm long. Nutlet subglobose, with tapering apex, not apiculate, black, 0.5–1 mm long, finely papillose but papillae not in longitudinal rows.

Zambia. W: Mwinilunga Dist., Matonchi Farm, fl., 22.x.1937, *Milne-Redhead* 2895 (BM K); Matonchi Farm, fl. & fr. 13.xi.1937, *Milne-Redhead* 3223 (BM K).

Also in Nigeria, Cameroon, Central African Republic, Gabon, Congo-Brazzaville, D.R. Congo, Angola and ?Tanzania. On hard lateritic soil in *Brachystegia* woodland; 1300–1400 m.

Conservation notes: Within the Flora area known only from these two collections at the same locality; possibly threatened. Widespread elsewhere in a very widespread non-threatened habitat.

I rather doubt the occurrence of this species in Tanzania. It is a long way outside the rest of the distribution area. The collection (*Newbould & Jefford* 2791) cited by Hoenselaar (2010) has a similar growth form, but has narrower spikelets and yellowish brown smaller glumes. It should probably be treated as a distinct infraspecific taxon or considered an abnormal specimen of *Pycreus nigricans*.

This is the only *Pycreus* species in the Flora area growing on dry soils in *Brachystegia* woodland.

24. **Pycreus richardsiae** Vollesen, sp. nov. Related to *Pycreus scaettae* and *P. smithianus*. *Pycreus scaettae* grows in dense tussocks, has filiform (not flat) leaves and has longer (3–3.5 not 2–3 mm long) glumes. *Pycreus smithianus* has narrower (2–3 not 3–4 mm wide) spikelets and shorter (1.5–2 not 2–3 mm long) glumes. Type: Mbala Dist., 24 km S of Mbala, Lunzawa, 25.xii.1961, *Robinson* 4743 (K holotype).

Stoloniferous perennial with thick fleshy rhizome; culms 1–2, erect, to 85 cm long an 0.5–1.5 mm wide, triangular, smooth. Leaves basal, sheaths to 7 cm long, pale brown, with narrow less than 0.5 mm wide mebranaceous ochrea-like flanges; blade flat, to 45 cm long and 2–3 mm wide, scabrid near tip. Inflorescence capitate or with a single stalked cluster with peduncle to 1 cm long, 2–3.5 × 2–3.5 cm, with c. 25–50 spikelets; involucral bracts leaf-like, 2, one much larger, largest to 20 cm long and c. 2 mm wide, scabrid near tip. Spikelets 5–15 × 3–4 mm, lanceolate to ovate; rachis straight in fruiting stage; glumes about 75% overlapping, appressed throughout, white or with a faint brown tinge, ovate, without hyaline margin, 2–3 × 1–1.5 mm, obtuse, keel pale green. Stamens (2)3; filaments 1–2 mm long; anthers 1–1.5 mm long. Stigma branches 2, 1–2 mm long. Nutlet broadly obovoid, truncate at apex, not apiculate, dark grey, 1 mm long, transversely ridged.

Zambia. N: Mbala Dist., Saisi, 1525 m, fl. 23.x.1954, *Richards* 2143 (K); Mbala Dist., 24 km S of Mbala, Lunzawa, fl. 25.xii.1961, *Robinson* 4743 (K holotype).

Not known elsewhere. Swampy grassland; 1525 m.

Conservation notes: Possibly threatened. Known only from these two collections which are in widespread and non-threatened habitats.

25. **Pycreus okavangensis** Podlech in Mitt. Bot. Staaatssaml. München **3**: 522 (1960); in Prodr. Fl. SW Afr. **165**: 42 (1967). Type: Namibia, Okavango R., Rundu (Runtu), 11.v.1939, *Volk* 1966 (M holotype, PRE).

    Cyperus okavangensis Reynders in Phytotaxa **166**: 41 (2014).

Single stemmed or tufted annual or perennial with short creeping rhizome; culms 1 to 7, erect, to 40 cm long and 2 mm wide, rounded, striate, smooth. Leaves basal or also along lower part of stem, sheaths straw-coloured to brown, to 8 cm long; blade flat, to 40 cm long and 1.5–2(3) mm wide, scabrid near tip. Inflorescence of 1–3 very dense sessile clusters of 5 to over 30 spikelets and with 1–4 stalked clusters with densely clustered spikelets, 1–5(7) × 1–4 cm, with erect branches; peduncles of stalked clusters to 5(6) cm long; involucral bracts leaf-like, 1–3 large and 0–1 small, largest to 13(27) cm long and 1.5–3 mm wide, scabrid near tip. Spikelets 3–7 × 1–2 mm (to 10 mm long in fruit), lanceolate; rachis straight in fruiting stage; glumes appressed at flowering, spreading later, falling at maturity, densely imbricate (about 50% overlapping), pale green to green, brown at very tip, the whole glume apart from veins almost hyaline and translucent, a single broad ribbon-like central vein on each side, 1–1.5 mm long, subacute to obtuse, keel prominent, rib-like, pale green. Stamens 3; filaments 1 mm long; anthers 0.25 mm long. Stigma branches 2, 1 mm long. Nutlet ellipsoid to obovoid, apex subacute to truncate, apiculate, olive brown to dark brown, 0.5–1 mm long, smooth.

Botswana. N: Ngamiland Dist., Boteti R, fl. & fr. 15.ii.1980, *P.A. Smith* 3030 (K, SRGH); Maun, Samedupe Bridge, 925 m, fl. 19.ii.2005, *Kabelo et al.* MSB164 (K). **Zambia**. B: Mongu, fl. & fr. 6.i.1966, *Robinson* 6779 (K).

Also in Namibia. Seasonally flooded grassland, wet dambos, mud at edge of rivers and waterholes, sometimes aquatic along edges of rivers; 900–1100 m.

Conservation notes: Fairly widespread distribution; Least Concern. In common and widespread habitats.

Annual plants have strictly basal leaves but perennial (sometimes semi-aquatic) plants can have 1–2 nodes along the lower part of the stem.

26. **Pycreus pauper** (A. Rich.) C.B. Clarke in Durand & Schinz, Consp. Afr. **5**: 540 (1894); in F.T.A. **8**: 291 (1901). —Hoenselaar in F.T.E.A., Cyperaceae: 279 (2010). Type: Ethiopia, Sana, Walcha, 6.viii.1841, *Schimper* III.1602 (P holotype, BR, GOET, H, HAL, K, UPS).

> *Cyperus pauper* A. Rich., Tent. Fl. Abyss. **2**: 478 (1850). —Haines & Lye, Sedges & Rushes E. Afr.: 286 (1983), in part. —Lye in Hedberg & Edwards, Fl. Ethiopia Eritrea **6**: 484, fig.212.147 (1997).

Slender single stemmed or tufted annual; culms 1 to 15 (sometimes appearing to be more when several plants grow closely together), erect, to 35 cm long and 0.5 mm wide, rounded, striate, smooth. Leaf sheaths pale brown to dark brown, to 6(8) cm long; blade filiform, to 15 cm long and 0.5 mm wide, scabrid near apex. Inflorescence to 2 × 2 cm, a single sessile cluster of 1–3 spikelets; involucral bract single and with a second smaller, sometimes two large, leaf-like, in small plants sometimes held as an extension of culm, making inflorescence appearing lateral, 3–10(15) cm long and 0.5–1 mm wide, scabrid near apex. Spikelets 5–15 × 3–4 mm, oblong; rachis zig-zag in fruiting stage; glumes appressed at flowering and fruiting, very tardily dehiscing, densely imbricate (about 75% overlapping), greyish brown to brown, with dark brown edges apically, apiculate tip pale, ovate to elliptic, without hyaline margin, 2–3 × 1.5–2 mm long, acute and apiculate, keel prominent, rib-like, pale brown. Stamens 2; filaments 1 mm long; anthers 0.5 mm long. Stigma branches 2, 1–2 mm long. Nutlet black, subglobose, apex truncate, not apiculate, 1 mm long, with conspicuous transverse ridges.

Zambia. N: Mbala Dist., 15 km from Mbala, 3 km from Kiwimbe, 1775 m, fl. 3.iii.1959, *McCallum Webster* C7 (K). W: Mwinilunga Dist., Matonchi, Kalenda Plain, 1400 m, fl. & fr. 16.iv.1960, *Robinson* 3620 (K).

Also in Nigeria, Cameroon, Central African Republic, Ethiopia and Tanzania. Seasonally inundated grassland, edges of waterholes and roadside ditches; 1400–1800 m.

Conservation notes: Restricted distribution in the Flora area but widespread outside; Least Concern. Probably undercollected. In common and widespread habitats.

The illustration of *Pycreus pauper* in Haines & Lye (1983) and in Lye (1997, fig.212.146) is of *P. melanacme*.

27. **Pycreus sp. B** of F.Z. (*Bingham* 12872).

Slender tufted annual; culms c. 15, erect, to 15 cm long and 0.5 mm wide, rounded, striate, smooth. Leaf sheaths brown, to 3 cm long; blade filiform, to 15 cm long and 0.5 mm wide, not scabrid. Inflorescence to 3 × 4 cm, consisting of a single subsessile cluster of 3–8 spikelets;

involucral bracts one large and leaf-like, a second much smaller, large one held as an extension of the culm, making inflorescence appear lateral, largest 2–7.5 cm long and 0.5 mm wide, not scabrid. Spikelets 10–22 × 2.5–4 mm, linear-lanceolate; glumes appressed at flowering, slightly spreading later, densely imbricate (about 75% overlapping), reddish brown, broadly ovate, without a hyaline margin, 2–2.5 × 1.5–2 mm, subacute, with a minute colourless apiculus, keel prominent, rib-like, white. Stamens 2; filaments 0.5 mm long; anthers 0.5 mm long. Stigma branches 2, 1–2 mm long. Immature nutlet oblong, apex truncate, slightly apiculate, 1 mm long, too young to see sculpturing.

Zambia. C: Serenje Dist., Kasanka Nat. Park, 1225 m, fl. 8.iii.2005, *Bingham 12872* (K). Not known elsewhere. Edge of small pond in lateritic pan; 1225 m.

Conservation notes: Possibly threatened but Data Deficient. Known only from this collection from a widespread habitat in a fairly well-collected area.

Related to *Pycreus capillifolius* from which it differs in the long spikelets with reddish brown glumes.

28. **Pycreus zonatissimus** Cherm. in Bull. Soc. Bot. France **74**: 605 (1928), replacement name. —Hoenselaar in F.T.E.A., Cyperaceae: 278 (2010). Type: Madagascar, Antsirabe, iii.1920, *Perrier de la Bathie 13061* (P holotype, BR, K).

　　Pycreus zonatus Cherm. in Bull. Soc. Bot. France **67**: 328 (1921), non *Cyperus zonatus* Kük. (1913). Type as for *P. zonatissimus*.
　　Cyperus zonatissimus (Cherm.) Kük. in Engler, Pflanzenr. **4**, 20(101): 395 (1936). —Haines & Lye, Sedges & Rushes E. Afr.: 285, fig.587 (1983).

Slender single stemmed or tufted annual; culms 1 to 15 (sometimes appearing more when several plants grow closely together), erect, to 30 cm long and 1 mm wide, rounded, striate, smooth. Leaf sheaths brown to reddish brown, to 2 cm long; blade flat (often wilting at time of flowering and appearing folded or filiform), to 15 cm long and 1–2 mm wide, scabrid near tip. Inflorescence to 2.5 × 3.5 cm, consisting of a single subsessile cluster of 2–10 spikelets; involucral bracts one large and leaf-like, a second much smaller, largest 2–10(15) cm long and 0.5–1.5 mm wide, scabrid near apex. Spikelets 5–18 × 2.5–4 mm, lanceolate; rachis slightly zig-zag in fruiting stage; glumes appressed at flowering, slightly spreading later, falling at maturity, densely imbricate (about 75% overlapping), dark brown, ovate, without a hyaline margin, 2–3 × 1.5–2 mm, subacute, sometimes with a minute colourless apiculus, keel prominent, rib like, white to pale brown. Stamens 2; filaments 0.5–1 mm long; anthers 0.5 mm long. Stigma branches 2, 1–2 mm long. Nutlet black, subglobose, apex truncate, slightly apiculate, 1 mm long, strongly reticulately ridged.

Zambia. N: Mbala Dist., Nkali dambo, fl. & fr. 21.iv.1962, *Robinson 5114* (K). W: Masaiti Dist., 16 km S of Ndola, 1300 m, fl. 2.iv.1960, *Robinson 3449* (K). C: Lusaka Dist., Mt Makulu, 1250 m, 15.iv.1956, *Robinson 1491* (K). S: Choma Dist., 5 km E of Choma, 1300 m, 26.iii.1955, *Robinson 1164* (K). **Zimbabwe**. C: Harare, 1425 m, fl. & fr. 12.iv.1932, *Brain 8893* (K). S: Masvingo Dist., Makaholi Exp. Station, fl. 10.iii.1978, *Senderayi 165* (K, SRGH).

Also in Tanzania and Madagascar. Seasonally inundated grassland, edges of waterholes, lateritic pans and roadside ditches; 1000–1750 m.

Conservation notes: Widespread distribution; Least Concern. In common and widespread habitats.

29. **Pycreus sanguineosquamatus** Van der Veken in Bull. Jard. Bot. État Bruxelles **25**: 145, fig.39 (1955). Type: D.R. Congo, Katanga, Lubumbashi (Elisabethville), Lake Mwero, Mwashia, v.1939, *Bredo 2750* (BR holotype, BM, C, K, NY, P, PRE).

Slender tufted annual; culms 1 to 25, erect, to 15 cm long and 1 mm wide, rounded, striate, smooth. Leaf sheaths brown to reddish brown, to 2 cm long; blade filiform, to 15 cm long and 0.5–1 mm wide, minutely scabrid near tip. Inflorescence to 2 × 3 cm, consisting of a single subsessile cluster of 1–4 spikelets; involucral bracts 1(2) large, leaf-like, with or without 1 small, the larger held as an extension of the culm making the inflorescence appearing lateral, largest 4–6 cm long and 0.5–1 mm wide, not scabrid. Spikelets 5–10 × 4–5 mm (to 25 mm long in fruit), lanceolate-ovate; rachis slightly zig-zag in fruiting stage; glumes appressed at flowering and fruiting stages, falling at maturity, densely imbricate (about 2/3 overlapping), reddish to purplish brown, broadly ovate, without hyaline margin, 3–4 × 3–4 mm, acute and minutely apiculate, with midrib reaching tip, keel prominent, rib-like, green. Stamens 2; filaments 1–1.5 mm long; anthers 0.5 mm long. Stigma branches 2, 2 mm long. Nutlet black, subglobose, apex truncate, not apiculate, 1.5 mm long, with numerous indistinct papillae in longitudinal rows.

Zambia. W: Chingola Dist., N of Chingola, banks of Kafue R., 1350 m, fr. 4.v.1960, *Robinson* 3705 (K).

Also in D.R. Congo. Dambo grassland; 1350 m.

Conservation notes. Known from only the cited locality and three collections from the same place in Congo; possibly threatened. Its habitat is widespread in N Zambia and S Congo.

The Zambian collection is clearly annual but one of the Congo collections seems to be perennial.

30. **Pycreus subtrigonus** C.B. Clarke in Durand & Schinz, Etud. Fl. Congo **1**: 282 (1896); in Mém. Couronnés Autres Mém. Acad. Roy. Sci. Belgique **53**: 282 (1896). —Clarke in F.T.A. **8**: 292 (1901). Types: D.R. Congo, Lutete, N'Tombi R., 27.ii.1888, *Hens* ser.A 251 (K syntype, BR); Equateurville, 8.v.1888, *Hens* ser.C 182 (K lectotype, BR), lectotypified here.

Tufted annual; culms (1)3–20, erect, to 30 cm long and 0.5–1.5 mm wide, triangular, striate, smooth. Leaf sheaths brown to reddish brown, to 5 cm long; blade flat, to 15 cm long and 1.5–3 mm wide, scabrid near tip. Inflorescence to 2.5 × 3.5 cm, consisting of a single subsessile cluster of (5)10 to over 40 spikelets; involucral bracts 2–3 large, leaf-like, also with 1–2 small ones, largest 3–12 cm long and 1.5–3.5 mm wide, scabrid near apex. Spikelets 5–10 × 1.5–2 mm (to 15 mm long in fruit), lanceolate; rachis zig-zag in fruiting stage; glumes appressed at flowering and fruiting stages, falling at maturity, densely imbricate (about 2/3 overlapping), olive brown to dark brown, ovate, without hyaline margin, 1.5–2 × 1 mm, acute, not apiculate, midrib reaching the tip, keel prominent, rib-like, green. Stamens 2; filaments 1 mm long; anthers 0.5 mm long. Stigma branches 2, 1–2 mm long. Nutlet dark brown, obtrigonous, apex retuse, apiculate, 0.5 mm long, with fine transverse 'crests' and very indistinct papillae in longitudinal rows.

Zambia. W: Mwinilunga Dist., 35 km W of Mwinilunga, 1400 m, fl. & fr. 16.iv.1960, *Robinson* 3648 (K); Matonchi, Chingabola dambo, 1350 m, fl. & fr. 16.ii.1975, *Hooper & Townsend* 103 (K).

Also in D.R. Congo and Congo-Brazzaville. Wet dambo grassland; 1350–1400 m.

Conservation notes. Data deficient, but probably Least Concern. Known only from two collections in the Flora area, but more common in neighbouring areas of Congo.

Material from the Flora area is of annual plants, but in the Congo short-lived perennial forms occur; one such form is *Hens* ser.A 251, one of the two syntypes. Inflorescence, glumes and nutlets are identical in annual and perennial specimens.

31. **Pycreus sp. C** of F.Z. (*Hooper & Townsend* 298).

Slender tufted annual; culms to 15 (appearing to be more when several plants grow closely together), erect, to 35 cm long and 0.5 mm wide, rounded, striate, smooth. Leaf sheaths brown to reddish brown, to 4 cm long; blade filiform, to 15 cm long and 0.5 mm wide, not scabrid. Inflorescence to 4 × 5 cm, consisting of a single subsessile cluster of 5–15 spikelets, usually also with 1 stalked cluster with 4–8 spikelets and peduncle 0.5–2 cm long; involucral bracts one large, leaf-like, sometimes also a second much smaller, large held as extension of culm, making inflorescence appear lateral, largest 1–8 cm long and 0.5 mm wide, not scabrid. Spikelets 5–10

× 1.5–2 mm (to 18 mm long in fruit), linear-lanceolate; rachis straight in fruiting stage; glumes appressed at flowering, slightly spreading later, falling at maturity, densely imbricate (about 75% overlapping), golden-yellow, ovate-elliptic, without a hyaline margin, 1.5–2 × 1 mm, subacute, with a colourless apex, keel prominent, rib-like, white. Stamens 2; filaments 0.5–1 mm long; anthers 0.5 mm long. Stigma branches 2, 1–2 mm long. Nutlet chestnut brown, glossy, oblong, apex truncate, apiculate, 1 mm long, with numerous indistinct papillae in longitudinal rows.

Zambia. W: Mwinilunga Dist., 14 km N of Kalene Hill towards Salujinga, 1250 m, fl. & fr. 21.ii.1975, *Hooper & Townsend* 298 (K).

Not known elsewhere. Edge of small pool on large granitic outcrop, sandy soils; 1250 m.

Conservation notes. Data Deficient, but possibly threatened. Only known from this collection which is in a widespread habitat.

32. **Pycreus capillifolius** (A. Rich.) C.B. Clarke in Durand & Schinz, Consp. Fl. Afr. 5: 535 (1894); in F.T.A. **8**: 300 (1901). —Hoenselaar in F.T.E.A., Cyperaceae: 284 (2010). Type: Ethiopia, Tigray, Kouaieta Village, n.d., *Quartin Dillon* s.n. (P holotype). FIGURE 14.**49**.

  Cyperus capillifolius A. Rich., Tent. Fl. Abyss. **2**: 475 (1850). —Kükenthal in Engler, Pflanzenr. **4**, 20(101): 357 (1936). —Haines & Lye, Sedges & Rushes E. Afr.: 287, figs.594,595 (1983).

Slender single stemmed or tufted annual; culms 1 to 15 (sometimes appearing to be more when several plants grow closely together), erect, to 35 cm long and 1 mm wide, rounded, striate, smooth. Leaf sheaths brown to reddish brown, to 2 cm long; blade linear, folded, to 12 cm long and to 0.5 mm wide (to 1 mm if flattened), not scabrid. Inflorescence of a single sessile cluster of (1)2 to over 30 spikelets (very rarely also with a single stalked cluster with peduncle to 2 cm long), 0.5 × 1–2.5 cm (to 3.5 × 4 cm in fruit); involucral bract single (or with a second much smaller), leaf-like, held as extension of culm making inflorescence appearing lateral, 1–10(12) cm long and 0.5–1 mm wide, not scabrid. Spikelets 5–10 × 1–2 mm (to 25 mm long in fruit), linear to lanceolate; rachis straight in fruiting stage; glumes appressed at flowering, slightly spreading later, falling at maturity, densely imbricate (about 50% overlapping), pale brown to almost black, ovate to elliptic, without or with a very narrow hyaline margin, 1–2 × 0.5–1 mm, rounded with midrib not reaching tip, keel prominent, rib-like, white to pale brown. Stamens 2; filaments 0.5–1 mm long; anthers 0.2 mm long. Stigma branches 2, 0.5–1 mm long. Nutlet dark brown, ellipsoid to obovoid or obtrigonous, apex truncate, apiculate or not, c. 1 mm long, with numerous papillae in longitudinal rows.

Zambia. N: Kasama Dist., Malole Rocks, 1275 m, fl. 2.iii.1960, *Richards* 12696 (K). W: Mwinilunga Dist., Matonchi, Kalenda Plain, 1400 m, fl. & fr. 16.iv.1960, *Robinson* 3609 (K). C: Serenje Dist., Kasanka Nat. Park, 1225 m, fl. & fr. 8.iii.2005, *Bingham* 12867 (K). S: Choma Dist., 18 km N of Choma, Muckle Neuk, 1275 m, fl. & fr. 28.ii.1954, *Robinson* 593a (K). **Zimbabwe**. W: Matobo Dist., Besna Kobila, 1450 m, fl. 14.iv.2015, *Browning* 965 (K). E: Mutare Dist., Vumba Mts, Globe Rock, fl. 19.iv.1993, *Browning* 571 (K, NU, SRGH). **Malawi**. N: Nyika Plateau, 2300 m, fl. 13.iii.1961, *Robinson* 4481 (K). S: Zomba Dist., Zomba Plateau, Chitinji stream, fl. 18.iii.1983, *Banda* 1810 (K, MAL).

Widespread in tropical Africa. In small pockets of soil on rocky outcrops, shallow sandy to gravelly soil in ironstone pans, edges of dambos and roadside ditches; 1200–2300 m.

Conservation notes: Widespread distribution; Least Concern. In common and widespread habitats.

Plants from lower altitudes in W Zambia often have paler glumes than plants from higher altitudes in E Zambia and Malawi. But the differences are not distinct and dark glumed specimens can also be found in the west.

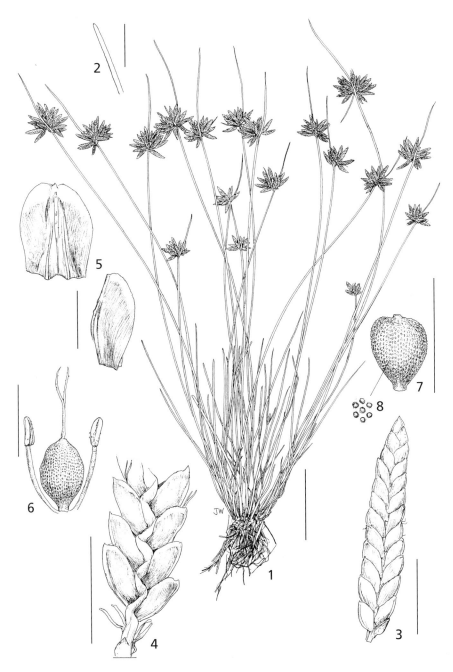

Fig. 14.**49**. PYCREUS CAPILLIFOLIUS. 1, habit; 2, tip of involucral bract; 3, spikelet; 4, detail of rachis; 5, glume, dorsal (left) and side view; 6, flower; 7, nutlet; 8, detail of surface. 1–3, 7, 8 from *Robinson* 4557; 4, 5 from *Richards* 12678; 6 from *Brummitt* 10352. Scale bars: 1 = 3 cm; 2–4 = 3 mm; 5–7 = 1 mm. Drawn by Juliet Williamson.

33. **Pycreus poikilostachys** Nelmes in Kew Bull. **6**: 320 (1952). Type: Zambia, Mwinilunga Dist., Matonchi Farm, 24.i.1938, *Milne-Redhead* 4311 (K holotype, BM, BR, PRE).

Pycreus heterochrous Nelmes in Kew Bull. **6**: 320 (1952). Type: Zambia, Mwinilunga Dist., Matonchi Farm, 24.i.1938, *Milne-Redhead* 4309 (K holotype, BM, BR, PRE).

Cyperus poikilostachys (Nelmes) Reynders in Phytotaxa **166**: 41 (2014).

Slender single stemmed or tufted annual; culms 1 to 10 (sometimes appearing to be more when several plants grow closely together), erect, to 25 cm long and 0.5 mm wide, rounded, striate, smooth. Leaf sheaths brown, to 4 cm long; blade filiform, to 12 cm long and 0.5 mm wide, not scabrid. Inflorescence to 2 × 35 cm (to 3 × 4.5 cm in fruit), of a single sessile cluster of 1–10(20) spikelets; involucral bract single (or also with a second much smaller), larger leaf-like, held as an extension of culm making inflorescence appear lateral, 1–12 cm long and 0.5 mm wide, not scabrid. Spikelets 5–20 × 2–3.5 mm (to 30 mm long in fruit), lanceolate; rachis strongly curved; glumes appressed at flowering and fruiting stages, falling at maturity, densely imbricate (about 2/3 overlapping), pale brown to reddish brown or dark brown, ovate to elliptic, with a broad hyaline margin, 2–2.5 × 1.5–2 mm, apex rounded with hyaline tip covering end of keel, keel prominent, rib-like, white to pale green. Stamens 2; filaments 0.5 mm long; anthers 0.5 mm long. Stigma branches 2, 1 mm long. Nutlet black, ellipsoid, apex truncate, shortly apiculate, 1–1.5 mm long, with numerous inconspicuous papillae in longitudinal rows.

Zambia. W: Mwinilunga Dist., Matonchi, Kalenda Plain, 1400 m, fl. & fr. 16.iv.1960, *Robinson* 3608 (K); Mwinilunga Dist., 6 km N of Kalene Hill, fl. 12.xii.1963, *Robinson* 5923 (K).

Not known elsewhere. On bare sandy soil in dambos, small lateritic pools and on granitic outcrops; 1300–1500 m.

Conservation notes. Known only from the Mwinilunga area where it seems to be fairly common; probably not threatened. Its habitats are widespread throughout N Zambia and into Angola and Congo.

The type of *Pycreus heterochrous* – collected on the same day from the same locality – differs only in having paler glumes, but not a bigger difference than seen in many other of the annual species in the Flora area. In my opinion it only shows variation within a single population.

34. **Pycreus atrorubidus** Nelmes in Kew Bull. **6**: 320 (1952). Type: Zambia, Mwinilunga Dist., Kalenda dambo, 14.ii.1938, *Milne-Redhead* 4565 (K holotype).

Slender single stemmed or tufted annual; culms 1 to 3, erect, to 15 cm long and 0.5 mm wide, rounded, striate, smooth. Leaf sheaths brown, to 3 cm long; blade filiform, to 8 cm long and to 0.25 mm wide, not scabrid. Inflorescence to 1.5 × 1.5 cm, of a single sessile cluster of 1–6 spikelets; involucral bract single (or with a second much smaller), larger leaf-like, not held as extension of culm, 1–3.5 cm long and 0.5 mm wide, not scabrid. Spikelets 5–10 × 2.5–3 mm, lanceolate; rachis becoming slightly zig-zag; glumes appressed at flowering, slightly spreading later, densely imbricate (about 75% overlapping), dark brown, ovate to elliptic, with a very narrow (to 0.2 mm) to almost absent hyaline margin, 2 × 1.5 mm, apex rounded with a brown tip which covers end of keel, keel prominent, rib-like, whitish. Stamens 2; filaments 0.5 mm long; anthers 0.2 mm long. Stigma branches 2, 1 mm long. Nutlet (immature) dark grey, ellipsoid, apex truncate, slightly apiculate, c. 0.5 mm long, with numerous inconspicuous papillae in longitudinal rows.

Zambia. W: Mwinilunga Dist., Kalenda dambo, fl., 14.ii.1938, *Milne-Redhead* 4565 (K).

Not known elsewhere. On bare sandy to peaty soil in dambos; 1350 m.

Conservation notes: Data Deficient, known only from the type collection. The habitat is widespread throughout N Zambia and into Angola and Congo.

Closely related to *Pycreus poikilostachys* from which it differs in the very narrow to almost absent hyaline glume margin and in the smaller nutlet. Also close to *P. capillifolius* which has narrower spikelets with smaller glumes.

35. **Pycreus demangei** J. Raynal in Kew Bull. **23**: 314 (1969). —Hoenselaar in F.T.E.A., Cyperaceae: 278 (2010). Type: Mali, Dogo, Soredina Plain, 12.ii.1966, *Demange* 3114 (P holotype, BR, K, WAG).

   Cyperus demangei (J. Raynal) Lye in Nord. J. Bot. **3**: 231 (1983). —Haines & Lye, Sedges & Rushes E. Afr.: 284, fig.584 (1983).

Slender single stemmed or tufted annual, terrestrial or aquatic with only inflorescence above water level; culms 1 to 15, erect or trailing, to 30 cm long and 1 mm wide, rounded, striate, smooth. Erect plants with a single internode (2 leaves per stem), aquatic or trailing plants with several internodes. Leaf sheaths brown, to 2 cm long; blade filiform in aquatic forms, flat in terrestrial forms, to 10 cm long (to 15 cm in aquatic forms) and 1 mm wide, scabrid near tip. Inflorescence to 5 × 7 cm, consisting of a single subsessile cluster of 1–10 spikelets, some or all also with 1–5 stalked clusters each with 1–5(7) spikelets and peduncle 0.5–4(6) cm long; involucral bracts 1–3, leaf-like or 1–2 smaller, largest 2–8(15) cm long and 0.5–1 mm wide, scabrid near apex. Spikelets 4–10 × 1.5–3 mm (to 15 mm long in fruit), lanceolate to oblong; rachis slightly zig-zag in fruiting stage; glumes appressed at flowering, spreading later, falling at maturity, densely imbricate (about 75% overlapping), pale brown to dark brown, broadly ovate, without hyaline margin, 1–1.5(2) × 1(1.5) mm, subacute to obtuse, keel prominent, rib-like, pale green. Stamens 2; filaments 1 mm long; anthers 0.5 mm long. Stigma branches 2, 1–2 mm long. Nutlet black, subglobose, apex truncate, slightly apiculate, 0.5–1 mm long, densely papillose with evenly (not in in longitudinal rows) distributed papillae.

Botswana. N: Okavango Delta, Chief's Is., 2 km SW of Chief's Camp, 975 m, fl. & fr. 31.x.2011, *A. & R. Heath* 2170 (K). **Zambia**. B: Kalabo Dist., 15 km S of Kalabo, fl. & fr. 2.viii.1962, *Robinson* 5450 (K). N: Mbala Dist., Uningi Pans, 1600 m, fl. 24.vi.1956, *Robinson* 1741 (K). W: Solwezi Dist., Solwezi, 1350 m, fl. 9.iv.1960, *Robinson* 3478 (K). S: Namwala Dist., Namwala–Ngoma road, Mulela Plain, fl. & fr. 19.iv.1963, *van Rensburg* 2066 (K). **Zimbabwe**. W: Matobo Dist., Besna Kobila Farm, 1450 m, fr. iv.1957, *Miller* 4345 (K, SRGH).

Also in Mali and Tanzania. Partly submerged in seasonal waterholes or on drying mud along streams and around waterholes, wet areas in dambos; 975–1600 m.

Conservation notes: Common from N Zambia into SW Tanzania; Least Concern. Inconspicuous, but in areas where it has been actively searched for it has proved to be common.

This can either be wholly aquatic as *Pycreus waillyi* with floating stems with several internodes and filiform leaves, or terrestrial with erect stems, 1–2 internodes and flat leaves. What seems to happen is that plants germinating early in the rains become aquatic while plants germinating later when the water retreats become erect terrestrials.

36. **Pycreus melanacme** Nelmes in Kew Bull. **10**: 91 (1955). —Hoenselaar in F.T.E.A., Cyperaceae: 280 (2010). Type: Zambia, Mbala Dist., Mpulungu–Mbala Road, Tsetse Control Station, 1.ii.1952, *Richards* 725 (K holotype). FIGURE 14.**50**.

   Cyperus pauper sensu Haines & Lye, Sedges & Rushes E. Afr.: 286 (1983), in part and for fig.589, non A. Rich.

Slender single stemmed or tufted annual; culms 1 to 10 (sometimes appearing to be more when several plants grow closely together), erect, to 35 cm long and 0.5 mm wide, rounded, striate, smooth. Leaf sheaths brown to reddish brown, to 6 cm long; blade filiform, to 15(20) cm long and 0.5–1 mm wide, not scabrid. Inflorescence to 1.5 × 1.5 cm, of a single sessile cluster of 1–5 spikelets; involucral bract single (sometimes also with a second much smaller), leaf-like, held as extension of culm making inflorescence appearing lateral, 1–12 cm long and 0.5–1 mm wide, not scabrid. Spikelets 5–15 × 3–5 mm, oblong; rachis straight in fruiting stage; glumes appressed at flowering, slightly spreading later, very tardily dehiscing, densely imbricate (about 75% overlapping), yellowish brown, sometimes tinged reddish, edges towards tip and tip dark chocolate brown to almost black, elliptic, without hyaline margin, 2–3 × 1.5–2 mm long, acute and apiculate, keel prominent, rib-like, white to pale brown. Stamens 2; filaments 0.5–1 mm long; anthers 1 mm long. Stigma branches 2, 1–2 mm long. Nutlet dark grey, broadly ellipsoid to subglobose, apex truncate or slightly subacute, apiculate or not, 1–1.5 mm long, with numerous evenly (not in in longitudinal rows) spread papillae.

Fig. 14.**50**. PYCREUS MELANACME. 1, habit; 2, inflorescence; 3, spikelet; 4, fruiting spikelet with nutlets; 5, glume, dorsal (left) and side view; 6, flower; 7, nutlet, with detail of sculpturing. All from *Bingham* 12884. Scale bars: 1 = 3 cm; 2–4 = 5 mm; 5–7 = 1 mm. Drawn by Juliet Williamson.

Zambia. N: Mbala Dist., Chilongwelo, Plain of Death, fl. 9.ii.1957, *Richards* 8127 (K). W: Mwinilunga Dist., 6 km N of Kalene Hill, Zambezi Rapids, 1350 m, fl. & fr. 22.ii.1975, *Hooper & Townsend* 306 (K). C: Serenje Dist., Kasanka Nat. Park, 1225 m, fl. & fr. 8.iii.2005, *Bingham* 12884 (K).

Also in Uganda and D.R. Congo. In small pockets of soil on rocky outcrops, shallow sandy to gravelly or peaty soil in ironstone pans, usually only appearing for a short period at the end of the rains; 1200–1650 m.

Conservation notes: Widespread distribution; Least Concern. In specialised habitats, but these are not threatened.

The illustration of *Pycreus pauper* in Haines & Lye (1983) and in Lye (1997, fig.212.146) is of this species.

37. **Pycreus melas** (Ridl.) C.B. Clarke in Durand & Schinz, Consp. Fl. Afr. **5**: 538 (1894); in F.T.A. 8: 302 (1901). —Hoenselaar in F.T.E.A., Cyperaceae: 279 (2010). Types: Angola, Pungo Andongo, Lombe, iii.1857, *Welwitsch* 6913 (BM syntype); Mutollo Sobata de Guinga, i.1857, *Welwitsch* 6914 (BM lectotype); Huíla, Morro de Lopollo to Empalanca, iv.1860, *Welwitsch* 6871 (BM syntype), lectotypified here.

    Cyperus melas Ridl. in Trans. Linn. Soc. London, Bot. **2**: 127 (1884). —Kükenthal in Engler, Pflanzenr. **4**, 20(101): 357 (1936). —Haines & Lye, Sedges & Rushes E. Afr.: 286, fig.590 (1983).

    Juncellus ater C.B. Clarke in Bull. Soc. Bot. France **54**, Mem.8: 26 (1907). Type: Central African Republic, Kago Do, near Demba, 26.x.1902, *Chevalier* 5886 (P holotype, BR, K).

    Pycreus ater (C.B. Clarke) Cherm. in Arch. Bot. Mém. 4(7): 11 (1931).

Slender tufted annual; culms 1 to over 30, erect, to 15(20) cm long andf 0.5 mm wide, rounded, striate, smooth. Leaf sheaths brown, to 2 cm long; blade linear, folded, to 6.5 cm long and 0.5 mm wide (to 1 mm if flattened), not or minutely scabrid near tip. Inflorescence of 1–3 sessile clusters of 2–15 spikelets, or also with 1–2 stalked clusters 0.5–1.5 × 1–2 cm (to 3 × 3 cm in fruit); peduncles of stalked clusters to 1 cm long; involucral bracts one large and 0–1(2) small, if only one then inflorescence appearing lateral, largest leaf-like, 5–60 × 0.5–1.5 mm, not scabrid. Spikelets 5–10 × 1–1.5 mm (to 20 mm long in fruit), linear-lanceolate; rachis slightly zigzag in fruiting stage; glumes appressed at flowering, slightly spreading later, falling at maturity, densely imbricate (about 50% overlapping), dark brown to almost black (rarely brown), ovate to obovate, apically with a hyaline margin, c. 1 × 0.5 mm, rounded with midrib not reaching the tip, keel prominent, pale brown. Stamens 2; filaments 0.5–1 mm long; anthers 0.2 mm long. Stigma branches 2, 0.5–1 mm long. Nutlet obovoid, apex truncate, apiculate, dark brown, c. 0.5 mm long, densely papillose with evenly (not in in longitudinal rows) distributed papillae.

Zambia. B: Mongu, fl., 5.iii.1966, *Robinson* 6869 (K). N: Mansa Dist., 90 km S of Mansa (Fort Rosebery), 1250 m, fr. 4.vi.1960, *Robinson* 3730 (K). W: Solwezi Dist., 85 km W of Chingola, 1350 m, fl. & fr. 9.iv.1960, *Robinson* 3501 (K). S: Choma Dist., 5 km E of Choma, 1300 m, fl. & fr. 27.iii.1955, *Robinson* 1196 (K). **Zimbabwe**. C: Harare, 1425 m, fl. 29.iv.1932, *Brain* 8945 (K).

Also in Ghana, Togo, Nigeria, Central African Republic, D.R. Congo, Burundi, Tanzania and Angola. Seasonally inundated grassland, edges of waterholes, roadside ditches, usually on sandy soils; 1000–1500 m.

Conservation notes: Widespread distribution; Least Concern. In common and widespread habitats.

38. **Pycreus micromelas** Lye in Nord. J. Bot. **1**: 617 (1981). Type: Zambia, Kasama Dist., Malole, 2.ii.1961, *Robinson* 4331 (EA holotype, K).

    Cyperus micromelas (Lye) Lye in Haines & Lye, Sedges & Rushes E. Afr., app.3: 2 (1982). —Haines & Lye, Sedges & Rushes E. Afr.: 287, figs.592,593 (1983).

Slender tufted annual; culms 1 to 10 (sometimes appearing to be more when several plants grow closely together), erect, to 10 cm long and 0.5 mm wide, rounded, striate, smooth. Leaf sheaths brown, to 0.5–1 cm long; blade linear, folded, to 4 cm long, and 0.5 mm wide (to 1 mm if flattened), not scabrid. Inflorescence a single sessile clusters of 2–15(25) spikelets or also with 1–2 stalked clusters, 0.5–2 × 1–1.5 cm (to 3 × 3 cm in fruit); peduncles of stalked clusters to 1.5 cm long; involucral bracts one large and 0–1(2) small, if only one then inflorescence appearing lateral, largest filiform, 0.5–2.5(3.5) cm long and 0.5 mm wide, not scabrid. Spikelets 5–12 × 0.5–1 mm (to 17 mm long in fruit), linear-lanceolate; rachis zig-zag in fruiting stage; glumes appressed at flowering, slightly spreading later, falling at maturity, loosely imbricate (about 33% overlapping), reddish brown to dark brown, ovate to obovate, without hyaline margin, c. 1 × 0.5 mm, rounded with midrib not reaching tip, keel prominent, wing-like, pale brown. Stamens 2; filaments 0.5–1 mm long; anthers 0.2 mm long. Stigma branches 2, 0.5–1 mm long. Nutlet obovoid, apex truncate, not apiculate, pale brown or grey, c. 0.5 mm long, with numerous minute papillae in longitudinal rows.

Zambia. N: Kasama Dist., 10 km E of Kasama, fl. & fr. 1.iii.1961, *Robinson* 4425 (K). W: Mwinilunga Dist., road to Salujinga, NW of Kalene Hill, 1250 m, fl. 21.ii.1975, *Hooper & Townsend* 302 (K). C: Serenje Dist., S of Kanona, Kundalila Falls, 1400 m, fl. 14.iii.1975, *Hooper & Townsend* 757 (K).

Also in D.R. Congo. On shallow sandy soil in small pans on rocky outcrops and on bare sandy soil in dambos; 1250–1500 m.

Conservation notes: Due to its inconspicuousness almost certainly very under-collected; probably not Vulnerable. Where it has been actively looked for it is fairly common. Its habitats are widespread in northern Zambia and neighbouring parts of D.R. Congo and Tanzania.

Excluded species.

Pycreus altus (Turrill) Lye in Nord. J. Bot. **1**: 622 (1981). Type: Angola, Benguella, Cuito, Tiengo R., 9.iii.1960, *Gossweiler* 2584 (K holotype, BM).

> *Juncellus altus* Turrill in Bull. Misc. Inform. Kew **1914**: 338 (1914).
>
> *Cyperus praealtus* Kük. in Engler, Pflanzenr. **4**, 20(101): 319 (1936). —Haines & Lye, Sedges & Rushes E. Afr.: 270 (1983), non *Cyperus altus* Nees.

This species is cited by Haines & Lye (1983) under *Cyperus* subgen. *Pycreus* (P. Beauv.) Miq. and recorded by them as occurring in Zambia. It has many-nerved glumes and a nutlet with the wider side facing the axis. Haines & Lye suggest it is related to *P. mundii*. Despite the superficial similarity with this species it is not a *Pycreus* but belongs in *Cyperus* subgen. *Juncellus* (C.B. Clarke) C.B. Clarke.

22. **QUEENSLANDIELLA** Domin [22]

Queenslandiella Domin in Biblioth. Bot. **85**: 415 (1915).

> *Mariscopsis* Cherm. in Bull. Mus. Natl. Hist. Nat. **25**: 60 (1919).
>
> *Cyperus* subgen. *Queenslandiella* (Domin) Govind. in Reinwardtia **9**: 194 (1975).

Tufted annuals with a pungent odour of curry or fenugreek, persistent in herbarium specimens; culms scapose. Leaves eligulate. Involucral bracts leaf-like. Inflorescences anthelate. Spikelets falling entire at maturity; glumes persistent on deciduous rachilla; each glume subtending a flower. Flowers bisexual. Stamens 2. Style 2-branched. Nutlets ± oblong in outline, laterally compressed.

A monotypic genus extending from the East African coast to N Queensland (Australia).

Larridon *et al.* in Bot. J. Linn. Soc. **172**: 106–126 (2013) placed the species of *Queenslandiella* in *Cyperus*, but it is kept separate here.

---

[22] by M. Xanthos

Queenslandiella hyalina (Vahl) Ballard in Hooker's Icon. Pl. **33**: t.3208 (1933). Type:
India, *Röttler* s.n. (C holotype). FIGURE 14.**51**.

    Cyperus hyalinus Vahl, Enum. Pl. **2**: 329 (1806). —Kükenthal in Engler, Pflanzenr. **4**,
20(101): 498 (1936). —Haines & Lye, Sedges & Rushes E. Afr.: 293, fig.608 (1983).
    Cyperus pumilus sensu Nees in Wight & Arnott, Contr. Bot. India: 74 (1834) in part excl.
synonym, non L.
    Pycreus pumilus sensu Nees in Linnaea **9**: 283 (1835) in part, non L.

Fig. 14.**51**. QUEENSLANDIELLA HYALINA. 1, habit (× ⅔); 2, spikelet (× 5); 3, glume (× 10);
4, flower (× 12); 5, young nutlet (× 10). 1, 3–5 from *Schlieben* 12140; 2 from *Kirika & Muthoka*
NMK 729. Drawn by Juliet Williamson. Reproduced from Flora of Tropical East Africa (2010).

Queenslandiella mira Domin in Biblioth. Bot. **85**: 416, t.11 figs.7–13 (1915). Type: Australia, Queensland, near Chillagoe, 1910, *Domin* 1598 (PR holotype).

Pycreus hyalinus (Vahl) Domin in Biblioth. Bot. **85**: 417 (1915), in annotation.

Mariscopsis suaveolens Cherm. in Bull. Mus. Natl. Hist. Nat. **25**: 61 (1919). Type: Madagascar, humid grassland in N, n.d., *Berrier* 33 (P syntype); Zanzibar, iv-v, *Boivin* s.n. (P syntype).

Mariscopsis hyalinus (Vahl) Ballard in Bull. Misc. Inform. Kew **1932**: 458 (1932).

Tufted annual, 6–15 cm tall; culms 2–6 cm long. Leaves basal, flat, 2–7 cm long, 1–1.5 mm wide, glabrous; sheaths reddish-brown. Involucral bracts 3, leaf-like, overtopping inflorescence, 1.5–6.5 cm long, 1.5 mm wide. Inflorescence a simple umbel or with 1 sessile spike and up to 4 stalked spikes; rays to 1.5 cm long; spikes 0.5–1.5 cm long, 6–7 mm wide, comprising 8–12 ovate to ovate-elliptic compressed spikelets, 4 × 2 mm. Glumes 4–5, yellow to greenish yellow with membranous margin, lanceolate-ovate to ovate, 2–2.5 mm long, 8–10 veined, with scabrid margins, green midrib excurrent with recurved tips. Nutlet dark grey or brown, 1.2 × 1 mm, laterally compressed, rounded in outline, truncate to somewhat emarginate at apex, finely densely papillate.

Mozambique. Unspecified locality ("P.E.A." = Portuguese East Africa), sandy flat, iv.1921, *Drummer* 4639 (K).

In Kenya, Tanzania, Madagascar, Mauritius and Mayotte; also in India, Maldives, Sri Lanka, Malaysia and Australia. In grassland, bushland on coral rag, also as a weed in coastal lawns, sisal plantations and under coconut palms; sea-level–30 m.

Conservation notes: Least Concern owing to its wide distribution.

The strongly aromatic odour sets this species apart from other related genera such as *Kyllinga* and *Pycreus*. Only one specimen of this species has been traced from the Flora region, but due to its extensive range in the Old World, it is probably found in other coastal parts of Mozambique.

23. **KYLLINGA** Rottb.[23]

Kyllinga Rottb., Descr. Icon. Rar. Pl.: 12, t.4 (1773), conserved name.
—Getliffe in J. S. African Bot. **49**: 261–304 (1983).
Cyperus L. subgen. *Kyllinga* (Rottb.) Valck. Sur., Cyperus Mal. Arch.: 42 (1898).
—Kükenthal in Engler, Pflanzenr. **4**, 20(101): 566 (1936). —Haines & Lye,
Sedges & Rushes E. Afr.: 224–250 (1983).

Annual or perennial herbs, often with rhizomes or stolons. Culms scapose to few-noded, usually triangular. Leaves sometimes reduced to sheaths only; ligule 0. Involucral bracts leaf-like. Inflorescence a single sessile ovoid or globose spike, or a complex head with smaller lateral spikes at base of main spike; laterals flower and fruit later than the main one. Spikelets narrowly ovoid, 1–6-flowered, with 2 sterile basal glumes and several larger fertile glumes; upper flowers sometimes male only; glumes in 2 rows, usually ovate and hooded, keel sometimes winged. Stamens 1–3, usually 3. Style with 2 branches. Nutlet oblong or ellipsoid, compressed laterally.

About 60 species, mostly in Africa but some in southern Asia and the Americas.

The genus *Kyllinga*, along with some other genera, has been subsumed in the genus *Cyperus* sensu lato by Larridon *et al.* in Phytotaxa **166**: 33–48 (2014). As none of the taxa below have been given new combinations, and to ensure consistency with F.T.E.A., I feel it would be premature to incorporate this change here.

Occasionally the spelling *Kyllingia* is encountered (as used by Kükenthal in Engler's Pflanzenreich and by Boeckeler, Chermezon, Kunth, Kuntze, Nees and Flora of Madagascar), but the original and correct spelling is without the 'i'.

Species from this genus are sometimes confused with those of others such as sessile-headed *Cyperus* (e.g. in section *Mariscus* (Vahl.) Griseb.). *Cyperus* has spikelets with

---

[23] by H. Beentje

many distichous glumes, while *Kyllinga* has just a few (but more than one), while *Ascolepis* has just a single glume per spikelet. The occasional *Cyperus* with very few glumes (such as *C. albopilosus* with 3) can be distinguished by 3-branched styles, rather than 2-branched styles in *Kyllinga*.

Note: In the absence of any published genus revision the sequence of species in this treatment is alphabetical.

1. Inflorescence head golden, yellow or yellow-green, red-brown or almost black. 2
– Inflorescence head white or cream 13
2. Leaves and sheaths in line with leaves pubescent; glumes hairy . . . **5.** *aureovillosa*
– Leaves at most scabrid on midrib and margins, otherwise glabrous; glumes not hairy 3
3. Flowering heads often with young plants growing out of them; glumes 4.5–7 mm long. . . . . **17.** *pauciflora*
– Flowering heads not viviparous; glumes 1.8–4.5 mm long (to 6.5 mm in *Kyllinga alba*) 4
4. Flattened leaves 0.8–2 mm wide 5
– Flattened leaves more than 2 mm wide 8
5. Culms tufted in tight groups, with swollen bases; glume apex ± recurved 6
– Culms solitary and spaced along rhizome (sometimes tightly); glume apex straight 7
6. Leaf width 0.8–1 mm; heads golden. . . . . **28.** *ugogensis*
– Leaf width 1–2.5 mm; heads yellow-green to black; rhizome fragrant . . .**14.** *nervosa*
7. Involucral bracts erect to spreading; spikelet 2–4-flowered; glumes uniformly straw-coloured or greenish . . . . **3.** *albiceps*
– One involucral bract erect, others spreading; spikelet 1–2-flowered; glumes yellow or orange with green keel. . . . . **6.** *brevifolia*
8. Glumes 3.5–6.5 mm long, winged and toothed; leaf 2–6 mm wide; rhizome fragrant. . . . . **2.** *alba*
– Glumes without clear wing, sometimes with a few small teeth; rhizome not fragrant (except in *K. nervosa*, which has no teeth) 9
9. Culm base not swollen (character not always clear); heads large, ovoid, red-brown to blackish brown; glumes 3–4.5 mm long with a few small teeth . . . . . . **18.** *peteri*
– Culm base swollen; heads yellow, golden or yellow-green (occasionally dark brown in *K. nervosa*, with fragrant rhizome); glumes 1.8–3.5 mm long. 10
10. Culms in dense tufts; leaf width 1–2.5 mm; culm base aromatic; glumes yellow with black apex . . . .**14.** *nervosa*
– Culms slightly spaced along creeping rhizome; leaf width 2–5 mm. 11
11. Inflorescence bracts 3–4; glumes smooth . . . . **10.** *erecta*
– Inflorescence bracts 3–9; some glumes with small teeth (not always) 12
12. Heads usually several; otherwise very close . . . .**20.** *polyphylla*
– Heads solitary; otherwise very close . . . .**11.** *melanosperma*
13. Culms tufted in tight groups. 14
– Culms solitary or (sometimes closely) spaced along creeping rhizome. 22
14. Glumes winged and with teeth on lower part 15
– Glumes smooth (or winged in *K. robinsoniana*, but then without teeth) 17
15. Head 9–18 mm across; glumes 3.5–6.5 mm long; culm base aromatic . . . . **2.** *alba*
– Single heads 4–9 mm across; glumes 1.5–2.9 mm long; culm base aromatic or not. 16
16. Whole plant sweet-scented; rhizome present; heads 1–3; glumes with few minute teeth . . . .**21.** *pumila*
– Plant not scented; rhizome absent; heads solitary; glumes with coarse hair-tipped teeth . . . . **25.** *squamulata*

17. Rhizome absent; heads 3–4 mm in diameter . . . . **1.** *afropumila*
–   Rhizome present 18
18. Glumes 4.2–4.8 mm long; rhizome fragrant . . . . **15.** *nigripes*
–   Glumes < 4 mm long; rhizome not fragrant (except in *Kyllinga odorata*) 19
19. Culm base swollen; leaf width 0.5–3 mm 20
–   Culm base not swollen; leaf width 1.5–7 mm 21
20. Flowering heads 5–12 mm wide; glumes 1–2.2 mm long, with green keel . . . . . .
    . . . . **27.** *tenuifolia*
–   Flowering heads 4–7 mm wide; glumes 2–2.4 mm long, with translucent keel . . .
    . . . . **23.** *robinsoniana*
21. Rhizome not fragrant; glumes 3–4 mm long, with whitish keel . . . . **19.** *platyphylla*
–   Rhizome fragrant; glumes 2–3 mm long, with green keel . . . . **16.** *odorata*
22. Glumes winged and with teeth or large marginal hairs; culms solitary or spaced
    23
–   Glumes smooth (or winged in *K. robinsoniana*, but then without teeth) 26
23. Leaf 1–2 mm wide; rhizome not fragrant; keel green; head 5–10 mm wide . . . . .
    . . . . **4.** *albogracilis*
–   Leaf 1.5–6 mm wide; rhizome fragrant; keel not green 24
24. Flowering heads 3–8 mm wide, occasionally viviparous; forest species . . . .
    . . . . **13.** *nemoralis*
–   Flowering heads 6–18 mm wide, never viviparous; usually from more open
    situations 25
25. Rhizome thin; glumes with large teeth (except one inland form) . . . . **2.** *alba*
–   Rhizome 3–5 mm thick; glumes with tiny teeth or dense minute hairs; coastal . .
    . . . . **8.** *cartilaginea*
26. Culms solitary or well-spaced 27
–   Culms closely spaced 30
27. Flowering heads 1–3 in number, 5–17 mm wide; glume keel green; leaf 1–5 mm
    wide 28
–   Flowering heads solitary, 4–8 mm wide; keel whitish; leaf 0.5–2.5 mm wide . . . 29
28. Long stolons/rhizomes present; flowering heads 5–17 mm wide; inflorescence
    bract to 10 cm long . . . . **7.** *bulbosa*
–   Rhizome absent or very short; flowering heads 6–10 mm wide; inflorescence bract
    to 3.3 cm long . . . . **12.** *mtotomwema*
29. Glumes 1.8–2.3 mm long; inflorescence bracts to 6.5 cm long . . **22.** *rhizomafragilis*
–   Glumes 2–3.5 mm long; inflorescence bracts to 4 cm long . . . . **26.** *tanzaniae*
30. Leaf base slightly swollen; leaves 10–38 cm long, 1.5–4.3 mm wide; heads solitary,
    involucral bracts (3)7–30 cm . . . . **9.** *crassipes*
–   Leaf base bulbous; leaves 3–15 cm long; heads 1–3, involucral bracts 1–10 cm   31
31. Heads 4–7 mm wide; glumes 2–2.4 mm long; stamens 3; altitudes 600–1900 m . .
    . . . . **23.** *robinsoniana*
–   Heads 6–12 mm wide; glumes 2.5–4.5 mm long; stamens 2 32
32. Involucral bracts up to 7 cm long; heads always solitary; usually below 1500 m . .
    . . . . **24.** *songeensis*
–   Involucral bracts to 3.3 cm long; heads 1–3; found at 1850–2300 m . . . .
    . . . . **12.** *mtotomwema*

1. **Kyllinga afropumila** Lye in Nordic J. Bot. **1**: 741 (1982). —Beentje in F.T.E.A.,
   Cyperaceae: 327 (2010). Type: Tanzania, Mbeya Dist., track from Kawetire to
   Mbeya Peak, 15.iii.1970, *Wingfield* 754 (DSM holotype, K).

   Cyperus afropumilus (Lye) Lye in Haines & Lye, Sedges & Rushes E. Afr, app.3: 2 and main
   work: 244, fig.489 (1983).

Rather slender perennial to 26 cm tall; base slightly swollen, coated with fibrous remains of old leaf sheaths. Culms in tight groups, 10–25 cm long, 0.5–1 mm wide, trigonous, glabrous. Leaves to 10 cm long; leaf sheath straw-coloured to brownish; leaf blade linear, 4–10 cm long, 1–2.5 mm wide, scabrid on margin. Involucral bracts leaf-like, spreading, 2–3, lowermost 2–13 cm long. Inflorescence of single central spike 5–8 × 3–4 mm, usually with 2 smaller lateral spikes; spikelets many per head, ellipsoid, 1.5–1.8 × 0.6–0.7 mm, 1-flowered. Glumes whitish, often with purplish dots, 1–1.5 mm long, keel with 3–4 veins on either side, without teeth or cilia, acute. Stamens not seen. Nutlet almost black, ellipsoid, 1.2–1.4 × c. 0.6 mm, minutely papillose.

Zambia. W: Mwinilunga Dist., Kalenda Plain S of Matonchi farm, 18.xii.1937, *Milne-Redhead* 3719 (K). E: Chipata Dist., Chipata (Fort Jameson), xii.1961, *Robinson* 4739 (K). **Malawi**. N: Nyika Plateau, Lake Kalume, 16.v.1970, *Brummitt* 10815 (K).

Also in Tanzania. Muddy grassland, shallow soil over laterite or rock; 1300–2350 m.
Conservation notes: Widespread; not threatened. Data Deficient in F.T.E.A. (2010).
Similar to *Kyllinga pumila* and *K. odorata* var. *cylindrica*, but with even smaller spikelets and glumes and a fibrous culm-base.

2. **Kyllinga alba** Nees in Linnaea **10**: 140 (1835). —Clarke in F.T.A. **8**: 271 (1901). —Podlech in Merxmüller, Prodr. Fl. SW Afr. **165**: 30 (1967). —Napper in J. E. Africa Nat. Hist. Soc. Natl. Mus. **28**(124): 20 (1971). —Gordon-Gray in Strelitzia **2**: 113 (1995). —Beentje in F.T.E.A., Cyperaceae: 331, fig.50 (2010). Type: South Africa, Cape, Zwarte Key R., probably *Drège* s.n. (B?, not found); Gordon-Gray (1995) says "*Ecklon* (S? holotype)". FIGURE 14.**52**.

   Kyllinga cristata Kunth, Enum. Pl. **2**: 136 (1837). Types: South Africa, Cape, Queenstown Dist., between Table Mt & Wildschutsberg, 27.xi.1832, *Drège* 3930 (B syntype, K, P); Tafelberg, Wildschutsberg, 11.xii.1832, *Drège* 7385 (B syntype, P).

   Kyllinga alata Nees var. *alba* (Nees) Kuntze, Rev. Gen. Pl. **3**(2): 335 (1895).

   Cyperus nigripes (C.B. Clarke) Kük. var. *grandiceps* Kük. in Engler, Pflanzenr. **4**, 20(101): 572 (1936). Types: Tanzania, Pare Dist., near Mkomazi, 4.vi.1915, *Peter* 10703 (B syntype, GOET, K, WAG); Pare Dist., near Mkomazi, 6.vi.1915, *Peter* 10831 (B syntype, GOET); Pangani Dist., N of Buiko, 30.v.1915, *Peter* 10426 (B syntype, GOET); Pangani Dist., N of Buiko, 30.v.1915, *Peter* 10466 (B syntype, GOET); S Pare, Buiko, 29.v.1915, *Peter* 10390 (B syntype); Lushoto Dist., W Usambaras between Manolo & Mtai, 30.v.1914, *Peter* 4188 (B syntype).

   Cyperus cristatus (Kunth) Mattf. & Kük. in Engler, Pflanzenr. **4**, 20(101): 609 (1936). —Haines & Lye, Sedges & Rushes E. Afr.: 248, fig.499 (1983).

   Cyperus cristatus var. *nigritanus* (C.B. Clarke) Kük. in Engler, Pflanzenr. **4**, 20(101): 610 (1936). Type: Nigeria, Nupe, 19.ii.1895, *Barter* 1588 (K holotype).

   Cyperus cristatus var. *exalatus* Merxm. in Trans. Rhodesia Sci. Assoc. **43**: 80 (1951). Type: Zimbabwe, Marondera (Marandellas), 12.xi.1942, *Dehn* 804 (M holotype).

   Kyllinga alba var. *exalata* (Merxm.) Podlech in Mitt. Bot. Staatssamml. München **3**: 525 (1960).

   Kyllinga alba var. *nigritana* (C.B. Clarke) Podlech. in Mitt. Bot. Staatssamml. München **3**: 525 (1960).

   Cyperus alatus (Nees) F. Müll. subsp. *albus* (Nees) Lye in Lidia **3**: 172 (1995).

Perennial 5–60 cm tall; with short creeping rhizome and aromatic base (smelling 'sweet' or of cedarwood). Culms spaced or in dense tussocks, 5–60 cm long, 0.7–1.8 mm wide, triangular, glabrous or with a few hairs just below inflorescence; culm-base not or slightly swollen, covered by old leaf sheaths. Leaves up to 37 cm long; leaf sheath pale brown to reddish, 1–9(14) cm long; leaf blade linear, flat or channelled, 5–25(37) cm long, 2–6 mm wide, scabrid on margins and midrib. Involucral bracts leaf-like, spreading or sharply reflexed, (2)3–4, lowermost 4–18 cm long, scabrid on margins and midrib. Inflorescence a single sessile whitish, yellow-green or pale yellow globose head (very rarely double-headed), (6)9–15 × (6)9–18 mm; spikelets many per spike, ovoid, 3.3–6.5 ×1–3 mm, 2-flowered. Glumes white or yellow and often with minute brown spots, ovate, 3.5–6.5 mm long, keel winged (wing to 1 mm wide) with fairly coarse hair-tipped

Fig. 14.**52**. KYLLINGA ALBA. 1, habit; 2, partial habit; 3, spikelet; 4, spikelet, two lower glumes removed, rest opened up; 5, 6, nutlet, side and transverse view. 1–4 from *Robinson* 55; 5, 6 from *Robinson* 2972. Scale bars: 1 = 250 mm; 2 = 40 mm; 3, 4 = 2 mm; 5 = 1 mm. Drawn by Jane Browning.

teeth in basal half, apex acute to acuminate (latter associated with slightly longer stamens). Stamens 3, 0.8–1.7 mm long; filaments to 4.7 mm. Nutlets 1–2, red-brown or black, ellipsoid, 1.3–1.8 × 0.6–0.8 mm, minutely papillose.

Glumes white (often with brown spots) . . . . .a) var. *alba*
Glumes yellow-green or golden yellow, often with green keel. . . . . .b) var. *alata*

a) Var. **alba**

Culms closely spaced or in dense tussocks, 5–60 cm long. Leaves 4–25(38) cm long, (2)3–5 mm wide, almost smooth. Involucral bracts (2)3–4. Inflorescence a single sessile whitish globose head (6)9–15 × (6)9–18 mm; spikelets 4–6.5 × 2–3 mm. Glumes white, often with minute brown spots, fading to pale brown, narrowly ovate, 3–5 mm long. Stamens yellow. Nutlet black, flattened ellipsoid, 1.5–1.8 × 0.7–0.8 mm, minutely papillose.

Botswana. N: Okavango Delta, Chief's Is., 29.xi.2010, *A. & R. Heath* 2061 (GAB, K). SE: South East Dist., 5 km N of Gaborone, 6.xi.1977, *Hansen* 3273 (C, GAB, K, PRE, SRGH, WAG). SW: Ghanzi/Kgalagadi Dist., 23°08'S 20°45'E 21.iii.1980, *Skarpe* 425 (K). **Zambia**. N: Mbala Dist., Mbala (Abercorn), 8.i.1954, *Siame* 239 (K). W: Ndola Dist., Ndola, 1.i.1960, *Robinson* 3270 (K). C: Lusaka Dist., Makeni Ranch, 5.ix.1931, *Trapnell* 456 (K). E: Katete Dist., Katete R., 8.i.1966, *Astle* 4312 (K). S: Kalomo Dist., 1.2 km S of Zimba, 22.xii.1963, *Bainbridge* 943 (K, NDO, SRGH). **Zimbabwe**. N: Makonde Dist., Great Dyke, Vanad Pass, 13.xii.1974, *Wild* 7971 (K, SRGH). W: Hwange Dist., Victoria Falls village, 17.i.1974, *Gonde* 48.73 (K, SRGH). C: Gweru Dist., Whitewaters Dam, Gweru (Gwelo), 20.i.1963, *Loveridge* 592 (K, SRGH). E: Chipinge Dist., border road 5 km S of Rusongo, 1.ii.1975, *Gibbs Russell* 2740 (K, SRGH). S: Masvingo Dist., S of Lake Kyle (Mutirikwe), 1.i.1995, *Wilkin* 746 (K, SRGH). **Malawi**. N: Mzimba Dist., 2.5 km SSW of Chikangawa, 5.xii.1978, *Phillips* 4356 (K, MO). C: Kasungu Dist., Kasungu Nat. Park, 23.xii.1970, *Hall Martin* 1362 (K, SRGH). S: Blantyre Dist., Limbe, football ground, 12.iii.1969, *Mshinkhu* 12 (K, SRGH). **Mozambique**. Z: Lugela Dist., Namagoa, n.d., *Faulkner* 138 (K). T: Mutarara Dist.?, between Lupata & Tete, ii.1859, *Kirk* s.n. (K). MS: Cheringoma Dist., 40 km N of Dando Camp to Inhaminga, 3.xii.1971, *Pope & Müller* 501 (K, LISC, SRGH). M: Namaacha Dist., Namaacha, 9.xii.1948, *Myre & Carvalho* 302 (K, LMA).

Also in Togo, Somalia, Uganda, Kenya, Tanzania, Angola, Namibia, Swaziland and South Africa. Grassland on sandy soil in mopane woodland, lake edges, dambo margins and streamsides on hardpan, heavy red soil or black clays, also on copper-bearing or alkaline soils and in thin soil over rock; 50–1800 m.

Conservation notes: Widespread; not threatened. Least Concern in F.T.E.A. (2010).

b) Var. **alata** (Nees) C.B. Clarke in Durand & Schinz, Consp. Fl. Afr. **5**: 526 (1894); in F.T.A. **8**: 272 (1901). —Beentje in F.T.E.A., Cyperaceae: 333 (2010). Types: South Africa, Uitenhage, Koegakamma ravine, Zwartkops R., *Ecklon* 883 (B syntype); near Pauli Maré, probably *Drège* s.n. (B? syntype).

Kyllinga alata Nees in Linnaea **10**: 139 (1835), mistakenly as *allata*. —Podlech in Merxmüller, Prodr. Fl. SW Afr. **165**: 30 (1967). —Napper in J. E. Africa Nat. Hist. Soc. Natl. Mus. **28**(124): 19 (1971). —Getliffe in J. S. African Bot. **49**: 280, fig.9 (1983). —Gordon-Gray in Strelitzia **2**: 112 (1995).

Cyperus alatus (Nees) F. Muell., Fragm. **8**: 272 (1874). —Kükenthal in Engler, Pflanzenr. **4**, 20(101): 611 (1936). —Haines & Lye, Sedges & Rushes E. Afr.: 249, fig.500 (1983).

Cyperus alatus var. *serratus* Peter & Kük. in Engler, Pflanzenr. **4**, 20(101): 611 (1936). Type: Tanzania, Dodoma Dist., Turu, E of Itigi near Bangayega, km 618, 1926, *Peter* 33888 (B holotype).

Plant with smell of ginger or camphor; culms 9–50 cm long. Leaves 5–37 cm long, 2.5–5 mm wide, scabrid on margins and midrib. Involucral bracts (2)3–4. Inflorescence an ovoid or globose white (elsewhere reported as yellow-green or pale yellow) head, 10–15 × 8–14 mm; spikelets 3.3–5.5 × 1–1.5 mm; glumes yellow-green or golden yellow, often with green keel. Nutlet 1.3 × 0.6 mm, minutely papillose.

Botswana. N: Okavango, Moremi Wildlife Reserve, Txatxanika campsite, 3.iii.1972, *Biegel & Russell* 3843 (K, SRGH). SW: Kgalagadi Dist., Kang, 300 km W of Gaborone, 4.iii.1975, *Mott* 1123 (K, ROML, SRGH). SE: Kweneng Dist., 16 km NW of Lephepe, Kalahari Pasture Research Station, iv.1969, *Kelaole* 574 (K, SRGH). **Zambia**. W: Mwinilunga Dist., 42 km E of Mwinilunga, 17.xii.1963, *Robinson* 6121 (K). C: Chibombo Dist., Kamaila Forest Resthouse, 36 km N of Lusaka, 9.ii.1975, *Brummitt et al.* 14281 (K). S: Choma Dist., Choma, 6.xii.1962, *van Rensburg* 1017 (K). **Zimbabwe**. W: Lupane Dist., Ngamo pans, (Gwaai Reserve), 16.xii.1956, *Paterson* 2 (K, SRGH). **Malawi**. S: Machinga Dist., Liwonde Nat. Park, Likwenu R., 24.ii.1985, *Dudley* 1483 (K, MAL).

Also in Somalia, Uganda, Kenya, Tanzania, Angola, Namibia and South Africa. Seasonally damp grassland or sandveld, often around rock outcrops; 400–1500 m.

Conservation notes: Widespread; not threatened. Least Concern in F.T.E.A. (2010).

I fully agree with Clarke (1895) that this is so close to *Kyllinga alba* that varietal status is better than specific. In the protologue, Nees already wrote "an var.?" when discussing the differences between *alata* and *alba*. Getliffe (1983) and Gordon-Gray (1995) disagree and say that "recent workers all accept that the morphological differences between them require specific ranking for both"; but their morphological differences (culms pubescent or glabrous/scabrid, a character not confirmed by our specimens; glumes 'golden' or 'pale/white' and with mucro 'often recurved' or straight) seem better to me at varietal level. I think the distinction should stand. In addition to the characters mentioned above, the glumes of var. *alata* seem to be more broadly ovate, often (not always) have a longer-acuminate apex, and slightly more pronounced teeth. All these characters seem more gradual than absolute, especially since the colour character does not seem to hold true for our specimens.

A group of specimens from Zambia, Zimbabwe and Botswana is close to *Kyllinga alba* sensu lato, but differs in having much smaller teeth than usual for var. *alba*, and a less pronounced wing. I do not feel happy treating this as a separate taxon; the whole *K. alba* complex needs more careful study. For the time being I prefer slotting it near *K. alba* as an unplaced taxon:

Kyllinga taxon A near **alba** var. **alba**

Perennial, 10–45 cm tall, densely tufted with slender root-system; apparently lacking rhizomes or stolons. Culms 10–45 cm long, 0.9–1.7 mm wide, sharply triangular, glabrous except for some minute spines/hairs near inflorescence. Leaves to 19 cm long; leaf sheath pale to mid-brown, 2–5 cm long; leaf blade linear, flat or slightly channelled, 5–19 cm long, (2)2.5–4.4 mm wide, scabrid on margins and midrib. Involucral bracts leaf-like, spreading or reflexed, 3–4, lowermost 2.5–20 cm long. Inflorescence a sessile white ovoid head of a single spike (6)8–17 × (6)9–15 mm; spikelets many, ovoid, 4–4.5 × 1.5–2 mm, (1)2-flowered. Glumes translucent to whitish or yellow, often with minute reddish dots, ovate, 3.6–4.2 mm long, keel slightly green and winged in lower 2/3, with small teeth (0.1 mm), apex acute to obtuse. Stamens 3, anthers yellow, 0.9–1.1 mm long. Nutlet flattened ellipsoid, blackish, 1.4–1.6 × 0.7–1.1 mm, minutely papillose.

Botswana. N: Ngamiland Dist., Khwebe Hills, 1.ii.1898, *Lugard* 139 (K). **Zambia**. N: Mpika Dist., Luangwa Valley near Mupamadzi R., 5.i.1966, *Astle* 4408 (K). C: Mpika Dist., S Luangwa Nat. Park, 05 road, 13°06'13"S 31°44'17"E, 535 m, 7.i.2008, *Bingham* 13345 (K, NDO). S: Siavonga Dist., Lusitu R. gorge, NE of Changa, 13.ii.1975, *Kornaś* 3277 (K). **Zimbabwe**. W: Hwange Dist., Hwange Nat. Park, Makalolo I pan, 11.xii.1968, *Rushworth* 1352 (K, LISC, SRGH). C: Kadoma Dist., Kadoma (Gatooma), Copper Pot Mine, 16.iii.1966, *Wild* 7557 (K, SRGH).

Not found elsewhere. Grassland in miombo zone on sand, sometimes in seasonally moist sites. May be frequent or locally dominant; 500–1740 m.

Conservation notes: Endemic to the Flora area. Widespread; not threatened.

3. **Kyllinga albiceps** (Ridl.) Rendle in Hiern, Cat. Afr. Pl. **2**: 106 (1899). —Clarke in F.T.A. **8**: 286 (1901). —Beentje in F.T.E.A., Cyperaceae: 335 (2010). Type: Congo, Lower Congo R., *C. Smith* 5 (BM holotype).

   Cyperus albiceps Ridl. in J. Bot. **22**: 16 (1884).

   Kyllinga merxmuelleri Podlech in Mitt. Bot. Staatssamml. München **3**: 525 (1960); in Merxmüller, Prodr. Fl. SW Afr. **165**: 29 (1967). —Mapaura & Timberlake, Checkl. Zimbabwe Pl.: 88 (2004). Type: Namibia, Okavango terrtitory 24 km E of Runtu, 12.iii.1958, *Merxmüller & Giess* 2136 (M holotype, PRE, WIN).

   Cyperus merxmuelleri (Podlech) Lye in Haines & Lye, Sedges & Rushes E. Afr., app.3: 2 and main work: 230, fig.464 (1983).

Slender perennial, 10–40 cm tall; culm base slightly thickened and long slender stolons 1 mm across, covered in fibrous bracts. Culms solitary, 20–40 cm long, triangular, smooth. Leaves few; leaf blade linear, 7–16 cm long, 0.5–2 mm wide, scabridulous near apex. Involucral bracts leaf-like, erect to spreading, 2–3, lowermost to 8 cm long. Inflorescence a single round to ovoid head, 8–11 × 5–8 mm, or 2–3 spikes confluent into a single head; spikelets many, lanceolate-ellipsoid, 3–3.5 mm long, 2–4-flowered. Glumes straw-coloured to greenish, ovate-lanceolate, 2.3–2.6 mm long, keel smooth, apex apiculate. Stamens 3; anthers 1–1.5 mm long. Nutlets 2, golden brown, ellipsoid, c. 1.2 mm long.

Zambia. N: Mbala Dist., near Mbala (Abercorn), 14.xii.1954, *Richards* 3644 (K). **Zimbabwe**. C: Harare Dist., 17 km NE of Harare, 27.i.1967, *Drummond* 7514 (K, SRGH). E: Nyanga Dist., Nyanga Nat. Park, Udu Dam, 29.xii.1974, *Burrows* 639 (K, SRGH).

Also in Togo, Nigeria, Cameroon, Tanzania, D.R. Congo, Angola and Namibia. In moist shallow soil on granite outcrops or in peaty pans; 1000–1700 m.

Conservation notes: Widespread; not threatened. Least Concern in F.T.E.A. (2010).

This taxon is close to *Kyllinga bulbosa* but distinct in the number of florets and leaf width. Hooper in F.W.T.A., ed.2 **3**: 304 (1968) has *K. albiceps* as a synonym of *K. bulbosa*.

4. **Kyllinga albogracilis** Lye in Nordic J. Bot. **1**: 742 (1982). —Beentje in F.T.E.A., Cyperaceae: 334 (2010). Type: Zambia, Mbala Dist., Old Mpulungu Road below Venning's Farm, 22.i.1955, *Richards* 4223 (K holotype).

   Cyperus albogracilis (Lye) Lye in Haines & Lye, Sedges & Rushes E. Afr., app.3: 2 and main work: 230, fig.465 (1983).

Slender perennial to 42 cm tall; short white rhizomes or stolons 2 mm thick, covered in pink-tinged scales. Culms solitary, arising from end of stolon or from several nodes, 5–40 cm long, 0.5–0.8 mm wide, trigonous, glabrous. Leaves up to 23 cm long; leaf sheath greyish green to pale reddish-brown, 6 cm long; leaf blade linear, flat or folded, 2–23 cm long, 1–2 mm wide, scabrid on margin and primary vein. Involucral bracts leaf-like, reflexed, 2–3, lowermost 3.5–18 cm long. Inflorescence white, ± spherical, 5–10 mm wide, of a single spike; spikelets lanceolate, 2.5–4.3 × 1–1.8 mm wide, flattened, mostly 2-flowered, but usually only one nutlet. Glumes whitish, ovate, 2–3.5 mm long, keel green in lower half, winged and minutely toothed in lower part, 2–3 veins on either side, apex acute. Stamens ?2, 0.9–1.4 mm long; filaments to 3 mm. Nutlet brown to almost blackish, ellipsoid, 1.7–2 × 0.6–1 mm, minutely papillose.

Zambia. N: Mporokoso Dist., Kipoma Falls, 3.2 km W of Mporokoso, 23.xii.1967, *Simon & Williamson* 1521 (K, SRGH). W: Ndola Dist., Ndola, 1.i.1960, *Robinson* 3269 (K).

Also in Tanzania. Miombo woodland on sandy soil, sometimes in moist sites; 1300–1800 m.

Conservation notes: Data Deficient, with less than 10 specimens and a restricted distribution. Least Concern in F.T.E.A. (2010).

5. **Kyllinga aureovillosa** Lye in Nordic J. Bot. **1**: 743 (1982). Type: Malawi, Nyika Plateau, Lake Kaulime, 4.i.1959, *Robinson* 3054 (K holotype).

Cyperus aureovillosus (Lye) Lye in Haines & Lye, Sedges & Rushes E. Afr., app.3: 2 (1983).

Perennial with short creeping rhizome 2–3 mm thick. Culms 15–50 cm long, triangular, glabrous or with scattered hairs in lower part, densely hairy just below inflorescence. Leaves from lower 4–7 cm only; sheaths green and tomentose on side of leaf, red-brown and almost glabrous on other side; blades 20–60 × 3–4 mm, pubescent to tomentose on both surfaces. Involucral bracts 2–3, reflexed or spreading, largest 10–29 × 2–3.5 mm. Inflorescence a globose golden or brown-green head 7–10 mm across; spikelets narrowly ellipsoid and pointed, 4–5 × 1–1.5 mm, 2–3-flowered. Glumes yellow to light brown, ovate-lanceolate, 3–4 mm long, with ± 3 veins on each side of slightly excurrent and hairy midrib. Stamens 3, 1.5–1.6 mm long. Stigmas 3. Nutlet black, 2–2.3 × 0.9 × 0.5 mm.

Malawi. N: Rumphi Dist., Nyika Plateau, 13 km N of M1, 23.xii.1977, *Pawek* 13311 (K, MAL, MO, SRGH).

Not known elsewhere. Upland grassland and miombo woodland; 1650–2400 m.

Conservation notes: Endemic to the Nyika Plateau and known from only 3 specimens; possibly threatened.

A similar but different specimen with longer, much less hairy leaves, white heads and a non-hairy glume keel, is recorded from the Zambian side of the Nyika Plateau (*Robinson* 2992 (K)).

6. **Kyllinga brevifolia** Rottb., Descr. Icon. Rar. Pl.: 13, t.4.3 (1773). —Clarke in F.T.A. **8**: 273 (1901). —Hooper in F.W.T.A., ed.2 **3**(2): 307 (1968). —Gordon-Gray in Strelitzia **2**: 113 (1995). —Beentje in F.T.E.A., Cyperaceae: 313, fig.47 1–4 (2010). Type: India, no locality, n.d., *König* s.n. (C holotype, BM, STB).

Cyperus brevifolius (Rottb.) Hassk., Cat. Hort. Bot. Bogor.: 24 (1844). —Kükenthal in Engler, Pflanzenr. **4**, 20(101): 600 (1936). —Haines & Lye, Sedges & Rushes E. Afr.: 236 (1983).

Kyllinga intricata Cherm. in Bull. Mus. Natl. Hist. Nat. **25**: 211 (1919). —Podlech in Merxmüller, Prodr. Fl. SW Afr. **165**: 29 (1967). —Getliffe in J. S. African Bot. **49**: 292, fig.14 (1983). —Gordon-Gray in Strelitzia **2**: 115 (1995). —Mapaura & Timberlake, Checkl. Zimbabwe Pl.: 88 (2004). Types: Madagascar, Imerina, xii.1880, *Hildebrandt* 3788 (K syntype, JE, P) is the only one Getliffe mentions, but the protologue also cites *Hildebrandt* 3740, *Prudhomme* 88, *Perrier* 2685, *d'Aleizette* 233, *Viguier & Humbert* 906, 1126 & 1740, *Goudot* s.n., *Campenon* s.n. & *Le Myre de Viler*s s.n., all from Madagascar (P syntypes).

Kyllinga erecta Schumach. var. *intricata* C.B. Clarke in Durand & Schinz, Consp. Fl. Afr. **5**: 529 (1894), invalid name.

Cyperus erectus (Schumach.) Mattf. & Kük. var. *intricatus* (Cherm.) Kük. in Engler, Pflanzenr. **4**, 20(101): 590 (1936).

Kyllinga colorata sensu Napper in J. E. Africa Nat. Hist. Soc. Natl. Mus. **28**(124): 20 (1971), non (L.) Druce.

Kyllinga aurata sensu Napper in J. E. Africa Nat. Hist. Soc. Natl. Mus. **28**(124): 21 (1971), non Nees.

Cyperus brevifolius subsp. *intricatus* (Cherm.) Lye in Haines & Lye, Sedges & Rushes E. Afr., app.3: 2 and main work: 236, fig.475 (1983).

Perennial with long, thin (1–2.5 mm) white or greenish-white creeping rhizome. Culms solitary and spaced along rhizome, 5–55 cm long, 0.5–1.2 mm wide, triangular, glabrous. Leaves up to 15 cm long; leaf sheath reddish to purplish, 1–7 cm long; leaf blade linear, grooved along midrib, 3–10(15) cm long, 1–2 mm wide, scabrid on margins and midrib. Involucral bracts leaf-like, 2–3, longest erect, others spreading or reflexed, lowermost 1.5–6.5 cm long. Inflorescence a greenish yellow, yellowish to dark brown or blackish small globose or ovoid head 4–10 × 4–8 mm; spikelets in a single spike, many and dense, narrowly ovoid, 2–3.5 × 0.6–1.3 mm, 1–2-flowered (when 2-flowered, lowermost is a ♂ with reduced stamens). Glumes yellow-green to golden or orange with green keel, ovate, 2–3.7 mm long, keel green, apex acuminate. Stamens 1–3; filaments to 2.6 mm, anthers 1–1.3 mm long. Nutlet black when mature, broadly (ob-)ovoid, 1.1–1.4 × 0.7–0.9 mm, minutely papillose.

Botswana. N: Ngamiland Dist., Samoqoma lediba, 22.x.1979, *P.A. Smith* 2861 (K, SRGH). **Zambia**. N: Mbala Dist., Mbala (Abercorn), 16.xi.1963, *Brown* 596b (EA, K). W: Mwinilunga Dist., Mwinilunga road between Kanyama turnoff and Samuteba, 25.ii.1975, *Hooper & Townsend* 388 (K). S: Namwala Dist., Namwala, 8.i.1957, *Robinson* 2037 (K, NRGH, NU, SRGH). **Zimbabwe**. W: Matobo Dist., Besna Kobila, xii.1955, *Miller* 2581 (K, SRGH). C: Gweru Dist., 13 km S of Gweru (Gwelo), 29.xii.1966, *Biegel* 1607 (K, SRGH). E: Nyanga Dist., Nyanga Nat. Park, Mare R., 21.x.1946, *Wild* 1439 (K). S: Masvingo Dist., Makoholi Expt. Station, 10.iii.1978, *Senderayi* 166 (K, SRGH). **Malawi**. N: Karonga Dist., Karonga, lakeside, 10.i.1959, *Robinson* 3138 (K). S: Nsessi R., "sand hills near N. Nyassa", uncertain locality, 5.ii.1896, *Scott* s.n. (K). **Mozambique**. M: Moamba Dist., Rikatla, xi.1917, *Junod* 69 (K, LISC, PRE).

Widespread in tropical Africa and South Africa; also in Indian Ocean Islands, S Asia, Australia and the Americas. Seasonally swampy grassland in floodplains or on riverbanks, dambos, lakeshores; can be locally common; 50–1900 m.

Conservation notes: Widespread; not threatened. Least Concern in F.T.E.A. (2010).

I have been unable to separate material on whether the involucral bracts are "shaped like a cross" (with one erect and two spreading, as *Kyllinga intricata*) or "more flaccid" (as in *K. brevifolia*); this is not easy in dried and arranged herbarium material. The colour differences between these two also seem more gradual than abrupt – 'paler' greenish or straw-coloured, or golden yellow. The protologue of *K. brevifolia* does not specify a colour of the inflorescence, apart from "flores grysei cum margine utroque viridissimus" (grey with green margins), which does not really help. Subspp. *brevifolia* and *intricata* were distinguished on greenish to straw-coloured and golden yellow inflorescences, respectively. I have merged these two in the F.T.E.A. and F.Z. treatments, but a final decision needs to be made on a world-wide basis.

A specimen from N Zambia (Luapula, Mbereshi, 12.i.1960, *Richards* 12343) is close to this taxon, but has a rather long lowermost involucral bract (to 7.5 cm) and very small heads (3–3.5 mm across, with very few spikelets). This is either a depauperate *Kyllinga brevifolia* or something new.

7. **Kyllinga bulbosa** P. Beauv., Fl. Oware **1**: 11, t.8.1 (1805). —Hooper in F.W.T.A., ed.2 **3**(2): 304 (1968). —Napper in J. E. Africa Nat. Hist. Soc. Natl. Mus. **28**(124): 24 (1971). —Beentje in F.T.E.A., Cyperaceae: 336 (2010). Types: Nigeria, Chama, n.d., *Palisot de Beauvois* s.n. (G syntype); "Oware et Benin", n.d., *Palisot de Beauvois* s.n. (G syntype).

    Kyllinga macrocephala A. Rich., Tent. Fl. Abyss. **2**: 491 (1850). —Clarke in F.T.A. **8**: 286 (1901). Type: Ethiopia, Tacazze, Tchélatchékanné, 1844, *Quartin Dillon* s.n. (P holotype).

    Cyperus richardii Steud., Syn. Pl. Glumac. **2**: 8 (1854). —Kükenthal in Engler, Pflanzenr. **4**, 20(101): 568 (1936). —Haines & Lye, Sedges & Rushes E. Afr.: 227, fig.457 (1983), new name for *Kyllinga macrocephala*.

    Kyllinga sphaerocephala Boeckeler in Flora **58**: 258 (1875). —Clarke in F.T.A. **8**: 274 (1901). Type: Zanzibar, xi.1863, *Speke & Grant* s.n. (K holotype).

    Kyllinga macrocephala var. *angustior* C.B. Clarke in Durand & Schinz, Consp. Fl. Afr. **5**: 529 (1894), invalid name; in Durand & Schinz, Etud. Fl. Congo **1**: 279 (1896). Type: D.R. Congo, Stanley-Pool, viii.1888, *Hens* 14 (BR holotype, K).

    Cyperus richardii var. *angustior* (C.B. Clarke) Kük. in Engler, Pflanzenr. **4**, 20(101): 568 (1936).

    Cyperus purpureoglandulosus Mattf. & Kük. in Engler, Pflanzenr. **4**, 20(101): 570 (1936), new name for *Kyllinga sphaerocephala*. —Haines & Lye, Sedges & Rushes E. Afr.: 228, fig.458 (1983).

Perennial 5–40 cm tall; stolons or rhizomes long, slender, whitish, 0.5–2 mm in diameter, covered at first with delicate sheaths, soon rotting leaving few short fibres to mark nodes with few roots from each node. Culms solitary, swollen at base, 5–40 cm long, 0.7–1.5 mm wide, trigonous, ridged, glabrous or with a few spine-like hairs below inflorescence. Leaves to 30 cm long, basal sheaths without blades; leaf sheath pale brown, 1–8 cm long, older ones darkening, covering base of culm; leaf blade linear, ± flat, 4–25 cm long, 2–5 mm wide, scabrid on margins and midrib, especially above. Involucral bracts leaf-like, spreading, 2–5, lowermost 2–10 cm long. Inflorescence a dense irregular or spherical to ovoid white head of one to several spikes, when several-spiked often triangular in outline, 5–15 × 5–17 mm (head may appear yellow at anthesis due to many stamens); spikelets narrowly ovoid, 2.5–4.5 × 1–1.7 mm, with 1–5 flowers per spikelet, usually 3 nutlets developing. Glumes whitish, occasionally with brownish dots, ovate, largest 2.5–3 mm long, keel green, with 2–4 veins on either side, apex acute. Stamens 2–3, filaments 2.4–3 mm, anthers 0.9–1.8 mm long. Nutlet pale-coloured to dark brown, ellipsoid to obovoid, 1–1.6 × 0.4–0.7 mm, minutely papillose.

Zambia. N: Mbala Dist., Mpulungu, 12.ii.1957, *Richards* 8183 (K). W: Mwinilunga Dist., no locality, n.d., *Milne-Redhead* 3601 (K). C: Mpika Dist., South Luangwa Nat. Park, Mfuwe, fl. 6.i.1970, *Astle* 5734 (K). S: Hwange Dist., Victoria Falls, 5.i.1955, *Robinson* 1429 (K). **Zimbabwe**. W: Matobo Dist., Matopos Nat. Park, Maleme Valley, 6.i.1963, *Wild* 5923 (K, SRGH). C: Chegutu Dist., Chegutu (Hartley), Poole Farm, 20.xii.1944, *Hornby* 2396 (K, SRGH). E: Chimanimani Dist., Rusitu R., 2 km W of Haroni confluence, 12.i.1969, *Biegel* 2804 (K, SRGH). **Malawi**. N: Karonga Dist., Karonga, 10.i.1959, *Robinson* 3133 (K). C: Nkhotakota Dist., Nkhotakota, near airfield, 16.vi.1970, *Brummitt* 11478 (K, LISC). S: Chikwawa Dist., Lengwe Nat. Park, 15.xii.1970, *Hall-Martin* 1158 (K).

Also in western, eastern and central Africa down to Zimbabwe. In grassland in damp sites, a weed in lawns, roadsides; 90–1700 m.

Conservation notes: Widespread; not threatened. Least Concern in F.T.E.A. (2010).

Haines & Lye (1983) felt *Kyllinga bulbosa* was very similar to *K. purpureoglandulosus* but "has more than one spike per inflorescence, thinner stolons and larger spikelets with more flowers; .... atypical plants of *purpureo-glandulosus* with several spikes can be identified by its spikelet producing one nutlet only". Napper thought *K. sphaerocephala* was a synonym, as do I.

8. **Kyllinga cartilaginea** K. Schum. in Engler, Pflanzenw. Ost-Afrikas **C**: 123 (1895). —Napper in J. E. Africa Nat. Hist. Soc. Natl. Mus. **28**(124): 20 (1971). —Beentje in F.T.E.A., Cyperaceae: 330 (2010). Type: Tanzania, Tanga, 31.i.1893, *Holst* 2082 (B holotype, K).

Cyperus cartilagineus (K. Schum.) Mattf. & Kük. in Engler, Pflanzenr. **4**, 20(101): 608 (1936). —Haines & Lye, Sedges & Rushes E. Afr.: 246, fig.493 (1983).

Perennial 15–60 cm tall; scale-covered stolons or rhizomes 3–5 mm thick, roots smelling of cinnamon or cough medicine. Culms spaced or dense, 15–50(73) cm long, 1.2–3 mm wide, triangular, glabrous. Leaves up to 55 cm long; lower leaf sheaths purple to red, 1–12 cm long; leaf blade linear, keeled, 20–55 cm long, 1.5–5 mm wide, scabrid on margins and midrib. Involucral bracts leaf-like, spreading, 4–5, lowermost 10–30(45) cm long. Inflorescence a sessile white globose or ovoid head, 8–18 × 9–14 mm; spikelets many and densely packed, narrowly ovoid, 4–5.8 × 1.1–1.3 mm, 2-flowered. Glumes white or off-white, ovate, 3.3–5 mm long, apex acute (lower) or almost tubular and obtuse (upper), keel with a few to many minute teeth. Stamens 3, anthers yellow, 2–2.5 mm long. Nutlet blackish, slighly obovoid, 1.8–2 × 0.7–0.8 mm, minutely papillose, containing oil (liquid even 50 years after collection).

Mozambique. MS: Marromeu Dist., "Zambesi", Kongone R. mouth, i.1861, *Kirk* s.n. (K).

Kenya, Tanzania and Madagascar. No habitat data given, but in East Africa found in lowland forest, coconut groves, beach crest, dunes, *Brachystegia* woodland, occasionally more inland in riverine situations; near sea level (to 200 m in East Africa).

Conservation notes: Widespread; not threatened. Least Concern in F.T.E.A. (2010).

9. **Kyllinga crassipes** Boeckeler in Flora **42**: 444 (1859); in Peters, Naturw. Reise Mossambique: 534 (1864), as *Kyllingia*. —Schumann in Engler, Pflanzenw. Ost-Afrikas **C**: 123 (1895). —Clarke in F.T.A. **8**: 275 (1901). —Napper in J. E. Africa Nat. Hist. Soc. Natl. Mus. **28**(124): 22 (1971). —Beentje in F.T.E.A., Cyperaceae: 342 (2010). Type: Tanzania, Zanzibar, ix.1843, *Peters* 23 (B holotype).

Cyperus bulbipes Mattf. & Kük. in Engler, Pflanzenr. **4**, 20(101): 587 (1936), new name for *Kyllinga crassipes*. —Lye in Haines & Lye, Sedges & Rushes E. Afr.: 247, fig.496 (1983).

Cyperus bulbipes var. *pallescens* Kük. in Engler, Pflanzenr. **4**, 20(101): 588 (1936). Types: Tanzania, many syntypes: *von Brehmer* 124, *Engler* 334, *Holst* 2018, *Peter* 3861, 4309, 6948, 31524, 31843 & 46565, *Prittwitz* 217, *Volkens* 67 (all B syntypes, only Peter ones seen).

Perennial; short creeping rhizome with faint sweet smell. Culms densely crowded along rhizome, 10–94 cm long, 0.8–1.5 mm wide, sharply triangular, glabrous or with a few hairs just below head, base slightly bulbous. Leaves up to 38 cm long; leaf sheath pinkish to purple-brown, 1.5–8 cm long; leaf blades several per culm, very short on basal sheaths, larger higher up, green and often with minute reddish dots, linear, slightly channeled, 10–38 cm long, 1.5–4.3 mm wide, scabrid on margins and midrib. Involucral bracts leaf-like, spreading or reflexed, 3–4, lowermost (3)7–16(30) cm long. Inflorescence a single pure white globose head 4–10 mm across, of one spike; spikelets many, narrowly ovoid, 1.8–3.8 × 0.5–0.7 mm, 2-flowered, the lower hermaphrodite, upper male. Glumes very pale brown with minute reddish dots, narrowly ovate, 1.8–3.3 mm long, smooth, apex acute to obtuse, several veins on each side. Stamens 3, filaments to 2 mm long, anthers yellow or white, 0.8–1.1 mm long. Nutlet dark grey, ellipsoid, 1.5–1.7 × 0.5–0.9 mm, minutely papillose.

Zambia. N: Kasama Dist., Chilubula, 12.ii.1961, *Robinson* 4380 (K). W: Ndola Dist., 9.i.1960, *Robinson* 3289 (K). C: Chongwe Dist., Chakwenga headwaters, 115 km E of Lusaka, 14.ii.1965, *Robinson* 6376 (K). S: Namwala Dist., Bambwe, 17.iv.1963, *van Rensburg* 2025 (K). **Zimbabwe**. E: Chimanimani Dist., Rusitu R., 2 km S of Haroni confluence, 12.i.1969, *Biegel* 2807 (K, SRGH). **Malawi**. N: Chitipa Dist., Nyika Plateau, N of Nganda, 3.iv.1997, *Patel et al.* 5069 (K, NHBG). C: Dedza Dist., Mua-Livulezi Forest Reserve, 9.iii.1966, *Banda* 809 (K, SRGH). S: Mulanje Dist., Mt Mulanje tea factory, 19.ii.1986, *Chapman & Chapman* 8304 (K, MO). **Mozambique**. N: Marrupa Dist., 20 km on Lichinga road from Marrupa, Missor area, 16.ii.1981, *Nuvunga* 529 (K, LISC, LMU). Z: Mopeia Dist., Cundine, 20.xii.1904, *Le Testu* 559 (K, P). MS: Báruè Dist., near Zimbabwe border opposite Gairesi Ranch, 19.xi.1956, *Robinson* 1949 (K).

Also in South Sudan, Uganda, Kenya, Tanzania, D.R. Congo and Angola. Seasonally wet grassland in miombo, plantations, clearings; 50–2500 m.

Conservation notes: Widespread; not threatened. Least Concern in F.T.E.A. (2010).

Often confused with *Kyllinga bulbosa* but distinct in the close-set culms on the rhizome; *K. bulbosa* has the culms distant on stolons.

10. **Kyllinga erecta** Schumach., Beskr. Guin. Pl.: 42 (1827). —Clarke in F.T.A. **8**: 274 (1901). —Hooper in F.W.T.A., ed.2 **3**(2): 307, fig.408 (1968). —Napper in J. E. Africa Nat. Hist. Soc. Natl. Mus. **28**(124): 22 (1971). —Gordon-Gray in Strelitzia **2**: 115 (1995). —Mapaura & Timberlake, Checkl. Zimbabwe Pl.: 88 (2004) as subsp. *erecta*. —Beentje in F.T.E.A., Cyperaceae: 322, fig.48 (2010). Type: Guinea, no locality, n.d., *Thonning* s.n. (C holotype).

Kyllinga aurata Nees in Linnaea **10**: 139 (1835). Types: South Africa, Zwartkopsrivier near Pauli Maré, no further details, possibly *Ecklon & Zeyher* s.n., 1803 (JE syntype); E Cape, mountains of van Stadensrivier / near Grahamstown, no further details; E Cape, Uitenhage, possibly *Zeyher* 30 (HAL syntype). Note: The protologue only gives localities without dates or collectors; choosing a lectotype should be done in South Africa.

Kyllinga erecta var. *intercedens* Kük. in Repert. Spec. Nov. Regni Veg. **12**: 91 (1913) as *Kyllingia*. Type: Malawi, Mt Zomba, 1200–1830 m, xii.1896, *Whyte* s.n. (K holotype).

Cyperus erectus (Schumach.) Mattf. & Kük. in Engler, Pflanzenr. **4**, 20(101): 588 (1936).
—Haines & Lye, Sedges & Rushes E. Afr.: 238, fig.478 (1983).
Cyperus erectus var. *auratus* (Nees) Kük. in Engler, Pflanzenr. **4**, 20(101): 589 (1936).
Cyperus erectus var. *intercedens* Kük. in Engler, Pflanzenr. **4**, 20(101): 589 (1936).
Kyllinga erecta subsp. *albescens* Lye in Nordic J. Bot. **1**: 745 (1982). Type: D.R. Congo,
Popokaba territory, 13.v.1959, *Pauwels* 3001 (BR holotype).
Cyperus erectus subsp. *albescens* (Lye) Lye in Haines & Lye, Sedges & Rushes E. Afr., app.3:
2 and main work: 238, fig.479 (1983).

Perennial with long creeping fleshy rhizome 2–5 mm thick, covered in large reddish or pink
scales. Culms single but densely set in a single row along rhizome, bases slightly bulbous (not
always clearly visible), 12–75 cm long, 0.8–2 mm wide, triangular, glabrous. Leaves up to 20
cm long; leaf sheath purplish red, basal ones without leaves, 0.5–9 cm long; leaf blade (rarely
completely absent) linear, flat or channeled, 2–20 cm long, 2–3.2 mm wide, scabrid on margins
and midrib. Involucral bracts leaf-like, spreading, 3–4, lowermost 2–7(11) cm long. Inflorescence
a solitary ovoid or subglobose greenish yellow head 5–12 × 5–9 mm; spikelets many, narrowly
ovoid, (1.8)2.5–3.5 × 0.8–1.5 mm, 1–2-flowered. Glumes golden yellow with green keel, 2–3.5
mm long, smooth, apex acuminate and somewhat recurved, 3–5 veins on each side. Stamens 3,
filaments to 1.8 mm long, anthers yellow, 1–1.2 mm long, apiculate. Nutlet 1, dark grey to dark
brown, ellipsoid, 1.1–1.4 × 0.6–0.7 mm, minutely papillose.

Botswana. N: Ngamiland Dist., Okavango R., 24.43°S 21.52°E, 17.ii.1979, *P.A. Smith*
2679 (K, MN, PRE, SRGH). SE: South East Dist., Kgale, 26.xii.1977, *Hansen* 3316 (BM,
C, GAB, K, PRE, SRGH, WAG). **Zambia**. B: Kaoma Dist., 15 km on Kaoma–Kasempa
road, 28.ii.1996, *Harder, Zimba & Luwiika* 3583 (K, MO). N: Nchelengwe Dist., E of
Lake Mweru, vi.1961, *Chongo* 8 (K). W: Ndola, 3.i.1960, *Robinson* 3280 (K). C: Serenje
Dist., Kundalila Falls, S of Kanona, 1300 m, 13.iii.1975, *Hooper & Townsend* 696 (K). E:
Petauke Dist., Great East Road between Nyimba & Petauke, 14.xii.1958, *Robinson* 949
(K). S: Namwala Dist., 3 km from Namwala, Kafue Nat. Park, 11.xii.1962, *van Rensburg*
1069 (K). **Zimbabwe**. N: Mutoko Dist., Mutoko (Mtoko), 160 km ENE of Harare,
6.xii.1946, *Weber* 288 (K). W: Hwange Dist., Victoria Falls 'rain forest', 4.iv.1956,
Robinson 1419 (K). C: Harare Dist., Lake Chivero (MacIlwaine), 13.x.1956, *Robinson*
1812 (EAR, IRLOS, K, LISC, NRGH, NU, SRGH). E: Chipinge Dist., Gungunyana
Forest Reserve v.1967, *Goldsmith* 68/67 (K, SRGH). **Malawi**. N: Mzimba Dist., Mzuzu,
Marymount, 11.iii.1973, *Pawek* 6489 (K, MAL). C: Lilongwe Dist., Dzalanyama Forest
Reserve, 7 km S of Forest Station, 5.iv.1989, *Iversen & Martinsson* 89/152 (K, UPS).
S: Mangochi Dist., Mangochi Mt, 30.i.1979, *Blackmore, Banda & Patel* 255 (K, MAL).
Mozambique. MS: Manica Dist., Bandula, 6.iv.1952, *Chase* 4537 (K, SRGH).

Widespread in tropical and South Africa, Mauritius, Madagascar. Seasonally flooded
grassland or sandy sites in miombo zone, dambos and lawns, may form almost pure
stands; 450–1800 m.

Conservation notes: Widespread; not threatened. Least Concern in F.T.E.A. (2010).

Lye separated his subsp. *albescens* on "fewer and shorter leaf blades and involucral
bracts as well as whitish glumes and spikelets". As regards the leaf size there is plenty
of variation, none of it discontinuous. Unlike in East Africa, I have seen several white-
headed specimens of this species from Malawi and Zambia.

11. **Kyllinga melanosperma** Nees in Wight, Contr. Bot. India: 91 (1834). —Clarke in
F.T.A. **8**: 277 (1901). —Hooper in F.W.T.A., ed.2 **3**(2): 307 (1968). —Napper in J.
E. Africa Nat. Hist. Soc. Natl. Mus. **28**(124): 22 (1971). —Getliffe in J. S. African
Bot. **49**: 296, fig.16 (1983). —Gordon-Gray in Strelitzia **2**: 115, fig.47 (1995). —
Beentje in F.T.E.A., Cyperaceae: 323, fig.49 (2010). Type: India, "Peninsula Ind.
orientalis", no locality, vii.1851, *Wight* 1851 (K holotype, E, NY). FIGURE 14.**53**.

Cyperus melanospermus (Nees) Valck. Sur., Cyperus Mal. Arch.: 50, t.2, fig.8 (1898). —
Kükenthal in Engler, Pflanzenr. **4**, 20(101): 583 (1936). —Haines & Lye, Sedges & Rushes
E. Afr.: 240 (1983).

Kyllinga imerinensis Cherm. in Bull. Mus. Natl. Hist. Nat. **25**: 210 (1919); in Humbert, Fl.
Madagascar, Cyperaceae: 13 (1937). Types: Madagascar, Manankazo, *Perrier de la Bathie* 2714
(P synonym); Vohimalaza Peak, *Viguier & Humbert* 1370 (P synonym), syn. nov.

Fig. 14.**53**. KYLLINGA MELANOSPERMA. 1, habit; 2, culm and inflorescence; 3, spikelet; 4,
part of spikelet opened out; 5, nutlet. All from *Browning* 240. Scale bars: 1, 2 = 40 mm; 3–5 = 1
mm. Drawn by Jane Browning. Reproduced from Strelitzia (1995) by kind permission of the
South African National Biodiversity Institute, Pretoria.

Var. **melanosperma**

Perennial with long creeping branched rhizome 3–8 mm across covered with short red-brown scales. Culms solitary from each node but set close, 12–100 cm long, 1.5–3 mm wide, often with a bulbous base, usually triangular, glabrous. Leaves few per culm, or one from uppermost sheath, or hardly any blade developed; leaf sheath reddish or purple, 1.5–17 cm long; leaf blade linear, flat, 2–17 cm long, 3–4 mm wide, scabrid on margins and midrib, apex acute. Involucral bracts leaf-like, usually reflexed, 3–7, largest 2–4 almost equal in size, lowermost 2.5–17 cm long, 2–7 mm wide. Inflorescence a green to golden yellow globose to ovoid head of a single spike, 7–15 × 6–11 mm; spikelets many, sessile, narrowly ovoid, 3–4 mm long (Getliffe says to 5.5 mm), 0.7–1.5 mm wide, (1)2-flowered. Glumes green to golden, ovate, 2.6–4 mm long, keel green and rarely with few small teeth, apex long-acuminate. Stamens 3, anthers 1.3–1.4 mm long. Nutlets 1(2), dark, almost black, obovoid, 1.1–1.3 × 0.5–0.8 mm, minutely papillose, apparently very few developing.

Zambia. N: Mbala Dist., Kali dambo, 15.i.1955, *Richards* 4125 (K). W: Mwinilunga Dist., Lunga R., 1.xii.1937, *Milne-Redhead* 3471 (K). **Zimbabwe**. N: Guruve Dist., Nyamunyeche (Nyamyetsi) Estate, 5.ix.1978, *Nyariri* 327 (K, SRGH). C: Harare (Salisbury), Prince Edward dam, 19.x.1956, *Robinson* 1815 (EAR, IRLOS, K, NRGH, NU, SRGH). E: Chimanimani Dist., Nyahodi R., 4.i.1995, *Wilkin* 756 (K). **Malawi**. N: Mzimba Dist., Mzimba R., 26.ii.1959, *Robinson* 1724 (BM, K, LISC). S: Chikwarwa Dist., Maperere Mission, 2.xi.1950, *Jackson* 246 (K). **Mozambique**. MS: Gorongosa Dist., Gorongosa Nat. Park (Game Reserve), W slopes of Cheringoma Plateau, 12.xi.1971, *Ward* 7444 (K, NH).

From Nigeria to D.R. Congo and south to the Cape; also in Madagascar, India, Sri Lanka, China and Malesia. Seasonally swampy grassland, lake and streamside grassland and dambos, forming dense stands near water; 100–1700 m.

Conservation notes: Widespread; not threatened. Least Concern in F.T.E.A. (2010).

Kyllinga imerinensis was said by Chermezon to differ in its less robust habit, smaller spikelets and golden-ferruginous glumes which are long-mucronate. In fact all these characters fall comfortably in the variability of *K. melanosperma* and I see no reason to uphold this taxon. There is a var. *hexalata* in Uganda with 6-angular stems.

12. **Kyllinga mtotomwema** Beentje, sp. nov. Related to *Kyllinga tanzaniae* and *K. songeensis* but lacks the obvious rhizome. It usually has more than one spike, with shorter involucral bracts, and occurs at much higher altitudes. Type: Malawi, Nyika Plateau, Chelinda Bridge, 3.i.1977, *Pawek* 12249 (K holotype, MO, MAL).

Perennial up to 38 cm tall; rhizomes short. Culms solitary or tightly spaced on rhizome, (8)18–38 cm long, 0.5–1.3 mm wide, rounded to subtriangular, scabridulous near apex; bases swollen, surrounded by dense old leaf fibres. Leaves to 13 cm long; leaf sheath light green to brownish, 1–4.5 cm long; leaf blade linear, 3–13 cm long, 1–4 mm wide, scabridulous near apex. Involucral bracts leaf-like, spreading, 2–3(4), lowermost 2–3.3 cm long, scabridulous near base. Inflorescence with light green to dirty white central spike, 5–9 × 6–8 mm, often with 2 smaller lateral spikes; spikelets many, narrowly ovoid, 2.9–3.9 mm long, 1-flowered. Glumes white, sometimes with small brown spots, ovate, 2.5–3.8 mm long, keel green and smooth, apex acute to apiculate. Stamens 2 (often no trace); anthers 0.6–0.8 mm long. Nutlet mid-brown, ovoid-oblong, 1.9–2 mm long, papillose.

Zambia. E: Isoka Dist., Mafinga Hills, 11.iii.1966, *Robinson* 2992 (K). **Malawi**. N: Chitipa Dist., 18 km SSE of Chisenga, 21.i.1989, *Thompson & Rawlins* 6088 (CM, K).

Not known elsewhere. Hilltop grassland and *Protea* scrubland; 1850–2300 m.

Conservation notes: Endemic to the Mafinga Mts and W Nyika plateau on the border of E Zambia and NW Malawi; possibly threatened.

I have named this species after Ndugu Kilian Mtotomwema, a Tanzanian sedge specialist with an interest in *Kyllinga* who died in 1992. Mtotomwema separated most of the specimens of this taxon under the manuscript name *K. mafingaensis*.

13. **Kyllinga nemoralis** (J.R. Forst. & G. Forst.) Hutch. & Dalziel in F.W.T.A. **2**: 487 (1936). —Hooper in F.W.T.A., ed.2 **3**(2): 307 (1968). —Napper in J. E. Africa Nat. Hist. Soc. Natl. Mus. **28**(124): 20 (1971). —Gordon-Gray in Strelitzia **2**: 117 (1995). —Beentje in F.T.E.A., Cyperaceae: 335 (2010). Type: no type indicated (see note); lectotype t.65 in J.R. & G. Forster, Char. Gen. Pl. (iconotype).

Thryocephalon nemorale J.R. Forst. & G. Forst., Char. Gen. Pl. ed.2: 130, t.65 (1776).
Cyperus kyllingia Endl. in Cat. Horti Vindob. **1**: 94 (1842). —Kükenthal in Engler, Pflanzenr. **4**, 20(101): 606 (1936). —Haines & Lye, Sedges & Rushes E. Afr.: 247, fig.497 (1983).

Perennial to 30 cm tall; stolons fleshy (called rhizomes by Napper), 1–10 cm long, 2 mm across, white and fragrant when fresh, with pink-brown scales and thin roots. Culms rather spaced along rhizome, sometimes dense, 8–24 cm long, 0.8–1.5 mm wide, triangular, glabrous. Leaves to 21 cm long; leaf sheath pale to mid-brown, 1–6 cm long; leaf blade dark green, linear, flat, 10–21 cm long, 2–5 mm wide, scabrid on margins and midrib. Involucral bracts leaf-like, spreading or reflexed, 3–4, lowermost 8–18 cm long. Inflorescence a globose or ovoid sessile head of a single spike (rarely with 1–2 smaller spikes) 3–8 mm across; spikelets many, narrowly ovoid, 2–2.5 (Getliffe says to 5 mm) × 0.6–1 mm, 1–2-flowered, sometimes viviparous. Glumes whitish, fading to pale red-brown, 2–2.5 mm long, keel winged (sometimes not very prominent), with minute teeth, apex acuminate, 2–4 veins each side of midrib. Stamens 3. Nutlet 1, red-brown turning black, oblong, 1–1.3 × 0.7–0.9 mm, minutely papillose.

Malawi. S: or **Mozambique**. T: "Shire & Shupanga, Zambezi", pre-1869, *Stewart* s.n. (K).

Guinée to Cameroon, Uganda, Tanzania, D.R. Congo and South Africa; also in Madagascar, Indian Ocean Is., India, Sri Lanka and SE Asia. No habitat indicated, but elsewhere in forest with open canopy or in forest clearings; c. 50 m.

Conservation notes: Widespread outside the Flora area; not threatened. Least Concern in F.T.E.A. (2010).

The Forsters' protologue has no details about the species at all, no specimen or origin mentioned, no description – but the name is validated by the "descriptio generico-specifica" (Code, art. 42.1) as the genus *Thryocephalon* is described by the Forsters and *nemorale* is the only species therein. I therefore chose the plate as the type in F.T.E.A. (2010). Gordon-Gray mentions a collection J.R. & G. Forster from the Society Is., which could serve as an epitype – she does not say in which herbarium.

Haines & Lye (1983) cite both *Kyllinga nemoralis* and *K. monocephala* Rottb. as synonyms of their *Cyperus kyllingia* Endl.; the latter name dates from 1842 and so has no priority. *Kyllinga monocephala* Rottb., however, dates back to 1773 (*K. monocephala* Rottb., Descr. Icon. Pl.: 13, t.4, fig.4 (1773). No type is mentioned in the protologue, which refers to several pre-Linnean names. The plate accompanying the protologue is not very diagnostic, so it does not help either; *K. monocephala* Rottb. will have to remain an unresolved taxon. There are no fewer than six other taxa named *K. monocephala*, although the Rottbøll name is the oldest.

14. **Kyllinga nervosa** Steud. in Flora **25**: 597 (1842). —Clarke in F.T.A. **8**: 279 (1901). —Napper in J. E. Africa Nat. Hist. Soc. Natl. Mus. **28**(124): 20 (1971). —Beentje in F.T.E.A., Cyperaceae: 317 (2010). Type: Ethiopia, Wadi Schoata, 24.vii.1838, *Schimper* 1375 (B† holotype, GOET, K, M, P, TUB, US).

Cyperus costatus Mattf. & Kük. in Engler, Pflanzenr. **4**, 20(101): 575 (1936), replacement name for *K. nervosa*.
Cyperus oblongus (C.B. Clarke) Kük. subsp. *nervosus* (Steud.) Lye in Haines & Lye, Sedges & Rushes E. Afr., app.3: 2 and main work: 235, fig.473 (1983).

Subsp. nervosa

Tufted perennial; rhizome short with dense root mass and swollen stem base, often surrounded by fibres from disintegrating leaf bases; stem base aromatic (smelling of eucalyptus, or ginger).

Culms tufted, 7–46 cm long, 0.8–1.5 mm wide, trigonous, glabrous. Leaves to 35 cm long; leaf sheath pale brown, more reddish near base, 1–6 cm long, lowermost sometimes leafless; leaf blade linear, flat or grooved, 7–35 cm long, 1–2.5 mm wide, scabrid near apex. Involucral bracts leaf-like, spreading or recurved, 1–3(4), lowermost 3–11 cm long. Inflorescence a single yellow-green conical-ovoid to cylindric spike, turning dark brown to black, 5–10(12) × 5–8 mm, rarely with a small subsidiary spike; spikelets many, 2.2–3.5 × 0.9–1 mm, 2–3-flowered. Glumes with yellow margins, green keel, blackish tips, ovate, 1.8–3.5 mm long, apex acuminate and slightly recurved, with distinct veins. Stamens 3, filaments 1.2–2.5 mm long, anthers 1.2 mm. Nutlet violet-black, 1–1.2 × 0.6 mm, minutely papillose.

Zambia. S: Choma Dist., 14.5 km E of Choma, 18.xii.1956, *Robinson* 1991 (IRLCS, K, NU, NRGH, SRGH). **Malawi**. N: Rumphi Dist., Nyika Plateau, E side, 5.i.1959, *Robinson* 3061 (K). S: Zomba, xi–xii.1899, *Cameron* 92 (K).

Also in Eritrea, Ethiopia, Somalia, Uganda, Kenya and Tanzania. Shallow soil over rock, seepage zones, seasonally swampy grassland, especially on black clay soils; 750–2150 m.

Conservation notes: Widespread; not threatened. Least Concern in F.T.E.A. (2010).

The other subspecies, subsp. *oblonga* (C.B. Clarke) J.-P. Lebrun & Stork, is restricted to East Africa.

A specimen from NC Mozambique (Z: Mt Namuli, Malema Valley, 1227 m, in very disturbed miombo woodland, 14.xi.2007, *Harris et al.* 306 (K, LMA)) keys out here but lacks the black apex. I am unable to place this specimen and prefer to leave it as 'cf. *nervosa*'.

Similarly a specimen from S Mozambique (M: Maputo, Jardin Vasco da Gama, 28.iv.1971, *Balsinhas* 1845 (K, LMU)) keys to *Kyllinga nervosa* but lacks the black apex. It is also from near sea level and at completely the wrong altitude, and a weed in a garden. It has smaller heads than the *Harris et al.* specimen mentioned above, but I am unable to place it; it does not resemble any of the South African specimens I have looked at.

Yet another specimen from Mozambique is different from both the above and *Kyllinga nervosa* itself, and different enough to warrant a description:

Kyllinga sp. near **nervosa**

Perennial to 40 cm tall; no rhizome observed. Culms tufted, 20–40 cm long, c. 1mm wide, rounded to triangular, glabrous, bases reddish, slightly swollen, surrounded by dense old leaf fibres; culm groups tightly packed. Leaves to 10 cm long; leaf sheath reddish brown to pale brown, lowermost (sterile) sheaths 7 mm, broadly ovate, a few leaf-bearing sheaths to 9 cm long; leaf blade linear, 1.5–10 cm long, 1.6–2.1 mm wide, glabrous. Involucral bracts leaf-like, 1 ?erect, others spreading, 3–4, lowermost 2.7–5.5 cm long, scabridulous near base. Inflorescence a solitary orange and green hemispherical or ovoid spike, 5–7 × 6–9 mm; spikelets many, narrowly ovoid, 3.5–4.5 mm long, 1–2-flowered. Glumes orange with green keel, ovate, 3.4–4.3 mm long, keel green and smooth, apex apiculate, slightly recurved. Stamens 3, anthers not present, just filaments 3.2–3.8 mm long. Nutlet only seen immature, ovoid-oblong, 1.5–1.9 mm long.

Mozambique. Z: Gurué Dist., Mt Namuli, Muretha Plateau, 19.xi.2007, *Harris et al.* 384 (K, LMA).

15. **Kyllinga nigripes** C.B. Clarke in F.T.A. **8**: 285 (1901). Type: Malawi, no locality, x.1892, *Buchanan* 1428 (K holotype, B).

    Cyperus nigripes (C.B. Clarke) Kük. in Engler, Pflanzenr. **4**, 20(101): 572 (1936).

Perennial to 50 cm tall; rhizome short, with dense almost contiguous tufts, smelling of sandalwood; roots fibrous, wiry. Culms 30–50 cm long, sharply angled in upper part, smooth, thickened at base by torn black basal sheaths 2–3.5 cm long. Leaves to 35 cm long; leaf sheath brownish, 2–8 cm long; leaf blade linear, folded, 7–34 cm long, 2.2–3.3 mm wide, scabrid in upper third. Involucral bracts 3–5, leaf-like, spreading or strongly reflexed, lowermost 5–11 cm long. Inflorescence head of 1 (sub)globose dense white to cream spike, 7–14 mm in diameter; spikelets 4–5.6 mm long, 2-flowered. Glumes elliptic-lanceolate, 4.2–4.8 × 1–1.3 mm, subacute to acuminate, ribbed on each side; keel wingless and smooth above, sometimes with short very contiguous cilia coalescing almost into a wing in lower half. Stamens 3, filaments 1.7–2.5 mm long, anthers 1.1–1.6 mm long. Nutlet black, ellipsoid, 2.1–2.2 × 0.7–0.8 mm.

Botswana. N: Okavango Delta, Shishikola Pan, 1.i.1993, *P.A. Smith* 330 (K, SRGH). **Zambia**. B: Kabompo Dist., 5 km towards Kabompo from Kabombo–Lukulu–Zambezi junction, 23.x.1986, *Linder* 3946 (K, PRE). N: Mbala Dist., Mbala (Abercorn), 31.xii.1958, *Robinson* 2968 (K). C: Chongwe Dist., Chakwenga headwaters, c. 115 km E Lusaka, 10.i.1964, *Robinson* 6175 (K). S: Namwala Dist., Kafue Nat. Park, Ngoma burning trials, 6.xii.1960, *Mitchell* 2/10 (K). **Zimbabwe**. N: Hurungwe Dist., Mensa Pan near Chirundu, 8.i.1958, *Goodier* 545 (K). C: Kadoma Dist., Ngezi Recreational Park, ii.1968, *Brits* 12 (K). E: Chipinge Dist., 1 km W of confluence of Musirizwi and Bwazi rivers, 30.i.1975, *Gibbs Russell* 2678 (K). **Malawi**. N: Malawi (Nyasaland), no locality, x.1892, *Buchanan* 1428 (B, K).

Not known elsewhere. Grassland or mopane woodland on Kalahari sand, sometimes in shallow soil over rock; 500–1600 m.

Conservation notes: Endemic to the Flora area, but widespread; not threatened.

16. **Kyllinga odorata** Vahl, Enum. Pl. **2**: 382 (1805). —Hooper in F.W.T.A., ed.2 **3**(2): 304 (1968). —Gordon-Gray in Strelitzia **2**: 117 (1995). —Beentje in F.T.E.A., Cyperaceae: 340 (2010). Types: 'America meridionalis', *von Rohr* s.n. & *Richard* s.n. (not located).

    Kyllinga cylindrica Nees in Wight, Contr. Bot. India: 91 (1834). —Clarke in F.T.A. **8**: 282 (1901). Type: Nepal, Himalaya, Kunawur, 1821, *Wallich* 3442 (B holotype).

    Kyllinga cylindrica var. *major* C.B. Clarke in Durand & Schinz, Consp. Fl. Afr. **5**: 528 (1894); in F.T.A. **8**: 283 (1901). Type: Tanzania, Kilimanjaro, 1884, *Johnston* 75 (K holotype).

    Cyperus sesquiflorus (Torr.) Mattf. & Kük. var. *cylindricus* (Nees) Kük. in Engler, Pflanzenr. **4**, 20(101): 593 (1936).

    Cyperus sesquiflorus var. *major* (C.B. Clarke) Kük. in Engler, Pflanzenr. **4**, 20(101): 594 (1936).

    Cyperus sesquiflorus subsp. *cylindricus* (Nees) T. Koyama in Bot. Mag. (Tokyo) **83**: 187 (1970). —Haines & Lye, Sedges & Rushes E. Afr.: 241, fig.484 (1983).

    Kyllinga odorata subsp. *odorata*. —Mapaura & Timberlake, Checkl. Zimbabwe Pl.: 88 (2004).

Tufted perennial; rhizome present or absent, spreading with basal buds or with fleshy aromatic rhizome. Culms in tight groups, 15–70 cm long, 1–2.3 mm across, triangular with blunt angles, glabrous, aromatic. Leaves to 50 cm long; leaf sheath pinkish to dark red, turning brown when dead, 1–10 cm long; leaf blade linear, 4–50 cm long, 1.5–6(7) mm wide, flat and grooved or channeled, scabrid on margins and midrib. Involucral bracts leaf-like, deflexed or spreading, 2–4, longest 3–18(30) cm long. Inflorescence a whitish or greenish-white, ovoid or ellipsoid head of 1–3 spikes, 5–22 × 4–18 mm; spikelets many, narrowly ovoid, 2–4.5 × 0.6–1.6 mm, 1–2-flowered. Glumes whitish with green keel, often with minute reddish dots, ovate, 2–3 mm long, short-acuminate; 2–5 veins on each side of midrib. Stamens 2, filaments to 3.8 mm long, anthers 0.7–1.3 mm. Nutlet brown to glossy black, ovoid, 1.2–1.7 × 0.8–1.1 mm, minutely papillose.

Botswana. N: Ngamiland Dist., Kwando Hunters Camp, 23.i.1978, *P.A. Smith* 2218 (K, LISC, SRGH). **Zambia**. B: Mongu, 12.i.1966, *Robinson* 6799 (K). N: Kasama Dist., Chibutubutu, 48 km S of Kasama, 23.ii.1960, *Richards* 12548 (K). W: Ndola Dist., 2 km from Ndola, 27.i.1963, *Symoens* 10021 (K). C: Kafue Dist., 1 km W of Kamaila Forest Station, 9.ii.1975, *Brummitt et al.* 14306 (K). S: Namwala Dist., Beambure–Kahumbwe road, 16.i.1964, *van Rensburg* 2783 (K). **Zimbabwe**. N: Mutoko Dist., 160 km ENE of Harare (Salisbury), 25.ii.1946, *Weber* 253 (K). C: Chegutu Dist., Chegutu (Hartley), Poole Farm, 11.ii.1945, *Hornby* 2394 (K, SRGH). E: Nyanga Dist., Nyanga Nat. Park,

6.ii.1974, *Burrows* 275 (K, SRGH). **Malawi**. N: Chitipa Dist., Misuku Hills, 21.i.1959, *Robinson* 3186 (K). C: Ntchisi Dist., between Chintembwe Mission and Nkotakota road, 21.ii.1959, *Robson & Steele* 1694 (BM, K). S: Zomba Dist., Zomba Mt, Chingwe's Hole, 11.ii.1970, *Brummitt & Banda* 8509 (K). **Mozambique**. Z: Gurué Dist., Mt Namuli, N Plateau, 16.xi.2007, *Harris et al.* 335 (K, LMA).

Sierra Leone to Kenya, Tanzania and D.R. Congo, South Africa, Madagascar, India, China, SE Asia, Australia and tropical America. Seasonally swampy and/or stony grassland in miombo zone, also in ruderal grassland; (600)1000–2100 m.

Conservation notes: Widespread; not threatened. Least Concern in F.T.E.A. (2010).

I am following Gordon-Gray and Getliffe here in uniting *Kyllinga cylindrica* and *K. odorata*, which Haines & Lye and Kükenthal kept apart at either subspecific or varietal level.

17. **Kyllinga pauciflora** Ridl. in Trans. Linn. Soc. London, Bot. **2**: 147, t.23 (1884). —Clarke in F.T.A. **8**: 273 (1901). —Napper in J. E. Africa Nat. Hist. Soc. Natl. Mus. 28(124): 21 (1971). —Gordon-Gray in Strelitzia **2**: 117 (1995). —Beentje in F.T.E.A. Cyperaceae: 319 (2010). Type: Angola, between Ferrão da Sola and Catumba, 1860, *Welwitsch* 6811 (BM holotype, LISU, colour painting by Ridley at K).

   Cyperus ridleyi Mattf. & Kük. in Engler, Pflanzenr. **4**, 20(101): 599 (1936), new name for *Kyllinga pauciflora*. —Haines & Lye, Sedges & Rushes E. Afr.: 237, fig.476 (1983).

Fairly robust perennial to 40 cm tall; rhizome slender, erect or creeping. Culms densely clustered, 10–60 cm long, 0.8–1.2 mm wide, triquetrous, glabrous, often decumbent when producing viviparous spikelets. Leaves to 15 cm long; leaf sheath purplish to reddish-brown, 1–12 cm long; leaf blade linear, flat, 4–12 cm long, 1–4 mm wide, scabrid on margin and primary vein. Involucral bracts leaf-like, erect to spreading, 4–8, lowermost 4–14 cm long. Inflorescence a solitary spike of 6–15 spikelets, 5–10 × 8–15 mm; spikelets narrowly ovoid, 4–7 × 1–1.5 mm, usually 2-flowered but often with leafy young plants arising. Glumes golden to straw-coloured, ovate, 4.5–7 mm long, keel green, often with a few small teeth, apex acuminate and slightly recurved. Stamens 3, filaments to 3 mm long, anthers 2.2–2.5 mm. Nutlet 1, yellowish-brown, obovoid, 2–3.5 mm.

Zambia. N: Mbala Dist., Sumbawanga road 13 km from Kawimbe, 30.i.1957, *Richards* 8010 (K). **Zimbabwe**. E: Mutare Dist., Lake Alexander, 7.i.1972, *Gibbs Russell* 1250 (K, SRGH). **Malawi**. N: Mzimba Dist., Mzuzu, Marymount, 14.vi.1973, *Pawek* 6863 (CAH, K, MAL, MO, UC).

Also in Tanzania, Angola and South Africa. Dambos, swamps, dam edges, streamsides, pools at base of rocks; 1250–2000 m.

Conservation notes: Widespread; not threatened. Least Concern in F.T.E.A. (2010).

Closely related to *Kyllinga brevifolia*, but distinct in its viviparous spikelets.

18. **Kyllinga peteri** (Kük.) Lye in Nordic J. Bot. **1**: 746 (1982). —Beentje in F.T.E.A. Cyperaceae: 317 (2010). Types: Tanzania, various localities (Saranda, Uyansi near Lake Tschaya, Ngulia, Unyanyembe at Malongwe), 1926, *Peter* 33394a, 34101, 34142, 34236, 34444, 34723, 34868, 45845 (B syntypes); Nhulu, E of Malongwe towards Tura, km 723, *Peter* 34723 (B lectotype), lectotypified by Beentje in F.T.E.A. (2010).

   Cyperus peteri Kük. in Engler, Pflanzenr. **4**, 20(101): 575 (1936). —Haines & Lye, Sedges & Rushes E. Afr.: 235, fig.474 (1983).

Fairly robust perennial to 62 cm tall; rhizome short, thick. Culms several closely together on short rhizome, 20–60 cm long, 1.3–2 mm wide, trigonous, glabrous. Leaves to 25 cm long; leaf sheath grey to brown, 1–6 cm long, many at base but only slightly splitting into fibres; leaf blade linear, flat or folded, 5–30 cm long, 2–4 mm wide, scabrid on margin in upper part. Involucral bracts leaf-like, spreading to reflexed, 3–4(12), lowermost 6–12 cm long. Inflorescence a cylindrical spike 10–16 × 6–10 mm; spikelets ovoid, 3–5.5 mm long, 1.5–2 mm wide, 2–3-flowered but perfecting 1–2 nutlets only. Glumes pale reddish-brown, blackish at apex, ovate, 3–4.5 mm long, keel slightly excurrent with a few teeth, 3 veins each side, apex acuminate, often slightly recurved. Stamens ?2, filaments 1–1.3 mm long, anthers 1.5–1.6 mm. Nutlet almost black, ellipsoid, c. 2 × c. 0.8 mm, minutely papillose.

Zambia. N: Kaputa Dist., Mweru-Wantipa, Salemani, Mwawe R., 17.xii.1960, *Vesey-Fitzgerald* 2830 (K). S: Choma Dist., Miller's Farm, 19 km N of Choma, 2.ii.1960, *White* 6728 (FHO, K). **Zimbabwe**. W: Matobo Dist., Shumba Shava, 7.i.1963, *Wild* 5953 (K, SRGH). C: Harare, Thorn Park, 25.xii.1956, *Paterson* 17 (K, SRGH).

Also in Tanzania. Floodplains and dambos; 900–1500 m.

Conservation notes: Widespread; not threatened. Uncertain status in F.T.E.A. (2010) as all specimens from there over 70 years old.

19. **Kyllinga platyphylla** K. Schum. in Bot. Jahrb. Syst. **30**: 270 (1901). —Napper in J. E. Africa Nat. Hist. Soc. Natl. Mus. **28**(124): 24 (1971). —Beentje in F.T.E.A. Cyperaceae: 339 (2010). Type: Tanzania, Mbeya Dist., Mbeya peak, 16.xi.1899, *Goetze* 1449 (B holotype, EA).

Kyllinga platyphylla var. *longifolia* Kük. in Repert. Spec. Nov. Regni Veg. **12**: 92 (1913). Type: Mozambique, Changara Dist., Luenha (Luia) near Luenha (Luia) R., i.1906, *Tiesler* 20 (B? holotype).

Cyperus ciliatopilosus Mattf. & Kük. in Engler, Pflanzenr. **4**, 20(101): 571 (1936), replacement name for *K. platyphylla*. —Haines & Lye, Sedges & Rushes E. Afr.: 245, fig.492 (1983).

Cyperus ciliatopilosus var. *rhizomatosus* Kük. in Engler, Pflanzenr. **4**, 20(101): 571 (1936). Type: Angola, between Ganda & Caconda, 1932, *Hundt* 272 (B? holotype).

Perennial to 65 cm tall; short rhizome present but stolons absent, with contiguous tufts. Culms densely tufted, bases covered in dark brown fibres, 15–65 cm long, 1.3–2 mm wide, sharply 3-angled, scabridulous. Leaves to 17(33) cm long; leaf sheath pale brown, darker at base, 1.5–8 cm long; leaf blade linear, recurved and channelled, 2–20(33) cm long, 4–7 mm wide, scabrid on margin and primary vein. Involucral bracts leaf-like, recurving, 3–5, lowermost 3–8(12.5) cm long. Inflorescence a hemispheric to subglobose white or cream (occasionally greenish yellow) head of 1–3 spikes, 8–15 mm wide; spikelets narrowly ovoid, 4–4.8 × 1.2–1.5 mm wide, 2-flowered. Glumes whitish, occasionally with reddish spots, narrowly ovate, 3–4 mm long, keel hairy to ciliate, sometimes appearing smooth, apex ± acuminate. Stamens 3, filaments 2–2.4 mm long, anthers 1.3–2.1 mm long. Nutlet dark grey (brown?), obovoid, 1.5–2.2 × 0.5–0.7 mm, minutely papillose.

Zambia. N: Mporokoso Dist., Chishimba Falls, 27.xi.1993, *Nkhoma et al.* 23 (K). W: Ndola, 1.i.1960, *Robinson* 3266 (K). C: Chongwe Dist., Chakwenga headwaters, 125 km E of Lusaka, 10.i.1964, *Robinson* 6154 (K). S: Mumbwa Dist., 48 km W of Kafue Hook pontoon on Mankoya road, 21.xi.1959, *Drummond & Cookson* 6735 (K). **Zimbabwe**. N: Guruve Dist., Nyamunyeche, Chiripanyanga area, 6.xii.1978, *Nyariri* 541 (K). C: "Marandellas and Sabi Dist.", xii.1931, *Myres* 639 (K). **Malawi**. N: Nyika Plateau, 19.xii.1975, *Philips* 646 (K). **Mozambique**. N: near Lake Nyasa, 1902, *Johnson* 493 (K). T: Changara Dist., Changara (Luenha, Luia) near Luenha (Luia) R., i.1906, *Tiesler* 20 (B?).

Also in Tanzania and Angola. Open miombo and other woodland types, grassland and dambos; 250–2200 m.

Conservation notes: Widespread; not threatened. Least Concern in F.T.E.A. (2010).

20. **Kyllinga polyphylla** Kunth, Enum. Pl. **2**: 134 (1837). —Clarke in F.T.A. **8**: 276 (1901). —Napper in J. E. Africa Nat. Hist. Soc. Natl. Mus. **28**(124): 22 (1971). —Gordon-Gray in Strelitzia **2**: 120 (1995). —Beentje in F.T.E.A., Cyperaceae: 318 (2010). Type: Mauritius, *du Petit Thouars* s.n. in Willd. Herb. 1441 (B-W holotype).

    Kyllinga aromatica Ridl. in Trans. Linn. Soc. London, Bot. **2**: 146 (1884). Type: Angola, Pungo Andongo, 23.ii.1857, *Welwitsch* 6801 (BM holotype, LISU).

    Cyperus aromaticus (Ridl.) Mattf. & Kük. in Engler, Pflanzenr. **4**, 20(101): 581 (1936). — Haines & Lye, Sedges & Rushes E. Afr.: 239, fig.480 (1983).

    Cyperus aromaticus var. *repens* Kük. in Engler, Pflanzenr. **4**, 20(101): 583 (1936). Type: Tanzania, ?Lushoto Dist., between Kalekwa & Gumbo, 3.xi.1916, *Peter* 18031 (B holotype, GOET, K, WAG).

    Cyperus aromaticus var. *brachyrhizomatosus* Kük. in Engler, Pflanzenr. **4**, 20(101): 583 (1936). Type: Tanzania, Morogoro Dist., Uluguru Mts, Fisigo valley, x.1913, *von Brehmer* 450, 453, 454, 455 (B syntypes).

    Kyllinga erecta K. Schum. var. *polyphylla* (Kunth) S.S. Hooper in Kew Bull. **26**: 580 (1972); in F.W.T.A., ed.2 **3**(2): 307 (1972).

Robust perennial to 92 cm tall; rhizome creeping, 2–8 mm diameter, covered in pinkish red to black scales with membranous margins. Culms green, densely set along rhizome, (4)20–90 cm long, 1–3.5 mm wide, triangular to almost winged, to 4 mm across, glabrous, swollen base covered with brownish or purplish membraneous sheaths. Leaves few, 2–4 fully developed, to 20 cm long; leaf sheath reddish brown to purplish, 2–10 cm long, lowermost without blades; leaf blade linear, flat, 2–20 cm long, 2–5 mm wide, scabrid on margins and midrib, apex acute. Involucral bracts leaf-like, spreading, (4)5–9, lowermost 6–21 cm long, 3–7 mm wide. Inflorescence a green to yellow-brown irregular hemispheric to ellipsoid head, 8–15 × 4–10 mm, with a central spike and often several smaller lateral spikes; spikelets many, olive green in flower, turning yellow-brown in fruit, narrowly ovoid, 2.5–4 × 0.8–1 mm, 1–2-flowered, when 2-flowered upper one male or bisexual. Glumes yellowish, golden yellow or straw-coloured, narrowly ovate, 2–3.5 mm long, keel green, frequently with dark brown dots or streaks, sometimes with 3–4 teeth, 2–5 veins on either side, apex shortly acuminate and slightly recurved. Stamens 3, filaments c. 1.5 mm long, anthers 1.4–1.6 mm long, yellow. Style white. Nutlet 1, dark red-brown to blackish, ellipsoid, 1–1.5 × 0.6–0.9 mm, minutely papillose.

a) Var. **polyphylla**

Culms densely set along the rhizome, glabrous. Head hemispheric to globose, 9–10 × 6–9 mm; glumes rarely with teeth.

Botswana. N: Ngamiland Dist., Okavango between Samocima & Red Cliffs, 30.iv.1975, *Gibbs Russell* 2869 (K, SRGH). **Malawi**. N: Karonga Dist., Kambwe lagoon, Karonga, 11.xi.1888, *Scott* s.n. (K). S: Machinga Dist., Ntangai R., 10.vii.1985, *Herb. Univ. Malawi* 1659 (K).

Widespread in W, C and NE Africa, down into Angola, South Africa and Mauritius. In moist sites by lakes and streams; 450–1100 m.

Conservation notes: Widespread; not threatened. Least Concern in F.T.E.A. (2010).

b) Var. **elatior** (Kunth) Kük. in Notizbl. Bot. Gart. Berlin-Dahlem **9**: 300 (1925) as *Kyllingia*. —Beentje in F.T.E.A., Cyperaceae: 319 (2010). Type: South Africa, between Cape and Durban (Port Natal), 1834, *Drège* 4384 (B holotype, K, P).

    Kyllinga elatior Kunth., Enum. Pl. **2**: 135 (1837). —Clarke in F.T.A. **8**: 275 (1901). — Hooper in F.W.T.A., ed.2 **3**(2): 305 (1968). —Napper in J. E. Africa Nat. Hist. Soc. Natl. Mus. **28**(124): 21 (1971). —Getliffe in J. S. African Bot. **49**: 290, fig.13 (1983). —Gordon-Gray in Strelitzia **2**: 115 (1995).

    Kyllinga pinguis C.B. Clarke in Bot. Jahrb. Syst. **38**: 131 (1906). Types: Tanzania, Kwai, xi.1899, *Stern* 235 (B syntype); Uganda, Entebbe, iii.1906, *E. Brown* 26 (K syntype); Kenya, Nairobi, *Linton* 7 (K syntype).

    Cyperus aromaticus (Ridl.) Mattf. & Kük. var. *elatior* (Kunth) Kük. in Engler, Pflanzenr. **4**, 20(101): 582 (1936).

Cyperus pinguis (C.B. Clarke) Mattf. & Kük. in Engler, Pflanzenr. **4**, 20(101): 583 (1936). —Haines & Lye, Sedges & Rushes E. Afr.: 239, fig.481 (1983).

Culms spaced along aromatic rhizome, scabrid or glabrous. Leaves lemon-scented when crushed. Head ellipsoid, central spike 8–15 × 4–8 mm; glumes often with teeth.

Zambia. N: Chinsali Dist., Shiwa Ngandu, 1.vi.1956, *Robinson* 1530 (K). **Malawi**. N: Chipata Dist., Misuku Hills, 11.i.1959, *Robinson* 3165 (K).

Also in Cameroon, Ethiopia, Rwanda, Uganda, Kenya, Tanzania, D.R. Congo, South Africa and Madagascar. Streamsides and moist grassland; 1500–2000 m.

Conservation notes: Widespread; not threatened. Least Concern in F.T.E.A. (2010).

The type of var. *elatior* seems to differ from that of var. *polyphylla* mainly in the slightly scabrid (not glabrous) culm, and in the more distant culms and a more ellipsoid head. Otherwise the two are similar. In most material with ellipsoid heads the culms are glabrous, unlike in the type. I here follow Kükenthal in varietal status, but more material might show the varieties are synonymous.

21. **Kyllinga pumila** Michx., Fl. Bor.-Amer. **1**: 28 (1803). —Clarke in F.T.A. **8**: 281 (1901). —Hooper in F.W.T.A., ed.2 **3**(2): 305 (1968). —Napper in J. E. Africa Nat. Hist. Soc. Natl. Mus. **28**(124): 22 (1971). —Beentje in F.T.E.A., Cyperaceae: 325 (2010). Type: North America, "in Shavanensium regione, ad amnem Scioto", *Michaux* s.n. (P-Michaux holotype).

Cyperus densicaespitosus Mattf. & Kük. in Engler, Pflanzenr. **4**, 20(101): 597 (1936). —Haines & Lye, Sedges & Rushes E. Afr.: 243, figs.487,488 (1983), new name for *Kyllinga pumila*.

Short-lived or annual, 8–47 cm tall; rather densely tufted on a short juicy brittle rhizome with a slender root system, whole plant sweet-scented. Culms 8–66 cm long, 0.7–1.2 mm wide, trigonous, ridged, glabrous, bases not swollen. Leaves up to 34 cm long; leaf sheath reddish or purple, upper ones green, 1–9 cm long, lower ones covering base of culm and bladeless; leaf blade linear, flat or channeled near midrib, 4–25 cm long, 2–3.2 mm wide, scabrid on midrib and margins. Involucral bracts leaf-like, erect to spreading, 3–5, lowermost 4–17 cm long. Inflorescence an irregular greenish white head consisting of a central ovoid spike, 5–8 × 4–6 mm, and 1–2(?3) smaller lateral spikes, spikelets on narrow receptacle; spikelets narrowly ovoid, 2–2.5 × 0.6–0.9 mm, 1-flowered. Glumes pale brown to transparent, narrowly ovate, 1.5–2.5 mm long, keel green and ± winged, sometimes with many minute reddish dots, 2–4 veins on either side, keel with a few minute teeth or cilia, apex acute. Stamens 1–2, filaments to 1.8 mm long. Nutlet pale to dark brown, ellipsoid, 1–1.2 × 0.6–0.7 mm, minutely papillose; endosperm liquid.

Zambia. B: Mongu Dist., Mongu, 11.iv.1966, *Robinson* 6932 (K). N: Kawambwa Dist., Mbereshi R. swamp, 20.iv.1957, *Richards* 9370 (K). W: Luanshya Dist., Masaiti Agric. Station, 13.iv.1961, *Angus* 2841 (K, OXF). E: Chipata Dist., Chipata (Fort Jameson), 6.vi.1954, *Robinson* 845 (K). **Zimbabwe**. E: Chimanimani Dist., junction of Haroni & Rusitu (Lushitu) roads, 29.v.1969, *Müller* 1159 (K, SRGH). **Malawi**. N: Chitipa Dist., Misuku Hills, 12.i.1959, *Robinson* 3174 (K).

Pantropical. Streamsides, ditches, boggy hollows, damp grassland, thin damp soil over granite; 300–2000 m.

Conservation notes: Widespread; not threatened. Least Concern in F.T.E.A. (2010).

22. **Kyllinga rhizomafragilis** Lye in Nordic J. Bot. **1**(6): 744 (1982). Type: Zambia, 27 km W of Mwinilunga, 17.iv.1960, *Robinson* 3660 (K holotype, BR).

Cyperus rhizomafragilis (Lye) Lye in Haines & Lye, Sedges & Rushes E. Afr., app.3: 2 and main work: 231 (1983).

Perennial to 40 cm high; rhizome crisp, 'tuberous', 1–4 mm thick, young parts covered by purplish scales. Culm 20–40 cm long, 0.5–0.8 mm wide when dry, ± triangular, glabrous, with slightly widened base. Leaves from lower 8 cm only; lower sheaths purplish, upper sheaths green, only uppermost 1–4 with lamina; leaf blade 0.5–15 cm long, 0.5–2.5 mm wide, scabrid on margin and major veins. Involucral bracts spreading, 3–4, largest to 6.5 cm long. Inflorescence a solitary white hemispherical head, 4–5 mm wide, of a single spike drying to cinnamon colour; spikelets 1-flowered, 2–2.2 × 1–1.2 mm. Glumes whitish with reddish lines and spots, ovate, 1.8–2.3 mm long, ± acute. Stamen 1. Nutlet grey-black, ellipsoid, 1.5–1.6 × 0.9–1.2 mm.

Zambia. W: Mwinilunga Dist., 27 km W of Mwinilunga, 17.iv.1960, *Robinson* 3660 (K, BR).

Not known elsewhere. Sandy plateau grassland in shade of trees; 1400 m.

Conservation notes: Known only from the type; possibly threatened.

Lye says it is related to *Kyllinga albogracilis* but differs in size and lacks the scabrid glume wing. A specimen from Zimbabwe C (50 km NE of Kwekwe, Iwaba Estate, 10.ii.2000, *Lye* 23783 (K)) is close but has green keels.

23. **Kyllinga robinsoniana** Mtot. in Nordic J. Bot. **9**: 637, fig.1 (1990). —Beentje in F.T.E.A., Cyperaceae: 325 (2010). Type: Zambia, Kasama Dist., Chishimba Falls, 12.ii.1961, *Robinson* 4367 (SRGH holotype, K?), not *Robinson* 4357 as in protologue.
    Cyperus robinsonianus (Mtot.) Lye in Lydia 7: 97 (2011).

Perennial to 45 cm tall; rhizomes short, brittle, with a pink apex. Culms densely tufted or just a few mm apart on rhizome, 3–43 cm long, 1–1.5 mm wide, subtriangular, strongly ridged; bases bulbous, surrounded by dense old leaf fibres. Leaves up to 13 cm long; leaf sheath light green to whitish, 1.5–3 cm long; leaf blade linear, 4–13 cm long, 0.5–3 mm wide. Involucral bracts leaf-like, spreading, usually 3, lowermost 2.5–10 cm long. Inflorescence with a central light green to dirty white spike, 5–12 × 4–7 mm, often with 2 smaller lateral spikes; spikelets many, ovoid, 1.2–2.8 mm long, 2-flowered, upper usually vestigial. Glumes ± translucent, ovate, 2–2.4 mm long, keel winged, ciliate to glabrous, not dentate, apex acute, 1–2-veined on each side. Stamens 3, sulphur yellow, anthers 0.8–1 mm long. Nutlet light brown, ovoid-oblong, 2.5–3.5 mm long, papillose.

Zambia. N: Kasama Dist., Kasama, 12.i.1961, *Robinson* 4254 (K). W: Mwinilunga Dist., 0.5 km S of Matonchi Farm, 6.xii.1937, *Milne-Redhead* 3519 (K). C: Kafue Dist., Lazy J Botanical Reserve, 26 km SE of Lusaka, 21.iii.1999, *Bingham* 11997 (K). S: Mazabuka Dist., Mazabuka to Magoye, 28.ii.1963, *van Rensburg* 1533 (K). **Zimbabwe**. N: Gokwe Dist., Gokwe, 12.v.1964, *Bingham* 1338 (K, SRGH). **Malawi**. S: Zomba Dist., Zomba Plateau, Chingwe's Hole, 11.ii.1970, *Brummitt & Banda* 8509a (K).

Also in Kenya, Tanzania and Angola. Miombo and mopane woodland and wooded grassland, also seasonally moist grassland; 600–1900 m.

Conservation notes: Widespread; not threatened. Least Concern in F.T.E.A. (2010).

This species is related to *Kyllinga pumila* from which it differs in small stature, densely fibrous bases, longer and narrower leaves, often densely ciliate glume keels, equal glumes and a 3–4 mm long prophyll, densely ciliate on keel.

A specimen from S Malawi (S: Liwonde Nat. Park, Chiunguni Hill, 28.xii.1987, *Dudley* 1406 (K, MAL)) is probably this species but has very small glumes – 1.4 mm long (or is possibly *Kyllinga afropumila*).

24. **Kyllinga songeensis** Lye in Bot. Not. **125**: 218 (1972). —Beentje in F.T.E.A., Cyperaceae: 343 (2010). Type: Tanzania, Songea Dist., R. Luhimba c. 28 km N of Songea, 6.v.1956, *Milne-Redhead & Taylor* 10106 (K holotype).
    Cyperus songeensis (Lye) Lye in Haines & Lye, Sedges & Rushes E. Afr., app.3: 2 and main work: 246, fig.494 (1983).

Perennial to 45 cm tall; rhizomes short, horizontal. Culms crowded on rhizome, bases bulbous, 18–45 cm long, 0.4–1 mm wide, terete except near apex where bluntly triangular, glabrous, basal parts covered by fibrous old leaf sheaths. Leaves to 15 cm long; leaf sheath pale to dark brown,

1–6 cm long; leaf blade linear, flat or folded, 5–15 cm long, 1–2 mm wide, scabrid on margin and midrib. Involucral bracts leaf-like, usually sharply reflexed, 3–4, lowermost 1–7 cm long. Inflorescence a single whitish subglobose head, 8–12 mm across; spikelets many, narrowly obovoid, 4–4.8 × 1–1.2 mm, 1–2-flowered. Glumes whitish, narrowly ovate to narrowly obovate, 3.2–4.5 mm long, unwinged and glabrous or with minute hairs, apex acute, c. 5-veined on each side of midrib. Stamens ?2, anthers 0.7–1 mm long. Nutlet only seen immature, brown, ellipsoid, 1.3 × 0.5 mm.

Zambia. N: Mbala Dist., Kalambo Falls, 7.xii.2006, *Bingham* 13210 (K). **Zimbabwe**. E: Chimanimani Dist., Chimanimani Mts., 31.xii.1964, *Robinson* 6346 (K).

Also in Tanzania. Upland grassland; 1200–1600 m.

Conservation notes: Disjunct distribution; probably not threatened. Data Deficient in F.T.E.A. (2010).

25. **Kyllinga squamulata** Vahl, Enum. Pl. **2**: 381 (1805). —Clarke in F.T.A. **8**: 270 (1901). —Hooper in F.W.T.A. ed.2 **3**(2): 304 (1968). —Napper in J. E. Africa Nat. Hist. Soc. Natl. Mus. **28**(124): 19 (1971). —Beentje in F.T.E.A., Cyperaceae: 326, fig.47.5–7 (2010). Type: Guinea, n.d., *Thonning* s.n. (possibly 547 but that is in different ink) (B holotype, C).

   Kyllinga metzii Steud., Syn. Pl. Glumac. **2**: 70 (1854). Type: India, Karnataka, Mangalor, n.d., *Hohenacker* 199 (B† holotype, JE).
   Cyperus metzii (Steud.) Mattf. & Kük. in Engler, Pflanzenr. **4**, 20(101): 612 (1935). —Haines & Lye, Sedges & Rushes E. Afr.: 250, fig.503 (1983).

Annual or short-lived herb, 6–37 cm tall; root-system slender, lacking rhizomes or stolons. Culms 2–36 cm long, 0.5–0.8 mm wide, trigonous or terete near base, glabrous. Leaves to 20 cm long; leaf sheath pinkish to red, 1–6 cm long; leaf blade linear, flat or slightly channelled, 4–20 cm long, 1–3 mm wide, scabrid on margins and midrib. Involucral bracts leaf-like, spreading or reflexed, 3(4), lowermost (2.5)6–15 cm long, with conspicuous transparent wings near base. Inflorescence a small sessile irregular or ovoid head of a single spike 6–10 × 6–9 mm; spikelets many, broadly ovoid, 2.5–4 × 1.3–1.8 mm, 1-flowered with 2 glumes and an elongated stalk with 2 minute bracts. Glumes translucent to whitish or yellow, often with minute reddish dots, keel green, ovate, 2–2.9 mm long, winged with coarse broad hair-tipped teeth to 0.7 × 0.3 mm, apex acuminate. Stamens 2. Nutlet dark brown, subcircular, 1.3–1.5 × 1.2–1.3 mm, minutely papillose.

Zambia. B: Mongu Dist., Mongu, 20.ii.1966, *Robinson* 6854 (K). N: Kasama Dist., Mungwi, 12.iv.1961, *Robinson* 4590 (K). W: Masaiti Dist., 40 km S of Ndola, 2.iii.1960, *Robinson* 3369 (K). C: Mpika Dist., Luangwa Valley, iv.1971, *Abel* 296 (K, SRGH). E: Katete Dist., Katete Hospital, 8.iii.1957, *Wright* 172 (K). S: Choma Dist., Mapanza Mission, 4.iii.1957, *Robinson* 2148 (K). **Zimbabwe**. E: Chiping Dist., Gungunyana Forest Reserve, v.1967, *Goldsmith* 65/67 (K, SRGH). **Malawi**. C: Kasungu Dist., Kasungu Nat. Park near Lifupa Lodge, 24.iii.1989, *Iversen & Martinsson* 89/009 (K, UPS). S: Blantyre Dist., Matenje road 2 km N of Limbe, 6.ii.1970, *Brummitt* 8453 (K). **Mozambique**. N: Malema Dist., Mutuali, 5.iv.1962, *Lemos & Marrime* 317 (IAM, K, LISC).

From Senegal to Ethiopia south to Tanzania and Mozambique; also in Madagascar, India and Vietnam. A weed of cultivation (maize, sorghum, yam), on streambanks, lakeshores and open grassland; 600–1750 m.

Conservation notes: Widespread; not threatened. Least Concern in F.T.E.A. (2010).

26. **Kyllinga tanzaniae** Lye in Bot. Not. **125**: 217 (1972). —Beentje in F.T.E.A., Cyperaceae: 343 (2010). Type: Tanzania, Ufipa Dist., Rukwa escarpment, Namwele, 28.xii.1961, *Robinson* 4783 (K holotype, K).

   Cyperus tanzaniae (Lye) Lye in Haines & Lye, Sedges & Rushes E. Afr., app.3: 2 and main work: 233, fig.469 (1983).

Medium-sized perennial to 40 cm tall; rhizome thickish, horizontal or curved. Culms solitary or somewhat crowded, 5–40 cm long, 0.4–0.8 mm wide, trigonous, glabrous. Leaves to 11 cm long; leaf blade linear, flat, 2–11 cm long, 1–2 mm wide, scabrid on margin and primary vein. Involucral bracts leaf-like, reflexed or spreading, 2–3, lowermost 2–4 cm long. Inflorescence a single white or cream ovoid to globose spike, 5–10 × 5–8 mm; spikelets 2.5–3.5 mm long, 1–2-flowered. Glumes whitish or cream, 2–3.5 mm long, keel unwinged, smooth, with 3–5 veins on either side, apex shortly acuminate. Stamens 3, anthers 1.2 mm long. Nutlet dark brown to blackish, obovoid, 1.8–2 × 0.9–1.2 mm, minutely papillose.

Zambia. W: Mwinilunga Dist., Kalenda Plain, N of Matonchi Farm, 8.xii.1937, *Milne-Redhead* 3559 (K). S: Choma Dist., Kabanga, Mapanza, 26.i.1956, *Robinson* 1331 (K). **Malawi**. S: Zomba, xii.1896, *Whyte* s.n. (K).

Also in Tanzania. Damp grassland, or shallow soil over laterite; 750–1350 m.

Conservation notes: Moderately widespread; not threatened. Data Deficient in F.T.E.A. (2010).

Lye says this species is related to *Kyllinga merxmuelleri* (= *K. albiceps*), but distinct in its thicker rhizome and absence of long stolons.

27. **Kyllinga tenuifolia** Steud., Syn. Pl. Glumac. **2**: 69 (1854). —Hooper in F.W.T.A., ed.2 **3**(2): 305 (1968). —Beentje in F.T.E.A., Cyperaceae: 329 (2010). Type: Senegal, no locality, n.d., *Leprieur* s.n. (P holotype).

Kyllinga triceps Rottb., Descr. Icon. Rar. Pl.: 14, t.4, fig.6 (1773), invalid name (see note). —Clarke in F.T.A. **8**: 280 (1901). —Podlech in Merxmüller, Prodr. Fl. SW Afr. **165**: 29 (1967). —Napper in J. E. Africa Nat. Hist. Soc. Natl. Mus. **28**(124): 22 (1971). Type: see note below.

Cyperus triceps (Rottb.) Endl. in Cat. Horti Vindob. **1**: 94 (1842). —Kükenthal in Engler, Pflanzenr. **4**, 20(101): 578 (1936). —Haines & Lye, Sedges & Rushes E. Afr.: 231, fig.466 (1983).

Kyllinga triceps var. ß *obtusiflora* Boeckeler in Linnaea **35**: 414 (1868). Type: Ethiopia, Matamma, *Schweinfurth* 2051 (B holotype (not seen), K).

Cyperus triceps var. *obtusiflorus* (Boeckeler) Kük. in Engler, Pflanzenr. **4**, 20(101): 579 (1936).

Cyperus tenuifolius (Steud.) Dandy in Exell, Cat. Vasc. Pl. S. Tomé: 363 (1944).

Tufted perennial; rhizome short or lacking; often in dense groups of tufts with swollen culm bases densely covered with old leaf sheaths, old bases persisting. Culms tufted, (2.5)5–40 cm long, 0.4–1 mm wide, triangular, glabrous. Leaves to 20 cm long; leaf sheath brown, more pinkish towards base, 1–8 cm long; leaf blade linear, flat, slightly channeled, often with small longitudinal purple marks, 5–20 cm long, 0.9–3 mm wide, scabrid on primary vein and margins. Involucral bracts 3–4, leaf-like, spreading to reflexed, 2–3, lowermost 2.5–12.5 cm long. Inflorescence an irregular white to cream head, often triangular, 5–14 × 5–12 mm, consisting of one to several (usually 3) spikes; spikelets many per spike, narrowly and sometimes asymmetrically ovoid, 1.5–2.7 × 0.7–0.8 mm, 1-flowered (very rarely with 2), slightly gaping at maturity. Glumes whitish or cream, with or without brown spots near midrib, ovate, 1–2.2 mm long, keel greenish or cream, 1–4 veins on either side, apex acute to hooded (upper) to slightly excurrent (lower). Stamens 1–3, filaments 1–2 mm long, anthers 0.8–1 mm. Nutlet brown, ellipsoid, 1–1.7 × 0.6–1.1 mm wide, minutely papillose.

a) Var. **tenuifolia**

Culms tufted, 5–40 cm long. Leaf sheath brown, more pinkish towards base, 1–8 cm long; leaf blade 1.5–3 mm wide. Inflorescence 6–14 × 6–12 mm. Glumes whitish or cream, with or without brown spots near midrib, 1.5–2.2 mm long, keel greenish and smooth. Stamens 1–3. Nutlet 1.5–1.7 × c. 1.1 mm, minutely papillose.

Zambia. N: Kasama Dist., Kasama, 29.xii.1962, *Wright* 320 (K). W: Solwezi Dist., Solwezi, 10.iv.1960, *Robinson* 3512 (K). C: Mpika Dist., Luangwa Valley, 4.8 km S of Lubi R. on 5° road, 2.ii.1967, *Prince* 97 (K). S: Kafue Dist., Kafue Gorge, 14.iii.1960, *Vesey-Fitgerald* 2718 (K). **Zimbabwe**. N: Hurungwe Dist., Hurungwe (Urungwe) Nat. Park, 1.5 km N of Marangora, 15.ii.1981, *Philcox et al.* 8614 (K). W: Hwange Dist., Victoria Falls,

4.iv.1956, *Robinson* 1426 (K). E: Chipinge Dist., Jersey Tea Estate, v.1967, *Goldsmith* 69/67 (K, SRGH). **Malawi**. N: Nyika Plateau, Chelinda camp, 3.ii.1978, *Pawek* 13744 (K).

Widespread in Africa and the Old World tropics, though not very common anywhere. Miombo or mopane woodland, dambo edges or streamsides, also in quite dry sites; 550–2250 m.

Conservation notes: Widespread; not threatened. Least Concern in F.T.E.A. (2010).

Kukkonen's proposal in Taxon **44**: 626 (1995) to conserve *Kyllinga triceps* was not approved (see Taxon **47**: 864, 1998). The protologue cited *Scirpus glomeratus* L. in synonymy, and therefore was both superfluous and illegitimate; the Linnean name to which Rottbøl's protologue refers was based either on Herb. Linn. 68.9 (= *Mariscus coloratus*) or on Herb. Linn. 69.1 (= *K. nemoralis*).

b) Var. **ciliata** (Boeckeler) Beentje in F.T.E.A., Cyperaceae: 330 (2010). Type: Mozambique, Tete, n.d., *Peters* s.n. (B holotype).

Kyllinga triceps Rottb. var. *ciliata* Boeckeler in Peters, Naturw. Reise Mossambique: 535 (1864).

Kyllinga blepharinota Boeckeler in Linnaea **35**: 414 (1868), illegitimate name as based on type of *Kyllinga triceps* var. *ciliata*.

Kyllinga welwitschii Ridl. in Trans. Linn. Soc. London, Bot. **2**: 147 (1884). —Hooper in F.W.T.A., ed.2 **3**(2): 305 (1968). —Napper in J. E. Africa Nat. Hist. Soc. Natl. Mus. **28**(124): 23 (1971). —Getliffe in J. S. African Bot. **49**: 277, fig.7 (1983). Type: Angola, Cuanza Norte (Pungo Andongo), near Calemba, iii.1857, *Welwitsch* 6779 (LISU holotype, BM).

Kyllinga controversa Steud. var. *subexalata* C.B. Clarke in F.T.A. **8**: 271 (1901). Types: "Senegambia", 1839, *Heudelot* 399 (P syntype); Niger, Nupe, 19.ii.1895, *Barter* 1588 (K syntype); Sudan, Kordofan, Gebel Kurbag, viii.1875, *Pfund* 359, 623 (both K syntypes); Ethiopia, Gursarfu, 9.viii.1854, *Schimper* 2201 (K syntype, JE); Eritrea, no locality, 1875, *Hildebrandt* 360 (syntype not found).

Cyperus triceps Rottb. var. *ciliatus* (Boeckeler) Kük. in Engler, Pflanzenr. **4**, 20(101): 579 (1936), in part.

Cyperus welwitschii (Ridl.) Lye in Haines & Lye, Sedges & Rushes E. Afr., app.3: 2 and main work: 232, fig.467 (1983).

Culms tufted, 2.5–20 cm long. Leaf sheath brown, 1–3 cm long; leaf blade 0.9–2 mm wide. Inflorescence 5–8 × 5–9 mm. Glumes translucent whitish or pale yellow, 1–1.8 mm long, keel ciliate. Stamens (1?)3. Nutlet 1–1.2 × 0.6 mm, minutely papillose.

Zambia. N: Kasama Dist., Chishimba Falls, 12.ii.1961, *Robinson* 4367 (K). W: Solwezi Dist., Mulenga Protected Forest Area, 20 km NW of Kanshanshi, 19.iii.1961, *Drummond & Rutherford-Smith* 7073 (K). **Mozambique**. MS: "Lower Zambesi", xi.1858, *Kirk* s.n. (K).

Also in Mauritania, Senegal, Ivory Coast, Burkino Faso, Nigeria, Chad, Cameroon, Ethiopia, Somalia, Uganda, Kenya, Tanzania, Angola and Namibia. In woodland on sand, sometimes in dambos; 10–1750 m.

Conservation notes: Widespread; not threatened. Least Concern in F.T.E.A. (2010).

28. **Kyllinga ugogensis** (Peter & Kük.) Lye in Bot. Not. **125**: 218 (1972). —Beentje in F.T.E.A., Cyperaceae: 320 (2010). Types: Tanzania, Dodoma Dist., Ugogo, between Bahi and Kitalalo steppe, 17.xii.1925, *Peter* 33267a (B, syntype, K, GOET); between Kitalalo & Tschali, 18.xii.1925, *Peter* 33327 (B syntype, GOET).

Cyperus ugogensis Peter & Kük. in Engler, Pflanzenr. **4**, 20(101): 572 (1936). —Haines & Lye, Sedges & Rushes E. Afr.: 237, fig.477 (1983).

Kyllinga ferruginea Peter in Engler, Pflanzenr. **4**, 20(101): 572 (1936), invalid name.

Small perennial to 20 cm tall; bulbous culm-base and thin creeping rhizome or stolon. Culms tufted, 6–20 cm long, 0.4–0.6 mm wide, triangular, glabrous. Leaves to 15 cm long; leaf sheath brownish, 0.8–2.5 cm long; leaf blade linear, flat or folded, 4–15 cm long, 0.8–1 mm wide, scabridulous on margins near apex. Involucral bracts 1–3, leaf-like, lowermost longest and ± erect, others (if present) spreading or reflexed, lowermost (1.3)1.8–7 cm long. Inflorescence a single ovoid or globose spike, 4.5–7 mm across; spikelets ovoid, 2–2.6 × 0.9–1.1 mm, 2–3-flowered. Glumes golden with green keel, ovoid, 1.8–2.6 mm long, smooth, apex mucronate and slightly recurved. Stamens 3, filaments to 3 mm long, anthers 1–1.2 mm long. Nutlet pale brown (immature?), ellipsoid, c. 1 × 0.4 mm, minutely papillose.

Zambia. C: Chongwe Dist., 85 km E of Lusaka, Chinyunyu hotspring, 27.ii.2000, *Bingham & Peters* 112143 (K). S: Mazabuka Dist., Kafue floodplain, Mazabuka, 6.ii.1963, *van Rensburg* 1348 (K). **Malawi**. E: Chitipa Dist., 5 km SE of Chitipa (Fort Hill), 11.iii.1961, *Robinson* 4437 (K).

Also in Tanzania. Floodplain and seasonally wet dambo edges; 950–1300 m.

Conservation notes: Widespread; not threatened. Data Deficient in F.T.E.A. (2010).

Excluded species.

Kyllinga buchananii C.B. Clarke in Fl. Cap. **7**: 155 (1897). Type: South Africa, KwaZulu-Natal, Umzula, n.d., *Mudd* s.n. (K).

In Flora Capensis Clarke says "Also in the Shire Highlands, British Central Africa", but does not cite any specimens. However, at K there is a specimen (*Buchanan* 159) from the Shire Highlands (now Malawi) with a note on it by C.B. Clarke dated 19.ii.1895 stating it is *Kyllinga buchanani* (sic) C.B. Clarke MS. I have determined *Buchanan* 159 as *K. crassipes*.

Kyllinga comosipes (Mattf. & Kük.) Napper in J. E. Africa Nat. Hist. Soc. Natl. Mus. **28**(124): 24 (1971). —Beentje in F.T.E.A., Cyperaceae: 338 (2010). Type: Tanzania, Uyanzi (Dodoma?) Dist., Mgunda Mkhali, 1240 m, 28.xii.1860, *Speke & Grant* s.n. (K holotype).

    Kyllinga leucocephala Boeckeler in Flora **58**: 257 (1875), illegitimate name. —Clarke in F.T.A. **8**: 287 (1901), not *Kyllinga leucocephala* Baldwin or *Cyperus leucocephala* Retz. Type as for *K. comosipes*.

    Cyperus comosipes Mattf. & Kük. in Engler, Pflanzenr. **4**, 20(101): 568 (1936), new name for *Kyllinga aurea*. —Haines & Lye, Sedges & Rushes E. Afr.: 228 (1983).

    Kyllinga chrysantha K. Schum. var. *comosipes* (Mattf. & Kük.) J.-P.Lebrun & Stork, Énum. Pl. Fleurs Afr. Trop. **3**: 191 (1995).

Found in Ethiopia, ?Somalia, Tanzania and South Africa.

Reported in F.T.E.A. (2010) to occur in Zambia and Zimbabwe. I could only find a Zambian specimen, which turned out to be *Kyllinga polyphylla*.

Kyllinga controversa Steud., Syn. Pl. Glumac. **2**: 70 (1854). —Clarke in F.T.A. **8**: 270 (1901). Type: Ethiopia, hills near Enderder, 16.vii.1838, *Schimper* 581 (B holotype, BM, M).

    Cyperus controversus (Steud.) Mattf. & Kük. in Engler, Pflanzenr. **4**, 20(101): 611 (1935).

The F.T.A. (1901) cites a specimen from Mozambique (Lupata, *Kirk* s.n.) under this species, a specimen I have since determined as *Kyllinga alba*.

Kyllinga pulchella Kunth, Enum. Pl. **2**: 137 (1837). —Clarke in F.T.A. **8**: 284 (1901). —Podlech in Merxmüller, Fl. S.W.A. **165**: 29 (1967). —Napper in J. E. Africa Nat. Hist. Soc. Natl. Mus. **28**(124): 23 (1971). —Getliffe in J. S. African Bot. **49**: 271, fig.5 (1983). —Gordon-Gray in Strelitzia **2**: 120 (1995). —Beentje in F.T.E.A., Cyperaceae: 316 (2010). Type: South Africa, Cape of Good Hope, 22.ii.1833, *Drège* 7384 (B holotype, K, P).

    Cyperus teneristolon Mattf. & Kük. in Engler, Pflanzenr. **4**, 20(101): 574 (1936). —Haines & Lye, Sedges & Rushes E. Afr.: 233, fig.470 (1983), new name for *Kyllinga pulchella*.

Cyperus transitorius Kük. in Engler, Pflanzenr. **4**, 20(101): 35 (1935), 574 (1936). Types: Tanzania, Dodoma Dist., Saranda, *Peter* 33396 & 33587 (B syntypes).

Also in Eritrea, Ethiopia, Kenya, Tanzania, Namibia, South Africa. Seasonally swampy grassland, seepage zone on rock, black clay soils and streamsides.

I did not find any specimens of this taxon from our area, although it occurs to the north and south.

24. **VOLKIELLA** Merxm. & Czech[24]

Volkiella Merxm. & Czech in Mitt. Bot. Staatssamml. München **8**: 317 (1953).

Plants inconspicuous, under or above ground, herbaceous to woody, usually rhizomatous (branched). Culms scapose, short. Leaves soon falling, few to absent, eligulate. Inflorescence terminal, capitate; primary bracts leaf-like, elongate, photosynthetic; partial inflorescences spikelet-like, of several distichous glume-like bracts each subtending a much-reduced spikelet of a solitary bisexual floret enveloped by a veined prophyll and hyaline, veinless glume. Stamens 2. Style short, 3-branched, long. Nutlets trigonous.

One species only, known from Zambia and Namibia. It is closely related to *Lipocarpha*, which is often frequent in localities where *Volkiella* is found.

Recently this genus has been subsumed into *Cyperus* sensu lato (Bauters *et al.* in Phytotaxa **166**: 1–32, 2014). However, it is treated separately here in order to ensure consistency with F.T.E.A.

Volkiella disticha Merxm. & Czech in Mitt. Bot. Staatssamml. München **8**: 318, t.1-2 (1953). —Hooper in Kew Bull. **41**: 945 (1986). Type: Namibia, Okavango, Rundu (Runtu), 7.v.1939, *Volk* 1815 (M holotype, BR, K, PRE). FIGURE 14.**54**.
  Cyperus distichus (Merxm. & Czech) Bauters in Phytotaxa **166**: 19 (2014).

Small, inconspicuous, tufted annual or perennial; rhizome short almost 'woody', adventitious roots bright red; more slender elongate rhizomes give rise to leaves (generally few and falling early) and stalked inflorescences. Culm short, less than or slightly exceeding 20 mm long. Inflorescence terminal, of 1–3(4) much-reduced short branches (spikes), one central, that may, or may not, branch laterally to form a somewhat flattened close head; each inflorescence subtended by one long inflorescence bract, extending above ground level and serving photosynthetically in plants otherwise totally buried; 2–3 much shorter inflorescence bracts present in all plants. Rachis of spike somewhat flattened (oval rather than cylindric in cross-section), carries short, subdistichously arranged spikelet bracts, 2.5–4 mm long, each subtending a spikelet consisting of a delicate rachilla with one bisexual floret closely enveloped by a veined prophyll and hyaline, veinless glume. (Without careful dissection and magnification the spikelet rachilla will not be observed, and indeed it may not be developed.) Perianth 0. Stamens 2, filaments exceeding glume length; anthers small, connective elongate. Style shorter than its 3 branches. Nutlet trigonous, cordate-obovoid, c. 1 mm long; surface finely puncticulate.

Zambia. S: Namwala Dist., between Namwala and Baambwe, 17.iv.1963, *van Rensburg* 2005 (K).

Known only from western parts of southern Africa (Zambia and Namibia). In Zambia, small above-ground caespitose plants are known from heavily grazed grassland on sand, while in Namibia it is recorded from periodically inundated areas; 1000–1100 m.

Conservation notes: Unknown status, but probably not threatened.

---

[24] by J. Browning and K.D. Gordon-Gray†

Fig. 14.**54**. VOLKIELLA DISTICHA. 1, 2, habit; 3, spike with bract and prophyll; 4, prophyll; 5, 6, spikelet bract, lateral and adaxial view; 7, spikelet, prophyll and glume displaced; 8, floret, glume removed; 9, immature style and ovary; 10, immature nutlet. All from *van Rensburg* 2005. Scale bars: 1, 2 = 10 mm; 3 = 5 mm; 4–9 = 2 mm; 10 = 1 mm. Drawn by Jane Browning.

It is probable that mature fruits and vegetative fragments are distributed by water and wind, growing aerially until inundated by flooding and submersed under deposited soil and detritus.

The reproductive organs differ in detail from those of *Lipocarpha*; the differences are explicable by soil pressure following inundation. It is debatable whether *Volkiella* should be maintained as separate from *Lipocarpha*.

25. **LIPOCARPHA** R. Br.[25]

Lipocarpha R. Br. in Tuckey, Narr. Exped. Zaire, app.5: 459 (1818). —Goetghebeur & Borre in Wageningen Agric. Univ. Pap. **89**(1): 1–87 (1989).

Annual herbs, more rarely caespitose or rhizomatous perennials. Culm erect, ± cylindrical, glabrous and smooth. Leaves basal, linear, with a rather short, closed leaf sheath, no ligule, flat or rarely ± inrolled, with a smooth glabrous leaf blade. Inflorescences terminal, head-like, with few to many spikes or reduced to a single spike, surrounded by 1 to few spike-bearing or empty bracts, lowermost bract(s) elongate, uppermost bracts smaller and resembling spikelet bracts; inflorescence sometimes pseudolateral due to the permanent upright position of lowermost spike bract. Spikes composed of a cylindrical axis with many well-developed spikelet bracts in a dense helicoidal arrangement, lateral spikes with an empty spike prophyll. Each spikelet bract is subtended by a highly reduced 1-flowered spikelet with 2 tiny hyaline scales, an empty adaxial prophyll, and a flower-bearing abaxial glume; either glume or both scales in some species can be partly to completely reduced. Spikelet bracts are ± obovate to spathulate, with a short to elongate apical part. The single bisexual flower has a 3- or bimerous ovary with a (very) short to medium-sized style and 2 or 3 stigmatic branches, 1–3 stamens at ribs of ovary, and no perianth. Fruits (broadly) obovate, ellipsoid to narrowly subcylindrical, often constricted at base and apically with a small stylebase remnant, rarely beaked.

A genus of 35 species from open, often weedy vegetation on temporarily moist or peaty soils; 14 species in the Flora area. The perennial species tend to grow more often under more or less permanently moist to wet conditions, whilst the annuals are found on seasonally wet sandy soils.

Recently this genus has been subsumed into *Cyperus* sensu lato (Bauters *et al.* in Phytotaxa **166**: 1–32, 2014). However, it is treated separately here in order to ensure consistency with F.T.E.A.

1. Rhizomatous perennial with a terminal, stellate, subspherical head 15–20 mm wide; spikelet bract 5–12 mm long, white apical part 3–5 times longer than violet body . . . . **5.** *comosa*
– Inflorescence head different; spikelet bract shorter than 3.9 mm. 2
2. Lower involucral bract always strictly erect; annuals. 3
– Lower involucral bract (obliquely) patent to reflexed, rarely erect in young inflorescence; annuals or perennials 6
3. Spikelet bract with short apical part, 0.4–0.7 mm wide; spikelet prophyll and glume always present; inflorescence of 1 spike; stamen single; stigmata 2 . . . . **7.** *hemisphaerica*
– Spikelet bract with longer apical part, at least ⅕ of basal part; spikelet prophyll and glume present, sometimes reduced or absent; inflorescence of 1 or more spikes 4

[25] by K. Bauters

4. Spikes 1–3; spikelet glume absent, prophyll present but shorter than fruit, often deeply bifid, rarely completely reduced; fruit thick, straight, 0.4–0.65 × 0.2–0.3 mm; spikelet bract 0.5–1.1 × 0.3–0.4 mm, shortly acuminate . . . .**9.** *micrantha*
–   Spike always single, horizontal-pseudolateral; no second involucral bract present (or very short); spikelet prophyll and glume present 5
5. Top of spikelet bract smooth and rounded; stigmata 2 . . . .**10.** *monostachya*
–   Top of spikelet bract with a few spinules; stigmata 3 . . . . **11.** *nana*
6. Apical part of spikelet bract more than ½ of bract 7
–   Apical part of spikelet bract less than ½ of bract 9
7. Spikelet prophyll and glume present, stigmas 3 . . . . **11.** *nana*
–   Spikelet prophyll and glume absent 8
8. Fruit conspicuously flattened, stigmas 2 . . . . **8.** *kernii*
–   Fruit rounded, trigonous in cross-section, stigmas 3 . . . .**13.** *rehmannii*
9. Stigmas 2 (rarely 3), fruit dorsiventrally flattened, annuals; spikes (1)3–5; spikelet bract 1.2–1.5 × 0.75–1.2 mm, apical part 0.1–0.25(0.4) mm long . . **12.** *prieuriana*
–   Stigmas 3, fruit rounded, trigonous to subterrete in cross-section 10
10. Perennial, with at least a few thick roots 11
–   Annual, with filiform roots 15
11. Spikes confluent in a pale, rarely purplish head; spikelet bract 1.8–3.9 × 0.5–1.1 mm, apical part 0.6–1.4 mm long . . . .**2.** *albiceps*
–   Spikes clearly individual, pale or dark; if pale then apical part of spikelet bract less than 1/3 of bract 12
12. Spike whitish to pale red brown; apical part of spikelet bract less than 1/3 of bract 13
–   Spike dark; apical part of spikelet bract longer, more than 1/3 of bract 14
13. Plant conspicuously rhizomatous with isolated stems; leaf blade inrolled to subterete, rather pungent; inflorescence with one single involucral bract . . . . . .
. . . .**14.** *robinsonii*
–   Plant not or slightly rhizomatous, stems tufted; leaf blade not or slightly inrolled, soft; involucral bracts 2–3(5) . . . .**4.** *chinensis*
14. Spikelet bract 1.3–1.6 mm wide, conspicuously and broadly shouldered, apical part abruptly narrowed . . . . **1.** *abietina*
–   Spikelet bract less than 1 mm wide, not or slightly shouldered, apical part gradually narrowed . . . . **3.** *atra*
15. Spikelet prophyll and glume dark; inflorescence extremely dark . . . . . .**6.** *echinus*
–   Spikelet prophyll and glume hyaline; apical part of spikelet bract conspicuously narrowed; spikes 2–12, cylindrical, 3–12 mm long . . . .*filiformis*

1. **Lipocarpha abietina** Goetgh. in Wageningen Agric. Univ. Pap. **89**(1): 19 (1989). —Beentje in F.T.E.A., Cyperaceae: 355 (2010). Type: Burundi, unclear locality, 6.vi.1952, *Michel* 2487 (BR holotype, K, MO, NY).

Lipocarpha triceps (Roxb.) Nees var. *latinux* Kük. in Repert. Spec. Nov. Regni Veg. **40**, Anhang: 123 (1936). Type: Tanzania, Buha Dist., Musosi (Mchaji) to Bugaga, *Peter* 37325 (B holotype).

Cyperus abietinus (Goetgh.) Bauters in Phytotaxa **166**: 18 (2014).

Tufted perennial; stem 45–80 cm tall, 1.25–1.5 mm across. Roots to 1 mm diameter. Leaves to 45 cm long, 1 mm wide, often inrolled. Involucral bracts 1–2, largest to 4.5 cm long. Inflorescence terminal, with 4–7 spikes, 2.5–10 × 2–4.5 mm, ovoidal to conical. Spikelet bract 1.5–2.1 × 1.3–1.6 mm, broadly obtrullate, conspicuously shouldered, apical part 0.6–0.7 mm long, acuminate, red brown, rarely dark brown, with pale top. Spikelet prophyll and glume 1.25–1.5 mm long. Stamens 2, anthers 0.7–0.75 mm long. Style 0.1–0.3 mm long, with 3 branches. Nutlet 0.9–1.1 × 0.3–0.5 mm, frontally obovate, rounded trigonous in cross-section.

Botswana. N: Okavango Swamps, Xobega lediba (Gobega lagoon), 7.iii.1972, *Biegel & Gibbs Russell* 3891 (K, LISC, MO, SRGH).

Widespread from Senegal to Central African Republic and Tanzania south to D.R. Congo, Angola and Botswana. On seasonally wet soils, mostly sandy grasslands; c. 1000 m.

Conservation notes: Widespread; not threatened.

The scar left by the deciduous spikelet bract on the spike axis has the form of a conspicuously winged V, whilst in *Lipocarpha atra* this V has only very small wings.

2. **Lipocarpha albiceps** Ridl. in Trans. Linn. Soc. London, Bot. **2**: 163 (1884). —Haines & Lye, Sedges & Rushes E. Afr.: 295 (1983). —Beentje in F.T.E.A., Cyperaceae: 354 (2010). Types: Angola, Pungo Andongo, Praesidium, xii.1855, *Welwitsch* 6785 (BM syntype); Pungo Andongo, Sansamande, iii.1857, *Welwitsch* 6786 (BM syntype); Pungo Andongo, Catete, iii.1857, *Welwitsch* 6786 (BM syntype, COI).

 Hypaelyptum albiceps (Ridl.) K. Schum. in Engler, Pflanzenw. Ost-Afrikas **C**: 127 (1895).

 Cyperus echinolepis T. Koyama in Bot. Mag. (Tokyo) **73**: 438 (1960). Type: Angola, Pungo Andongo, Praesidium, xii.1855, *Welwitsch* 6785 (BM syntype); Pungo Andongo, Sansamande, iii.1857, *Welwitsch* 6786 (BM syntype); Pungo Andongo, Catete, iii.1857, *Welwitsch* 6786 (BM syntype, COI).

Shortly rhizomatous or tufted perennial; stem 10–60 cm tall, 1–2(3) mm across. Rhizome to 3 mm in diameter, covered by red-brown cataphylls. Leaves to 20 cm long, 1.5 mm wide, ± inrolled. Involucral bracts (1)2–3(4), largest to 9(13) cm long. Inflorescence terminal, with 1–6 confluent spikes, terminal spike 3–10(15) × 3–7 mm, lateral spikes 2–6 × 1.5–4 mm, ovoidal to conical. Spikelet bract 1.8–3.9 × 0.5–1.1 mm, obtrullate, apical part 0.6–1.4 mm long, apiculate, body and top creamy, or body (partly) purplish; spikelet prophyll and glume 1.35–2.2 mm long. Stamens 3, anthers 0.9–1.4 mm long. Style 0.5–1.4 mm long, with 3 branches. Nutlet 0.8–1.2 × 0.3–0.75 mm, obovate in frontal view with small style base remnant, rounded trigonous in cross-section.

Zambia. B: Senanga Dist., Mashi R. by Angola border, 30.ix.1964, *Verboom* 1769 (K). N: Kawambwa Dist., Nchelenge–Luapula road, 900 m, 1.xii.1961, *Richards* 15465 (K). W: Mwinilunga Dist., Kalenda plain, N of Matonchi Farm, 8.xii.1937, *Milne-Redhead* 3552 (BM, BR, K, S). C: Chongwe Dist., Chakwenga headwaters, 10.i.1964, *Robinson* 6180 (K, M, NY, SRGH). E: Katete Dist., Katete, St Frances' Hospital, 1070 m, 12.i.1958, *Wright* 217 (K). S: Choma Dist., Muckle Neuk, 19.3 km N of Choma, 13.i.1954, *Robinson* 454 (K). **Zimbabwe**. C: Goromonzi Dist., Bromley, Liemba Farm, Carter Estate, 3.ii.1999, *Goetghebeur* 9167 (GENT). E: Nyanga Dist., Nyamaropa, Regina Coeli Mission, 16.i.1967, *Biegel* 1747 (K, MO, SRGH). **Malawi**. N: Mzimba Dist., 10 km NW of Mzuzu, 7.iii.1976, *Pawek* 10888 (K, MO, SRGH, WAG).

Senegal and Chad to East Africa and south to Angola and Zimbabwe. Open woodlands and grassland on sandy and lateritic soils, dambos and along rivers; 800–1600 m.

Conservation notes: Widespread; not threatened.

The notable differences in colour between typical *Lipocarpha albiceps* (creamy) and *L. purpureolutea* Ridl. (purplish-violet) are not matched by any other morphological feature. The dark specimens are centred in Angola and the D.R. Congo, but many transitional forms blur the picture.

3. **Lipocarpha atra** Ridl. in Trans. Linn. Soc. London, Bot. **2**: 162 (1884). —Clarke in F.T.A. **8**: 472 (1902). —Haines & Lye, Sedges & Rushes E. Afr.: 297 (1983). —Beentje in F.T.E.A., Cyperaceae: 355 (2010). Types: Angola, Huíla, Lake Ivantala, ii.1860, *Welwitsch* 6961 (BM syntype, LISU); Huíla, Lake Ivantala, 1860, *Welwitsch* s.n. (BM syntype).

Lipocarpha barteri C.B. Clarke in F.T.A. **8**: 472 (1902). Type: Nigeria, Nupe, 1857–1859, *Barter* 1585 (K holotype, GS, S).

Lipocarpha atra var. *barteri* (C.B. Clarke) J. Raynal in Adansonia, n.s. **7**: 85 (1967).

Lipocarpha atra var. *atra*. —Mapaura & Timberlake, Checkl. Zimbabwe Pl.: 88 (2004).

Cyperus lipoater Goetgh. in Phytotaxa **166**: 20 (2014), new name, non *Cyperus ater* Vahl (1805).

Shortly rhizomatous or tufted perennial with a short rhizome, rarely elongated; stem 20–65 cm long, 1–1.5 mm across. Roots thick, to 1 mm diameter. Leaves to 20 cm long, 1.5 mm wide, often inrolled. Involucral bracts 1–2, largest up to 4.5 cm long. Inflorescence terminal, with (3)4–10 spikes, 3–13 × 2–4 mm, ovoidal to conical. Spikelet bract 1.2–1.75 × 0.65–1 mm, obovate to obtrullate, apical part 0.5–0.7 mm long, acuminate, smooth to slightly scaberulous at top, dark red-brown, top often pale; spikelet prophyll and glume 1–1.2 mm long. Stamens 2, anthers 0.25–0.35 mm long. Style 0.05 mm long or shorter, with 3 branches. Nutlet 0.7–1.0 × 0.2–0.45 mm, in front view obovate with small style base remnant, rounded trigonous in cross-section.

Zambia. B: Zambezi Dist., Zambezi (Balovale), 8.vii.1963, *Robinson* 5578 (K). N: Kasama Dist., off Luwingu road, 1275 m, 29.ii.1960, *Richards* 12653 (K). W: Mwinilunga Dist., banks of Luao R., 27.xii.1937, *Milne-Redhead* 3841 (BM, BR, K, L, MO, NY, P). C: Lusaka Dist., Munali school, 13.iv.1956, *Robinson* 1463 (BR, K, NY, SRGH). S: Chongwe Dist., Chakwenga headwaters, 8.ix.1963, *Robinson* 5640 (K). **Zimbabwe**. W: Matobo Dist., Besna Kobila farm, iv.1957, *Miller* 4335 (K, NY, SRGH). C: Marondera Dist., Grasslands Research Station, i-ii.1948, *Colvile* 61 (K, SRGH). **Mozambique**. N: Ribáuè Dist., Ribáuè, 615 m, 18.iii.1942, *Tinoco* 33 (LISC).

Also in Tanzania, D.R. Congo and Angola. In moist grasslands, drier dambos, sandy ditches of dry rivers, along rivers and streams; 500–1500 m.

Conservation notes: Widespread; not threatened.

4. **Lipocarpha chinensis** (Osbeck) J. Kern in Blumea, suppl. **4**: 167 (1958). —Hooper in F.W.T.A., ed.2 **3**: 328 (1972). —Haines & Lye, Sedges & Rushes E. Afr.: 296 (1983). —Gordon-Gray, Cyperaceae Natal: 121 (1995). —Beentje in F.T.E.A., Cyperaceae: 354 (2010). Type: China, no locality, n.d., *Osbeck* s.n. (S holotype). FIGURE 14.**55**.

Scirpus chinensis Osbeck, Dagb. Ostind. Resa: 220 (1757).

Scirpus senegalensis Lam., Tab. Encycl. **1**: 140 (1791). Type: Senegal, no locality, n.d., *Rousillon* s.n. coll. 1789 (P-LA holotype, G, P).

Lipocarpha argentea (Vahl) R. Br. in Tuckey, Narr. Exp. Zaire, app.5: 477 (1818). —Rendle in Hiern, Cat. Afr. Pl. **2**: 129 (1899). —Clarke in F.T.A. **8**: 469 (1902). —Peter, Fl. Deutsch-Ostafr.: 383 (1937), superfluous name.

Lipocarpha senegalensis (Lam.) Durand & H. Durand, Syll. Fl. Congol.: 619 (1909).

Cyperus lipocarpha T. Koyama in Bot. Mag. (Tokyo) **73**: 438 (1960). Type as for *Lipocarpha chinensis*.

Tufted perennial; stem 15–80 cm long, 0.5–2 mm across, round or obscurely angled. Leaves glaucous or pale green, to 40 cm long, 4 mm wide, ± flat or sometimes inrolled. Involucral bracts 2–3(5), largest to 13 cm long. Inflorescence terminal, with 2–12 subequal spikes, 3–13 × 1.5–5 mm, ovoidal to conical. Spikelet bract 1.5–2.35 × 0.45–1 mm, white to creamy with red stripes; spikelet white to pale yellowish brown, often with red dots and green midnerve; spikelet prophyll and glume 1.25–2 mm long, often with reddish stripes. Stamens 1(2), anthers 0.8–1 mm long. Style 0.2–0.8 mm long, with 3 branches. Nutlet 0.8–1.15 × 0.25–0.4 mm, in frontal view oblong to narrowly obovate with small style base remnant, (rounded) trigonous in cross-section.

Botswana. N: Ngamiland Dist., Santantadibe R., 19°24.2'S 23°15.4E, 11.i.1978, *P.A. Smith* 2183 (K, SRGH). **Zambia**. B: Kaoma Dist., Mangango Mission, Luena R., 17.iv.1964, *Verboom* 1718 (K). W: Mwinilunga Dist., Chingabola dambo near Matonchi, 1350 m, 16.ii.1975, *Hooper & Townsend* 105 (K). N: Nchelenge Dist., E

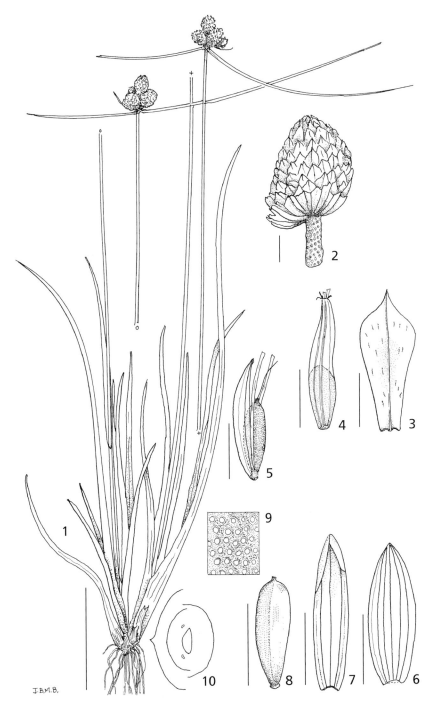

J.B.M.B.

Fig. 14.**55**. LIPOCARPHA CHINENSIS. 1, habit; 2, spike; 3, bract; 4, spikelet; 5, spikelet and glume, prophyll removed; 6, prophyll, 7, glume; 8, 9, nutlet and surface detail; 10, diagram of spikelet. All from *Robinson* 1559. Scale bars: 1 = 40 mm; 2–8 = 1 mm. Drawn by Jane Browning.

of Lake Mweru, 1525 m, vi.1961, *Chongo* 18 (BR, K, SRGH, WAG). C: Lusaka Dist., Nyanshishi R. near Chinhuli, 15°16'S 28°36'E, 1100 m, 10.xii.1972, *Kornaś* 2796 (GENT). E: Lundazi Dist., Nsefu game camp, Luangwa R. bed, 750 m, 15.x.1958, *Robson* 117 (K, LISC). S: Choma Dist., 8 km E of Choma, 26.iii.1955, *Robinson* 1171 (K). **Zimbabwe**. N: Gokwe Dist., Chimvuri vlei, 2 km SW of Gokwe, 12.xi.1963, *Bingham* 907 (K, SRGH). W: Matobo Dist., Quaringa Farm, 1460 m, iv.1957, *Miller* 4305 (K, SRGH). C: Harare Dist., S side of Cleveland dam c. 10 km W of Harare, 1500 m, 13.xii.1991, *Laegaard* 15792 (GENT). E: Chipinge Dist., Chipinga Expt. Farm, ropadside near Nyokari R., 15.ii.1962, *Chase* 7616 (K, SRGH). S: Masvingo Dist., Morgenster Mission, 40 km SE of Masvingo (Fort Victoria), 5.iii.1970, *Mavi* 1069 (K, SRGH). **Malawi**. N: Mzimba Dist., Mzuzu, Marymount dambo, 19.vii.1973, *Pawek* 7208 (K, MO, SRGH, WAG). C: Mchinji Dist., Mchinji, banks of Bua R. on road to Zambia, 30.iii.1970, *Brummitt* 9547 (K, LISC). S: Machinga Dist., Kawinga, 8 km NE of Ntaja by Mikoko bridge, 16.ii.1979, *Blackmore, Brummitt & Banda* 450 (K, MAL). **Mozambique**. N: Murrupula Dist., between Murrupla and Chefe Amoceta, 24.i.1961, *Carvalho* 461 (K, LMA). MS: Manica Dist., Bandula, 6.iv.1953, *Chase* 4544 (K, LISC, MO, SRGH, UPS).

Widespread in tropical Africa, Asia and Australia, also in Angola and South Africa (KwaZulu-Natal). On moist soils in grasslands, dambos, along rivers and streams; 400–1800 m.

Conservation notes: Widespread; not threatened.

The involucral bracts are sometimes deciduous; the smaller ones are shed first, the largest (= lowermost) last of all, conspicuously so on several inflorescences of *Kornaś* 2796 (GENT).

5. **Lipocarpha comosa** J. Raynal in Bull. Mus. Natl. Hist. Nat., ser.2 **41**: 974 (1970). — Haines & Lye, Sedges & Rushes E. Afr.: 295 (1983). —Burrows & Willis, Pl. Nyika Plat.: 301 (2005). —Beentje in F.T.E.A., Cyperaceae: 347 (2010). Type: Zambia, Chongwe Dist., Chakwenga headwaters, 14.ii.1965, *Robinson* 6380 (P holotype, K, NY).

Cyperus lipocomosus Goetgh. in Phytotaxa **166**: 21 (2014), new name, non *Cyperus comosus* Sm. (1806).

Rhizomatous perennial; stem 15–55 cm long, 1–1.5 mm across. Rhizome, slender, 1–3 mm diameter, covered by small red-brown cataphyll. Leaves to 20 cm, 1.5 mm wide, inrolled to subterete. Involucral bracts 2–3, largest 4–8 cm long. Inflorescence terminal, capitate, with several confluent spikes, 15–20 mm wide. Spikelet bract 5–12 × 0.4–1 mm, body obovate, yellowish to violet, apical part 4–10 mm long, long acuminate, creamy; spikelet prophyll and glume 1–1.7 mm long. Stamens 3, anthers 1 mm long. Style 0.2 mm long with 3 branches. Nutlet 1–1.2 × 0.5–0.6 mm, obovate in frontal view with small style base remnant, (slightly rounded) trigonous in cross-section.

Zambia. N: Isoka Dist., 20 km SE of Tunduma, 1.i.1959, *Robinson* 2977 (K, NY, PRE). W: Ndola Dist., Bwana Mkubwa, 31.xii.1907, *Kassner* 2273 (BM, BR, K, P, Z). C: Mkushi Dist., Fiwila, 5.i.1958, *Robinson* 2626 (K, NY, P, SRGH). S: Chongwe Dist., Chakwenga headwaters, 100–129 km E of Lusaka, 14.ii.1965, *Robinson* 6380 (K, NY, P). **Malawi**. N: Rumphi Dist., Nyika Plateau, 13 km N of M1, 1690 m, 23.xii.1977, *Pawek* 13286 (K, MAL, MO). C: Dedza Dist., Chongoni Forest, 8.ii.1961, *Adlard* 402 (K, SRGH).

Also in D.R. Congo and Tanzania. In seasonally moist grasslands and miombo woodland; 1000–1800 m.

Conservation notes: Widespread; not threatened.

6. **Lipocarpha echinus** J. Raynal in Adansonia, n.s. **13**: 159, pl.6 (1973). Type: Zambia, Chinsali Dist., Shiwa Ngandu, 2.vi.1956, *Robinson* 1550 (NY holotype, K).

    Cyperus echinus (J. Raynal) Bauters in Phytotaxa **166**: 19 (2014).

Tufted annual with thin roots; stem 5–20 cm long, 0.4–0.6 mm across. Leaves to 7 cm long, 0.5 mm wide, inrolled. Involucral bracts 1–3, largest to 5 cm long. Inflorescence terminal, with 3–5 spikes, 3–7 × 2–3 mm, ovoidal to conical. Spikelet bract 1.3–1.7 mm × 0.8 mm, obovate, apical part 0.25–0.45 mm long, long acuminate, scabrid at top, dark violet, with pale to greenish midnerve; spikelet prophyll and glume 0.7–0.8 mm long, dark violet. Stamens 3. Style 0.1 mm or shorter, with 3 branches. Nutlet 0.75–0.8 × 0.5–0.6 mm, frontally obovate with small style base remnant, subtriquetrous in cross-section.

Zambia. N: Chinsali Dist., Shiwa Ngandu, 2.vi.1956, *Robinson* 1550 (K, NY). Not found elsewhere. In seasonally wet grasslands and swamps; c. 1500 m.
Conservation notes: Known only from the type; probably threatened.

7. **Lipocarpha hemisphaerica** (Roth) Goetgh. in Wageningen Agric. Univ. Pap. **89**(1): 37 (1989). —Gordon-Gray, Cyperaceae Natal: 122 (1995). —Beentje in F.T.E.A., Cyperaceae: 348 (2010). Type: India ('India orientale'), no locality, *Heyne* s.n. (B† holotype, mixed collection). FIGURE 14.**56**.

    Scirpus hemisphaericus Roth, Nov. Pl. Sp.: 29 (1821).
    Isolepis hemisphaerica (Roth) A. Dietr., Sp. Pl. **2**: 109 (1833).
    Hemicarpha isolepis Nees in Edinburgh New Philos. J. **17**: 263 (1834). Type: India ("Peninsula India orientalis"), n.d., *Wight* 1856 (C, G, GE, K, LE, NY; not clear which is holotype).
    Scirpus isolepis (Nees) Boeckeler in Linnaea **36**: 498 (1870). —Clarke in F.T.A. **8**: 459 (1902). —Hooper in F.W.T.A., ed.2 **3**: 310 (1972).
    Lipocarpha isolepis (Nees) R.W. Haines in Bot. Not. **124**: 476 (1971). —Goetghebeur in Adansonia, n.s. **19**: 303 (1979). —Haines & Lye, Sedges & Rushes E. Afr.: 300 (1983). —Mapaura & Timberlake, Checkl. Zimbabwe Pl.: 88 (2004).
    Cyperus isolepis (Nees) Bauters in Phytotaxa **166**: 19 (2014).

Tufted annual with thin roots; stem 2–15(25) cm long, 0.25–0.5 mm across. Leaves to 6 cm long, 0.6 mm wide. Involucral bract 1, to 10(16) cm long, erect. Inflorescence pseudolateral, with 1 spike, 1–7(10) × 1–3 mm, spherical, ovoidal to conical. Spikelet bract 0.5–0.95 × 0.4–0.7 mm, broadly obovate, apical part 0.2–0.5 mm long, shortly acuminate to obtuse, smooth at top, brown to red-brown, often with pale to greenish midnerve; spikelet prophyll and glume 0.5–0.7 mm long. Stamen 1, anthers 0.25 mm long. Style 0.1 mm or shorter, with 2 branches. Nutlet 0.5–0.75 × 0.25–0.3 mm, frontally obovate with small style base remnant, rounded to elliptical in cross-section.

Botswana. N: Ngamiland Dist., Okavango Delta, 3 km NW of Duba Plains Camp, 19°00.053'S 22°40.669'E, 9.iv.2013, *Heath* 2644 (K). **Zambia**. N: Mbala Dist., Niamkola, near Lake Tanganyika, 11.vi.1961, *Robinson* 4710 (K). W: Ndola, Itawa dambo, 1300 m, 3.iv.1960, *Robinson* 3456 (K). C: Chongwe Dist., E of Lusaka, 29.v.1958, *Robinson* 2869 (K, M, NY, SRGH). S: Choma Dist., Muckle Neuk, 19.3 km N of Choma, 14.i.1954, *Robinson* 467 (K). **Zimbabwe**. W: Hwange Dist., Victoria Falls, 10.iii.1932, *Brain* 8728 (BR, K, LISC, MO, S, SRGH). C: Harare, Old Charter Road, 27.iii.1952, *Wild* 3790 (K, LISC, NY, S, SRGH). **Malawi**. N: Chitipa Dist., 15 km E of Chitipa at Kaseye Mission, 18.iv.1976, *Pawek* 11080 (BR, K, MO, SRGH, WAG). S: Mulanje Dist., 16 km NW of Likabula Forestry Depot, 700 m, 15.vi.1962, *Robinson* 5355 (K).

Widespread from Senegal to Ethiopa, Tanzania and Kenya south to Angola, Namibia and South Africa (KwaZulu-Natal); also in India and Thailand. On moist sandy soils, dambos, alluvial plains along rivers, disturbed areas and on wet rocks; 1000–1600 m.
Conservation notes: Widespread; not threatened.

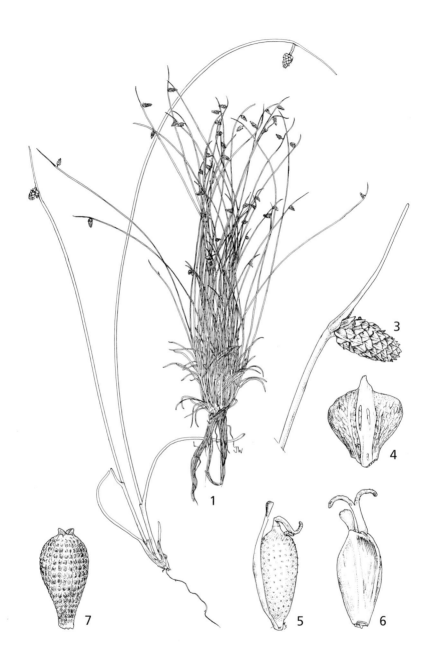

Fig. 14.**56**. LIPOCARPHA HEMISPHAERICA. 1, habit (× 1); 2, small habit (× 2); 3, spikelet (× 8); 4, glume (× 50); 5, floret (× 50); 6, floret (× 50); 7, nutlet (× 50). 1, 4, 7 from *Milne-Redhead & Taylor* 9944; 3, 5, 6 from *Richards* 19108. Drawn by Juliet Williamson. Reproduced from Flora of Tropical East Africa (2010).

8. **Lipocarpha kernii** (Raymond) Goetgh. in Wageningen Agric. Univ. Pap. **89**(1): 42 (1989). —Beentje in F.T.E.A., Cyperaceae: 352 (2010). Type: Senegal, Simiti, 12.i.1954, *Berhaut* 4692 (MT holotype, P).

> *Scirpus squarrosus* sensu C.B. Clarke in F.T.A. **8**: 458 (1902) in part for *Schweinfurth* 2572 & 3003, non L.
>
> *Scirpus kernii* Raymond in Naturaliste Canad. **86**: 230 (1959). —Raynal in Adansonia, n.s. **8**: 95 (1968). —Hooper in F.W.T. A., ed.2 **3**: 310 (1972).
>
> *Rikliella kernii* (Raymond) J. Raynal in Adansonia, n.s. **13**: 155 (1973). —Haines & Lye, Sedges & Rushes E. Afr.: 301 (1983).
>
> *Cyperus kernii* (Raymond) Bauters in Phytotaxa **166**: 20 (2014).

Tufted annual with thin roots; stem 2–40 cm long, 0.5–1.5 mm across. Leaves to 16 cm long, 2 mm wide. Involucral bracts 2–5, largest to 15 cm long. Inflorescence terminal, with (1)2–8 spikes, 2–8 × 1.5–5 mm, ovoidal. Spikelet bract 1.45–2.6 × 0.4–0.55 mm, elliptical to obovate, apical part 0.7–1.5 mm long, (very) long acuminate, scaberulous at top, body yellowish green to pale brown with red dots, top yellow; spikelet prophyll and glume absent. Stamen 1, anthers 0.2–0.25 mm long. Style 0.1 mm or shorter, with 2 branches. Nutlet 0.45–0.65 × 0.3–04 mm, frontally obovate with very small style base remnant, dorsiventrally flattened and elliptical to narrowly rhombic in cross-section.

Zambia. N: Mbala Dist., Lake Tanganyika, Namkolo, 22.iii.1960, *Richards* 12768 (K). **Zimbabwe**. N: Omay Dist., Matusadonha Nat. Park, Tashinga Camp, 11.iv.1978, *Mushori* 97 (SRGH, not found). **Malawi**. N: Karonga Dist., Kondowe to Karonga, vii.1896, *Whyte* s.n. (K). C: Nkhotakota Dist., Nkhotakota (Kota-Kota), 2.v.1963, *Verboom* 110S (K).

From Senegal and Ethiopia south to Zimbabwe. On seasonally wet sandy soils, in swamps and grasslands on shallow soil; 400–1000 m.

Conservation notes: Widespread; not threatened.

Taxonomic placement of this species has always been difficult. Raynal (1968) suggested that *Scirpus kernii* had originated from *Ascolepis* or *Lipocarpha* by the loss of their hypogynous scales. In 1973 he created a new genus, *Rikliella*, for this species, also including *Scirpus squarrosus* L. and *Scirpus rehmannii* Ridl. This new genus was supposed to be clearly related to either *Ascolepis* or *Lipocarpha*, depending on interpretation of the remaining scale either as the spikelet glume or spikelet bract. Raynal himself was inclined to accept *Ascolepis* as the closer relative, and thus interpreted the remaining scale as a spikelet glume. Later on, Goetghebeur (1980) argued that a derivation from *Lipocarpha* was much more probable due to the presence of a spike prophyll in *Rikliella* and *Lipocarpha* and the absence of this in *Ascolepis*. Furthermore, the description of *Rikliella australiensis* J. Raynal made *Rikliella* polyphyletic – *Rikliella rehmannii* (Ridl.) J. Raynal and *R. kernii* are closely related while *R. squarrosa* (L.) J. Raynal has a morphology intermediate between *Lipocarpha microcephala* and *R. kernii*. *Rikliella australiensis* is no more than a prophyll- and glumeless form of *Lipocarpha microcephala* (R. Br.) Kunth with no other differences (Wilson 1981). Both 'species' (from two 'genera') were repeatedly collected in one population, leading Goetgebeur and Van den Borre (1989) to include *Rikliella* in *Lipocarpha*. However, recent molecular phylogenetic analysis has shown that *L. kernii* and *L. rehmannii* are not related to other *Lipocarpha* species, suggesting a reinstatement of the genus *Rikliella*. The position of the clade including former *Rikliella* species suggests a different interpretation of the inflorescence found in *Lipocarpha* (Bauters *et al.*, submitted). The remaining scales are spirally arranged, one-flowered glumes, forming a dense headlike spikelet. In *Lipocarpha*, the spikelets are one-flowered and spirally arranged along the rachis, thus forming a spike of spikelets, a so-called pseudospikelet.

9. **Lipocarpha micrantha** (Vahl) G.C. Tucker in J. Arnold Arbor. **68**: 410 (1987). —
Gordon-Gray, Cyperaceae Natal: 122 (1995). Type: French Guiana, no locality,
n.d., *Richard* s.n. (P holotype, C).

> *Scirpus micranthus* Vahl, Enum. Pl. **2**: 254 (1805).
> *Isolepis micrantha* (Vahl) Roem. & Schult., Syst. Veg. **2**: 110 (1817).
> *Hemicarpha micrantha* (Vahl) Pax in Engler & Prantl, Nat. Pflanzenfam. **2**(2): 105 (1888),
> (January 1888).
> *Hemicarpha micrantha* (Vahl) Britton in Bull. Torrey Bot. Club **15**: 104 (1888), superfluous
> combination (April 1888).

Tufted annual with thin roots; stem 2–20 cm long, 0.3 × 0.5 mm across. Leaves to 10 cm long,
0.5 mm wide. Involucral bracts 1–2(3), largest to 3–5 cm long, erect. Inflorescence pseudolateral,
with 1–2(3) spikes, 1–5 × 1–2 mm, ovoidal to spherical. Spikelet bract 0.5–1(1.1) × 0.3–0.4 mm,
obovate to obtrullate, apical part 0.1–0.25 mm long, acuminate, scaberulous at top, yellowish to
red-brown, with greenish midnerve; spikelet prophyll reduced, often deeply bifid, or virtually
absent; spikelet glume absent with small, almost invisible remnant near base. Stamen 1, anthers
0.15 mm long. Style 0.1 mm or shorter, with 2 branches. Nutlet 0.4–0.65 × 0.2–0.3 mm, frontally
obovate to subelliptical with small style base remnant, rounded in cross-section.

Zimbabwe. N: Gokwe Dist., Sengwa, 23.iii.1964, *Jarman* BM22 (SRGH). N: Hurungwe
Dist., Zambezi Valley, Mhenza (Menswa) Pan, 27.ii.1953, *Wild* 4022 (K, MO, SRGH).

Also in N & C America, Namibia and South Africa (KwaZulu-Natal). Banks along
lakes and rivers, on wet sandy soils; c. 600 m.

Conservation notes: Widespread; not threatened.

This species is widespread in N, C and S America but rather rare in Africa. Some
minor variations are found between American and African specimens, but overall
they appear to be the same.

10. **Lipocarpha monostachya** R. Gross & Mattf. in Notizbl. Bot. Gart. Berlin-Dahlen
14: 189 (1938). —Haines & Lye, Sedges & Rushes E. Afr.: 300 (1983). —Beentje
in F.T.E.A., Cyperaceae: 348 (2010). Type: Tanzania, Lindi Dist., 150 km SW of
Lindi, near Masasi, 25.iv.1935, *Schlieben* 6399 (B holotype, BM, BR, G, GENT, M,
P, Z).

> *Cyperus lipomonostachyus* Goetgh. in Phytotaxa **166**: 21 (2014), new name, non *Cyperus*
> *monostachyus* Link (1825).

Tufted annual with thin roots; stem 2–18 cm long, 0.2–0.3 mm across. Leaves to 3.5 cm, 0.5
mm across. Involucral bract 1, to 1.5 cm long. Inflorescence pseudolateral, with 1 spike, 1.5–5.5
× 1–3 mm, ovoidal to conical. Spikelet bract 1–1.6 × 0.4–0.6, obovate, apical part 0.5–1.1 mm
long, acuminate, more or less recurved, smooth at top, yellowish to reddish brown, with mostly
greenish midnerve; spikelet prophyll and glume 0.3–0.5 mm long. Stamen 1, anthers 0.15 mm
long. Style 0.1 mm long or shorter, with 2 branches. Nutlet 0.35–0.65 × 0.2–0.25 mm, frontally
obovate with very small style base remnant, rounded in cross-section.

Zimbabwe. C: Kwekwe Dist., 50 km NE of Kwekwe town, Iwaba estate, 30°08'E
18°50'S, 1200 m, 13.ii.2001, *Lye* 24553 (K). **Malawi.** N: Chitipa Dist., 5 km SE of Chitipa
(Fort Hill), 16.iii.1961, *Robinson* 4530 (M, MO, NY, SRGH). C: Kasungu Dist., 10 km S
of Kasungu, 7.iv.1978, *Pawek* 14342b (K, MAL, MO).

Also in Tanzania and D.R. Congo. In seasonally wet sandy places along lakes, rivers
and streams; 1000–1500 m.

Conservation notes: Widespread; not threatened.

Specimens of this species are often identified as *Lipocarpha nana*, which it closely
resembles. However, *L. monostachya* is easily recognized by the smooth tip of the
spikelet bract and the 2-branched style. Furthermore, its inflorescence is "uniformly
monocapitate over a wide range of habitat" (note on *Robinson* 4530, M), which is
rarely the case in *L. nana*.

11. **Lipocarpha nana** (A. Rich.) Cherm. in Bull. Soc. Bot. France **71**: 142 (1924). —
Hooper in F.W.T.A., ed.2 **3**: 328 (1972). —Haines & Lye, Sedges & Rushes E. Afr.:
299 (1983). —Gordon-Gray, Cyperaceae Natal: 123 (1995). —Beentje in F.T.E.A.,
Cyperaceae: 351 (2010). Type: Ethiopia, Shire, Kouaitea, n.d., *Quartin-Dillon &*
Petit s.n. (P holotype). FIGURE 14.**57**.

 Fuirena nana A. Rich., Tent. Fl. Abyss. **2**: 497 (1850).

 Lipocarpha pulcherrima Ridl. in Trans. Linn. Soc. London, Bot. **2**: 162 (1884). —Clarke
in F.T.A. **8**: 473 (1902). Types: Angola, Catete, xii.1856, *Welwitsch* 6774 (BM syntype, COI);
Quilange, ii.1857, *Welwitsch* 6774 (BM syntype, COI); Huíla, v.1860, *Welwitsch* 6775 (BM
syntype); Catete, ii.1857, *Welwitsch* 6785 in part (BM syntype).

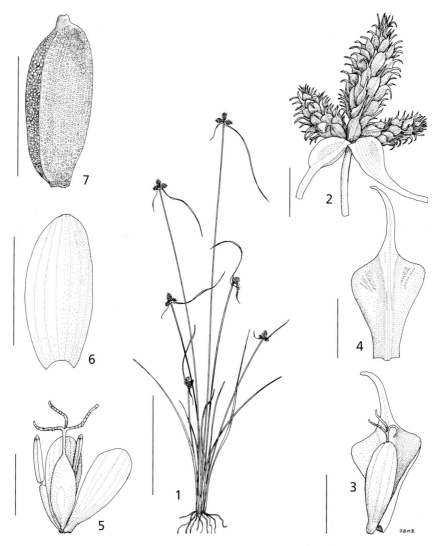

Fig. 14.**57**. LIPOCARPHA NANA. 1, habit; 2, inflorescence; 3, spikelet with bract; 4, bract; 5,
floret, prophyll and glume displaced; 6, glume, flattened; 7, nutlet. All from *Gordon-Gray* s.n.
Scale bars: 1 = 40 mm; 2 = 2 mm; 3–7 = 0.5 mm. Drawn by Jane Browning. Reproduced from
Strelitzia (1995) by kind permission of the South African National Biodiversity Institute, Pretoria.

Lipocarpha atropurpurea Boeckeler, Beitr. Cyper. **1**: 21 (1888). Type: Malawi, Shire Highlands, vii.1885, *Buchanan* 69 (B holotype, K, LE, NY, P).

Lipocarpha tenera Boeckeler, Beitr. Cyper. **1**: 21 (1888). Type: Malawi, Shire Highlands, vii.1885, *Buchanan* 63 (B holotype, K).

Lipocarpha pulcherrima Ridl. forma *luxurians* Merxm. in Mitt. Bot. Staatssamml. München **1**(5): 164 (1952). Type: Mozambique, Manica Dist., Bandula, iv.1952, *Schweickerdt* 2319 (M holotype).

Cyperus persquarrosus T. Koyama in Bot. Mag. (Tokyo) **73**: 438 (1960), new name, non *Cyperus pulcherrimus* Kunth (1837).

Tufted annual with thin roots; stem 3–40 cm long, 0.3–0.75 mm across. Leaves to 7 cm long, 1.2 mm wide. Involucral bracts (1)2(3), largest to 4 cm long. Inflorescence terminal (rarely pseudolateral), with 1–9 spikes, 2–8 × 1.5–4 mm, ovoidal to conical. Spikelet bract 0.9–1.65(2.2) × 0.25–0.8 mm, obtrullate, apical part 0.45–0.8(1.3) mm long, (very) long acuminate, recurved, scabrid at top, body red-brown to black, with pale midnerve and yellowish to greenish apical part; spikelet prophyll and glume 0.3–0.8 mm long. Stamens 1–2, anthers 0.35 mm long. Style 0.15 mm or shorter, with 3 branches. Nutlet 0.55–0.8 × 0.2–0.4 mm, frontally obovate with short style base remnant, rounded trigonous in cross-section.

Zambia. B: Mongu Dist., no locality, 5.iii.1966, *Robinson* 6870 (K, NY, P). N: Mbala Dist., Chilongowelo, Plain of Death, 9.ii.1957, *Richards* 8124 (BR, K, SRGH). W: Mwinilunga Dist., 2 km E of Matonchi Farm, 1350 m, 18.ii.1975, *Hooper & Townsend* 203 (K). C: Serenje Dist., Kundalila Falls, S of Kanona, 1400 m, 12.iii.1975, *Hooper & Townsend* 692 (K). E: Chipata Dist., Lunkwaka dambo, 24.iv.1963, *Verboom* 85S (K). S: Choma Dist., 11 km SW of Mapanza, 20.iii.1955, *Robinson* 1138 (K). **Zimbabwe**. W: Matobo Dist., Besna Kobila farm, iv.1955, *Miller* 2753a (K, LISC, SRGH). C: Harare, 1460 m, 2.v.1931, *Brain* 3744 (K). E: Nyanga Dist., 8 km N of Nyanga village, 28.iv.1967, *Rushworth* 900 (K, L, P, SRGH, WAG). S: Chivi Dist., near Madziviri Dip, 6.4 km N of Runde R., 3.v.1962, *Drummond* 7920 (K, SRGH). **Malawi**. N: Nkhata Bay Dist., 56 km SW of Mzuzu, 27.iii.1976, *Pawek* 10934 (BR, K, MO, SRGH, WAG). C: Mchinji Dist., c. 7 km W of Namitete, 33°16'E 14°01'S, 29.iv.1989, *Radcliffe-Smith et al.* 5787 (K). S: Shire Highlands, n.d., *Buchanan* 69 (K, LE, NY, P). **Mozambique**. MS: Manica Dist., Dororo near Bandula, *Fisher & Schweickerdt* 249 (BM, M).

From Guinea to Sudan and Ethiopia south to Angola and South Africa (KwaZulu-Natal); also in Madagascar. In seasonally wet grasslands, sandy ditches and on rocky outcrops; 600–1800 m.

Conservation notes: Widespread; not threatened.

12. **Lipocarpha prieuriana** Steud., Syn. Pl. Glumac. **2**: 130 (1855). —Clarke in F.T.A. **8**: 471 (1902). —Hooper in F.W.T.A., ed.2 **3**: 328 (1972). —Haines & Lye, Sedges & Rushes E. Afr.: 298 (1983). —Beentje in F.T.E.A., Cyperaceae: 353 (2010). Type: Senegal, no locality, n.d., *Leprieur* s.n. (P holotype, G, L).

Lipocarpha prieuriana Steud. var. *prieuriana*. —Raynal in Adansonia, n.s. **7**: 86 (1967). — Mapaura & Timberlake, Checkl. Zimbabwe Pl.: 88 (2004).

Cyperus prieurianus (Steud.) T. Koyama in Bot. Mag. (Tokyo) **73**: 438 (1960).

Tufted annual with thin roots; stem 5–50 cm long, 0.5–1 mm across. Leaves to 25 cm long, 2 mm wide. Involucral bracts 2, largest to 20 cm long. Inflorescence terminal, with (1)3–5 spikes, 3–10 × 2–4 mm, ovoidal. Spikelet bract 1.2–1.5 × 0.75–1.2 mm, broadly obovate, apical part 0.1–0.25(0.4) mm long, rather short, rounded, smooth at top, ± hooded, pale brown to red-brown with green midnerve; spikelet prophyll and glume 1–1.2 mm, red-brown spotted. Stamen 1, anthers 0.3–0.4 mm long. Style 0.1–0.15 mm long, with 2 branches (3 branches in some specimens). Nutlet 1–1.1 × 0.65–0.7 mm, in frontal view broadly obovate with small style base remnant, flattened trigonous in cross-section.

Zambia. N: Mbala Dist., old road to Cascalawa from Chemba village, 1090 m,

16.ii.1960, *Richards* 12481 (K). C: Mumbwa Dist., 62 km W of Lusaka, on Mumbwa road, 20.iii.1965, *Robinson* 6457 (K, M, NY). S: Namwala Dist., Kabulamwanda, 110 km N of Choma, 1000 m, 21.iv.1955, *Robinson* 1286 (K, NY, SRGH). **Zimbabwe**. W: Hwange Dist., Hwange Nat. Park, Tshabema Pan, 24 km W of Main Camp, 10.xii.1973, *Rushworth* 2564 (SRGH). **Malawi**. C: Nkhotakota Dist., Nkhotakota airport, 2.v.1963, *Verboom* 101S (K). S: Machinga Dist., Liwonde Nat. Park, near Mvuu Camp, 14.iv.1985, *Dudley* 1529 (K, MAL).

From Senegal to Sudan and Ethiopia south to Zimbabwe. On damp sand and in swamps; 500–1500 m.

Conservation notes: Widespread; not threatened.

13. **Lipocarpha rehmannii** (Ridl.) Goetgh. in Wageningen Agric. Univ. Pap. **89**(1): 64. —Gordon-Gray, Cyperaceae Natal: 123 (1995). —Beentje in F.T.E.A., Cyperaceae: 352 (2010). Types: South Africa, KwaZulu-Natal, Griffin's Hill, near Estcourt, 1875–1880, *Rehmann* 7305, 7315 (K syntype, Z); Angola, Huíla, Lopollo, v.1860, *Welwitsch* 6771 (BM syntype).

> *Scirpus rehmanni* Ridl. in Trans. Linn. Soc. London, Bot. **2**: 159 (1884). —Raynal in Adansonia, n.s. **8**: 97 (1968).
>
> *Isolepis rehmannii* (Ridl.) Lye in Bot. Not. **124**: 479 (1971).
>
> *Rikliella rehmannii* (Ridl.) J. Raynal in Adansonia, n.s. **13**: 155 (1973). —Haines & Lye, Sedges & Rushes E. Afr.: 301 (1983).
>
> *Scirpus hystricoides* B. Nord. in Dinteria **11**: 55 (1974). Type: Namibia, Omaruru Dist., Brandberg, c. 1900 m, 31.v.1963, *Nordenstam* 2836 (S holotype, M).
>
> *Cyperus hystricoides* (B. Nord.) Bauters in Phytotaxa **166**: 19 (2014).

Tufted annual with thin roots; stem 2–25 cm long, 0.5–1.5 mm across. Leaves to 7 cm long, 1.5 mm wide. Involucral bracts (3)4–9, largest to 12 cm long. Inflorescence terminal, with 3–12 spikes, 3–10 × 3–5 mm, ± confluent. Spikelet bract 1.2–3.15(3.45) × 0.2–1 mm, elliptical, ovate or obovate, apical part 0.6–1.9(2.2) mm long, (very) long acuminate, recurved, scaberulous at top, body red-brown, top yellow; spikelet prophyll and glume absent. Stamen 1, anthers 0.25–0.35 mm long. Style 0.1–0.15 mm long, with 3 branches. Nutlet 0.5–0.6 × 0.2–0.4 mm, frontally obovate with small style base remnant, rounded trigonous in cross-section.

Botswana. SE: Gaborone, Aedume Park, 24°42'S 25°54'E, 1050 m, 15.iii.1978, *Hansen* 3377 (K, C, GAB, PRE, SRGH, UPS, WAG). **Zambia**. N: Kasama Dist., 100 km E of Kasama, 18.ii.1962, *Robinson* 4943 (K). W: Solwezi Dist., 100 km W of Solwezi, 1350 m, 15.iv.1960, *Robinson* 3549 (K). C: Kabwe Dist., Kabwe (Broken Hill), 1200 m, v.1909, *Rogers* 8103 (K, P, SRGH). E: Chipata Dist., Chipata (Fort Jameson), Ngoni area, ii.1962, *Robinson* 5003 (K). S: Choma Dist., 20 km N of Choma, Pengelly's Dam, 1280 m, 14.i.1954, *Robinson* 465 (K, NY). **Zimbabwe**. N: Gokwe Dist., Sengwa Research Station, near Ntaba Manque, 2.v.1969, *Jacobsen* 644 (K, SRGH). W: Hwange Dist., Matetsi Safari Area game fence, LK845724, 950 m, 22.v.1978, *Gonde* 211 (K, MO, SRGH). C: Harare, no locality, 1460 m, 20.iv.1932, *Brain* 8936 (K, MO, SRGH). E: Mutare Dist., NW corner Marange (Maranke) communal land, 48 km SW of Odzi, 11.vi.1969, *Plowes* 3212 (K, SRGH). S: Gutu Dist., 37.5 km peg from Chivhu on Gutu road, 12.v.1993, *Browning* 637b (NU, SRGH, GENT, PRE). **Malawi**. N: Chitipa Dist., Chitipa (Fort Hill), n.d., *Whyte* s.n. (K). S: Blantyre, Nyambadwe Hill, 1120 m, 6.v.1980, *Townsend* 2135 (K). **Mozambique**. MS: Manica Dist., Bandula, 700 m, 6.iv.1952, *Chase* 4549 (K, SRGH).

Kenya and Tanzania south to D.R. Congo, Angola, Namibia and South Africa (KwaZulu-Natal). On moist sandy soils and along rivers; 700–1800 m.

Conservation notes: Widespread distribution; not threatened.

The species was included in the genus *Rikliella*; see note under *Lipocarpha kernii*.

14. **Lipocarpha robinsonii** J. Raynal in Adansonia, n.s. **7**: 81, pl.1 (1967). Type: Zambia, 40 km NE of Mongu, 12.xii.1965, *Robinson* 6739 (P holotype, K, M, NY).

> *Cyperus liporobinsonii* Goetgh. in Phytotaxa **166**: 21 (2014), new name, non *Cyperus robinsonii* Podlech (1961).

Rhizomatous perennial; stems 50–90 cm long, 1.5–3 mm across, isolated on a long-creeping, reddish, somewhat fleshy rhizome, covered by distant cataphylls of c. 3 cm long. Leaves to 50 cm long, 3 mm wide, rather thick and pungent. Involucral bract 1, 1–10 cm long. Inflorescence terminal with 3–9 subequal spikes, 3–7 × 2–4 mm, ovoidal. Spikelet bract 2.2–3 × 0.5–1 mm, obtrullate, apical part 0.3 mm long, white to pale yellowish; spikelet prophyll and glume 2.5–3 mm long. Stamens 3, anthers 0.75–1.25 long. Style 0.75–1.8 mm long, with 3 branches. Nutlet 1–1.1 × 0.7 mm, obovate in frontal view with small style base remnant, flattened trigonous in cross-section.

Zambia. B: Mongu Dist., 40 km NE of Mongu, 12.xii.1965, *Robinson* 6739 (K, M, NY, P). N: Kasama Dist., Mungwi, 23.ii.1962, *Robinson* 5053 (M, NY). S: Namwala Dist., Kafue, Mutenda Swamp, 22.x.1962, *Vesey-FitzGerald* 3814 (K, NY).

Also in Angola. In permanent wet swamps; 800–1500 m.

Conservation notes: Fairly widespread; probably not threatened.

Excluded species.

Lipocarpha filiformis (Vahl) Kunth, Enum. Pl. **2**: 267 (1837). Type: Ghana, no locality, n.d., *Thonning* s.n. (C holotype, LE).

> *Hypaelyptum filiforme* Vahl, Enum. Pl. **2**: 284 (1805).

Tufted annual to 50 cm long; leaves to 15 cm. Inflorescence terminal with (2)3–7(12) spikes, recognized by its spikelet bract, 1.1–1.5 × 0.4–0.6 mm, obovate to obtrullate, apical part 0.2–0.4 mm long, shortly acuminate, smooth to minutely scaberulous at top, pale red-brown with pale midnerve. Nutlet frontally obovate to subelliptical with small style base remnant, (rounded) trigonous in cross-section, slightly constricted at base.

Mainly western Sudano-Zambesian from Senegal to Chad, and along the west coast to Angola. In grasslands, savannas, near ponds, in mangrove swamps, on wet sandy to dry lateritic soils.

While this species is not yet recorded from the Flora region, it is very likely to occur in W Zambia.

Lipocarpha leucaspis J. Raynal in Bull. Mus. Natl. Hist. Nat., sér.2 **41**(4): 978 (1969).

Known from D.R. Congo, Burundi and W Tanzania, with one unconfirmed record from W Zambia: Mwinilunga Dist., Zambezi rapids, 20.ii.1975, *Hooper & Townsend* 241 (K). This may possibly be a mis-identification for *Lipocarpha nana*.

26. RHYNCHOSPORA Vahl[26]

Rhynchospora Vahl, Enum. Pl. **2**: 229 (1805), conserved name.
—Robinson in Kirkia **1**: 32–43 (1961).

Perennial, sometimes short-lived or annual herbs, tall or short, erect or tufted, sometimes geniculate rooting from lower culm nodes or rhizomatous. Culms strict or branching, noded or nodeless, internodes trigonous or irregularly ridged. Leaves radical or radical and cauline; sheaths closed, ligule minute or 0. Inflorescence compound, a fasciculate corymbose cyme or a solitary, capitate head. Bracts leaf-like or reduced. Spikelets with glumes spirally imbricate or distichous, few to many (*R. candida*); central florets bisexual, with reduction of lowest few to sterility and/or upper florets staminate (variable with species). Perianth of (2–3)6(7) rigid, scabrid bristles, sometimes reduced or lacking. Stamens 3(2–1). Style branches 2, long, short

---

[26] by J. Browning and K.D. Gordon-Gray†

(hardly visible), or 0; base persistent, epaulette-like over nutlet apex (variable in form with species). Nutlets biconvex, obovoid, elliptic or square; surface smooth to faintly to markedly transversely ridged, or glabrous and punctute, or partially or completely hairy.

The key below is a practical one. As nutlet details were found to be distinctive, species are arranged as seen in the nutlets illustration.

1. Spikelet white; glumes numerous per spikelet . . . . **8.** *candida*
– Spikelet golden, straw coloured or brown; glumes not very numerous per spikelet 2
2. Inflorescence a solitary compact head, terminal on culm; nutlet almost completely clothed by delicate hair-like outgrowths . . . . **9.** *rubra*
– Inflorescence compound, of more than one compact terminal head; nutlet with few hairs on shoulders, or lacking such outgrowths 3
3. Inflorescence a central compact head, sessile or almost so, accompanied by (0)5 primary branches each terminating in a similar head, with or lacking secondary branches . . . . **3.** *holoschoenoides*
– Inflorescence of a terminal and several lateral inflorescences forming an irregularly branched structure 4
4. Robust perennial plant, usually rhizomatous or soboliferous; with whitish scales (modified leaves) on rhizome; spikelets 6–10 mm long; style unbranched (or slightly bifid) 5
– Slender annual or short-lived perennial, rhizomes short or lacking; spikelets generally less than 6 mm long; style with 2 long branches 6
5. Spikelets in large clusters on stout spreading pedicels; leaves keeled, often wider than 10 mm . . . . **1.** *corymbosa*
– Spikelets 1 to few together in small clusters on long slender erect pedicels; leaves often less than 10 mm wide. . . . . **2.** *triflora*
6. Spikelets dark brown; nutlet transversely rugose; style base conical, attached throughout width. 7
– Not this combination of characters 8
7. Stem trigonous, usually scabrous; perianth bristles shorter than nutlet and style base together, variable in number (rarely absent) . . . . **5.** *brownii*
– Stem rounded, glabrous; perianth bristles 5–6, usually 6, up to and beyond apex of crown; nutlet 4.1–4.4. . . . . **4.** *angolensis*
8. Nutlet shortly obovate to almost orbicular in outline; surface smooth, shining, cellular swellings on either side near base. **10.** *brevirostris*
– Nutlet shape in outline different; surface not smooth shining, but transversely wavy or faintly to strongly rugose 9
9. Spikelets densely clustered, mostly sessile; nutlet strongly rugose, style base markedly apiculate in centre . . . . **6.** *perrieri*
– Spikelets mostly pedicelled in looser inflorescence; nutlet not strongly rugose, with apiculate style base 10
10. Pedicel stout, erect or spreading; leaves 1–5 mm wide, rachilla not sinuous . . . . **7.** *eximia*
– Pedicel slender; leaves filiform, up to 1 mm wide, rachilla markedly sinuous . . . . **11.** *gracillima*

1. **Rhynchospora corymbosa** (L.) Britton in Trans. New York Acad. Sci. **11**: 84 (1892). —Robinson in Kirkia **1**: 35 (1961). —Haines & Lye, Sedges & Rushes E. Afr.: 313, fig.643 (1983). —Gordon-Gray in Strelitzia **2**: 150 (1995). —Hoenselaar in F.T.E.A., Cyperaceae: 360 (2010). Type: India, no locality, *Herb. Linn.* 71.48 (LINN holotype). FIGURE 14.**58A**.

Scirpus corymbosus L., Cent. Pl. **II**: 7 (1756).

Rhychospora spectabilis C. Krauss in Flora **28**: 760 (1845). Type: South Africa, KwaZulu-Natal, ponds near Umlaas R., xii.1839, *Krauss* 210 (K holotype, BM, MO, TUB).

Perennial robust herb to 185(200) cm tall, strongly to slenderly tufted, glabrous; rhizome short, linking shoot bases of tufts. Leaves several, radical, cauline, gradually reduced upwards; sheaths entire, mouth broadly U-shaped or lipped; no ligule. Culm erect, nodose, internodes trigonous. Inflorescence compound, a terminal corymbose cyme plus one to several lateral cymes developed singly from upper culm nodes, corymbs fairly compact. Bracts leaf-like. Spikelets numerous per corymbose cyme, 6–10 mm long; lowest floret hermaphrodite, 1 nutlet maturing, upper florets male. Glumes 6–7 per spikelet, spirally imbricate, apex acute to apiculate. Perianth bristles 3 + 3, usually uniform within a floret, not exceeding nutlet length plus persistent style base. Stamens 3. Style entire or minutely bifid. Nutlet obovoid, ± lenticular (biconvex), 4–4.5 (excluding style base) × 1.9–2.1 mm, each face smooth but with 1–2 longitudinal invaginations, surface eventually faintly papillose, crown (persistent style base) conical, 4–5 mm long.

Botswana. N: Chobe Dist., near main Linyati channel, upstream of No 2 camp, 18°23'S 23°47.2'E, 8.xi.1974, *P.A. Smith* 1175 (K). **Zambia**. B: Kalabo Dist., 14.x.1962, *Robinson* 5483 (K). N: Kaputa Dist., Chishi Lake, swamp forest, 900 m, 24.ix.1956, *Richards* 6277 (K). W: Mwinilunga Dist., Hillwood Wildlife Reserve Campsite, 11°16'S 24°19'E, 1380 m, 5.i.2009, *Bingham* 13608 (K). C: Kabwe Dist., Mpunde, 53 km NW of Kabwe, 14°06'S 28°06'E, 1170 m, 1.x.1972, *Kornaś* 2224 (K). **Zimbabwe**: E: Chipinge Dist., Chirinda road to Espungabera, 1160 m, 23.x.1947, *Rattray* 1203 (K, SRGH). **Malawi**. N: Mzimba Dist., Mzuzu, Marymount dambo, 1370 m, 20.xi.1947, *Pawek* 7511 (K). **Mozambique**. MS: Mossurize Dist.?, Mt Marume, 1070 m, 13.ix.1906, *Swynnerton* 901 (K). GI: Jangamo Dist., between Inhambane and Jangamo (Yangamo), 15.ix.1948, *Myre & Cavalho* 208 (K).

Widely distributed in the tropics and subtopics; also in Uganda, Tanzania, South Africa (KwaZulu-Natal). Coastal swamps, river margins, lakes and shallow pools; 20–1500 m.

Conservation notes: Widespread; not threatened.

2. **Rhynchospora triflora** Vahl, Enum. Pl. **2**: 232 (1805). —Robinson in Kirkia **1**: 35 (1961). —Haines & Lye, Sedges & Rushes E. Afr.: 314 (1983). Type: India, "Habitat in India oriental", n.d., unknown collector (C-Vahl holotype). FIGURE 14.**58B**.

Schoenus triflorus (Vahl) Poir. in Lamarck, Encycl., suppl. **2**: 248 (1811).

Perennial herb 58–70 cm tall; rhizome 3–6 mm thick, creeping, horizontal. Leaves basal, cauline, c. 8, sheaths eligulate, blade linear, to 9 mm wide, narrower when folded, margins and midribs adaxially scabrous. Culm closely spaced, erect, (1)1.5–4.5 mm wide, trigonous, smooth proximally to antrorsely scabrous distally, often closely ribbed and channelled on one face. Inflorescence terminal, compound, of a central and 1–2 lateral open corymbose panicles, each subtended by a leaf-like bract; primary rays to 25 cm long, tipped by 3–9 hemispherically arranged spikelets, 6–9 × 1.7–2.1 mm long. Glumes close-packed, 3–7 per spikelet, 5.4–7.5 × 3–4 mm, lowest bisexual, central staminate, uppermost few generally sterile (perhaps young with undeveloped stamens). Stamens 3, anthers appendaged apically. Perianth bristles 6, sparsely antrorsely spinous (spicules not barbs). Style unbranched, long exserted from glumes at maturity. Nutlet biconvex, obovoid, 3–4.1 (excluding style base) × 2.5–2.8 mm, surface transversely rugulose often with prickle hairs on shoulders, cells linear, vertically arranged; style base subulate-lanceolate, laterally compressed, becoming unequally 4-angled upwards, 4–5.5 mm long, × 0.4–0.6 mm basally across nutlet apex, tip exserted beyond subtending glume.

Zambia. B: Mongu Dist., Sefula, JICA rice project, 15°22.9'S 23°10.5'E, 1000 m, fl. 11.ii.1999, *Bingham & Luwiika* 11827 (K). N: Kawambwa Dist., Chishinga Ranch near Luwingi, 1400 m, fl. 28.iv.1961, *Astle* 577 (K). W: Mwinilunga Dist., 7 km N of Kalene Hill, fl. 17.iv.1965, *Robinson* 6622 (K). **Malawi**. C: Nkhota-Kota Dist., Nkhotakota (Kota-

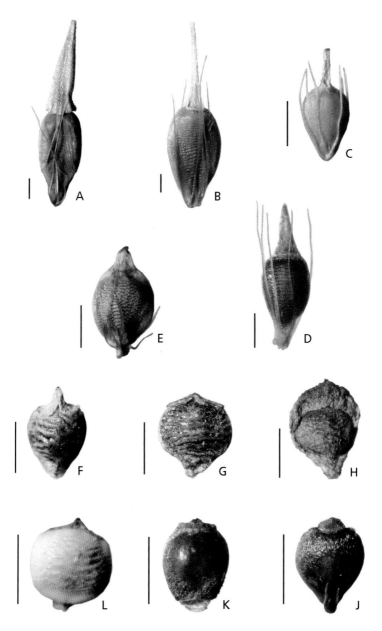

Fig. 14.**58**. RHYNCHOSPORA nutlets. A. —RHYNCHOSPORA CORYMBOSA, from *Bingham* 13608. B. —RHYNCHOSPORA TRIFLORA, from *Bingham & Luwiika* 11827. C. — RHYNCHOSPORA HOLOSCHOENOIDES, from *Robinson* 4699. D. —RHYNCHOSPORA ANGOLENSIS, from *Robinson* 4314. E. —RHYNCHOSPORA BROWNII, from *Thompson & Rawlins* 5711. F. —RHYNCHOSPORA PERRIERI, from *Robinson* 3698. G. —RHYNCHOSPORA EXIMIA, from *Verboom* 114. H. —RHYNCHOSPORA CANDIDA, from *Renvoize* 5687. J. — RHYNCHOSPORA RUBRA subsp. AFRICANA, from *Schlecter* 12090. K. —RHYNCHOSPORA BREVIROSTRIS, from *Robinson* 2850. L. —RHYNCHOSPORA GRACILLIMA subsp. SUBQUADRATA, from *Browning* 356. Scale bar = 1 mm. Photomicrographs by Jane Browning.

Kota), rice swamp, fl. 2.v.1963, *Verboom* 1185 (K). **Mozambique**. MS: Muanza Dist., Nyamarua Camp, halfway between camp and road junction to Chinizaia Lighthouse, Cheringoma coastal area, fl. v.1973, *Tinley* 2900 (K, LISC).

In West Africa and Angola; also in tropical C and S America, Sri Lanka, Indo-China and Malesia. Rare in Africa but associated with rice-growing; possibly an introduction. Wet mud of floodplains; 10–1400 m.

Conservation notes: Probably introduced; not threatened.

The nutlet surface is distinctly transversely rugulose.

3. **Rhynchospora holoschoenoides** (Rich.) Herter in Revista Sudamer. Bot. **9**: 157 (1953). —Haines & Lye, Sedges & Rushes E. Afr.: 314, fig.645 (1983). —Gordon-Gray in Strelitzia **2**: 150 (1995). —Hoenselaar in F.T.E.A., Cyperaceae: 358, fig.53 (2010). Type: French Guiana, Cayenne, 1792, *LeBlonde* 36 (P holotype). FIGURES 14.**58C**, 14.**59**.

> *Schoenus holoschoenoides* Rich. in Actes Soc. Hist. Nat. Paris **1**: 106 (1792).
> *Rhynchospora cyperoides* Mart. in Denkschr. Königl. Akad. Wiss. München **6**: 149 (1824). —Clarke in F.T.A. **8**: 479 (1902). Type: uncertain.
> *Rhynchospora mauritii* Steud., Syn. Pl. Glumac. **2**: 149 (1855). —Robinson in Kirkia **1**: 34 (1961). Type: Mauritius, "ex herbario Urville decta in Ins. Mauritii"; possibly 1824, *Chauvin* 1274 (P).
> *Rhynchospora arechavaletae* Boeckeler, Beitr. Cyper. **1**: 24 (1888). —Podlech in Mitt. Bot. Staatssamml. München **4**: 118 (1961); in Merxmüller, Prodr. Fl. SW Afr. **165**: 44 (1967). Type: Uruguay, near Montevideo, other details uncertain.

Perennial herb to 110(180) cm tall, glabrous; rhizome horizontal or oblique, to c. 4 cm long, scales imbricate, whitish becoming naked in age. Leaves radical, cauline, latter reduced upwards; sheath entire, mouth broadly U-shaped or lipped; ligule very small or lacking. Culm erect, eventually recurving under weight of mature fruits, nodose; internodes trigonous to triquetrous. Inflorescence variable, usually compound of a terminal contracted globose head (cyme) 15 mm diameter plus several pedicelled heads forming an anthelate cluster, sometimes also branched from lower nodes or reduced to 2 or 1 shortly pedicelled heads terminating a culm. Bracts leaf-like, unequal in length. Spikelets numerous per globose head, 3–5 mm long; lower 1–2 florets sterile, 3 bisexual, upper 4–5 male or sterile, glumes c. 5 per spikelet, spirally imbricate. Perianth of 3 + 3 bristles, antrorsely barbed, uniform or of varying lengths, not exceeding length of nutlet plus style base. Stamens 3. Style entire or minutely bifid, surface scabrid. Nutlet elliptic in outline, biconvex 2.5–2.7 (including style base) × 1.1–1.2 mm; finely transversely rugulose with few coarse scabrid outgrowths on shoulders; style base persistent, minutely scabrid, attached to nutlet by small central area only, with thickened rims running to shoulders.

Zambia. B: Sesheke Dist., Mashi, Kwando R., dambo, 30.x.1964, *Verboom* 1777 (K). N: Luwingu Dist., edge of Lake Bangweulu, 22.v.1961, *Robinson* 4699 (K). W: Ndola Dist., Itawa dambo, 1300 m, 29.iii. 1060, *Robinson* 3438 (K). S: Livingstone Dist., banks of Zambezi above Boat Club, 7.iv.1956, *Robinson* 1456 (K). **Zimbabwe**. W: Hwange Dist., Victoria Falls rainforest, 18.xi.1949, *Wild* 3190 (K, SRGH). **Mozambique**. GI: Inhassoro Dist., Bazaruto Is., Ponta Mulderga, 7.xi.1958, *Mogg* 28845 (K).

Also in Central and South America, Greater and Lesser Antilles, West Africa (Senegal, Ivory Coast, Gabon), Tanzania, Namibia, South Africa (Eastern Cape, KwaZulu-Natal), Madagascar and the Mascarene Is. Wet sandy areas in grasslands and swamps, stream and riverbanks; 10–1300 m.

Conservation notes: Widespread; not threatened.

Van Laren (MSc thesis, Univ. Natal, 1979) gave special consideration to the opinions of Kükenthal (Bot. Jahrb. **74**: 375–509, 1949) and Podlech (1961) in their maintenance of one-headed, longer-nutted examples of *Rhynchospora holoschoenoides* as a distinct species, *R. arechavaletae*. It was found that one-headed inflorescences were not correlated with longer nutlets. So variable were the characters used by Podlech

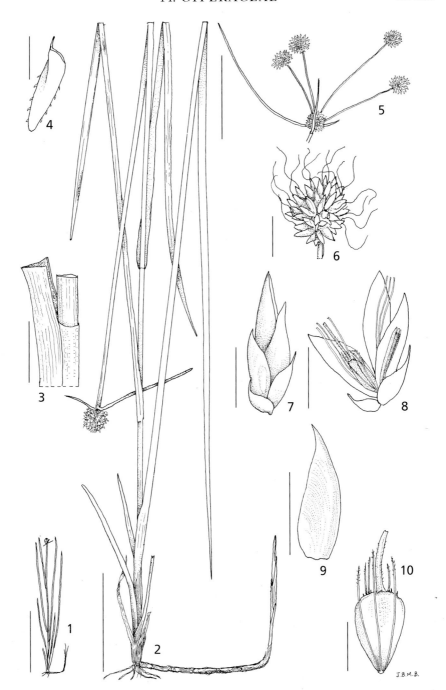

Fig. 14.**59**. RHYNCHOSPORA HOLOSCHOENOIDES. 1, 2, habit; 3, leaf sheath; 4, leaf apex; 5, inflorescence; 6, head of spikelets; 7, 8, spikelet, complete and opened to show florets; 9, glume, lateral view; 10, nutlet. 1–4 from *Robinson* 6123; 5, 7, 9 from *Robinson* 1048; 6 from *Robinson* 1456; 10 from *Renvoize* 5598. Scale bars: 1 = 250 mm; 2, 5 = 40 mm; 3, 6 = 5 mm; 7–9 = 2 mm; 4, 10 = 1 mm. Drawn by Jane Browning.

(1967) to distinguish *R. arechavaletae* and *R. mauritii* that no satisfactory limits could be applied to maintain them. Thus all southern African specimens are regarded as constituting a single species, not distinct at specific level from the American plants.

Rhynchospora holoschoenoides is distinguished from other species in southern Africa by its development of shortly elongated rhizomes when young clothed with whitish imbricate scales that fall away with ageing. In Zambia there is a species allied to *R. corymbosa*, namely *R. triflora* which also displays these scales on the rhizome.

4. **Rhynchospora angolensis** Turrill in Bull. Misc. Inform. Kew **1914**: 136 (1914). — Haines & Lye, Sedges & Rushes E. Afr.: 318, figs.655,656 (1983). —Hoenselaar in F.T.E.A., Cyperaceae: 360 (2010). Type: Angola, Benguella, country of Ganguellas and Ambuellas, n.d., *Gossweiler* 3268 (K holotype). FIGURE 14.**58D**.

   Rhynchospora africana Cherm. in Arch. Bot. Mém. **4**(7): 44 (1931). —Robinson in Kirkia **1**: 37 (1961).

Perennial herb, slender to tufted, erect, 120 cm tall; rhizome contracted, almost lacking. Leaves radical, blade to 50 × 1–3 cm, several cauline, size reduced upwards. Culm glabrous, rounded with longitudinal ridges. Inflorescence a compound, corymbose cyme with several to 1 lateral cymes from upper culm nodes; corymbs loose or compact, sometimes reduced to quasi-terminal heads. Bracts several, leaf-like. Spikelets few to many per cyme, 5–8(12) × 1.8–3.5(5) mm, each maturing 2–4 nutlets; florets bisexual, glumes spiralled, rachilla straight. Perianth bristles 5–6 (usually 6), of uniform length, 3–3.5 mm long, generally up to and beyond apex of crown. Stamens 2. Style deeply bifid. Nutlet 4.1–4.4 (including style base) × 1.2–1.4 mm, elliptic to ovoid, surface transversely rugose to rugulose; style base conical, 1.2–1.7 mm long, attached throughout width.

Zambia. B: Mongu Dist., Kande Plain, E of Mongu, 20.vii.1961, *Angus* 3001 (K). N: Chinsali Dist., Shiwa Ngandu, 28.i.1961, *Robinson* 4314 (K, LISC, LMA). W: Mwinilunga Dist., 6 km N of Kalene Hill, 12.xii.1963, *Robinson* 5908 (K). C: Chongwe Dist., Chakwenga headwaters, 100–129 km E of Lusaka, 10.i.1964, *Robinson* 6171 (K). **Zimbabwe**. C: Marondera Dist., Marondera (Marandellas), Grasslands Research Station, 1520 m, 3.i.1949, *Corby* 332 (K, SRGH). E: Chimanimani Mts, E side of upper Bundi plain, 31.xii.1957, *Goodier & Phipps* 231 (SRGH).

Also in Cameroon, Central African Republic, D.R. Congo, Uganda, Tanzania, Angola and Madagascar. Marshes, near streams and in boggy grassland; 1000–1600 m.

This species differs from the variable *Rhynchospora brownii* in its glabrous rounded culms and larger spikelets and nutlets. Examination of bristle length on the nutlets is needed to distinguish *R. angolensis* from *R. brownii*.

Rhynchospora angolensis has numerous 'transparent hairs' attached at the base of the perianth segments (bristles); these are sometimes present in *R. brownii*. It may not be possible to easily distinguish the two species in the field, and both may be collected together, as in *Robinson* 2225a,b.

Robinson 3572 (K) from Mwinilunga in NW Zambia is similar to *R. angolensis*, but thought to be new (see note under Uncertain species).

5. **Rhynchospora brownii** Roem. & Schult., Syst. Veg. **2**: 86 (1817). —Robinson in Kirkia **1**: 36 (1961). —Haines & Lye, Sedges & Rushes E. Afr.: 317, fig.653 (1983). —Gordon-Gray in Strelitzia **2**: 150, fig.64 (1995). —Hoenselaar in F.T.E.A., Cyperaceae: 361 (2010). Type: Australia (New Holland), Endeavour R., 1770, *Banks & Solander* s.n. (BM lectotype), lectotypified by Browning (1989). FIGURES 14.**58E**, 14.**60**.

   Rhynchospora juncea Kunth, Enum. Pl. **2**: 298 (1837).
   Rhynchospora glauca sensu C.B. Clarke in F.T.A. **8**: 482 (1902), non Vahl.
   Rhynchospora rugosa sensu Mapuara & Timberlake, Checkl. Zimbabwe Vasc. Pl.: 89 (2004), non (Vahl) Gale.

Perennial slender to tufted herb, 80(115) cm tall; rhizome contracted, almost lacking. Leaves radical, blade to 50 cm long by 1–2 mm wide; several cauline, size reduced upwards. Culm trigonous with shallow longitudinal ridges, occasionally scabrid. Inflorescence a compound, corymbose cyme with several to 1 lateral cymes from upper culm nodes; corymbs loose or compact, sometimes reduced to quasi-terminal heads. Bracts several, leaf-like. Spikelets few to many per cyme, (3)5 mm long, each maturing 1–5 nutlets; florets bisexual, glumes spiralled,

Fig. 14.**60**. RHYNCHOSPORA BROWNII. 1, habit; 2, spikelet; 3, glume, lateral view; 4, nutlet. All from *Browning* 153. Scale bars: 1 = 40 mm; 2–4 = 2 mm. Drawn by Jane Browning. Reproduced from Strelitzia (1995) by kind permission of the South African National Biodiversity Institute, Pretoria.

rachilla straight. Perianth bristles variable, (0)6(7), of uniform length or not, seldom exceeding nutlet + style base length. Stamens 3(1). Style deeply bifid. Nutlet 2.5 (including style base) × 1.3–1.6 mm, elliptic to ovoid, surface transversely rugose to rugulose; style base conical, attached throughout width.

Botswana. N: Okavango swamps, Mboroga R., 19°15.25'S 23°09.7'E, 21.i.1980, *P.A. Smith* 3013 (K). **Zambia**. B: Mongu Dist., 40 km NE of Mongu, 12.xii.1965, *Robinson* 6725 (K). N: Mansa Dist., Mansa (Fort Roseberry), Chipili dambo, 18.vi.1957, *Robinson* 2300 (K). W: Ndola Dist., Itawa dam, 3.vi.1957, *Robinson* 2225 (K). C: Serenje Dist., Kundalila Falls, S of Kanona, 1300 m, 13.iii.1975, *Hooper & Townsend* 707 (K). E: Isoka Dist., Nyika plateau, 2300 m, 2.i.1959, *Robinson* 3006 (K). **Zimbabwe**. C: Marondera Dist., Marondera, Digglefold, 16.xi.1949, *Corby* 546 (K, SRGH). E: Chimanimani Dist., Chimanimani Mts, 31.xii.1964, *Robinson* 6347 (K, SRGH). **Malawi**. N: Chitipa Dist., 18 km SSE of Chisenga, Jembya Forest Reserve, 10°08'S 33°27'E, 1870 m, 26.xii.1988, *Thompson & Rawlins* 5711 (K). **Mozambique**. Z: Gurué Dist., Mt Namuli, Muretha Plateau, 15°23'49.8"S 37°02'42.9"E, 15.xi.2007, *Harris* 316 (K). MS: Gorongosa Dist., Mt Gorongosa, Gogogo summit area, iii.1973, *Tinley* 2424 (K, SRGH).

Also in West Africa (Ivory Coast, Nigeria), Cameroon, Ethiopia, Kenya, Uganda, Tanzania, Rwanda, D.R. Congo, Angola, South Africa (Eastern and Western Cape, Gauteng, KwaZulu-Natal, Limpopo, Mpumalanga, North West), Lesotho and Swaziland; also in Madagascar, Asia and Pacific. Wet grassland, stream margins, edges of dambos and amongst rocks; 800–2300 m.

Conservation notes: Widespread; not threatened.

Here *Rhynchospora brownii* is accepted as a distinct species from the closely related *R. rugosa* (Vahl) Gale, which is confined to C & S America. This follows the status given this taxon by Hoenselaar in F.T.E.A., Cyperaceae (2010) and does not follow the infraspecific level as subsp. *brownii* (Roem. & Schult.) T. Koyama (Enum. Fl. Pl. Nepal **1**: 118, 1978) given by Govaerts & Simpson in the World Checklist of Cyperaceae (2007).

Perianth bristles are noted to be variable. In *Robinson* 1637 (K) and *P.A. Smith* 3013 (K) from W Zambia and Botswana no bristles were found. Robinson (1961) gave the name *R. juncea* Kunth to his collection, but this is here considered to be a synonym.

In several populations nutlets on plants at high altitudes are noted to be smaller than those from coastal habitats.

6. **Rhynchospora perrieri** Cherm. in Bull. Soc. Bot. France **69**: 721 (1923). — Robinson in Kirkia **1**: 40 (1961). —Haines & Lye, Sedges & Rushes E. Afr.: 319, fig.657 (1983). —Gordon-Gray in Strelitzia **2**: 152, fig.65b (1995). —Hoenselaar in F.T.E.A., Cyperaceae: 362 (2010). Type: Madagascar, Berizoka, ix.1897, *Perrier de la Bâthie* 305 (P lectotype); Madagascar, Ile Mahakamby, *Waterlot* 565 (P syntype). FIGURE 14.**58F**.

    Rhynchospora deightonii Hutch. in F.W.T.A. **2**: 468 (1936). Types: Sierra Leone, Malema, *Deighton* 326 (?); Newton, *Deighton* 1449 (?); Southern Province between Gboyama and Bendu, *Deighton* 1633 (K syntype).

Annual or short-lived perennial, slender, sparsely tufted to 30–45 cm tall, or loosely spreading and rooting from geniculate nodes; rhizome inconspicuous. Radical leaves few, cauline more numerous, reduced upwards, blades mostly 1–2 mm wide. Culms nodose, erect or decumbent, subterete, often filiform, glabrous. Inflorescence compound, a terminal corymbose cyme, generally with 3–4 lateral cymes single from culm nodes, each corymb loose. Bracts leaf-like. Spikelets 1–10 per corymb, (2)3.5–5 × 0.7–1.5 mm; florets bisexual, uppermost male, glumes spiralled, 4–5 per spikelet, 2.1–2.5 × 1.3–1.5 mm, apex acuminate to awned. Perianth bristles absent. Stamens 2. Style branches 2, long, base attached to nutlet epaulette-like over shoulders, central peak 0.25–0.7 mm long. Nutlet obovoid, 1.5–1.7 (including style base) × 1.2–1.4 mm, surface strongly transversely corrugated (rugose).

Botswana. N: Okavango, Mojei, Dindinga footpath, 19°07.1'S 23°05.2'E, 28.iii.1975, *P.A. Smith* 1324 (K). **Zambia**. N: Kasama Dist., Chishimba Falls, 2.iv.1961, *Robinson* 4567 (K, LMA). W: Mwinilunga Dist., Zambezi Falls, 6.4 km N of Kalene Mission, 1200 m, 10.xi.1962, *Richards* 17135 (K). **Zimbabwe** W: Hwange Dist., Victoria Falls rainforest, 4.iv.1956, *Robinson* 1418 (MPR). **Mozambique**. MS: Cheringoma Dist., Cheringoma section, river terrace, 40 m, 13.vii.1972, *Ward* 7886 (NU).

Also in West Africa (Guinea, Ivory Coast, Nigeria, Senegal), D.R. Congo, Kenya, Tanzania, South Africa (KwaZulu-Natal), Swaziland and Madagascar. Sometimes a pioneer establishing as an inconspicuous creeping herb in wet to moist disturbed areas; 20–1400 m.

Conservation notes: Widespread; not threatened.

With it slender culms and inflorescences almost concolourous with vegetative organs plants are often inconspicuous.

Confusion with *Rhynchospora gracillima* subsp. *subquadrata* is possible, but the sinuous rachilla and squareness of nutlet of the latter are distinctive. *R. perrieri* may also be confused with *R. brevirostris*.

7. **Rhynchospora eximia** (Nees) Boeckeler in Linnaea **37**: 601 (1873). —Haines & Lye, Sedges & Rushes E. Afr.: 316, fig.650 (1983). —Hoenselaar in F.T.E.A., Cyperaceae: 363 (2010). Types: Mexico, Hacienda de la Laguna, *Schiede* 864 (B syntype); Panama, *Seemann* s.n. (B syntype, K). FIGURE 14.**58G**.

Spermodon eximius Nees in Seemann, Bot. Voy. Herald: 222 (1854).

Annual or short-lived perennial herb, slender to sparsely tufted, (10)20–60 cm tall; rhizome undeveloped. Leaves often exceeding inflorescence; blade flat, 1–5 mm wide. Culm spreading or erect, obtusely triangular, glabrous, longitudinally ridged. Inflorescence terminal, usually with axillary branches below, carrying cluster of 1(5) spikelets sessile or on stiff ascending pedicels. Bracts leaf-like, mostly inconspicuous, occasionally exceeding inflorescence. Spikelets (5)6–10 mm long, apices distinctly acuminate. Glumes many per spikelet, c. 5 mm long, 1-veined, midribs mostly excurrent. Perianth 0. Stamens 2. Style bifid. Nutlet broadly ovate to almost orbicular, 1.5–1.6 (including style base) × 1.3–1.4 mm; style base low, capping apex of dark clearly transversely rugose body, edges (ridges) of which are lined by contiguous raised cells.

Malawi. C: Nkhotakota Dist., Nkhotakota (Kota-Kota), rice swamp, 2.v.1963, *Verboom* 114 (K).

Also in West Africa, Tanzania (one specimen only) and tropical America. Said to be present from sea level to 1000 m altitude in tropics, but rare in Africa; presently known only from Malawi, where it may be a chance introduction.

Conservation notes: Probable introduction.

8. **Rhynchospora candida** (Nees) Boeckeler in Linnaea **37**: 605 (1873). —Robinson in Kirkia **1**: 34 (1961). —Haines & Lye, Sedges & Rushes E. Afr.: 319, fig.659 (1983). —Hoenselaar in F.T.E.A., Cyperaceae: 357 (2010). Type: Guyana (British Guyana), 1838, *Schomburgk* 685 (B† holotype, BM, G, K). FIGURE 14.**58H**.

Psilocarya candida Nees in Martius, Fl. Bras. **2**(1): 117 (1842).

Rhynchospora adscendens C.B. Clarke in Durand & Schinz, Consp. Fl. Afr. **5**: 652 (1894), illegitimate name.

Perennial herb, stolons slender, scale leaves straw-coloured, striate; rhizome base thickened. Leaves mostly cauline, blade with scattered villous hairs; sheaths closed; ligule 0. Culm c. 50 cm tall, slender, distantly noded, usually solitary, less often 2–3 clustered; internodes round becoming trigonous upwards. Inflorescence simple or compound, corymbose with 2–6 primary branches. Spikelets solitary, terminating branches, 5.5–12 × 3–5.5 mm. Glumes imbricate, to ± 50 per spikelet, white only, or with small brown dots; lowest 6–8 sterile, central 30 ± larger, bisexual, upper sometimes staminate showing reduction; rachilla with rounded, elongated projections.

Perianth lacking. Stamens 3. Style bifid. Nutlet 2–2.2 (including style base) × 1.4–1.5 mm; surface transversely wavy, crown covering apex spongy, fawn coloured; pedicel well-developed.

Botswana. N: Okavango swamps, 19°11.4'S 23°15.7'E, in 10–15 cm of water, 10.xii.1979, *P.A. Smith* 2917 (K, LISC). **Zambia**. B: Senanga Dist., 16 km N of Senanga, 1040 m, 31.vii.1952, *Codd* 7303 (K). N: Mbala Dist., Lucheche R. cattle crossing 2.4 km from Mbala (Abercorn), 1520 m, 28.iv.1952, *Richards* 1536 (K). W: Mwinilunga Dist., source of Matonchi dambo, 16.xi.1937, *Milne-Redhead* 3258 (K). C: Mkushi Dist., Chinshinshi dambo, Fiwila, 1400 m, 9.i.1958, *Robinson* 2702 (K). **Zimbabwe**. N: Sebungwe, Kanyangwe Tsetse, 15.xi.1958, *Phipps* 1473 (SRGH). W: Matobo Dist., Besna Kobila, 1460 m, i.1957, *Miller* 4039 (K, SRGH). C: Marondera Dist., Marondera (Marandellas), 1650 m, 14.v.1931, *Brain* 4357 (K, SRGH). E: Chimanimani Dist., Chimanimani Mts, Stonehenge plateau bog, 1675 m, 1.ii.1955, *Phipps* 360 (K, SRGH). **Malawi**. N: Rumphi Dist., 37 km N of Rumphi, 1.6 km S of Nchena Nchena, 1250 m, 22.ii.1976, *Pawek* 13818 (K). C: Nkhota Kota Dist., Nkhota Kota (Kota Kota), Sani Road, 22.iii.1953, *Jackson* 1095 (K). **Mozambique**. N: Marrupa Dist., c. 20 km from Naboina, 17.ii.1981, *Nuvunga* 567 (K, LISC). MS: Muanza Dist., Cheringoma Section, Chinizuia sawmills, 13.vii.1972, *Ward* 7894 (K).

Also in West Africa, (Ivory Coast, Nigeria, Senegal), Gabon, Uganda, Tanzania and Angola; also in Madagascar and S tropical America. In wet grassland; 20–1700 m.

Conservation notes: Widespread; not threatened.

Usually only few of the many florets per spikelet mature to produce a full-sized, viable fruit.

9. **Rhynchospora rubra** (Lour.) Makino in Bot. Mag. (Tokyo) **17**: 180 (1903).
   Schoenus ruber Lour., Fl. Cochinch.: 52 (1790).

Subsp. **africana** J. Raynal in Adansonia, n.s. **7**: 323 (1967). —Haines & Lye, Sedges & Rushes E. Afr.: 314, fig.646 (1983). —Hoenselaar in F.T.E.A., Cyperaceae: 358 (2010). Type: Tanzania, Mafia Is., 7.viii.1936, *Vesey Fitzgerald* 5217 (K holotype). FIGURE 14.**58J**.
   Chaetospora madagascariensis Steud., Syn. Pl. Glumac. **2**: 161 (1855).
   Rhynchospora minor Nelmes in Kew Bull. **11**: 533 (1957), illegitimate name.

Annual or short lived perennial herb, to 35 cm tall, tufted, glabrous; rhizome almost absent. Radical leaves numerous, sheaths splitting early; cauline leaves few from low culm nodes; sheath mouth truncate to V-shaped; ligule absent. Culm erect to suberect, filiform, nodeless except at extreme base and apex. Inflorescence a single bracteate head. Bracts 4–7, variable to 3 cm long, bases hairy. Spikelets numerous per head, 2–3 mm long, maturing one nutlet if any; basal florets female, upper male. Glumes numerous per spikelet, spirally imbricate, apices acute. Perianth absent, or 3 + 3 bristles, uniform or variable, not exceeding mature nutlet length, hairy below, scabrid above. Stamens 2. Style minutely bifid or entire. Nutlet elliptic to obovoid, biconvex, 1.4 (including style base) × 1 mm, almost black at maturity, surface minutely bristly, hairs upward pointing, scabrid; style base attached to nutlet apex by small central area only.

Mozambique. MS: Muanza Dist., Cheringoma coast, Nyamaruza dambo, halfway between camp and road junction to Chiniziua lighthouse, v.1973, *Tinley* 2901 (K). GI: Inhambane, "Jangamo Dist.", 1.ii.1898, *Schlechter* 12090 (K, PRE).

Also in West Africa (Ivory Coast, Liberia, Senegal), Tanzania, South Africa (KwaZulu-Natal) and Madagascar. Swampy brackish and fresh water grassland; 20–100 m.

Conservation notes: Widespread; not threatened.

Kükenthal in Bot. Jahrb. **74**: 375–509 (1949) recognised *Rhynchospora rubra* as Asiatic and absent from Africa, re-establishing the African plants as *R. parva* (Nees) Kunth [= *R. madagascariensis* (Steud.) Cherm.]. Raynal's revision resulted in the recognition of one species, with five subspecies to accommodate the differing geographical races.

Only subsp. *africana* is southern African, identified by the length of hairs at the base of the perianth bristles, the size of nutlets and style base. Subsp. *rubra* occurs in Asia and is not present in Africa.

10. **Rhynchospora brevirostris** Griseb., Cat. Pl. Cub.: 246 (1866). —Robinson in Kirkia **1**: 39 (1961). —Haines & Lye, Sedges & Rushes E. Afr.: 316, fig.652 (1983). —Hoenselaar in F.T.E.A., Cyperaceae: 362 (2010). Type: Cuba, 1860–1864, *Wright* 3413 (GOET holotype). FIGURE 14.**58K**.

    Rhynchospora barteri C.B. Clarke in Durand & Schninz, Consp. Fl. Afr. **5**: 653 (1894), invalid name; in F.T.A. **8**: 482 (1902). Type: Nigeria, Nupe, 1858, *Barter* 1010 (K holotype, P).

Slender annual herb to 20 cm tall. Leaves mostly cauline, linear, blade 0.3–1.2 mm wide. Culm trigonous, glabrous. Inflorescence a panicle of 2–4 scattered groups, each of a few sessile and shortly pedicelled spikelets, 3.8–4.3 × 1–1.4 mm. Glumes 2.3–3.7 × 1.2–1.4 mm, keel projecting to form a scabrid awn. Perianth lacking. Stamens 2. Style bifid. Nutlet broadly ovoid, 1.3–1.4 (including style base) × 0.9–1 mm, surface smooth with 2 cellular areas on each side at base; style base small, conical, set within a raised margin that shrinks and alters appearance on drying.

Botswana. N: Chobe Dist., Kwando R., 18°09.7'S 23°23.4'E, 24.i.1978, *P.A. Smith* 2246 (K). **Zambia**. N: Kawambwa Dist., 1340 m, 21.vi.1957, *Robinson* 2316 (K). W: Mwinilunga Dist., Matonchi, Kalenda dambo, 1400 m, 16.iv.1960, *Robinson* 3587 (K). S: Choma Dist., Mapanza, Pangama, 1070 m, 27.iv.1958, *Robinson* 2850 (K).

Also in West Africa (Guinea, Nigeria, Senegal), Cameroon, D.R. Congo, Tanzania, Mexico and tropical America. Often in wet mud fringing rivers, waterholes and in damp grasslands; 1000–1400 m.

Conservation notes: Infrequently collected, but probably not threatened.

This species is imperfectly known in Africa, requiring further field study. Morphologically it is said to be similar to *Rhynchospora perrieri*, but differs in its smooth nutlet carrying an inconspicuous style base and bearing slightly smaller spikelets.

11. **Rhynchospora gracillima** Thwaites & Hook., Enum. Pl. Zeyl.: 435 (1864). Type: Sri Lanka (Ceylon), S of island, *Thwaites* 3818 (BR holotype, G, P, US).

Subsp. **subquadrata** (Cherm.) J. Raynal in Adansonia, n.s. **7**: 321 (1967). —Haines & Lye, Sedges & Rushes E. Afr.: 315, fig.648 (1983). —Gordon-Gray in Strelitzia **2**: 150, fig.63F (1995). —Hoenselaar in F.T.E.A., Cyperaceae: 364 (2010). Types: Madagascar, Firingalava, viii.1898, *Perrier de la Bâthie* 920 (P syntype); Tamatave, 27.ix.1912, *Viguier & Humbert* 397 (P syntype). FIGURE 14.**58L**.

    Rhynchospora subquadrata Cherm. in Bull. Soc. Bot. France **69**: 720 (1923). —Robinson in Kirkia **1**: 40 (1961).

    Rhynchospora testui Cherm. in Arch. Bot. Mém. **4**(7): 42 (1931). Types: Central African Republic, Haut-Obangui, Pawa, 90 km W of Yalinga, 23.viii.1921, *Le Testu* 3123 (P syntype, BR); Haut-Obangui, Mangapou, 75 km W of Yalinga, 26.viii.1926, *Le Testu* 3158 (P syntype, BR); R. Kpalato, near Bambari, 14.viii.1921, *Tisserant* 411 (P syntype).

Annual or short-lived perennial herb to 70 cm tall, delicate slenderly tufted, glabrous; rhizome almost lacking, shortly linking shoot bases. Radical leaves few, blade to 1 mm wide, cauline leaves numerous, size reduced upwards; sheath leaf-like; ligule minute, sometimes a distinct ridge. Culm erect or supported by other vegetation, filiform, nodose, internodes subterete. Inflorescence compound, a terminal corymbose cyme with several branches single, from lower culm nodes, cymes loose. Bracts leaf-like, reduced. Spikelets few per cyme, 5–7 mm long, usually each maturing 2 nutlets; upper florets sometimes male. Glumes spiralled, apices acute, rachilla sinuous. Perianth bristles 0. Stamens 2. Style deeply bifid. Nutlet (including style base) 1.2–1.3 × 1.2–1.3 mm, square in outline, biconvex, surface transversely corrugated; style base fully attached to nutlet crown, shorter than nutlet width.

Zambia. B: Mongu Dist., Mongu, 17.xi.1962, *Robinson* 5499 (K). N: Kasama Dist., Chibutubutu, Lukulu R., 48 km S of Kasama on Mpika road, 1320 m, 23.ii.1960, *Richards* 12533 (K). W: Mwinilunga Dist., between Kalene and Sakatwala, 1400 m, 22.ii.1975, *Hooper & Townsend* 320 (K). C: Lusaka Dist., Chakwenga headwaters, 100–129 km E of Lusaka, 27.iii.1965, *Robinson* 6545 (K). **Zimbabwe**. W: Matobo Dist., Besna Kobila, iv.1958, *Miller* 5243 (SRGH). E: Mutare Dist., Vumba Mts, Eastern Beacon, 3.iii.1991, *Browning* 356 (GENT, K, NU, SRGH). **Malawi**. N: Karonga Dist., Kayelekera, 28 km WSW of Karonga, 950 m, ix.1989, *Collinson & Davy* s.n. (K).

Also in West Africa (Guinea, Ivory Coast, Liberia, Senegal), D.R. Congo, Ethiopia, Uganda, Tanzania, South Africa (KwaZulu-Natal), Swaziland and Madagascar. In seasonally wet areas, usually inconspicuous among other vegetation; 900–1400 m, not coastal except further south.

Conservation notes: not threatened.

This subspecies is recognised by the square outline shape of the nutlet, the sinuous spikelet rachilla and the plant's fragility.

Uncertain species.

Robinson (1961: 38) cites specimen *Robinson* 3572 (K) as a possible new species from Mwinilunga in NW Zambia. I have examined specimens at K and agree with his statement regarding the affinities with *Rhynchospora brownii* and *R. angolensis*. The nutlet in particular resembles that of *R. angolensis* but is smooth. No other collections similar to *Robinson* 3572 have been located, and until more material is available this cannot be described as a new species.

27. **ACTINOSCHOENUS** Benth.[27]

Actinoschoenus Benth. in Hooker's Icon. Pl. **14**: 33 (1881).

Perennial herbs, densely tufted, rhizomatous or stoloniferous; culms scapose, ± 4-angular, bases covered by closed leaf sheaths, blades reduced. Inflorescences terminal, a shortly bracteated head of 3–4 closely aligned, linear-lanceolate spikelets each with a veinless prophyll and (4)5(7) glumes, distichous, successively larger upwards, uppermost enclosing a single bisexual floret. Stamens 3. Style trifid, base enlarged, soon deciduous, moving upward to project beyond enveloping glume. Nutlet ovate in outline, trigonous, surface faintly cellular, becoming puncticulate.

Traditionally, a genus of 3 species – *Actinoschoenus repens* (endemic to the Flora area), *A. thouarsii* (Kunth) Benth. (W-C tropical Africa, W Indian Ocean, Sri Lanka, S China, Indochina to W & C Malesia and N Australia) and *A. yunnanensis* (C.B. Clarke) Y.C. Tang (India to China). Four new species and one new combination under the genus have been recognized from W Australia by Rye *et al.* in Nuytsia **26**: 167–184 (2015).

Actinoschoenus repens J. Raynal in Adansonia, n.s. **7**: 92 (1967). Type: Zambia, Mwinilunga, source of Zambezi R., 13.xii.1963, *Robinson* 5977 (K holotype, P). FIGURE 14.**61**.

Rhizomatous densely tufted perennial, appearing leafless, but bases enveloped by closed, short leaf sheaths with ligules c. 1.3 mm long. Culms 50–60 cm × 0.5–0.8 mm, curved basally, becoming erect or almost so. Inflorescence terminal, a shortly bracteated head of (3)4 closely aligned, linear-lanceolate spikelets c. 10 mm long, each comprising a short, sometimes long-mucronate bract, a short ± ovate, veinless prophyll with lateral edges infolded, and 5 glumes with emarginate, mucronate apices, distichous, successively larger upwards, the basal sterile, topmost

---

[27] by J. Browning and K.D. Gordon-Gray†

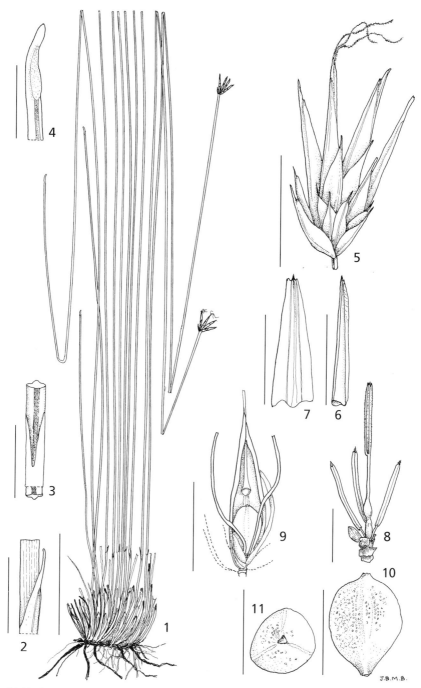

Fig. 14.**61**. ACTINOSCHOENUS REPENS. 1, habit; 2, 3, leaf sheath; 4, culm apex, inflorescence undeveloped; 5, inflorescence; 6, 7, apical glume, folded and opened; 8, 9, floret, developing and mature within glume; 10, 11, nutlet, abaxial and apical views. 1–4, 6, 7, 9–11 from *Robinson* 3643; 5, 8 from *Milne Redhead* 3107. Scale bars: 1 = 40 mm; 2–4, 8, 10, 11 = 2 mm; 5–7, 9 = 5 mm. Drawn by Jane Browning.

closely enveloping a single floret. Stamens 3; anthers 2.8–3 mm long, connective prolonged, red; at anthesis filaments extruding beyond enfolding, enwrapping glume. Style elongate, exceeding uppermost glume in length, base expanded into a 'funnel' covering apex of nutlet, soon detaching, moving upwards through enwrapping glume to project as a dark bristle; branches 3, fimbriate, soon drying off. Nutlet ovate in outline, trigonous, c. 2.5 × 1.7 mm, cream to brownish; surface topography faintly cellular, cells isodiametric, eventually papillate.

Zambia. W: Mwinilunga Dist., Chingobola dambo near Matonchi, 1350 m, 17.ii.1975, *Hooper & Townsend* 142 (K); Mwinilunga Dist., just E of Kasompa R. in *Cryptosepalum* woodland, 1.ii.1938, *Milne-Redhead* 4433 (K).

Monotypic in the Flora area. Almost exclusively on sand in open parts of *Cryptosepalum* woodland; 1300–1400 m.

Conservation notes: Endemic to the Mwinilunga area; unlikely to be threatened.

Affinity of this species appears to be with *Abildgaardiae* and/or *Fimbristylideae*, but requires verification. Morphological support for the first is given by (a) breaking glumes resulting in ragged spikelet rachilla, (b) reduction in number of spikelets per inflorescence within the genus, (c) the expanded style base, and (d) reduced leaf blades. The persistent style is probably effective in fruit dissemination.

28. SCHOENUS L. [28]

Schoenus L., Sp. Pl.: 42 (1753).

Tufted, tussocky, leaf-bearing perennials; without stolons. Culms scapose. Leaves ligulate, basal, sometimes reduced to a sheath. Inflorescence terminal, capitate. Bracts sheathing basally, apices elongate. Spikelets numerous. Lower florets bisexual, the remainder functionally male. Glumes c. 6, distichous, size increasing upwards, perfecting (1)3 nutlets. Bristles 0–6, rarely laminar, usually deciduous with fruit. Stamens 3. Style branches 3. Nutlets obovoid, faintly trigonous, surface smooth.

About 108 species worldwide. The distribution of the genus is very wide, and has possibly increased into the New World by introduction.

Schoenus nigricans L., Sp. Pl.: 43 (1753). —Clarke in Fl. Cap. **7**: 272 (1898); in F.T.A. **8**: 484 (1902). —Schönland in Mem. Bot. Surv. S. Africa **3**: 57 (1922). —Gordon-Gray in Strelitzia **2**: 171, 174, figs.76,77 (1995). Type: Europe, "In Europae paludibus aestate, exsiccatis", *Herb. Linn.* 68.5, 68.6 (LINN). FIGURE 14.**62**.

Schoenus aggregatus Thunb., Phytogr. Blatt. **1**: 5 (1803).
Chaetospora nigricans (L.) Kunth, Enum. Pl. **2**: 323 (1837).

Densely tufted perennial, 60–100 cm tall; roots 1.2–1.5 mm wide; leaf bases 7–8 cm long, dark brown to black shiny. Leaves shorter than culms; blade c. 40 cm long × 0.5 mm wide, stiff, scabrid on margins towards apex. Culms, scapose, 1–2 mm wide, nodeless, trigonous. Inflorescence terminal, 13–15 × 10–20 mm, a head of closely-packed, dark spikelets, subtended by 2 bracts, their bases sheathing, apices elongate, lower 35–90 mm long, exceeding upper in length. Spikelets c. 10 mm long, bisexual, blackish, occasionally brown. Glumes distichous, usually 6, size increasing upwards to 6–8 mm long, keels minutely scabrous. Perianth 3, minute scales. Stamens 3; anthers crested. Style branches 3, ± style length. Nutlet (1)3-perfected, white, ovoid, faintly trigonous, c. 1.5 mm long, surface topography smooth.

Zambia. S: Choma Dist., Muckle Neuk, 20 km N of Choma, 1280 m, 10.x.1954, *Robinson* 903 (K). **Zimbabwe**. N: Binga Dist., Chitiwatata Hot Springs, 480 m, 7.xi.1958, *Phipps* 1388 (K, NU, SRGH).

---

[28] by J. Browning and K.D. Gordon-Gray†

Cosmopolitan: North Africa (Algeria, Libya, Tunisia), Ethiopia, Somalia, D.R. Congo (Katanga), South Africa (Eastern Cape, KwaZulu-Natal, Free State, Mpumalanga, Northwest, Western Cape), but with a disjunct distribution. Inland in damp grassland bordering hot springs or at coast in ± sands of undulating dunes in coastal heath, or estuarine to riverine in fissures of wet rock outcrops; 300–1300 m.

Conservation notes: Not threatened.

Fig. 14.**62**. SCHOENUS NIGRICANS. 1, partial habit; 2, 3, inflorescence; 4, bract; 5, spikelet, two lower glumes removed, remainder displaced; 6, floret, part of stigma removed; 7, nutlet. 1–5 from *Strey* 7739; 6, 7 from *Turrill & Montford* 552. Scale bars: 1, 2 = 25 mm; 3 = 10 mm; 4, 5 = 2 mm; 6, 7 = 1 mm. Drawn by Jane Browning. Reproduced from Strelitzia (1995) by kind permission of the South African National Biodiversity Institute, Pretoria.

29. **CLADIUM** P. Browne[29]

Cladium P. Browne, Civ. Nat. Hist. Jamaica: 114 (1756).

Robust perennials; stem base, rhizome and stolon swollen; aerial stem sparsely noded, internodes hollow. Leaves cauline, V-shaped to flat, midribs and margins scabrid, eligulate. Inflorescence paniculate, a series of short, club-shaped branches, bearing numerous short stalked and sessile spikelets comprising prophyll, 3–5 sterile, (1)2–3 fertile glumes, florets bisexual, bristles absent. Stamens 2(3). Style branches 2(3), base corky, persistent. Nutlets rounded, apex beaked, surface smooth, shining.

A genus of 2 or 3 species, only one from Africa that Kükenthal (1938) named *Cladium mariscus* (L.) Pohl subsp. *jamaicense* (Crantz) Kük. as there is resemblance in fruit size and shape to the American *C. jamaicense*, and in spikelet size and prolongation of anther connective to the European *C. mariscus*.

Cladium mariscus (L.) Pohl, Tent. Fl. Bohem. **1**: 32 (1809). Type: Europe, "Habitat in Europae paludibus", *Herb. Linn.* no.68.1 (LINN lectotype), lectotypified by Kukkonen (Taxon **53**: 179, 2004).

> *Schoenus mariscus* L., Sp. Pl.: 42 (1753).

Subsp. **jamaicense** (Crantz) Kük. in Repert. Spec. Nov. Regni Veg. Beih. **40**(1): 523 (1938). —Haines & Lye, Sedges & Rushes E. Afr.: 323 (1983). —Gordon-Gray in Strelitzia **2**: 44, 45, figs.18,19 (1995). —Hoenselaar in F.T.E.A., Cyperaceae: 364, 366, fig.54 (2010). Type: Jamaica, no further details. FIGURE 14.**63**.

> *Cladium jamaicense* Crantz, Inst. Rei Herb. **1**: 362 (1766). —Clarke in F.T.A. **8**: 484 (1902).
> *Cladium mariscus* R. Br. sensu Clarke in Fl. Cap. **7**: 291 (1898).

Perennial herb to 4 m high; rhizomes and stolons developed. Culms rounded or somewhat trigonous, noded, basally covered by scales and thickened leaf-bases. Leaves large, 200 × 5 cm, dark green, margins and abaxial midrib scabrid; upper leaves narrower with tubular sheaths, uppermost indistinguishable from involucral bracts, bearing inflorescence branches. Inflorescence paniculate, a series of inverted triangular, club-shaped branches (longest side uppermost) with numerous dark brown, closely-packed, short-stalked sessile spikelets in groups of 3–15. Spikelets comprising a lowest branch scale (prophyll) 1–3 mm long, with 2 distinct nerves and 3–5 sterile glumes, then by one with bisexual floret, uppermost glumes reduced, seldom fertile, largest glumes 2–3.5 mm long, glabrous, apices obtuse. Perianth lacking. Stamens 2. Style 3-branched, long; style base darkening with age, persistent. Nutlet 2–3 × 2 mm, lanceolate to ovoid, rounded, never angled, pale to dark brown, surface smooth, becoming faintly longitudinally ridged with wrinkles between ridges in age; usually water-distributed.

Botswana. N: Chobe Dist., Zibadianja lagoon, 18°30'S 23°10'E, 1000 m, 19.x.1972, *Gibbs Russell* 2153 (K). **Zambia**. W: Ndola Dist., Chichele Forest Reserve, Ndola, 1300 m, 17.i.1960, *Robinson* 3301 (K). N: Kawambwa Dist., Lake Lusenga, S of Baka-baka, 11.xi.1969, *Verboom* 2739 (K). C: Serenje Dist., Kasanka Nat. Park, Fibwe, 12°35'S 30°17'E, 1300 m, 14.xi.1998, *Bingham* 11749 (K). S: Mumbwa Dist., Kafue Nat. Park, 26.ix.1964, *van Rensburg* KBS 2968 (K). **Zimbabwe**. N: Kariba Dist., Chirundu Hot Spring, 19.i.1967, *Wild* 7592 (K, SRGH). **Mozambique**. GI: Zavala Dist., Lake Marangue, 46 m, 29.x.1935, *Lea* 127 (K). M: Record from Inhaca Is. (Macnae & Kalk, Nat. Hist. Inhaca Is.: 141, 1969).

Also in West Africa, Ethiopia, Kenya, Uganda, Tanzania, Angola, Namibia, South Africa (Eastern, Western and Northern Cape, Gauteng, KwaZulu-Natal, Limpopo, Mpumalanga, North West), Madagascar and in the Old and New World Tropics. Mostly

---

[29] by J. Browning and K.D. Gordon-Gray†

fringing permanent water bodies (rivers, lakes, occasionally swamps), localised stands blend with grasses and other vegetation of equal height; 10–1400 m.

Conservation notes: Widespread; not threatened.

Cladium mariscus subsp. *jamaicense* is characterized by the very serrated and sharp edges to the leaves and bracts. It bears some resemblance to *Rhynchospora corymbosa*, which has a less compact inflorescence with larger, paler spikelets.

Fig. 14.**63**. CLADIUM MARISCUS subsp. JAMAICENSE. 1, habit; 2, inflorescence; 3, plant base, longitudinal section; 4, detail of leaf; 5, spikelet; 6, nutlet and upper unisexual floret with 2 filaments and rudimentary gynoecium. All from *Ward* 9082. Scale bars: 1 = 250 mm; 2, 3 = 50 mm; 4 = 10 mm; 5, 6 = 2 mm. Drawn by Jane Browning. Reproduced from Strelitzia (1995) by kind permission of the South African National Biodiversity Institute, Pretoria.

30. **CARPHA** R. Br.[30]

Carpha R. Br., Prodr. Fl. Nov. Holland.: 230 (1810). —Kuekenthal in Feddes Repert. **47**: 101–119, 209–211 (1939).

Asterochaete Nees in Linnaea **9**: 300 (1834).

Oreograstis K. Schum. in Engler, Pflanzenw. Ost-Afrikas **C**: 127 (1895).

Perennial herbs with a short rhizome. Culms erect, trigonous to ± circular in cross-section. Leaves basal and cauline. Inflorescence a narrow panicle with slender branches, sometimes head-like, spikelets loosely arranged to densely clustered. Spikelets compressed, flowers in each spikelet 1–2, bisexual, lower flower male in *Carpha schlechteri*. Glumes 4–6(7), distichous, lower ones empty, persistent, ± half length of upper glumes, uppermost glume sometimes linear, empty. Hypogynous perianth with 6 bristles, upwardly scabrous or plumose in lower half, usually with inner whorl much longer than outer, falling with nut. Stamens 3; anthers mostly greyish rarely yellow, with a pyramidal to conical apical appendage. Style slender, continuous with ovary, persistent; stigmas 3. Nutlet trigonous, ornamentation reticulate or punctulate.

A genus of c. 16 species with 11 species in eastern and southern Africa, Madagascar and the Mascarenes. Most species in Africa are relatively narrow endemics but *Carpha capitellata*, the only species in the Flora area, is widespread (South Africa, Zimbabwe and Madagascar).

While Zhang *et al.* (in Austral. Syst. Bot. **19**: 438–465, 2006 & in Austral. Syst. Bot. **20**: 93–106, 2007) found species of *Carpha* assignable to *Asterochaete* (all from Africa) to be sister species of *Carpha sensu stricto* (from Australia, Malesia and South America), Viljoen *et al.* (in Amer. J. Bot. **100**: 2494–2508, 2013) and Bruhl *et al.* (unpublished data) did not recover the reciprocal monophyly of *Carpha* and *Asterochaete*. Hence here we maintain the recognition of *Carpha* in a broad sense over *Asterochaete*.

Carpha capitellata (Nees) Boeckeler in Linnaea **38**: 266 (1874). —Clarke in Durand & Schinz, Consp. Fl. Afr. **5**: 656 (1894); in Fl. Cap. **7**: 270 (1898); in Illustr. Cyp.: t.76.2 (1909). —Pfeiffer in Repert. Spec. Nov. Regni Veg. **29**: 179 (1931). — Kükenthal in Repert. Spec. Nov. Regni Veg. **47**: 116 (1939). Type: South Africa, Cape, *Forbes* s.n. (CGE). FIGURE 14.**64**.

Asterochaete capitellata Nees in Linnaea **9**: 300 (1834); in Linnaea **10**: 194 (1835). —Kunth, Enum. Pl. **2**: 312 (1837). —Steudel, Syn. Pl. Glumac. **2**: 155 (1855).

Asterochaete tenuis Kunth, Enum. Pl. **2**: 312 (1837). —Steudel, Syn. Pl. Glumac. **2**: 155 (1855). Types: South Africa, Cape, Zuureberg Range, 4.xi.1829, *Drège* 2031 (P? syntype); Zuurberg Range, *Drège* 1840 (K? syntype); Cape, Roodeberg, 10.xi.1830, *Drège* 2449 (P? syntype).

Carpha bracteosa C.B. Clarke in Durand & Schinz, Consp. Fl. Afr. **5**: 656 (1894); in Fl. Cap. **7**: 270 (1898). —Pfeiffer in Repert. Spec. Nov. Regni Veg. **29**: 179 (1931). Types: South Africa, W Cape, Somerset East, Boschberg, *MacOwan* 1616 (SAM syntype, K); Boschberg, *MacOwan* 2187 (K syntype); Baines' Kloof, x.1873, *Bolus* 2867 (K syntype, BOL).

Erect perennial 23–70 cm high; rhizome short to long. Culms erect with 0–2 nodes, each with a single leaf. Leaves crowded at culm base; blade straight, 16–36 cm long, 1.5–4(5) mm wide, V-shaped in cross-section. Inflorescence with 2–5 short internodes, 5–38 cm long; spikelets 50–100, in 2–8 ovoid heads. Involucral bracts several, leaf-like or ovate, to 25 cm long and 5.8 mm wide, shorter or slightly longer than heads of spikelets. Spikelets 5–7.5 mm long, with (4)5 distichous glumes; glumes straw-coloured to brownish, lowest 3 (rarely 2) empty, persistent, upper 2 usually fertile (uppermost sometimes sterile), deciduous, lowest shorter than upper empty one and two adjacent fertile ones, basal fertile glume 4.4–6.2 × 1.5–2.5 mm, second fertile glume 4.2–6.5 × 1–2.3 mm; 'rachilla' elongated above basal fertile glume, attached to base. Flowers 1–2 per spikelet, bisexual; bristles 6, scabrous over whole length or nearly smooth

---

[30] by J. Bruhl, K. Wilson and X. Zhang

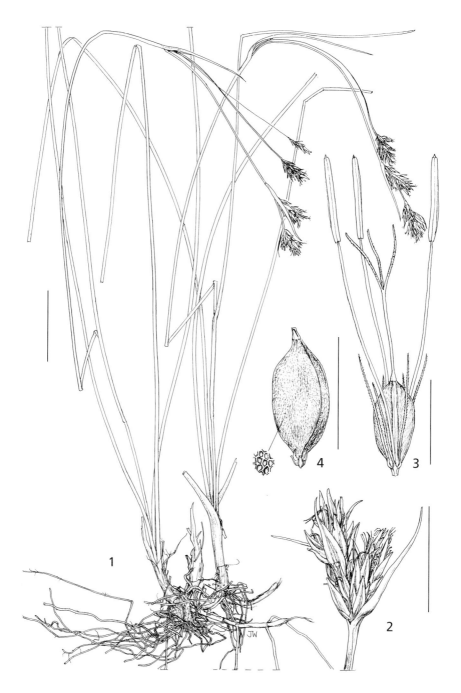

Fig. 14.**64**. CARPHA CAPITELLATA. 1, habit; 2, part of the inflorescence; 3, floret; 4, nutlet. All from *Robinson* 1976. Scale bars: 1 = 3 cm; 2 = 1 cm; 3, 4 = 2 mm. Drawn by Juliet Williamson.

at base, in 2 whorls; outer 3 bristles shorter than inner 3, inner 1.5–3.5 mm long. Stamens 3; anthers 2–2.9 mm long, greyish, with apical appendage 0.1–0.3 × c. 0.1 mm. Style branches 3; style base persistent, 0.3–1 × 0.15–0.3 mm. Nutlets 1–2 per spikelet, ellipsoid, trigonous in cross-section, 2–2.7 × 0.8–1.4 mm, dark brown at maturity, surface reticulate.

Zimbabwe. E: Mutasa Dist., Nyanga, Gairezi Ranch, near Mozambique border, 21.xi.1956, *Robinson* 1976 (IRLCS, K, LISC, NRGH, NU, SRGH).

Also in Madagascar and South Africa. In marshy places; c. 1900 m (lower altitudes in South Africa).

Conservation notes: Known only from one specimen in the Flora area; probably Vulnerable, but not threatened globally.

Carpha capitellata is a variable species that needs more detailed study. The only Zimbabwe specimen seen, *Robinson* 1976, is particularly slender, especially its inflorescence branches. It is similar to *C.* sp. 'Uluguru Mts' (Zhang, Bruhl & Wilson, pers. comm.), but differs in significant characters such as the inflorescence arrangement and fruit ornamentation. A molecular study of populations is needed to understand variation in the *C. capitellata* complex, other species limits within the genus, and the limits of the genus.

31. **TETRARIA** P. Beauv.[31]

Tetraria P. Beauv. in Mém. Cl. Sci. Math. Inst. Natl. France **13**(2): 54 (1816).

Perennial herbs, tufted, or with short seriate rhizomes. Culms scapose or few-noded, terete or faintly 3–4 angled. Leaves basal or basal and cauline, about inflorescence height, grass-like, sometimes reduced to sheaths, ligule when present cylindric, dark brown, conspicuous banding culms at intervals. Inflorescence paniculate, open (interrupted) or contracted; primary bracts leaf-like, subtending inflorescence branches. Spikelets several to many. Glumes subdistichous, of increasing length upwards, margins of proximal ones somewhat serrate, sterile; distal ones (1)2–3(4), subtending a floret; all glumes closely enwrapped; 1–2(3) lower florets usually functionally male, upper 1–2 florets bisexual (rarely male). Perianth bristles (0)3–6, or absent. Stamens 3–6, clearly crested prior to anthesis. Style 3-branched, occasionally 4–9, thickened, robust. Nutlets rounded, trigonous to 4-angled, beaked or not, surface often smooth.

Genus of c. 55 species, predominantly South African with a few outliers at high altitudes in E. tropical Africa, Malesia and Australia.

Tetraria mlanjensis J. Raynal in Adansonia, n.s. **12**: 213 (1972). Type: Malawi, Mt Mulanje, Castle Rock, 27.vii.1956, *Newman & Witmore* 251 (WAG holotype, BR, NY, P, SRGH). FIGURE 14.**65**.

Tufted perennial, often with deep red/purple colouration at base. Leaves 0.5 mm wide, wiry, scabrid, usually not exceeding inflorescence height. Inflorescence a narrow panicle, 50–150 mm long, with branches in series along culm, usually 2 per node subtended by a single bract. Spikelets 4–5 × 1.5 mm, ovate, cylindrical. Glumes 7–10, distichous to spiral, closely involute, lowest short, lengths increasing towards apex, c. 6 empty; uppermost 2 mucronate to awned, each subtending a single unisexual or bisexual floret (total number of florets per spikelet variable, only one usually perfecting a fruit). Stamens 3; anther connective flattened, produced sometimes beyond anther length. Style base slightly pyramidal, persistent after style shed, or deciduous with style leaving apical depression, 3-branched. Nutlet obovoid, trigonous, surface topography faintly finely cellular.

Zimbabwe. E: Nyanga Dist., no locality, 23.xi.1930, *Fries et al.* 3160 (K). **Malawi**. S: Mt Mulanje massif, summit ridge of Mchesi Mt, 2150 m, 11.vii.1987, *J.D. & E.D. Chapman*

---

[31] by J. Browning and K.D. Gordon-Gray†

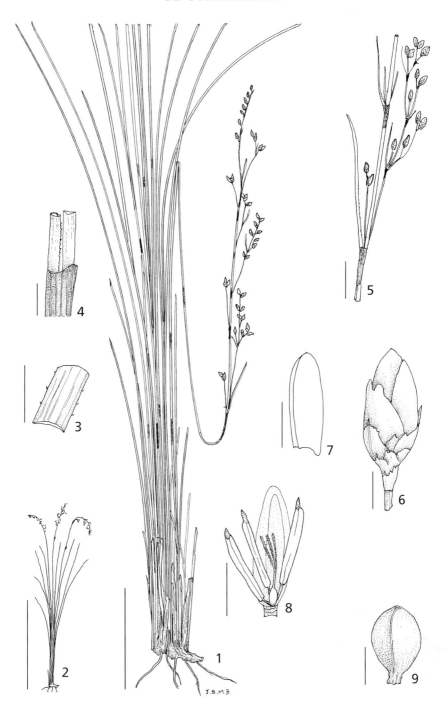

Fig. 14.**65**. TETRARIA MLANJENSIS. 1, 2, habit; 3; part of leaf; 4, sheath; 5, part of panicle; 6, spikelet; 7, apical glume, lateral view; 8, floret, two glumes displaced; 9, nutlet. 1–8 from *Hilliard* and *Burtt* 6391; 9 from *Reid* 1755. Scale bars: 1 = 40 mm; 2 = 250 mm; 5 = 10 mm; 3, 4, 6–9 = 1 mm. Drawn by Jane Browning.

8690 (MO, K); Mt Mulanje, Tuchila to Great Ruo Basin, 2225 m, 4.xi.1971, *Hilliard & Burtt* 6391 (K).

Also in South Africa (Mpumalanga). Clumps interspersed in montane grassland and in ericaceous zone, rooted in dry sand among rocks; 1800–2250 m.

Conservation notes: Not threatened.

Mature nutlets were not present on material at Kew.

Tetraria mlanjensis from E Zimbabwe has a close relationship to *T. usambarensis* K. Schum. in Engler, Pflanzenw. Ost-Afrikas **C**: 128 (1895). It is possible that they are the same species, the difference in size being due to latitude and environmental factors. If so, the name *T. usambarensis* has precedence.

32. **COSTULARIA** C.B. Clarke[32]

Costularia C.B. Clarke in Fl. Cap. **7**: 274 (1898).

Tufted perennial herbs; rhizome woody, laterally branched. Culms erect, densely clothed by leaves; radical with poorly defined sheaths, cauline enveloping up to ½ internode length; margins scabrid, eligulate. Inflorescence paniculate, branching from nodes. Spikelets numerous, solitary or paired. Glumes subdistichous, 6–12, proximal 3–9 sterile. Florets 2, lower usually functionally male with rudimentary gynoecium; perianth bristles 3 + 3, fimbriate proximally, ciliate distally deciduous with fruit. Stamens 3. Style trifid, base often thickened, persistent. Nutlets ovate to oblong, rounded to 3-ribbed, somewhat stipitate, beaked, smooth to rugulose.

A genus of medium-sized to tall perennials; some species may appear as dwarf shrubs with a caudex. One species in Africa and about 25 species from Madagascar, other West Indian Islands and New Caledonia. Species favour rather dry, sandy or rocky places on mountain slopes.

Costularia natalensis C.B. Clarke in Fl. Cap. **7**: 274 (1898). —Schönland in Mem. Bot. Surv. S. Africa **3**: 58 (1922). —Burtt in Notes Roy. Bot. Gard. Edinburgh **45**: 77 (1988). —Browning & Gordon-Gray in S. African J. Bot. **62**: 155–159 (1996). Types: South Africa, KwaZulu-Natal, possibly Noodsberg, iv.1883, *Buchanan* 152 (K syntype); possibly Noodsberg, n.d., *Buchanan* 354 (K syntype, NH) – see Burtt (1988). FIGURE 14.**66**.

Tetraria natalensis (C.B. Clarke) T. Koyama in J. Fac. Sci. Univ. Tokyo, Sect. 3, Bot. **8**: 75 (1961). —Gordon-Gray in Strelitzia **2**: 190, figs.87C,88 (1995).

Tufted, erect perennial; rhizome 1–1.5 mm diameter, woody, with lateral branches increasing tuft size. Leaf bases persistent, up to 15 mm wide; basal leaves spirodistichously arranged, grading into numerous radical and cauline foliage leaves; radical leaves with poorly defined sheaths, 0.3–0.6 m long, tapering to 3–4 mm at mid-length, tough, glabrous, margins scabrid; cauline leaves enveloping up to half internode length, mouth V-shaped, eligulate; base as for foliage leaves but reduced in size to form smaller bracts at reproductive nodes. Culms erect, 0.6–1.8 m high, nodes dark to chestnut brown, internodes subtrigonous, greenish, with numerous floral branches. Inflorescence a panicle, bracteated axillary branches from nodes with clustered spikelets. Spikelets solitary or somewhat paired, pedicelled or subsessile, oblong, 6–9 × 1.8–2 mm, dark brown. Glumes subdistichous, 6–12, proximal 3–9 empty; lowest 3 with attenuate apices, remainder increasing in length upwards, apices acute or acuminate; next 3 largest, 6–7 × 3 mm, boat-shaped, margins ciliate, uppermost enclosed within two preceding. Florets 2, lower functionally male, usually with rudimentary gynoecium, upper bisexual. Perianth (glumellae) 3 + 3, 6–7 mm long, villous distally, prickle hairs basally. Stamens 3, filaments persistent after anthesis. Style 3-branched; lower portion persistent as short beak on fruit. Nutlet globose, narrowed basally into a funnel-shaped extension, 5–3 mm in total length, faintly 3-ridged longitudinally, whitish to pale fawn, surface faintly transversely rugose.

---

[32] by J. Browning and K.D. Gordon-Gray†

Zimbabwe. E: Mutare Dist., Bvumba Mts, Castle Beacon, c. 1800 m, 4.ii.1990, *Browning* 296 (NU). **Malawi**. S: Mulanje Dist., Mt Mulanje, above hut, 1874 m, 28.v.2005, *Chapman et al.* 253 (K). **Mozambique**. MS: Báruè Dist., serra de Choa, 21 km from Catandica (Vila Gouveia), 1400 m, 28.iii.1966, *Torre & Correia* 15439 (BR, LD, LISC, LMA, MO, UPS, WAG).

Fig. 14.**66**. COSTULARIA NATALENSIS. 1, habit and inflorescence; 2, spikelet; 3, upper florets opened, glumes displaced. All from *Hilliard & Burtt* 14504. Scale bars: 1 = 40 mm; 2, 3 = 2 mm. Drawn by Jane Browning. Reproduced from Strelitzia (1995) by kind permission of the South African National Biodiversity Institute, Pretoria.

Also in South Africa (KwaZulu-Natal, Mpumalanga, Limpopo) and Swaziland. Rocky slopes with coarse grasses, often with restiads and short grasses; 1400–1900 m.

Conservation notes: Scarce in most localities with little active regeneration observed. In KwaZulu-Natal plantations of alien tree species have reduced its habitat extent; possibly Near Threatened.

33. **COLEOCHLOA** Gilly[33]

Coleochloa Gilly in Brittonia **5**: 12 (1943). —Nelmes in Kew Bull. **8**: 373–381 (1953). —Kativu in Kirkia **15**: 33–37 (1994).

Eriospora A. Rich. in Tent. Fl. Abyss. **2**: 508 (1850), illegitimate name.
Catagyna sensu Hutchinson in F.W.T.A. **2**: 490 (1936), non Lestib. (1819).

Loosely or densely tufted perennials forming tussocks. Culms erect or slightly wanting, ± scapose, basal part of culms covered by remains of old leaf sheaths. Leaves 2-ranked, grass-like; leaf sheaths open on ventral side with a ligule at sheath-blade junction represented by a line of hairs; leaf blades inrolled, conduplicate or ± flat, deciduous in unfavourable seasons. Inflorescence paniculate with clusters of spikes on the tips of long peduncles, or appearing sessile. Spikes of dense clusters of spikelets subtended by glume-like bracteoles. Spikelets bisexual of 4–5 distichous unisexual glumes, the lowest two empty and upper 2–3 contain 1–2 male flowers and 1–2 female flowers; male flowers with 3 stamens, mucronate at apex; female flowers with soft-textured utricle-like nutlet overtopping glumes when mature, obscurely trigonous, long-beaked, scabrid or smooth on angles and beak. Style 3-branched, continous with nutlet, style tips easily breaking off. Hypogynous bristles slender or less so, colourless or tinged red in the top half, antorsely barbellate or smooth, deciduous with nutlet or detaching in groups. Seed brown, oblong, caryopsis-like, loosely enveloped by nutlet.

A genus of seven species in Africa and Madagascar.

1. Hypogynous bristles smooth, colourless; angles of nutlet scabrid; anthers without a crest of white setae on apex; leaves conduplicate to ± flat, often shorter than or scarcely exceeding inflorescence . . . . **3.** *pallidior*
– Hypogynous bristles antorsely barbellate, tinged red; angles of nutlet smooth or scabrid; anthers with a crest of white setae on apex; leaves filiform, convolute-cylindric, often much curved above and exceeding inflorescence 2
2. Spikes lax and individually peduncled; culms and abaxial leaf surfaces sparsely to densely villous, rarely with lanate adaxial leaf surface . . . . **1.** *setifera*
– Spikes aggregated and clustered together, sessile and appearing secund; culms and abaxial leaf surface glabrous, adaxial leaf surface minutely hispidulous (Malawi, Mt Mulanje only) . . . . **2.** *virgata*

1. **Coleochloa setifera** (Ridl.) Gilly in Brittonia **5**: 14 (1943). —Haines & Lye, Sedges & Rushes E. Afr.: 363 (1983). Type: Madagascar, no locality, n.d., *Hilsenberg & Bojer* s.n. (BM lectotype), lectotypified by Nelmes (1953). FIGURE 14.**67**.

Fintelmannia setifera Ridl. in J. Linn. Soc., Bot. **20**: 337 (1883).
Trilepis oliveri Boeckeler, Beitr. Cyper. **1**: 18 (1888). Type: Malawi, Shire Highlands, vii.1885, *Buchanan* 25 (K holotype).
Eriospora oliveri (Boeckeler) C.B. Clarke in Durand & Schinz, Consp. Fl. Afr. **5**: 676 (1894); in F.T.A. **8**: 513 (1902).
Eriospora villosula C.B. Clarke in Trans. Linn. Soc. London, Bot. **4**: 54 (1894). —Clarke in Durand & Schinz, Consp. Fl. Afr. **5**: 676 (1894); in F.T.A. **8**: 513 (1902). —Schumann in Engler, Pflanzenw. Ost-Afrikas **C**: 128 (1895). Type: Malawi, Mt Mulanje, x.1891, *Whyte* 68 (K holotype, BM).

[33] by M. Xanthos

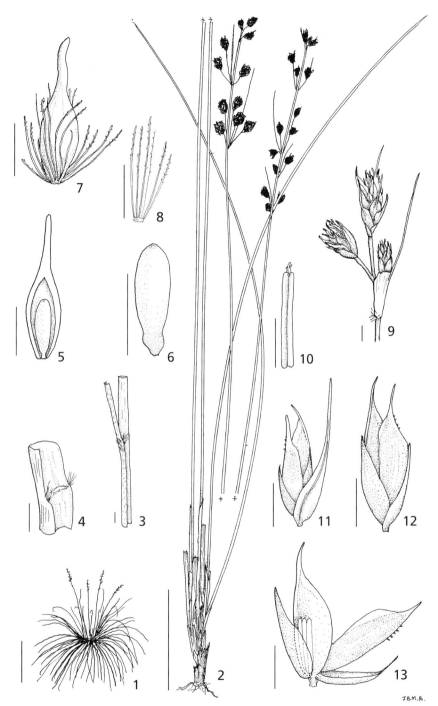

Fig. 14.**67**. COLEOCHLOA SETIFERA. 1 habit, entire plant; 2, habit; 3, part of culm and leaf sheath; 4, leaf sheath opened to show ligule; 5, nutlet split open, seed enclosed; 6, seed; 7, nutlet, some bristles removed; 8, bristles detached from base of nutlet; 9, three fascicles of spikelets; 10, mature anther; 11, spikelet and bract, abaxial view; 12, spikelet, bract removed, adaxial view; 13, spikelet, glumes displaced, adaxial view. 1–4, 9–13 from *Browning* 560; 5–8 from *Pawek* 13626a. Scale bars: 1 = 250 mm; 2 = 40 mm; 3–13 = 1 mm. Drawn by Jane Browning.

Eriospora rehmanniana C.B. Clarke in Fl. Cap. **7**: 297 (1898). Type: South Africa, Mpumulanga, Houtbosch, 1875–1880, *Rehmann* 5624 (K holotype, BM).

Caespitose perennial to 70 cm tall. Culms densely tufted, cylindric, 0.3–1 mm wide, sparsely to densely villous. Leaves to 60 cm long, sheaths 1–5 cm long, usually blackened basally and covering basal part of culm; blade narrowly linear, convolute-cylindrical, 60 cm long by 1 mm wide, curving towards apex and often overtopping inflorescence, abaxial surface sparsely to densely villous, adaxial surface usually glabrous. Inflorescence of pedunclcd spikes emerging from upper leaf sheaths, peduncles villous. Spike narrowly ellipsoid, becoming somewhat obconical in maturity, 4–7 mm long. Spikelets of 4–5 distichously arranged glumes, many per spike, lanceolate to elliptic, 2–4 mm long, laterally compressed. Glumes 2–3 mm long, ovate to lanceolate, maroon to purplish-red sometimes with patches of yellow near base, glabrous to hispidulous on midrib and flanks, apex muticous to aristate. Stamens 3, anthers 0.7–1.5 mm long, with crest of white setae on apex. Style 3-branched, continuous with nutlet beak. Nutlet utricle-like, 3–4 × 0.6–0.9 mm, glabrous, stramineous often with red patches, obscurely trigonous, beak straight or slightly falcate. Hypogynous bristles 2/3 to as long as nutlet, detaching in groups or deciduous with nutlet, antrorsely barbellate, colourless to reddish. Seeds oblong to ellipsoid, 1–1.25 mm long.

Zambia. N: Kasama Dist., Mwela Rocks, 5 km E of Kasama, 4.xii.2006, *Bingham* 13174 (K). W: Chingola, 26.xi.1965, *Fanshawe* 9427 (K, NDO, SRGH). C: Mkushi Dist., Mkushi, Fwila, 3.i.1958, *Robinson* 2579 (K, SRGH). **Zimbabwe**. W: Matobo Dist., Gordon Park, 7.xi.1952, *Plowes* 1527 (K, PRE). C: Goromonzi Dist., Chinamora communal land, Domboshawa, 16.ii.1947, *Wild* 1660 (K, SRGH). E: Mutare Dist., Vumba Mts, Castle Beacon summit, 18.iv.1993, *Browning* 560 (GENT, K, NU, PRE). S: Mberengwa Dist., Mt Buhwa, top of ridge, 2 km E of summit, 30.x.1973, *Biegel, Pope & Gordon* 4329 (K, PRE, SRGH). **Malawi**. N: Mzimba Dist., Hola Mt, 19.i.1971, *Pawek* 13262a (K, MO, SRGH, UC). S: Blantyre Dist., Ndirande Mt, NW side of summit, 3.iii.1970, *Brummitt* 8853a (K, LISC). **Mozambique**. N: Mecula Dist., Niassa Nat. Reserve, Mbatamila camp, 11.vi.2003, *Boane* 46 (K, LMU). Z: Lugela Dist., Mt Mabu, 17.x.2008, *Timberlake* 5426 (K, LMA). T: Cahora Bassa Dist., Songo, Bairro da Zango, 17.iv.1972, *Macêdo* 5206 (LISC, LMA, SRGH). MS: Sussundenga Dist., Makurupini Falls, S foothills of Chimanimani Mts, 25.xi.1967, *Simon & Ngoni* 1323 (K, LISC, SRGH).

Also in D.R. Congo, Tanzania, Swaziland, South Africa and Madagascar. In shallow soil over exposed rock slopes and in rock crevices; 800–2300 m.

Conservation notes: Widely distributed; Least Concern.

The most common species in our area. The degree of hairiness on the whole plant, particularly the culms and lower leaf surfaces, can be very variable. Some specimens collected on Mt Mulanje (e.g. *Chapman* 8112, K) have glabrous culms and lanate upper leaf surfaces, which may indicate hybridisation with *Coleochloa virgata*. The peduncled spikes and pubescence are generally enough to distinguish it.

Records of *Coleochloa abyssinica* (A.Rich.) Gilly from the Flora area, such as from S Zimbabwe in Mapaura & Timberlake, Checkl. Zimbabwe Pl. (2004), are probably referrable to this species.

2. **Coleochloa virgata** (K. Schum.) Nelmes in Kew Bull. **8**: 381 (1953). —Haines & Lye, Sedges & Rushes E. Afr.: 363 (1983). Types: Tanzania, Arusha Dist., Mt Meru, n.d., *Fischer* 624 (B† holotype); Malawi, Mt Mulanje, Lichenya Plateau, 8.vii.1946, *Brass* 16737 (K lectotype, BM, PRE, SRGH), lectotypified here.

Eriospora virgata K. Schum. in Engler, Pflanzenw. Ost-Afrikas **C**: 128 (1895).

Caespitose perennial to 80 cm tall, culms densely tufted, cylindric, glabrous. Leaves to 60 cm long, sheaths 3–15 cm long, blackened basally becoming yellow higher up culm; blades narrowly linear, convolute-cylindrical, 30–55 cm long by 0.7–1.4 mm wide, abaxial surface glabrous, adaxial surface densely minutely hispidulous. Inflorescence of interrupted groups of sessile or very shortly pedicelled spikes clustered together, sometimes appearing second. Spikes obovoid, oblong-ellipsoid, 4–7 mm long; spikelets of 4–5 distichously arranged glumes, many per spike, lanceolate-elliptic, distinctly aristate, 3.5–4 mm long, laterally compressed. Glumes dark red to purplish red, often with a yellow base, 2–3 mm long (excluding awn), scurfy-hispidulous; awn 1–2 mm long. Stamens 3, anthers 1.6–2.4 mm long with crest of white setae on apex. Style and nutlet not seen.

Malawi. S: Mulanje Dist., Mt Mulanje, Lichenya Plateau, 7.vi.1962, *Robinson* 5296 (K). Known only from Mt Mulanje in Malawi and (possibly) Mt Meru in N Tanzania. Shallow rocky soil or in rock crevices; 1800–2300 m.

Conservation notes: From herbarium specimens this appears to be a Mt Mulanji endemic. The type from Tanzania may have been named in error but this is now impossible to verify. Probably Near Threatened.

Unless good fruiting material is seen it remains unclear whether this species is sufficiently distinct from *Coleochloa setifera* to be maintained. The aggregated nature of the inflorescence appears to hold up and I consider it a more consistent taxonomic character than the glabrous culms. Previous literature states the type was destroyed in Berlin, although Kativu (1994) erroneously cites it in the Kew herbarium. The lectotype was chosen from amongst specimens cited in Nelmes (1953).

3. **Coleochloa pallidior** Nelmes in Kew Bull. **8**: 378 (1953). —Haines & Lye, Sedges & Rushes E. Afr.: 363 (1983). Type: South Africa, Soutspansberg Dist., Greefswald Farm, 2.4 km W of Mapungubwe by Limpopo R., 15.iii.1948, *Codd & Dyer* 3835 (PRE holotype, K, SRGH).

Caespitose perennial to 70 cm tall. Culms loosely tufted, cylindric, glabrous, usually fibrous at base. Leaves 35–70 cm long, sheaths 5–13 cm long, stramineous yellowish-brown; blades 27–56 cm and 1–2.5 mm wide, glabrous or lanate on adaxial surface, conduplicate to flat, rarely convolute-cylindric, curving towards apex. Inflorescence of peduncled spikes emerging from upper leaf sheaths. Spikes 4–8 mm long, lanceolate-elliptic. Spikelets of 4–5 distichously arranged glumes, many per spike, lanceolate-elliptic, 2.5–3 mm long. Glumes stramineous, sometimes with red patches, 2–3 mm long, ovate to lanceolate, glabrous or sometimes sparsely hairy on flanks. Stamens 3, anthers 1–1.5 mm long with a red blunt mucro. Style 3-branched, continuous with nutlet beak. Nutlet utricle-like, 2.5–3.5 × 0.6–1 mm, stramineous, prominently angled, with a straight beak, hispid on angles and beak. Hypogynous bristles 2/3 to as long as nutlet, detaching in groups or deciduous with nutlet, colourless and smooth. Seeds stramineous to brown, oblong to ellipsoid, 1–1.25 mm long.

Zimbabwe. N: Mutoko Dist., 10 km SW of Mutoko on road to Nyadiri, 23.iii.1999, *Williamson* 243 (SRGH). W: Matobo Dist., Besna Kobila Farm, i.1958, *Miller* 5023 (LISC, SRGH). C: Kadoma Dist., Ngezi Recreational Park, ii.1968, *Brits* 6 (K, SRGH). E: Mutare Dist., 18 km S of Mutare, 4.xi.1956, *Chase* 6229 (K, SRGH). S: Masvingo Dist., Kyle Nat. Park, Chembira Hill, 21.v.1971, *Ngoni* 89 (K, PRE, SRGH). **Malawi**. S: Machinga Dist., Makulu-kulu Hill, T.A. Nkhokwe, 9.vii.1983, *Seyani & Tawakali* 1127 (K, MAL). **Mozambique**. N: Muecate Dist., Imala on road to Mecubúri, 21.xi.1936, *Torre* 1062 (LISC). Z: Milanje Dist., Sabelua, 1.vii.1972, *Bowbrick* 72 (LISC, SRGH). MS: Gondola Dist., near Chimoio, slopes of Mt Zembe, 15.vi.1959, *Leach* 9113 (K, SRGH).

Also in South Africa. Shallow soil on granitic rocks and in rock crevices; 500–1500 m.

Conservation notes: Widespread; not threatened.

This species is easily distinguished by its nutlet characters and the absence of setae on the anther apex. The culms are also more robust and the leaves generally wider than in the other species. Haines & Lye (1983) stated that *Coleochloa virgata* is the only species in the genus with hairy upper leaf surfaces, but this character has also been found in a specimen of *C. pallidior* from Zimbabwe (*Beigel* 601, PRE).

34. **SCLERIA** P.J. Bergius[34]

Scleria P.J. Bergius in Kongl. Vetensk. Acad. Handl. **26**: 142 (1765). —Robinson in
Kew Bull. **18**: 487–551 (1966). —Hennessy in Bothalia **15**: 505–530 (1985).

Annual or perennial herbs of variable size and habit, from erect plants 15–20 cm tall up to
scramblers attaining 5 m or more. Stems leafy, 3-angled, the stem angles and leaf margins
sometimes very finely dentate, teeth silicified and thus extremely sharp. Leaves with closed
sheaths, sometimes ligulate, the throat margin sometimes extending as a tongue. Inflorescence
either paniculate with a terminal panicle and usually one to many lateral panicles from upper
leaf axils, or glomerulate-spicate and ebracteate with clusters of sessile spikelets. Spikelets with
2–3 basal sterile glumes; bisexual ones with lowest flower female and upper ones male; male
ones similar but lacking the basal female flower; female ones with a single female flower and 0–3
empty glumes above. Flowers unisexual, in axils of spirally or distichously arranged glumes. Male
flowers with 1–3 stamens. Female flowers with 3-branched, sometimes persistent style. Nutlets
ovoid-ellipsoid to depressed spherical, somewhat trigonous, smooth or variously ornamented,
often shiny and sometimes hairy, pericarp silicified at maturity therefore extremely brittle,
stipitate with a thick trigonous or cylindrical stipe, or a very short obpyramidal stipe expanded
at apex into a persistent discoid, triangular, trilobed, zoniform or cupulate hypogynium (see
below) with entire, fimbriate or ciliate margin.

A large genus of about 250 species in the tropical and subtropical zones of the Old
and New Worlds.

This account follows that of Edward Robinson (1966) in almost all respects; I have
used his descriptions, often almost unaltered and, with just minor modifications, his
key. I have also drawn on the late Bernard Verdcourt's account for F.T.E.A. Robinson
made excellent and extremely thorough collections of the genus from throughout
Zambia, with some excursions into Mozambique, Malawi and Zimbabwe. Without these
collections, and Robinson's subsequent work on them, our understanding of the genus
in the Flora area would be extremely poor (see Bingham in Kirkia **18**: 111–115, 2002 for
an account of Robinson's collections). The only major subsequent relevant study of the
genus is that of E.F. Franklin Hennessy (in Bothalia **15**: 505–530, 1985), who dealt with
the genus in southern Africa, discussed spikelet morphology in detail, and provided
useful illustrations of the hypogynium and of nutlet ornamentation. She has gone
through this account extremely thoroughly and made numerous valuable comments.

Robinson's account has been departed from only once – I have recognised *Scleria
lateritica*, which he placed in the synonymy of *S. flexuosa*, with reasons given under the
former taxon. I have also added *S. sobolifer*, described from KwaZulu-Natal and found
in S Mozambique after Robinson's account was written.

Robinson compared a number of the species with material from both the New
World and southern Asia, and suggested some possible synonymies. These have not
been pursued here. The question of intercontinental synonymy needs detailed study,
using both morphology and molecular methods, before any definite conclusions can
be drawn.

Collections of most *Scleria* species are difficult or impossible to name without ripe
fruits and the basal parts and rhizomes. The ripe fruits often fall easily and should
be collected separately into a carefully labelled small packet. Unripe fruits, which
are often white and empty, may not show the normal surface pattern, and certainly
will not show the mature colour. They are also easily crushed. Would be collectors of
Scleria should be aware that the stems and leaves can be extremely sharp. Robinson
remarked graphically on this when he described *S. porphyrocarpa*, but many of the
larger species share this feature.

---

[34] by J.M. Lock

Robinson made much use of the characters of the hypogynium. This structure, which lies immediately below the nutlet, may be absent or take two forms. In some species it is cup-shaped, reaching its greatest development in Section *Ophyroscleria* (*Scleria racemosa*) where it forms a very obvious cup enclosing the base of the nutlet. More commonly, the cup is small and ± divided into three lobes of varying shape. The base of the nutlet (i.e. the pedicel) may also take the form of a thickened stipe developed to varying degrees. The question of whether these two structures are really homologous is not addressed here.

While most rhizome systems are obvious and easy to collect, some species produce tubers on the end of slender rhizomes, which break off unless particularly care is taken when collecting.

Without seeing the plants in their natural habitat it is very difficult to come to a meaningful conservation assessment. The notes given under each species are based almost entirely on distribution and whether a species has been collected more than once or in more than just a very few localities. Many species occur in seasonally or perennially wet places (dambos), which are liable to be drained (often unsuccessfully) for agriculture, or flooded because of deforestation of the interfluves and consequent greater run-off into the valley dambos (M. Bingham, pers. comm.).

1. Hypogynium ciliate on upper margin, forming a cup-like structure surrounding base of nutlet at maturity; nutlet with a prominent persistent style-base at apex . . . . **60.** *racemosa*
–   Hypogynium absent or, if present, not ciliate on upper margin, forming a shallow cup-like, obscurely or distinctly lobed structure at base of nutlet; style-base absent or represented by small apical projection 2
2. Fertile spikelets predominantly bisexual; hypogynium absent. 3
–   Fertile spikelets predominantly unisexual, or with only rudimentary staminate flowers; hypogynium generally well-developed 34
3. Stems branched repeatedly from near base; plants procumbent or only weakly erect, forming dense semiprostrate masses in permanently wet places. . . . .**31.** *procumbens*
–   Stems unbranched or with only very few branches; erect. 4
4. Inflorescence a single terminal spike or panicle. 5
–   Inflorescence with terminal and lateral panicles, the latter subtended by leafy bracts 32
5. Plant annual. 6
–   Plant perennial 13
6. Inflorescence clearly branched, branches to 10 cm or more long. . . . . . .**2.** *glabra*
–   Inflorescence simply spicate, or only shortly branched towards base 7
7. Glumes hairy . . . .**9.** *melanotricha*
–   Glumes glabrous, or hairy only on midrib. 8
8. Mature nutlets dark red, with 3 longitudinal ribs of ± transparent tissue on angles . . . . **6.** *delicatula*
–   Mature nutlets grey, brown or black, without such transparent tissue. 9
9. Stem and leaves hairy; inflorescence a simple unbranched spike . . . . **7.** *zambesica*
–   Stem, and often leaves, glabrous; inflorescence rarely completely unbranched 10
10. Glumes hispid-spinulose on midrib; nutlets reticulate . . . . **5.** *hispidula*
–   Glumes entirely glabrous; nutlets reticulate or tuberculate 11

11. Plant small, stems rarely exceeding 20 cm in height, inflorescence rarely more than 3 cm long; nutlets less than 1 mm broad, tuberculate . . . .**3.** *pulchella*
– Plant with stems usually more than 20 cm in height; nutlets at least 1 mm broad 12
12. Inflorescence simply spicate or with very short branches; glumes reddish black with a green midrib; nutlets tuberculate . . . .**1.** *pergracilis*
– Inflorescence usually shortly branched towards base, branches up to 1.5 cm long, rarely simply spicate; glumes pale straw-coloured with a green midrib, often with a reddish black margin; nutlets reticulate . . . .**4.** *calcicola*
13. Nutlets hairy. . . . . **27.** *fulvipilosa*
– Nutlets glabrous. 14
14. Stems bulbous and ± woody at base . . . .**12.** *bulbifera*
– Stems not bulbous at base. 15
15. Stems arising in small groups from a slender elongated underground stem 1–1.5 mm in diameter; coastal Mozambique (and KwaZulu-Natal) only. . . . **21.** *sobolifer*
– Stems and underground parts not as above. 16
16. Stems arising in a ± straight line from a stiff horizontally elongated rhizome at least 2 mm thick. 17
– Stems and rhizomes not as above 24
17. Glumes hairy 18
– Glumes glabrous 21
18. Nutlets at least 2 mm wide, pale to dark brown at maturity; with a conspicuous black stipe . . . . **16.** *longispiculata*
– Nutlets less than 2 mm wide; stipe not black. 19
19. Glomerules reflexed at maturity; at least some hairs on glumes black . . **8.** *distans*
– Glomerules not reflexed; hairs on glumes not black 20
20. Leaves 0.75–2 mm wide; glumes sparsely hispidulous; nutlets strongly ornamented . . . .**14.** *lateritica*
– Leaves 2–6 mm wide; glumes densely hairy; nutlets smooth . . . . .**15.** *erythrorrhiza*
21. Inflorescence 6–25 cm long; spikelets 4.5–7 mm long . . . .**17.** *welwitschii*
– Inflorescence generally less than 10 cm long; spikelets less than 5 mm long . . 22
22. Inflorescence not more than 4 cm long (usually less than 3 cm); nutlets fulvous or pale brown at maturity, with a stipe the same colour . . . .**19.** *paupercula*
– Inflorescence more than 4 cm long; nutlets not as above 23
23. Rhizome not more than 2.5 mm thick; stipe darker in colour than nutlet at maturity; leaves setaceous . . . .**22.** *angustifolia*
– Rhizome 3–4 mm thick; stipe not darker than nutlet at maturity; leaves 1–3(4) mm wide. . . . .**18.** *rehmannii*
24. Glumes glabrous 25
– Glumes hairy or hispidulous. 28
25. Geophytic plant perennating by means of one or more slender rhizomes that become tuberous; not caespitose 26
– Hemicryptophytic or phanerophytic plant, usually caespitose or, if rhizomatous, with very short rhizomes 27
26. Plant producing a single rhizome each year . . . . **24.** *woodii*
– Plant producing 2–4 slender rhizomes simultaneously. . . . .**25.** *polyrrhiza*
27. Inflorescence generally a ± crowded spike, occasionally shortly branched; nutlets smooth or papillose. . . . .**20.** *dregeana*
– Inflorescence always branched, with distant glomerules; nutlets lightly to strongly trabeculate . . . . **26.** *richardsiae*
28. Inflorescence compound, with many slender compound branches . . . **23.** *pooides*
– Inflorescence either spicate or, if branched, only simply so. 29

29. Inflorescence clearly branched, with slender branches; glomerules distant . . . . .
. . . . **26.** *richardsiae*
– Inflorescence usually spicate, occasionally with a few ± stout branches in lower part; glomerules ± crowded 30
30. Plant geophytic, perennating by means of a ± spherical or extended tuber produced at the end of a slender rhizome (so that imperfect specimens often appear to be annuals); glumes castaneous to dark reddish, with inconspicuous hairs . . . . **13.** *flexuosa*
– Plant not geophytic; hairs on glumes generally conspicuous, often black 31
31. Glomerules reflexed at maturity; inflorescence simply spicate; nutlets smooth; ± slender rhizomes produced from stem base . . . . **10.** *catophylla*
– Glomerules not reflexed; inflorescence spicate or shortly branched; nutlets strongly trabeculate-reticulate; plant caespitose, without rhizomes . . . .
. . . . **11.** *veseyfitzgeraldii*
32. Nutlets purple; scrambling plant to 4–5 m . . . . **30.** *porphyrocarpa*
– Nutlets not purple; erect plant of smaller stature 33
33. Leaves not more than 5 mm wide . . . . **28.** *lithosperma*
– Leaves 10 mm or more wide . . . . **29.** *lacustris*
34. Stems branched repeatedly from near base; plants procumbent or only weakly erect, forming dense semi-prostrate masses in permanently wet bogs; leaves and stems citrus-scented when crushed 35
– Stems not or very rarely branched, erect; leaves and stems not citrus-scented . 36
35. Stems not swollen at base but rising from a ± horizontal tuberous rhizome; nutlets only obscurely trigonous; hypogynium soft and spongy when fresh, even when dried at least half nutlet length . . . . **32.** *bequaertii*
– Stems swollen at base and either forming ± uniseriate rows or massed into knots; nutlets strongly trigonous; hypogynium inconspicuous, much smaller that nutlet
. . . . **33.** *laxiflora*
36. Hypogynium fimbriate on upper margin . . . . **59.** *angusta*
– Hypogynium not fimbriate 37
37. Plant annual 38
– Plant perennial 50
38. Lateral panicles all single at nodes 39
– Lateral panicles two or more on at least one node 48
39. Nutlets quite smooth . . . . **34.** *schimperiana*
– Nutlets not smooth 40
40. Peduncles of lateral panicles erect 41
– Peduncles of lateral panicles pendulous 45
41. Nutlets not more than 2 mm long including hypogynium; plant small with stems 3–25 cm tall . . . . **40.** *patula*
– Nutlets at least 2.5 mm long including hypogynium; stems at least 15 cm tall, usually at least 30 cm 42
42. Male spikelets shorter than their pedicels; nutlets glabrous but not shiny . . . . . .
. . . . **36.** *mikawana*
– Male spikelets as long or longer than their pedicels; nutlets hairy or, if glabrous, shiny 43
43. Nutlets globose, hairy or glabrous . . . . **37.** *tessellata* var. *sphaerocarpa*
– Nutlets ovoid or ellipsoid-cylindric, glabrous 44
44. Nutlets ovoid, irregularly alveolate-lacunose, smoother towards apex, which is always quite smooth . . . . **35.** *foliosa*
– Nutlets ellipsoid-cylindric, tessellate-lacunose, the surface relief uniform from base to apex . . . . **37.** *tessellata*

45. Nutlets globose, hairy . . . .**38.** *globonux*
– Nutlets ovoid to subglobose, hairy or glabrous 46
46. Nutlets hairy or glabrous, grey at maturity; male spikelets as long as or longer than their pedicels; hypogynium distinctly 3-lobed. . . . .**39.** *bambariensis*
– Nutlets glabrous, grey or black at maturity; male spikelets shorter than their pedicels; hypogynium not or only obscurely 3-lobed 47
47. Nutlets distinctly striate-lacunose, grey at maturity. . . . . **41.** *chlorocalyx*
– Nutlets lightly and indistinctly lacunose, shining black at maturity. . . . .
. . . . **42.** *lucentinigricans*
48. Nutlets smooth. . . . .**45.** *gracillima*
– Nutlets not smooth 49
49. Nutlets ovoid to subglobose, hairy or glabrous; hypogynium distinctly 3-lobed . .
. . . . **43.** *parvula*
– Nutlets oblong-ellipsoid, glabrous; hypogynium only faintly 3-lobed .**44.** *clathrata*
50. Inflorescence terminal, ebracteate. . . . . **58.** *poiformis*
– Inflorescence terminal and lateral, the lateral panicles subtended by leafy bracts
. . . . 51
51. Leaves strongly serrate-ciliate on margins; inflorescence lax and copious, composed of hundreds of spikelets with a very high proportion being male . . . .
. . . . **57.** *greigiifolia*
– Leaves not serrate-ciliate on margins; inflorescence not as above 52
52. Nutlets violet towards apex . . . . **53.** *iostephana*
– Nutlets without any violet colouring 53
53. Nutlets glabrous, or with very sparse hairs proximally, usually 4–5 mm long . . 54
– Nutlets hairy, usually less than 4 mm long. 56
54. Lateral panicles pendulous; inflorescence much elongated; leaves 7–20 mm wide. . . . . **56.** *melanomphala*
– Lateral panicles erect; inflorescence not much elongated; leaves less than 11 mm wide. 55
55. Rhizome much elongated and horizontally creeping; male spikelets 5–6 mm long; stems to 2 m tall . . . . **54.** *pachyrrhyncha*
– Rhizome short or absent, the stems forming a ± caespitose mass; male spikelets 8–9 mm long; stems not more than 1 m tall . . . . **55.** *induta*
56. Nutlets regularly and distinctly tessellate-lacunose 57
– Nutlets only faintly lacunose, or completely smooth. 58
57. Male spikelets 5–8 mm long; leaves 2–5(7) mm wide . . . . **47.** *nyasensis*
– Male spikelets 4–5 mm long; leaves 1.5–3.5(4) mm wide . . . . **48.** *unguiculata*
58. Nutlets completely smooth 59
– Nutlets not completely smooth. 60
59. Leaves, sheaths (except sometimes at mouth) and stems glabrous . . .**49.** *lagoensis*
– Leaves, sheaths and stems covered with short appressed hairs so plant appears pale green when dried. . . . . **50.** *adpressohirta*
60. Male spikelets 7–9 mm long; lateral panicles always single at nodes . . **46.** *achtenii*
– Male spikelets not more than 6 mm long; lateral panicles single or more than one at each node. 61
61. Lateral panicles on erect or stiffly outward-curving peduncles, always single at nodes . . . . **52.** *arcuata*
– Lateral panicles on pendulous panicles, single or (usually) more than one at each node 62
62. Nutlets globose or depressed-globose, 2.8–3 mm wide. . . . . **51.** *xerophila*
– Nutlets ovoid, not more than 2.5 mm wide, usually less . . . . **49.** *lagoensis*

1. **Scleria pergracilis** (Nees) Kunth, Enum. Pl. **2**: 354 (1837). —Clarke in F.T.A. **8**: 495 (1902); in Illustr. Cyper.: t.121 (1909). —Nelmes in Kew Bull. **10**: 445 (1955). — Robinson in Kew Bull. **18**: 494 (1966). —Haines & Lye, Sedges & Rushes E. Afr.: 342, figs.705,706 (1983). —Hennessy in Bothalia **15**: 521 (1985). —Gordon-Gray in Strelitzia **2**: 184, fig.84 (1995). —Verdcourt in F.T.E.A., Cyperaceae: 395, fig.60 (2010). Type: India, Silhet, n.d., *Wallich* 3406 (B† holotype, K, PH).

> *Hypoporum pergracile* Nees in Edinburgh New Philos. J. **17**: 267 (1834).
> *Scleria ustulata* Ridl. in Trans. Linn. Soc. London, Bot. **2**: 168 (1884). —Clarke in Durand & Schinz, Consp. Fl. Afr. **5**: 625 (1894); in F.T.A. **8**: 497 (1902). Type: Angola, Pungo Andongo, banks of R. Cuanza, 12.iii.1857, *Welwitsch* 7134 (LISU holotype, K).
> *Scleria pergracilis* var. *brachystachys* Nelmes in Kew Bull. **10**: 446 (1955). Type: Zimbabwe, Harare (Salisbury), 25.iv.1931, *Brain* 3710 (K holotype, LISC, PRE, SRGH).

Annual herb, 15–40 (60) cm tall, entirely glabrous except for leaf sheaths, which are sometimes sparsely hairy. Leaves 1–2 mm wide. Inflorescence simply spicate or very shortly branched near base, (2)4.5–6(18) cm long, with many erect or spreading glomerules of 1–3, 4–5 mm long dark red spikelets; bracteoles erect, 3–9 mm long, acuminate at base. Glumes pale brown to dark reddish black, paler at base and midrib pale green in female, 3–4(5) mm long, acuminate to shortly mucronate. Nutlets ± globose, 1–1.5(1.9) × 1–1.5(1.7) mm wide, tuberculate, with 3 longitudinal smooth bands, grey, brownish or blackish, raised parts lighter in colour.

Zambia. N: Mbala Dist., Saisi R., Issa Ranch, 2.v.1961, *Richards* 15093 (K). C: Mpika Dist., Mutinondo Wilderness area, inselberg near falls, 7.iv.2003, *P.A. Smith & Chishala* 1845 (K). S: Kafue Dist., Kafue Gorge, 14.iii.1960, *Vesey-Fitzgerald* 2708 (K). **Zimbabwe**. N: Murewa Dist., Shavanoe R., 14.iii.1954, *Whellan* 781 (K, SRGH). C: Kwekwe Dist., c. 50 km N of Kwekwe Town, Iwaba Estate, c. 10 km SE of camp, 13.ii.2001, *Lye* 24550 (K). S: Masvingo Dist., Makoholi Expt. Station, 13.iii.1978, *Senderayi* 218 (K, SRGH). **Malawi**. N: Chitipa Dist., 5 km SE of Chitipa (Fort Hill), 11.iii.1961, *Robinson* 4441 (K). C: Kasungu Dist., 48 km N of Kasungu, 3.vi.1962, *Robinson* 5256 (K). S: Machinga Dist., near Mvuu Camp, Liwonde Nat. Park, 14.iv.1985, *Dudley* 1535 (K, MAL). **Mozambique**. T: Ulónguè Dist., 30 km from Dedza along Zomba road, 4.vi.1962, *Robinson* 5259 (K).

Senegal to Nigeria, Sudan, Ethiopia, Angola and South Africa (KwaZulu-Natal); also in India, Sri Lanka and New Guinea. In drier parts of seasonally wet dambos, streamsides and seasonally wet shallow soil over rock; 500–1500 m.

Conservation notes: Widespread; not threatened.

There is an unconfirmed record from Zimbabwe W: in Mapaura & Timberlake (Checkl. Zimbabwe Pl.: 89, 2004).

Robinson felt that Nelmes's var. *brachystachys* might merit recognition as a geographical subspecies. These plants, found mainly in south and central Africa, have shorter and denser spikes and sometimes longer bracts, although similar plants occur in West Africa. Further study is needed.

2. **Scleria glabra** Boeckeler, Beitr. Cyper. **1**: 35 (1888). —Clarke in F.T.A. **8**: 497 (1902). —Nelmes in Kew Bull. **10**: 435 (1955). —Robinson in Kew Bull. **18**: 495 (1966). —Haines & Lye, Sedges & Rushes E. Afr.: 322, figs.707,708 (1983). — Verdcourt in F.T.E.A., Cyperaceae: 397 (2010) Type: Malawi, Shire Highlands, vii.1885, *Buchanan* 2 (B†? holotype, E, K, P). FIGURE 14.**68**.

Glabrous annual herb, erect stems 0.25–0.8(1.2) m tall. Leaves 1.5–4(6) mm wide. Inflorescence paniculate, 4–25 cm long, often twice branched from lower part, the branches slender and often compound with many glomerules of 2–16 blackish, 3.5–5 mm long spikelets, upper glomerules closely spaced, the lower up to 4 cm apart. Glumes dark reddish black, paler at base, all muticous or shortly mucronate, female glumes 2.5–3 mm long with green midrib, male glumes 2.5–4 mm

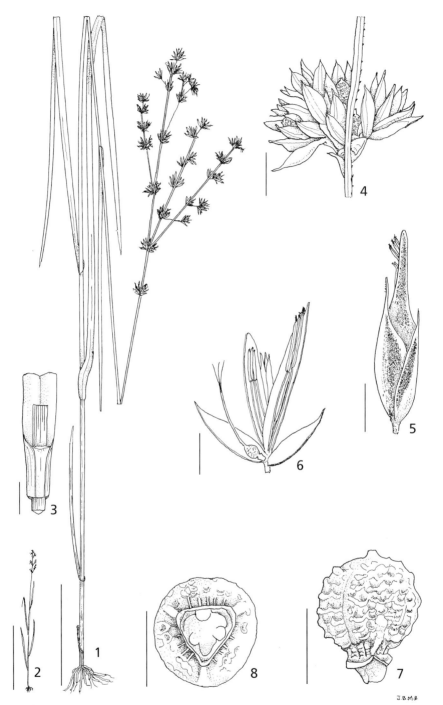

Fig. 14.**68**. SCLERIA GLABRA. 1, habit, entire plant; 2, habit, entire plant with culm and branched inflorescence; 3, junction of lamina and sheath; 4, part of inflorescence and culm; 5, androgynous spikelet, prophyll removed; 6, androgynous spikelet, prophyll removed and remainder displaced to show lower female and upper male florets; 7, nutlet, lateral view; 8, nutlet, basal view. All from *Robinson* 4658. Scale bars: 1 = 40 mm; 2 = 250 mm; 3, 4 = 2 mm; 5–8 = 1 mm. Drawn by Jane Browning.

long. Nutlets ± globose, 1.2–1.5 mm long, c. 1 mm wide, trabeculate-tuberculate or shallowly pitted or transversely wrinkled, with 3 longitudinal smooth bands, shortly apiculate, grey or black but raised parts appearing lighter in colour.

Zambia. N: Mbala Dist., Mbala Kawimbe, Lumi R., 30.v.1961, *Richards* 15176 (K). W: Kasempa Dist., 55 km NW of Kasempa, 7.vii.1963, *Robinson* 5540 (K). E: Lundazi Dist., Tumbuka area, iii.1962, *Robinson* 5085 (K). **Malawi**. N: Karonga Dist., 5 km S of Chitipa (Fort Hill), 22.v.1962, *Robinson* 5218 (K, M, SRGH). S: Mulanje Dist., 16 km NW of Likabula Forestry Depot, 15.vi.1962, *Robinson* 5357 (K). **Mozambique**. T: Ulónguè Dist., 30 km from Dedza along Zomba road, 4.vi.1962, *Robinson* 5262 (K).

Also in Tanzania, D.R. Congo (Katanga) and Burundi. In seasonally or permanently wet bogs and dambos; 700–1750m.

Conservation notes: Widespread; not threatened.

Robinson regarded this species as "curiously uncommon" in spite of the wide range of habitat and rainfall that it tolerates. Plants from drier habitats are small, with an almost simple inflorescence while those from wetter places are more robust with a larger inflorescence.

3. **Scleria pulchella** Ridl. in Trans. Linn. Soc. London, Bot. **2**: 168 (1884). —Clarke in F.T.A. **8**: 495 (1902). —Nelmes in Kew Bull. **10**: 442 (1955). —Robinson in Kew Bull. **18**: 496 (1966). —Haines & Lye, Sedges & Rushes E. Afr.: 341, figs.703,704 (1983). —Verdcourt in F.T.E.A., Cyperaceae: 394 (2010). Type: Angola, Huíla, Empalanca, xii.1859, *Welwitsch* 7141 (LISU holotype, BM, K).

   Scleria suaveolens Nelmes in Kew Bull. **10**: 442 (1955). Type: Zambia, Mbala, Uninji Pans, 29.ii.1952, *Richards* 847 (K holotype).

Annual herb, 5–25 cm tall, almost entirely glabrous. Leaves 1–2 mm wide, glabrous or very sparsely hairy. Inflorescence simple, 1–3(–4) cm long, some lower branches to 1.5 cm long, with erect or spreading glomerules of 1–8 dark red 2–4 mm long spikelets, upper glomerules crowded and ± touching, lower up to 1 cm apart. Glumes blackish red or green sometimes with reddish streaks, midrib green in female, c. 2 mm long, shortly awned. Nutlets globose, c. 1 × 0.8 mm wide, faintly to strongly transversely ridged or pitted, grey, brownish or blackish, sometimes with 3 longitudinal dark stripes.

Zambia. N: Mbala Dist., Uninji Pans, 24.vi.1956, *Robinson* 1738 (K). C: Mpika Dist., Mutinondo Wilderness Area, inselberg, 7.iv.2003, *Chishala & P.A. Smith* 29 (K). E: Lundazi Dist., Nyika Plateau, Kangampande Mt, 8.v.1952, *White* 2809 (FHO, K). **Zimbabwe**. C: Goromonzi Dist., Chinamora communal land (Reserve), Domboshawa, Ngomakurira, 14.iii.1965, *Loveridge* 1371 (K, SRGH). E: Mutare Dist., Vumba Mts, Eastern Beacon slope, 17.iv.1993, *Browning* 551 (K, NU). **Malawi**. N: Nyika Plateau, Sengula Rock, 21.iv.2006, *Chapama et al.* 444 (K). S: Zomba Dist., Zomba Plateau, 31.v.1946, *Brass* 16108 (K, NY).

Also in Tanzania, D.R. Congo (Katanga) and Angola. In wet bogs and dambos, shallow soil over granite or sandstone; 1450–2300 m.

Conservation notes: Widespread; not threatened.

The species is strongly lemon-scented in the field, but the smell disappears on drying, according to *Richards* 847.

4. **Scleria calcicola** E.A. Rob. in Kew Bull. **18**: 496, fig.1 (1966). —Haines & Lye, Sedges & Rushes E. Afr.: 340 (1983). Type: Zambia, Ndola, Itawa dambo, 23.iv.1960, *Robinson* 3676 (K holotype, EA, M, MTJB, NU, SRGH, YBI).

Annual herb to 70 cm talll, almost glabrous. Leaves 1–2.5 mm wide, glabrous, almost flat; sheaths glabrous but sparsely pilose around mouth. Inflorescence 1–4.5 cm long, simple or with some lower branches to 2 cm long; bracts 5–22 mm long, thin, glabrous but hispid on margins towards apex; glomerules 5–7 mm wide, of 5–9 spikelets each 3.5–4 mm long. Male glumes c. 3 mm long, dark reddish; female glumes 2.5–3.5 mm long, ovate-lanceolate, glabrous, midrib and central part green, margins dark reddish, elsewhere straw-yellow or scarious. Nutlets globose, 1.5–1.7 × 1.2–1.4 mm wide, reticulate-tessellate, grey brownish or dark grey; beak short.

Zambia. W: Ndola Dist., Ndola, Itawa dambo, 23.iv.1960, *Robinson* 3676 (EA, K, M, MTJB, NU, SRGH, YBI). C: Lusaka Dist., 8 km E of Lusaka, Munali, little dambo, 14.vi.1955, *Robinson* 1298 (K).

Not known elsewhere. Calcareous dambos; 1100–1300 m.

Conservation notes: Only known from two localities in C Zambia in an unusual habitat; possibly Vulnerable.

There is an unconfirmed record from Zimbabwe N in Mapaura & Timberlake, Checkl. Zimbabwe Pl.: 89 (2004).

According to Robinson (1966) the stems of this species remain green and juicy long after the leaves and inflorescence have withered.

5. **Scleria hispidula** A. Rich., Tent. Fl. Abyss. **2**: 511 (1850). —Clarke in F.T.A. **8**: 497 (1902). —Nelmes in Kew Bull. **10**: 436 (1955). —Napper in J. E. Africa Nat. Hist. Soc. Coryndon Mus. **24**(109): 32 (1964). —Robinson in Kew Bull. **18**: 498 (1966). —Haines & Lye, Sedges & Rushes E. Afr.: 340, figs.699,700 (1983). —Verdcourt in F.T.E.A., Cyperaceae: 394 (2010). Type: Ethiopia, Tigre, Guendepta (Gafta), 12.ix.1838, *Schimper* II 1277 (P holotype, BM, K, UPS).

Slender glabrous or hairy annual herb, 5–40(60) cm tall, stems less than 1 mm wide. Lower leaf sheaths glabrous or hairy; blades to 20 cm long, 1–3 mm wide, glabrous or sparsely hairy, scabrid on margins towards apex. Inflorescence simply spicate, 3–11 cm long, or with 1–2 lower lateral branches 2 cm long, with 2–10 sessile glomerules of 2–6, 3–4 mm long spikelets; bracteoles 3–9 mm long, hispidulous. Female glumes medium to blackish red with green midrib, 2–4 mm long, scabrid on midrib but otherwise glabrous, awned; male glumes similar but darker and not awned. Nutlets obovoid or globose, 1.2–1.6 × 1–1.2 mm wide, finely reticulate, sometimes with 3 longitudinal smooth bands, greyish white.

Zambia. C: Lusaka Dist., Munali School, 11.iv.1956, *Robinson* 1459 (K, MRSC, MTJB, SRGH); Chibombo Dist., Chisamba, 50 km N of Lusaka, 15.v.1959, *Robinson* 3260 (EA, GC, K).

Also in Tanzania, Ethiopia and Eritrea. Seasonally damp places on calcareous soils, sometimes closely associated with *Scleria calcicola*; 1200–1250 m.

Conservation notes: Widespread, but in the Flora area only known from C Zambia; possibly Near Threatened.

Similar in appearance to *Scleria pergracilis* but the glomerules are well separated and the nutlet pattern is reticulate rather than tuberculate.

6. **Scleria delicatula** Nelmes in Kew Bull. **10**: 448 (1955). —Napper in J. E. Africa Nat. Hist. Soc. Coryndon Mus. **24**(109): 31 (1964). —Robinson in Kew Bull. **18**: 498 (1966). —Haines & Lye, Sedges & Rushes E. Afr.: 342, figs.701,702 (1983). —Verdcourt in F.T.E.A., Cyperaceae: 394 (2010). Type: Zambia, Mbala Dist., Chilongowelo, Plain of Death, 5.ii.1952, *Richards* 600 (K holotype).
    Scleria spondylogona Nelmes in Kew Bull. **10**: 448 (1955). Type: Zambia, Mbala Dist., Lunzua R., banks below falls, 18.iv.1950, *Bullock* 2871 (K holotype).

Slender annual herb, 12–45 cm tall. Leaves c. 1 mm wide, glabrous or sparsely hairy. Inflorescence simply spicate, 3–13 cm long with glomerules of 1–8 spikelets. Spikelets 2–5 mm long, pale or reddish brown; glumes pale chestnut, often spotted with red. Nutlets oblong-ellipsoid to broadly ovoid, c. 1mm long, muricate-trabeculate, dark red with raised parts translucent, often with 3 longitudinal ridges of semi-translucent tissue on angles.

Zambia. N: Kasama Dist., 13 km NE of Kasama, 27.iv.1961, *Robinson* 4621 (K). W: Mwinilunga Dist., Kalenda dambo, 14.ii.1938, *Milne-Redhead* 4569 (K). **Mozambique**. N: Nampula Dist., 60 km from Ribáuè to Nampula, 24.iii.1964, *Torre & Paiva* 11381 (BR, LISC, LMA, LMU, MO, WAG).

Also in S Tanzania. Seasonally damp soils on sandstone or laterite outcrops where it is often locally dominant; sometimes with *S. melanocarpa*; 800–1650 m.

Conservation notes: Widespread; not threatened.

According to Robinson the species is common and often locally abundant in suitable locations in the Northern Province of Zambia and at Mwinilunga.

7. **Scleria zambesica** E.A. Rob. in Kew Bull. **18**: 499, fig.2 (1966). —Haines & Lye, Sedges & Rushes E. Afr.: 341 (1983). Type: Zambia, Mwinilunga Dist., 7 km N of Kalene Hill, Zambezi Rapids, 16.iv.1965, *Robinson* 6578 (K holotype, B, M, MRSC, NY, SRGH).

Erect annual herb; stems 0.25(0.3) m tall, hairy. Leaves c. 1 mm wide, hairy. Inflorescence spicate, 3.5–10 cm long; glomerules sessile, 1 cm apart below, contiguous above, of 2–5 spikelets subtended by hairy bracts 3–5 mm long. Spikelets bisexual, 3–4 mm long, ± erect; glumes usually glabrous (but with a few hairs on lower part of keel), lower ones scarious, reddish brown with a green keel, upper ones (male) reddish with dark red spots, or dark red. Nutlets c. 1 × 0.8–1 mm wide, glabrous, trigonous, broadly obovoid to subglobose, surface lightly reticulate, grey or brown; hypogynium darker, beak short or absent.

Zambia. W: Mwinilunga Dist., c. 6 km N of Kalene Hill, 20.ii.1975, *Hooper & Townsend* 260 (K).

Not known elsewhere. Damp peaty places in grassy areas over granite, wet flushes; 1250 m.

Conservation notes: Apparently endemic to the Mwinilunga area; possibly Vulnerable.

Although determined as *Scleria zambesica*, the specimen *Brummitt* 9135 from S Malawi is actually *S. flexuosa*.

8. **Scleria distans** Poir. in Lamarck, Encycl. **7**: 4 (1806). —Haines & Lye, Sedges & Rushes E. Afr.: 330 (1983). —Gordon-Gray in Strelitzia **2**: 183 (1995). —Verdcourt in F.T.E.A., Cyperaceae: 383 (2010). Type: Puerto Rico, *Ledru* 110 (P holotype). FIGURE 14.**69**.

Scleria nutans Kunth, Enum. Pl. **2**: 351 (1837). —Napper in J. E. Africa Nat. Hist. Soc. Coryndon Mus. **24**(109): 41 (1964). —Robinson in Kew Bull. **18**: 502 (1966). —Napper in F.W.T.A., ed.2 **3**: 344 (1972). —Hennessy in Bothalia **15**: 522 (1985). Type: Venezuela, Cumana, ?1799, *Humboldt* s.n. in *Herb. Willd.* 17336 (B-W holotype).

Scleria cenchroides Kunth, Enum. Pl. **2**: 352 (1837). Type: South Africa, Cape of Good Hope, 4.v.1832, *Drège* 4365 (?B holotype, K, P).

Scleria hirtella Sw. var. *tuberculata* C.B. Clarke in Fl. Cap. **7**: 294 (1898). Type: South Africa, Magaliesberg, n.d., *Burke* 62 (K holotype).

Scleria hirtella auct., non Sw.

Slender perennial herb 20–90 cm tall; many stems at 2–20 mm intervals from a creeping rhizome, 2–4 mm thick, 10 cm or more long, hairy or ± glabrous; stem bases sometimes swollen and bulbous. Leaves to 18 cm long, 1–3 mm wide; lower leaf sheaths brown or pale reddish brown to purple, without blades; ligule an indistinct rim, throat with dense rim of hairs. Inflorescence a lax spike 5–10 cm long, with 4–9 sessile drooping glomerules 5–6 × 4–10 mm. Spikelets (1)2–8(10), densely crowded, bisexual, a solitary female flower below upper male flowers; glumes reddish brown to blackish, 3–6 mm long, the outer ending in a long awn densely set with spreading reddish brown hairs. Nutlet 1.4–1.5 × 0.9–1.2 mm wide, smooth or with strong transverse wrinkles or tubercles, white, greyish or pale-violet-tinged; stipe reddish or yellowish brown, trigonous, 0.3–0.6 mm long.

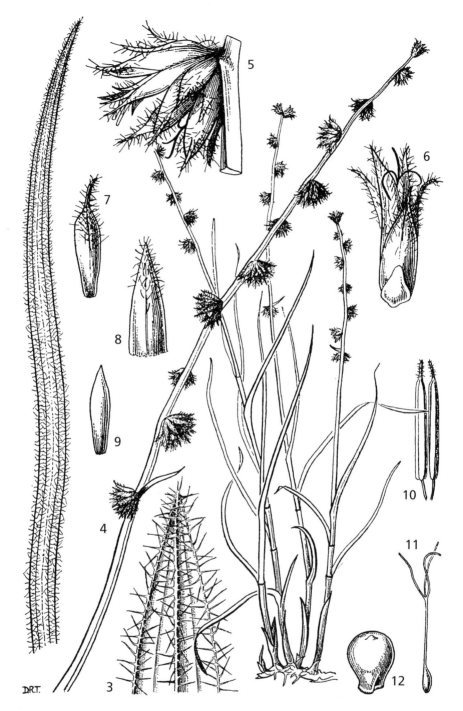

Fig. 14.**69**. SCLERIA DISTANS. 1, habit (× ⅔); 2, leaf blade (× 4); 3, tip of leaf blade, lower surface (× 12); 4, inflorescence (× 1½); 5, group of spikelets (× 9); 6, spikelet detail (× 9); 7, sterile glume (× 6); 8, female glume (× 6); 9, male glume (× 9); 10, stamens (× 20); 11, ovary and stigma (× 9); 12, nutlet (× 9). All from *Greenway* 3282. Drawn by Dorothy Thompson. Reproduced from Flora of West Tropical Africa, ed. 1 (1936), as *Scleria nutans*.

Botswana. N: Okavango Swamps, floodplain swamp off the Khianadianavhu R., 19°01.9'S 22°55.5'E, 16.v.1979, *P.A. Smith* 2749 (K, SRGH). **Zambia**. N: Mbala Dist., Lake Chila, 13.ii.1959, *Richards* 10858 (K). **Malawi**. S: Chikwawa Dist.?, Elephant Marsh, 'Mkyusa's Country', 15.xii.1887, *Scott* s.n. (K). **Mozambique**. N: Mecula Dist., Mbatamila airstrip, 12°10.12S 37°32.12E, 17.vi.2003, *Golding et al.* 123 (K, LMA).

Also in Nigeria, Cameroon, Ethiopia, D.R. Congo, Burundi, Uganda, Kenya, Tanzania and South Africa, also Madagascar and tropical America. In permanently wet bogs; 400–1600 m (2500 m in East Africa).

Conservation notes: Widespread; not threatened.

In East Africa there are three varieties – *distans, chondrocarpa* (Nelmes) Lye and *glomerulata* (Oliv.) Lye. All material seen from the Flora area is var. *distans*.

9. **Scleria melanotricha** A. Rich., Tent. Fl. Abyss. **2**: 511 (1850). —Clarke in F.T.A. **8**: 495 (1902).—Nelmes in Kew Bull. **10**: 452 (1955). —Napper in J. E. Africa Nat. Hist. Soc. Coryndon Mus. **24**(109): 33 (1964). —Robinson in Kew Bull. **18**: 501 (1966). —Napper in F.W.T.A., ed.2 **3**: 346 (1972). —Haines & Lye, Sedges & Rushes E. Afr.: 338, figs.696,697 (1983). —Verdcourt in F.T.E.A., Cyperaceae: 392 (2010). Type: Ethiopia, Tigray, Guendepta (Gafta, Gapdia), vii.1843, *Schimper* II 830 (P lectotype, BM, K), lectotypified indirectly by Robinson (1966), see F.T.E.A.

   Scleria grata Nelmes in Kew Bull. **10**: 453 (1955). —Napper in J. E. Africa Nat. Hist. Soc. Coryndon Mus. **24**(109): 33 (1964). Type: Zambia, Mbala (Abercorn), above Chilongowelo, 15.v.1952, *Richards* 1688 (K holotype).

   Scleria melanotricha var. *grata* (Nelmes) Lye in Nordic J. Bot. **3**: 243 (1983). —Haines & Lye, Sedges & Rushes E. Afr.: 339 (1983).

Annual herb, (6)10–50(60) cm tall with erect stems, hairy on angles. Leaf blades to 30 cm long, 1–3 mm wide, densely hairy, particularly on veins. Inflorescence a simple spike 3–20 cm long with 4–12(15) sessile or very shortly stalked spreading or usually reflexed glomerules of 1–9(12) dark 5–19 mm long spikelets. Glumes pale green or reddish, 2–3 mm long (but 4–7 mm in Ethiopia), with a mixture of thicker brown or black and thinner pale hairs, prominently awned. Nutlets ovoid, obovoid or subglobose, obtusely trigonous, 1–1.2 × 0.8–1 mm broad, tuberculate or trabeculate, grey or yellow brown, sometimes with raised parts paler, with 3 darker longitudinal bands and stipe darker in colour, shortly apiculate.

Zambia. N: Kasama Dist., 95 km E of Kasama, 3.iv.1961, *Robinson* 4578 (K). W: Mwinilunga Dist., Zambezi Rapids, c. 6 km N of Kalene Hill, 20.ii.1975, *Hooper & Townsend* 243 (K). C: Serenje Dist., Kundalila Falls, S of Kanona, 14.iii.1975, *Hooper & Townsend* 758 (K). **Malawi**. C: Ntchisi Dist., Ntchisi Forest Reserve road near Chitembwene, 13.iv.1991, *Radcliffe-Smith* 5965 (K).

Also in Guinea, Ivory Coast, Mali, Ghana, Nigeria, Ethiopia, D.R. Congo, Burundi, Rwanda and Tanzania. Shallow sandy or gravelly soil over rock outcrops, often with *Scleria delicatula*; 750–1550 m.

Conservation notes: Widespread; not threatened.

10. **Scleria catophylla** C.B. Clarke in Durand & Schinz, Consp. Fl. Afr. **5**: 670 (1894); in F.T.A. **8**: 498 (1902); Illustr. Cyper.: t.122, fig.1–4 (1909). —Napper in J. E. Africa Nat. Hist. Soc. Coryndon Mus. **24** (109): 33, fig.3 (1964). —Robinson in Kew Bull. **18**: 501 (1966). —Haines & Lye, Sedges & Rushes E. Afr.: 329, figs.671,672 (1983). —Verdcourt in F.T.E.A., Cyperaceae: 381, fig.58 (2010). Type: Nigeria, *Barter* 1561 (K holotype).

   Scleria hirtella Sw. var. *aterrima* Ridl. in Trans. Linn. Soc. London, Bot. **2**: 166 (1884). Type: Angola, iii-iv.1860, *Welwitsch* 7143 (BM, K).

   Scleria aterrima (Ridl.) Napper in Kew Bull. **25**: 445 (1971); in F.W.T.A., ed.2 **3**: 344 (1972). —Hennessy in Bothalia **15**: 522 (1985). —Gordon-Gray in Strelitzia **2**: 181, fig.82 (1995). —Mapaura & Timberlake, Checkl. Zimbabwe Pl.: 89 (2004).

Perennial herb 0.2–1.2 m tall; glabrous or hairy stems slightly swollen at base, producing up to 4 slender ± fleshy stolons 1–2(3) mm thick. Leaves mostly produced at or near base of stem, 10–25 mm long, 2–5 mm wide, glabrous to densely hairy; ligule a dense rim of short hairs. Inflorescence simply spicate, 6–18 cm long, with many reflexed glomerules of 2–7 dark, 4–6 mm long bisexual spikelets. Glumes 3–5 mm long, reddish brown to blackish, the outer ending in a long awn densely covered with almost black hairs; hairs below awn black, often white or reddish brown. Nutlet obovoid, (1.2)1.5–1.7 × 1–1.2 mm wide, smooth, white, grey or grey-brown, stipe minute.

Zambia. B: Mongu Dist., 32 km NE of Mongu, 10.xi.1958, *Drummond & Cookson* 6315 (K, SRGH). N: Kasama Dist., small marsh off Luwingu road, 29.ii.1960, *Richards* 12642 (K). W: Mwinilunga Dist., 18 km SW of Kalene Hill, Kafweko Forest Res. near Lisombo R., 22.ii.1975, *Hooper & Townsend* 308 (K). **Zimbabwe**. E: Chimanimani Dist., Chimanimani Mts, 1.5 km NE of Mt Hut on Bundi Plain, 6.iv.1969, *Kelly* 77 (K, SRGH). **Mozambique**. N: Muecate Dist., Imala, 36 km from Imala towards Mecubúri (Mocuburi), 16.i.1963, *Torre & Paiva* 10022 (BR, LISC, LMA, LMU, MO, WAG). Z: Milange Dist., Belua, Milange–Quelimane road, v.1972, *Bowbrick* J10B (LISC, SRGH). MS: Cheringoma Dist., 110 km from Beira to Inhaminga, 26.v.1942, *Torre* 4223 (LISC, LMA, LMU, PRE, WAG).

Senegal to Nigeria, Uganda, Tanzania, Angola and South Africa (KwaZulu-Natal). In dambos, bogs and marshes, usually on sandy or peaty ground; 950–1700 m.

Conservation notes: Widespread; not threatened.

11. **Scleria veseyfitzgeraldii** E.A. Rob. in Kew Bull. **18**: 503, fig.3.1–9 (1966). —Haines & Lye, Sedges & Rushes E. Afr.: 332, figs.679,680 (1983). —Hennessy in Bothalia **15**: 518 (1985). —Verdcourt in F.T.E.A., Cyperaceae: 386 (2010). Type: Zambia, Chililabombwe Dist., banks of Kafue R., 11 km N of Chingola, 3.i.1961, *Robinson* 4220 (K holotype, B, EA, GC, M, MPR, MTJB, NY, PRE, SRGH).

Perennial, usually caespitose herb; stems robust, glabrous, to 1 m tall, 1–3 mm in diameter, thickened and covered with dead leaf sheaths at base. Leaves 2–7 mm wide, almost glabrous to densely hairy. Inflorescence spicate, usually simple, 5–12(15) cm long, of 4–8 sessile glomerules, occasionally with glomerule-bearing branchlets towards base; glomerules dense, of many spikelets, to 14 mm broad. Glumes chestnut-brown or blackish brown, with green keels, awned, ± pilose, hairs black or pale; female glumes 5–6 mm long, including awn. Nutlet 2 × 1.3–1.6 mm wide, acutely trigonous, broadly obovoid, distinctly reticulate-trabeculate, grey.

Caprivi. Kavango R., Singalamwe, 31.xii.1958, *Killick & Leistner* 3218 (K, M, PRE). **Zambia**. B: Mongu Dist., Mongu, 14.xii.1962, *Robinson* 5520 (B, K, LISC, SRGH). W: Lufwanyama Dist., 80 km W of Chingola, 18.xii.1963, *Robinson* 6127 (B, K, M, MRSC, SRGH). N: Kasama Dist., Chambeshi Flats, 50 km SE of Kasama, 23.i.1961, *Robinson* 4300 (EA, GC, K, M, MRSC, MTJB, NU, PRE, SRGH).

Also in Tanzania. Dambos, usually seasonal but sometimes permanently wet; 1000–1400 m.

Conservation notes: Widespread; not threatened.

There is an unconfirmed record for Botswana in Setshogo (Prelim. Checkl. Pl. Botswana: 121, 2005).

A rather robust-looking species with compact glomerules subtended by small bracts that are less obvious than in the drawing accompanying the original description. Named after Desmond Vesey-Fitzgerald (1909–1974), an ecologist in Zambia and Tanzania.

12. **Scleria bulbifera** A. Rich., Tent. Fl. Abyss. **2**: 510 (1850). —Clarke in F.T.A. **8**: 500 (1902); Illustr. Cyper.: t.122, fig.56 (1909). —Nelmes in Kew Bull. **10**: 438 (1955). —Robinson in Kew Bull. **18**: 503, fig.3 (1966). —Napper in F.W.T.A., ed.2 **3**: 344 (1972). —Haines & Lye, Sedges & Rushes E. Afr.: 331, figs.677,678 (1983). —

Hennessy in Bothalia **15**: 518 (1985). —Gordon-Gray in Strelitzia **2**: 181, figs.82,83 (1995). —Verdcourt in F.T.E.A., Cyperaceae: 386 (2010). Type: Ethiopia, Mt Scholoda, 5.viii.1841, *Schimper* 1557 (P holotype, BM, K). FIGURE 14.**70**.

Scleria mechowiana Boeckeler in Bot. Jahrb. Syst. **5**: 510 (1884). —Clarke in F.T.A. **8**: 498 (1902). Type: Angola, Malange, *Mechow* 345 (B† holotype).

Scleria buchananii Boeckeler, Beitr. Cyper. **1**: 33 (1888). —Clarke in F.T.A. **8**: 499 (1902). Type: Malawi, Shire Highlands, vii.1885, *Buchanan* 32 (B† holotype, K).

Scleria verdickii De Wild. in Rev. Zool. Bot. Africaines, suppl. Bot. **14**: 26 (1926). Type: D.R. Congo, Katanga, Lukafu, iii.1900, *Verdick* 398 (BR holotype).

Scleria schliebenii R. Gross in Notizbl. Bot. Gart. Berlin-Dahlem **11**: 657 (1932); in Peters, Fl. Deutsch-Ostafrika **1**: 531 (1938). Type: Tanzania, Iringa Dist., Upper Ruhudje, Lupembe, Likanga, 28.iii.1931, *Schlieben* 484 (B holotype?).

Scleria thomasii Piérart in Bull. Soc. Roy. Bot. Belgique **83**: 405 (1951). Type: D.R. Congo, Kundelungu Plateau, v.1923, *Thomas* 1202 (BR holotype).

Perennial herb with woody bulbous stem base to 12 mm wide, contiguous or connected by slender rhizomes. Stems 20–120 cm tall, glabrous or hairy. Leaves 1–10 mm wide, glabrous or hairy. Inflorescence with main axis 2–20 cm long, usually simply spicate, 2 lowest glomerules may produce branches to 4 cm long. Spikelets 4–6.5 mm long; glumes glabrous, or hairy on midrib. Nutlets obovoid to subglobose, 1.5–2 × 1–1.5 mm broad, smooth to lightly to strongly tuberculate or trabeculate, grey or pale brown.

Zambia. B: Kabompo Dist., 21 km WSW of Kabompo on Zambezi (Balovale) road, 24.iii.1961, *Drummond & Rutherford-Smith* 7308 (K, LISC, SRGH). N: Mbala Dist., Ndundu, 1.xii.1960, *Richards* 13641 (K). W: Mwinilunga Dist., just E of Dobeka bridge, 7.xii.1937, *Milne-Redhead* 3539 (K). C: Serenje Dist., S of Kanona, Kundalila Falls, 18.iii.1975, *Hooper & Townsend* 709 (K). E: Chipata Dist., Ngoni area, ii.1962, *Robinson* 5001 (K). S: Kalomo Dist., 1 km S of Zimba Township, 22.xii.1963, *Bainbridge* 941 (K). **Zimbabwe**. N: Murewa Dist., Mangwende communal land (Reserve), 14.i.1958, *Cleghorn* 346 (K, SRGH). W: Hwange Dist., Fuller Siding, i.1928, *Brain* 826 (K). C: Harare Dist., near Harare, 9.xii.1936, *Eyles* 8837 (K, SRGH). E: Hurungwe Dist., 6 km N of Msuora, 1.iii.1958, *Phipps* 996 (K, LISC, SRGH). **Malawi**. N: Mzimba Dist., Mzuzu, Marymount, 31.i.1969, *Pawek* 1672 (K). S: Mangochi Dist., Mwawa Valley, 22.xii.1956, *Jackson* 1427 (K). **Mozambique**. N: Marrupa Dist., c. 20 km from Marrupa, Matikite, 18.ii.1982, *Jansen & Boane* 7846 (K, LISC, WAG). Z: Alto Molócuè Dist., 67 km on Gilé road, 19.xii.1967, *Torre & Correia* 16604 (BR, LISC, LMA, LMU, MO, WAG). MS: Sussundenga Dist., Moribane, 17.xi.1942, *Salbany* 90 (LISC).

Widespread throughout tropical Africa from Senegal to Ethiopia and South Africa, also in Madagascar. Open woodlands and grasslands, sometimes in seasonally damp but not permanently waterlogged dambos; 0–2200 m.

Conservation notes: Widespread; not threatened.

A widespread species. The bulbous base is distinctive, and there are more glomerules in the inflorescence than in *Scleria veseyfitzgeraldii*, which is more vigorous. The species varies greatly in stature and leaf width and it is possible that with detailed analysis it could be subdivided. Robinson comments that the ornamentation of the nutlets is very variable and not correlated with other characters.

For a fuller synonymy see Verdcourt (2010).

13. **Scleria flexuosa** Boeckeler, Beitr. Cyper. **1**: 33 (1888). —Nelmes in Kew Bull. **10**: 431 (1955). —Napper in J. E. Africa Nat. Hist. Soc. Coryndon Mus. **24**(109): 36 (1964). —Robinson in Kew Bull. **18**: 505 (1966), in part. —Napper in Kew Bull. **25**: 444 (1971). —Haines & Lye, Sedges & Rushes E. Afr.: 332, figs.681,682 (1983). —Verdcourt in F.T.E.A., Cyperaceae: 387 (2010). Type: Malawi, Shire Highlands, vii.1885, *Buchanan* 60 (B?† holotype, E, K).

Fig. 14.**70**. SCLERIA BULBIFERA. 1, habit showing rhizome; 2, junction of lamina and sheath; 3, androgynous spikelet with lower prophyll and one glume removed; remainder displaced to show lower female floret and upper male florets; 4, glomerule of three spikelets; 5, nutlet. 1–4 from *Hilliard & Burtt* 13394; 5 from *Hilliard & Burtt* 18036. Scale bars: 1 = 40 mm; 2–5 = 2 mm. Drawn by Jane Browning. Reproduced from Strelitzia (1995) by kind permission of the South African National Biodiversity Institute, Pretoria.

Perennial herb; stems 15–65 cm tall. Rhizome c. 1 mm diameter, descending, pliable or brittle, ending in a single fleshy tuber, varying from ± spherical and 5–8 mm in diameter to elongate to 4 cm long and 2–4 mm thick. Tubers easily detached or missed; the plant may then appear to be annual. Leaves 1–2 mm wide, glabrous or often hairy. Inflorescence 3–10 cm long, simply spicate, occasionally with 1–2 short branches in lower part. Spikelets 3–5 mm long; glumes hairy, chestnut-brown, often dark reddish in upper half. Nutlets 1.1–1.3 × c. 1 mm broad, trigonous, broadly ovoid to subglobose, strongly tuberculate-trabeculate, blackish red, raised parts paler.

Zambia. N: Mbala Dist., Mbala, Mningi Pans, 22.ii.1959, *Richards* 10965 (K). W: Mpongwe Dist., 25 km S of Luanshya, 12.iii.1960, *Robinson* 3388 (K). C: Mpika Dist., Mutinondo Wilderness Area, inselberg near falls, 12°27.065'S 31°17.38'E, 7.iv.2003, *P.A. Smith & Chishala* 1846 (K). S: Choma Dist., 15 km E of Choma, 18.xii.1956, *Robinson* 1992 (K). **Zimbabwe**. W: Matobo Dist., Besna Kobila Farm, xii.1954, *Miller* 2566 (K, LISC, SRGH). E: Chimanimani Dist., Chimanimani Mts where Bundi leaves Upper Bundi Plain, 1.ii.1957, *Phipps* 327 (K, SRGH). **Malawi**. S: Zomba Dist., Zomba Plateau, Chingwe's Hole Nature Trail, 13.ii.1982, *Brummitt* 15869 (K). **Mozambique**. Z: Gúruè Dist., top of serra do Gúruè, 5 km from waterfall, 21.xi.1966, *Torre & Correia* 14770 (LISC, LMU, LUA, MO, WAG).

Also in Tanzania and possibly in West Africa. Moist places in upland grasslands; 1200–2300 m.

Conservation notes: Widespread; not threatened.

There are unconfirmed records from Zimbabwe N & C in Mapaura & Timberlake (Checkl. Zimbabwe Pl.: 89, 2004).

Robinson placed *Scleria lateritica* in synonymy under this species, but material at Kew shows that *S. lateritica* has a relatively thick elongated rhizome rather than tubers at the end of slender rhizomes, and smaller spikelets than in most material of *S. flexuosa*. I treat it here as a good species. Robinson also placed *S. dieterlenii* Turrill from Lesotho and KwaZulu-Natal as a synonym of *S. flexuosa*. Like *S. flexuosa*, *S. dieterlenii* has tubers on the end of slender rhizomes, but its spikelets are much hairier than those of *S. flexuosa* and the plants are generally smaller; interpretation is difficult because the type of *S. flexuosa* is very poor. A future revision may find it useful to designate an interpretive epitype.

The tubers of *Scleria flexuosa* are easily detached and are often missing from dried material.

14. **Scleria lateritica** Nelmes in Kew Bull. **10**: 432 (1955). Type: Zambia, Mwinilunga Dist., Kalenda Plain, 8.xii.1937, *Milne-Redhead* 4568 (K holotype, PRE).

   Scleria flexuosa sensu Robinson in Kew Bull. **18**: 505 (1966), in part, non Boeckeler.

Perennial herb. Rhizome horizontal, short, woody, producing 0.3–0.65 m tall stems at short intervals. Leaves 0.75–2 mm wide, mostly sparsely villous. Inflorescence simply spicate, 6–10 cm long, upper glomerules contiguous or not more than twice their length apart, lower ones more widely spaced, each with 2–6 spikelets 3–4 mm long;. Female glumes 2.2–2.5 mm long, sparsely hispidulous, streaked and splashed dark chestnut; midrib pale; male glumes 3–4 mm long, more deeply coloured. Nutlets quadrate-subglobose, obtusely trigonous, 1.5–1.75 × 1–1.1 mm broad, ± verrucose-lacunose or verrucose-trabeculate, pale to light brownish or greying, shortly beaked.

Zambia. N: Mporokoso Dist., Kasanshi dambo, 55 km ESE of Mporokoso, 29.ix.1962, *Robinson* 4953 (K). W: Mwinilunga Dist., Matonchi Farm, Kalenda dambo, 16.iv.1960, *Robinson* 3573 (K).

Also in Angola. In seasonally wet places, sometimes over laterite; 1200–1300 m.

Conservation notes: Moderately widespread; not threatened.

Robinson treated this species as a synonym of *Scleria flexuosa*, but the horizontal rhizomes and smaller spikelets distinguish it. Some of Robinson's collections mention tubers but they are not apparent on the specimens; one wonders if this species sometimes produces small tubers as well as horizontal rhizomes, or whether Robinson saw tubers belonging to another species growing together with this.

15. **Scleria erythrorrhiza** Ridl. in Trans. Linn. Soc. London, Bot. **2**: 167 (1884). — Clarke in F.T.A. **8**: 499 (1902). —Nelmes in Kew Bull. **10**: 437 (1955). —Napper in J. E. Africa Nat. Hist. Soc. Coryndon Mus. **24**(109): 32 (1964). —Robinson in Kew Bull. **18**: 506 (1966). —Haines & Lye, Sedges & Rushes E. Afr.: 333, fig.683 (1983). —Verdcourt in F.T.E.A., Cyperaceae: 387 (2010). Type: Angola, Huíla, Morro de Ferrão da Sola, i.1860, *Welwitsch* 7136 (LISU holotype, BM).

Perennial herb. Rhizome 3–6 mm thick, woody, straight, producing erect 0.3–1 m tall stems at short intervals. Leaves 2–6 mm wide, glabrous, ligulate, hairy or hispid. Inflorescence simply spicate, 3–14 cm long, with many glomerules of 3–16 dark, 4–6 mm long spikelets, occasionally a single glomerule produced on a short stalk in the axil of a leaf-like bract well below main spike. Glumes 3–5 mm long, chestnut, often dark red towards apex, with a green midrib, sometimes awned, densely white-hairy. Nutlets broadly ovoid, 2 × 1–1.2 mm wide, smooth, apiculate, pale brown or pale grey with darker interangular stripes, sometimes tinged violet at apex; stipe clearly differentiated, white, ± spongy in texture when fresh.

Zambia. B: Kalabo Dist., 5 km N of Kalabo, 16.xi.1959, *Drummond & Cookson* 6531 (K, LISC, SRGH). N: Kawambwa Dist., 18 km W of Kawambwa, 7.iv.1961, *Angus* 2748 (K, OXF). W: Mwinilunga Dist., 45 km E of Mwinilunga, 15.iv.1960, *Robinson* 3566 (K). **Malawi**. N: Mzimba Dist., 10 km NW of Mzuzu town, Lupaso, 7.iii.1976, *Pawek* 10890 (K, MO, SRGH). **Mozambique**. N: Lichinga Dist., Lichinga (Vila Cabral), road to Maniamba, 25.i.1935, *Torre* 728 (LISC).

Also known from Tanzania, Angola and D.R. Congo (Katanga). In peaty, permanently wet dambos, usually towards the damp margins and avoiding permanently wet parts, and in upland grassland; 1200–1820 m.

Conservation notes: Widespread; not threatened.

Resembles *Scleria veseyfitzgeraldii* in its leaves, but has a creeping rhizome, more conspicuous inflorescence bracts, and smooth, not patterned, nutlets.

16. **Scleria longispiculata** Nelmes in Kew Bull. **13**: 150 (1958). —Napper in J. E. Africa Nat. Hist. Soc. Coryndon Mus. **24**(109): 32 (1964). —Robinson in Kew Bull. **18**: 506 (1966). —Podlech in Merxmüller, Prodr. Fl. SW Afrika **165**: 52 (1967). —Hennessy in Bothalia **15**: 518 (1985). —Haines & Lye, Sedges & Rushes E. Afr.: 333, figs.684,685 (1983). —Verdcourt in F.T.E.A., Cyperaceae: 388 (2010). Type: Tanzania, Songea Dist., Kitai, 16.iv.1956, *Milne-Redhead & Taylor* 9739 (K lectotype), lectotypified by Hennessey (1985).

Perennial herb; stems erect, 0.6–0.9(1.2) m tall. Rhizome long, straight, creeping, 4–6 mm in diameter, woody tawny, stems at 0.5–2 cm intervals; stem bases slightly swollen. Leaves 2–5 mm wide, hairy or hispid. Inflorescence simply spicate, 5–10 cm long with many glomerules 1–2 cm apart, each of 2–6 greenish or pale brown minutely hispidulous spikelets 8–9 mm long. Glumes appressed-hispidulous-pubescent, chestnut with green midrib, darker at maturity. Nutlets ovoid, 3.5–4 × 2–2.5 mm wide, smooth, brown or light brown with interangular stripes of darker brown; stipe black.

Zambia. N: Chinsali Dist., Manshya Dambo, Shiwa Ngandu, xii.1963, *Symoens* 10785 (K). C: Mkushi Dist., Fiwila, 8.i.1958, *Robinson* 2688 (K, M, MRSC, MTJB, SRGH, YBI). **Malawi**. N: Mzimba Dist., 9 km NW of Mzuzu towards Lupaso, 7.iii.1976, *Pawek* 10890 (K, MAL, MO, SRGH).

Also in S Tanzania and NE Namibia. In moist grassland and at sandy dambo margins, often in relatively dry situations for the genus; 1200–1400 m.

Conservation notes: Widespread; not threatened.

The patterning on the nutlet is only visible on mature specimens, which are rarely seen.

17. **Scleria welwitschii** C.B. Clarke in Durand & Schinz, Consp. Fl. Afr. **5**: 675 (1894). —Clarke in F.T.A. **8**: 501 (1902). —Nelmes in Kew Bull. **10**: 423 (1955). —Napper in J. E. Africa Nat. Hist. Soc. Coryndon Mus. **24**(109): 30 (1964). —Robinson in Kew Bull. **18**: 506 (1966). —Haines & Lye, Sedges & Rushes E. Afr.: 334, fig.686 (1983). —Hennessy in Bothalia **15**: 517 (1985). —Gordon-Gray in Strelitzia **2**: 186, fig.85 (1995). —Verdcourt in F.T.E.A., Cyperaceae: 388 (2010). Type: Angola, Huíla, Catumba, iii.1860, *Welwitsch* 7138 (LISU lectotype, BM), lectotypified by Nelmes (1955).

> *Scleria junciformis* Ridl. in Trans. Linn. Soc. London, Bot. **2**: 168 (1884), illegitimate name, non Kunth (1837).

Slender perennial herb 0.3–1 m tall. Rhizome ± straight, woody, reddish, 3–4 mm wide, with weakly erect stems at 0.5–1.5 cm intervals. Leaves 1.5–3 mm wide, glabrous or hairy, ligulate. Inflorescence sparingly branched, to 25 cm long, branches to 10 cm long, lax and ± drooping at maturity with glomerules of 2–6, 4.5–7 mm long spikelets. Glumes dark chestnut or reddish, 3–4 mm long, long-acuminate, glabrous or hairy. Nutlets ovoid to ellipsoid, 1.5–1.8 × 1–1.2 mm wide, not or very shortly apiculate, smooth, grey.

Zambia. N: Mansa Dist., 30 km S of Mansa (Fort Rosebery), 4.i.1961, *Robinson* 4237 (K). S: Namwala Dist., Ngoma, 7.xii.1960, *Mitchell* 2/51 (MRSC). **Zimbabwe**. C: Marondera Dist., Grasslands Research Station, 17.x.1948, *Corby* 155 (K, SRGH). E: Mutasa Dist., Nuza Plateau, 1934, *Gilliland* K1299 (K). **Malawi**. N: Nyika Plateau, Lake Kaulime, 4.i.1959, *Robinson* 3027 (K). **Mozambique**. N: Malema Dist., 16 km from Malema (Entre-Rios) towards Ribáuè, 3.ii.1964, *Torre & Paiva* 10407 (COI, LISC, LMA, LMU, LUA, PRE).

Also in Angola and South Africa (KwaZulu-Natal). In upland bogs and dambos; 1600–2300 m.

Conservation notes: Widespread; not threatened.

Robinson states that this species is "not uncommon in suitable localities", so it is surprising that there are only three collections by him at Kew, with none from Zambia C, where he mentions that it occurs. Perhaps he regarded it as too distinctive to warrant frequent collection.

18. **Scleria rehmannii** C.B. Clarke in Fl. Cap. **7**: 295 (1898); in F.T.A. **8**: 501 (1902). —Nelmes in Kew Bull. **10**: 425 (1955). —Napper in J. E. Africa Nat. Hist. Soc. Coryndon Mus. **24**(109): 30, fig.11 (1964). —Robinson in Kew Bull. **18**: 507 (1966). —Podlech in Merxmüller, Prodr. Fl. SW Afr. **165**: 52 (1967). —Haines & Lye, Sedges & Rushes E. Afr.: 334, fig.687 (1983). —Hennessy in Bothalia **15**: 517 (1985). —Verdcourt in F.T.E.A., Cyperaceae: 388 (2010). Type: South Africa, Houtbosch, 1875, *Rehmann* 5626 (K holotype).

Slender perennial herb, 0.3–1.5 m tall. Rhizome long, straight, woody tawny, 3–4 mm in diameter, with erect stems 0.5–1.5 cm apart. Leaves 1–3(4) mm wide, ± hairy, ± ligulate. Inflorescence a stiffly erect simple panicle 4–12(15) cm long with branches to 7 cm, sometimes simply spicate with glomerules of 2–6 dark red, 3.5–5 mm long spikelets. Glumes shortly acuminate, glabrous, dark brown to reddish brown. Nutlets irregularly globose or broadly ovoid, 1.5–1.8 × c. 1 mm wide, apiculate, generally smooth but sometimes slightly papillose or strongly tuberculate in transverse lines, grey or pale brown.

Zambia. B: Kalabo Dist., 5 km S of Kalabo, 16.xi.1959, *Drummond & Cookson* 6554 (K, LISC, SRGH). N: Chinsali Dist., 27 km SW of Isoka, 13.i.1957, *Robinson* 3197 (K). W: Mwinilunga Dist., Chingabola dambo near Matonchi, 16 ii.1975, *Hooper & Townsend* 102 (K). C: Mkushi Dist., Fiwila, 3.i.1958, *Robinson* 2596 (K). S: Namwala Dist., Namwala, 8.i.1957, *Robinson* 2044 (GH, IRLCS, K, NR, NU, SRGH). **Zimbabwe**. N: Hurungwe Dist., Mwami (Miami), K34 Farm, 4.iii.1947, *Wild* 1754 (K, SRGH). W: Matobo Dist., Farm Besna Kobila, xi.1957, *Miller* 4763 (K, LISC, SRGH). C: Chegutu Dist., Msengeri Expt. Farm, Makwire, xi.1955, *Conradine* 12 (K, LISC, SRGH). E: Nyanga Dist., Gairesi Ranch, 9 km N of Troutbeck, 20.xi.1956, *Robinson* 1964 (K, LISC). **Malawi**. N: Karonga Dist., 20 km W of Karonga, Kayelekera, ix.1989, *Collinson & Davy* s.n. (K). C: Dedza Dist., Kachere Village, 7.i.1959, *Jackson* 2267 (K). S: Shire Highlands, 8.iv.1908, *Adamson* 109 (K). **Mozambique**. Z: Maganja da Costa Dist., 64 km from Maganja da Costa on road to Mocuba, 9.i.1968, *Torre & Correia* 17003 (LISC, LMA, WAG). T: Moatize Dist., 16 km from Zóbuè towards Metengobalama, 11.i.1966, *Correia* 383 (LISC).

Also in Uganda, Tanzania, D.R. Congo, Angola, Namibia and South Africa. In seasonally or permanently wet dambos; 950–1750 m.

Conservation notes: Widespread; not threatened.

There is an unconfirmed record from Botswana in Setshogo, Prelim. Checkl. Pl. Botswana: 121 (2005).

This is similar to the previous species but with smaller spikelets. Robinson regarded it as tolerant of a wider range of conditions than *Scleria welwitschii* and he discusses two forms – one more slender with stems to 2 mm in diameter and leaves to 3 mm wide, and another with stems less than 1 mm in diameter and leaves little more than 1 mm wide. The first occurs in sites that dry out completely in the winter, and the second in permanently wet sites. However, intermediate forms occur in intermediate sites and the full range of variation may be seen along a transect across the fringe of a wetland.

19. **Scleria paupercula** E.A. Rob. in Kew Bull. **18**: 508, fig.4 (1966). —Haines & Lye, Sedges & Rushes E. Afr.: 335, figs.688,689 (1983). —Verdcourt in F.T.E.A., Cyperaceae: 389 (2010). Type: Zambia, 8 km E of Kasama, 12.xii.1961, *Robinson* 4723 (K holotype, B, EA, GC, M, MPR, MTJB, NY, PRE, SRGH).

Slender perennial herb; stems erect, slender, 0.5–1 mm wide, 20–40(50) cm tall, close or up to 1 cm apart. Rhizome long, creeping, ± straight juicy (woody when dry), 2–3 mm thick, entirely glabrous except for mouths of leaf sheaths, glumes sometimes slightly hairy. Leaves less than 1 mm wide with incurved margins. Inflorescence 1.5–2.5 cm long (6 cm in F.T.E.A. area), spicate or shortly branched with glomerules of 2–6, 3–4 mm long spikelets. Glumes dark brown, keel and margins often paler. Nutlets irregularly ovoid to ovoid-globose, 1.2–1.4 × 0.7–0.8 mm wide, smooth, dark to pale brown; stipe trigonous, brown, 0.4 mm long.

Zambia. N: Mbala Dist., Lunzua Swamp near Mbala, 17.i.1962, *Richards* 15920 (K). C: Lusaka Dist., Chakwenga headwaters, 100–129 km E of Lusaka, 27.x.1963, *Robinson* 5767 (EA, K, MRSC, NY, PRE, SRGH). **Zimbabwe**. C: Marondera Dist., Grassland Res. Station, 2.ii.1948, *Colville* 74 (K, SRGH).

Also in Tanzania. In permanently wet bogs and dambos; 1400–1550 m.

Conservation notes: Widespread; not threatened.

Resembles *Scleria rehmannii* but altogether a more delicate plant with narrower leaves, thinner stems, a less complex inflorescence and dark brown glumes.

20. **Scleria dregeana** Kunth, Enum. Pl. **2**: 354 (1837). —Clarke in Fl. Cap. **7**: 295 (1898); in F.T.A. **8**: 499 (1902). —Nelmes in Kew Bull. **10**: 426 (1955). —Napper in J. E. Africa Nat. Hist. Soc. Coryndon Mus. **24**(109): 32 (1964). —Robinson in Kew Bull. **18**: 510 (1966). —Haines & Lye, Sedges & Rushes E. Afr.: 336, figs.690,691

(1983). —Hennessy in Bothalia **15**: 519 (1985). —Gordon-Gray in Strelitzia **2**: 183, figs.81,82 (1995). —Verdcourt in F.T.E.A., Cyperaceae: 389 (2010). Type: South Africa, Cape of Good Hope (Kat Berg), Stockenstrom Division, xi.1840, *Drège* s.n. (under *Clarke* 3934) (B† holotype, K).

    Scleria caespitosa Ridl. in Trans. Linn. Soc. London, Bot. **2**: 167 (1884). Type: Angola, Pungo Andongo, Condo, iii.1857, *Welwitsch* 7135 (LISU holotype, K).

    Scleria setulosa Boeckeler, Beitr. Cyper. **1**: 33 (1888). Type: Malawi, Shire Highlands, vii.1885, *Buchanan* 36 (B† holotype, K).

    Scleria bulbifera A. Rich. var. *hirsuta* sensu Fl. Deutsch Ost-Afrika **1**: 531 (1938) as regards *Peter* 38937, non Peter & Kük.

Perennial herb, usually caespitose or sometimes with a rhizome 1–2 mm thick, with closely placed slender stems 0.2–1 m tall. Leaves 1–2(3) mm wide, glabrous or slightly hairy. Inflorescence 3–10 cm long, simply spicate or sparsely to strongly branched, branches 3–5 cm long; glomerules close-set of 2–9 blackish, 4.5–6 mm long spikelets. Glumes usually dark to pale brown or blackish with conspicuous green midrib, glabrous or hairy. Nutlets ovoid, 1.4–2 × 1.1–1.3 mm wide, apiculate, smooth or slightly or distinctly tuberculate towards apex, distinctly beaked, whitish to olive brown, often darker towards apex.

Botswana. N: Okavango Swamps, Dassakao Is. on Monachira R., 19°12'S 23°19.7'E, 10.ii.1979, *P.A. Smith* 2635 (K, PRE, SRGH). **Zambia**. C: Mkushi Dist., Fiwila dambo, 5.i.1958, *Robinson* 2643 (MRSC). S: Choma Dist., Choma, 27.i.1957, *Robinson* 2121 (K, MRSC). **Zimbabwe**. N: Gokwe Dist., Sengwa Research Station, 19.i.1975, *Guy* 2274 (K, SRGH). C: Marondera Dist., Grasslands Research Station, 19.i.1948, *Colville* 38 (K, SRGH). E: Nyanga Dist., Nyanga Nat. Park, Udu Dam, 29.xii.1974, *Burrows* 636 (K, SRGH). **Malawi**. N: Karonga Dist., Chisenga, 23.v.1962, *Robinson* 5229 (K). C: Kirk Range, Kanyaloma, 16.xi.1950, *Jackson* 287 (K). S: Blantyre Dist., Manganja Hills, near Soche Hill, 8.iii.1862, *Kirk* s.n. (K).

Also in South Africa northwards to Angola, D.R. Congo (Katanga) and Tanzania. In permanently wet dambos; 900–1500 m (50–2100 m in South Africa).

Conservation notes: Widespread; not threatened.

There is an unconfirmed record from Zimbabwe S in Mapaura & Timberlake, Checkl. Zimbabwe Pl.: 89 (2004).

The form of the inflorescence is very variable and Kunth based three names on Drège's material; Robinson (1966: 510) was happy to place all of these under *Scleria dregeana*.

See Robinson (1966: 508) for a note on the type of *Scleria setulosa* Boeckeler.

21. **Scleria sobolifer** E.F. Franklin in Kew Bull. **38**: 33 (1983); in Bothalia **15**: 521, fig.6 (1985). —Gordon-Gray in Strelitzia **2**: 185, fig.84 (1995). Type: South Africa, KwaZulu-Natal, 10.ix.1965, *Ward* 5128 (K holotype, PRE, NH, NU, UD-W). FIGURE 14.**71**.

Perennial herb; stems 20–100 cm tall, glabrous or glabrescent. With slender underground stems, 1–1.5 mm thick, hard, whitish with wine-red blotches, scales purple-red. Leaves 1.1–2.6 mm broad, glabrous adaxially, sparsely hirsute abaxially; mouth of sheath truncate, hirsute. Inflorescence unbranched, 20–65 mm long; glomerules 2–6, of 2–6 sessile spikelets. Bracts glabrous, the margins sparsely hirsute, shortly awned, pale stramineous with red striae. Spikelets c. 4 mm long; glumes 1.7–2.6 mm long, glabrous or sparsely hirsute, midrib stramineous with red striae. Nutlet subglobose, 1.5–1.8 × 1.2 mm, glabrous, undulate-tuberculate, grey, with a series of deep horizontal troughs separated by horizontal and vertical ridges at junction of stipe and body.

Mozambique. M: Matutuine Dist., Ponta do Ouro, 19.vi.1981, *de Koning, Hiemstra & Nuvunga* 8823 (K, LMU).

Also known from South Africa (KwaZulu-Natal). By pools among old coastal dunes; 0–50 m.

Conservation notes: Widespread; probably not threatened across its range but possibly Vulnerable in the Flora area.

Not previously recorded from the Flora area, although Hennessy suggested that it would probably be found in S Mozambique. The thin elongated rhizomes ('soboles') are unlike those of any other species in the area.

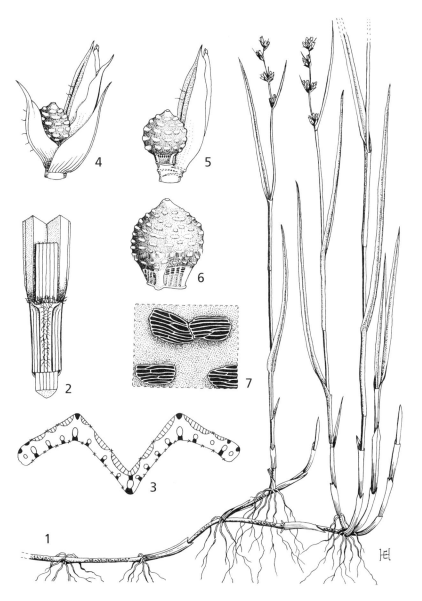

Fig. 14.**71**. SCLERIA SOBOLIFER. 1, habit (× 2/3); 2, junction of lamina and sheath (× 6); 3, plan of transverse section of mid-lamina (× 20); 4, androgynaceous spikelet (× 9); 5, androgynaceous spikelet with lower glume removed to reveal nutlet (× 9); 6, nutlet (× 17); 7, nutlet surface pattern, from SEM (× 142). 1–5 from *Ward* 5218; 6, 7 from *Ward* 8851. Drawn by E.H. Franklin Hennessy. Reproduced from Bothalia (1985).

22. **Scleria angustifolia** E.A. Rob. in Kirkia **3**: 8 (1962); in Kew Bull. **18**: 510, fig.5 (1966). —Haines & Lye, Sedges & Rushes E. Afr.: 336 (1983). Type: Zambia, Kasama Dist., Mungwi, 17.xi.1960, *Robinson* 4079 (K holotype, EA, GC, M, MTJB, PRE, SRGH).

Perennial herb; stems arising 3–5 mm apart, up to 80 cm tall, less than 1 mm wide. Rhizome 2–2.5 mm thick, rigid, ± straight, horizontal. Leaves setaceous, glabrous. Inflorescence 3–9 cm long, branched; branches to 2.5(3) cm long, simple or more rarely compound. Spikelets 3–4 mm long; glumes glabrous, reddish-brown or blackish red with paler midrib. Nutlets 1–1.2 × 0.7–0.9 mm broad, irregularly ovoid to globose, almost smooth to lightly or strongly papillose or tuberculate, pale reddish brown or reddish grey; stipe darker reddish brown.

Zambia. N: Kasama Dist., 8 km N of Kasama, 22.i.1961, *Robinson* 4286 (K, M, SRGH). W: Chingola Dist., Chingola, 16.iii.1960, *Robinson* 3406 (K, LISC, MTJB, SRGH). C: Lusaka Dist., Leopard's Hill Road, 41 km from Lusaka, 21.x.1972, *Kornaś* 2405 (K).

Known only from Zambia. According to Robinson (1962) in "Bogs which remain wet for all or most of the year, where it may be found more or less closely associated with any of the following species: *Scleria rehmannii, S. bequaertii, S. procumbens, S. laxiflora, S. erythrorrhiza, S. greigiifolia, S. pooides, S. unguiculata* or *S. welwitschii*"; 1200–1500 m.

Conservation notes: Endemic to Zambia but widespread; not threatened.

A very delicate species. The stems arise from the rhizome at very short intervals.

23. **Scleria pooides** Ridl. in Trans. Linn. Soc. London, Bot. **2**: 170 (1884) as *poaeoides*. —Clarke in F.T.A. **8**: 502 (1902) as *poaeoides*. —Nelmes in Kew Bull. **10**: 433 (1955) as *poaeoides*. —Napper in J. E. Africa Nat. Hist. Soc. Coryndon Mus. **24**(109): 30 (1964) as *poaeoides*. —Robinson in Kew Bull. **18**: 512 (1966). —Napper in F.W.T.A., ed.2 **3**: 343 (1972). —Haines & Lye, Sedges & Rushes E. Afr.: 336, figs.16b,692 (1983). —Verdcourt in F.T.E.A., Cyperaceae: 390 (2010). Type: Angola, Huíla, near Quilebe, xi.1859, *Welwitsch* 7142 (LISU holotype, K).

   Scleria multispiculata Boeckeler, Beitr. Cyper. **1**: 36 (1888). —Clarke in Durand & Schinz, Consp. Fl. Afr. **5**: 673 (1895) as *multispiculosa*. —Schumann in Engler, Pflanzenw. Ost-Afrikas C: 129 (1895). —Clarke in F.T.A. **8**: 501 (1902). Type: Malawi, Shire Highlands, vii.1885, *Buchanan* 1 (B† holotype, E, K).

   Scleria prophyllata Nelmes in Kew Bull. **10**: 434 (1955). Type: Angola, Moxico Dist., Mumbala R., 8.i.1938, *Milne-Redhead* 3995 (K holotype).

Slender glabrous tufted perennial herb, 30–80 cm tall, or with a short creeping rhizome c. 1 mm thick (particularly if growing in water or very wet sites). Leaves 1–2 mm wide. Inflorescence a spreading compound panicle 3–15(20) cm long, branches slender, compound; spikelets dark red axillary and pedicellate, up to 170 per stem, 3–4(5) mm long. Glumes arranged ± distichously on spikelet; male glumes 2–3 mm long, female glumes 1.5–2 mm, hispidulous. Nutlets ovoid to globose, 1–1.5 × 0.8–1 mm wide, densely tuberculate to trabeculate, grey.

Zambia. N: Kasama Dist., Mungwi, 6.ii.1961, *Robinson* 4362 (K). W: Mwinilunga Dist., Zambezi Rapids c. 6 km N of Kalene Hill, 20.ii.1975, *Hooper & Townsend* 263 (K). C: Serenje Dist., Kundalila Falls, 4.ii.1973, *Strid* 2873 (K). **Zimbabwe**. W: Matobo Dist., Besna Kobila Farm, i.1960, *Miller* 7045 (K, LISC, SRGH). C: Harare (Salisbury), 19.ii.1927, *Eyles* 4732 (K, SRGH). E: Nyanga Dist., Nyanga (Inyanga) road, *Gilliland* K.1033 (K). **Malawi**. N: Mzimba Dist., Mzuzu waterworks, 15.vi.1974, *Pawek* 8714 (K, MO, SRGH). C: Lilongwe Dist., Dzalanyama Forest Reserve, above Chiunjiza road, c. 5 km SE of Chaulongwe Falls, 23.iii.1970, *Brummitt* 9294 (K, LISC). S: Shire Highlands, vii.1895, *Buchanan* s.n. (K).

Also in Nigeria, D.R. Congo, Angola, Kenya, Tanzania and Madagascar. In permanently wet bogs, where it may be dominant; 1200–1700 m.

Conservation notes: Widespread; not threatened.

The delicate, open, grass-like inflorescence is distinctive. Not to be confused with the similarly named *Scleria poiformis* which is coastal and rare in the Flora area.

24. **Scleria woodii** C.B. Clarke in Fl. Cap. **7**: 295 (1898). —Clarke in F.T.A. **8**: 501 (1902). —Nelmes in Kew Bull. **10**: 428 (1955). —Napper in J. E. Africa Nat. Hist. Soc. Coryndon Mus. **24**(109): 31 (1964). —Robinson in Kew Bull. **18**: 512 (1966). —Haines & Lye, Sedges & Rushes E. Afr.: 336 (1983). —Hennessy in Bothalia **15**: 516 (1985). —Gordon-Gray in Strelitzia **2**: 186, fig.85 (1995). Type: South Africa, Zululand, Inyoni R., 4.iv.1888, *Wood* 3994 (K lectotype, BOL, NH), lectotypified by Gordon-Gray (1995).

Perennial herb; stems 25–75 cm tall, irregularly spaced on the rhizome. Rhizome single, obliquely descending, scented white or pink, soft and tuberous, 5–8(10) cm long, 2–3 mm thick, with slightly swollen internodes; eventually produces a new plant at tip, up to 10 cm from parent, while the latter dies after flowering. Leaves 1–2.5(3.5) mm wide, glabrous or hairy. Inflorescence a simple or compound panicle 5–14 cm long, 2–5 cm wide with short erect or extended branches up to 6(10) cm long, slender and nodding with 10–30 glomerules of (1)2–6, 4–6 mm long spikelets; glumes 3–4(5) mm long, glabrous, pale to blackish brown. Nutlets ovoid to globose, 1.5–1.7(2) mm long, 1–1.4 mm wide, usually smooth, or faintly striate-tessellate to papillose or strongly tuberculate even on the same plant, grey with darker longitudinal stripes; stipe dark brown with whitish border 0.5 × 0.3 mm wide, sometimes persistent.

Robinson considered this species to be very variable and did not divide it into two species or recognise two varieties. The two varieties described here certainly grow together; Robinson collected both as consecutive numbers from near Choma (S Zambia). More recently, Napper (Kew Bull. **25**: 442, 1971) has treated *Scleria striatonux* as a good species, and both Haines & Lye and Verdcourt distinguished two varieties. Perversely, it is the taxon with smooth unornamented nutlets that bears the name *ornata*.

a) Var. **woodii**. —Napper in J. E. Africa Nat. Hist. Soc. Coryndon Mus. **24**(109): 31 (1964). —Haines & Lye, Sedges & Rushes E. Afr.: 336, fig.693 (1983). —Verdcourt in F.T.E.A., Cyperaceae: 391 (2010).

    Scleria striatonux De Wild. var. *lacunosa* Piérart in Lejeunia Mém. **13**: 30 (1953). Type: Kenya, Mt Elgon, v.1931, *Lugard* 667b (K lectotype), lectotypified by Robinson in F.T.E.A.

Nutlet strongly tuberculate or with transverse ridges. Inflorescence with several simple branches from lower clusters.

Botswana. N: Chobe Nat. Park, Zweizwe flats, 18.v.1977, *P.A. Smith* 2033 (K). **Zambia**. N: Mbala Dist., track to Goddard's Farm, 5.ii.1957, *Richards* 8073 (K). S: Choma Dist., 9 km E of Choma, 27.i.1957, *Robinson* 2124 (K). **Zimbabwe**. C: Chegutu Dist., Chegutu (Hartley), Poole Farm, 15.ii.1946, *Hornby* 2435 (K).

Widespread in tropical Africa and South Africa. In seasonally wet grassland; 900–2050 m.

Conservation notes: Widespread; not threatened.

b) Var. **ornata** (Cherm.) W. Schultze-Motel in Willdenowia **2**: 504 (1960). —Napper in J. E. Africa Nat. Hist. Soc. Coryndon Mus. **24**(109): 31 (1964). —Haines & Lye, Sedges & Rushes E. Afr.: 337, fig.694 (1983). —Verdcourt in F.T.E.A., Cyperaceae: 391 (2010). Type: D.R. Congo, Kabare, W of Lake Kivu, 20.xi.1929, *Scaetta* 2373 (BR holotype, NY, P).

    Scleria striatinux De Wild. in Rev. Zool. Bot. Africaines, suppl. Bot. **14**: 22, fig.5 (1926) as *striatonux*. —Nelmes in Kew Bull. **10**: 429 (1955) as *striatonux*. —Robyns & Tournay, Fl. Parc Nat. Albert **3**: 277, t.39 (1955) as *striatonux* (corrected according to Article 73 of code). —Napper in Kew Bull. **25**: 442 (1971); in F.W.T.A., ed.2 **3**: 343 (1972). Types: D.R. Congo, Beni, 3.iv.1914, *Bequaert* 3357; Beni, 6.iv.1914, *Bequaert* 3428; Rutshuru, 11.ix.1914, *Bequaert*

5640; Rutshuru, 19.x.1914, *Bequaert* 6048; Rutshuru, 29.x.1914, *Bequaert* 6098; and D.R. Congo, xi.1916, *Vanderyst* 6245 (all BR syntypes).

    Scleria rehmannii C.B. Clarke var. *ornata* Cherm. in Bull. Jard. Bot. État Bruxelles **13**: 283 (1935).

    Scleria lelyi Hutch. in F.W.T.A. **2**: 493 (1936). Types: Nigeria, Jos Plateau, v.1930, *Lely* P292 (K syntype); Naraguta, 21.vi.1921, *Lely* P299 K (K syntype), invalid name, English description only.

Nutlets quite smooth or almost so. Inflorescence more open, sometimes branched twice.

Zambia. N: Mbala Dist., Mbala, Lefhandside Pan, 11.ii.1957, *Richards* 8162 (K). W: Ndola Dist., Chichele Forest Reserve, 17.i.1960, *Robinson* 3302 (K). S: Namwala Dist., Namwala, 8.i.1956, *Robinson* 2043 (K). **Zimbabwe**. W: Matobo Dist., Besna Kobila Farm, ii.1959, *Miller* 5771 (K, SRGH).

Also in West Africa, D.R. Congo, Sudan, Uganda, Kenya and Tanzania. In seasonally wet grassland; 1000–1400 m.

Conservation notes: Widespread; not threatened.

Some specimens cannot be attributed to variety as they have no ripe fruit, e.g. N Malawi, Nyika Plateau, Western Valleys, 6.i.1959, *Robinson* 3088 (K).

Both varieties were lumped in the Zimbabwe Sabonet checklist (Mapaura & Timberlake, 2004), and recorded from Zimbabwe N, W, C, E.

25. **Scleria polyrrhiza** E.A. Rob. in Kew Bull. **18**: 513, fig.6 (1966). —Haines & Lye, Sedges & Rushes E. Afr.: 338 (1983). Type: Zambia, 55 km ESE of Mporokoso, Kasanshi dambo, 4.ii.1962, *Robinson* 4921 (K holotype, B, EA, M, MRSC, MTJB, NY, PRE, SRGH).

Perennial herb; stems to 65 cm long, 1 mm thick at base, glabrous or sparsely hairy. Rhizomes 2–4, produced from base each year, initially c. 1 mm thick, flexible, pale, covered with brown-striate scales, later the internodes thickening to 6 mm and alternately swollen and contracted (like an intestine), fairly rigid, fragile, rather sappy. Leaves 0.5–1.5 mm broad, often exceeding inflorescence, upper surface sparsely hairy, glabrous beneath; sheaths pilose, mouth truncate. Inflorescence 4–6.5 cm long, simply branched, nodding at maturity; branches to 2.5 cm long, very slender, bearing 1–2 glomerules. Glomerules of 1–6 spikelets, axillary ones often reduced to a prophyll. Spikelets androgynous and male, 4–5 mm long. Glumes glabrous, dark brown (keel and margins often paler or pale-spotted), female ones broadly ovate, aristate, male ones ovate-lanceolate. Nutlet 1.2–1.5 × c. 1 mm wide, broadly ovoid or subglobose, obscurely trigonous, obtusely beaked, lightly and indistinctly papillose or striate, grey with 3 darker interangular lines.

Zambia. N: Mporokoso Dist., Kasanshi dambo, 55 km ESE of Mporokoso, 22.ii.1962, *Robinson* 4952 (EA, K, M, MTJB, PRE, SRGH). W: Solwezi Dist., Solwezi, 9.iv.1960, *Robinson* 3485 (K).

Not known elsewhere. In ± permanently wet bogs; associated with *Scleria flexuosa* in the type locality, elsewhere with *S. glabra* and *S. pergracilis*; 1300–1500 m.

Conservation notes: Endemic to the northern parts of Zambia, but probably not threatened.

Very similar to *Scleria woodii* and herbarium specimens often do not show the rhizome characters. The pattern on the nutlets described and illustrated by Robinson (1966: 513) is not always apparent in herbarium specimens and may be best developed only in really mature nutlets.

26. **Scleria richardsiae** E.A. Rob. in Kirkia **3**: 9 (1962). —Napper in J. E. Africa Nat. Hist. Soc. Coryndon Mus. **24**(109): 31 (1964). —Robinson in Kew Bull. **18**: 515, fig.7 (1966). —Haines & Lye, Sedges & Rushes E. Afr.: 338, fig.695 (1983). —Verdcourt in F.T.E.A., Cyperaceae: 392 (2010). Type: Zambia, Nyika Plateau, near rest house, 13.iii.1961, *Robinson* 4473 (K holotype, EA, M, MTJB, SRGH).

Perennial herb, 0.6–1 m tall with rhizome formed from a knotty mass of hard fleshy (when fresh) stem-bases each c. 3 mm thick; stems slender, weakly erect, 60–100 cm tall. Leaves 1(2) mm wide, glabrous or sparsely hairy. Inflorescence a simple condensed to very lax panicle 9–20 cm long, branches to 6 cm long; glomerules 1–2 cm apart, more crowded towards apex, each with 2–4, 4–5 mm long spikelets; axes often with long hairs. Glumes glabrous or minutely hairy, dark reddish brown to blackish. Nutlets ovoid, 1.4–1.8 × c. 1 mm wide, lightly to strongly trabeculate-reticulate, transversely wrinkled or pitted, sometimes with raised cubic crystal-like tubercles, often cuspidate, grey or whitish with darker interangular stripes.

Zambia. E: Nyika Plateau, 6.i.1959, *Robinson* 3099 (K, LISC). **Malawi**. N: Nyika Plateau, 5.i.1959, *Robinson* 3078 (K, LISC).

Also in Tanzania. Grassy upland streamsides; 2000–2500 m.

Conservation notes: A montane species of restricted distribution, but probably not threatened.

A delicate upland species (Robinson regarded it as the only truly montane species in Africa); the strongly patterned nutlets are distinctive. The specific name commemorates Mary Richards (1885–1977), a prolific collector of plants in Zambia and Tanzania.

27. **Scleria fulvipilosa** E.A. Rob. in Kew Bull. **18**: 515, fig.7 (1966). —Haines & Lye, Sedges & Rushes E. Afr.: 338 (1983). Type: Zambia, Mbala Dist., Nondo, 80 km S of Mbala, 1.x.1961, *Robinson* 4712 (K holotype, B, EA, M, MPR, MTJB, NY, PRE, SRGH).

Perennial herb, erect, almost rigid. Rhizome absent except for thickened culm bases, aggregated into a sub-woody mass so the plant forms a tussock. Stems to 75 cm tall, c. 2 mm thick at base, pilose below, subglabrous near inflorescence. Leaves 2–2.5(3) mm wide, subrigid, pilose. Inflorescence mostly terminal, simply branched; rarely with 1–2 lateral small simple panicles at nodes, inflorescence measures up to 29 cm from basal panicle-bearing node to apex. Spikelets androgynous and male, 9–11 mm long, glabrous, in compact glomerules of 3–6 spikelets. Glumes glabrous, lanceolate; male ones pale brown or brown-spotted; female pale green, shiny, green-keeled. Nutlet oblong-ovoid, 1.9–2.1 × 1 mm wide, distinctly trigonous, dark brown to black at maturity except for the white hypogynium, with a dense brown villous indumentum; beak absent.

Zambia. N: Chinsali Dist., Shiwa Ngandu, 21.xii.1964, *Robinson* 6324 (B, K, M, MRSC, NY, SRGH); Mbala Dist., 80 km S of Mbala (Abercorn), 12.xi.1960, *Robinson* 4064 (EA, GC, K, M, MTJB, SRGH).

Not known elsewhere. In permanently wet dambos; associated with *Scleria rehmannii*, *S. angustifolia*, *S. paupercula*, *S. procumbens*, *S. bequaertii*, *S. greigiifolia* and *S. laxiflora* in the type locality; 1400–1500 m.

Conservation notes: Endemic to N Zambia; possibly Vulnerable.

The hairy nutlets are distinctive. Robinson suggests that the lack of an underground rhizome may make this species intolerant of fires and explain its relative rarity in apparently suitable habitats.

28. **Scleria lithosperma** (L.) Sw., Prodr. Veg. Ind. Occ.: 18 (1788). —Clarke in Durand & Schinz, Consp. Fl. Afr. **5**: 672 (1894); in F.T.A. **8**: 502 (1902); in Illustr. Cyp.: t.123 (1909). —Nelmes in Kew Bull. **10**: 421 (1955). —Napper in J. E. Africa Nat. Hist. Soc. Coryndon Mus. **24**(109): 30 (1964). —Robinson in Kew Bull. **18**: 503 (1966). —Napper in F.W.T.A., ed.2 **3**: 343 (1972). —Haines & Lye, Sedges & Rushes E. Afr.: 343, fig.709 (1983). —Verdcourt in F.T.E.A., Cyperaceae: 397 (2010). Type: India, Rheede, Hort. Mal. **12**: 89, t.48 (1693) (iconotype).

Scirpus lithospermus L., Sp. Pl.: 51 (1753).

Perennial herb with short rhizomes, forming clumps; stems 1–2.5 mm wide, 30–90 cm tall, minutely scabrid. Lower leaf sheaths brown, upper green, blades to 20 cm long by 2–5 mm wide, scabrid on margin at least near tip, otherwise glabrous or hairy. Inflorescence of 1 terminal and 2–3 lateral panicles, the latter arising singly from leafy bracts. Spikelets 4–5 mm long, all androgynous; glumes straw-coloured to pale green or light brown, female ones often with green midrib, scabrid on midrib and margins, otherwise glabrous. Nutlets ovoid or obovoid, 2.5–3 × 1.5–2 mm wide, ± trigonous, smooth, glabrous, apiculate with 3 depressions near base, olive-grey or olive-brown or pearly white; stipe represented by an unlobed disc at base of nutlet.

Zambia. S: Choma Dist., Mapanza, Choma, 26.i.1958, *Robinson* 2754 (GC, K, M, MRSC, MTJB, NU, PRE, SRGH, YBI). **Mozambique**. MS: Dondo Dist., Dondo (25 Miles Station), 10.iv.1898, *Schlechter* 12241 (BM, K, PRE).

Also in Ivory Coast, Ghana, Nigeria, D.R. Congo, Ethiopia, Kenya and Tanzania, also in tropical and subtropical Asia, Australia and America. Found in well drained evergreen riverside thicket in the Flora area, elsewhere in shady and open places in evergreen forest, woodland, thickets and termite mound thickets; 50–1350 m.

Conservation notes: Widespread; not threatened.

This is the first species in a group in which the nutlets are very prominent and relatively large. It has fine silicified teeth on the midribs and leaf margins so that leaves can cut the skin.

29. **Scleria lacustris** Sauvalle in Anales Acad. Ci. Méd. Habana **8**: 152 (1871). — Nelmes in Kew Bull. **10**: 422 (1955). —Robinson in Kew Bull. **18**: 517 (1966). —Haines & Lye, Sedges & Rushes E. Afr.: 343 (1983). —Hennessy in Bothalia **15**: 523 (1985). Type: Cuba, near Pinar del Río, 1869, *Wright* s.n. (K).

   Scleria aquatica Cherm. in Bull. Soc. Bot. France **77**: 279 (1930). Type: Gabon, Haute-Ngounyé, 22.xii.1925, *Le Testu* 5845 (CN holotype, BR, P).

Annual herb, entirely glabrous, but with very small sharp silicified teeth on angles of stems, leaves and sheaths. Stems partly hollow, up to 180 cm tall by 12 mm thick, with many adventitious roots at lower nodes where submerged in water. Leaves to 15 mm wide; mouth of sheath produced into an oval membranous tongue or contraligule 5–10 mm long with a narrow, blackish hispid zone in angles at its base. Inflorescence of 1 terminal and 1–2 lateral panicles, the latter rising singly in axils of leafy bracts, to 1 m long. Spikelets 4–6 mm long, predominantly androgynous or male, though some female ones may be found in which the male part is much reduced or ± vestigial; female glumes apiculate or aristate, dark reddish brown. Nutlets 3–3.5 × 2–2.5 mm broad, ovoid, smooth, grey, grey-brown, fulvous brown or dark brown (much variation in colour occurs on a single plant), with darker interangular stripes; hypogynium small and very narrow, but clearly differentiated from nutlet and obscurely 3-lobed.

Botswana. N: Okavango Swamps, floodplain of Mboroga R., 19°13.7'S 23°08.1'E, 19.vi.1979, *P.A. Smith* 2796 (K). **Zambia**. N: Luwingu Dist., Nsombo, edge of Lake Bangweulu, 22.v.1961, *Robinson* 4700 (K, MRSC).

Also in Sierra Leone, Gabon, Central African Republic and Madagascar, and in the Caribbean and South America. In swamps and on lake margins; 950–1200 m.

Conservation notes: Widespread; not threatened.

There is only a single record from Zambia but there are several collections from the Okavango Delta in Botswana. This species generally grows with the base submerged. The nutlets lack an apiculus but taper somewhat towards the apex.

30. **Scleria porphyrocarpa** E.A. Rob. in Kew Bull. **18**: 519, fig.9 (1966). —Haines & Lye, Sedges & Rushes E. Afr.: 356 (1983). Type: Zambia, Kasama Dist., Mungwi, 21.xii.1961, *Robinson* 4742 (K holotype, B, EA, GC, M, MRSC, MTJB, NU, NY, PRE, SRGH). FIGURE 14.**72**.

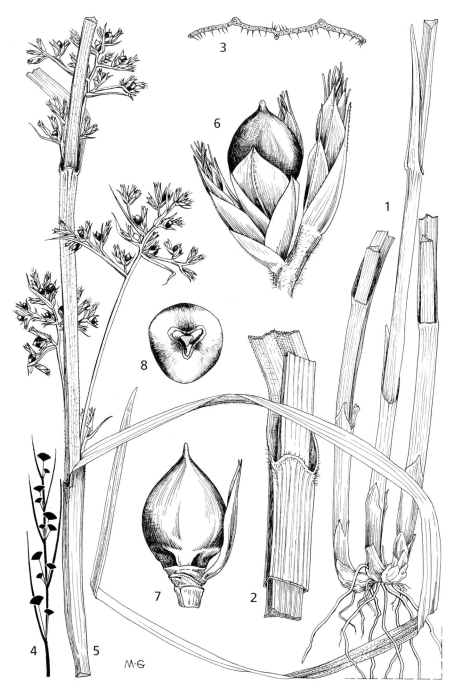

Fig. 14.**72**. SCLERIA PORPHYROCARPA. 1, basal part of plant showing rhizome (× ²⁄₃); 2, junction of lamina and sheath (× 2); 3, transverse section of leaf (× 4); 4, diagram of entire inflorescence; 5, part of inflorescence (× ²⁄₃); 6, single glomerule (× 6); 7, nutlet, lateral view with empty male glumes (× 6); 8, nutlet, basal view (× 6). All from *Robinson* 3849. Drawn by Mary Grierson. Reproduced from Kew Bulletin (1966).

Perennial herb, either erect or scrambling over trees and bushes. Stems 5–6 mm thick at base, to 5 m long, smooth, triquetrous; retrorse-scabrid on angles above. Rhizome robust, 3–4 mm thick, with short internodes. Leaves 5–24 mm wide, sheaths covering greater part of stems, shortly and sparely hairy, margins markedly scabrid, lower leaves reduced to sheaths or with a short lamina; lower sheaths smooth or scabrid, reddish, upper ones scabrid on angles and shortly hairy; mouth of sheath produced into a short rounded tongue 1.5–4 mm long with a brown-pilose margin. Inflorescence up to 90 cm long; panicles sometimes corymbose, often broadly pyramidal, lower branches spreading or reflexed, 1.5–4 × 2–6.5 cm; peduncles compressed, 1–1.7 mm wide, scabrid or hairy, almost erect, exserted 2–16 cm from sheaths. Spikelets androgynous and male; in androgynous ones, male flower often reduced to empty glumes; male spikelets 6–7 mm long; glumes of androgynous spikelets 4–5 mm long, ovate, acuminate, glabrous or minutely hispidulous on median nerve towards apex, keel green, sides purple or brown-purple. Nutlets ovoid, 4.8–6 × 2.7–3.7 mm, smooth, glabrous, shiny, purple, paler or white where covered by glumes, obscurely trigonous, markedly and acutely beaked, beak gradually narrowing from nutlet; stipe distinctly trigonous but not entirely differentiated from nutlet.

Zambia. N: Kasama Dist., Mungwi, 21.ix.1960, *Robinson* 3849 (EA, GC, K, LISC, M, MRSC, MTJB, PRE, SRGH). W: Mwinilunga Dist., Chingabola dambo near Matonchi, 17.ii.1975, *Hooper & Townsend* 143 (K). C: Mkushi Dist., Fiwila, 3.i.1958, *Robinson* 2591 (K, M, MTJB, SRGH).

Also in D.R. Congo. In relict evergreen swamp forest (mushitu) on permanently damp soil; 1300–1500 m.

Conservation notes: Found in a somewhat threatened habitat; possibly Near Threatened.

According to Robinson (1966), "In this association *Scleria porphyrocarpa* often forms a dense tangle of erect or trailing rough-edged stems which make progress almost impossible and generally bloody." The teeth on the angles are minute and hard to see but clearly large enough to cut the skin.

The following three species are very similar in appearance and, according to Robinson, often grow together.

31. **Scleria procumbens** E.A. Rob. in Kew Bull. **18**: 521, fig.10 (1966). —Haines & Lye, Sedges & Rushes E. Afr.: 352 (1983). Type: Zambia, Kasama Dist., Mungwi, 16.ii.1961, *Robinson* 4387 (K holotype, SRGH).

Perennial herb, similar to *Scleria bequaertii* De Wild. but differs in narrower leaves 1–2.5(3) mm wide with deflexed hairs on sheaths, in the androgynous spikelets including staminiferous flowers, in the shorter spikelets (5–6(7) mm long), in the shorter nutlets (1–1.2 mm long) that are smooth or more lightly papillose or lacunose, and in the stipe that equals or is longer than the nutlet, which is smooth and 1.1–1.4 mm long before drying.

Zambia. N: Mbala Dist., Lunzua R., Kambole Bridge, 26.ii.1961, *Robinson* 4420 (K, LISC); Kawambwa Dist., Kawambwa, 22.vi.1956, *Robinson* 2364 (K, M, MPR, MTJB, SRGH). C: Mpika Dist., Mutinondo Wilderness Area, Kabulu dambo, 12°23'59"S 31°19'06"E, *Bauters* 2015-171 (GENT, UZL).

Not known elsewhere. In permanently wet bogs; 1350–1500 m.

Conservation notes: Apparently endemic to NE Zambia; probably Near Threatened.

This species is similar to both *Scleria bequaertii* and *S. laxiflora* in its weakly erect habit, numerous short leaves, branched stems, and in its aromatic stems and leaves. It is variable in hairiness, but the relatively large stipe is characteristic. Robinson warns that duplicates of his numbers of this species, collected before he was fully aware of the differences between it and *S. bequaertii*, may be mixed. However, the collections listed in his 1966 paper (p.521) are all of this taxon.

32. **Scleria bequaertii** De Wild. in Rev. Zool. Bot. Africaines, suppl. Bot. **14**: 17 (1926). —Nelmes in Kew Bull. **11**: 101 (1956). —Robinson in Kew Bull. **18**: 523 (1966). —Haines & Lye, Sedges & Rushes E. Afr.: 352 (1983). Types: D.R. Congo, Kinshasa (Leopoldville), 1915, *Bequaert* 7158, 7335, 7597 (BR syntypes); Kinshasa, 15.v.1915, *Bequaert* 7597 (BR lectotype), lectotypified by Nelmes (1956).

    Scleria bequaertii De Wild. var. *laevis* Piérart in Lejeunia Mém. **13**: 38 (1953). Type: Angola, Moxico, near R. Chibamba, 16.i.1963, *Milne-Redhead* 4157 (K).

Perennial herb; stems weakly erect or semi-prostrate, less than 1 mm wide at base, up to 1 m long, branching at many nodes to form a ± bushy growth. Rhizome 2–5 mm thick, hard, horizontal, ± straight, occasionally branched and often forming (with other plants of same species) a thickly knotted mass. Leaves 1–3.5 mm wide, sometimes glabrous but more often hairy. Inflorescences terminal and lateral; lateral panicles single at nodes in axils of leafy bracts, on erect peduncles; panicles generally spicate but sometimes branched shortly in the lower part. Spikelets 7–9(10) mm long, in effect all unisexual, although in the female ones there is always an abortive male flower reduced to empty glumes; glumes hairy, pale reddish-brown or straw-coloured. Nutlets 1.2–1.5 mm long excluding the hypogynium, 0.8–1 mm broad, glabrous, obscurely trigonous, oblong-cylindric, shortly beaked, dark grey or dark brown, surface densely and minutely trabeculate, the trabeculae arranged in ± straight horizontal rows; hypogynium (when fresh) soft, white and spongy, 0.8–1 mm long, cylindrical, almost as broad as nutlet.

Zambia. N: Chinsali Dist., Shiwa Ngandu, 2.vi.1956, *Robinson* 1534 (K). C: Mkushi Dist., Fiwila, 4.i.1958, *Robinson* 2618 (K, MRSC).

Also in Angola and D.R. Congo. In permanently wet bogs, often with *Scleria procumbens*; 1300–1500 m.

Conservation notes: Widespread; not threatened.

Apart from the differences in the fruit, the leaves are broader than in *Scleria procumbens* and usually more hairy; Robinson regards the variation in indumentum as being of no significance. He also regarded the mature nutlets as very distinctive, "the horizontal arrangement of the fine trabeculae on the surface of the nutlet gives an effect reminiscent of a maize cob."

33. **Scleria laxiflora** R. Gross in Notizbl. Bot. Gart. Berlin-Dahlem **11**: 658 (1932). —Napper in J. E. Africa Nat. Hist. Soc. Coryndon Mus. **24**(109): 28 (1964). —Robinson in Kew Bull. **18**: 524 (1966). —Haines & Lye, Sedges & Rushes E. Afr.: 352, fig.729 (1983). —Verdcourt in F.T.E.A., Cyperaceae: 406 (2010). Type: Tanzania, Njombe Dist., Lupembe, Ruhudji R., iv.1931, *Schlieben* 782 (B holotype).

Perennial herb; stems weakly erect or semi prostrate, to 1 m tall and 1 mm wide, branched at several nodes, stem bases 3–4 mm wide, packed closely together to form a caespitose or ± straight row. Leaves 0.5–1.5(2) mm wide, glabrous or sparsely hairy. Inflorescences terminal and lateral, elongated so that apex may be 50 cm from lowest panicle-bearing node, lateral panicles usually single at nodes, on slender erect peduncles exserted up to 8 cm from leaf sheaths; panicles spicate or very shortly branched near base. Spikelets 6–8(9) mm long, in effect all unisexual although the females always contain an aborted male flower; glumes straw-coloured, reddish brown or pale reddish. Nutlets dark, oblong to ellipsoid, distinctly trigonous, 2 × 0.7–1 mm wide, minutely transversely rugulose, ± papillate, grey or dark brown with raised parts of surface lighter; hypogynium inconspicuous scarcely more than a dark ring, 3-lobed.

Zambia. N: Chinsali Dist., Shiwa Ngandu, 25.v.1962, *Robinson* 5243 (K); Kasama Dist., Mungwi, 3.iv.1962, *Robinson* 5075 (K, LISC). **Mozambique**. Z: Mocuba Dist., rocky hill, 32 km from Mocuba on road to Mugeba, 5.iii.1966, *Torre & Correia* 15039 (LISC, LMU, PRE, WAG).

Also in Tanzania and D.R. Congo (Katanga, Kundelungu Plateau). In perennially wet bogs, often with *Scleria bequaertii* and *S. procumbens*; 1300–1500 m.

Conservation notes: Widespread; not threatened.

Very similar to the last two species, but differs from *Scleria procumbens* in the very short hypogynium, and from *S. bequaertii* in lacking an elongated rhizome and in the minute trabeculae on the nutlet. Robinson (1966) states that it is "locally abundant in the Abercorn [Mbala], Kasama and Chinsali Districts".

The holotype at Berlin is probably still extant as Robinson cites it and he generally only cited specimens he had seen.

The following twelve species (33–44) are all annuals and can easily be confused. They form the section *Tessellatae* of Robinson (Kirkia **2**: 173–192, 1961). Careful examination of the nutlets and the hypogynium is needed for identification. Other useful characters mentioned by Robinson include the number of lateral panicles arising from each node, the nature of the lateral peduncles, and the relative length of the male spikelets and their pedicels.

34. **Scleria schimperiana** Boeckeler in Linnaea **38**: 466 (1874). —Clarke in F.T.A. **8**: 504 (1902). —Robinson in Kirkia **2**: 176 (1961). —Napper in J. E. Africa Nat. Hist. Soc. Coryndon Mus. **24**(109): 29 (1964). —Robinson in Kew Bull. **18**: 524 (1966). —Napper in F.W.T.A., ed.2 **3**: 343 (1973) as *schimperana*. —Haines & Lye, Sedges & Rushes E. Afr.: 344, fig.710 (1983). —Verdcourt in F.T.E.A., Cyperaceae: 399 (2010). Type: Ethiopia, Begemeder, Senka Berr, 1863-68, *Schimper* 1235 (B† holotype, BM, K, M, P).

    Scleria schimperiana var. *hypoxis* (Boeckeler) C.B. Clarke in F.T.A. **8**: 504 (1902).

Loosely tufted annual herb, (20)35–70(90) cm tall; stems 1–3 mm wide, glabrous or sparsely hairy above. Leaf sheaths glabrous or hairy; blades up to 40 cm long, 3–8 mm wide, glabrous or hairy on margins and ribs. Inflorescence of one terminal and 1–2 lateral panicles, always solitary from leaf sheaths, 2–5 × 1.5–2.5 cm, usually pendulous on slender hairy peduncles to 8 cm long. Male spikelets 4–5 mm long with dark reddish brown glumes; female spikelets (5)6–8 mm long; glumes pale or dark reddish brown with green midrib. Nutlets globose or depressed globose, 2.5–3 × 3–3.5 mm, smooth and glabrous, pale duck-egg blue or green when fresh but whitish when dried, without an apiculus; hypogynium with 3 short yellow-brown rounded lobes.

Botswana. N: Chobe Dist., near Nata–Kazungula road, 18°18'7"S 25°27'3"E, 14.iv.1983, *P.A. Smith* 4309 (K, SRGH). **Zambia**. W: Ndola Dist., Ndola, Itawa dambo, 3.iv.1960, *Robinson* 3458 (K, LISC). C: Kabwe Dist., Kabwe (Broken Hill), 30.iv.1961, *Robinson* 4629 (K). S: Choma Dist., Muckleneuk, 16 km N of Choma, 23.iii.1957, *Robinson* 2167 (K, NU, SRGH). **Zimbabwe**. N: Gokwe Dist., 15 km N of Gokwe on road to Chinyenyetu, 12.iii.1963, *Bingham* 513 (K, SRGH). W: Hwange Dist., Matetsi Safari Area, unit L, LK 775715, 6.iv.1978, *Gonde* 139 (K, SRGH).

Also in Nigeria, D.R. Congo, Ethiopia, Uganda and Tanzania. In seasonally wet dambos, perhaps confined to basic or neutral soils; 900–1400 m.

Conservation notes: Widespread; not threatened.

There is an unconfimed record from Zimbabwe C in Mapaura & Timberlake, Checkl. Zimbabwe Pl.: 90 (2004).

Robinson reckoned this species was always found in association with *Scleria foliosa*. The smooth nutlets with rounded apices and no apiculus are distinctive.

35. **Scleria foliosa** A. Rich., Tent. Fl. Abyss. **2**: 509 (1850). —Ridley in Trans. Linn. Soc. London, Bot. **2**: 170 (1884). —Clarke in F.T.A. **8**: 503 (1902). —Nelmes in Kew Bull. **11**: 102 (1956). —Robinson in Kirkia **2**: 177 (1961). —Napper in J. E. Africa Nat. Hist. Soc. Coryndon Mus. **24**(109): 29 (1964). —Robinson in Kew Bull. **18**: 525 (1966). —Napper in F.W.T.A., ed.2 **3**: 343 (1972). —Haines & Lye, Sedges & Rushes E. Afr.: 344, figs.711,712 (1983). —Hennessy in Bothalia **15**: 525 (1985). —Verdcourt in F.T.E.A., Cyperaceae: 400 (2010). Type: Ethiopia, Guendepta

(Gafta), 18.ix.1838, *Schimper* II 1232 (P lectotype, BM, BR, E, K, L), lectotypified by Lye (Fl. Ethiopia, 1997).

Scleria dumicola Ridl. in Trans. Linn. Soc. London, Bot. **2**: 169 (1884). Type: Angola, Pungo Andongo, between Quilanga and Pedras de Quinga, i.1857, *Welwitsch* 7122 (LISU holotype, BM).

Annual herb, loosely or densely tufted, 0.2–2 m tall. Stems 1–4 mm in diameter, glabrous, or scabrid on angles. Mouth of leaf sheath extended into a tongue (contraligule) with dark brown margin; leaf blades 6–40 cm long, 2–8 mm wide, glabrous but scabrid on margin and veins beneath. Inflorescence of a terminal and 1–3 lateral panicles, always single at nodes of upper leaves, with mostly stiffly erect peduncles which may become pendulous after maturity, to 2 cm long; panicles 1–6 × 1–2 cm; bracteoles within panicles rigid and erect, 1–4 cm long, giving a characteristic prickly look to fresh inflorescence but easily broken off when dry. Male spikelets 4–5 mm long, dark blackish red; female glumes 3–5 mm long, glabrous but midrib usually scabrid, green to blackish red. Nutlets ovoid, 3.5–4 × 2–2.5 mm, glabrous, lightly to moderately verrucose-lacunose but smooth at apex, white, grey or sometimes dark; hypogynium with 3 whitish to pale yellowish brown stiff rounded lobes.

Zambia. B: Kabompo Dist., 65 km W of Kabompo, 7.vii.1963, *Robinson* 5567 (K). N: Mbala Dist., Chilongwelo, Plain of Death, 5.v.1955, *Richards* 5530 (K). W: Mwinilunga Dist., Zambezi Rapids, c. 6 km N of Kalene Hill, 20.ii.1975, *Hooper & Townsend* 245 (K). E: Lundazi Dist., Rukuzi R., iii.1962, *Robinson* 5086 (K). S: Choma Dist., 8 km E of Choma, 27.iii.1955, *Robinson* 1197 (K). **Zimbabwe**. N: Darwin Dist., vlei of Nyamalengeni R. near Chirenga Mission, 12.iv.1972, *Loveridge* 1851 (K, LISC, SRGH). W: Hwange Dist., Kazuma Range, Forest Area boundary road, 9.v.1972, *Russell* 1847 (K, SRGH). C: Chegutu Dist., Poole Farm, 3.v.1945, *Hornby* 2413 (K, SRGH). S: Masvingo Dist., Makoholi Expt. Station, 13.iii.1978, *Senderayi* 219 (K, SRGH). **Malawi**. N: Karonga Dist., 5 km S of Chitipa (Fort Hill), 22.v.1962, *Robinson* 5219 (K). **Mozambique**. N: Nampula Dist., 21 km from Nampula towards Nametil, 1.iv.1964, *Torre & Paiva* 11565 (BR, LISC, LMU, MO, PRE, WAG). Z: Mocuba Dist., Namagoa, vi.1947, *Faulkner* 151 (K, PRE).

Widespread in tropical Africa from Senegal to South Africa, but absent from moist forest regions. In seasonal dambos and wet places at roadsides and beside dams; 900–1400 m in the Flora area (sea level to 2050 m elsewhere).

Conservation notes: Widespread; not threatened.

Also recorded from Botswana (Setshogo, Prelim. Checkl. Pl. Botswana: 121, 2005) and from Zimbabwe E (Mapaura & Timberlake, Checkl. Zimbabwe Pl.: 89, 2004), but not confirmed.

Often growing with the previous species. The nutlet is more ovoid than in *Scleria schimperiana* and distinctly apiculate, and its surface is weakly lacunose although smooth at the apex; the male spikelets are paler and not as strongly patterned.

36. **Scleria mikawana** Makino in Bot. Mag. (Tokyo) **27**: 57 (1913). —Nelmes in Kew Bull. **11**: 107 (1956). —Robinson in Kirkia **2**: 185 (1961); in Kew Bull. **18**: 525 (1966). —Napper in F.W.T.A., ed.2 **3**: 342 (1972). —Haines & Lye, Sedges & Rushes E. Afr.: 345, figs.713,714 (1983). —Verdcourt in F.T.E.A., Cyperaceae: 400 (2010). Types: Japan, Aichi Prefecture (Mikawa Prov.), Hosoya-mura and near Futakawa in Atsumi-gôri, 28.x.1894, *Makino* s.n. (MAK syntype); Aichi, Takashi, 28.x.1894, *Makino* s.n. (MAK syntype); Aichi, 29.ix.1896, *Nagura* s.n. (MAK syntype).

Scleria glabroreticulata De Wild., Pl. Bequaert. **4**: 230 (1927). Type: D.R. Congo, Wombali, x.1913, *Vanderyst* 2232 (BR holotype).

Annual herb, loosely to densely tufted, to 0.5–1.2(2) m tall; stems 2–3 mm in diameter. Leaves to 20 cm long, 3.7 mm wide, glabrous; lower leaf sheaths purplish without developed blades. Inflorescence of a terminal panicle and 2–3 laterals, single at nodes, 2–5 × 1–2 cm, on erect peduncles often little exserted from sheaths. Male spikelets (3)4–5(6) mm long, usually pale chestnut with dark reddish pedicels 4–14 mm long; glumes straw-coloured with green midrib, glabrous. Nutlets broadly ovoid to globose, 2.8–3.2 × 2.2–2.4 mm, dull, glabrous but appearing minutely hairy due to many minute brownish glands, with rather regular lacunae arranged in longitudinal lines, white or cream or grey to pale brown with 3 darker longitudinal stripes; hypogynium whitish with 3 well separated acutely to obtusely pointed lobes.

Zambia. B: Mongu Dist., Mongu, 5.iii.1966, *Robinson* 6862 (K). N: Kasama Dist., 103 km E of Kasama, Kalungu dambo, 3.iv.1961, *Robinson* 4583 (K). W: Mwinilunga Dist., Matonchi, Kalenda dambo, 16.iv.1960, *Robinson* 3582 (K).

Also in Senegal, Sierra Leone, Ivory Coast, Gabon, Uganda, D.R. Congo, Burundi and Angola; widespread in Asia extending N to Japan. In seasonally flooded areas in higher rainfall areas of Zambia, often in shallow standing water; 1250–1400 m.

Conservation notes: Widespread; not threatened.

The nutlet resembles a slightly irregularly patterned golf ball and has a dark apiculus.

37. **Scleria tessellata** Willd., Sp. Pl. **4**: 315 (1805). —Nelmes in Kew Bull. **11**: 108 (1956). —Robinson in Kirkia **2**: 178 (1961). —Napper in J. E. Africa Nat. Hist. Soc. Coryndon Mus. **24**(109): 29 (1964). —Napper in F.W.T.A., ed.2 **3**: 343 (1972). —Haines & Lye, Sedges & Rushes E. Afr.: 348, figs.720,721 as *tesselata* (1983). Type: India, ?König, *Herb. Willdenow* 17323 (B-W lectotype).

Annual herb, densely tufted, 0.15–1 m tall. Leaves plicate, 2–6 mm wide, glabrous or more rarely shortly hairy. Inflorescence of lateral panicles borne singly at nodes on short erect peduncles, not or scarcely exserted from leaf sheaths. Male spikelets 4–5 mm long, sessile or with pedicels 1–2 mm long, pale green or chestnut; female glumes pale green or ± scarious with a green midrib. Nutlets cylindric-ellipsoid to globose, 2.2–3.5 × 1.2–2.5 mm, deeply lacunose-tessellate or striate-tessellate, glabrous or hairy, grey or olive-grey with irregular often interrupted longitudinal lines of darker colour; hypogynium yellow, 3-lobed, lobes brown, squarish at apex.

a) Var. **tessellata**. —Robinson in Kew Bull. **18**: 526 (1966). —Verdcourt in F.T.E.A., Cyperaceae: 403 (2010).

   Scleria glandiformis Boeckeler in Linnaea **38**: 458 (1874). —Clarke in Durand & Schinz, Consp. Fl. Afr. **5**: 671 (1894); in F.T.A. **8**: 503 (1902). —Hutchinson in F.W.T.A. **2**: 493 (1936) in part. Type: Nigeria, Nupe, *Barter* 1042 (K holotype).

Nutlets cylindric-ellipsoid, 3.2–3.8 × 1.5–2 mm, shiny, glabrous, lightly striate-lacunose.

Zambia. C: Mumbwa Dist., 62 km W of Lusaka on Mumbwa road, 20.iii.1965, *Robinson* 6455 (K). **Mozambique**. N: Ribáuè Dist., from Ribáuè to Lalaua, 24.iv.1937, *Torre* 1435 (LISC).

Common in West Africa but rarer elsewhere in Africa, from Senegal to Nigeria, Sudan and Tanzania, also in Madagascar and India. In seasonally damp grassland; 600–1000 m.

Conservation notes: Widespread; not threatened.

C.B. Clarke excluded the Rumphius reference cited by Willdenow.

b) Var. **sphaerocarpa** E.A. Rob. in Kew Bull. **18**: 526 (1966). —Verdcourt in F.T.E.A., Cyperaceae: 403 (2010). Type: Zambia, 100 km E of Kasama, 17.iv.1962, *Robinson* 5080 (K holotype, B, EA, M, MPR, MTJB, NU, NY, PRE, SRGH).

   Scleria glandiformis sensu Hutchinson in F.W.T.A. **2**: 493 (1936) in part, non Boeckeler
   Scleria globonux sensu Nelmes in Kew Bull. **11**: 105 (1956) in part, non C.B. Clarke.
   Scleria ? *tessellata* × *globonux*, Robinson in Kirkia **2**: 181 (1961).
   Scleria sphaerocarpa (E.A. Rob.) Napper in Kew Bull. **25**: 441 (1971); in F.W.T.A., ed.2 **3**: 342 (1972). —Haines & Lye, Sedges & Rushes E. Afr.: 348 (1983).

Nutlets grey, globose, 2.7–3 × 2.3–2.5 mm, glabrous or shortly hairy, deeply lacunose-tessellate.

Zambia. B: Zambezi Dist., Zambezi (Balovale), 8.vii.1963, *Robinson* 5575 (K, PRE, SRGH). N: Kasama Dist., 105 km E of Kasama, 6.v.1962, *Robinson* 5157 (K, M, MTJB, PRE, SRGH). C: Mumbwa Dist., 62 km W of Lusaka on Mumbwa road, 20.iii.1965, *Robinson* 6456 (B, K, M, MPR, NY, SRGH).

Also in Senegal, Ivory Coast, Ghana, Cameroon, Sudan, Tanzania and Angola. In ± permanently wet dambos; 1000–1400 m.

Conservation notes: Widespread; not threatened.

Napper raised this taxon to specific status, but Verdcourt followed Robinson in treating it as a variety. The pits on the nutlet surface tend to be in longitudinal lines; they are much more prominent in var. *sphaerocarpa*. Var. *tessellata* has a slight apiculus to the nutlet; in var. *sphaerocarpa* it is less prominent and the whole nutlet is more rounded.

38. **Scleria globonux** C.B. Clarke in F.T.A. **8**: 504 (1902). —Nelmes in Kew Bull. **11**: 104 (1956). —Robinson in Kirkia **2**: 179 (1961). —Napper in J. E. Africa Nat. Hist. Soc. Coryndon Mus. **24**(109): 30, figs.1,5 (1964). —Robinson in Kew Bull. **18**: 527 (1966). —Napper in F.W.T.A., ed.2 **3**: 343 (1972). —Haines & Lye, Sedges & Rushes E. Afr.: 348, fig.719 (1983). —Verdcourt in F.T.E.A., Cyperaceae: 402 (2010). Type: Sudan, Bahr al Ghazal, 23.x.1869, *Schweinfurth* 2560 in part (K holotype, P), see Robinson (1961) for discussion.

Scleria glandiformis sensu Hutchinson in F.W.T.A. **2**: 493 (1936) in part, non Boeckeler.

Annual herb to 0.5–1 m tall. Leaves up to 40 cm long, 3–9 mm wide, flat or plicate, scabrid on margins and ribs. Inflorescence a terminal panicle and 2–3 lateral panicles occurring singly at nodes on long flexuous hairy peduncles. Male spikelets 3–5(6) mm long, dark reddish black, on pedicels of the same length or slightly shorter; female spikelets 7–8 mm long; glumes straw-coloured, with or without reddish streaks, glabrous or hispidulous on midrib. Nutlets globose, 2.8–3.2 × 2.5–2.7(3) mm, deeply lacunose-tessellate, ridges with pale to bright ferruginous short hairs, white or ferruginous; hypogynium white or brownish, deeply 3-lobed, lobes rounded at apex.

Zambia. N: Kasama Dist., 100 km E of Kasama, 17.v.1961, *Robinson* 4657 (K). W: Mwinilunga Dist., Matonchi, Kalenda dambo, 16.iv.1960, *Robinson* 3581 (K).

Widely distributed in West and Central Africa, Senegal, Guinea-Bissau, Sierra Leone, Liberia, Ghana, Nigeria, D.R. Congo, Sudan and Uganda. In dambos, sometimes in disturbed areas; 1200–1500 m.

Conservation notes: Widespread; not threatened.

Robinson (1966) states "Common in Zambia in areas of >1000 mm annual rainfall, and frequently recorded in the Northern and Western Provinces".

The pits on the nutlet are irregular, similar to those of *Scleria mikawana* but much deeper.

39. **Scleria bambariensis** Cherm. in Arch. Bot. Mém. **4**(7): 48 (1931). —Robinson in Kirkia **2**: 182 (1961); in Kew Bull. **18**: 527 (1966). —Haines & Lye, Sedges & Rushes E. Afr.: 346, fig.716 (1983). —Verdcourt, F.T.E.A., Cyperaceae: 401 (2010). Type: Central African Republic, Yanguya, 40 km SE of Bambari, 21.vii.1928, *Tisserant* 2693 (P lectotype); 7 km S of Ippy, x.1928, *Tisserant* 2694 (P syntype), lectotypified by Haines & Lye (1983).

Annual herb, densely tufted, (20)45–90(100) cm tall. Leaves few, 1.5–4(7) mm wide, hairy on both surfaces or sometimes glabrous. Inflorescence a terminal and 1–3 lateral panicles, usually single at nodes on slender pendulous peduncles, well exserted from sheaths. Male spikelets 3–4 mm long, pale greenish to dark reddish brown, on 1–4 mm long pedicels; female spikelets 4–6 mm long, glumes pale green to dark reddish or almost scarious with green midrib. Nutlets ovoid to subglobose or oblong-ellipsoid, 2–2.5(3.2) × 1.6–2.3 mm, with moderate to deep lacunae arranged in straight rows, white, grey or blackish; hypogynium yellowish to brownish, deeply 3-lobed, lobes rounded but sometimes with a whitish apical part which can be erose.

a) Var. **bambariensis**.

Zambia. B: Seen but not collected according to Robinson (1966). W: 30 km S of Ndola, 21.v.1960, *Robinson* 3719 (K). N: Kasama Dist., R bank of R. Lukulu, 40 km W of Kasama, 7.v.1961, *Robinson* 4635 (K). S: Namwala Dist., Kabulamwanda, 50 km N of Choma, 13.ii.1955, *Robinson* 1103 (K). **Zimbabwe**. W: Matobo Dist., Farm Besna Kobila, iv.1957, *Miller* 4329 (K, SRGH). **Malawi**. N: Chitipa Dist., 5 km S of Chitipa, 22.v.1962, *Robinson* 5221 (K). S: Mulanje Dist., 16 km N of Likabula, 15.vi.1962, *Robinson* 5360 (K). **Mozambique**. MS: Muanza Dist., Cheringoma Section, Lower Chinizuía sawmill, 13.vii.1972, *Ward* 7896 (K, NU).

Also in Central African Republic, Kenya, Tanzania, Madagascar and (probably) America. In seasonal swamps and seasonally wet grassland; 1000–1400 m.

Conservation notes: Widespread; not threatened.

b) Var. **B** of Robinson (1966). —Mapaura & Timberlake, Checkl. Zimbabwe Pl.: 89 (2004) as *var. B* sensu Haines & Lye. —Verdcourt, F.T.E.A. Cyperaceae: 402 (2010).

Larger in all its parts than typical *Scleria bambariensis*. Stems up to 1 m; leaves 2–6 mm wide. Male spikelets 3–4 mm long. Nutlets grey or nearly black, 3–3.2 × 2–2.3 mm, glabrous or hairy.

Zambia. W: Ndola, Itawa dambo, 18.v.1960, *Robinson* 3715 (K). N: Kasama Dist., 25 km W of Kasama, 7.v.1961, *Robinson* 4634 (K). E: Chipata Dist., 45 km W of Chipata (Fort Jameson), 23.vi.1962, *Robinson* 5406 (K). S: Choma Dist., Choma–Namwala road, 17 km N of Choma, 17.iv.1956, *Robinson* 1369 (K). **Zimbabwe**. Recorded from N, W, C in Mapaura & Timberlake (2004), but unconfirmed. **Malawi**. C: Kasungu Dist., 48 km N of Kasungu, 3.vi.1962, *Robinson* 5255 (K). S: Zomba Dist., W of Lake Chila, 26.vi.1962, *Robinson* 5401 (K). **Mozambique**. T: Ulónguè Dist., 30 km from Dedza along Zomba road, 4.vi.1962, *Robinson* 5258 (K).

Also in Guinea, Sierra Leone, Ivory Coast, Kenya, Tanzania, and from tropical Asia north to Japan and (probably) Tropical America. Habitat as for var. *bambariensis*; 1000–1400 m.

Conservation notes: Widespread; not threatened.

According to Robinson (1966) it is common and widespread throughout Zambia (except B), and fairly frequently recorded from Zimbabwe and Malawi.

The two varieties occur together and Robinson collected both from the same site on the same day (e.g. Kasama dambo, 25 km W of Kasama, 7.v.1961 – *Robinson* 4635 is var. *bambariensis* and *Robinson* 4634 is var. *B*).

Robinson reckoned *Scleria bambariensis* to be the same as the American species for which the name, when he wrote in 1966, was not settled, but now appears to be *S. reticularis* Michx. Haines & Lye used the name *S. bambariensis* and included America in the distribution. However, Flora of North America (**23**: 251, 2003) states that *S. reticularis* does not occur outside the United States. As mentioned in the introduction, this highlights the importance of studies of species of *Scleria* on a worldwide basis. In view of this uncertainty, neither Robinson nor Verdcourt saw fit to give this variety a name, and I feel this approach is best followed pending a full revision of the group.

40. **Scleria patula** E.A. Rob. in Kirkia **2**: 186 (1961); in Kew Bull. **18**: 528, fig.11 (1966). —Haines & Lye, Sedges & Rushes E. Afr.: 346 (1983). Type: Zambia, Chingola Dist., 11 km N of Chingola, 4.v.1960, *Robinson* 3696 (K holotype, B, EA, GC, M, MPR, MRSC, MTJB, NU, NY, PRE, SRGH, YBI).

Annual herb; stems 3–25 cm long, spreading. Leaves 1–2 mm wide, glabrous. Lateral panicles single at nodes, on erect peduncles little exserted from sheaths. Male spikelets c. 2 mm long, straw-coloured or pale reddish, on pedicels 1.5–3 mm long; female glumes glabrous. Nutlets 1.8–2 × 1.2–1.4 mm broad, ovoid or subglobose, lacunose-tessellate, hairy or glabrous, grey with darker longitudinal stripes; hypogynium 3-lobed, pale greenish or yellowish; lobes rounded at apex, often edged with a ± scarious margin which may be 1- to 3-apiculate.

Zambia. W: Mwinilunga Dist., Matonchi, Kalenda dambo, 4.v.1960, *Robinson* 3684 (EA, K, LISC, M, MTJB, SRGH).
Not known elsewhere. In seasonally damp places; 1350–1450 m.
Conservation notes: Local distribution, only known from these two records; possibly Near Threatened.
Of small stature. The nutlets are apiculate, with clearly square or rectangular cavities.

41. **Scleria chlorocalyx** E.A. Rob. in Kirkia **2**: 184 (1961); in Kew Bull. **18**: 530, fig.12 (1966). —Haines & Lye, Sedges & Rushes E. Afr.: 347 (1983). Type: Mporokoso Dist., Chiwala, 25.v.1961, *Robinson* 4681 (GC, K, M, MTJB, NU, PRE, SRGH).

Annual herb; stems 12–60 cm long. Leaves 2–3 mm wide, hairy. Lateral panicles single at nodes, on pendulous peduncles exserted up to c. 2.5 cm from sheaths. Male spikelets 2.5–3 mm long, pale chestnut-brown, on pedicels 3–6 mm long, becoming spreading or replexed at maturity; female glumes glabrous. Nutlets 2.5–2.7 × 1.5–1.7 mm, ellipsoid to ovoid-ellipsoid, glabous, lightly lacunose-tessellate, grey with darker longitudinal stripes; hypogynium yellowish to greenish, upper margin fringed with ± scarious extension obscuring division between the 3 lobes, so lower $^1/_4$–$^1/_3$ of nutlet is held as in a cup.

Zambia. N: Mporokoso Dist., Chiwala, 25.v.1961, *Robinson* 4681 (GC, K, M, MTJB, PRE, SRGH). W: Mwinilunga Dist., Matonchi, Kalenda Plain, 16.iv.1960, *Robinson* 3603 (K holotype, GC, M, MTJB, PRE).
Not known from elsewhere. On shallow soils over sandstone or laterite which are wet for at least 8 months in the year (Robinson 1966); 1300–1400 m.
Conservation notes: Known from only four specimens. Endemic to N Zambia but habitat not endangered; not threatened.

42. **Scleria lucentinigricans** E.A. Rob. in Kew Bull. **18**: 530, fig.13 (1966). —Haines & Lye, Sedges & Rushes E. Afr.: 347, fig.716b (1983). Type: Zambia, Kasama Dist., 25 km W of Kasama, 7.v.1961, *Robinson* 4644 (K holotype, B, EA, GC, M, MRSC, MTJB, NU, PRE, SRGH). FIGURE 14.**73**.

Annual herb, erect, almost entirely glabrous; stems to 85 cm long, glabrous. Leaves 1.5–3(5) mm wide, subrigid, plicate so there are 2 marked nerves on upper surface, glabrous or with sparse long hairs; sheaths glabrous, mouth forming a shortly convex splitting and usually glabrous tongue. Inflorescence interrupted so lowest panicle-bearing node is up to 53 cm from apical panicle; lateral panicles single at nodes 1–3, pendulous, exserted 5(9) cm from sheaths on slender glabrous peduncles. Male spikelets 3–3.5(4) mm long, pedicels 6(9) mm long, pale or black; female glumes 4–5 mm long, ovate, acuminate, glabrous, straw-yellow, distinctly green – or brown-keeled. Nutlet ellipsoid or ovoid, 3.3–3.6 × 1.9–2.1 mm, glabrous, shiny, black, surface lightly and indistinctly lacunose; hypogynium obtusely and obscurely 3-lobed, joined to nutlet and forming a single terete ellipsoid shape.

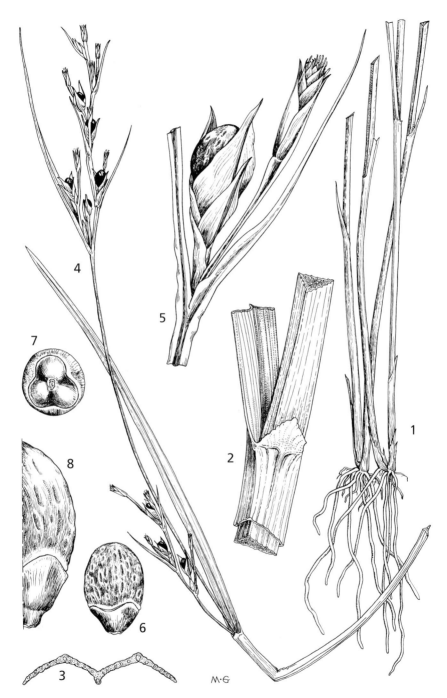

Fig. 14.**73**. SCLERIA LUCENTINIGRICANS. 1, base of plant (× 1); 2, junction of lamina and sheath (× 6); 3, transverse section of leaf (× 8); 4, inflorescence (× 1); 5, glomerule (× 6); 6, nutlet, lateral view (× 6); 7, nutlet, basal view (× 6); 8, nutlet, enlarged portion showing surface pattern (× 12). All from *Robinson* 5148. Drawn by Mary Grierson. Reproduced from Kew Bulletin (1966).

Zambia. N: Isoka Dist., Kalungu dambo, 100 km E of Kasama, 6.v.1962, *Robinson* 5148 (EA, K, M, MTJB, NU, SRGH); same locality, 17.v.1961, *Robinson* 4660 (K, NU, PRE).

Not known elsewhere. In seasonally wet dambos; c. 1400 m.

Conservation notes: Known only from the broad Kasama area in NE Zambia; possibly Near Threatened.

Robinson 4660 and 5148 show contrasting states in successive seasons, the first with a wet rainy season and the second with a relatively dry one.

43. **Scleria parvula** Steud., Syn. Pl. Glumac. **2**: 174 (1855). —Nelmes in Kew Bull. **11**: 105 (1956) in part. —Robinson in Kirkia **2**: 190 (1961); in Kew Bull. **18**: 532 (1966). —Napper in Kew Bull. **25**: 442 (1971); in F.W.T.A., ed.2 **3**: 343 (1972). —Haines & Lye, Sedges & Rushes E. Afr.: 347, figs.717,718 (1983). —Verdcourt in F.T.E.A., Cyperaceae: 402 (2010). Type: India, Nilgherri Hills (Nilagiri), n.d., *Hohenacker* 1295 (B† holotype, BM, K, M, P, UPS).

Annual herb, 30–60 cm tall, with simple stem or occasionally branched near base. Leaves 2–5 mm wide, sparsely hairy. Inflorescence with 2–3(5) lateral panicles (only single in small immature specimens) on pendulous peduncles. Male spikelets 3–4(5) mm long, pale or dark brown, on pedicels 1–3 mm long; female glumes glabrous. Nutlets ovoid to subglobose, 2–2.3× 1.5–1.7 mm, glabrous or minutely papillose (sometimes hairy), distinctly tessellate-lacunose, grey or grey-brown; hypogynium 3-lobed, lobes ± rounded but shape obscured by scarious margins.

Zambia. N: Mbala Dist., 120 km ESE of Mbala (Abercorn), on Nakonde road, 2.vi.1962, *Robinson* 5253 (K, M, MTJB, SRGH).

Also known from Sierra Leone and Guinea Republic. In wet places; c. 1400 m.

Conservation notes: According to Robinson (1966) apparently only known from 3 records in Africa (N Zambia, Sierra Leone, Guinea); possibly Near Threatened, or best considered Data Deficient.

Verdcourt (2010) follows Haines & Lye in accepting records of *Scleria parvula* from Kenya and Tanzania and discusses the apparent confusion in the naming of some East African specimens.

The nutlets are apiculate, with rectangular pits arranged in vertical lines. The margins of the hypogynium are irregularly toothed.

44. **Scleria clathrata** A. Rich., Tent. Fl. Abyss. **2**: 510 (1850). —Clarke in F.T.A. **8**: 502 (1902). —Nelmes in Kew Bull. **11**: 104 (1956). —Robinson in Kirkia **2**: 189 (1961); in Kew Bull. **18**: 534 (1966). —Haines & Lye, Sedges & Rushes E. Afr.: 349, fig.722 (1983). —Verdcourt in F.T.E.A., Cyperaceae: 404 (2010). Type: Ethiopia, Walcha, 10.viii.1841, *Schimper* 1603 (P holotype, BM, M).

Slender annual herb, 5–60 cm tall, glabrous or minutely scabrid or shortly hairy above. Leaves 3–40 cm long, 2–6 mm wide, flat, scabrid to shortly hairy on margins and main ribs near apex. Inflorescence with a terminal panicle and several lateral panicles, usually 2–3 per node on very unequal slender pendulous peduncles. Male spikelets reddish brown, 4–6 mm long; female spikelets 8–10 mm long; glumes glabrous, pale to red brown with prominent green keel. Nutlets ovoid- or oblong-ellipsoid, 2–3 (3.4–3.8 according to Robinson) × 1.6–1.8 mm, lightly striate-lacunose, variable in colour, light grey, pinkish grey, light brown or red to almost black; hypogynium yellow brown, 3-angled with 3 very indistinct lobes, base drying to form a cylindrical stipe.

Zambia. N: Mbala Dist., Kawimbe, 21.v.1962, *Robinson* 5211 (K, LISC). W: Solwezi Dist., Solwezi, 9.iv.1960, *Robinson* 3481 (K). **Mozambique**. T: Ulóngùe Dist., 30 km from Dedza along Zomba road, 4.vi.1962, *Robinson* 5257 (K).

Also known from Ethiopia, Kenya and Tanzania. In perennially wet dambos; 1200–1400 m.

Conservation notes: Widespread; not threatened.

Robinson comments that this is an inconspicuous species that may well be more common than present records suggest.

45. **Scleria gracillima** Boeckeler in Flora **62**: 570 (1879). —Clarke in F.T.A. **8**: 505 (1902). —Nelmes in Kew Bull. **11**: 10 (1956). —Robinson in Kirkia **2**: 188 (1961). —Napper in J. E. Africa Nat. Hist. Soc. Coryndon Mus. **24**(109): 29 (1964). —Robinson in Kew Bull. **18**: 534 (1966). —Napper in F.W.T.A., ed.2 **3**: 343 (1972). —Haines & Lye, Sedges & Rushes E. Afr.: 349, fig.723 (1983). —Verdcourt in F.T.E.A., Cyperaceae: 404 (2010). Type: Sudan, Jur Ghattas, 20.x.1869, *Schweinfurth* III 189 (B?† holotype, K).

Slender annual herb, 25–55(70) cm tall, entirely glabrous. Leaves 1–2 mm wide. Inflorescence with lateral panicles, single or paired at each node, few-flowered; peduncles filiform, pendulous. Male spikelets 3–4 mm long, straw-coloured, pedicels to 4 mm long; female glumes tinged with red. Nutlets oblong-cylindric, (2.2)2.9–3.3 × 1.7–1.8 mm, smooth and shiny, whitish or grey with darker grey or blackish longitudinal stripes; hypogynium pale greenish yellow, with or without a dark reddish brown margin, not or only faintly lobed.

Zambia. N: Kasama Dist., 100 km E of Kasama, 17.iv.1962, *Robinson* 5078 (K). W: Kapiri Mposhi Dist., 50 km N of Kabwe (Broken Hill), 29.iv.1961, *Robinson* 4626 (K). S: Choma Dist., Simansunda, Mapanza, 13.iv.1958, *Robinson* 2840 (K). **Malawi**. N: Karonga Dist., 5 km S of Chitipa (Fort Hill), 22.v.1962, *Robinson* 5220 (K). S: Mulanje Dist., 16 km SW of Likabula Forestry Depot, 15.vi.1962, *Robinson* 5359 (K).

Also in Senegal, Ghana, Ivory Coast, S Nigeria, Central African Republic, Sudan and Tanzania. In seasonal dambos; 700–1400 m.

Conservation notes: Widespread; not threatened.

46. **Scleria achtenii** De Wild. in Rev. Zool. Bot. Africaines, suppl. Bot. **14**: 16, fig.1 (1926); Pl. Bequaert. **4**: 219, fig.1 (1927). —Napper in J. E. Africa Nat. Hist. Soc. Coryndon Mus. **24**(109): 28 (1964). —Robinson in Kew Bull. **18**: 534 (1966). —Napper in F.W.T.A., ed.2 **3**: 342 (1972). —Haines & Lye, Sedges & Rushes E. Afr.: 350, figs.725,726 (1983). —Hennessy in Bothalia **15**: 526 (1985). —Gordon-Gray in Strelitzia **2**: 181, fig.82 (1995). —Verdcourt in F.T.E.A., Cyperaceae: 405 (2010). Type: D.R. Congo, Kinshasa (Leopoldville), viii.1915, *Achten* 97 (BR holotype).

Scleria substriato-alveolata De Wild. in Rev. Zool. Bot. Africaines, suppl. Bot. **4**: 23, 33, fig.6 (1926); Pl. Bequaert. **4**: 240, fig.8 (1927). Types: D.R. Congo, Wombali, vi.1913, *Vanderyst* 1060, 1890 (BR syntypes); Kimpako, ii.1909, *Vanderyst* s.n. (BR syntype).

Scleria subintegriloba De Wild., Pl. Bequaert. **4**: 238 (1927). Type: D.R. Congo, Katchaka, xii.1913, *Vanderyst* 2839 (BR holotype).

Scleria nyasensis sensu Nelmes in Kew Bull. **11**: 86 (1956) in part, non C.B. Clarke.

Perennial herb, 0.4–1.3(2) m tall. Rhizome usually straight, 3–6 mm wide, swollen bases of hairy stems arising up to 1 cm apart. Leaves 30–60 cm long, 2.5–5 mm wide, glabrous above, hairy on 5 principal veins beneath. Inflorescence reddish, elongate, 20–85 cm long, terminal panicle to 7.5 cm, lateral panicles single at 2–3(4) upper nodes, 2.5–4(5) cm long, on pendulous hairy peduncles exserted up to 18 cm from sheaths. Male spikelets 7–9 mm long, sessile on short pedicels, glumes straw-coloured or reddish; female spikelets 5–7 mm long, glabrous, with straw-coloured or reddish glumes. Nutlets obovoid to subglobose, 2.6–2.9 × 1.9–2.1 mm, lightly irregularly pitted or transversely ridged, hairy on ridges, sometimes smooth and hairless near apex, with microscopic papillae, grey, brownish-grey or violet-grey; hypogynium greyish white, with 3 lobes each terminating in a semi-scarious ligulate extension to 1 mm long appressed to nutlet base.

Zambia. N: Kasama Dist., Malole, 15.v.1961, *Robinson* 4656 (K). W: Mwinilunga Dist., Chingabola dambo near Matonchi, 16.ii.1975, *Hooper & Townsend* 94 (K). **Mozambique**. MS: Muanza Dist., Cheringoma Section, Lower Chinizíua R., 1 km from Gano, 13.vii.1972, *Ward* 7879 (K, NU).

Also in Congo, Uganda, Kenya(?), Tanzania and South Africa (KwaZulu-Natal). Perennially damp but not waterlogged ground; 40–1350 m.

Conservation notes: Widespread; not threatened.

According to Robinson, this species is frequent in the Mbala, Kasama, Mporokoso, Mansa and Mwinilunga districts of Zambia.

Scleria subintegrifolia De Wild., listed as a synonym in F.T.E.A., is a misprint for *S. subintegriloba*.

47. **Scleria nyasensis** C.B. Clarke in F.T.A. **8**: 504 (1902). —Nelmes in Kew Bull. **11**: 86 (1956), in part. —Napper in J. E. Africa Nat. Hist. Soc. Coryndon Mus. **24**(109): 28 (1964). —Robinson in Kew Bull. **18**: 535 (1966). —Haines & Lye, Sedges & Rushes E. Afr.: 351, figs.727,728 (1983). —Verdcourt in F.T.E.A., Cyperaceae: 406 (2010). Types: Malawi, Zomba Mt, xii.1896, *Whyte* s.n. (K syntype); Zomba Dist., Mt Malosa, xii.1896, *Whyte* s.n. (K syntype); Blantyre, near Mt Sochi, *Kirk* s.n. (K syntype).

Erect perennial herb, (0.45)0.8–1.5(2) m tall. Rhizome short, reddish brown, crowded stems with swollen bases to 5 mm wide, usually joined to form an irregular woody mass or clump up to 1 m across. Leaves 20–30 cm long, 2–5(7) mm wide, glabrous or hairy but sharply scabrid on margin and veins. Inflorescence 25–50(100) cm long overall, lateral panicles rarely single, usually 3–5(6) at 2–5 nodes, on slender pendulous glabrous or hairy peduncles exserted to 20(30) cm from sheaths. Male spikelets 5–7(8) mm long with pedicels to 11 mm long, usually shorter than spikelets; female spikelets 7–9 mm long, glumes straw-coloured to pale brown, sometimes with dark purple patches and green keels. Nutlets ovoid-ellipsoid, obovoid or subglobose, 2.5–3.6 × 1.8–2.2 mm wide, distinctly regularly finely pitted with longitudinal rows of pits and ridges with white hairs, white or greyish to brown or reddish brown; hypogynium greyish white with 3 broadly triangular white or yellowish lobes usually extended into semiscarious, sometimes bicuspidate, tips adpressed to nutlet base.

Zambia. N: Kasama Dist., Mungwi, 21.vi.1960, *Robinson* 3765 (K). W: Mwinilunga Dist., Samahina, NE of Ikelenge, 17.iv.1969, *Robinson* 6613 (K). C: Mkushi Dist., Fiwila, Chinshinshi dambo, 9.i.1958, *Robinson* 3707 (K). **Zimbabwe**. E: Nyanga Dist., 10 km N of Troutbeck, Gairesi Ranch, 12.xi.1956, *Robinson* 1865 (K, LISC, NU, SRGH). **Malawi**. N: Nyika Plateau, Lake Kaulime, 4.i.1959, *Robinson* 3031 (K). S: Mt Mulanje, Lichenya Plateau, 6.vi.1962, *Robinson* 5280 (K, LISC). **Mozambique**. MS: Probably present as occurs on border in Zimbabwe E.

Also in Burundi, D.R. Congo, Uganda and Tanzania. In perennially wet places, usually in shade, often at edges of mushitus; 1200–2200 m.

Conservation notes: Widespread; not threatened.

There is an unconfirmed record for Zimbabwe C in Mapura & Timberlake, Checkl. Zimbabwe Pl.: 89 (2004).

Similar to the previous species, but distinguished by the shorter rhizomes and more strongly patterned nutlets.

48. **Scleria unguiculata** E.A. Rob. in Kew Bull. **18**: 536, fig.14 (1966). —Haines & Lye, Sedges & Rushes E. Afr.: 352, figs.730,731 (1983). —Hennessy in Bothalia **15**: 525 (1985). —Verdcourt in F.T.E.A., Cyperaceae: 407 (2010). Type: Zambia, Luwingu Dist., Luena Mission, 28.iii.1962, *Robinson* 5056 (K holotype, M, MTJB, PRE, SRGH).

Tufted perennial herb; stems erect, to 1.3 m tall and 1–2 mm wide, thickened bases 3–4(5) mm wide. Rhizome reduced to connections between these to form a subwoody mass. Leaves 1.5–4 mm wide, glabrous or sparsely hairy. Inflorescence to 70 mm long with (1)2–4(5) panicles at each node, each 1.5–3 cm long on a pendulous peduncle to 26 cm long. Male spikelets 4–5 mm long, sometimes to 10 mm long but usually shorter than spikelets; female glumes 3.5–5 mm long, acuminate, glabrous, straw-coloured or brown with green keel. Nutlets ovoid to globose, 2–2.8 × 1.7–2 mm, striate-lacunose with pits in longitudinal lines, shortly beaked, with white or yellowish hairs, grey or pale brown with darker longitudinal lines; hypogynium brownish with 3 triangular lobes.

Botswana. N: Okavango Swamps, near Dindinga Is. on banks of Mboroga R. headwaters, 19°06'8"S 23°05'2"E, 18.vi.1979, *P.A. Smith* 2790 (K, PRE, SRGH). **Zambia**. N: Mporokoso Dist., Chiwala, 53 km W of Mporokoso, 21.v.1961, *Robinson* 4683 (EA, GC, K, MRSC, MTJB, SRGH). W: Mwinilunga Dist., Dobeka Bridge near Matonchi, 17.ii.1975, *Hooper & Townsend* 146 (K). C: Mpika Dist., Mutinondo Wilderness Area, Musamfushi dambo, 1420 m, 12°27'41"S 31°17'09"E, *Bauters et al.* 2015-011 (GENT, UZL).

Also in Togo, Central African Republic and Tanzania. In perennially wet bogs; 950–1400 m.

Conservation notes: Widespread; not threatened.

Forma **B**. Robinson in Kew Bull. **18**: 536 (1966).

Robinson (1966) notes: "from typical examples of that species (*S. unguiculata*) they differ in having slightly narrower leaves (1–3 mm wide), a somewhat stricter inflorescence, nutlets which are ovoid-ellipsoid or cylindric-ellipsoid and covered with fulvous hairs, and a smaller hypogynium with 3 short triangular lobes".

Zambia. B: Kalabo Dist., Kalabo, 14.x.1962, *Robinson* 5486 (B, K, LISC, MTJB, SRGH). N: Kasama Dist., 25 km W of Kasama, Chilibula, 11.v.1961, *Robinson* 4563 (K, M, MPR, MTJB, PRE, SRGH). W: Chingola Dist., 22 km W of Chingola, 9.vii.1963, *Robinson* 5588 (K). C: Mkushi Dist., Fiwila, 4.i.1958, *Robinson* 2621 (K, SRGH).

Conservation notes: Seemingly widespread, although endemic to N Zambia; not threatened.

Robinson even gave this taxon a manuscript name (*Scleria densicaespitosa*, ms.) which appears on several sheets at K, but he later changed his mind and treated it as above.

49. **Scleria lagoensis** Boeckeler in Vidensk. Meddel. Dansk Naturhist. Foren. Kjøbenhavn **1869**: 151 (1869). —Robinson in Kew Bull. **18**: 538 (1966). —Napper in F.W.T.A., ed.2 **3**: 342 (1972). —Hennessy in Bothalia **15**: 526 (1985). —Verdcourt in F.T.E.A., Cyperaceae: 407 (2010). Type: Brazil, Lagoa Santa, *Warming* s.n. (C lectotype), see extensive notes on typification in Robinson (1966: 539).

    Scleria canaliculatotriquetra Boeckeler in Flora **62**: 573 (1879). —Clarke in F.T.A. **8**: 505 (1902). —Hutchinson in F.W.T.A. **2**: 493 (1936). —Nelmes in Kew Bull. **11**: 84 (1956). Type: Sudan, Jur Ghattas, 12.x.1869, *Schweinfurth* 2474 (B† holotype, K).

    Scleria cervina Ridl. in Trans. Linn. Soc. London, Bot. **2**: 171 (1884). Type: Angola, between Mutollo and Candumba, iii.1857, *Welwitsch* 7127 (LISU holotype, BM).

    Scleria vanderystii De Wild. in Rev. Zool. Bot. Africaines, suppl. Bot. **14**(2): 25 (1916). Types: D.R. Congo, Wombali, ii.1914, *Vanderyst* 3471 (BR syntype) and 5 other syntypes.

    Scleria canaliculatotriquetra var. *clarkeana* Piérart in Lejeunia Mém. **13**: 49, t.2, figs.20,21 (1953). Types: D.R. Congo, 10 syntypes from widely separated localities (BR syntypes).

    Scleria lagoensis subsp. *canaliculatotriquetra* (Boeckeler) Lye in Nordic J. Bot. **3**: 243 (1983). —Haines & Lye, Sedges & Rushes E. Afr.: 353, fig.732 (1983).

Perennial herb; stems 50–150(200) cm tall, 2–3 mm wide, scabrid. Stem-bases swollen, to 5 mm wide, forming a ± shapeless knotty mass or sometimes extended into a ± straight row; true rhizome lacking, roots becoming cylindrical and tuberous 3–10 cm from stem base. Leaves 20–50 cm long, 5–12 mm wide, usually sharply scabrid on margins and ribs but otherwise glabrous.

Inflorescence of 1 terminal and 3–6 lateral panicles, sometimes single but mostly 2–3 per node, 3–8(12) cm long, on erect or pendulous glabrous or scabrid peduncles to 6 cm long. Male spikelets 5–6 mm long with straw-coloured glumes, sessile or pedicels very short; female spikelets 6–7 mm long; glumes straw-coloured or green, often strongly speckled with deep reddish brown, with green midrib. Nutlets narrowly ovoid to ovoid-subglobose, 2.7–4.2 × 2–2.5 mm wide, smooth or faintly striate-lacunose, hairy, more so towards base, ± glabrous above, green turning white or grey to brown; hypogynium yellowish brown with narrowly acuminate lobes, sometimes almost or completely absent.

Zambia. N: Kasama Dist., Chibutubutu/Lukulu R., 50 km S of Kasama on Mpika road, 23.ii.1960, *Richards* 12554 (K). W: Masaiti Dist., 60 km S of Ndola, 2.iii.1960, *Robinson* 3368 (K). C: Lusaka Dist., 8 km E of Lusaka, Munali School, Little Dam, 11.iv.1956, *Robinson* 1457 (K). E: Chipata Dist., Ngeni area, ii.1962, *Robinson* 5002 (K). S: Mapanza Dist., Muckle Neuk, 18 km N of Choma, 14.i.1954, *Robinson* 468 (K). **Zimbabwe**. N: Guruve Dist., Nyamunyeche Estate, 5.iii.1979, *Nyariri* 741 (K, SRGH). E: Mutare Dist., Nyamkwarara (Nyamquarara) Valley, ii.1935, *Gilliland* K1497 (K). **Malawi**. N: Nkata Bay Dist., Kandoli, 20.ii.1961, *Vesey-Fitzgerald* 2995 (K). S: Blantyre Dist., near Soche Hill, 'Manganja's Country', 8.iii.1862, *Kirk* s.n. (K). **Mozambique**. N: Muecate Dist., Imala, 36 km from Imala to Mecubúri, 16.i.1964, *Torre & Paiva* 10016 (BR, LISC, LMU, MO, PRE, WAG).

Senegal to Nigeria, Cameroon, D.R. Congo, Sudan, Ethiopia, Angola, Uganda, Kenya, Tanzania and Swaziland, also in Madagascar, Comoro Is., Brazil, Columbia and Venezuela. In seasonally or (more usually) permanently wet swamps or grassland; 700–1350 m (elsewhere near sea level).

Conservation notes: Widespread; not threatened.

The hairs on the nutlet can be difficult to see on a white nutlet as they are themselves white.

50. **Scleria adpressohirta** (Kük.) E.A. Rob. in Kirkia **3**: 10 (1962). —Napper in J. E. Africa Nat. Hist. Soc. Coryndon Mus. **24**(109): 29 (1964). —Robinson in Kew Bull. **18**: 540 (1966). —Haines & Lye, Sedges & Rushes E. Afr.: 353, fig.733 (1983). —Verdcourt in F.T.E.A., Cyperaceae: 408 (2010). Type: Tanzania, Kigoma Dist., Ujiji, between Kandega and Lake Tanganyika, 21.iii.1926, *Peter* 38957a (B† holotype, EA, K frag.).

    Scleria canaliculatotriquetra Boeckeler var. *adpressohirta* Kük. in Repert. Spec. Nov. Regni Veg. **40**(1), Anh.: 142 (1938).

Perennial herb; stems 50–90 cm tall. Rootstock a knotty mass of swollen stem-bases, roots becoming cylindrical and tuberous 3–8 cm from stem bases. Leaves 4–7 mm wide, hairy, appearing pale green when dried. Inflorescence a terminal panicle 2–8 cm long, 1–2 cm wide, and 1–2 lateral panicles at each of 1–3 nodes; peduncles exserted 0.5–10 cm from sheaths, shortly hairy. Male spikelets 4–5.5 mm long, straw-coloured to chestnut, hispidulous, sessile or pedicels 1–3 mm long; female glumes 4.5–5 mm long, shortly hairy near apex and on keel, otherwise glabrous, straw-coloured with chestnut markings or entirely chestnut with green keel. Nutlets broadly ovoid-globose, globose or depressed globose, 2.5–2.8 × 2.4–2.8 mm, smooth, shortly hairy below, glabrous above, grey, yellowish- or brownish-grey; hypogynium with 3 shortly acuminate lobes, yellowish, light brown or chestnut.

Zambia. N: Kasama Dist., 10 km E of Kasama, 6.iv.1961, *Robinson* 4587 (K). W: Mwinilunga Dist., Samahina, NE of Ikelenge, 17.iv.1965, *Robinson* 6612 (K).

Otherwise known only from the type in Tanzania. Perennially damp ground on fringes of bogs; 1250–1400 m.

Conservation notes: Moderately widespread although known from few specimens; probably not threatened.

51. **Scleria xerophila** E.A. Rob. in Kew Bull. **18**: 540, fig.15 (1966). —Haines & Lye, Sedges & Rushes E. Afr.: 354 (1983). Type: Zambia, 30 km W of Mwinilunga, 16.iv.1960, *Robinson* 3650 (K holotype, M, MTJB, NU, SRGH).

Perennial herb; stems erect, to 70 cm long and 1 mm thick, glabrous, triquetrous, striate. Rhizome c. 3 mm thick, woody, scarcely visible between thickened clump-forming stem-bases. Leaves 2–3 mm wide, sparsely hairy on upper surface on midrib, glabrous beneath. Inflorescence elongate, lower panicle-bearing node up to 60 cm from terminal one; panicles compact, terminal to 3 mm long, laterals to 2 mm, pendulous, single or two together at 1–2 nodes, on slender mostly glabrous peduncles (very shortly pilose below each panicle), exserted 3–20 cm from sheaths; bracteoles to 4 mm long from a pilose red-brown glume-like base. Male spikelets 4–5 mm long, on 1–8 mm long peduncles; female glumes glabrous, broadly ovate or somewhat rounded, with a 1 mm long excurrent arista, straw-coloured at base, elsewhere red brown except for green keel, sometimes red-brown throughout. Nutlet globose or quadrate-globose, 3–3.4 × 2.8–3 mm, very lightly tessellate-striate, shortly and sparsely pilose, pale grey or pale brown. Hypogynium c. 0.6 mm broad, dark brown (when dried before maturity brown or straw-coloured), like shiny polished wood, lobes very short (c. 0.5 mm), narrowly apiculate.

Zambia. W: Mwinilunga Dist., 30 km W of Mwinilunga, 16.iv.1960, *Robinson* 3650 (K, M, MTJB, NU, SRGH).

Not known elsewhere. Sandy well-drained grassland with scattered trees; c. 1400 m.

Conservation notes: Known only from the type specimen; possibly Vulnerable.

The species grows in much drier situations than most other species of the genus and may well have been overlooked.

52. **Scleria arcuata** E.A. Rob. in Kew Bull. **18**: 541, fig.16 (1966). —Haines & Lye, Sedges & Rushes E. Afr.: 354 (1983). Type: Zambia, Kasama Dist., 80 km S of Kasama, 1 km N of Old Chambeshi Pontoon, 29.iv.1962, *Robinson* 5136 (K holotype, B, EA, M, MRSC, MTJB, NU, PRE, SRGH).

Perennial herb, at first erect, later spreading; stems to 40 cm long, rigid, curved so inflorescence sometimes pointing downwards, glabrous or pilose, striate. Rhizome creeping, firm, almost straight, ± completely covered by red-streaked scales, white inside, 4–6 mm thick, fleshy and acrid-tasting when fresh, becoming woody and somewhat constricted when dry. Leaves 1–3.5 mm wide, equal or longer than culms, stiff, markedly keeled, sparsely pilose. Inflorescence interrupted; lateral panicles usually solitary (occasionally another at a second node), on erect flattened glabrous or pilose peduncles, not or very shortly exserted from sheaths (occasionally to 4 cm). Male spikelets 4.5–6 mm long, sessile or shortly pedicellate; female glumes 5–6 mm long, ovate, acuminate, glabrous, straw-coloured, green-keeled, sometimes red-tinged or red-spotted. Nutlet 3.3–3.6 × 1.8–2.4 mm, ovoid, broadly ovoid or subglobose, very lightly tessellate-lacunose, sparsely hairy, grey, dark grey or dark brown, colour varying so that upper parts are paler; hypogynium 3-lobed, lobes shortly to long-acuminate, rarely bicuspid at apex, pale, straw-coloured, brown, or blackish brown.

Zambia. N: Kasama Dist., banks of Chambeshi R., 16.iv.1961, *Robinson* 4607 (EA, GC, K, M, MTJB, NU, SRGH). W: Mwinilunga Dist., 27 km W of Mwinilunga, 17.iv.1960, *Robinson* 3658 (K, M, MTJB, SRGH).

Not known elsewhere. On black sandy soil, probably damp for 7–8 months of the year, but not fully saturated for long, with very sparse grass cover; 1200–1400 m.

Conservation notes: Moderately widespread but in a limited habitat; possibly Near Threatened.

Robinson points out that the distinctive habit, with the stems bending stiffly outwards at maturity so that the apex of the inflorescence points downwards, is obvious in the field but much less so after the specimen has been pressed and dried.

53. **Scleria iostephana** Nelmes in Kew Bull. **11**: 94 (1956). —Napper in J. E. Africa Nat. Hist. Soc. Coryndon Mus. **24**(109): 28 (1964). —Robinson in Kew Bull. **18**: 544 (1966). —Napper in F.W.T.A., ed.2 **3**: 342 (1972). —Haines & Lye, Sedges & Rushes E. Afr.: 354, figs.734,735 (1983). —Verdcourt in F.T.E.A., Cyperaceae: 409 (2010). Type: Uganda, Mengo, Kyewaga Forest, 7.ix.1949, *Dawkins* 365 (K holotype, EA, ENT).

Perennial herb; stems 0.6–2.1 m tall, 2–4 mm wide, ± glabrous below, hairy and usually with sharp scabrid angles. Stem bases swollen to 5–8 mm, forming an irregularly shaped knot or (rarely) stems in a straight line. Leaves 30–60 cm long, 5–8 mm wide, scabrid on margins and ribs, sparsely to densely hairy beneath. Inflorescence a terminal and 2–3 lateral panicles placed singly at nodes, 3–7 cm long, 1.5–4 cm wide, on stiffly erect pubescent peduncles exserted to 3 cm from sheaths. Male spikelets 4–5 cm long, glumes straw-coloured with dark reddish brown sides, hairy particularly on margin; female spikelets 5–7 mm long, dark reddish brown glumes with midrib and area near it straw-coloured or greenish, hairy on margin and midrib. Nutlets broadly ovoid or ellipsoid, 3–4.2 × 2.5–2.8 mm, smooth, greyish white or whitish below, dark above or sometimes all over; hypogynium yellowish to reddish brown, with 3 distinct lobes with recurved margins.

Zambia. N: Mwense Dist., Chipili, 27.vi.1956, *Robinson* 1752 (K, MRSC). W: Mwinilunga Dist., source of R. Kanyamwana, just W of Dobeka Plain, 8.ii.1938, *Milne-Redhead* 4497 (K). C: Mkushi Dist., Mwendashi Hills, David Moffat's Farm, 19.ix.1993, *Bingham & Nkhoma* 9678 (K). **Mozambique**. Z: Pebane Dist., Pebane, 43 km on road to Mualama, 15.i.1968, *Torre & Correia* 17143 (LISC, LMU). MS: Muanza Dist., Cheringoma Plateau, source of R. Mueredzi, 18°50'S 34°46'E, 11.vii.1972, *Ward* 7814 (K, NU)·

Also in Nigeria, D.R. Congo, Central African Republic, Uganda and Tanzania. Shady places along streams or in dense vegetation at margins of relict evergreen swamp forest (mushitu); 100–1400 m.

Conservation notes: Widespread; not threatened.

Robinson remarks that this species is very close to *Scleria barteri* Boeckeler, occurring in West Africa from Senegal to D.R. Congo. *Scleria barteri* is often a scrambler, sometimes attaining 5 m or more in height, and has ovoid hairy nutlets. It was treated by Verdcourt as a synonym of *S. boivinii* Steud., described from Madagascar and recorded from East Africa (Kenya, Uganda, Tanzania), and should be looked for in coastal Mozambique.

54. **Scleria pachyrrhyncha** Nelmes in Kew Bull. **11**: 99 (1956). —Napper in J. E. Africa Nat. Hist. Soc. Coryndon Mus. **24**(109): 28 (1964). —Robinson in Kew Bull. **18**: 544 (1966). —Haines & Lye, Sedges & Rushes E. Afr.: 354, fig.736 (1983). — Verdcourt in F.T.E.A., Cyperaceae: 409 (2010). Type: Tanzania, Uluguru Mts, Bunduki, 27.i.1935, *Bruce* 669 (K holotype).

Perennial herb; stems to 2 m long, 3–5 mm in diameter, stem-bases swollen to 7 mm wide. Rhizome horizontally creeping, 3–5 mm thick, covered with red striate scales, with glabrous stems at intervals of 3–7 cm. Leaves 5–11 mm wide, glabrous above, sparsely hairy beneath; sheaths hairy. Inflorescence a simple terminal panicle or terminal and lateral panicles together to 35 cm overall, terminal panicle 6–9 cm long, 5–7(9) cm wide, laterals 2.5–5 cm long, 1–4 cm wide, single or paired at a node, on erect peduncles exserted to 10(14) cm from sheath. Male spikelets 5–6 mm long, sessile, hispidulous, brown; female spikelets hairy, greenish or brown. Nutlets broadly ovoid, 4–5 × 2.5–2.8 mm wide, obtusely trigonous with angles often white, smooth, strongly beaked, yellowish, pinkish or greyish brown or green, becoming blue at tip; hypogynium reduced to a narrow brown collar, barely 3-lobed.

Zimbabwe. E: Nyanga Dist., E of Mt Nyangani, near Nyamingura R., Eastern Highlands Tea Estate, 22.iv.1964, *Robinson* 6235 (B, K, M, NY, SRGH). **Mozambique**. MS: Sussundenga Dist., S tip of Chimanimani Mts., slopes above Haroni–Makurupini Forest, 31.v.1969, *Müller* 1256 (K, LISC, SRGH).

Also in Tanzania. Open places in rain forest; 600–1000 m (to 1850 m in East Africa). Conservation notes: Restricted forest distribution, but probably not threatened.

55. **Scleria induta** Turrill in Bull. Misc. Inform. Kew **1914**: 137 (1914). —Nelmes in Kew Bull. **11**: 98 (1956). —Robinson in Kew Bull. **18**: 545 (1966). —Haines & Lye, Sedges & Rushes E. Afr.: 355 (1983). Types: Angola, Benguella (Ganguellas & Ambuellas), n.d., *Gossweiler* 3658 (K lectotype); Benguella, *Gossweiler* 3757 (K syntype), lectotypified by Nelmes (1956).

Scleria angolensis Turrill in Bull. Misc. Inform. Kew **1914**: 136 (1914). Type: Angola, Benguella, *Gossweiler* 4115 (K).

Scleria duvigneaudii Piérart in Bull. Séances Inst. Roy. Colon. Belge **22**: 351 (1951) Type: D.R. Congo, no locality, n.d., *Duvigneaud* 54 (BRLU).

Perennial herb; stems 50–90 cm long, 2–3 mm in diameter, rigid, bases swollen to 6–7 mm thick, either contiguous or joined by a stout woody rhizome to form a knotty mass. Leaves 2–6 mm wide, rather rigid, glabrous or sometimes hairy beneath. Inflorescence of single lateral panicles at 2–3 nodes, on rigid peduncles shortly exserted from sheaths. Male spikelets 8–9 mm long, very numerous, spreading at maturity and then a conspicuous feature, anthers emerging from top in a brush-like tuft, mostly sessile; female glumes lanceolate, to 14 mm long, hispidulous on midrib. Nutlets 4.2–4.6 × c. 3 mm, ovoid, smooth, shining, pale grey, pale brown or pale violet-brown; hypogynium without lobes, when fresh white and ± spongy, 1–1.5 mm long.

Zambia. B: Kalabo Dist., 5 km S of Kalabo, 17.xi.1957, *Drummond & Cookson* 6564 (K, LISC, SRGH). W: Mwinilunga Dist., c. 6 km S of Mwinilunga, 3.xii.1937, *Milne-Redhead* 3502 (K). N: Kasama Dist., 80 km S of Kasama, 29.iv.1962, *Robinson* 5128 (K).

Also in Angola and D.R. Congo. In well-drained shady woodland in areas with over 900 mm rainfall; 1200–1400 m.

Conservation notes: Widespread; not threatened.

Robinson (1966) states: "Immature specimens in which the male spikelets have not fully developed, so that the panicles are in consequence much more compact, have a very different appearance; such is the type of *Scleria angolensis* Turrill. The white unlobed hypogynium is unusual and recalls that of *S. erythrorrhiza* Ridl. In this connection it may be significant that abortive male flowers are always present within the female glumes of *S. induta*".

56. **Scleria melanomphala** Kunth, Enum. Pl. **2**: 345 (1837). —Clarke in Fl. Cap. **7**: 296 (1898); in F.T.A. **8**: 506 (1902). —Piérart in Lejeunia Mém. **13**: 26, t.1, figs.26,31 (1951). —Nelmes in Kew Bull. **11**: 88 (1956). —Napper in J. E. Africa Nat. Hist. Soc. Coryndon Mus. **24**(109): 27 (1964). —Robinson in Kew Bull. **18**: 546 (1966). —Napper in F.W.T.A., ed.2 **3**: 340 (1972). —Haines & Lye, Sedges & Rushes E. Afr.: 356, figs.741,742 (1983). —Hennessy in Bothalia **15**: 526 (1985). —Gordon-Gray in Strelitzia **2**: 184, figs.81,82 (1995). —Verdcourt in F.T.E.A., Cyperaceae: 411 (2010). Types: South Africa, Cape of Good Hope, E Coast, *Drège* s.n. (B† holotype); 'Tembuland', between Baskee R. & Morley, 1840, *Drège* s.n. under *Clarke* 4369 (K lectotype, OXF).

Robust, tussock-forming perennial herb; stems 0.6–2.4 m tall to 1 cm wide across leaf sheaths, glabrous but minutely to strongly scabrid on angles. Rhizome thick, 4–6 mm wide with very short internodes. Leaves 20–60 × 0.7–1.8(2) cm, glabrous or hairy, scabrid on margin and ribs. Inflorescence of 1 terminal and 5–9 lateral often drooping narrowly ovate panicles, 2–10 × 1–3 cm, borne singly or 2–3 at nodes on pendulous peduncles exserted to 30 cm from sheaths. Male spikelets 8–13 mm long, ± sessile, the glumes straw-coloured with usually dark reddish brown margins and green scabrid produced midrib; female spikelets often even darker in middle, glumes 10–12 mm long, hispidulous or hairy on midrib. Nutlets ovoid, 3.7–4(5) × 2.7–3.2 mm, smooth, glabrous or proximally sparsely hairy, white and shining, usually with blue-black apex; hypogynium white or yellowish brown, irregular but unlobed.

Botswana. N: Okavango Swamps, banks of Mboroga R. near Dindinga Is., 19°06'8"S 23°05'2"E, 18.vi.1979, *P.A. Smith* 2789 (K, SRGH). **Zambia**. B: Mongu Dist., Sefula, JICA Rice Project, 11.ii.1999, *Bingham & Luwiika* 11825 (K). N: Kawambwa Dist., Kawambwa, 21.vi.1957, *Robinson* 2331 (K). W: Mwinilunga Dist., by R. Lunga, 1.xii.1937, *Milne-Redhead* 3473 (K). C: Serenje Dist., S of Mankone, Kundalila Falls, 13.iii.1975, *Hooper & Townsend* 700 (K). E: Katete Dist., Katete Hills, 10.vi.1963, *Verboom* 138 (K). **Zimbabwe**. N: Gokwe Dist., 40 km from Gokwe, Lutope R., 18.iii.1963, *Bingham* 536 (K, SRGH). E: Mutare Dist., Mutare (Umtali), streamside commonage, 23.xi.1954, *Chase* 5337 (K, LISC, SRGH). S: Masvingo Dist., Great Zimbabwe, Mzero Farm, 19.iv.1972, *Chiparawacha* 443 (K, LISC, SRGH). **Malawi**. N: Chitipa Dist., N end of Nyika Plateau, near Lower Mondwe R., 13.viii.1972, *Brummitt & Synge* WC232 (K). C: Dedza Dist., Chongeni, Fala Village, 6.iii.1970, *Salubeni* 1462 (K, SRGH). S: Zomba Dist., Thondwe, Mpita Tobacco Farm, 5.xii.1985, *Balaka & Nachamba* 1304 (K, MAL). **Mozambique**. N: Ngauma Dist., Massangulo, iii.1935, *Gomes e Sousa* 1264 (K). Z: Milange Dist., Metolola, 23.vi.1972, *Bowbrick* JA607 (LISC, SRGH). MS: Sussundenga Dist., c. 28 km W of Dombe, base of SE slopes of Chimanimani Mts, 25.iv.1974, *Pope & Müller* 1303 (K, LISC, SRGH). M: Marracuene Dist., near Marracuene (Vila Luiza), 9.x.1940, *Torre* 1741 (BR, LISC, LMA, LMU).

Widespread in Tropical Africa from Guinea to South Africa; also in Madagascar and South America. Seasonally or permanently wet grasslands, dambos and swamps; 900–1700 m.

Conservation notes: Widespread; not threatened.

57. **Scleria greigiifolia** (Ridl.) C.B. Clarke in F.T.A. **8**: 509 (1902) as *griegifolia*. — Robinson in Kew Bull. **18**: 546 (1966). —Haines & Lye, Sedges & Rushes E. Afr.: 358, figs.744,745 (1983). —Hennessy in Bothalia **15**: 527 (1985). —Gordon-Gray in Strelitzia **2**: 183, fig.82 (1995). —Verdcourt in F.T.E.A., Cyperaceae: 412 (2010). Type: Angola, Huíla, marshes by R. Cacolobar, near Lake Ivantala, ii.1860, *Welwitsch* 6959 (LISU holotype, BM).

    Acriulus greigifolius Ridl. in J. Linn. Soc., Bot. **20**: 336 (1883); in Trans. Linn. Soc., Bot. **2**: 166, t.22 (1884). —Clarke in Consp. Fl. Afr. **5**: 676 (1895). —Schumann in Engler, Pflanzenw. Ost-Afrikas **C**: 128 (1895). —Rendle in Cat. Afr. Pl. Welw. **2**: 132 (1899).

    Acriulus madagascariensis Ridl. in J. Linn. Soc., Bot. **20**: 336 (1883). —Clarke in Durand & Schinz, Consp. Fl. Afr. **5**: 676 (1894). —Schumann in Engler, Pflanzen. Ost-Afrikas **C**: 128 (1895). Types: Madagascar, Ambatolampy, x.1882, *Baron* 1870 (K syntype); Andrangaloaka, xi.1880, *Hildebrandt* 3751 (K syntype).

    Scleria acriulus C.B. Clarke in F.T.A. **8**: 509 (1902), new name, non *Scleria madagascariensis* Boeckeler (1884).

    Acriulus titan C.B. Clarke in Bull. Misc. Inform. Kew, add. ser. **8**: 62 (1908). Type: D.R. Congo, Mandimba, Djuma valley, vii.1902, *Gentil* s.n. (BR holotype).

    Scleria friesii Kük. in Fries, Wiss. Ergebn. Schwed. Rhodesia-Kongo-Exped. 1911–1912, Erg.: 9 (1921). Type: Zambia, Lake Bangweulu, Mano, 22.ix.1911, *Fries* 743 (UPS holotype).

Perennial herb; stems 1–2 m tall, 2–6 mm wide, triangular in cross-section, 2–6 mm wide, glabrous but scabrid on angles; plant either forming loose clumps or from a long creeping rhizome 6–10 mm thick, loosely covered with broad striate brown scales. Leaves 50–80 cm long, 8–12 mm wide, stiff, closely imbricate below, roughly hairy or almost glabrous, margins coarsely serrate, ribs scabrid. Inflorescence of one terminal and many lax lateral copious panicles, 4–7 at each node, on slender pendulous scabrid straw-coloured to dark reddish peduncles to 20 cm long. Male spikelets 6–8 times as numerous as female, straw-coloured or chestnut above but reddish brown to almost blackish below, 4.5–5.5 mm long; female spikelets with some straw-coloured obtuse glumes below and 3 mostly reddish black acuminate glumes above; upper with scabrid and ciliate margin and midrib, with stiff hairs on upper half of inner surface. Nutlets broadly ovoid, 4–5 × 2.8–3.8 mm, strongly apiculate, glabrous, smooth, white with pinkish, often with dark violet blotches; hypogynium pale orange-brown, unlobed and disc-like or angular with obscure lobes.

Zambia. N: Luapula Dist., Kalasa Mukoso Flats, Lwame dambo, 26.ii.1996, *Renvoize* 5690 (K). W: Mwinilunga Dist., c. 6.5 km from Lwawu Mission, 24.ii.1995, *Nawa et al.* 127 (K, MO). C: Chongwe Dist., Great East Road, between Undaunda and Rufunsa, 128 km N of Lusaka, 2.i.1972, *Kornaś* 781 (K). **Zimbabwe**. N: Guruve Dist., Nyamunyeche Estate, 21.iii.1979, *Nyariri* 777 (K, SRGH). C: Chirumanzu Dist., Mvuma (Umvuma), 15.ii.1971, *Chiparawasha* 349 (K, SRGH). E: Nyanga Dist., Nyanga (Inyanga) Road, *Gilliland* K1029 (K). **Malawi**. N: Mzimba Dist., 5 km S of Mzuzu, Katoto Estate, 23.ii.1969, *Pawek* 1730 (K, MO). C: Lilongwe Dist., Dzalanyama Forest Reserve, valley NW of Kazuzu Hill, 24.ii.1982, *Brummitt et al.* 16106 (K). **Mozambique**. N: Lichinga Dist., Lichinga (Vila Cabral), 23.v.1934, *Torre* 202 (LISC).

Also in Uganda, Tanzania, D.R. Congo, Angola, South Africa (KwaZulu-Natal) and Madagascar. In permanetly wet bogs, often in association with *Scleria bequaertii*, *S. laxiflora* and *S. procumbens*; 950–1450 m.

Conservation notes: Widespread; not threatened.

58. **Scleria poiformis** Retz., Observ. Bot. **4**: 13 (1786) as *poaeformis*. —Fischer in Bull. Misc. Inform. Kew **1931**: 265 (1931); in Bull. Misc. Inform. Kew **1932**: 70 (1932), all as *poaeformis*. —Nelmes in Kew Bull. **11**: 110 (1956) as *poaeformis*. —Napper in J. E. Africa Nat. Hist. Soc. Coryndon Mus. **24**(109): 27 (1964) as *poaeformis*. —Robinson in Kew Bull. **18**: 547 (1966) as *poaeformis*. —Haines & Lye, Sedges & Rushes E. Afr.: 357, fig.743 (1983). —Hennessy in Bothalia **15**: 527 (1985). —Gordon-Gray in Strelitzia **2**: 184, fig.84 (1995). —Verdcourt in F.T.E.A., Cyperaceae: 412 (2010). Type: India, *König* s.n. (LD holotype, K, LZ).

> *Scleria coriacea* G. Bertol. in Mem. Reale Accad. Sci. Ist. Bologna **5**: 34 (1854); Ill. Piant. Mozamb. Dissert. **IV**: 14, t.5, figs.1–4 (1855), illegitimate name, non Liebm. (1851). Type: Mozambique, no locality, ?1842, *Fornasini* s.n. (?BOLO holotype, P).

Stout perennial herb; stem 3-angled, 1.2–2.1 m tall, 3–10 mm wide, rooting from submerged nodes. Rhizome creeping, 5–10 mm thick. Leaves tough, to 70 cm or more long, 1–4 cm wide, glabrous but scabrid on veins and with saw-edged margins. Inflorescence a single terminal ellipsoid panicle, 10–20 × 5–12 cm with compound branches bearing very many spikelets; male spikelets 3.5–4.5 mm long; female glumes 3.5–5 mm long, glabrous or hispidulous, straw-coloured or brown. Nutlets broadly ovoid to subglobose, 3.5 × 2.5–2.8 mm, smooth, glabrous, whitish; hypogynium small with 3 short triangular lobes.

Mozambique. Z: Maganja da Costa Dist., Maganja, Raraga, near beach, 14.ix.1944, *Mendonça* 2076 (LISC, LMU, MO, WAG). MS: Muanza Dist., Cheringoma, 80 km from Beira on road to Inhaminga, 21.v.1942, *Torre* 4152a (LISC, LMA, LMU). GI: Inhassoro Dist., Bazaruto Is., E of Ponta Gengarene, 6.xi.1958, *Mogg* 28890 (K, LISCK). M: Maputo, 20.ii.1952, *Myre & Carvalho* 1147 (NU).

Also in South Africa, Tropical Asia and Australia. In shallow freshwater lakes near the sea; 0–50 m.

Conservation notes: Widespread; not threatened.

According to Art. 60.8 of the Code compounds contrary to Rec. 60 G must be corrected so that *poaeformis* used by nearly all authors has to be changed (Verdcourt 2010).

Both Robinson and Haines & Lye (1983) suggest the type was at Leipzig and destroyed. It was seen by C.E.C. Fischer (Bull. Misc. Inform. Kew **1932**: 70, 1932) who borrowed the whole König collection from Lund. His meticulous paper has been ignored by most people dealing with Retzius. Gordon-Gray gives it correctly (see note by Verdcourt in F.T.E.A.).

59. **Scleria angusta** Kunth, Enum. Pl. **2**: 346 (1837). —Clarke in Fl. Cap. **7**: 296 (1898). —Nelmes in Kew Bull. **11**: 73 (1956). —Robinson in Kew Bull. **18**: 548 (1966). —Hennessy in Bothalia **15**: 528 (1985). —Gordon-Gray in Strelitzia **2**: 181, fig.82

(1995). Type: South Africa, KwaZulu-Natal (Pondoland), between Umtentu & Umzimkulu rivers, ii.1840, *Drége* s.n. [under *Clarke* 4246] (B† holotype, K).

Stoutly erect perennial herb; stems to 250 cm tall, 3–6 mm thick, glabrous below, hairy above. Rhizome 4–5 mm thick, ± straight, stems rising at intervals of 1.5–2 cm. Leaves at least 45 cm long, 20 mm wide, with lateral flanges ending unequally towards apex ('praemorse' of some authors; the specific epithet means 'narrowed'). Inflorescence interrupted, lateral panicles single at 3–7 nodes, on erect, ± shortly exserted peduncles, panicles rather compact, 3–4(6) cm long. Male spikelets 4 mm long; female glumes glabrous or hispidulous, pale brown with darker reddish streaks. Nutlets ovoid to ovoid-globose, 2.3–3.5 × 1.8–2.3 mm, smooth, glabrous, dark purple (violet or partly white if dried before maturity); hypogynium barely 3-lobed, fimbriate on upper margin.

Mozambique. MS: Cheringoma Dist., Lower Zuni drainage, Ntambani swamp forest, v.1973, *Tinley* 2830 (K, SRGH). GI: Xai-Xai Dist., Chongoéne, 24.x.1947, *Barbosa* 502 (K). M: Matutuine Dist., Bela Vista, Zitundo, 3.x.1968, *Balsinhas* 1344 (LISC).

Also in South Africa (KwaZulu-Natal) and Madagascar. Coastal swamp forests; 0–100 m. Conservation notes: Widespread; not threatened.

60. **Scleria racemosa** Poir. in Lamarck, Encycl. **7**: 6 (1806). —Oliver in Trans. Linn. Soc. London **29**: 169, t.111 (1875). —Clarke in F.T.A. **8**: 50 (1902); in Illustr. Cyper.: t.131, fig.5 (1909). —Piérart in Lejeunia Mém. **13**: 58, t.3, figs.11,12 (1953). —Nelmes in Kew Bull. **11**: 76 (1956). —Napper in J. E. Africa Nat. Hist. Soc. Coryndon Mus. **24**(109): 27, fig.7 (1964). —Robinson in Kew Bull. **18**: 540 (1966). —Haines & Lye, Sedges & Rushes E. Afr.: 358, figs.746,747 (1983). — Verdcourt in F.T.E.A., Cyperaceae: 413 (2010). Type: Madagascar, no locality, *du Petit-Thouars* (P holotype, B-W 17319).

     Scleria ciliolata Boeckeler in Flora **65**: 31 (1882). Type: Madagascar, Nosy Be (Nossi-bé), iv.1879, *Hildebrandt* 2921 (B† holotype, K).
     Scleria verrucosa sensu Clarke in F.T.A. **8**: 509 (1902) in part, non Willd. (see note).

Perennial herb; stems 1–3.5(4) m and 4–7 mm thick at base, glabrous, with razor-sharp basal sheaths. Rhizome creeping, horizontal, ± straight, 4–6 mm thick. Leaves up to c. 60 cm, 1–2.5(3.5) cm wide, with razor-sharp margins. Inflorescence of one terminal and 3–6 lateral panicles, single or double at nodes, elliptic to lanceolate in outline on erect minutely hairy peduncles. Male spikelets 5–6 mm long, sessile or with very short pedicels, glumes minutely hairy, straw-coloured with reddish brown dots and dashes; female spikelets 7–9 mm long, glumes minutely hairy on margin and obscure midrib, straw-coloured or pale brown with dark reddish marks. Nutlet ovoid, 4–5 mm long (excluding style base and hypogynium), 3.5–4.5 mm wide, smooth, glabrous, white with pinkish brown tinge; style base persistent, dark brown, woody when dried, 1–1.5 mm long; hypogynium yellowish brown above, dark reddish brown below, cupular, 3–4 × 4–5 mm wide, smooth or wrinkled, corky, margin with many close-set whitish, yellow or reddish brown hairs.

Zambia. N: Mbala Dist., Chianga dambo, 23.vi.1956, *Robinson* 1716 (K). W: Mwinilunga Dist., between Kalene & Sakatwala, 22.ii.1975, *Hooper & Townsend* 318 (K). C: Luangwa Dist., 8 km N of Katondwe Mission, Chitapa, 28.ix.1964, *Angus* 3363 (FHO, K). E: Petauke Dist., Petauke, 15.xii.1958, *Robson* 953 (BM, K). **Zimbabwe**. E: Chimanimani Dist., banks of Rusitu R. near junction with Haroni R., 25.xi.1955, *Drummond* 5025 (K, LISC, SRGH). **Malawi**. N: Mzimba Dist., Mzuzu, Marymount, 11.vii.1976, *Pawek* 11486 (K, MAL, MO). C: Mchinji Dist., Mchinji, banks of Bua R. by road to Zambia, 30.iii.1970, *Brummitt* 9536 (K, LISC). S: Mulanje Dist., Mulanje Forest Reserve, Fort Lister, Phalombe R., 19.viii.1983, *Seyani & Balaka* 1225 (K, MAL). **Mozambique**. N: Sanga Dist., Macaloge (Mecaloja de Rovuma), 2.viii.1934, *Torre* 203 (LISC). Z: Between Quelimane and Marral, 7.ix.1941, *Torre* 3441 (COI, LISC, LMA, LMU, LUA, PRE, WAG). T: Marávia Dist., Fíngoé, 2.iv.1942, *Mendonça* 389 (BR, LISC, LMU, MO, WAG). MS: Mossurize Dist., 12 km E on Gogi Mission Road from Frontier Barrier on Espungabera road, L bank of Busi R. above bridge, 16.viii.1961, *Chase* 7522 (K, LISC, SRGH).

Also in Sudan, Ethiopia, D.R. Congo, Uganda, Kenya, Tanzania, Angola and Madagascar. In swamp forests and streamsides; 300–1400 m.

Conservation notes: Widespread; not threatened.

The fruit is distinct from all other species in the genus in its persistent hypogynium that forms a cupule at the nutlet base. Another species, *Scleria verrucosa* Willd., with warty, not smooth, nutlets, occurs in West Africa from Senegal to W Uganda and D.R. Congo; it could occur in N Zambia. The nutlets of *S. racemosa* can appear lumpy if immature when collected and dried, but the warts of *S. verrucosa* nutlets are large, deflexed, and usually crowned with a tuft of brownish hairs.

35. **DIPLACRUM** R. Br.[35]

Diplacrum R. Br., Prodr. Fl. Nov. Holland.: 241 (1810).

Slender, inconspicuous herb, annual; stem slender, under 150 mm tall. Leaves alternate, simple, linear-lanceolate, eligulate, often reddish purple abaxially. Inflorescence a pedunculate spikelet cluster axillary to a leaf, lowest 1–2 leaves sterile. Spikelets unisexual, female solitary, terminal between 2 distichous, acuminate glumes; male 0–2 below female. Stamens 3–1. Style branches 3. Nutlet round, longitudinally ribbed, grey to blackish, hypogynium white eventually deciduous.

A genus of about 6 species widely distributed in the tropical and subtropical Old World, with 2 species in Africa, one in West Africa and one in the Flora area.

Diplacrum africanum (Benth.) C.B. Clarke in F.T.A. **8**: 510 (1902); in Consp. Fl. Afr. **5**: 668 (1894), invalidly published. —Haines & Lye in Sedges & Rushes E. Afr.: 360, fig.750 (1983). —Beentje in F.T.E.A., Cyperaceae: 415, fig.62 (2010). Type: Nigeria, Nupe, 1857-59, *Barter* 1041 (K holotype). FIGURE 14.**74**.

    Scleria africana Benth. in Bentham & Hooker, Gen. Pl. **3**: 1071 (1883).

Slender inconspicuous herb, rooting system suggesting an annual. Stems ribbed, less than 150 mm tall; leaves numerous, alternate, simple, 20–40 × 2–3 mm, with 6–7 veins, linear, narrowing abruptly to acuminate, scabrid apex, often purple-red abaxially, ligule lacking. Inflorescence a series of shortly pedunculate spikelet clusters each axillary to a leaf sheath, except lowest 1–2 sterile leaves. Spikelets unisexual, lanceolate, 2–3 mm long; female solitary, appearing terminal between 2 lower strongly acuminate glumes with receptacle (short axis) produced forming a pair of clips between glumes holding the ripe nutlet; male spikelets 0–2, below female; florets few per spikelet, enclosed by strongly acuminate glumes. Stamen 1–2. Style branches 3. Nutlets 0.5–0.8 × 0.5–0.6 mm, ovate in outline, faintly trigonous, longitudinally ribbed, grey to almost black; hypogynium c. 1 mm deep, smooth, white, eventually falling from clips.

Zambia. W: Chingola Dist., banks of Kafue R., 11 km N of Chingola, 1350 m, 4.v.1960, *Robinson* 3701 (K). S: Choma Dist., Pangama, Mapanza, 1067 m, 27.iv.1958, *Robinson* 2851 (K).

Found in West Africa (Guinea, Sierra Leone, Ivory Coast and Nigeria), Central African Republic, Sudan, Kenya, Uganda, Tanzania and Madagascar; also in S India and tropical America (Surinam). In seasonally wet places, sometimes locally dominant on bare ground surrounding dams, or on shallow soil over rock outcrops; 1000–1400 m.

Conservation notes: Widespread; not threatened.

Closely related to *Scleria* and previously sometimes included as a section or subgenus within it. In the field the species has the appearance of a small dicotyledonous plant due to many lanceolate leaves on the short stem, and is easily hidden below the surrounding *Polygonum* growing at the margins of dams.

---

[35] by J. Browning and K.D. Gordon-Gray†

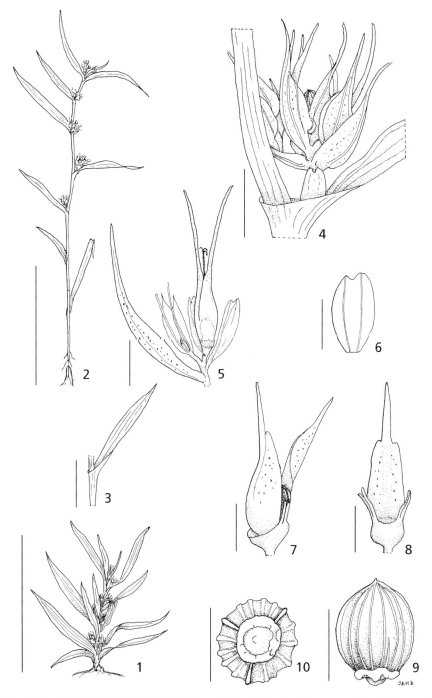

Fig. 14.**74**. DIPLACRUM AFRICANUM. 1, 2, habit; 3, leaf; 4, inflorescence; 5, spikelets, male and female with bract and prophyll; 6, prophyll; 7, 8, female spikelet, with and without nutlet; 9, 10, nutlet, lateral and basal view. 2–4, 7–10 from *Robinson* 2851; 1, 5, 6 from *Bidgood et al.* 3967. Scale bars: 1, 2 = 25 mm; 3 = 10 mm; 4–8 = 1 mm; 9, 10 = 0.5 mm. Drawn by Jane Browning.

36. **SCHOENOXIPHIUM** Nees[36]

Schoenoxiphium Nees in Linnaea **7**: 531 (1832). —Kukkonen in
Bothalia **14**: 819–823 (1983).

Robust or slender rhizomatose or tufted perennials or slender grass-like plants (easily confused with them). Culms scapose. Leaves ligulate. Involucral bracts leaf-like or short. Inflorescence a slender to large panicle with unisexual flowers or reduced to a single spike; main axis with a succession of bracts, each of lowest subtending utricles enclosing a female floret; rachilla with apical partial spikelet of several glumes each subtending a male floret; upper bracts of main axis also subtending male florets to form a terminal male spikelet; rachilla may not be developed into a male spikelet but reduced to flattened scabrid remnant protruding from utricle mouth, or so reduced it does not protrude but is shorter than ovary, or entirely lacking. Male flowers with 3 stamens. Female flowers with trifid style. Nutlet trigonous, often beaked.

A small genus close to *Carex* of c. 12 mainly montane species, with its greatest diversity in South Africa but extending north to Ethiopia; 2 species in Madagascar. Recently this genus has been subsumed into *Carex* sensu lato (Global Carex Group in Bot. J. Linn. Soc. **179**: 1–42, 2015). However, it is treated separately here in order to ensure consistency with F.T.E.A.

Species delimitation in this genus is difficult; the descriptions and illustrations in Gordon-Gray (1995) are valuable. Descriptions used here and the key are largely taken from Verdcourt (2010).

1. Main inflorescence branches with 20–40 female flowers; leaves 6–9 mm wide . . .
   . . . . **1.** *ludwigii*
 – Main inflorescence branches with 2–10 female flowers; leaves 1–5 mm wide . . . 2
2. Plant slender with lax slender inflorescence; utricles stipitate, 4–6 mm long, beak
   1.5–2.0 mm long . . . . **4.** *lehmannii*
 – Plant more rigid with more rigid compact inflorescence; utricles 2–3 mm long,
   beak under 1 mm. 3
3. Largest glumes subtending utricles 6–7 mm long including 3–4 mm awn . . . . . .
   . . . . **2.** *sparteum*
 – Largest glumes subtending utricles 10–12 mm long including 7–9 mm awn . . . .
   . . . . **3.** *caricoides*

1. **Schoenoxiphium ludwigii** Hochst. in Flora **28**: 764 (1845). —Gordon-Gray in Strelitzia **2**: 168 (1995). —Verdcourt in F.T.E.A., Cyperaceae: 420 (2010). Type: South Africa, Cape peninsula ("Cap. b. Spei"), no locality, 1837, *Ludwig* s.n. (TUB holotype).

   Carex ludwigii (Hochst.) Luceño & Martín-Bravo in Bot. J. Linn. Soc. **179**: 27 (2015).

Robust perennial herb 60–80 cm tall. Leaves 20–36 cm long by 6–9 mm wide, scabrid particularly towards apex on margins and midribs. Inflorescence a rather narrow panicle from each of upper leaf sheaths, c. 5 × 2.5 cm wide with 3–12 cm peduncles, fairly dark brown; spikelets 10 × 2.5 mm, female glumes brown, 3–4 mm long with long scabrid awn 0.5–1.7(3) mm long; male part of spikelet 6 mm long, glumes 3.5 mm long, glossy coppery-brown, with hyaline margin. Utricle greenish, 4.5–5.5(6) mm long without distinct beak. Nutlet pale 2.5–3 × 2 mm without an invagination on plane face, with a symmetrical apex.

Zimbabwe. E: Nyanga Dist., 10 km N of Troutbeck on Mozambique border, Gairesi Ranch, 16.xi.1956, *Robinson* 1909 (K, LISC).
Also in Tanzania and South Africa (KwaZulu-Natal), probably also in Mozambique. In peaty moorland bog; 1850 m.

---

[36] by M. Lock

Verdcourt used this name for East African material but noted that there had been confusion with *Schoenoxiphium rufum* Nees; he stated that *S. ludwigii* occurs in Zimbabwe. Gordon-Gray kept these two species separate, with the main difference being the greater size of *S. ludwigii*, which also tends to have curved rather than straight utricles. *Robinson* 1909 has no mature utricles and the determination must therefore remain in doubt. *Robinson* 1909 has also been named as *S. ?buchananii* by Kukkonen, but the type from KwaZulu-Natal (not Malawi as are many *Buchanan* specimens), does not appear to be the same.

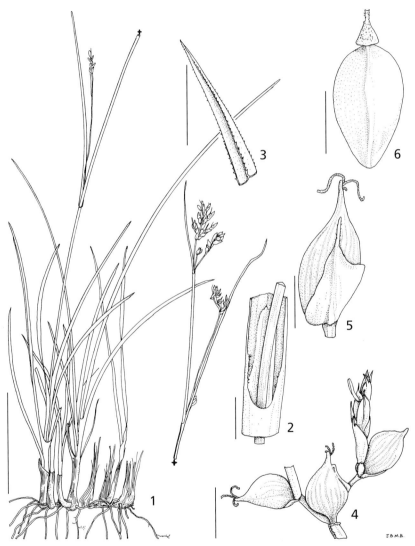

Fig. 14.**75**. SCHOENOXIPHIUM SPARTEUM. 1, habit; 2, junction of leaf sheath and blade; 3, tip of involucral bract; 4, part of inflorescence; 5, fertile utricle and glume; 6, immature nutlet. All from *Browning* 227. Scale bars: 1 = 40 mm; 2, 4 = 2 mm; 3, 5, 6 = 1 mm. Drawn by Jane Browning. Reproduced from Strelitzia (1995) by kind permission of the South African National Biodiversity Institute, Pretoria.

This species with its abundant dark brown spikelets, looks completely different from the other *Schoenoxiphium* taxa in the Flora area.

2. **Schoenoxiphium sparteum** (Wahlenb.) C.B. Clarke in Bull. Misc. Inform. Kew, add. ser. **8**: 67 (1908). —Kükenthal in Engler, Pflanzenr. **4**, 20(38): 31 (1909). —Haines & Lye, Sedges & Rushes E. Afr.: 367, fig.759 (1983). —Gordon-Gray in Strelitzia **2**: 171, fig.75 (1995). —Verdcourt in F.T.E.A., Cyperaceae: 418 (2010). Type: South Africa, no locality, n.d., *Thunberg* s.n. (UPS holotype). FIGURE 14.**75**.

    Carex spartea Wahlenb. in Kongl. Vetensk. Akad. Nya Handl. **24**: 149 (1803). —Clarke in Fl. Cap. **7**: 304 (1897).

    Carex schimperiana Boeckeler in Linnaea **40**: 373 (1876). —Engler, Hochgebirgsfl. Afrika: 152 (1892). —Clarke in Durand & Schinz., Consp. Fl. Afr. **5**: 690 (1894); in F.T.A. **8**: 548 (1902). Type: Ethiopia, Tigray, Debra Tabor, 27.viii.1863, *Schimper* 1318 (B† holotype, BM, E, K).

    Schoenoxiphium schimperianum (Boeckeler) C.B. Clarke in Bull. Misc. Inform. Kew, add. ser. **8**: 67 (1908).

    Schoenoxiphium sparteum var. *schimperianum* (Boeckeler) Kük. in Engler, Pflanzenr. **4**, 20(38): 32 (1909).

    Kobresia spartea (Wahlenb.) T. Koyama in J. Fac. Sci. Univ. Tokyo, sect. 3, Bot. **8**: 80 (1961).

Erect tufted slender perennial 25–80 cm tall, with short base covered with fibrous leaf remains; stems ridged and scabrid. Leaves to 40 cm long by 1–4 mm wide, flat, scabrid on margins and some ridges; leaf sheaths 1–4 cm long, all subtending pedunculate inflorescences, green or whitish on one face, often purple dotted at base; ligule with a distinct whitish or violet-dotted rim. Inflorescence with two main branches at different lengths, peduncle 1–7 cm long, very scabrid; glumes yellow-green or tinged ochre-brown, green at edges, flowers crowded, a few males at branch tips, female beneath; upper female glumes pale brown, sometimes with dark dots and green or white midrib, broadly triangular, 2–8 mm long, acute or subulate, scabrid; lower female glumes 6–7 mm long including 3–4 mm scabrid awn to twice length of utricle, male glumes shorter and narrower. Utricles brown with green stripe on two sides, obtusely triangular with 3 flat ridges, 1.3 × (2)2.5–3 mm including 0.5–0.8 mm beak, distinctly ridged. Nutlet yellowish, 2 mm long, slightly stipitate; style branches papillate, projecting from beak; reduced male axis slightly protruding, enclosed part very scabrid.

Zimbabwe. E: Chimanimani Mts, Long Gulley, 30 xii.1959, *Goodier & Phipps* 361 (K, SRGH). **Malawi**. N: Nyika Plateau, E side, 5.i.1959, *Robinson* 3068 (K, PRE). C: Ncheu Dist., Gochi, Kirk Range, 14.xi.1950, *Jackson* 276 (K). S: Thyolo Dist., Bvumbwe (Cholo Tung) Research Station, 24.xi.1950, *Jackson* 302 (K).

Also in Ethiopia, Sudan, Uganda, Kenya, Tanzania, D.R. Congo and South Africa. In upland grassland and on forest margins; 1100–2300 m.

Almost all the Kew material comes from the Eastern Highlands of Zimbabwe.

3. **Schoenoxiphium caricoides** C.B. Clarke in Bull. Misc. Inform. Kew, add. ser. **8**: 67 (1908). —Haines & Lye, Sedges & Rushes E. Afr.: 367, fig.760 (1983). —Gordon-Gray in Strelitzia **2**: 166 (1995). —Verdcourt in F.T.E.A., Cyperaceae: 420 (2010). Types: South Africa, W Cape, Swellendam, near Swellendam R., n.d, *Zeyher* 4440 (K syntype); E Cape, Murberg Range, Alexandria Division, n.d, *Drège* (K syntype); E Cape, Thembuland, *Baur* 744 (K syntype).

Erect shortly rhizomatous tussock-forming perennial herb 15–40 cm tall. Leaves yellow-green, 10–30 cm long by 2.5 mm wide, flat, densely scabrid on margins and ribs; leaf sheaths 2.4 cm long; ligule with distinct rib. Inflorescence branches usually borne singly at each node; partial units pyramidal with conspicuous bracts; branches 0.5–6 mm long, flattened, densely scabrid; the upper entirely hidden in leaf sheaths; each spike with a few male flowers at top and 4–10 female flowers below; glumes greenish or light brown with a greenish midrib with or without dark reddish brown dots, and lines; largest female glumes 10–12 mm long including 7–9 mm densely scabrid awn, but upper glumes with much shorter awn; male glumes 3–4 mm long with awn only 0.5–1 mm long. Utricles light reddish brown, ellipsoid, 2.7–2.8 × 1.3–2 mm, not distinctly ridged; style branches dark reddish, 2.5–3 mm long. Nutlet reddish brown, densely papillose; male axis remnant flattened, scabrid or ciliate, shorter than nutlet (but sometimes reduced or undeveloped – Gordon-Gray).

Zambia. W: Mwinilunga Dist., just S of Matonchi Farm, 12.xi.1937, *Milne-Redhead* 3217 (K). C: Serenje Dist., Kaudinia Falls, S of Kanona, 13.iii.1975, *Hooper & Townsend* 703 (K).

Also in Uganda, Tanzania and South Africa. *Brachystegia* woodland and grassland at dambo margins; 1300–1400 m.

These plants agree with *Robinson* 4765 and 4874 from SW Tanzania which both Verdcourt and Haines & Lye accepted as *Schoenoxiphium caricoides*. Unfortunately, Gordon-Gray does not seem to have examined material from outside South Africa. Kukkonen determined *Milne-Redhead* 3217 as *S. sparteum*.

4. **Schoenoxiphium lehmannii** (Nees) Steud., Syn. Pl. Glumac. **2**: 245 (1855). — Haines & Lye, Sedges & Rushes E. Afr.: 366, figs.756,758 (1983). —Gordon-Gray in Strelitzia **2**: 168 (1995). —Verdcourt in F.T.E.A., Cyperaceae: 417 (2010). Type: South Africa, E slope of Table Mt, near Constantia, n.d., *Ecklon & Zeyher* s.n. (S lectotype), lectotypified by Kukkonen (1983). FIGURE 14.**76**.

Uncinia lehmannii Nees in Linnaea **10**: 206 (1835).

Carex uhligii C.B. Clarke in Bot. Jahrb. Syst. **38**: 136 (1906); in Bull. Misc. Inform. Kew, add. ser. **8**: 73 (1908).

Schoenoxiphium sparteum (Wahlenb.) C.B. Clarke var. *lehmannii* (Nees) Kük. in Engler, Pflanzenr. **4**, 20(38): 32 (1909). —Peter, Fl. Deutsch Ost-Afrikas **1**: 533 (1929).

Kobresia lehmannii (Nees) T. Koyama in J. Fac. Sci. Univ. Tokyo, sect. 3, Bot. **8**: 80 (1961).

Slender greenish yellow perennial herb 30–90 cm tall with slightly swollen base covered with fibrous remains of old leaf-bases. Leaves 15–40 cm long by 2–5 mm wide, flat, scabrid on margin and some veins; sheaths pale green to reddish, 1–2 cm long; ligule with distinct brownish or white rib. Inflorescences borne at most nodes with 1–2 branches, 6–20 mm long, axes very scabrid; some male flowers at each branch tip with 2–5 female flowers below; lower female glumes 6 mm long including 2–3 mm long awn, the upper 4–5 mm long equal to or shorter than utricle; male glumes smaller. Utricles brown with a green stripe on at least 2 faces, triangular, 4–6 mm long including 1.5–2 mm beak, with very distinct longitudinal ridges, containing a nutlet smooth, 3 mm long, yellowish triangular, with flattened green scabrid axis which sometimes develops into a narrow club-like male spikelet to 10 mm long.

Zimbabwe. E: Chipinge Dist., Chirinda Forest, 13.ix.1947, *Whellan* 146 (K, SRGH); Chipinge Dist., Mt Selinda, Chipete Forest, 28.iv.1976, *Müller* 2823 (K, SRGH).

Also in Ethiopia, Uganda, Kenya and South Africa. Upland forest, sometimes at margins or in clearings; 1100–1800 m.

Within the Flora area this species is only recorded from Chipinge District in the Eastern Highlands of Zimbabwe, with 6 collections. In general appearance these specimens are laxer, taller and have slightly broader leaves than *Schoenoxiphium sparteum*.

Fig. 14.**76**. SCHOENOXIPHIUM LEHMANNII. 1, 2, habit; 3, ligule apex, leaf displaced; 4, leaf underside; 5, partial inflorescence; 6, 7, female floret, with and without utricle; 8, rachilla; 9, utricle of female floret with extended rachilla bearing terminal male spikelet; 10, male spikelet, glumes *in situ*; 11, male floret with one stamen, a filament and rachilla, glumes displaced; 12, 13, nutlet, adaxial and apical view. 5–11 from *Robinson* 6352; 1–4, 12, 13 from *Whellan* 413. Scale bars: 1 = 40 mm; 2 = 250 mm; 3–13 = 2 mm. Drawn by Jane Browning.

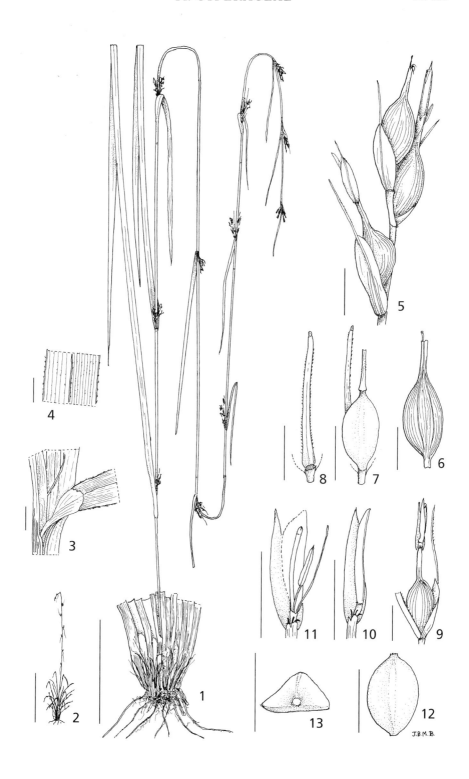

J.B.M.B.

37. CAREX L.[37]

Carex L., Sp. Pl.: 972 (1753). —Gehrke in Bot. J. Linn. Soc. **166**: 51–99 (2011).

Perennial herbs of permanent wetlands or damp forest margins, tufted or rhizomatous, sometimes producing 'tiller clumps'. Culms 3-angled, occasionally rounded, mostly nodose, a few scapose. Leaves sheathing at base, ligule present; blade linear, elongated, V-shaped or flanged usually with prominent abaxial keel. Inflorescences (in the Flora area) either a panicle with all branches sessile or sub-sessile (foxtail-like), an interrupted panicle with partial long-stalked panicles or a raceme of spikes; lowermost branches frequently distant. Inflorescence bracts prominent, leaf-like or reduced, bristle-like; higher order branches in paniculate inflorescences with inner (axillary) bract frequently inflated, nervose and utricle-like. Spikelets unisexual, 1-flowered, solitary in axils of glume-like bracts, male spikelet naked, female spikelet (sometimes together with vestigial rachilla) surrounded by modified prophyll (utricle or perigynium); utricle bottle-shaped with apex drawn out into short or long rostrum through which style branches protrude, rostrum mostly 2-toothed. Perianth 0. Stamens 3. Style 2- or 3-branched. Nutlet elliptic to obovate, biconvex or 3-angled, often stipitate, beaked, papillose, remaining within utricle at dispersal.

A very large genus of 1500–2000 species occurring worldwide in moist to wet places, with a concentration in the Northern Hemisphere. 9 species in the Flora area.

I differ from Verdcourt (2010) in the interpretation of the spikelet – in my view both male and female spikelets in *Cariceae* comprise a single flower.

In his monograph, Kükenthal in Engler, Pflanzenr. **4**, 20(38): 67–767 (1909), recognized four subgenera with 69 sections, largely corresponding to inflorescence morphology. With the impact of cladistic theory and molecular techniques Kükenthal's classification has become obsolete, but there is as yet no satisfactory replacement and it still serves as a valuable starting point for further study.

Three of Kükenthal's subgenera occur in the Flora area:

Subgenus *Vignea* (T. Lestib.) Peterm.; basionym (genus) *Vignea* T. Lestib. (species 1).

Subgenus *Vigneastra* (Tuck.) Kük.; basionym (genus) *Vigneastra* Tuck. (subgenus *Indocarex* (Baill.) Kük. (species 2–5).

Subgenus *Carex* (subgenus *Eu-Carex* sensu Kükenthal, 1909) (species 6–9).

These subgenera are further divided into numerous sections, but they are not particularly useful in the African context, where relatively few species occur.

Key: mature inflorescences and utricles should always be examined, along with several utricles per inflorescence because the relative lengths of utricle body and rostrum vary greatly, even within one inflorescence.

1. Culm scapose; inflorescence dense, paniculate, foxtail-like; branches ± sessile; main inflorescence bracts reduced, bristle-like . . . . **1.** *lycurus*
 – Culm nodose; inflorescence not dense or foxtail-like; partial panicles or spikes stalked (stalks short in some racemose-spicate taxa); main inflorescence bracts well developed, leaf-like 2
2. Inflorescence of several branched partial inflorescences 3
 – Inflorescence a raceme of spikes 6
3. Higher order inflorescence branches hardly spreading at anthesis; inner (axillary) bract of ultimate branches not inflated or utricle-like; nutlets clawed . **2.** *steudneri*
 – Higher order inflorescence branches spreading at c. 90° at anthesis; inner (axillary) bract of ultimate branches inflated, utricle-like; nutlets not clawed . . 4
4. Rostrum nearly filiform, long, nearly equalling utricle body . . . . . . **3.** *chlorosaccus*
 – Rostrum thicker than above, shorter to much shorter than utricle body 5
5. Utricle body 3.5 mm long . . . . **4.** *macrophyllidion*

[37] by C. Archer

1. **Carex lycurus** K. Schum. in Engler, Pflanzenw. Ost-Afrikas **C**: 129 (1895). —Clarke in F.T.A. **8**: 517 (1902). Type: Tanzania, Usambara, Heboma, iii.1893, *Holst* 2554 (K lectotype, M, Z), lectotypified by Lye (1983).

    Carex robinsonii Podlech in Mitt. Bot. Staatssamml. München **4**: 122 (1961). Type: Zambia, Nyika Plateau, no locality, c. 2100 m, 2.i.1959, *Robinson* 3000 (M holotype, K, NU, PRE, SRGH).
    Carex conferta A. Rich. var. *lycurus* (K. Schum.) Lye in Nordic J. Bot. **3**: 244 (1983). —Haines & Lye, Sedges & Rushes E Afr.: 373, fig.769 (1983). —Burrows & Willis, Pl. Nyika Plateau: 298 (2005).
    Carex lycurus subsp. *lycurus*. —Verdcourt in F.T.E.A., Cyperaceae: 429 (2010).

Perennial herb 0.3–2 m tall, with stout tufts from a creeping rhizome, base stout and 3-angled. Leaves to 60 cm long, (5.5)7–12.5 mm wide, slightly scabrid to smooth; basal sheaths pale brownish. Inflorescence green and brown, (3)6–11.5 × 1–3 cm, interrupted and ± lobulate, stalk scabrid below inflorescence; lower bracts sometimes 2–4 cm long but often not developed; individual spikes 1.5–2.5 cm long, mainly female, some males at top; female spikelet bracts chestnut with green margins and keel, thin, ovate, lanceolate, 3 mm long. Utricle greenish, often dark or black with age, ovoid-lanceolate, c. 4 mm long, drawn out into narrow rostrum for ± half length of utricle, convex side with 6–9(14) prominent continuous ribs, rostrum margin ± scabrid. Style branches 2. Nutlets dark, 2 mm long including persistent style base.

Zambia. W: Mwinilunga Dist., by R. Lunga, 1.xii.1937, *Milne-Redhead* 3460 (K). N: Isoka Dist., Nyika Plateau, c. 2180 m, 2.i.1959, *Robinson* 3000 (K, M, NU, PRE, SRGH). **Zimbabwe**. E: Mutare Dist., Sheba Forest Estate, c. 1800 m, 23.i.1967, *Jacobsen* 3073 (PRE). **Malawi**. N: Nyika Plateau, 5.ii.2005, *Gehrke & Patel* 312 (MAL, Z). C: Dedza Dist., Dedza Mt, edge of stream, 2.ii.1967, *Salubeni* 529 (K, NU, SRGH).

Also in Kenya and Tanzania. In upland grassland in wet partly shaded places along streams; 1300–2200 m.

Conservation notes: Widespread; not threatened.

Related to several similar-looking species of high altitude wetlands in N & E Africa. The foxtail-shaped inflorescence is immediately distinctive. The 2-branched style, flattened utricle and nutlet occur throughout the subgenus, but are also present in Sect. *Praelongae* Kük. of subgenus *Carex*. The sole example from the Flora area is *Carex rhodesiaca* which, however, has a panicle of pendulous spikes and a papillate surface to the utricle.

2. **Carex steudneri** Boeckeler in Linnaea **40**: 364 (1876). —Clarke in F.T.A. **8**: 520 (1902). —Haines & Lye, Sedges & Rushes E Afr.: 376, fig.775 (1983). —Verdcourt in F.T.E.A., Cyperaceae: 435 (2010). —Gehrke in Bot. J. Linn. Soc. **166**: 83 (2011), excluding *C. zuluensis*. Type: Ethiopia, Semien, Ghaba, *Steudner* 931 (B† holotype, K lectotype), lectotypified by Gehrke (2011).

    Carex zuluensis C.B. Clarke in Bull. Misc. Inform. Kew, add. ser. **8**: 74 (1908). —Nelmes in Mem. New York Bot. Gard. **9**: 100 (1954). —Gordon-Gray in Strelitzia **2**: 43 (1995). —Mapaura & Timberlake, Checkl. Zimbabwe Pl.: 87 (2004). —Gehrke in Bot. J. Linn. Soc. **166**: 83 (2011). Type: South Africa, Baziya, Tembuland, 610 m, n.d., *Baur* 1156 (K lectotype, BOL), lectotypified by C. Reid on sheet.

Carex huttoniana Kük. in Engler, Pflanzenr. **4**, 20(38): 271 (1909).
Carex condensata C.B. Clarke in Fl. Cap. **7**: 305 (1898); in F.T.A. **8**: 521 (1902), invalid name, non Nees (1834).

Tufted rhizomatous perennial herb 0.4–1 m tall with stiff stems. Leaf blades 5–30 cm long, 3–12 mm wide, flat, scabrid on margins and midrib. Inflorescence ± of several narrow, dense, often pendulous panicles, 1–2 from each of uppermost leaf sheaths; spikes brown, lanceolate, 6 × 1–2 mm, bisexual, male above and few female flowers below; glumes red-brown with pale midrib, 3–6 mm long (upper always 5–6 mm). Utricle greenish to red-brown to dark brown, sometimes with green ridges, lanceolate, 5–6(7) mm long (including 2 mm rostrum), pubescent or densely scabrid at least on beak and major ribs, or ± glabrous except for marginal or scattered scabrid hairs; beak with short erect teeth. Nutlets long-clawed.

Zimbabwe. E: Nyanga Dist., Mt Nyangani, 2590 m, 9.vi.1993, *Browning* 626 (MICH, NU). **Malawi**. N: Nyika Plateau, no locality, 5.ii.2005, *Gehrke & Patel* 309 (MAL, Z) S: Mt Mulanje, Lichenya Plateau, along Lichenya R. below Forestry Rest-House, 1740 m, 31.xii.1988, *Chapman & Chapman* 9461 (MO, PRE). **Mozambique**. Z: Gurué Dist., R. Malema, Marope, 22 km from Gurué, 1140 m, 1.viii.1979, *de Koning* 7521 (K, LISC, LMU).

Also in Ethiopia, Kenya, Tanzania, Swaziland and South Africa. On mountain slopes, frequently amongst rocks, often shaded; 1100–2600 m.

Conservation notes: Widespread; not threatened.

Plants with mature inflorescences cannot be confused with any other species in the subgenus owing to the narrow, dense, rusty-red panicles.

I have decided to follow Verdcourt (2010) in treating *Carex zuluensis* as synonymous with *C. steudneri*. Gehrke (2011) maintained the two species as separate on the basis of nutlet shape, with that of *C. steudneri* oblong and that of *C. zuluensis* obovate. However, the inflorescences of all material I have examined from the region are not fully mature. South African material has proved to be rather variable in nutlet shape but is generally narrowly to broadly elliptical rather than obovate. On the basis of gross inflorescence morphology there may be some merit in retaining *C. zuluensis* at subspecific level, but I hesitate to make further changes until sufficient mature material has been examined.

3. **Carex chlorosaccus** C.B. Clarke in J. Linn. Soc., Bot. **34**: 298 (1899). —Clarke in F.T.A. **8**: 519 (1902). —Haines & Lye, Sedges & Rushes E Afr.: 375, fig.773 (1983). —Burrows & Willis, Pl. Nyika Plateau: 298 (2005). —Verdcourt in F.T.E.A., Cyperaceae: 432, fig.65 (2010). Type: Equatorial Guinea, Bioko (Fernando Po), Clarence Peak, 2500 m, xii.1860, *Mann* 653 (K lectotype, K), lectotypified by Gehrke (2011). FIGURE 14.**77**.
     Carex echinochloe Kunze var. *chlorosaccus* (C.B. Clarke) Kük. in Engler, Pflanzenr. **4**, 20(38): 271 (1909).
     Carex brassii Nelmes in Mem. New York Bot. Gard. **9**: 100 (1954). —Burrows & Willis, Pl. Nyika Plateau: 298 (2005). —Gehrke in Bot. J. Linn. Soc. **166**: 85 (2011). Type: Malawi, Mt Mulanje, Luchenya Plateau, 1890 m, 7.vii.1946, *Brass* 16714 (K holotype, BM, BR, PRE, SRGH).

Perennial tussock herb 0.3–1(1.2) m tall with woody rhizome; basal leaf sheaths dark red. Leaf blades 50–90 cm long, 5–9 mm wide, scabrid on margins and veins. Inflorescence green to brownish, slender, much branched panicles 8–50 cm long, often with one short and one longer branch at lower nodes; inflorescence axis and side branches densely pubescent; upper bracts 2 mm long with 3–8 mm awn, lower ones leaf-like; spikes 6–20 × 3–5 mm, few male spikelets above and 5–10 female spikelets below; glumes pale green turning pale brown or greyish, pubescent, 4–5 mm long including 1.5–2 mm awn. Utricle green becoming pale brown, 4–5.5 long including 1.5–2.2 mm glabrous or slightly scabrid rostrum, often ± curved with several distinct ribs on each side, with short spine-like hairs below beak.

Fig. 14.**77**. CAREX CHLOROSACCUS. 1, habit (× ⅔); 2, spikelet (× 3); 3, male floret and tip of spikelet (× 8); 4, glume from female floret (× 12); 5, utricle (× 12); 6, beak details (× 16). 1, 7 from *F. Rose* 1156; 2 from *Trelawny* AB4381; 3–5 from *Napper* 831; 6 from *Lye* 9125. Drawn by Juliet Williamson. Reproduced from Flora of Tropical East Africa (2010).

Zambia. N: Mbala Dist., Lunzua Falls near Mbala, 1500 m, 2.ii.1959, *Richards* 10844 (K). C: Kapiri Mposhi Dist., 48 km E of Kapiri Mposhi where Great North Road crosses R. Lusenfwa, 1220 m, 17.i.1959, *Robinson* 3234 (K). E: Nyika Plateau, Chowo Forest, grid ref. 575 8829, 2145 m, 17.v.1970, *Brummitt* 10857 (K, LISC, PRE). **Malawi**. N: Nyika Nat. Park, Juniper Forest, 10°45'06"S 33°53'15"E, 2176 m, 5.iv.2000, *Smook* 10904 (MAL, PRE). C: Nchisi Dist., Nchisi Mt., 6.v.1963, *Verboom* 1295 (K). S: Mt Mulanje, Luchenya Plateau, stream bank, 1890 m, 7.vii.1946, *Brass* 16714 (BM, BR, K, PRE, SRGH). **Mozambique**. Z: Gurué Dist., Mt Namuli, Cascades de Namuli, 15°24'39"S 36°58'38"E, 1055 m, 25.xi.2007, *Mphamba et al.* 32 (K, LMA, MAL).

Also in Equatorial Guinea, D.R. Congo, Rwanda, Sudan, Ethiopia, Uganda, Kenya and Tanzania. Along streams in forest; 1000–2200 m.

Conservation notes: Widespread; not threatened.

Neither Verdcourt (2010) nor Gehrke (2011) list the species for the Flora Zambesiaca area.

The species is unmistakable with its (usually) greenish to brownish inflorescence and strongly falcate utricles with fine needle-like rostrums projecting in all directions. *Carex castanostachya* Kük. is probably referable to this species, although until now it has been segregated on the basis of its dark chestnut-coloured female spikelet bracts. *C. brassii* is included here because the type specimen has the same slender panicle and nearly glabrous utricles with bodies of the same narrowly elliptical shape and filiform rostrums, although the utricles in the type are not falcate. A similar situation occurs in *C. vallis-rosetto* with which *C. cyrtosaccus* was synonymised by Gehrke (2011); the former has falcate utricles and the latter not.

4. **Carex macrophyllidion** Nelmes in Bull. Misc. Inform. Kew **1940**: 161 (1940). —Haines & Lye, Sedges & Rushes E Afr.: 376, fig.774 (1983). —Verdcourt in F.T.E.A., Cypeaceae: 434 (2010). Type: Angola, Moxico Dist., by R. Mfumbu, 7.i.1938, *Milne-Redhead* 3971 (K holotype, BR).

Slender perennial herb to 60 cm tall, forming large tufts from a short woody rhizome. Leaves V-shaped in section or ± flat, 5–60 cm long, 2.5–4.5 mm wide, upper leaves overtopping the inflorescence, long-attenuate, scabrid, with long smooth basal sheaths becoming blackish and fibrous. Panicle interrupted, 1.5–4 × 1–2 cm; secondary panicles 5–6, the lower single, upper paired, subpyramidal; lower long-peduncled, upper more shortly; rachis glabrous excepting scabrid angles; bracts leaf-like overtopping inflorescence; spikes dense-flowered with male part shorter than female, 7–10 mm long; bracteoles long-aristate with hispidulous margins; female spikelet bracts pale greenish brown, ovate-lanceolate, 5–6 mm long with scabrid arista. Utricles yellow-green with green veins, equal to or longer than glumes, 5.5–6 mm long, glabrous, narrowed into a long bidentate ± scabrid rostrum to 1.5 mm.

Zambia. N: Chinsali Dist., 42 km S of Chinsali, 14.i.1959, *Robinson* 3202 (K, PRE). W: Mwinilunga Dist., 0.8 km S of Matonchi Farm, 6.xii.1937, *Milne-Redhead* 3511 (K, PRE). S: Choma Dist., Choma, Beckett's Farm, 1300 m, 16.i.1959, *Robinson* 3236 (K, PRE, SRGH). **Zimbabwe**. E: Nyanga Dist., N of Nyanga Expt. Station, Nyanga Nat. Park, 17.vi.1975, *Burrows* 741 (PRE, SRGH). **Malawi**. N: Mzimba Dist., Mzuzu, Marymount, c. 1370 m, 18.ii.1974, *Pawek* 8121 (K, MO, PRE, SRGH, UC).

Also in Angola and Tanzania. In *Brachystegia* woodland; 1300–1900 m.

Conservation notes: Widespread; not threatened.

The combination of narrow leaves, sparsely-branched inflorescence, spikes with few spikelets and large utricles makes this species unmistakable in the subgenus.

5. **Carex spicatopaniculata** C.B. Clarke in Fl. Cap. **7**: 304 (1898). —Nelmes in Bull. Misc. Inform. Kew **1940**: 160 (1940). —Haines & Lye, Sedges & Rushes E Afr.: 374, fig.771 (1983). —Gordon-Gray in Strelitzia **2**: 43 (1995). —Burrows &

Willis, Pl. Nyika Plateau: 298 (2005). —Verdcourt in F.T.E.A., Cyperaceae: 431 (2010). Type: South Africa, KwaZulu-Natal, Inanda, 7.i.1881, *Medley Wood* 1190 (K lectotype, BOL, BM, BR, NH, SAM), lectotypified by Nelmes (1940).

> *Carex nyasensis* C.B. Clarke in F.T.A. **8**: 519 (1902). —Nelmes in Bull. Misc. Inform. Kew **1940**: 162 (1940). —Mapaura & Timberlake, Checkl. Zimbabwe Pl.: 88 (2004) as *nyassensis*. Type: Malawi, Zomba plains, n.d., *Whyte* s.n. (K lectotype), lectotypified by Nelmes (1940).

> *Carex echinochloe* Kunze var. *nyasensis* (C.B. Clarke) Kük. in Engler, Pflanzenr. **4**, 20(38): 271 (1909).

> *Carex tricholepis* Nelmes in Bull. Misc. Inform. Kew **1940**: 160 (1940). —Mapaura & Timberlake, Checkl. Zimbabwe Pl.: 87 (2004). —Gehrke in Bot. J. Linn. Soc. **166**: 85 (2011). Type: Zambia, Mwinilunga Dist.; just S of Matonchi Farm, 18.ii.1938, *Milne-Redhead* 3686a (K holotype).

> *Carex angolensis* Nelmes in Bull. Misc. Inform. Kew **1940**: 162 (1940). —Gehrke in Bot. J. Linn. Soc. **166**: 83 (2011). Type: Angola, Moxico, between R. Monu and R. Kampashi, 19.i.1938, *Milne-Readhead* 4222 (K holotype, BR, K, PRE).

> *Carex echinochloe* subsp. *nyasensis* (C.B. Clarke) Lye in Nordic J. Bot. **3**: 244 (1983). —Haines & Lye, Sedges & Rushes E. Afr.: 375 (1983). —Verdcourt in F.T.E.A., Cyperaceae: 431 (2010). —Gehrke in Bot. J. Linn. Soc. **166**: 88 (2011).

Perennial tussock-forming herb 0.5–1.2 m tall with short rhizome. Leaves 20–50 cm long, 5–13 mm wide, flat or plicate, scabrid on veins and margins. Panicles green and brown with densely pubescent axes and branchlets, to 8 × 3–5 cm; spikes 5–10 × 4–5 mm; female spikelet bracts reddish brown, 3.5–4 mm long, awn 1–1.5 mm, scabrid. Utricle ellipsoid, 3.5–4 mm long (including rostrum), up to 30 pale ribs; rostrum 1.5–2 mm, bifid, lobes linear; rostrum shortly hairy, hairs extending down to 'shoulders' of utricle.

Zambia. N: Kasama Dist., Chishimba Falls, 12.ii.1961, *Robinson* 4374 (K). W: Mwinilunga Dist., just W of Matonchi Farm, 7.xii.1937, *Milne-Redhead* 3543 (K, PRE). C: Chongwe Dist., Chakwenga headwaters, 100–129 km E of Lusaka, 10.i.1964, *Robinson* 6207 (K). S: Mazabuka Dist., Nachibanga stream, 9.ii.1960, *White* 6915 (FHO, K). **Zimbabwe**. N: Mazowe Dist., upper reaches of Wengi R. where crosses Mutoroshanga Concession road, 12.iv.1959, *Drummond* 6048 (K, LISC, PRE, SRGH). W: Matobo Dist., Besna Kobila Farm, 1525 m, iii.1962, *Miller* 8199 (K, NU, PRE, SRGH). C: Goromonzi Dist., Dombashawa, 1370 m, 21.vii.1962, *Moll* 271 (NU). E: Mutare Dist., Vumba Mts, Thordale, 19.i.1990, *Browning* 264 (GENT, NU, PRE, SRGH). S: Mberengwa Dist., Bukwa Mt, SW gully between main and secondary ridges on W side, c. 1200 m, 1.v.1973, *Pope, Biegel & Simon* 1036 (K, PRE, SRGH). **Malawi**. N: Nyika Nat. Park, towards Sangula kopje, 10°39'54"S 33°45'06"E, 1.iv.2000, *Smook* 10855 (LMA, MAL, PRE, SRGH, UZL). C: Nkhotakota Dist., Ntchisi Mt, 1400 m, 26.vii.1946, *Brass* 16948 (K, NY, PRE). S: Mulanje Mt, foot of Great Ruo Gorge between hydroelectric station and dam, grid ref. 784 8232, 870–1060 m, 18.iii.1970, *Brummitt & Banda* 9218 (K, PRE). **Mozambique**. Z: Gurué Dist., Gurué, 1100 m, 30.vii.1979, *de Koning* 7477 (K, LISC, LMU). T: Angónia Dist., Ulónguè, NE of serra Dómue, 19.xii.1980, *Macuácua* 1491 (K, LISC, LMA, PRE). MS: Sussundenga Dist., 20 km W of Dombe, foothills at SE end of Chimanimani Mts., 24.iv.1974, *Pope & Müller* 1289 (K, LISC, SRGH).

Also in Ethiopia, D.R. Congo, Kenya, Tanzania, South Africa and Swaziland and possibly the Comoros and Mascarene Is. Damp places between boulders or on forest margins in shade or half-shade, 750–1860 m.

Conservation notes: Widespread; not threatened.

Subgenus *Vigneastra* and especially the *Carex spicatopaniculata* complex is an exceptionally difficult group that cannot be resolved in a short space of time. Some characters used by previous authors, like leaf width, inflorescence size, shape and colour, are not entirely reliable. Utricle characters including indumentum also seem to be highly variable, and it is clear that much more field and herbarium work is necessary before a completely satisfactory treatment of the complex can be produced.

6. **Carex rhodesiaca** Nelmes in Bull. Misc. Inform. Kew **1939**: 159 (1939). Type: Zambia, Mwinilunga Dist., by R. Lunga just E of Mwinilunga, 27.xi.1937, *Milne-Redhead* 3422 (K lectotype, B, BR, K, PRE), lectotypified by Gehrke (2011).

 Carex austroafricana (Kük.) Raymond in Naturaliste Canad. **91**: 126 (1964). Type: South Africa, Mooi R., xii.1890, *Medley Wood* 1690 (BM lectotype, Z), lectotypified here. *Medley Wood* 1690 (BM) is not the holotype as stated by Gehrke (2011).
 Carex cernua Boott var. *austroafricana* Kük. in Engler, Pflanzenr. **4**, 20(38): 354 (1909).

Plants 25–60 cm tall, usually tufted, occasionally with short rhizomes. Leaves slightly glaucous. Largest basal leaf blade 30–48 cm long, 3–7 mm wide, flat or plicate, distally minutely scabrid. Inflorescence a raceme of spikes, 5–31 cm long. Spikes 4–6, pendulous, largest 25–70 × 7–11 mm. Peduncles 2–15 cm long. Apical spike usually male, sometimes bisexual. Bracts of female spikelets ovate, straw-coloured, with broad keel extending into awn; awn 2–4.2 mm, margin scabrid. Utricle 3–4 × 1.5–1.8 mm, usually reddish brown with whitish rostrum, surface papillate, inconspicuously nerved; rostrum very short, 0.3–0.5 mm, apex truncate. Stigmas 2. Nutlet obovate, 1.9–2.2 × 1–1.7 mm wide, dark brown with lighter angles.

Zambia. W: Mwinilunga Dist., by R. Lunga just E of Mwinilunga, c. 1300 m, 27.xi.1937, *Milne-Redhead* 3422 (B, BR, K, PRE). C: Serenje Dist., Kasanka Nat. Park, in mushitu near Fibwe Hide, c. 0.5 km N of carpark, 12°35'20"S 30°15'07"E, 1170 m, 19.xi.1993, *Harder et al.* 1958 (MO, PRE). **Zimbabwe**. C: Harare, Highlands Park, 30.xi.1947, *Whellan* 290 (K, SRGH). E: Nyanga Dist., Nyanga Nat. Park, Rhodes Hotel drift, 1370 m, 30.xii.1954, *Whellan* 814 (K, SRGH).

Also in South Africa. In seasonally wet areas and along streams in light shade to full sun; 1150–1300 m.

Conservation notes: Widespread: not threatened.

The panicle of pendulous spikes, papillate utricle and 2-branched style make the section unmistakable. In Tanzania (and probably D.R. Congo, Gehrke 2011) this taxon is represented by the similar-looking *Carex papillosissima* Nelmes, which differs in the awn of the female spikelet bract being shorter than the utricle and the utricle surface being more densely papillate with larger papillae.

7. **Carex cognata** Kunth, Enum. Pl. **2**: 502 (1837). —Haines & Lye, Sedges & Rushes E. Afr.: 384 (1983), as var. *cognata*. —Gordon-Gray in Strelitzia **2**: 39, fig.14 (1995). —Verdcourt in F.T.E.A., Cyperaceae: 446 (2010), excluding var. *drakensbergensis*. —Archer & Balkwill in Bothalia **42**: 190 (2012). Type: South Africa, Zwellendam, George, *Mundt* (B† holotype); Eastern Cape, Kentani Dist., 22.ix.1910, *Pegler* 151 (PRE neotype), neotypified by Archer & Balkwill (2012). FIGURE 14.**78**.

 Carex drakensbergensis C.B. Clarke in Fl. Cap. **7**: 309 (1898). —Gehrke in Bot. J. Linn. Soc. **166**: 74 (2011). Type: South Africa, KwaZulu-Natal (East Griqualand), near Kokstad, Vaalbank Farm, 18.xii.1889, *Wood* 4201 (K lectotype, BOL, NH), lectotypified by Gehrke (2011).
 Carex cognata var. *drakensbergensis* (C.B. Clarke) Kük. in Engler, Pflanzenr. **4**, 20(38): 699 (1909). —Mapaura & Timberlake, Checkl. Zimbabwe Pl.: 87 (2004). —Verdcourt in F.T.E.A., Cyperaceae: 447 (2010).
 Carex congolensis Turrill in Bull. Misc. Inform., Kew **1912**: 240 (1912). —Gehrke in Bot. J. Linn. Soc. **166**: 75 (2011). Type: D.R. Congo, Lubumbashi, 21.xi.1911, *Rogers* 10082 (K holotype, BOL).
 Carex cognata var. *congolensis* (Turrill) Lye in Nordic J. Bot. **3**: 244 (1983). —Haines & Lye, Sedges & Rushes E Afr.: 384, fig.791 (1983). Type: D.R. Congo, Katanga, Lubumbashi (Elisabethville), 11°37'S 27°24'E, 1150 m, 21.x.1911, *Rogers* 10082 (K lectotype, BOL), lectotypified by Gehrke (2011).
 Carex pseudosphaerogyna Nelmes in Bull. Misc. Inform. Kew **1937**: 473 (1937). Type: Uganda, Kigezi Dist., Virunga Mts, NW end of Lake Bunyonyi, 27.xi.1934, *Taylor* 2146 (BM holotype).

Tufted perennial from creeping rhizome with short scaly rhizomes; stems 0.3–1 m tall, glabrous. Leaf-blades 10–80(120) cm long, 3–7(10) mm wide, plicate, scabrid on margins at least at tips; some transverse ribs occur between longitudinal veins. Main inflorescence bracts leaf-like, 3–6 times as long as spikes, bracts successively smaller upwards. Inflorescence of 4–6 sub-erect, sessile or pendulous, pedunculate spikes; spikes 2–3, 10–40 × 2–4 mm, terminal spike male, lateral spikes female; spikelet bracts pale golden brown, dark brown or red-brown, 3–5 mm long, acuminate, densely scabrid on margin and veins. Utricles pale greenish, golden brown or red-brown, ovoid, 3–4.5 mm long (including 1–1.5 mm rostrum), many-veined;

Fig. 14.**78**. CAREX COGNATA. 1, habit; 2, glume and utricle; 3, nutlet, style branches shrivelled. All from *Browning* 256. Scale bars: 1 = 40 mm; 2, 3 = 2 mm. Drawn by Jane Browning. Reproduced from Strelitzia (1995) by kind permission of the South African National Biodiversity Institute, Pretoria.

rostrum glabrous, strongly 2-toothed. Nutlet yellowish white or dark with pale edges, with long persistent curving style.

Caprivi. Linyanti R. swamp, Shaile, 12.x.1975, *Edwards* 4391 (K, PRE). **Botswana**. N: Ngamiland Dist., Quangwa in Thaoge R. bed, 19°30.25'S 22°17.05'E, just below junction with Wedamo R., 5.xi.1975, *P.A. Smith* 1531 (K, PRE, SRGH). **Zambia**. W: Mwinilunga Dist., by R. Lunga at Mwinilunga, 1.xi.1937, *Milne-Redhead* 3464 (K, PRE). N: Kaputa Dist., Mweru Wantipa, Lake Chisi, 1000 m, 13.xii.1960, *Richards* 13668 (K); Isoka Dist., Nyika Plateau, c. 2180 m, 3.i.1959, *Robinson* 3023 (PRE). **Zimbabwe**. N: Mhangura Dist., Dichwe Farm, Lemon Forest, c. 1170 m, 16.xi.1969, *Jacobsen* 4057 (PRE). C: Harare Dist., Mount Pleasant, edge of golf course, 1495 m, xii.1959, *Drummond* 6774 (K, LISC, PRE, SRGH). E: Nyanga Dist., Nyanga Nat. Park, Nyanora R. at bridge on road to campsite, c. 1920 m, 5.i.1972, *Gibbs Russell* 1184 (K, PRE, SRGH). **Malawi**. N: Nyika Plateau, no locality, 5.ii.2005, *Gehrke* 310 (MAL, Z). **Mozambique**. M: Delagoa Bay, 1893, *Junod* 414 (K).

Also in Namibia (Waterberg Plateau), Angola, D.R. Congo, Kenya, Tanzania, Swaziland, South Africa and Lesotho. Wetlands, wet places on margins of watercourses or in wet hillside flushes in grassland, 1000–3000 m.

Conservation notes: Widespread; not threatened.

Carex cognata is a well-marked species with its yellow-green foliage and conspicuous transverse venation, glabrous utricles and sharply-pointed rostrum teeth. In *C. cognata* *sensu stricto* the inflorescence comprises several short, ± sessile, sub-erect, clustered yellowish-greenish spikes, while its high-altitude form, *C. drakensbergensis* has longer, penduncled, pendulous spikes in which the female spikelet bracts are frequently rusty-red. Both forms occur in South Africa but an extensive field- and herbarium study (Archer & Balkwill, 2012) showed that *C. drakensbergensis* has no taxonomic merit.

8. **Carex petitiana** A. Rich., Tent. Fl. Abyss. **2**: 513 (1850). —Clarke in F.T.A. **8**: 522 (1902). —Haines & Lye, Sedges & Rushes E Afr.: 382, fig.789 (1983). — Verdcourt in F.T.E.A., Cyperaceae: 439, fig.66 (2010). Type: Ethiopia, mountain in 'Ouodgerate province' (probably Amhara), n.d., *Petit* s.n. (P lectotype, K frag.), lectotypified by Gehrke (2011). FIGURE 14.**79**.

   Carex longipedunculata K. Schum. in Engler, Pflanzenw. Ost-Afrikas **C**: 130 (1895). Type: Tanzania, Kilimanjaro, near Nobolu cave, 3200 m, 1893, *Volkens* 2015 (B lectotype, K), lectotypified by Gehrke (2011).
   Carex longipedunculata subsp. *cuprea* Kük. in Notizbl. Bot. Gart. Berlin-Dahlem **9**: 315 (1925).
   Carex cuprea (Kük.) Nelmes in Bull. Misc. Inform. Kew **1938**: 247 (1938). Type: Kenya, Aberdare Mts, near West Kenya Forest Station, 4.i.1922, *Fries & Fries* 734 (UPS lectotype, BR, K), lectotypified by Gehrke (2011).
   Carex aethiopica sensu Burrows & Willis, Pl. Nyika Plateau: 298 (2005), non Schkuhr.
   Carex fischeri sensu Verdcourt in F.T.E.A., Cyperaceae: 439 (2010), non K. Schum.

Perennial tussock-forming herb 0.4–1.2 m tall and to 90 cm wide with short to long creeping rhizome, glabrous. Leaf-blades 10–60 cm long, 3–10 mm wide, flat or plicate, scabrid at least on margins or ± smooth; leaf-sheaths orange-brown or purplish. Inflorescence of 4–8 suberect or pendulous greenish or brown spikes, arising singly at nodes, 20–70 × 4–7(10) mm; terminal spike entirely male or to half female spikelets above male spikelets, lateral spikes female with only few male spikelets at base or entirely female; peduncles 1–5(14) cm long; glumes pale brown, coppery or reddish brown with broad pale or green midribs, 3–6 mm long including awn. Utricles green with dense dark speckling, (3.5)4–6 mm long including 1–1.5 mm rostrum, glabrous or scabrid; rostrum deeply bifid with distinct divaricate spine-like teeth.

Malawi. N: Nyika Plateau, Lake Kaulime, c. 2330 m, 4.i.1959, *Robinson* 3028 (PRE, UZL). **Zimbabwe**. E: Nyanga Dist., Nyanga Nat. Park, Mare Dam, c. 1960 m, 6.i.1972, *Gibbs Russell* 1208 (K, PRE, SRGH).

Fig. 14.**79**. CAREX PETITIANA. 1, habit (× ²/₃); 2, spike (× 1.5); 3, male floret (× 8); 4, glume from female floret (× 8); 5, utricle and stigmas (× 8); 6, beak detail (× 16); 7, nutlet (× 12). 1 from *Fries & Fries* 405; 2–7 from *Greenway & Kanuri* 13666. Drawn by Juliet Williamson. Reproduced from Flora of Tropical East Africa (2010).

Also in West Africa (Cameroon and Nigeria), Ethiopia, Kenya, Uganda and Tanzania. Along watercourses, on wet forest margins or in swamps, often on sloping ground; 1700–2400 m.

Conservation notes. Widespread; not threatened.

Due to its bisexual spikes *Carex petitiana* is a distinctive species that shows little variation over its wide range.

9. **Carex vallis-rosetto** K. Schum. in Engler, Pflanzenw. Ost-Afrikas **C**: 130 (1895). —
    Clarke in F.T.A. **8**: 521 (1902). —Haines & Lye, Sedges & Rushes E Afr.: 381,
    fig.787 (1983). —Verdcourt in F.T.E.A., Cyperaceae: 443 (2010). Type: Tanzania,
    Usambaras, Rosetto valley, n.d., *Holst* 3823 (B lectotype, K frag.), lectotypified by
    Gehrke (2011).

    Carex cyrtosaccus C.B. Clarke in F.T.A. **8**: 524 (1902). —Verdcourt in F.T.E.A., Cyperaceae:
    444 (2010). Type: Malawi, Mt Mulanje, 1891, *Whyte* s.n. (K lectotype), lectotypified by
    Haines & Lye (1983); not holotype as stated by Gehrke (2011).

Tufted perennial with thick rhizome, 0.4–2 m tall; basal leaf sheaths dark purple or spotted purple. Leaf-blades 20–80 cm long, 4–13 mm wide, flat or plicate, scabrid at least at apex. Inflorescence medium brown, of 8–20 long sub-erect to pendulous spikes arising in pairs from leaf sheaths (occasionally in 3s or solitary), often branched with 1–6 much smaller spikes at base of main spike; main spikes 20–200 × 4–8 mm, upper 1–3 spikes usually entirely male, the lower usually female with short male area at tip, longest peduncles 3–7 cm long; smaller spikes at base of main spikes to 1.5 cm; female spikelet bracts brownish red or speckled, with broad green midrib, obovate to ovate-lanceolate with excurrent scabrid midrib, 4–6 mm long. Utricles greenish to brown and often speckled dark reddish brown or purplish, 4–6 mm long including 1–1.5(2) mm smooth or scabrid rostrum, distinctly longitudinally ribbed, in our area usually falcate; rostrum distinctly toothed.

Malawi: N: Nkhata Bay Dist., 2.8 km ENE of Chikangawa, c. 1780 m, 13.x.1978, *Phillips* 4081 (MO, PRE). C: Dedza Dist., forests on Dedza Mt., 6.ix.1950, *Jackson* 159 (K). S: Mulanje Dist., Mt Mulanje, 1891, *Whyte* s.n. (K). **Mozambique**. Z: Gurué Dist., Mt Namuli, S part of plateau, 15°23'06"S 37°02'12"E, 1880 m, 26.xi.2008, *Patel* 23 (K, LMA).

Also in Kenya and Tanzania. High altitude swamps and margins of watercourses, 1700–2300 m.

Conservation notes. Widespread; not threatened.

A very distinctive large species, only occurring in the Flora area on the Nyika Plateau and on Mt Mulanje. I agree with Gehrke (2011) that *Carex cyrtosaccus* cannot be adequately separated from *C. vallis-rosetto*.

Omitted species.

Carex johnstonii Boeckeler in Bot. Jahrb. Syst. **7**: 278 (1886).

This species is stated by several authors to occur in Malawi, e.g. Haines & Lye: 377 (1983), Verdcourt (2010) and Gehrke (2011), all possibly due to Kükenthal (1909: 593) who cites "Nyassaland: Rungwe-Stock, hochste Spitze in der Bambuszone, 2500 m, *W. Goetze* 1305". However, this locality is in Tanzania, not Malawi. It is a very distinctive species due to its profusely branched inflorescence and long narrow utricles.

Carex echinochloe Kunze (sensu stricto).

I have not seen any specimens that I can identify with certainty as this N and E African species, but *Milne-Redhead* 3494 (PRE) and 3686 (PRE) from W Zambia (both cited as *Carex tricholepis* in Nelmes, Bull. Misc. Inform. Kew **1940**: 160, 1940) have the rather broad leaves and very dense, greenish straw-coloured inflorescences of *C. echinochloe* and may represent the species. However, the inflorescences lack mature utricles and are here excluded from the very broadly defined *C. spicatopaniculata*.

INDEX TO BOTANICAL NAMES

Accepted names in roman, synonyms in *italic*. Bold page numbers indicate main entries of accepted names (names with description and in the keys,) and illustrations.

alba sensu lato 314
albiceps (Ridl.) Rendle 309, **315**
albogracilis Lye 310, 315, 330
aromatica Ridl. 328
aurata Nees 319
aurata sensu Napper 316
aureovillosa Lye 309, **316**
blepharinota Boeckeler 333
brevifolia Rottb. 309, **316**, 317, 326
  subsp. brevifolia 317
  subsp. intricata 317
buchananii C.B. Clarke **334**
bulbosa P. Beauv. 310, 315, **317**, 319
cartilaginea K. Schum. 310, **318**
chrysantha K. Schum. var. *comosipes* (Mattf. & Kük.)
    J.-P.Lebrun & Stork 334
colorata sensu Napper 316
comosipes (Mattf. & Kük.) Napper **334**
controversa Steud. **334**
  var. *subexalata* C.B. Clarke 333
crassipes Boeckeler 310, **319**
cristata Kunth 311
cylindrica Nees 325, 326
  var. *major* C.B. Clarke 325
cyperoides Roxb. 252
elatior Kunth. 328
erecta Schumach. **319**
  subsp. *albescens* Lye 320
  var. *polyphylla* (Kunth) S.S. Hooper 328
  var. *intercedens* Kük. 319
  var. *intricata* C.B. Clarke 316
ferruginea Peter 333
imerinensis Cherm. 321, 322
intricata Cherm. 316, 317
leucocephala Boeckeler 334
macrocephala A. Rich. 317
  var. *angustior* C.B. Clarke 317
mafingaensis 322
melanosperma Nees 309, **320**, **321**
  var. hexalata Lye 322
  var. melanosperma **322**
merxmuelleri Podlech 315
metzii Steud. 331
microcephala Steud. 145
monocephala Rottb. 323
mtotomwema Beentje 310, **322**
nemoralis (J.R. Forst. & G. Forst.) Hutch. & Dalziel
    310, **323**
nervosa Steud. 309, **323**
  subsp. nervosa **323**
  subsp. oblonga (C.B. Clarke) J.-P. Lebrun &
    Stork 324
nigripes C.B. Clarke 310, **324**
odorata Vahl 310, **325**, 326
  subsp. *odorata* 325
  var. cylindrica 311
pauciflora Ridl. 309, **326**
peteri (Kük.) Lye 309, **326**
pinguis C.B. Clarke 328
platyphylla K. Schum. 310, **327**
  var. *longifolia* Kük. 327
polyphylla Kunth 309, **328**, 334
  var. elatior (Kunth) Kük. **328**, 329
  var. polyphylla **328**
pulchella Kunth **334**
pumila Michx. 309, 311, **329**, 330
rhizomafragilis Lye **329**
robinsoniana Mtot. 310, **330**
songeensis Lye 310, 322, **330**
sphaerocephala Boeckeler 317

sp. near nervosa **324**
squamulata Vahl 309, **331**
sumatrensis Retz. 242
tanzaniae Lye 310, 322, **331**
taxon A near alba var. alba 314
tenuifolia Steud. 310, **332**
  var. ciliata (Boeckeler) Beentje **333**
  var. tenuifolia **332**
triceps Rottb. 332
  var. *ciliata* Boeckeler 333
  var. ß *obtusiflora* Boeckeler 332
ugogensis (Peter & Kük.) Lye 309, **333**
welwitschii Ridl. 333
KYLLINGIELLA R.W. Haines & Lye 3, **144**, 180
  microcephala (Steud.) R.W. Haines & Lye 144,
    145, **146**, 147
  polyphylla (A. Rich.) Lye **144**
  simpsonii Muasya 144, **147**
Limnochloa limosa (Schrad.) Nees 51
LIPOCARPHA R. Br. 1, 2, 335, **337**
  abietina Goetgh. **338**
  albiceps Ridl. 338, **339**
  argentea (Vahl) R. Br. 340
  atra Ridl. 338, **339**
    var. *atra* 340
    var. *barteri* (C.B. Clarke) J. Raynal 340
  atropurpurea Boeckeler 348
  barteri C.B. Clarke 340
  chinensis (Osbeck) J. Kern 338, **340**, **341**
  comosa J. Raynal 337, **342**
  echinus J. Raynal 338, **343**
  filiformis (Vahl) Kunth 338, **350**
  hemisphaerica (Roth) Goetgh. 337, **343**, **344**
  isolepis (Nees) R.W. Haines 343
  kernii (Raymond) Goetgh. 338, **345**, 349
  leucaspis J. Raynal **350**
  micrantha (Vahl) G.C. Tucker 338, **346**
  microcephala (R. Br.) Kunth 345
  monostachya R. Gross & Mattf. 338, **346**
  nana (A. Rich.) Cherm. 338, 346, **347**
  prieuriana Steud. 338, **348**
    var. *prieuriana* 348
  pulcherrima Ridl. 347
    forma *luxurians* Merxm. 348
  purpureolutea Ridl. 339
  rehmannii (Ridl.) Goetgh. 338, **349**
  robinsonii J. Raynal 338, **350**
  senegalensis (Lam.) Durand & H. Durand 340
  tenera Boeckeler 348
  triceps (Roxb.) Nees var. *latinux* Kük. 338
Loudetia 192
Ludwigia 228
Mariscopsis Cherm. 306
  hyalinus (Vahl) Ballard 308
  suaveolens Cherm. 308
Mariusculus Goetgh. 255
  peteri (Kük.) Goetgh. 258
Mariscus Gaertn. 147, 164, 166, 236
  assimilis (Steud.) Podlech 250
  alpestris (K. Schum.) C.B. Clarke 233
  amauropus (Steud.) Cufod. 193
  angularis Turrill 241
  aristatus (Rottb.) Cherm. var. *atriceps* (Kük.)
    Podlech 166
  bequaertii Cherm. 248
  chersinus N.E. Br. 238
  chrysocephalus K. Schum. 187
  coloratus (Vahl) Nees 199, 333
  congestus (Vahl) C.B. Clarke 182
  cylindristachyus Steud. 237

pelophilus (Ridl.) C.B. Clarke 271, **286**
permutatus (Boeckeler) Napper 292
poikilostachys Nelmes 270, **302**
polystachyos (Rottb.) P. Beauv. 270, 271, **293**
    var. *laxiflorus* (Benth.) C.B. Clarke 293
propinquus Nees 280
pumilus (L.) Nees 270, **287**
    subsp. *patens* (Vahl) Podlech 287
    var. *patens* (Vahl) Kük. 287
rehmannianus C.B. Clarke 277, 280
richardsiae Vollesen 269, **296**
robinsonii Vollesen 271, **280**
sanguineosquamatus Van der Veken 269, **298**
sanguinolentus (Vahl) Nees 270, 271, **273**
scaettae Cherm. 269, **295**, 296
    var. *katangensis* Cherm. 295
segmentatus C.B. Clarke 290
smithianus (Ridl.) C.B. Clarke 269, **294**, 296
sp. A 271, **283**
sp. B 269, **297**
sp. C 270, **299**
spissiflorus C.B. Clarke 276
subtrigonus C.B. Clarke 270, **299**
sulcinux (C.B. Clarke) C.B. Clarke 286
sumbawangensis Hoenselaar 292, 293
tremulus (Poir.) C.B. Clarke 284
umbrosus Nees 275
    var. *grantii* (C.B. Clarke) C.B. Clarke 275
unioloides (R. Br.) Urb. 270, 271, **276**, 280, 293
waillyi Cherm. 269, **274**, 303
xantholepis Nelmes 269, 270, 271, **283**
zonatissimus Cherm. 270, **298**
zonatus Cherm. 298
QUEENSLANDIELLA Domin 1, 3, **306**
hyalina **307**
mira Domin 308
RHYNCHOSPORA Vahl 2, 180, **350**
adscendens C.B. Clarke 359
africana Cherm. 356
angolensis Turrill 351, **353**, 356
arechavaletae Boeckeler 354, 356
barteri C.B. Clarke 361
brevirostris Griseb. 351, **353**, 359, **361**
brownii Roem. & Schult. 351, **353**, 356, **357**, 358
    subsp. brownii (Roem. & Schult.) T. Koyama
    358
bulbocaulis Boeckeler 204
candida (Nees) Boeckeler 351, **353**, **359**
corymbosa (L.) Britton 351, **353**, 367
cyperoides Mart. 354
deightonii Hutch. 358
erinacea (Ridl.) C.B. Clarke 253
eximia (Nees) Boeckeler 351, **359**, 353
glauca sensu C.B. Clarke 356
gracillima Thwaites & Hook. 351, **361**
    subsp. subquadrata (Cherm.) J. Raynal **353**,
    359, **361**
holoschoenoides (Rich.) Herter 351, **353**, **354**,
    355, 356
juncea Kunth 356, 358
madagascariensis (Steud.) Cherm. 360
mauritii Steud. 354, 356
minor Nelmes 360
ochrocephala Boeckeler 188
parva (Nees) Kunth 360
perrieri Cherm. 351, **353**, 358, 361
rubra (Lour.) Makino 351, **360**
    subsp. africana J. Raynal **353**, **360**
rugosa (Vahl) Gale 358
rugosa sensu Mapuara & Timberlake 356

spectabilis C. Krauss 352
subquadrata Cherm. 361
testui Cherm. 361
triflora Vahl 351, **352**, **353**
Rikliella 345, 349
australiensis J. Raynal 345
kernii (Raymond) J. Raynal 345
rehmannii (Ridl.) J. Raynal 345, 349
Scaevola plumieri 190
SCHOENOPLECTIELLA Lye 2, 4, **27**, 37
articulata (L.) Lye **36**, 37, 39
erecta (Poir.) Lye 28, **41**, 43
    subsp. erecta **41**, 42, 43
    subsp. raynalii 42, 43, 45
    subsp. raynalii (Schuyler) Beentje **42**
hooperae (J. Raynal) Lye 28, **40**, 44
juncea (Willd.) Lye 28, **40**
lateriflora (J.F. Gmel.) Lye 28, **42**, 43, 45
    subsp. lateriflora 43
    subsp. *laevinux* (Lye) Beentje 41
leucantha (Boeckeler) Lye 28, **44**, 45
microglumis (Lye) Lye 28, **43**
patentiglumis (Hayas.) Hayas. 36, 37
proxima (Steud.) Lye 28, 40, **44**, 45
    var. *botswanensis* (Hayas.) Hayas. 44
roylei (Nees) Lye 28, **39**
senegalensis (Steud.) Lye 28, **37**, **38**, 39
SCHOENOPLECTUS (Rchb.) Palla 2, 4, **27**
articulata 28
articulatus (L.) Palla 36
brachyceras (A. Rich.) Lye 32
confusus (N.E. Br.) Lye 28, **32**, 33, 34
    subsp. confusus **33**
    subsp. natalitius Browning 33
    subsp. *natalitius* Browning sensu Mapuara &
    Timberlake 33
    var. confusus **33**, 34
    var. rogersii (N.E. Br.) Lye **33**, 34
corymbosus (Roem. & Schult.) J. Raynal 28, **30**,
    31, 34
    var. *brachyceras* (A. Rich.) Lye 32
    var. *corymbosus* 32
cernuus (Vahl) Hayek 137
erectus (Poir.) J. Raynal
    subsp. *erectus* 41, 42
    subsp. *raynalii* (Schuyler) Lye 42
hooperae J. Raynal 40
inclinatus (Delile) Lye 32
junceus (Willd.) J. Raynal 40
lateriflorus (J.F. Gmel.) Lye 42, 43
    subsp. *laevinux* Lye 41
leucanthus (Boeckeler) J. Raynal 44
littoralis (Schrad.) Palla var. *pterolepis* (Nees) C.C.
    Towns 29
maritimus (L.) Lye 25
microglumis Lye 43
mucronatus (L.) A. Kern. 28, **30**
muricinux (C.B. Clarke) J. Raynal 27, 28, **34** 36, 43
muriculatus (Kük.) Browning 27, 28, 34, **35**, **36**
patentiglumis Hayas. 36
proximus 44
proximus (Steud.) J. Raynal 44
    var. *botswanensis* Hayas. 44
rhodesicus (Podlech) Lye 27, **28**
rogersii (N.E. Br.) Lye 33
roylei (Nees) Ovcz. & Czukav. 28, **39**
scirpoides (Schrad.) Browning 28, **29**
senegalensis (Steud.) Palla 37
subulatus sensu Lye 29
SCHOENOXIPHIUM Nees 1, 2, **427**

caricoides C.B. Clarke 427, **429**, 430
lehmannii (Nees) Steud. 427, **430**, **431**
ludwigii Hochst. **427**, 428
rufum Nees 428
schimperianum (Boeckeler) C.B. Clarke 429
sparteum (Wahlenb.) C.B. Clarke 427, **428**, **429**, 430
  var. *lehmannii* (Nees) Kük. 430
  var. *schimperianum* (Boeckeler) Kük. 429
SCHOENUS L. 2, **364**
  aggregatus Thunb. 364
  erinaceus Ridl. 253
  erraticus Hook. f. 89, 90
  holoschoenoides Rich. 354
  junceus Willd. 40
  lagoensis subsp. *canaliculatotriquetra* (Boeckeler) Lye 417
  mariscus L. 366
  nigricans L. 364, **365**
  pilosus Willd. 85
  ruber Lour. 360
  triflorus (Vahl) Poir. 352
Scirpidium nigrescens Nees 58
SCIRPOIDES Ség. 1, 4, **140**
  dioeca (Kunth) Browning 140, **141**, **142**
  varia Browning **140**
Scirpus L. 140
  sect. *Vaginaria* Koyama 4
  subgen. *Fuirena* Kuntze 4
  [no rank] *Bolboschoenus* Asch. 25
  acutangulus Roxb. 48
  angolensis C.B. Clarke 126
  articulatus L. 36
  atropurpureus Retz. 57
  atrosanguineus Boeckeler 121
  aureiglumis S.S. Hooper 40
  bisumbellatus Forssk. 74
  bivalvis Lam. 70
  boeckelerianus Schweinf. 93
  brachyceras A. Rich. 30
  caducus Delile 54
  cernuus Vahl 137
  chaetarius Spreng. 60
  chinensis Osbeck 340
  chlorostachyus auct. 137
  ciliaris L. 19
  cinnamomeus Boeckeler 89
  collinus Boeckeler var. *boeckelerianus* (Schweinf.) Schweinf. 93
  complanatus Retz. 64
  confervoides Poir. 61
  confusus N.E. Br. 32
  corymbosus L. 352
  costatus (A. Rich.) Boeckeler 136
  cubensis Poepp. & Kunth 142
  cyperoides L. 242
  densus Wall. 117
  dichotomus L. 73, 74
  dioecus (Kunth) Boeckeler 142
  diphyllus Retz. 73
  equitans Kük. 129
  erectus Poir. 41
  ferrugineus L. 70
  fistulosus 48
  fluitans L. 134
  geniculatus L. 55
  glaucus Lam. 25
  glomeratus L. 333
  gracilis Poir. 131
  griquensium C.B. Clarke 137

hemisphaericus Roth 343
hispidulus Vahl 99
hystricoides B. Nord. 349
hottentotus L. 10
isolepis (Nees) Boeckeler 343
jacobii C.E.C. Fisch. 37
kernii Raymond 345
kyllingioides (A. Rich.) Boeckeler 145
lateriflorus J.F. Gmel. 42
leucanthus Boeckeler 44
leucocoleus K. Schum. 133
limosus Schrad. 51
littoralis Schrad. var. *pterolepis* (Kunth) C.B. Clarke 29
littoralis sensu C.B. Clarke 29
lugardii 164
macer Boeckeler 136
maritimus L. 25
michelianus L. 169
micranthus Vahl 346
microcephalus (Steudel) Dandy 145
miliaceus L. 68
mucronatus L. 30
muricinux C.B. Clarke 34
muriculatus Kük. 36
mutatus L. 49
natans Thunb. 139
parvinux (C.B. Clarke) K. Schum. 86
praelongatus sensu Cufod. 37
prolifer Rottb. 136
pubescens (Poir.) Lam. 13
quinquangularis Vahl 68
quinquefarius Boeckeler 39
raynalii Schuyler 42
rehmannii Ridl. 345, 349
retroflexus Poir. 59
rhodesicus Podlech 28
rivularis (Schrad.) Boeckeler 139
rogersii N.E. Br. 33
schinzii Boeckeler 142
schoenoides Retz. 76
schweinfurthianus Boeckeler 116
senegalensis Lam. 340
spadiceus (Lam.) Boeckeler var. *ciliatus* Ridl. 126
squarrosus L. 345
squarrosus sensu C.B. Clarke 345
squarrosus (Vahl) Poir. 77
steudneri Boeckeler 144
supinus L.
  var. *leucosperma* C.B. Clarke 44
  var. *uninodis* (Delile) Asch. & Schweinf. 41
tenerrimus Peter 43
uninodis (Delile) Boiss. 41
uninodis (Delile) Coss. & Durieu 41
ustulatus Podlech 128
variabilis 140
variegatus Poir. 51
varius C.B. Clarke 140
SCLERIA P.J. Bergius 1, 2, **378**, 425
  achtenii De Wild. 382, **415**
  acriulus C.B. Clarke 422
  adpressohirta (Kük.) E.A. Rob. 382, **418**
  africana Benth. 425
  angolensis Turrill 421
  angusta Kunth 381, **423**
  angustifolia E.A. Rob. 380, **399**, 402
  aquatica Cherm. 403
  arcuata E.A. Rob. 382, **419**
  aterrima (Ridl.) Napper 389
  bambariensis Cherm. 382, **410**

FAMILIES OF VASCULAR PLANTS REPRESENTED IN THE FLORA ZAMBESIACA AREA

PTERIDOPHYTA
(Flora Zambesiaca families and family number. Published 1970)

Actiniopteridaceae		Grammitidaceae	20	see Adiantaceae	18
see Adiantaceae	18	Hymenophyllaceae	15	Polypodiaceae	21
Adiantaceae	18	Isoetaceae	4	Psilotaceae	1
Aspidiaceae	27	Lindsaeaceae	19	Pteridaceae	
Aspleniaceae	23	Lomariopsidaceae	26	see Adiantaceae	18
Athyriaceae	25	Lycopodiaceae	2	Salviniaceae	12
Azollaceae	13	Marattiaceae	7	Schizaeaceae	10
Blechnaceae	28	Marsileaceae	11	Selaginellaceae	3
Cyatheaceae	14	Oleandraceae		Thelypteridaceae	24
Davalliaceae	22	see Davalliaceae	22	Vittariaceae	17
Dennstaedtiaceae	16	Ophioglossaceae	6	Woodsiaceae	
Dryopteridaceae		Osmundaceae	8	see Athyriaceae	25
see Aspidiaceae	27	Parkeriaceae			
Equisetaceae	5				
Gleicheniaceae	9				

GYMNOSPERMAE
(Flora Zambesiaca families and family number. Volume 1(1) 1960)

Cupressaceae	3	Cycadaceae	1	Podocarpaceae	2

ANGIOSPERMAE
(Flora Zambesiaca families, volume and part number and year of publication)

Acanthaceae			Balanitaceae	2(1)	1963	
tribes 1–5	8(5)	2013	Balanophoraceae	9(3)	2006	
tribes 6–7	8(6)	-	Balsaminaceae	2(1)	1963	
Agapanthaceae	13(1)	2008	Barringtoniaceae	4	1978	
Agavaceae	13(1)	2008	Basellaceae	9(1)	1988	
Aizoaceae	4	1978	Begoniaceae	4	1978	
Alangiaceae	4	1978	Behniaceae	13(1)	2008	
Alismataceae	12(2)	2009	Berberidaceae	1(1)	1960	
Alliaceae	13(1)	2008	Bignoniaceae	8(3)	1988	
Aloaceae	12(3)	2001	Bixaceae	1(1)	1960	
Amaranthaceae	9(1)	1988	Bombacaceae	1(2)	1961	
Amaryllidaceae	13(1)	2008	Boraginaceae	7(4)	1990	
Anacardiaceae	2(2)	1966	Brexiaceae	4	1978	
Anisophylleaceae			Bromeliaceae	13(2)	2010	
see Rhizophoraceae	4	1978	Buddlejaceae			
Annonaceae	1(1)	1960	see Loganiaceae	7(1)	1983	
Anthericaceae	13(1)	2008	Burmanniaceae	12(2)	2009	
Apocynaceae	7(2)	1985	Burseraceae	2(1)	1963	
Aponogetonaceae	12(2)	2009	Buxaceae	9(3)	2006	
Aquifoliaceae	2(2)	1966	Cabombaceae	1(1)	1960	
Araceae	12(1)	2011	Cactaceae	4	1978	
Araliaceae	4	1978	Caesalpinioideae			
Aristolochiaceae	9(2)	1997	see Leguminosae	3(2)	2006	
Asclepiadaceae	-	-	Campanulaceae	7(1)	1983	
Asparagaceae	13(1)	2008	Canellaceae	7(4)	1990	
Asphodelaceae	12(3)	2001	Cannabaceae	9(6)	1991	
Avicenniaceae	8(7)	2005	Cannaceae	13(4)	2010	

Capparaceae	1(1)	1960		Hemerocallidaceae	12(3)	2001
Caricaceae	4	1978		Hernandiaceae	9(2)	1997
Caryophyllaceae	1(2)	1961		Heteropyxidaceae	4	1978
Casuarinaceae	9(6)	1991		Hyacinthaceae	-	-
Cecropiaceae	9(6)	1991		Hydnoraceae	9(2)	1997
Celastraceae	2(2)	1966		Hydrocharitaceae	12(2)	2009
Ceratophyllaceae	9(6)	1991		Hydrophyllaceae	7(4)	1990
Chenopodiaceae	9(1)	1988		Hydrostachyaceae	9(2)	1997
Chrysobalanaceae	4	1978		Hypericaceae		
Colchicaceae	12(2)	2009		see Guttiferae	1(2)	1961
Combretaceae	4	1978		Hypoxidaceae	12(3)	2001
Commelinaceae	-	-		Icacinaceae	2(1)	1963
Compositae				Illecebraceae	1(2)	1961
tribes 1–5	6(1)	1992		Iridaceae	12(4)	1993
Connaraceae	2(2)	1966		Irvingiaceae	2(1)	1963
Convolvulaceae	8(1)	1987		Ixonanthaceae	2(1)	1963
Cornaceae	4	1978		Juncaceae	13(4)	2010
Costaceae	13(4)	2010		Juncaginaceae	12(2)	2009
Crassulaceae	7(1)	1983		Labiatae		
Cruciferae	1(1)	1960		see Lamiaceae, Verbenacaeae		
Cucurbitaceae	4	1978		Lamiaceae		
Cuscutaceae	8(1)	1987		Viticoideae, Pingoideae	8(7)	2005
Cymodoceaceae	12(2)	2009		Lamiaceae		
Cyperaceae	14	2020		Scutellaroideae-		
Dichapetalaceae	2(1)	1963		Nepetoideae	8(8)	2013
Dilleniaceae	1(1)	1960		Lauraceae	9(2)	1997
Dioscoreaceae	12(2)	2009		Lecythidaceae		
Dipsacaceae	7(1)	1983		see Barringtoniaceae	4	1978
Dipterocarpaceae	1(2)	1961		Leeaceae	2(2)	1966
Dracaenaceae	13(2)	2010		Leguminosae,		
Droseraceae	4	1978		Caesalpinioideae	3(2)	2007
Ebenaceae	7(1)	1983		Mimosoideae	3(1)	1970
Elatinaceae	1(2)	1961		Papilionoideae	3(3)	2007
Ericaceae	7(1)	1983		Papilionoideae	3(4)	2012
Eriocaulaceae	13(4)	2010		Papilionoideae	3(5)	2001
Eriospermaceae	13(2)	2010		Papilionoideae	3(6)	2000
Erythroxylaceae	2(1)	1963		Papilionoideae	3(7)	2002
Escalloniaceae	7(1)	1983		Lemnaceae		
Euphorbiaceae	9(4)	1996		see Araceae	12(1)	2011
Euphorbiaceae	9(5)	2001		Lentibulariaceae	8(3)	1988
Flacourtiaceae	1(1)	1960		Liliaceae sensu stricto	12(2)	2009
Flagellariaceae	13(4)	2010		Limnocharitaceae	12(2)	2009
Fumariaceae	1(1)	1960		Linaceae	2(1)	1963
Gentianaceae	7(4)	1990		Lobeliaceae	7(1)	1983
Geraniaceae	2(1)	1963		Loganiaceae	7(1)	1983
Gesneriaceae	8(3)	1988		Loranthaceae	9(3)	2006
Gisekiaceae				Lythraceae	4	1978
see Molluginaceae	4	1978		Malpighiaceae	2(1)	1963
Goodeniaceae	7(1)	1983		Malvaceae	1(2)	1961
Gramineae				Marantaceae	13(4)	2010
tribes 1–18	10(1)	1971		Mayacaceae	13(2)	2010
tribes 19–22	10(2)	1999		Melastomataceae	4	1978
tribes 24–26	10(3)	1989		Meliaceae	2(1)	1963
tribe 27	10(4)	2002		Melianthaceae	2(2)	1966
Guttiferae	1(2)	1961		Menispermaceae	1(1)	1960
Haloragaceae	4	1978		Menyanthaceae	7(4)	1990
Hamamelidaceae	4	1978		Mesembryanthemaceae	4	1978

Family			Family		
Mimosoideae			Restionaceae	13(4)	2010
see Leguminosae	3(1)	1970	Rhamnaceae	2(2)	1966
Molluginaceae	4	1978	Rhizophoraceae	4	1978
Monimiaceae	9(2)	1997	Rosaceae	4	1978
Montiniaceae	4	1978	Rubiaceae		
Moraceae	9(6)	1991	subfam. Rubioideae	5(1)	1989
Musaceae	13(4)	2010	tribe Vanguerieae	5(2)	199
Myristicaceae	9(2)	1997	subfam.Cinchonoideae	5(3)	2003
Myricaceae	9(3)	2006	Rutaceae	2(1)	1963
Myrothamnaceae	4	1978	Salicaceae	9(6)	1991
Myrsinaceae	7(1)	1983	Salvadoraceae	7(1)	1983
Myrtaceae	4	1978	Santalaceae	9(3)	2006
Najadaceae	12(2)	2009	Sapindaceae	2(2)	1966
Nesogenaceae	8(7)	2005	Sapotaceae	7(1)	1983
Nyctaginaceae	9(1)	1988	Scrophulariaceae	8(2)	1990
Nymphaeaceae	1(1)	1960	Selaginaceae		
Ochnaceae	2(1)	1963	see Scrophulariaceae	8(2)	1990
Olacaceae	2(1)	1963	Simaroubaceae	2(1)	1963
Oleaceae	7(1)	1983	Smilacaceae	12(2)	2009
Oliniaceae	4	1978	Solanaceae	8(4)	2005
Onagraceae	4	1978	Sonneratiaceae	4	1978
Opiliaceae	2(1)	1963	Sphenocleaceae	7(1)	1983
Orchidaceae	11(1)	1995	Sterculiaceae	1(2)	1961
Orchidaceae	11(2)	1998	Strelitziaceae	13(4)	2010
Orobanchaceae			Taccaceae		
see Scrophulariaceae	8(2)	1990	see Dioscoreaceae	12(2)	2009
Oxalidaceae	2(1)	1963	Tecophilaeaceae	12(3)	2001
Palmae	13(2)	2010	Tetragoniaceae	4	1978
Pandanaceae	12(2)	2009	Theaceae	1(2)	1961
Papaveraceae	1(1)	1960	Thymelaeaceae	9(3)	2006
Papilionoideae			Tiliaceae	2(1)	1963
see Leguminosae	-	-	Trapaceae	4	1978
Passifloraceae	4	1978	Turneraceae	4	1978
Pedaliaceae	8(3)	1988	Typhaceae	13(4)	2010
Periplocaceae			Ulmaceae	9(6)	1991
see Asclepiadaceae	-	-	Umbelliferae	4	1978
Philesiaceae			Urticaceae	9(6)	1991
see Behniaceae	13(1)	2008	Vacciniaceae		
Phormiaceae			see Ericaceae	7(1)	1983
see Hemerocallidaceae	12(3)	2001	Vahliaceae	4	1978
Phytolaccaceae	9(1)	1988	Valerianaceae	7(1)	1983
Piperaceae	9(2)	1997	Velloziaceae	12(2)	2009
Pittosporaceae	1(1)	1960	Verbenaceae	8(7)	2005
Plantaginaceae	9(1)	1988	Violaceae	1(1)	1960
Plumbaginaceae	7(1)	1983	Viscaceae	9(3)	2006
Podostemaceae	9(2)	1997	Vitaceae	2(2)	1966
Polygalaceae	1(1)	1960	Xyridaceae	13(4)	2010
Polygonaceae	9(3)	2006	Zannichelliaceae	12(2)	2009
Pontederiaceae	13(2)	2010	Zingiberaceae	13(4)	2010
Portulacaceae	1(2)	1961	Zosteraceae	12(2)	2009
Potamogetonaceae	12(2)	2009	Zygophyllaceae	2(1)	1963
Primulaceae	7(1)	1983			
Proteaceae	9(3)	2006			
Ptaeroxylaceae	2(2)	1966			
Rafflesiaceae	9(2)	1997			
Ranunculaceae	1(1)	1960			
Resedaceae	1(1)	1960			